STP 989

Performance of Protective Clothing: Second Symposium

S. Z. Mansdorf, Richard Sager, and Alan P. Nielsen, editors

ASTM
1916 Race Street
Philadelphia, PA 19103

ASTM Publication Code Number (PCN: 04-989000-55)
ISBN: 0-8031-1167-3
ISSN: 1040-3035

NOTE

The Society if not responsible, as a body,
for the statements and opinions
advanced in this publication.

Peer Review Policy

Each paper published in this volume was evaluated by three peer reviewers. The authors addressed all of the reviewers' comments to the satisfaction of both the technical editor(s) and the ASTM Committee on Publications.

The quality of the papers in this publication reflects not only the obvious efforts of the authors and the technical editor(s), but also the work of these peer reviewers. The ASTM Committee on Publications acknowledges with appreciation their dedication and contribution of time and effort on behalf of ASTM.

Printed in Baltimore, MD
December 1988

Foreword

The papers in this publication, *Performance of Protective Clothing—Second Symposium,* have been selected from those presented at the Second International Symposium on the Performance of Protective Clothing, which was held in Tampa, Florida, during 19–21 January, 1987. This meeting was sponsored by the ASTM Committee F-23 on Protective Clothing and cosponsored by the American Industrial Hygiene Association Committee on Protective Devices and the Royal Institute of Technology of Stockholm, Sweden. This symposium was the second in a series of symposia held to bring together internationally known experts to discuss the emerging issues related to worker protection through the use of protective clothing.

The symposium chairmen were S. Z. Mansdorf, S. Z. Mansdorf & Associates, Inc., and Richard Sager, Sager Corporation. Additional support was provided by Alan Nielsen of the U.S. Environmental Protection Agency who was largely responsible for the overwhelming success of the pesticides sessions. These key individuals also served as editors of this publication.

Contents

Cut Protection

PROTECTION FROM INDUSTRIAL CHEMICAL STRESSORS

Dermal Toxicology

Chemical Breakthrough Parameters

NEW MATERIALS AND TECHNOLOGIES

Overview

The performance of protective clothing has become a significant concern of the health and safety community over the last ten years. This has been due in large part to the development of standard test methods by the F-23 Committee of ASTM and others which demonstrated significant limitations to previously considered "safe" uses of this equipment. Secondly, increased use of personal protective equipment as an apparent cost effective alternative to engineering controls and for those operations where engineering controls are not feasible has been evidenced.

The F-23 Committee

The F-23 Committee was originally organized in 1977 as the Chemical Protective Clothing Committee under the ASTM organizational umbrella. This committee was formed as a direct result of the recognized need by manufacturers and users for uniform standards for chemical protective clothing.

The first official F-23 standard test method [Resistance of Protective Clothing Materials to Permeation by Hazardous Liquid Chemicals (F 739-81)] was published in 1981. Following the publication of this test method and others, a considerable amount of data were generated that indicated most chemical protective clothing was not an absolute safeguard as was once commonly believed.

The F-23 Committee has grown to become the major recognized force in the protective clothing arena with over 160 active members representing protective clothing users, manufacturers, government, and academia.

Symposia

To help satisfy the growing interest of the general health and safety community in the activities of F-23 and the use of protective clothing, a symposium was sponsored in Raleigh, North Carolina in 1984. The Symposium was an overwhelming success in terms of participation and the transfer of technical knowledge through an ASTM Special Technical Publication (STP 900), *Performance of Protective Clothing,* containing 48 peer-reviewed technical papers.

The Second International Symposium on Protective Clothing was held three years later in Tampa, Florida. It was the most comprehensive and well attended symposium ever held on the subject of protective clothing. Over 150 papers encompassing a number of broad areas related to protective clothing were presented by an internationally recognized roster of experts. The international scope of the subject was evidenced by the fact that two of the four plenary speakers were from outside of the United States.

This publication, as the second volume to the original STP 900 contains 84 papers selected from the original presentations. These cover issues and areas of concern related to protective clothing selection, use, and testing. All of the manuscripts have undergone extensive peer review in accordance with ASTM requirements.

Organization

This STP is divided into six general topic areas. These sections contain a diverse range of papers on protective clothing which are characteristic of this type of symposium and the breadth of the subject. They represent the current issues of interest in this emerging field and thus will be of value to readers desiring both an overview and specific information on the latest research in protective clothing.

The major topic areas of the book are the voluntary standards process, human factors, protection from physical stressors, protection from industrial chemical stressors, protection from pesticides, and new materials and technologies.

The first topic area covering voluntary standards includes a history of the function and purpose of the ASTM Committee F-23 on Protective Clothing and a paper on the need for international cooperation for the development of standard test methods. The second topic area is human factors which contains four papers addressing the proper fit and testing of protective clothing. The third topic area, containing nine papers on protection from physical stressors, has a major emphasis on thermal performance and testing but also contains a paper on cut resistance of protective leggings. The fourth topic area on protection from industrial chemical stressors is one of the larger sections of the book with 38 papers. It is subdivided into sections containing papers on dermal toxicology, permeation theory and testing of protective clothing (including an expert roundtable discussion), new laboratory test methods, field test methods and the application of their data, field experiences, decontamination issues, selection and use of chemical protective clothing, emergency response and military applications, and the performance of full ensembles. The fifth topic area covering protection from pesticides contains 28 papers. It is divided into three major sections. These are field performance, laboratory test methods for materials resistance and decontamination, and user attitudes and work practices. The final topic areas of the book contains three papers on new materials and technologies.

Significance

This publication in combination with STP 900 contains the most comprehensive body of knowledge on the subject of protective clothing currently available. It spans the range of thermal protection to human factors and as such should be a valuable resource for those interested or responsible for the selection, use, or testing of protective clothing.

It is the hope of the editors that this book will encourage protective clothing research and subsequently lead to advancements in its selection, safe use and testing. As stated best by

John Moran of NIOSH at the plenary session, ". . . Protective clothing is clearly the last line of defense . . ."

S. Z. Mansdorf
S. Z. Mansdorf & Associates, Inc.
Cuyahoga Falls, OH 44223

Richard Sager
Sager Corporation
Prospect Heights, IL 60070

Alan Neilsen
US EPA
Office of Pesticide Programs
Washington, DC 20460

The Voluntary Standards Process

Norman W. Henry III[1]

A Decade of Protective Clothing Standards Development

REFERENCE: Henry, N. W., III, **"A Decade of Protective Clothing Standards Development,"** *Performance of Protective Clothing: Second Symposium, ASTM STP 989,* S. Z. Mansdorf, R. Sager, and A. P. Nielsen, Eds., American Society for Testing and Materials, Philadelphia, 1988, pp. 3–6.

ABSTRACT: ASTM Committee F23 on protective clothing will complete a decade of standards development activities this year. Since its original formation as an official ASTM committee in 1977, it has grown in size from 36 to 170 members, produced three standard methods, held one successful international symposium on protective clothing, and published one Special Technical Publication. Charged with a responsibility to develop standard test methods and standard terminology, classifications, and performance specifications for clothing used to protect against occupational hazards, Committee F23 has become the worldwide focal point of protective clothing standards development. This presentation reviews the history and development of Committee F23, looking at highlights of its past, present and future activities.

KEY WORDS: protective clothing, history, activities and accomplishments, standards writing organizations

ASTM Committee F23 will complete a decade of standards development this year. The committee was originally organized on 4 Oct. 1977 in Valley Forge, Pennsylvania. The intent was to meet a need expressed by members of industry for protective clothing standards for workers exposed to various chemical formulations. Both producers and users of protective clothing were present at this organizational meeting. They agreed that standards in this area were needed and unanimously approved a title, scope, and committee structure. Since that time, Committee F23 has seen the development of three standard test methods, broadened its subcommittee structure to include a molten metals group, changed its original title from *chemical* protective clothing to *protective* clothing, and increased its membership almost five-fold. It also has held a successful international symposium on protective clothing in Raleigh, North Carolina, in July 1984 and produced its first Special Technical Publication. Therefore, it is appropriate to review Committee F23's past, present, and future activities as it steps into its second decade of standards development.

The original planning session for the development of Committee F23 was held in ASTM headquarters in Philadelphia, Pennsylvania on 15 June 1977. It was the consensus of the 18 attendees that a new ASTM committee on chemical protective clothing should be organized. A proposed scope, title, and subcommittee structure were developed for submission at an organizational meeting scheduled for 4 Oct. 1977 in Valley Forge, Pennsylvania. At this organizational meeting, 36 attendees unanimously approved the following title, scope, and subcommittee structure.

---

[1] Research chemist, E. I. du Pont de Nemours and Company, Inc., Haskell Laboratory for Toxicology and Industrial Medicine, Newark, DE 19714, and chairman of F23.

Title—Chemical Protective Clothing

Scope—The development of standard test methods, terminology, classifications, and performance specifications for clothing used to protect against chemical hazards.

Subcommittee structure—

F23.1 Hazard Identification and Classification

F23.2 Test Methods

F23.3 Garment and Materials Performance

F23.4 Research

F23.5 Definitions

F23.6 Liaison

At the first committee meeting held in January 1978 at ASTM headquarters in Philadelphia, subcommittee assignments were developed and task groups formed. Allan Sisson, vice chairman of Committee F23 also reported to the members that he would give a presentation of Committee F23 goals and objectives at the American Industrial Hygiene Conference scheduled for May in Los Angeles, California, to encourage AIHA members to participate in the work of Committee F23. Jim Thomas was also introduced as the first ASTM staff manager assigned to the committee. The second annual meeting was held in Boston, Massachusetts, where committee bylaws were accepted by membership. Jerry Hess from Goodyear Tire and Rubber was the chairman of Committee F23 during its first two years.

In 1979, the committee met twice. The first meeting in January in Orlando, Florida, resulted in a draft of the permeation test method. An additional highlight was a tour of the Kennedy Space Center. The second meeting was held in San Francisco, California, where new officers were elected and an initial evaluation of interlaboratory permeation test results were reviewed. At this meeting, Jerry Coletta became the second chairman of the committee. Under his direction the committee was reorganized into new subcommittees in October 1979. They were as follows:

F23.10 Planning

F23.20 Physical Properties

F23.30 Chemical Resistance

F23.40 Classification

F23.90 Executive

F23.91 Terminology

F23.92 Liaison and Publicity

In 1980, the committee met in Fort Lauderdale, Florida, and formally adopted the new subcommittee structure of four technical and three administrative committees. Committee goals were established and there was discussion about a new subcommittee to be formed on clothing used to protect against molten substances. The committee next met in June in Chicago, Illinois. There were approximately 60 members on the committee at this time.

Committee F23 held two meetings in 1981. The first was in New Orleans, Louisiana, where approximately 120 members attended and formally welcomed a new subcommittee, F23.80, on Molten Metals. At this meeting, the concept of management by objective (MBO) was introduced and emphasis placed on long-range planning. The second meeting was in Philadelphia at ASTM headquarters where there were technical presentations on molten metals, hazardous material response, and the use of chemical protective clothing (CPC). The National Institute for Occupational Safety and Health (NIOSH) was also represented at the meeting by Steve Berardinelli. He reported on NIOSH's desire to coordinate its research with Committee F23 and industry. It was also reported that the liquid permeation test method had passed committee ballot. A new staff manager was introduced, Drew Azzara, to replace Jim Thomas who became vice president of standards development.

On 25 Sept. 1981, Committee F23's first standard, Test Method for Resistance of Protective Clothing Materials to Permeation by Hazardous Liquid Chemicals (F 739-81) was approved by Society Ballot. Draft standards were also being developed for methods to liquid penetration, degradation, and thermal resistance to materials exposed to molten substances.

In 1982, Committee F23 met twice, first in January in Houston, Texas, and then in June in Toronto, Canada. There were approximately 120 members on the committee. Most of the activity at these meetings focused on methods development and comparison of interlaboratory test results. Zack Mansdorf reported on AIHA's protective equipment committee and encouraged liaison with ASTM Committee F23.

In 1983, the committee met twice. At the first meeting in Atlanta, Georgia, the new publication of Committee F23's official newsletter, *The Clothesline* was passed out to members. It was also reported that Committee F23 would sponsor a panel discussion on protective clothing at the AIHA conference in May in Philadelphia, Pennsylvania. At the second meeting in Kansas City, Missouri, about 100 members were present. At this meeting an international guest, Krister Forsberg from Sweden, participated for the first time to report on European and Scandinavian developments on protective clothing standards.

The first Committee F23 service award was also given at this meeting to Jerry Coletta. This award recognizes those committee members for their distinguished service to Committee F23 for the cause of voluntary standardization. A Committee F23 Fund was also established at this meeting to support those activities not covered by dues.

In 1984, the committee met on the West Coast in San Diego, California, to plan for the first international symposium on protective clothing to be held in July in Raleigh, North Carolina. At this meeting, another new subcommittee on Ensembles (totally encapsulated suits) was formed and given the designation F23.50. In July, Committee F23 sponsored the First International Symposium on Protective Clothing. About 250 attendees were present for this symposium that was held in Raleigh, North Carolina. The keynote speaker was Dr. John Millar, Director of NIOSH. Papers presented at this symposium were peer reviewed and published in Committee F23's first Special Technical Publication (STP). The second service award was also given at the symposium to Norm Henry.

In 1985, Committee F23 met twice. The first meeting was in Reno, Nevada, where it was announced that the penetration test method, ASTM Test Method for Resistance of Protective Clothing Materials to Penetration by Liquids (F 903-84) had passed society ballot and was now a published ASTM standard. Membership at this point was at 137. A U.S. Environmental Protection Agency (EPA) representative, Alan Neilson, participated at this meeting expressing interest in protective clothing standards for agricultural workers.

The second meeting was held in Washington, D.C., where 149 members were in attendance. At this meeting, three new task forces in F23.30 were formed. They were F23.30.5 on Standard Test Chemicals, F23.30.7 on Particulates, and F23.30.8 on Pesticides. One new subcommittee, F23.51 on Human Factors, was also formed to address clothing sizing, comfort, and stress. Bobby Sasser was also recognized and given Committee F23's third service award. On 27 Sept. 1985, a new standard test method, ASTM Evaluating Heat Transfer Through Materials for Protective Clothing Upon Contact With Molten Substances (F 955-85) passed committee ballot and became the third method developed by Committee F23.

In 1986, Committee F23 met twice. The first meeting was in Cocoa Beach, Florida, where new officers took over the committee's activities and plans were agreed on the Second International Protective Clothing Symposium to be held in 1987. An additional activity included a tour of laboratories at Kennedy Space Center. At the next meeting that was held in Louisville, Kentucky, in June, plans for the symposium were formalized and a schedule set for Tampa, Florida, in January 1987. The fourth annual service award was given to Art

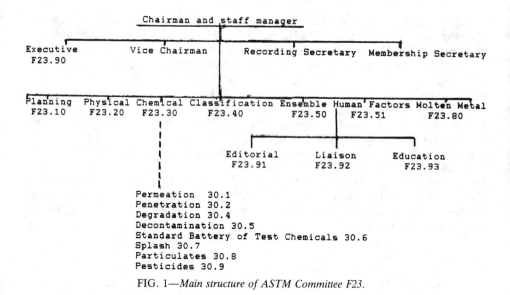

FIG. 1—*Main structure of ASTM Committee F23.*

Schwope, and Anne McKlindon was introduced as the new staff manager to replace Drew Azzara.

In January 1987, Committee F23 held its Second International Symposium on Protective Clothing. The symposium was sponsored by ASTM Committee F23 on Protective Clothing, the Royal Institute of Technology Stockholm, Sweden, and the Personal Protective Devices Committee of the American Industrial Hygiene Association (AIHA). It was an overwhelming success with over 270 attendees, 150 technical presentations, and 25 different sessions. Papers presented at the symposium were submitted for publication in Committee F23's second Special Technical Publication.

Currently, Committee F23 is organized into several technical and support committees. Figure 1 shows a block diagram of the structure of Committee F23 and gives an overall picture of the committee's organization. As of February 1987, the total membership of Committee F23 was 164.

Michael Hougaard[1]

The Need for International Cooperation Regarding Approvals

REFERENCE: Hougaard, M., **"The Need for International Cooperation Regarding Approvals,"** *Performance of Protective Clothing: Second Symposium, ASTM STP 989,* S. Z. Mansdorf, R. Sager, and A. P. Nielsen, Eds., American Society for Testing and Materials, Philadelphia, 1988, pp. 7–14.

ABSTRACT: Research shows that chemical protective equipment should be permeation tested against the exact chemical it is expected to protect against. Because more than 10 000 different chemicals are used today, the amount of permeation testing is immense.

Public reaction to and the legal ramifications of inadequate protection are discussed. A negotiation model, including the four major interests to be fulfilled by an international approval or certification system, is established.

Standards and certification concepts are defined as well as the roles of the organizations involved. Due to the amount of testing, if a national approval or certification system for chemical protective clothing is created, then it should be designed with the prospect of becoming international at a later stage.

A data presentation form that can be implemented immediately by manufacturers of chemical protective equipment is proposed.

KEY WORDS: protective equipment, permeation testing, standards, certification, approval, conformity, organizational relationships, data presentation, protective clothing

In this paper, various topics related to approval and certification of protective equipment are presented. Special reference is made to chemical protective equipment.

It is estimated that more than 100 000 chemical products with very different toxicological properties are in use in about 100 countries.

As knowledge of risks from hazardous chemicals is becoming more recognized, many countries are expected to regulate the use and supply of personal chemical protective equipment.

Supply regulations could include the requirements that the equipment be tested, pass the tests, be approved, (certified), and presented to the customers in a standardized way.

The legal ramifications and liability questions, giving employers, manufacturers, and suppliers a very specific interest in third party approval or conformity certification, will be mentioned.

Central Problems

New Information

A number of years ago, it was generally believed that permeation was sufficiently described by degradation data and that permeation data of one chemical could be derived from permeation data of a chemical in the same chemical group. It is now recognized that these

---

[1] Technical officer, Danish Standards Association, Aurehoejvej, DK-2900, Hellerup, Denmark.

beliefs are not correct. Nevertheless, some manufacturers still use degradation data in their advertising.

Consequently, the new information shows that a certain protective item should be permeation tested against the exact chemical (or mixture of chemicals) it is expected to protect against. This means that a great amount of permeation testing should be carried out.

Many scientists are trying to establish relevant data regarding permeation and to find underlying laws of nature from which permeation properties can be derived.

The Chemical Industry

Situations where protective equipment may be needed are shown in Fig. 1.

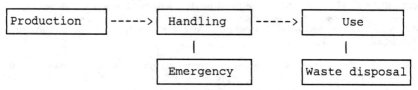

FIG. 1—*Chemical use situations.*

In almost any situation, it is of interest to know the type of personal protection and other safeguards to apply.

The manufacturer of the chemical should give relevant information about handling, use, etc., of the product. Relevant information includes statements such as "use glove type XYZ" instead of "use gloves," when the latter might be understood by a nonprofessional user as ordinary throw-away plastic gloves or perhaps even cotton gloves.

There are cases where the manufacturer or his agent have had the opportunity to sell the protection along with the chemical. No doubt this will be more common in the future, because the manufacturer of the chemical should be able to make the best choice of protective equipment, for example, gloves.

This particular solution is not expected to be commonplace. Probably, glove manufacturers will market gloves that can be used for several chemicals, and companies using a variety of chemicals, especially fire brigades supposed to handle emergencies, will prefer to select several glove types or protective gear types so that a wide range of chemicals are covered. Therefore, we should ask the suppliers of chemicals to give exact information about the type of personal protection to apply in specific cases.

Protective Equipment Industry

The manufacturer of protective equipment is up against a number of difficulties. Some of the problems are:

   (*a*) many thousands of chemicals are used in industry;
   (*b*) in many countries, the exact content of a given chemical product container cannot be derived from any source of information (laws in different countries vary considerably with respect to labeling);
   (*c*) even when the container is labeled and content identified in percentages, there may be minor amounts of unknown chemicals not stated, but of importance regarding the selection of protection;
   (*d*) when selling protective equipment, the manufacturer is often asked to recommend the best protection;
   (*e*) if the manufacturer gives advice, his legal position changes, and suddenly he can find himself made responsible in a lawsuit.

The manufacturer of protective wear must have a carefully designed internal control system. Minor changes in the chemical composition of raw materials may be followed by a change in protective properties causing the resources laid down in testing to become ineffective. To some extent, he can protect his own business by understating the limits of protection, but such a policy may be detrimental to his business.

It is interesting to study the legal situation between the supplier of the chemicals and the supplier of protective equipment. If a person working with chemicals has been disabled by the chemicals, he can try to obtain compensation from the employer, the supplier of the chemical, and the supplier of protective equipment. In this case, it appears that the supplier of the chemicals and supplier of the protective equipment have opposing interests.

On the other hand, the chemical and the protective equipment suppliers and the employer could benefit from an approval or certification system relieving them from part of the liability burden.

Public Reaction

During the last 10 years, the public has witnessed a number of large-scale accidents in chemical plants and the discovery of many contaminated hidden dumps inside city boundaries. A number of workers became disabled due to many years of exposure to various chemicals that were believed to have only limited risks. As a consequence, the public has become more interested in the chemical industry.

With regard to personal protection, the problem now is to teach users how to select the proper equipment and to adhere to safety instructions. The next step will be to ask to what degree does the equipment protect, that is, what is the breakthrough time. Professional people should realize that in the future such technical terms as "permeation rate" will become part of the language of many nonprofessional people. The public—including representatives of users—will ask questions about protective equipment quality and, if this quality appears to be insufficient, a political reaction can be expected. A possible political reaction to public pressure could be the creation of a national approval or certification system based on any principle that comes to hand.

International Cooperation

National and international resources should be combined in such a way that political and technical solutions can be found.

It is important to develop international standards for protective clothing. It is suggested that the United States join the international committee and take part in this work. Also, representatives of government institutions should discuss the various possibilities for international cooperation regarding approval and certification systems in order that we do not suddenly find ourselves testing the same equipment for the same chemicals in many countries at the same time.

The Negotiation Model

If an international approval or certification system is to be created, several interests must be considered. They include the following.

1. A number of countries will insist on having approval or certification systems (not necessarily national).
2. Many countries will want to establish accredited national testing facilities.
3. There will be a general agreement that duplication of testing in many countries will be a waste of resources.

4. Each type of equipment should be tested against as many chemicals as possible in order to cover a wide range of situations.

Writing down a list of interests in a given case always involves decisions as to which ones should be included and excluded. One could argue that "limited cost" should appear on the list. However, when considering the cost of alternative solutions to personal protection, where the proper protection can be a question of life and death, it was deleted.

The Tools

Standards

International standards for personal protection are developed in the International Organization for Standardization (ISO) where Technical Committee (TC) 94 is responsible for personal protection. It has several subcommittees (SC) of which ISO/TC94/SC13 covers personal protective clothing. The working program includes:

(a) revision of ISO 6530-1980, Limited protection against dangerous liquid chemicals—Resistance to penetration—Marking;
(b) development of ISO DP (draft proposal) 6529, Protective clothing—Protection against liquid chemicals—Determination of the rate of permeability through air-impermeable materials. This document is similar to ASTM Test Method for Resistance of Protective Clothing Materials to Permeation by Liquids or Gases (F 739–85);
(c) development of a standard for degradation of materials;
(d) development of a range of standards covering requirements, test methods, identification, and instruction for protective gloves against chemicals. In ISO the word "marking" is used instead of "identification"; and
(e) development of a document for selection of chemical protective equipment.

The *ISO/IEC Guide 7* [1] must be followed if the standards are to become a basis for an international certification scheme.

Certification Concepts

Here, the most important concepts, namely, approval, certification, and the manufacturer's declaration of conformity with standard, will be defined. It appears that the difference between approval and certification is:

(a) approval is an authority's statement, whereas certification relates to a non-authority institution; and
(b) approval is a declaration whereas certification involves signing a contract.

Approval—Approval can be defined as a declaration by a body vested with the necessary authority that a set of published criteria has been fulfilled. This definition is found in the first edition of *ISO Guide 2,* [2] but has been deleted in the second edition.

Certification—Certification can be defined as a contract between a supplier and an independent institution allowing the manufacturer to mark his product with the mark of that institution on the condition that conformity with a standard or other document is assured by impartial testing.

Manufacturer's Declaration of Conformity—This term is used when a company claims that a product is manufactured according to a standard, but when the claim is not backed by an independent testing institution. This type of declaration is often termed "self-certification."

Organizations Involved

The following organizations will be described briefly: governments, standards organizations, testing laboratories, and insurance companies.

Governments—In relation to protective equipment, governments have the power to regulate:

(*a*) use of the equipment, and
(*b*) sale and marketing of the equipment, including establishing an approval system or mandatory certification scheme.

Sometimes, governments agree to accept approvals mutually. Governments can establish mandatory approval systems, give various certification systems mandatory status, or encourage the establishment of certification schemes and testing procedures. As long as there is no public pressure, the government has considerable freedom to act as it chooses. It is advocated that governments should try to establish a system with "checks and balances," and with defined legal and liability positions. This could be in the form of delegating the approval/certification procedure to other organizations, with the government maintaining a supervising role.

Standards Organizations—Many standards bodies throughout the world offer certification service. For instance, the Danish Standards Association has operated certification schemes since 1929. Standards bodies cooperate in international organizations like ISO and CEN (Comite Europeen de Normalisation). The cooperation is not confined to standards development, but includes international certification schemes. In ISO and CEN, this type of cooperation is termed CASCO (ISO Committee on Conformity Assessment) and CENCER (CEN's certification body). The work in CASCO and CENCER is still in its infancy.

Testing Laboratories—Testing laboratories are organized differently from country to country. International accreditation of testing laboratories is being developed by the International Laboratory Accreditation Conference (ILAC). Nevertheless, anyone receiving a test report from an unknown testing laboratory must evaluate not only the results, but also the laboratory. Sometimes testing laboratories offer national certification schemes. It is important that the testing laboratory is impartial, that is, the person doing the testing should have no interest in arriving at a specific result. Consequently, the testing laboratory should not be part of a production department and the certifying body should not be too associated to a specific testing laboratory. Also, it is argued that the laboratory should not be part of the certifying body, as this will, taking it by and large, bind the certifying body to this particular laboratory.

Insurance Companies—For accident or disablement cases, the insurance companies will be involved. Their economic risk will be reduced if a third party inspection system is established (approval or certification) and if all protective equipment are tested against many different chemicals.

Relationships between National Organizations

National organizations can cooperate in various ways according to their powers and abilities. The government can choose to establish an approval system, ask somebody to operate a certification system, or do nothing. When establishing an approval system, the government cooperates with manufacturers (and agents for manufacturers). If a government chooses to do nothing, manufacturers will apply their own standards (self-certification) in a nonstandardized form making it difficult for buyers and users to select the correct product. If the government wants a certification system, an operator must be selected. In this case, the following relationships must be established:

(*a*) agreement between government and certification body,
(*b*) agreement between certification body and one or more testing laboratories, and
(*c*) contracts between manufacturers and certification bodies regulating legal responsibilities, use of certification marks, and external and internal inspection procedures.

The public should be informed of the principles in these agreements and contracts in order to be able to evaluate the system. Figure 2 shows the certification relationships.

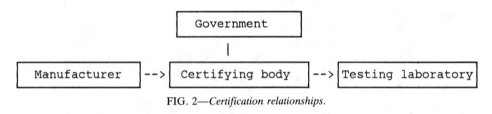

FIG. 2—*Certification relationships.*

Relationships between International Organizations

Governments have their own way of negotiating with each other, and the rules for establishing agreements are well known.

Standards organizations are connected on a regional and world-wide basis. When they act as certifying bodies, they can establish certification schemes covering many countries. Standards organizations have established a number of basic rules for international certification schemes. The *ISO/IEC Guide 42*, [3] outlines a step-by-step process for international certification. It should be read only in connection with a number of other ISO/IEC Guides that are collected in a compendium [4]. *Guide 42* does not take into account, however, that many countries want to have national testing facilities.

Solutions

Combining the Interests and Tools

If we are to find solutions, we must

(*a*) understand the goals,
(*b*) use the tools, and
(*c*) fulfill the interests.

The *goal* is to ensure that the protective equipment meets performance requirements.

The *tools* are standards, certification concepts, standards organizations, and their relationships.

The *interests* include the establishment of approval and certification systems, the desire for national testing facilities, avoidance of duplicate testing, and the testing of the equipment for many different chemicals.

National Systems

The easiest solution is that each country establishes its own system. This can be carried out by mandatory government standards or by a voluntary system established by certification bodies. Because personal protection from chemicals is a serious problem, a standard initiated by a national organization should correspond to systems in other countries.

A purely national system will only fulfill the first two interests. The last two interests, namely, avoidance of duplicate testing and testing of each type of equipment against many chemicals, will not be fulfilled. One can expect that only a few chemicals will be tested, resulting in an approval or certification label that is extended beyond its limits, unfortunately.

In Denmark, we have written a series of standards covering single-use gloves [5]. A certification scheme is established and described in the Special Rules for Certification [6]. According to these rules, it is possible to accept test results from other countries under certain conditions.

It is recommended that national rules be written so they can become international at a later time. Those organizations that already have close international bonds should be chosen.

CASCO (ISO)

Recently, the fundamental rules for certification, designated CASCO, within ISO were approved. Basically, the country of origin is responsible for testing and surveillance of a product. No products have passed through this international system as yet, because it is very new and no method is expected to be implemented in the near future.

Under the CASCO system, the country of origin is the testing country. This implies that countries without national production are not expected to do any testing at all and that countries with national production are not supposed to check imported goods. It is clearly a political wish to abolish trade barriers, but many occupational safety and health organizations want to check imported protective wear as well as national products. Only on a long-term basis can CASCO be expected to play a central role in the certification of protective equipment.

Suggestions to Manufacturers

The following proposals to manufacturers are:

1. The manufacturers realize the interests: that (*a*) a system should be established with (*b*) many national testing facilities (*c*) without duplicate testing, and should (*d*) include the testing with many chemicals.
2. They take steps to ensure that their products are tested in many countries, but against different chemicals in different countries.
3. They present their results according to the following proposal.

Proposal for Presentation of the Data

A proposal for the data is shown in Table 1. The "use recommendation" is expected to be defined in the ISO standards. They would be based on test results for a chemical.

TABLE 1—*Proposal for short-form presentation.*

Manufacturer:

Equipment designation:

Test Chemical	Test Results and Use Recommendations	Certifying or Approving Body/Testing Laboratory
1		USA/...
2		AFNOR/...
3		DIN/...
4		DS/Dantest
5		BSI/BSI test house

Some products under certain circumstances will be used only once against certain chemicals and several times against others. Sometimes, special decontamination and regeneration procedures must be followed. The idea here is that each test result and its "use recommendation" can be added to the list.

By accepting test results from other laboratories while still requiring national testing, information about equipment will be accumulated. An international data base can be established that would help those who give advice on selecting the proper equipment.

This proposed system can be started immediately by the manufacturers. If the manufacturers start now, before many countries write national rules, they will make it difficult for countries to establish barriers to trade under the pretext of "safety."

However, if manufacturers miss this opportunity, there will soon be a number of independent national systems. This means that we risk a lot of duplicate testing with a few chemicals, and can impede the development of international systems for years to come.

References

[1] *ISO/IEC Guide 7, Requirements for Standards Suitable for Product Certification,* International Organization for Standardization, Geneve, Switzerland, 1982.
[2] *ISO/IEC Guide 2,* General terms and their definitions concerning standardization certification and testing laboratory accreditation, International Organization for Standardization, Geneve, Switzerland, 1986.
[3] *ISO/IEC Guide 42,* Guidelines for a step-by-step approach to an international certification system, International Organization for Standardization, Geneve, Switzerland, 1984.
[4] ISO: Compendium of certification documents, International Organization for Standardization, Geneve, Switzerland, 1985.
[5] DS/R 2321, Protective gloves against chemicals (Part 0: General information; Part 1: Terminology; Part 2: Classification, requirements, instruction and marking (re. single use gloves); Part 3: Test method for protection against chemicals; Part 4: Test methods for mechanical properties, etc.), Danish Standards Association, Hellerup, Denmark, 1985.
[6] Special rules for certification of single use gloves against chemicals, SBC 232, Danish Standards Association, Hellerup, Denmark, 1985 (in Danish).

Human Factors

Bjarne W. Olesen[1] and Francis N. Dukes-Dobos[2]

International Standards for Assessing the Effect of Clothing on Heat Tolerance and Comfort

REFERENCE: Olesen, B. W. and Dukes-Dobos, F. N., **"International Standards for Assessing the Effect of Clothing on Heat Tolerance and Comfort,"** *Performance of Protective Clothing: Second Symposium, ASTM STP 989,* S. Z. Mansdorf, R. Sager, and A. P. Nielsen, Eds., American Society for Testing and Materials, Philadelphia, 1988, pp. 17–30.

ABSTRACT: Three standards recently developed by the International Organization for Standardization (ISO) Working Group for Thermal Environments for assessing the thermal load of workers are examined for how they deal with the effect of clothing on the worker's feeling of thermal comfort or heat tolerance. In the standard for assessing thermal comfort, only the thermal insulation of the clothing worn is taken into account, but in the standard for analytical determination of heat stress, values for both the insulation and vapor permeability of the working uniform and protective garments are needed for estimating the worker's heat load. However, data on these thermal properties for protective garments are not available. In the third standard examined, the assessment of heat stress is based on measurement of environmental heat and estimation of metabolic heat, while clothing insulation and permeability are considered to be constant.

KEY WORDS: clothing insulation, vapor permeability, heat stress, thermal comfort, discomfort, heat stress standards, heat stress index, International Organization for Standardization (ISO), heat tolerance, Wet-Bulb Globe-Temperature (WBGT) index, Predicted Mean Vote (PMV) comfort index, Predicted Percentage of Dissatisfaction (PPD), climatic conditions

Recently, a series of international standards for assessing and keeping within safe limits the thermal load of workers at their job sites have been adopted by the Working Group for Thermal Environments that was convened by the International Organization for Standardization (ISO). These standards include methods to be used in moderate thermal environments (ISO 7730) and in hot environments (ISO 7243 and ISO DIS 7933).

The ISO 7730 describes how to determine whether a living or work space is thermally comfortable. This is important because thermal comfort is conducive to optimal performance in mental, skilled, and heavy physical work. Unfortunately, the cost is prohibitive when maintaining comfortable conditions in large work spaces, such as industrial plants, particularly during the winter and summer seasons and where the work is connected with heat generating or cooling processes. In cold environments, workers can protect themselves by wearing heavier clothing, up to the point where the garments start to interfere with the performance of necessary skills. In hot environments, however, the worker may tolerate

---

[1] Research associate, Technical University of Denmark, Laboratory of Heating and Air Conditioning, Copenhagen, Denmark.
[2] Adjunct professor of Occupational Health and Ergonomics, University of South Florida, College of Public Health, Tampa, FL 33612-4799.

more heat even after shedding all unnecessary garments, because sweat evaporation may provide enough cooling to maintain an acceptable body temperature. Such conditions, however, far excede thermal comfort, and the main concern is to protect the workers from the harmful consequences of excessive thermal exposures. The ISO 7243 and ISO DIS 7933 standards describe methods for assessing the heat load of jobs where the workers may be exposed to more heat than what can be safely tolerated. These standards also include safe exposure limits. While clothing is only one of several factors that have an impact on the heat load to which a person is exposed, all three of the ISO standards deal in some way with the effect of clothing on human heat tolerance.

Purpose

The purpose of this paper is to show how the effect of clothing is accounted for in these standards and how thermal comfort and heat tolerance is influenced by clothing in terms of these standards.

Methods for Including the Clothing Factors in the Standards

The Standard for Assessing Thermal Discomfort and Comfort (ISO 7730)

To quantify the degree of thermal discomfort, an index has been devised by Fanger [1] that gives the predicted mean vote (PMV) of a large group of subjects according to the following psycho-physical scale:

+3 hot
+2 warm
+1 slightly warm
 0 neutral
−1 slightly cool
−2 cool
−3 cold

The ISO standard, Moderate Thermal Environments—Determination of the Predicted Mean Vote (PMV) and Predicted Percentage of Dissatisfaction (PPD) Indices and Specification of the Conditions for Thermal Comfort (ISO 7730), contains an equation that combines all the six factors (air temperature, mean radiant temperature, air velocity, humidity, level of physical activity, and clothing) determining human response to the thermal environment.

$$PMV = (0.303e^{-0.036M} + 0.028)[(M - W) - 3.05 \cdot 10^3\{5733 - 6.99(M - W) - p_a\}$$

$$- 0.42\{(M - W) - 58.15\} - 1.7 \cdot 10^{-5}M(5867 - p_a) - 0.0014M(34 - t_a)$$

$$- 3.96 \cdot 10^{-8}f_{cl}\{(t_{cl} + 273)^4 - (\bar{t}_r + 273)^4\} - f_{cl}h_c(t_{cl} - t_a)]$$

$$t_{cl} = 35.7 - 0.028(M - W)$$

$$- 0.155I_{cl}[3.96 \cdot 10^{-8}f_{cl}\{(t_{cl} + 273)^4 - (\bar{t}_r + 273)^4\} + f_{cl}h_c(t_{cl} - t_a)]$$

$$h_c = \begin{cases} 2.38(t_{cl} - t_a)^{0.25} & \text{for } 2.38(t_{cl} - t_a)^{0.25} > 12.1\sqrt{V_{ar}} \\ 12.1\sqrt{V_{ar}} & \text{for } 2.38(t_{cl} - t_a)^{0.25} < 12.1\sqrt{V_{ar}} \end{cases}$$

$$f_{cl} = \begin{cases} 1.00 + 0.2I_{cl} \text{ for } I_{cl} < 0,5 \text{ clo} \\ 1.05 + 0.1I_{cl} \text{ for } I_{cl} > 0,5 \text{ clo} \end{cases}$$

where

M = metabolic rate, W/m²;
W = external work, W/m²;
p_a = water vapor pressure, Pa;
t_a = air temperature, °C;
f_{cl} = clothing area factor, nondimensional;
t_{cl} = mean clothing surface temperature, °C;
$\bar{t}_r$ = mean radiant temperature, °C;
h_c = convective heat transfer coefficient, W/m²°C;
I_{cl} = thermal insulation of clothing, clo (1 clo = 6.45 W/m²°C); and
V_{ar} = relative air velocity, m/s.

From these equations, the PMV can be calculated for different combinations of clothing as well as metabolic rate, air temperature, mean radiant temperature, air velocity, and air humidity. The equations for t_{cl} and h_c are solved by iteration. The I_{cl} values can be obtained from Tables 1 and 2. This standard also includes a computer program for calculating the PMV and the PPD values and some reference tables for PMV values when the operative temperature, the relative air velocity, the activity level of the workers, and I_{cl} value of their clothing are known.

The PMV index predicts the mean value of the thermal votes of a large group of people exposed to the same environment, and individual votes are scattered around this mean value. If the PMV value equals 0 or is close to 0, that means that the majority of the exposed persons will consider the environment comfortable. For estimating the number of people likely to feel uncomfortably warm or cool, the PPD index has been developed to predict the percentage of a large group of people likely to feel thermally uncomfortable; that is, voting hot ($+3$), warm ($+2$), cool (-2), or cold (-3) on the psycho-physical sensation scale.

When the PMV value has been determined, the PPD can be found from Fig. 1, or determined by use of the equation

$$PPD = 100 - 95 \times e^{-(0.33\,53 \times PMV^4 + 0.217\,9 \times PMV^2)}$$

Although the ideal PMV value is 0 because it is equivalent to thermal neutrality, in ISO 7730, the recommended limit for an acceptable thermal environment is 0 ± 0.5. As shown in Fig. 1, within these limits, the PPD will not exceed 10%.

For the purpose of keeping the standard as simple as possible, the assumption is made that all evaporation from the skin will be transported through the clothing to the environment and that the relative humidity of the ambient air is 50%. Therefore, this index is not applicable for hot or very humid conditions or both. Actually, it is valid only in the range between PMV values of -2 and $+2$, that is, in a moderate thermal environment where sweating is minimal.

TABLE 1—*Thermal insulation of work clothing and daily wear clothing ensembles (I_{cl}).*

Work Clothing	I_d clo	I_d m²°C/W
Underpants, boiler suit, socks, shoes	0.70	0.110
Underpants, shirt, trouser, socks, shoes	0.75	0.115
Underpants, shirt, boiler suit, socks, shoes	0.80	0.125
Underpants, shirt, trousers, jacket, socks, shoes	0.85	0.135
Underpants, shirt, trousers, smock, socks, shoes	0.90	0.140
Underwear with short sleeves and legs, shirt, trousers, jacket, socks, shoes	1.00	0.155
Underwear with short legs and sleeves, shirt, trousers, boiler suit, socks, shoes	1.10	0.170
Underwear with long legs and sleeves, thermo jacket, socks, shoes	1.20	0.185
Underwear with short sleeves and legs, shirt, trousers, jacket, thermojacket, socks, shoes	1.25	0.190
Underwear with short sleeves and legs, boiler suit, thermojacket + trousers, socks, shoes	1.40	0.220
Underwear with short sleeves and legs, shirt, trousers, jacket, thermojacket and trousers, socks, shoes	1.55	0.225
Underwear with short sleeves and legs, shirt, trousers, jacket, heavy quilted outer jacket and overalls, socks, shoes	1.85	0.285
Underwear with short sleeves and legs, shirt, trousers, jacket, heavy quilted outer jacket and overalls, socks, shoes, cap, gloves	2.00	0.310
Underwear with long sleeves and legs, thermo jacket + trousers, outer thermo jacket + trousers, socks, shoes	2.20	0.340
Underwear with long sleeves and legs, thermo jacket + trousers, Parca with heavy quilting, overalls with heavy quilting, socks, shoes, cap, gloves	2.55	0.395

Daily Wear Clothing	I_d clo	I_d m²°C/W
Panties, T-shirt, shorts, light socks, sandals	0.30	0.050
Panties, petticoat, stockings, light dress with sleeves, sandals	0.45	0.070
Underpants, shirt with short sleeves, light trousers, light socks, shoes	0.50	0.080
Panties, stockings, shirt with short sleeves, skirt, sandals	0.55	0.085
Underpants, shirt, light weight trousers, socks, shoes	0.60	0.095
Panties, petticoat, stockings, dress, shoes	0.70	0.105
Underwear, shirt, trouser, socks, shoes	0.70	0.110
Underwear, track suit (sweater + trousers) long socks, runners	0.75	0.116
Panties, petticoat, shirt, skirt, thick kneesocks, shoes	0.80	0.120
Panties, shirt, skirt, roundneck sweater, thick kneesocks, shoes	0.90	0.140
Underpants, singlet with short sleeves, shirt, trousers, V-sweater, socks, shoes	0.95	0.145
Panties, shirt, trousers, jacket, socks, shoes	1.00	0.155
Panties, stockings, shirt, skirt, vest, jacket	1.00	0.155
Panties, stockings, blouse, long skirt, jacket, shoes	1.10	0.170
Underwear, singlet with short sleeves, shirt, trousers, jacket, socks, shoes	1.10	1.170
Underwear, singlet with short sleeves, shirt, trousers, vest, jacket, socks, shoes	1.15	0.180
Underwear with long sleeves and legs, shirt, trousers, V-sweater, jacket, socks, shoes	1.30	0.200
Underwear with short sleeves and legs, shirt, trousers, vest, jacket, coat, socks, shoes	1.50	0.230

TABLE 2—*Thermal insulation of individual garments* (I_{clu}). (*To obtain* I_{cl} *values for clothing ensembles add up* I_{clu} *values.*)

Garment Description	Thermal Insulation, clo (I_{clu})	Garment Description	Thermal Insulation, clo (I_{clu})
Underwear		Thin sweater	0.20
Panties	0.03	Sweater	0.28
Underpants with long legs	0.10	Thick sweater	0.35
Singlet	0.04	Jackets	
T-Shirt	0.09	Light summer jacket	0.25
Shirt with long sleeves	0.12	Jacket	0.35
Panties + Bra	0.03	Smock	0.30
Shirts—Blouses		High Insulative, Fiber-pelt	
Short sleeves	0.15	Boiler suit	0.90
Light weight, long sleeves	0.20	Trousers	0.35
Normal, long sleeves	0.25	Jacket	0.40
Flannel shirt, long sleeves	0.30	Vest	0.20
Lightweight blouse, long sleeves	0.15	Outdoor clothing	
Trousers		Coat	0.60
Shorts	0.06	Down jacket	0.55
Light weight	0.20	Parca	0.70
Normal	0.25	Fiber-pelt overalls	0.55
Flannel	0.28	Sundries	
Dresses—Skirts		Socks	0.02
Light skirt (summer)	0.15	Thick ankle socks	0.05
Heavy skirt (winter)	0.25	Thick long socks	0.10
Light dress, short sleeves	0.20	Nylon stockings	0.03
Winter dress, long sleeves	0.40	Shoes (thin soled)	0.02
Boiler suit	0.55	Shoes (thick soled)	0.04
Sweaters		Boots	0.10
Sleeveless vest	0.12	Gloves	0.05

The Standard for Estimating Heat Stress Based on the WBGT Index (ISO 7243)

The ISO standard, Hot Environments—Estimation of the Heat Stress on Working Man, Based on the Wet Bulb-Globe Temperature (WBGT) Index (ISO 7243), is based on the work of Dukes-Dobos and Henschel [2] whose recommendations for safe exposure limits for heat stress were adopted by the American Conference of Governmental Industrial Hygienists (ACGIH) as a Threshold Limit Value (TLV) for heat stress in 1972 [3]. Table 3 shows the safe exposure limits of this ISO standard, which differ only slightly from the ACGIH TLV. What makes this standard attractive for practical use is its simplicity. Its application requires the assessment of the work metabolism and the measurement of only two environmental factors when used indoors or outdoors in shade, that is, the natural wet-bulb temperature (t_{nw}) and the globe temperature (t_g). If it is used for establishing the heat stress exposure for a person working outdoors in the sun, then it is necessary to measure the dry-bulb temperature (t_a) as well. The equations for calculating the WBGT index are: for inside buildings and outside without solar load

$$WBGT = 0.7t_{nw} + 0.3t_g$$

and for outside buildings with solar load

$$WBGT = 0.7t_{nw} + 0.2t_g + 0.1t_a$$

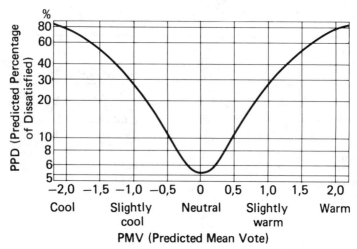

FIG. 1—*The relationship between the predicted percentage of thermally dissatisfied persons (PPD) and the predicted mean vote (PMV) indices.*

Since the natural wet-bulb thermometer is sensitive to air velocity, it is not necessary to measure air velocity.

As far as the effect of clothing on heat tolerance is concerned, there is a statement in the text of this standard that limits the use of this standard to conditions where the workers wear customary workers' uniforms that have an insulation value of 0.6 clo and are permeable to air and vapor. Furthermore, the standard stipulates that, if clothing is impermeable to water vapor, the reference values have to be lowered; and when wearing reflective clothing during exposure to radiant heat, the reference values should be increased. Because these instructions are very vague, the risk of over-exposure to heat may become significant when the reference values, no matter how slightly, are exceeded. The standard, therefore, advises that when the workers' heat exposure exceeds the reference values or when protective garments are worn, a more detailed analysis of the heat stress is necessary for determining whether a worker's exposure is within the safe limits.

The Standard for Analytical Determination of Thermal Stress (ISO DIS 7933)

The analytical approach to the development of a heat stress index requires the quantitative assessment of all important factors that influence the heat exchange between man and the environment, similar to the case of the PMV comfort index just described. However, under conditions of heat stress, the exposed person will sweat heavily. Therefore, the vapor permeability of the clothing and the vapor uptake capacity of the ambient air must be included in the calculation of total heat load. The index of required sweat (SW_{req}) established by Vogt, et al. [4] was developed by the ISO Working Group for Thermal Environments into an analytical heat stress standard, because it gives the cooling efficiency of sweating (η) and vapor resistance of the clothing (R_e) the necessary emphasis.

For calculating SW_{req} (W/m²), one has to assess quantitatively the activity level of the exposed person (M and W), the thermal insulation (I_{cl}) and vapor resistance (R_e) of the clothing, the four environmental parameters (t_a, $\bar{t}_r$, p_a, and V_{ar}), and the saturated vapor pressure at the skin ($p_{sk,s}$).

The M, W, and I_{cl} can be taken from tables available both in ISO 7243 and ISO/DIS 7933.

TABLE 3—*Safe exposure limits of the ISO 7243 heat stress standard, based on the wet-bulb globe-temperature (WBGT) index.*

Metabolic Rate Class	Metabolic rate, M — Related to a Unit Skin Surface Area, W/m²	Metabolic rate, M — Total (for a mean skin surface area of 1.8 m²), W	Reference Value of WBGT — Person Acclimatized to Heat, °C	Reference Value of WBGT — Person Not Acclimatized to Heat, °C
0 (resting)	M < 65	M < 117	33	32
1	65 < M < 130	117 < M < 234	30	29
2	130 < M < 200	234 < M < 360	28	26
3	200 < M < 260	360 < M < 468	26 (sensible air movement), 25 (no sensible air movement)	23 (sensible air movement), 22 (no sensible air movement)
4	M > 260	M > 468	25 (sensible air movement), 23 (no sensible air movement)	20 (sensible air movement), 18 (no sensible air movement)

The R_e can be calculated by the following equations

$$R_e = 1/16.7h_cF_{pcl}$$

$$F_{pcl} = 1/(1 + 0.92I_{cl})$$

where F_{pcl} is the moisture permeation efficiency factor [5]. The four environmental parameters must be measured where the worker performs the job. If the worker moves around between several sites that have different heat conditions, an hourly time-weighted average value should be calculated.

The value of $p_{sk,s}$ is dependent on the mean weighted skin temperature ($\bar{t}_{sk}$). However, the measurement of $\bar{t}_{sk}$ is a tedious procedure and requires special instrumentation. To avoid this measurement, this standard is based on the assumption that workers exposed to hot environments have a $\bar{t}_{sk}$ of 36°C that corresponds with a $p_{sk,s}$ value of 5.9 kPa.

The value of SW_{req} index is calculated by the following equations

$$SW_{req} = E_{req}/\eta$$

where E_{req} is the evaporation required for the maintenance of the body's thermal equilibrium.

$$E_{req} = M - W \pm C \pm R$$

where C and R are the heat exchanges by convection and radiation, respectively.

$$\eta = 1 - 0.51^{-6.6(1-w_{req})}$$

where w_{req} = required skin wettedness = E_{req}/E_{max}.

$$E_{max} = (p_{sk,s} - p_a)/R_e$$

where E_{max} is the maximum vapor uptake capacity of the ambient air.

$$R = h_rF_{cl}(\bar{t}_{sk} - \bar{t}_r)$$

$$C = h_cf_{cl}(\bar{t}_{sk} - t_a)$$

where F_{cl} is the clothing insulation factor and is calculated by the following equation

$$F_{cl} = 1/[(h_c + h_r)I_{cl} + 1/f_{cl}]$$

where

h_c = convective heat exchange coefficient, W/m²°C;
h_r = radiant heat exchange coefficient, W/m²°C;
I_{cl} = thermal insulation of clothing, m²°C/W; and
$\bar{t}_{sk}$ = mean weighted skin temperature, 36°C.

The values for h_c can be obtained by the following equations [6]

$$h_c = 3.5 + 5.2V_{ar}$$

if V_{ar} is less than 1 m/s.

$$h_c = 8.7\ V_{ar}^{0.6}$$

if V_{ar} is more than 1 m/s.

For calculating h_r, an equation is given in the standard that gives a linearized approximation

$$h_r = 4\sigma \cdot \epsilon_{sk} \cdot A_r/A_{DU}[(\bar{t}_r + \bar{t}_{sk})/2 + 273]^3$$

where

σ = universal radiation constant ($5.67.10^{-8}$ $Wm^{-2}K^{-4}$),

ϵ_{sk} = skin emissivity (0.97), and

A_r/A_{DU} = fraction of skin surface involved in radiant heat exchange: 0.77 standing, 0.70 seated, and 0.67 crouching.

To facilitate the practical application of this standard, it includes a computer program for a hand-held computer. Also included are recommended limit values in terms of maximum sweat rate (SW_{max}), skin wettedness (w_{max}), dehydration (D_{max}), and heat storage (O_{max}). Table 4 shows these limits with the addition of corresponding changes of rectal and skin temperature.

The Influence of Clothing on Thermal Comfort

Using the equation for calculating PMV, a diagram has been prepared (Fig. 2) that shows the optimal operative temperature[3] and the acceptable temperature range as a function of activity and clothing. Using this diagram, it is possible to find the influence of different

TABLE 4—*Recommended heat exposure limit values of the ISO heat stress draft standard, ISO DIS 7933, based on the required sweat rate (SW_{req}) index.*

Functions	Symbols	Units[a]	Not-acclimatized		Acclimatized	
			Warning	Danger	Warning	Danger
Maximum sweat rate	SW_{max} (resting)	W/m² (g/h)	100 (260)	150 (390)	200 (520)	300 (780)
	(working)	W/m² (g/h)	200 (520)	250 (650)	300 (780)	400 (1040)
Skin wettedness	W_{max}		0.85		1.00	
Dehydration	D_{max}	Wh/m² (g)	1000 (2600)	1250 (3250)	1500 (3900)	2000 (5200)
Heat storage	Q_{max}	Wh/m²	50	60	50	60
Increase rectal and skin temperature	t_{re}	°C	0.8	1	0.8	1
	t_{sk}	°C	2.4	3	2.4	3

[a] Estimation of sweat and dehydration in g/h and g, respectively, are based on an average person with 1.8 m² surface area.

[3] Operative temperature is the mean of air temperature and mean radiant temperature, weighted by h_c and h_r, respectively.

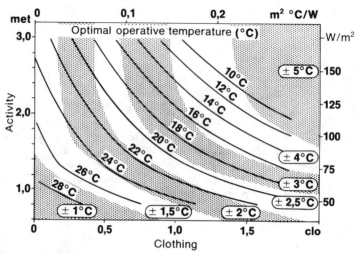

FIG. 2—*The optimal operative temperature (°C) (curved lines) corresponding to PMV = 0, as a function of activity and clothing levels. The shaded areas indicate the ±0.5 comfort range (Relative humidity 50% and no artificial air movement). The ± temperatures inserted in this diagram show the acceptable range of environmental temperatures.*

clothing ensembles on the optimal temperature (PMV = 0). By increasing insulation, that is, raising the clo value of the clothing, the optimal environmental temperature for comfort decreases, while the acceptable range of environment temperatures increases at a given activity level. For instance, in order to maintain thermal comfort at rest in the sitting position (activity level 1.0 met = 58 W/m²) where the operative temperature is 26°C, one has to wear a garment with an I_{cl} value of 0.5 clo corresponding to light summer clothing. If the operative temperature decreases to 21°C, one has to wear a garment with an I_{cl} value of 1.5 clo, corresponding to a heavy winter suit in order to maintain thermal comfort.

If the activity level increases to 2.0 mets, corresponding to walking at leisurely speed or performing light arm work in sitting position, one can feel thermal comfort in a 0.5 clo garment at an operative temperature of 21°C whereas in a 1.5 clo suit, one can still feel comfortable at 12°C. This means that an increase of clothing insulation from 0.5 clo to 1.5 clo will require a reduction of operative temperature from 26 to 21°C at 1 met activity level, and a reduction from 21°C to 12°C at 2 met activity level, in order to maintain a person's thermal comfort. However, the lower the environmental temperature, the higher will be the acceptable range. For instance, at 28°C, the range is ±1°C whereas at 14°C, it is ±4°C.

The Influence of Clothing on Heat Load

Using the equations for calculating SW_{req}, tables can be prepared showing the required sweat rates and skin wettedness values for different levels of clothing insulation at various work intensities, levels of relative humidity, operative temperature, and air velocity. Such tables can be used to determine the influence of clothing insulation on the heat load when working in hot environments.

One example is Table 5 that shows SW_{req} and w_{req} values for I_{cl} values from 0.1 to 1.0 clo at light work, under relative humidities from 20 to 80%, at operative temperatures from 25 to 35°C, and at air velocities from still air (less than 0.1 m/s) to 2.0 m/s. From this table, it becomes apparent that, in terms of total heat load (SW_{req}), reducing the insulation of the

TABLE 5—*Required sweat rate* (SW$_{req}$) *and required skin wettedness* (W$_{req}$) *values for light activity* (M = 115 W/m²) *at different combinations of clothing insulation* (I$_{cl}$), *relative humidity, operative temperature* (t$_o$) *and air velocity* (V$_a$).

Relative Humidity, %	Operative Temperature, t$_o$ (°C)	Air velocity, V$_a$, m/s, SW$_{req}$ (w$_{req}$)				
		<0.1	0.2	0.5	1.0	2.0
		CLOTHING, I$_{cl}$ = 0.1				
20	25	16 (0.04)	12 (0.02)			
20	30	61 (0.14)	58 (0.12)	52 (0.09)	41 (0.05)	26 (0.03)
20	35	106 (0.26)	106 (0.24)	105 (0.19)	103 (0.14)	100 (0.10)
20	40	153 (0.40)	155 (0.37)	159 (0.31)	165 (0.25)	176 (0.20)
20	50	296 (0.83)	286 (0.78)	284 (0.67)	300 (0.57)	333 (0.48)
20	60	* (1.00)	* (1.00)	* (1.00)	* (1.00)	* (1.00)
50	25	16 (0.04)	12 (0.03)			
50	30	61 (0.19)	59 (0.16)	52 (0.12)	41 (0.07)	26 (0.03)
50	35	107 (0.40)	106 (0.36)	105 (0.29)	103 (0.22)	100 (0.16)
50	40	174 (0.79)	169 (0.74)	165 (0.62)	168 (0.50)	177 (0.40)
80	25	16 (0.06)	12 (0.04)			
80	30	61 (0.28)	59 (0.25)	52 (0.18)	41 (0.11)	26 (0.05)
80	35	135 (0.87)	122 (0.80)	110 (0.64)	104 (0.49)	101 (0.36)
		CLOTHING, I$_{cl}$ = 0.5				
20	20	21 (0.06)	19 (0.05)	12 (0.03)		
20	25	50 (0.14)	49 (0.13)	44 (0.10)	37 (0.07)	29 (0.05)
20	30	80 (0.23)	79 (0.22)	76 (0.18)	73 (0.14)	68 (0.11)
20	35	110 (0.34)	109 (0.32)	109 (0.27)	108 (0.23)	107 (0.19)
20	40	141 (0.46)	141 (0.44)	142 (0.38)	144 (0.32)	147 (0.28)
20	50	250 (0.86)	238 (0.82)	226 (0.72)	226 (0.63)	232 (0.55)
20	60	* (1.00)	* (1.00)	* (1.00)	* (1.00)	* (1.00)
50	20	21 (0.07)	19 (0.06)	12 (0.03)	2 (0.00)	
50	25	50 (0.17)	49 (0.16)	44 (0.12)	37 (0.09)	29 (0.06)
50	30	80 (0.31)	79 (0.29)	76 (0.24)	73 (0.19)	68 (0.15)
50	35	111 (0.52)	111 (0.49)	110 (0.42)	109 (0.35)	103 (0.29)
50	40	203 (0.93)	179 (0.88)	157 (0.76)	151 (0.65)	151 (0.55)
80	20	21 (0.08)	19 (0.06)	12 (0.03)		
80	25	50 (0.22)	49 (0.20)	44 (0.15)	37 (0.11)	29 (0.07)
80	30	81 (0.47)	79 (0.43)	77 (0.36)	73 (0.29)	68 (0.22)
80	35	* (1.00)	* (1.00)	154 (0.92)	121 (0.77)	112 (0.64)
		CLOTHING, I$_{cl}$ = 1.0				
20	20	48 (0.16)	46 (0.15)	43 (0.12)	39 (0.10)	35 (0.08)
20	25	69 (0.24)	68 (0.23)	66 (0.19)	63 (0.16)	60 (0.14)
20	30	90 (0.33)	90 (0.31)	88 (0.27)	87 (0.24)	85 (0.21)
20	35	112 (0.43)	112 (0.41)	111 (0.36)	111 (0.32)	110 (0.28)
20	40	136 (0.55)	135 (0.53)	135 (0.47)	136 (0.42)	136 (0.37)
20	50	269 (0.94)	240 (0.90)	210 (0.82)	199 (0.73)	197 (0.66)
20	60	* (1.00)	* (1.00)	* (1.00)	* (1.00)	* (1.00)
50	20	48 (0.18)	46 (0.17)	43 (0.14)	39 (0.11)	35 (0.09)
50	25	69 (0.29)	68 (0.27)	66 (0.24)	63 (0.20)	60 (0.17)
50	30	91 (0.44)	90 (0.41)	89 (0.36)	87 (0.32)	85 (0.27)
50	35	117 (0.66)	116 (0.63)	114 (0.56)	112 (0.49)	111 (0.44)
50	40	* (1.00)	* (1.00)	202 (0.94)	162 (0.84)	149 (0.75)
80	20	48 (0.22)	46 (0.20)	43 (0.17)	39 (0.13)	35 (0.10)
80	25	69 (0.37)	68 (0.35)	66 (0.30)	63 (0.26)	60 (0.22)
80	30	94 (0.66)	93 (0.62)	90 (0.55)	88 (0.48)	86 (0.41)
80	35	* (1.00)	* (1.00)	* (1.00)	* (1.00)	179 (0.96)

worker's clothing will lower the total heat load to a much greater extent at high air velocities than at lower ones. Furthermore, an increase in air velocity does not reduce the total heat load when the operative temperature goes beyond skin temperature if the I_{cl} is 0.1 clo. This is reversed when I_{cl} is 0.5 clo or more. In other words, increasing the air velocity when the air temperature is above 36°C, will not reduce the worker's heat load when wearing shorts only but will be beneficial when a worker's clothing has an I_{cl} value of 0.5 or more.

Increasing the insulation of clothing does not always result in higher total heat load. For instance, when I_{cl} is 0.1 at a relative humidity of 20%, a t_0 of 50°C, and a V_a of 2 m/s, the value of SW_{req} will be 333, as can be seen in Table 5. Increasing the value of I_{cl} to 1.5 at the same environmental conditions, SW_{req} will be only 197, that is, the total heat load decreased by 136 W/m². Apparently, in this instance the higher clothing insulation has a protective effect against the heat uptake from the environment that is much hotter than the skin temperature. At the same time, the increased clothing insulation does not interfere with sweat evaporation due to high air velocity and low humidity.

Discussion

The most significant feature of ISO 7730 is the PMV index. To calculate this index one has to perform quite complex calculations. However, the diagram shown in Fig. 2, which is included in this standard, makes it easy to determine what the most comfortable temperature is for a given activity level and for a given level of clothing insulation. This diagram, when used together with Table 2, enables one also to predict how efficiently workers could compensate for a somewhat higher or lower thermostat setting than what is considered to be optimal, by putting on or taking off certain pieces of clothing, for example, a sweater or a jacket. This feature makes this standard very practical for finding the proper setting of the thermostat in an air-conditioned office or workshop during summer and winter to make sure that most employees will feel comfortable. However, for work spaces that are not air conditioned, this standard cannot be used most of the time because for the purpose of keeping this standard simple, its validity was limited to environments where the relative humidity is 50% and the air is not circulated artificially. To expand the utility of this standard, it would be necessary to modify its equation so that relative humidity would also be treated as a variable and the diagrams, such as Fig. 2, would have to be prepared for several levels of humidity and air velocity.

To promote practical applicability, the equations for ISO 7243 were kept very simple, but the validity of the reference values given in Table 3 are limited to conditions where the workers wear light ($I_{cl} = 0.6$) and vapor-permeable clothing. However, in many hot shops, workers have to wear protective garments (against splashes of molten metal or harmful chemical substances) that have a much higher I_{cl} value than 0.6 and they are not, or only slightly, permeable to water vapor, thus preventing sweat evaporation. In principle, ISO DIS 7933 makes it possible to overcome this problem because its equations for calculating the SW_{req} values for hot jobs take into consideration not only the work intensity and the climatic conditions but also clothing insulation (from Table 4, one can determine whether these SW_{req} values are within the safe limit). To make the calculations easier, ISO DIS 7933 also includes a computer program. Furthermore, the publication of more tables, such as Table 5 for various work intensities, makes it possible to establish the SW_{req} value of a job without performing the calculations, if the climatic conditions and the clothing insulation values are known.

Unfortunately, the insulation values and water vapor permeability of the protective garments are unknown, therefore, SW_{req} values cannot be calculated for jobs where such garments are worn. The ISO Working Group for Thermal Environments is now in the process of drafting yet another standard that will specify all the necessary measurements and calculations for assessing clothing insulation and water vapor permeability. The measurements and calculations for this purpose are quite complex and the equipment, an electrically heated copper manikin, is quite expensive. Such manikins are available only in a few specialized laboratories in the United States and abroad. Since there is no simple and inexpensive methodology known at present for those measurements, it is unlikely that many industries will install laboratories for assessing the thermal properties of protective garments worn by their workers in the near future. The ideal solution would be if the manufacturers of protective garments would measure these properties and mark these values on the garments' labels. Some manufacturers have already taken steps in this direction.

Conclusions

1. There are three international standards dealing with the thermal effects of clothing on the wearer.
2. The ISO 7730 standard describes a method for identifying conditions that make people feel thermally comfortable. It provides easily accessible information on the effect of clothing insulation on thermal comfort and includes tables presenting clothing insulation values for single garments as well as clothing ensembles. However, the validity of this standard is limited to conditions where the relative humidity is 50% and the clothing worn is water vapor permeable.
3. The ISO 7243 and ISO DIS 7933 standards describe methods for assessing the heat load of workers exposed to hot environments and provide safe exposure limits. The difference between these two standards is that ISO 7243 avoids the need for measuring clothing insulation while ISO DIS 7933 provides equations to accommodate values of clothing insulation and water vapor permeability. As a consequence, the applicability of ISO 7243 is limited to conditions where workers wear light and vapor-permeable garments, whereas ISO DIS 7933 can be used, in principle, for all combinations of clothing insulation and permeability.
4. Tables of insulation values for commonly worn garments are available, but such information is not available for protective garments. Values for the vapor permeability of these garments are also unavailable. As a result, ISO DIS 7933 has limited utility at the present time.
5. An ISO standard is now under development for the assessment of clothing insulation and water vapor permeability. However, the available methods for this purpose are complex and require expensive instrumentation. The ideal solution would be for protective garments to be labeled as to their thermal properties.
6. Data presented in terms of ISO 7730 and ISO DIS 7933 show that clothing insulation has a significant impact both on the feeling of thermal comfort and on the total heat load of workers exposed to hot environments.

References

[1] Fanger, P. O., "Thermal Comfort, Analyses and Applications," *Environmental Engineering*. Danish Technical Press, Copenhagen, 1970.
[2] Dukes-Dobos, F. N. and Henschel, A., "Development of Permissible Heat Exposure Limits for Occupational Work," *Journal,* American Society of Heating, Refrigeration and Air Conditioning Engineers, Vol. 15, No. 9, Sept. 1973, pp. 57–62.
[3] "Heat Stress," *Threshold Limit Values for Chemical Substances and Physical Agents in the Work Environment.* American Conference for Governmental Industrial Hygienist, published annually since 1972.

[4] Vogt, S. S., Candas, V., Libert, S. P. and Daull, F., "Required Sweat Rate as an Index of Thermal Strain in Industry," *Bioengineering, Thermal Physiology and Comfort,* K. Cena and S. A. Clark, Eds., Elsevier, Amsterdam, 1981.

[5] Nishi, Y. and Gagge, A. P., "Moisture Permeation of Clothing—A Factor Governing Thermal Equilibrium and Comfort," *Transactions,* American Society of Heating, Refrigeration and Air Conditioning Engineers, Vol. 76, Part 1, 1970, pp. 137–145.

[6] Missenard, F. A., "Coefficient d'echange de chaleur du corps humanin par convention, en fonction de la position, de l'activité du sujet et de l'environment," *Archives des Sciences Physiologiques,* Vol. 27, 1973, pp. A45–A50.

Sarah L. Cowan,[1] *Rosser C. Tilley,*[1] *and Mary Ellen Wiczynski*[1]

Comfort Factors of Protective Clothing: Mechanical and Transport Properties, Subjective Evaluation of Comfort

REFERENCE: Cowan, S. L., Tilley, R. C., and Wiczynski, M. E., "**Comfort Factors of Protective Clothing: Mechanical and Transport Properties, Subjective Evaluation of Comfort,**" *Performance of Protective Clothing: Second Symposium, ASTM STP 989*, S. Z. Mansdorf, R. Sager, and A. P. Nielsen, Eds., American Society for Testing and Materials, Philadelphia, 1988, pp. 31–42.

ABSTRACT: The purpose of this research was to determine the mechanical and transport properties contributing to fabric comfort for a series of nonwoven fabrics known to resist spray penetration and to relate those properties to the subjective evaluation of comfort.

Seven fabrics that may be used in protective apparel were evaluated for their comfort properties. One hundred percent cotton woven chambray with and without a fluorocarbon finish served as control fabrics. Five nonwoven test fabrics known to resist spray penetration were selected. Three spun-laced or spun-bonded test fabrics contained 100% olefin fibers with either a fluorocarbon finish or a polyethylene coating. Both woodpulp, polyester-blended nonwovens had a fluorocarbon finish. One woodpulp, polyester fabric also contained a flame retardant finish.

For each fabric, standard ASTM test methods were used to measure air and water vapor permeability, fabric weight, thickness, flexural rigidity, breaking load and elongation, and bursting and tear strengths. The Kawabata Evaluation System (KES) was used to provide an objective evaluation of hand characteristics. A panel of judges subjectively ranked the seven fabrics on smoothness, stiffness, stretchiness, thickness, weight, and hand.

The woodpulp, polyester nonwoven fabric with a commercial finish was found to be the nonwoven fabric having comfort characteristics closest to those of the chambrays. Among the nonwovens, it was ranked the highest for hand by the panel of judges while the olefin nonwoven fabric with the polyethylene coating was ranked as having the least desirable hand. The woodpulp, polyester nonwoven with a commercial finish was comparable to the chambray fabrics in terms of fabric stiffness. Both woodpulp, polyester fabrics had better air and water vapor transmission properties than did the other nonwoven fabrics evaluated.

According to the subjective evaluations, the stiffer and smoother the fabric, the less desirable the hand. The olefin nonwoven fabric with the polyethylene coating strongly exhibited both of these characteristics. In the subjective evaluation, it was ranked as having the least desirable hand. Both spun-bonded olefins had the lowest air and water vapor transport properties of the fabrics evaluated. Lack of these properties makes the olefin nonwovens a poor choice for comfortable protective apparel.

The nonwoven fabrics had significantly weaker tensile and bursting strength properties compared to the two chambray fabrics. There were no significant differences between wet and dry strengths of the nonwoven fabrics. Therefore, the presence of moisture would not be expected to cause premature fabric failure. The spun-bonded olefins were the most durable of the nonwoven fabrics evaluated.

KEY WORDS: comfort, disposable garments, hand, nonwoven fabrics, protective clothing, permeation, physical properties

[1] Assistant professor, graduate research, and graduate research assistant, respectively, Department of Clothing and Textiles, University of North Carolina, Greensboro, NC 27412.

Although the use of protective clothing by agricultural workers exposed to pesticides has been mandated by the U.S. Environmental Protection Agency [1,2], most agricultural workers are reluctant to wear apparel that offers adequate protection. Instead, workers prefer to wear woven chambray shirts and denim jeans [3] that are inadequate in preventing dermal exposure to pesticides [4,5]. Use of apparel that provides adequate protection to dermal pesticide exposure would probably increase if comfortable and affordable protective clothing was readily available. Disposable garments of nonwoven fabrics are potential alternatives.

An individual's perception of comfort is a subjective response to physical, physiological, and psychological variables [6]. Physical variables include not only environmental factors, but also fabric and garment characteristics such as air and water vapor permeability, durability, and fit. A key physiological variable affecting an individual's comfort status is level of physical activity. Psychological factors include the individual's general state of mind [7] and perceptions not only of style or fashion, but also of the appropriateness of the fabric for the intended end use [6].

Ideal apparel from a comfort standpoint facilitates vapor transmission. When sweating occurs, fabrics perceived as comfortable transfer moisture away from the skin and do not feel wet.

The properties associated with comfort of a cotton polyester blend were evaluated in a series of papers [8–10]. The woven fabric was subjected to a wide range of physical tests and subjective evaluations shown to measure components of comfort.

Although the literature on the properties associated with comfort of woven fabrics is extensive, a limited amount of work has focused on these properties in nonwoven fabrics. Branson, DeJonge, and Munson [11] examined these properties for a spun-bonded olefin by having human subjects wear coverall-type garments of the test fabric under conditions that simulated the typical temperature and relative humidity of a Michigan summer day.

The surface properties of a fabric are an important component of an individual's perception of comfort. The Kawabata method of hand evaluation [12] correlates objective data with subjective data to provide a quantitative method of evaluation. Other research efforts to correlate objective and subjective hand evaluations have been reported by Barker and Scheininger [13].

The woven fabrics widely used by agricultural workers as "protective" clothing are permeable to spray. Alternatives to such woven fabrics should be evaluated for properties associated with comfort. This research was designed to determine the mechanical and transport properties contributing to fabric comfort for a series of nonwoven fabrics known to resist spray penetration. A second objective of the research was to relate those properties associated with comfort to wearer perception of comfort. Hereafter, in this paper, those properties will be referred to as comfort properties or comfort descriptors.

Both objective and subjective fabric properties were determined. Standard test methods of textile evaluation were used. A panel of judges was selected to rank the fabrics on a series of fabric comfort descriptors. Correlation coefficients were developed comparing the objective and the subjective measurements.

Procedure

Fabrics

The five nonwoven test fabrics included:

1. Spun-laced woodpulp and polyester blend with both flame retardant and fluorocarbon finishes (woodpulp, polyester/FR, FC).

2. Spun-laced woodpulp and polyester blend with commercially applied fluorocarbon finish (woodpulp, polyester/COM).
3. Spun-bonded, melt-blown polypropylene fibers with fluorocarbon finish (polypropylene/FC).
4. Spun-bonded olefin with fluorocarbon finish (olefin/FC).
5. Spun-bonded olefin with polyethylene coating (olefin/PC).

The two control fabrics were 100% cotton chambray (65 by 42) with and without a fluorocarbon finish (chambray and chambray/FC) (see Table 1).

The nonwoven test fabrics were thinner and weighed less than the two chambray control fabrics (Table 2). The density of the woodpulp, polyester/COM fabric is closest to that of the chambrays. The least dense fabric was polypropylene/FC, and the fabric with greatest density was the olefin/PC.

Finishes

The fluorocarbon finish was applied to the chambray/FC, woodpulp, polyester/FR, FC, and polypropylene/FC fabrics by an area textile manufacturer using plant spraying and drying equipment. Since the olefin/FC fabric did not tolerate the elevated drying temperatures, a fluorocarbon finish in spray cans was applied by hand and allowed to air dry. The woodpulp, polyester/COM nonwoven fabric has a fluorocarbon finish that was applied by the textile converter who markets the product. The flame retardant finish was already on the woodpulp, polyester/FR, FC fabric when it was donated for the project. The polyethylene coating (olefin/PC) also was applied by the supplier of the textile fabric to the research project.

Test Methodology

Physical and mechanical testing procedures followed standard American Society for Testing and Materials (ASTM) test methods. Air and water vapor permeability were assessed according to ASTM Test Method for Air Permeability of Textile Fabrics (D 737-75) and ASTM Test Methods for Water Vapor Transmission of Materials in Sheet Form (E 96-66). Fabric weight, thickness, and flexural rigidity were determined in accordance with the ASTM

TABLE 1—*Description of fabrics.*

Designation	Fiber Content	Method of Fabrication	Finish
	CONTROL FABRICS		
Chambray	cotton	woven	none
Chambray/FC	cotton	woven	fluorocarbon
	TEST FABRICS		
Woodpulp, polyester/ FR, FC	woodpulp, polyester	spun-laced	flame retardant, fluorocarbon
Woodpulp, polyester/ COM	woodpulp, polyester	spun-laced	fluorocarbon
Polypropylene/FC	melt-blown polypropylene	spun-bonded	fluorocarbon
Olefin/FC	olefin	spun-bonded	fluorocarbon
Olefin/PC	olefin	spun-bonded	polyethylene coating

TABLE 2—*Fabric thickness, weight, and density.*

| | Thickness, mm | Weight,[a] | | Density, kg/m³ |
		g/m²	(oz/yd²)	
Chambray	0.35	130	(3.840)	362
Chambray/FC	0.30	122	(3.612)	395
Woodpulp, polyester/FR, FC	0.23	99	(2.926)	431
Woodpulp, polyester/COM	0.20	76	(2.240)	380
Polypropylene/FC	0.25	51	(1.509)	205
Olefin/FC	0.10	48	(1.417)	480
Olefin/PC	0.13	82	(2.423)	632

[a] Oz/yd² is equivalent to 0.0295 g/m².

Test Methods for Mass per Unit Area (Weight) of Woven Fabric (D 3776-79), the ASTM Test for Measuring Thickness of Textile Materials (D 1777-64), and the ASTM Test Methods for Stiffness of Fabrics (D 1388-64). The Kawabata Evaluation System (KES) [12] was used to measure fabric properties identified as components of fabric hand. The wet and dry strength characteristics of the fabrics were determined using the ASTM Test Method for Hydraulic Bursting Strength of Knitted Goods and Nonwoven Fabrics: Diaphragm Bursting Strength Tester Method (D 3786-80a) and ASTM Test Methods for Breaking Load and Elongation of Textile Fabrics (D 1682-64).

Analysis of variance was used to evaluate the mechanical and transport properties. Results were considered significant at or above the 95% confidence level. Utilization of the *post hoc* Tukey's honestly-significant-difference test located significant differences (at the 0.05 level) among fabric types.

A panel of judges composed of 20 college students participated in a blind hand evaluation of the fabrics. Students were blindfolded and asked to rank the fabrics on smoothness, stiffness, stretchiness, weight, thickness, and hand. The order of fabric presentation was strictly randomized. For each characteristic, the rank order ranged from one for the fabric that was smoothest, most flexible, most stretchy, lightest, thinnest, or had the most preferred hand to seven for the fabric that was the roughest, most stiff, least stretchy, heaviest, thickest, or had the least preferred hand.

Results and Discussion

Tensile Strength

The evaluation of fabric strength characteristics is important since rupture of the fabric would facilitate spray penetration. Because protective apparel may become wet during use due to perspiration, inclement weather, or an accidental spill, the wet strength characteristics are as important as the dry strength characteristics in evaluating the appropriateness of the fabrics for use in protective apparel. Only the chambrays demonstrated a difference between wet and dry tensile strengths, with the wet fabrics being significantly stronger than the dry fabrics (Table 3). The dry tensile strength of the chambray fabrics was at least three times greater than that of the nonwoven fabrics in the warp or machine direction. The wet tensile strength of the chambrays was four times greater.

The tensile strength differential among all fabrics tested was less marked but nevertheless significant in the filling or cross-machine direction. Among the nonwovens, the woodpulp, polyester/FR, FC was the strongest in the machine direction while the two olefin fabrics were strongest in the cross-machine direction. If the nonwovens are designed for single use only, the observed lack of durability may not be an important factor.

Bursting Strength

The bursting strength of the olefin/FC, both wet and dry, was superior to the bursting strength of the other nonwoven fabrics (Table 3). The polyproplyene/FC fabric had the weakest bursting strength. There were no statistically significant differences between the wet and the dry bursting strength characteristics of each fabric evaluated.

Permeability

The ability to transfer air and water vapor is an important component of a comfortable fabric (Table 4). Neither olefin nonwoven fabric allowed passage through the fabric of a measurable amount of air. Among the nonwovens, the woodpulp, polyester fabrics permitted the greatest passage of air. However, it was less than one-half the amount of air that passed through the chambrays under the same conditions.

There was no statistically significant difference between the water vapor permeabilities of the chambrays and either the woodpulp, polyester/COM or the polypropylene/FC nonwoven (Table 4). The chambrays and the woodpulp, polyester fabrics contain hygroscopic cellulose fibers that could facilitate the movement of water vapor. Because the density of the polypropylene/FC was the lowest of all fabrics evaluated (Table 2), the spaces between individual fibers may have facilitated water vapor movement.

The woodpulp, polyester/FR, FC fabric allowed the greatest water vapor movement of any fabric evaluated. The presence of hygroscopic cellulose fibers along with the flame retardant finish appear to have improved the transfer of water vapor. The two olefin fabrics had the lowest water vapor permeability. The olefin/PC allowed only a negligible amount of water vapor to pass through.

Flexibility

The flexibility of a fabric makes a significant contribution to comfort. The chambrays and the nonwovens had low levels of percent elongation in the warp or machine direction (Table 5). However, in the cross-machine direction, the extensibility of both woodpulp, polyester blends increased more than two-fold over the extensibility in the machine direction. This cross-machine extensibility may contribute to a comfortable fit for protective clothing at points of stress such as the knee or elbow.

Only the woodpulp, polyester/COM fabric had a flexural rigidity value comparable to that of the chambrays (Table 6). The olefin/PC fabric was at least twice as stiff as the other nonwovens and more than four times stiffer than the chambrays and the woodpulp, polyester/COM.

Kawabata Evaluation System

The Kawabata Evaluation System (KES) was developed in Japan for the purpose of objectively measuring fabric hand. Tensile, bending, shear, compression, and surface properties are measured with the KES precision instrumentation. There is an important difference between this system of measurement and other similar methods. The KES properties are measured at low degrees of deformation. Such low deformation levels are believed to duplicate deformations produced when judging the hand of a fabric.

The tensile, bending, and surface KES properties were selected for evaluation. Each is related to the perception of comfort.

KES mean deviation of surface roughness was said to characterize the fabric surface. Surface roughness is far greater for the chambrays than it is for any of the nonwovens (Table 7).

TABLE 3—Fabric wet and dry strength characteristics.

| | Breaking Load, kg (lbs)[a] | | | | Bursting Strength, kg/cm² (psi) | |
| | Warp[b] | | Filling[b] | | | |
	Dry	Wet	Dry	Wet	Dry	Wet
Chambray	26.56 (59.02)	31.96 (71.02)	12.70 (28.22)	14.28 (31.74)	7.04 (102.0)	7.57 (109.7)
Chambray/FC	25.53 (56.74)	28.30 (62.88)	13.06 (29.02)	15.08 (33.50)	8.00 (116.0)	8.62 (124.9)
Woodpulp, polyester/FR, FC	7.80 (17.34)	6.69 (14.86)	3.19 (7.08)	2.86 (6.36)	3.38 (49.0)	3.47 (50.3)
Woodpulp, polyester/COM	6.44 (14.30)	6.05 (13.44)	2.68 (5.96)	2.75 (6.10)	3.04 (44.0)	3.14 (45.5)
Polypropylene/FC	3.47 (7.72)	3.33 (7.40)	3.32 (7.38)	3.47 (7.70)	2.70 (39.2)	2.72 (39.4)
Olefin/FC	4.45 (9.88)	4.53 (10.06)	5.09 (11.30)	4.55 (10.12)	4.93 (71.5)	5.20 (75.3)
Olefin/PC	4.94 (10.98)	5.44 (12.08)	4.99 (11.08)	4.99 (11.08)	3.74 (54.2)	4.70 (68.1)

[a] The breaking load values for 1-in. widths of fabric.
[b] In the case of nonwovens, the machine direction is synonymous with the warp direction; the cross-machine direction, with the filling direction.

TABLE 4—*Fabric air and water vapor permeability.*

	Air		Water Vapor		
	$cm^3/cm^2 \cdot s$	Tukey's Multiple Range Grouping[a]	$g/m^2/24\ h$	Tukey's Multiple Range Grouping[a]	
Chambray	77.724	A	5.85	A	B
Chambray/FC	105.766	B	5.93	A	B
Woodpulp, polyester/ FR, FC	32.197	C	6.32	C	...
Woodpulp, polyester/ COM	35.865	C	6.06	C	B
Polypropylene/FC	18.451	D	5.72	A	...
Olefin/FC	<0.442	E	4.51	D	...
Olefin/PC	<0.442	E	0.14	E	...

[a] Means with the same letter are not significantly different at the 0.05 level.

The bending properties (Table 7) were evaluated and compared with the ASTM test measurements of flexural rigidity (Table 6). With the exception of the woodpulp, polyester/ COM fabric, the nonwovens as a group were much stiffer (had a greater bending rigidity) than the chambrays. The woodpulp, polyester/COM is the most flexible of the nonwovens but is stiffer than the chambrays. In the ASTM evaluation of fabric stiffness, the woodpulp, polyester/COM fabric was comparable to the chambrays.

The KES tensile characteristics show differences among the fabrics (Table 8). The work of tensile deformation was greater for the chambray fabrics that also displayed the greatest percent elongation. The weakest fabric was the olefin/PC that also had the lowest percent elongation.

Among all the fabrics tested, the olefin/PC and polypropylene/FC fabrics had the greatest percent resilience in the machine direction. In the cross-machine direction, the olefin/PC had the greatest percent resilience. The least resilient fabric in both the machine and the cross-machine directions was the woodpulp, polyester/COM fabric.

Subjective Evaluations

A panel of judges selected from the student body at the University of North Carolina at Greensboro was asked to rank the seven fabrics on a series of attributes. These characteristics

TABLE 5—*Fabric percent elongation.*

	Elongation, %			
	Warp[a]		Filling[a]	
	Dry	Wet	Dry	Wet
Chambray	22.4	25.0	16.8	18.0
Chambray/FC	14.8	19.4	13.2	19.5
Woodpulp, polyester/FR, FC	27.1	35.4	84.7	94.1
Woodpulp, polyester/COM	28.2	31.9	92.4	98.5
Polypropylene/FC	22.7	23.7	28.8	30.5
Olefin/FC	10.9	15.2	12.9	14.1
Olefin/PC	15.0	14.8	20.2	18.7

[a] For nonwovens, the machine direction is synonymous with the warp direction; the cross-machine direction, with the filling direction.

TABLE 6—*Fabric flexural rigidity.*

	Overall Flexural Rigidity, g · cm	Tukey's Multiple Range Grouping[a]
Chambray	75.61	A
Chambray/FC	71.41	A
Woodpulp, polyester/FR, FC	210.73	B
Woodpulp, polyester/COM	56.47	A
Polypropylene/FC	163.55	B
Olefin/FC	222.21	B
Olefin/PC	410.66	C

[a] Means with the same letter are not significantly different at the 0.05 level.

(smoothness, stiffness, stretchiness, thickness, and weight) are believed to contribute to fabric hand. All judges were enrolled at the University in the School of Home Economics. Although many panelists had taken courses in clothing and textiles, they are believed to be more representative of the typical consumer than of an expert in the field of fabric hand evaluation. Consequently, their evaluations should be a fair representation of the evaluation a typical consumer might give.

Participants first ranked the fabrics for each characteristic. Then each student was asked to rank the fabrics according to overall fabric hand. They were instructed that the fabrics might be used in apparel and told to rate the fabrics according to their preference for the hand or how it feels. The fabric hand most preferred was assigned a one by the test administrator and the fabric hand least preferred was assigned a seven, with the other fabrics ranked in between. For each characteristic, the mean scores were computed per fabric and then the fabric mean scores were ranked from 1 to 7.

In terms of hand, the chambrays were the fabrics most preferred and the olefin/PC fabric was least preferred (Table 9). The woodpulp, polyester/COM fabric was thought to be most nearly like the chambrays in terms of hand and stretchiness while the polypropylene/FC was ranked closest to the two chambrays in terms of stiffness and smoothness.

Of the fabrics evaluated, the olefin/PC was judged to be the least like the two chambrays. For example, the chambrays were ranked 1 and 2 as the fabrics having the most preferred hand, the most flexibility, and the greatest degree of stretch while the olefin/PC was ranked as the fabric having the least preferred hand, the greatest stiffness, and the least degree of

TABLE 7—*Fabric stiffness and surface characteristics (Kawabata descriptors).*[a]

	Bending Rigidity, gf · cm²/cm (B)[b]	Surface Roughness, μm (SMD)[b]
Chambray	0.0311	9.87
Chambray/FC	0.0473	11.14
Woodpulp, polyester/FR, FC	0.2504	3.54
Woodpulp, polyester/COM	0.0629	2.71
Polypropylene/FC	0.1229	3.83
Olefin/FC	0.2006	3.38
Olefin/PC	0.2985	1.78

[a] All values represent the mean of five warp (machine direction) and five filling (cross-machine direction) measurements.
[b] Kawabata reference terminology.

TABLE 8—*Fabric strength characteristics (Kawabata descriptors).*[a]

	Tensile Energy, gf · cm/cm² (WT)[b]		Elongation, % (EMT)[b]		Resilience, % (RT)[b]	
	Warp	Filling	Warp	Filling	Warp	Filling
Chambray	0.3500	0.3925	3.725	3.200	62.13	66.21
Chambray/FC	0.1825	0.2375	1.175	1.375	72.53	68.40
Woodpulp, polyester/ FR, FC	0.0825	0.1475	2.988	2.450	72.86	60.98
Woodpulp, polyester/ COM	0.0975	0.3225	1.788	3.675	58.89	52.73
Polypropylene/FC	0.1000	0.1400	1.463	2.013	87.36	71.19
Olefin/FC	0.0775	0.1100	1.613	1.725	84.18	66.17
Olefin/PC	0.0575	0.0750	1.088	1.163	87.26	82.26

[a] For nonwovens, the machine direction is synonymous with the warp direction; the cross-machine direction, with the filling direction.
[b] Kawabata reference terminology.

stretch. The olefin/PC was judged the smoothest fabric; the chambrays, the roughest. The chambrays were ranked as the most flexible and stretchy while the olefin/PC was ranked at the opposite end of the scale on these characteristics. The olefin/FC was judged to be the thinnest; the woodpulp, polyester/FR, FC, the thickest.

These subjective rankings are in agreement with the objective evaluations. The olefin/PC had the greatest overall flexural rigidity (Table 6) and had the least surface roughness (Table 7) of the fabrics evaluated.

Correlations

Pearson correlation coefficients were computed for all of the subjective hand evaluations (Table 10). Smoothness, stiffness and stretchiness all had a significant correlation with hand. The respondents did not equate smooth hand with preferred hand: the fabrics that were perceived to be smoothest were not the fabrics having the preferred hand. For example, the olefin/PC fabric was ranked as the smoothest fabric (rank of 1) while the two chambrays were ranked as the roughest fabrics (ranks of 6 and 7). However, the chambrays were the

TABLE 9—*Mean score—subjective ranking of fabrics.*[a]

	Hand	Smoothness	Stiffness	Stretchiness	Thickness	Weight
Chambray	1	6	1	1	5	5
Chambray/FC	2	7	2	2	3	4
Woodpulp, polyester/ FR, FC	5	4	5	5	7	6
Woodpulp, polyester/ COM	3	3	4	3	2	3
Polypropylene/FC	4	5	3	4	6	1
Olefin/FC	6	2	6	6	1	2
Olefin/PC	7	1	7	7	4	7

[a] A low numeral score = most preferred hand, smoothest, most flexible, stretchiest, thinnest, or lightest in weight. The mean scores were ranked in order from one to seven. The rank order value is reported.

TABLE 10—*Pearson correlation coefficients—subjective hand evaluations.*

	Hand	Smoothness	Stiffness	Stretchiness	Thickness	Weight
Hand	1.00	...	...	...	...	...
Smoothness	−0.84[a]	1.00	...	...	...	...
Stiffness	0.99[a]	−0.89[a]	1.00	...	...	...
Stretchiness	0.96[a]	−0.84[a]	0.94[a]	1.00	...	...
Thickness	−0.23	0.26	−0.19	−0.24	1.00	...
Weight	0.31	−0.35	0.39	0.26	0.47	1.00

[a] Significant at the 0.05 level.

fabrics with the most preferred hand, and the olefin/PC had the least preferred hand. The paired characteristics of smoothness and stiffness, smoothness and stretchiness, and stiffness and stretchiness had significant correlation coefficients.

Comparisons were made between the subjective hand evaluations and specific objective physical and mechanical measures of hand (Table 11). The comparisons were between single subjective and objective measurements. At no time were two objective measurements submitted simultaneously for correlation analysis. The subjective rankings of hand and smoothness had significant correlations with the Kawabata measure of geometrical roughness. Both hand and stiffness had significant correlations with flexural rigidity as measured by the ASTM standard test method. However, there was no corresponding level of significance for the correlation coefficients between the hand and stiffness subjective rankings and the Kawabata

TABLE 11—*Pearson correlation coefficients—subjective versus objective evaluations.*

	Subjective Evaluation	
Objective Evaluation	Hand	Other Subjective Evaluations
		Smoothness
Smoothness		
Kawabata, geometrical roughness	−0.83[a]	0.80[a]
		Stiffness
Stiffness		
ASTM, flexural rigidity	0.89[a]	0.90[a]
Kawabata, bending rigidity	0.95[a]	0.95[a]
		Stretchiness
Stretchiness		
ASTM, % elongation W[b]	−0.36	−0.52
ASTM, % elongation F[b]	0.02	−0.19
		Thickness
Thickness		
ASTM	−0.93[a]	0.52
Kawabata	−0.93[a]	0.20
		Weight
Weight		
ASTM	−0.61	0.52
Kawabata	−0.65	0.46

[a] Significant at the 0.05 level.
[b] In the case of the nonwovens, the machine direction is synonymous with the warp direction; the cross-machine direction, with the filling direction.

bending rigidity values. The thickness and weight subjective rankings did not have a significantly high correlation with either the ASTM or Kawabata data on thickness and weight.
The overall hand preference ranking and ASTM and Kawabata data on thickness had
significant correlation coefficients. Hand preference ranking also correlated well with the
Kawabata method of evaluating weight. The statistical analysis suggests that subjective
rankings of fabric weight and thickness are not good predictors of actual fabric weight and
thickness. In both cases, the correlation coefficients were not significant. Fabric weight and
thickness (determined objectively) do have a strong relationship with subjective overall hand
preference ranking.

The panel of judges was unable to differentiate between the fabrics on the attributes of
stretchiness, thickness, and weight. The correlation coefficients between the subjective and
objective evaluations of these attributes were not significant (Table 11). However, the
subjective evaluation of hand *per se* correlates well with objective measures of smoothness,
stiffness, and thickness.

Conclusions

The nonwoven fabrics were significantly weaker in strength characteristics than were the
chambray fabrics. Because the nonwovens would be expected to be single-use items, this
lack of strength may not be critical. Of the nonwovens evaluated, the spun-bonded olefins
were the most durable in the cross-machine direction while the woodpulp, polyester/FR,
FC was the strongest in the machine direction. No differences in wet and dry strengths were
observed for each nonwoven fabric. Consequently, the nonwoven fabrics in the study would
not be expected to fail more readily when subjected to moisture from perspiration, inclement
weather, or a liquid spill.

The olefin fabrics allowed no measurable amount of air to permeate the fabric substrate
and allowed the least amount of water vapor to pass through. Since both air and water
vapor permeability are important components of apparel comfort, the olefin fabrics may
not be comfortable to wear.

The woodpulp, polyester fabrics are expected to be the most comfortable nonwoven fabrics
in terms of air and water vapor permeability. The woodpulp and polyester fabrics, while
not as air permeable as the chambrays, had greater air permeability than did the other
nonwoven fabrics. The polypropylene and woodpulp, polyester fabrics had higher water
vapor permeability compared to the other nonwoven fabrics.

The woodpulp, polyester/COM fabric stiffness was comparable to that of the chambrays.
The olefin fabrics were significantly stiffer than the other nonwovens and the chambrays.
In the subjective rankings by the panel of judges, the stiffer the fabric the less desirable the
fabric was in terms of hand preference.

According to the Kawabata evaluations, the nonwoven fabrics are smoother and stiffer
than are the chambrays. The differences in both characteristics may be contributing factors
in the general reluctance to wear nonwoven fabric protective apparel. If the properties of
chambray are assumed to be preferable, nonwovens designed to be less stiff and smooth
may be more acceptable in the marketplace.

The chambray fabrics were the most preferred in terms of fabric hand while the olefin/
PC was the least preferred fabric. Among the nonwoven fabrics, the woodpulp, polyester/
COM fabric had the best hand and was most like the chambrays in terms of stretchiness.
The polypropylene/FC was judged to be most nearly like chambray for stiffness and smooth-
ness. The olefin/PC was thought to be least like chambray in the subjective testing.

The attributes of smoothness and stiffness are important components in the subjective
evaluation of hand. Nonwoven fabrics that are similar to chambray in degree of smoothness

and stiffness would be expected to be more acceptable to the individual who is required to wear protective apparel.

Acknowledgments

This study was supported by the North Carolina Agricultural Research Services. Test fabrics were donated by Cone Mills Corporation, E. I. DuPont de Nemours, Durafab, and Surgikos. Burlington Industries allowed the authors access to testing equipment. Culp Industries donated and applied the fluorocarbon finish.

This paper is registered as Paper No. 11406 of the Journal Series of the North Carolina Agricultural Research Services, Raleigh, NC 27695-7601.

References

[1] *Federal Register,* Vol. 39, 1974, pp. 16888–16891.
[2] Ezell, D. A., "Pesticide Protection of Workers Reviewed," *The Packer,* 23 Feb 1985, p. 3a.
[3] DeJonge, J. O., Vredevoogd, J., and Henry, M. S., *Clothing and Textiles Research Journal,* Vol. 2, 1983–1984, pp. 9–14.
[4] DeJonge, J. O., "Clothing as a Barrier to Pesticide Exposure," presented at the National Meeting of the Chemical Health and Safety Division, American Chemical Society, March 1983.
[5] Hobbs, N. E., Oakland, B. G., and Hurwitz, M. D., in *Performance of Protective Clothing, ASTM STP 900,* R. L. Barker and G. C. Coletta, Eds., American Society for Testing and Materials, Philadelphia, 1986, pp. 151–161.
[6] Pontrelli, G. J., "Partial Analysis of Comfort's Gestalt." *Clothing Comfort,* N. R. S. Hollies and R. F. Goldman, Eds., Ann Arbor Science, Ann Arbor, 1977, pp. 71–80.
[7] Vanderpoorten, A., "Use of Clothing and Perceptions of Thermal Comfort in the Home Environment," *Dissertation Abstracts International,* Vol. 42, 1981, p. 4759B; University Microfilms No. 8209007.
[8] DeMartino, R. N., Yoon, H. N., and Buckley, A., *Textile Research Journal,* Vol. 54, 1984, pp. 602–613.
[9] Yoon, H. N. and Buckley, A., *Textile Research Journal,* Vol. 54, 1984, pp. 289–298.
[10] Yoon, H. N., Sawyer, L. C., and Buckley, A., *Textile Research Journal,* Vol. 54, 1984, pp. 357–365.
[11] Branson, D. D., DeJonge, J. O., and Munson, D., *Textile Research Journal,* Vol. 56, 1986, pp. 27–34.
[12] Kawabata, S., *The Standardization and Analysis of Hand Evaluation,* 2nd ed., The Textile Machinery Society of Japan, Osaka, 1980.
[13] Barker, R. L. and Scheininger, M. M., *Textile Research Journal,* Vol. 52, 1982, pp. 615–620.

Eva Mauritzson-Sandberg[1] and Lennart Sandberg[1]

Evaluation of Psychological Reactions in Children When Using Respiratory Protective Devices

REFERENCE: Mauritzson-Sandberg, E. and Sandberg, L., **"Evaluation of Psychological Reactions in Children When Using Respiratory Protective Devices,"** *Performance of Protective Clothing: Second Symposium, ASTM STP 989,* S. Z. Mansdorf, R. Sager, and A. P. Nielsen, Eds., American Society for Testing and Materials, Philadelphia, 1988, pp. 43–49.

ABSTRACT: Three experiments were carried out with the aim of evaluating psychological reactions in children (0 to 7 years) when using respiratory protective devices (PD). Psychological reactions like uncertainty, anxiety, claustrophobia, and physiologic distress are linked with clothing, design, and technical conditions. From a psychological point of view, aspects connected with the possibility of satisfying psychosocial requirements such as communication, supervision, and nursing are equally important as clothing, design, and technique when evaluating the effectiveness of the PD. Results show that reliability and confidence in the PD are important aspects when parents (adults) chose PDs for children. If two protective devices are technically equal, the one that is shown to be most reliable will be chosen.

KEY WORDS: psychological reaction, protective devices, respiratory protection, personal protective equipment, clothing, evaluation, children, protective clothing

The use of protective devices (PD), particularly respiratory protective devices, highlights a number of technical and design problems as well as psychological and personal problems. These problems, taken individually or together, may create negative conditions (that is, anxiety, claustrophobia, and hyperventilation) that can thwart the use of the PD even though the PD is close to technically perfect. Furthermore, the problems may be of a more specific nature in that some problems are related to a particular age group. One can not assume that the juvenile population is comparable to the adult population. What will evoke emotional reactions in a child may be completely unimportant to an adult. Since the juvenile population constitutes a vital part of the total population and must be protected against environmental hazards, it is of great importance to evaluate what aspects associated with PD will create reactions of uncertainty and fear. In an attempt to study some of these problems, the Swedish Civil Defense Administration and The University of Umeå, Department of Psychology, have for the last five years cooperated in research studying and evaluating psychological reactions associated with the use of PDs for children.

Theoretical Background

The theoretical foundation [1] of this research is built on the insights of developmental psychology made by cognitive and behavioral theories represented by Piaget [2] and Bowlby [3] and on psycho-dynamic theories like Freud's [4]. All these theories point out the critical

---

[1] Psychologists, Department of Psychology, University of Umeå, Sweden.

periods in the cognitive and emotional development of the child. One such critical period takes place by the age of four when separation from the mother (caretaker) is completed. In the cognitive area, these theories show that logical thinking and reasoning starts to develop in preschool age and reaches its full development in adolescence. This means that one can not motivate a child to wear a PD on purely logical grounds. The PD must be accepted on emotional grounds to be accepted and used.

The theoretical foundation may be summarized as follows:

(*a*) a child is not an adult in miniature,
(*b*) a child has a limited capacity to understand logical reasons,
(*c*) children are extremely sensitive to emotional reactions (their own as well as others), and
(*d*) children need straight and undisturbed communication with parents. This is particularly important in an actual emergency.

The main purpose of the research was to evaluate what kinds of emotional reactions different age groups of children will experience when confronted with different types of PD. Two different situations were used:

(*a*) the dressing procedure can generate emotional reactions from the sight of the PD and from imagining what it is like to wear the particular PD, and
(*b*) wearing the PD can generate emotional reactions including design and technical aspects such as the clothing itself, fit, and restriction of communication.

FIG. 1—*Protective Jacket 36.*

TABLE 1—*The total number of children in different age groups who accept and refuse to wear the PD.*

| | Age Group | | | | | | Total, |
| | 2 to 3, | | 3 to 4, | | 4 to 7, | | |
	n	*(%)*	*n*	*(%)*	*n*	*(%)*	*n*
Accepts	3	(9)	9	(36)	44	·(69)	56
Refuses	30	(91)	16	(64)	20	(31)	66
Totals	33		25		64		122

Empirical Results

The children participating in the experiments were chosen from an ordinary Swedish day-care center.

The testing procedure was as follows:

(*a*) an introduction, where the children could look at and play with the PDs;
(*b*) determining the developmental level for each child;
(*c*) the test phase, where the children were dressed in the PD;
(*d*) during the test, the children were observed by a trained psychologist and were video recorded; and
(*e*) after the test, the children and participating adults (day-care staff and parents) were interviewed.

Accepting behavior was defined as wearing the PD through the whole test phase (about 20 min), while rejecting behavior was defined as:

(*a*) the child refuses to wear the PD,
(*b*) the child tries the PD on but takes it off immediately, and
(*c*) the testing procedure was terminated due to psychological reaction.

Study 1—Determination of the Age Limit for Use of Protective Jacket 36 [5]

In this study, 122 children participated. The protective jacket is made for children between 2 and 4 years old (see Fig. 1). The jacket has a hood covering the whole head and is supported with air by a fan. The clothing is made of a rather stiff nylon fabric covered with

TABLE 2—*A comparison between the original protective jacket and the modified protective jacket showing the total number of children who accept and refuse to wear the PD.*

| | Age Group (2 to 4 years) | | | | Total, |
| | Original Jacket, | | Modified Jacket, | | |
	n	*(%)*	*n*	*(%)*	*n*
Accepts	12	(21)	24	(47)	36
Refuses	46	(79)	27	(53)	73
Totals	58		51		109

rubber. The dressing procedure required the children to crawl into the PD, because the only opening was at the bottom of the jacket. The sound level in the hood is 72 dB(A) which makes the child almost completely deaf. Table 1 shows the total number of children who accepted or refused to wear the PD.

The age distribution of the children who accepted the PD was tested by CHI-square and was found statistically significant ($p < 0.01$) with older children (ages 4 to 7) accepting the PD to the greatest degree.

The jacket was modified in four ways:

(a) the clothing was changed to a more supple material (nylon fabric covered with PVC),
(b) the field of vision was made larger,
(c) a zipper was placed in the back of the jacket, and
(d) the sound level was lowered to a maximum of 65 dB(A).

Table 2 shows the result of the modification in terms of the accepting and refusing rates for children ages 2 to 4.

A comparison between the results from Table 1 and Table 2 shows a significant difference ($p < 0.01$) in favor of the modified jacket (CHI-square), where acceptance was increased by approximately twofold.

The acceptance of the modified jacket was also compared with another modified jacket that was totally transparent. Statistical analysis indicated there was no significant difference in behavior between the two designs. However, when parents estimated the reliability of the PD, the transparent jacket was judged less reliable than the modified jacket (the transparent jacket looked "cheaper") ($p < 0.001$).

Study 2—Examining the Gas-Mask 33 on Children Ages 3 to 7 Years [6]

This study examined an alternative PD to the protective Jacket 36. The protective mask (see Fig. 2) is an ordinary full-face mask for use by civilians. Eighty-six children, 3 to 7 years of age, participated. The results indicated a much higher acceptance level for the protective mask compared to the protective jacket, particularly for group 3 to 4 years old. In Table 3, a comparison is made between the two PDs [7].

The difference between the accepting and refusing behavior of the two PDs was tested by CHI-square and was found statistically significant ($p < 0.01$). The most likely explanation of this finding is that the protective mask will leave the hearing undisturbed (the protective

FIG. 2—*The protective mask (Gas Mask 33) for civilians.*

TABLE 3—*A comparison between the original protective jacket and the gas-mask showing the total number of children who accept and refuse to wear the PD.*

| | Age Group (3 to 4 years) | | | | |
| | Gas Mask, | | Original Jacket, | | Total, |
	n	(%)	*n*	(%)	*n*
Accepts	15	(79)	9	(36)	24
Refuses	4	(21)	16	(64)	20
Totals	19		25		44

jacket will not), allowing the child to have a far better interface with the environment and situation, creating less uncertainty and anxiety, than when dressed in the protective jacket.

Study 3—Empirical Evaluation of PD Prototypes for Small Children [8]

Four different PD prototypes were tested (see Fig. 3):

(*a*) a transparent plastic bag placed inside a baby carriage,
(*b*) a transparent plastic bag placed in a topless corrugated cardboard box,
(*c*) a rucksack in which the child is carried on the back of an adult, and
(*d*) a protective suit covering both child and adult.

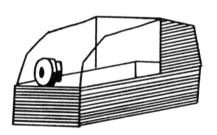

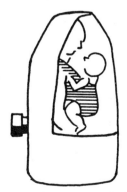

FIG. 3—(a) *Transparent plastic bag inside a baby carrier,* (b) *transparent plastic bag in a cardboard box,* (c) *the rucksack, and* (d) *the protective suit.*

TABLE 4—*Mean estimation of the possibility to satisfy psychosocial requirements (adult-child communication, comfort, supervision, and nursing) and rank (1 = maximal possibilities/highest rank; 4 = no possibilities/lowest rank).*

Variables	Protective Devices			
	Plastic bag	Corrugated Box	Rucksack	Protective Suit
Communication	2.8	2.5	3.4	1.0
Comfort	1.5	1.9	2.0	1.2
Supervise	1.9	1.1	3.0	1.0
Nursing	3.0	2.6	4.0	1.5
Rank	2	3	4	1

All prototypes have a fan-assisted air support.

In Phase One of the experiment, 37 parents with children participated. The adults were to evaluate the PDs using four variables (communication, comfort, supervision, and nursing) and then to rank the devices. Table 4 shows the mean estimations and rank (the lower the value, the more favorable the evaluation).

In Phase Two, 44 adults (without children) were to do the same evaluations. The results showed no significant differences between the parent group in Phase One and the adult group in Phase Two in estimates or rank. Note the similar results between the plastic bag and the corrugated box. The difference in rank was statistically significant ($p < 0.05$), which mirrors the lack of reliance in the corrugated box.

Conclusions

From a psychological point of view, considerations like reliability, dressing and wearing comfort, and the possibility of satisfying psychosocial demands are of vital importance when trying to obtain acceptable behavior in an emergency, particularly in situations where children are involved. These psychological considerations are as equally important as technique, clothing, and design. If two protective devices are technically equal, the one that is judged more reliable will be chosen. In addition to these psychosocial requirements, internal and external motivation are relevant aspects of successful behavior. In a situation that is extremely boring and with an ill-fitting PD, the decision to continue to wear the PD is exclusively a result of internal motivation. In an actual emergency where the internal motivation should be high, it may be necessary for an adult to counteract (for example, through external motivation) this boredom, especially in younger children.

References

[1] Mauritzson-Sandberg, E. and Sandberg, L., "Theories and Methods for Evaluation of Respiratory Protective Devices for Children," Report C40200-C2, National Defense Research Institute, July 1984.

[2] Piaget, J. and Inhelder, B., "The Psychology of the Child," Routledge & Kegan Paul, London, 1969.

[3] Bowlby, J., "Modern och barnets själsliga hälsa Stockholm," *Natur & Kultur*, 1954.

[4] Spitz, R. A., *The First Years of Life, a Psychoanalytic Study of Normal and Deviant Development of Object Relations*, International University Press Inc., New York, 1965.

[5] Sandberg, L. and Mauritzson-Sandberg, E., "Determination of the Age Limit for Use of Protective Jacket 36," Report C40163-C2, National Defense Research Institute, Nov. 1982.

[6] Mauritzson-Sandberg, E. and Sandberg, L., "Determination of Lower Age Limit for the Use of a Protective Mask for Children," Report C40182-C2, National Defense Research Institute, Sept. 1983.

[7] Jönsson, P-G., Sandberg, L., and Sundqvist, R., "Respiratory Protection for Children, 3–6 Years of Age," Report C40197-C2, National Defense Research Institute, May 1984.

[8] Technical Report to the Swedish Civil Defense Administration, Nov. 1984.

Cay A. Ervin[1]

A Standardized Dexterity Test Battery

REFERENCE: Ervin, C. A., **"A Standardized Dexterity Test Battery,"** *Performance of Protective Clothing: Second Symposium, ASTM STP 989,* S. Z. Mansdorf, R. Sager, and A. P. Nielsen, Eds., American Society for Testing and Materials, Philadelphia, 1988, pp. 50–56.

ABSTRACT: Though numerous investigators have studied the effects of various impediments on manual performance, few if any standardized test procedures have been developed even for particular conditions such as protective glove wear, climatic extremes, or drug use. The use of different dexterity tests, minor modifications of the same tests, and changes in experimental design result in frequent duplication of very similar experimentation. Standardized test batteries would streamline future research in this area by decreasing redundancy and making it possible to compare results from parallel studies. In developing a standardized test battery, several factors must be considered: availability and durability of equipment, control of learning effects, ease of scoring and administration, ability to discriminate between conditions being tested, and reliability. These factors, which were addressed in an investigation of the effects of chemical defense gloves on manual performance for the U.S. Air Force, are discussed here.

KEY WORDS: manual dexterity, standardized test battery, chemical defense gloves

Modern technology and the proliferation of hazardous materials have made the use of protective clothing increasingly common. Because most job functions require the use of the hands, reducing the limiting effects of protective gloves is useful in maintaining efficiency, quality control, and cost-effectiveness. Determining if newly developed gloves provide maximum protection with minimum hindrance is achievable through dexterity testing. The testing itself can be made more efficient by developing a standardized battery that could be used for evaluating new prototypes and comparing results with those obtained in previous tests of current hand wear. Factors to be considered in developing a test battery for assessing the effects of gloves on manual performance are outlined in this paper and include: selecting tests which can be standardized and are sensitive to glove types as well as applicable to the tasks for which the gloves will be worn; reviewing the tests for administrative or scoring modifications or both; and developing an experimental design in which several tests can be used to evaluate more than one condition and can be administered in such a way as to provide maximum control for learning effects. These factors were worked out through a series of studies of chemical defense (CD) gloves conducted under the direction of Kathleen Robinette of the Armstrong Aerospace Medical Research Laboratory, Wright-Patterson Air Force Base, Ohio.

Test Selection

A major factor to be considered when selecting dexterity tests for a battery is their sensitivity to different types of hand wear. The tests selected should be complex enough to

---

[1] Research associate, Anthropology Research Project, Inc., 503 Xenia Avenue, Yellow Springs, OH 45387.

reveal differences in hand wear, yet not be so difficult as to be virtually impossible to perform while wearing gloves. The only way to discover which tests meet this criterion is through pilot testing. At first glance, for instance, the Minnesota Rate of Manipulation-Turning Test (MRMTT), which involves turning over pegs in a specific order, did not appear to show much promise for discriminating between glove types because the pieces are relatively large in comparison with other tests. However, when the MRMTT was used to evaluate several types of CD gloves, analysis proved that it was able to reveal differences between 0.3175-mm (12.5 mil) (1 mil = 1/1000 of an inch) and 0.3556-mm (14 mil) CD gloves [1]. By comparison, the Crawford Small Parts Dexterity Test–Screws (CSPDTS), which involves using a small screwdriver to turn a screw through a threaded plate, was not useful for detecting any such differences, apparently because the test was too difficult to perform while wearing gloves. Because it had not proved useful in detecting differences between glove types in a previous study, the CSPDTS was dropped from the battery.

A second major factor in reviewing candidate tests is repeatability. That is, the test design should be sufficiently specific to enable all subjects to perform the test in exactly the same manner. This ensures that any differences found in performance are due to the conditions being evaluated rather than to differences in subjects' techniques. One of the tests that posed problems in this regard was the Bennett Hand Tool Dexterity Test, which requires subjects to use tools to disassemble and reassemble nut and bolt combinations of various sizes. We found that subjects' scores were affected by how tightly the nut and bolt combinations were assembled, by whether or not subjects dropped parts, and by subjects' familiarity with tools. We investigated the use of the torque wrench to control the tightness of the assemblies, but determined that a costly custom-made wrench would be required. (An objective of this study was not to develop new equipment but to find suitable, readily available equipment.) In short, too extensive modifications were required to make it impervious to extraneous factors. Tests which met this requirement but did not have standardized administrative procedures were accepted and the problems corrected by modifying the tests.

In the assembly of a dexterity test battery, the objective should be to test as wide a variety of relevant skills as possible with the smallest number of non-repetitive tests. Tests which add no new information could increase boredom and affect subjects' scores. Portions of some tests can be eliminated to achieve a battery which minimizes redundancy. The Pennsylvania Bi-Manual Worksample Assembly Test (PBWAT), for instance, has two parts: Assembly and Disassembly. Though the relationship between the two parts was not statistically analyzed, they both require the same type of hand movements, so the Disassembly portion was eliminated from the CD glove-testing battery.

Test Modification

Once tests pass the criteria of sensitivity and repeatability, the scoring procedures should be reviewed. Often the tests can be shortened or the scoring simplified. The PBWAT, which requires the assembly of 100 nut and bolt combinations, is scored, in its original version, by the amount of time needed to complete 80 of the 100 assemblies. While this may not take long bare-handed, a subject wearing gloves can take up to 15 min too much time when added to the number of practices required to control for learning effects. Changing scoring procedures so that a specific time limit is set (for example, 2 min) enables the investigator to schedule subjects more efficiently and reduces frustration and boredom for subjects. Of the five tests used to evaluate CD gloves, three were modified in this manner. The other two tests were relatively short in their original version and so were not changed.

An example of scoring simplifications is the Roeder Manipulative Aptitude Test-Rods and Caps, for which the number of rods and the number of caps placed are scored separately.

This was simplified by scoring the total number of pieces placed. Because the pieces are placed in an alternating fashion (rod, cap, rod, cap, rod, . . .) no information was lost.

Other tests were not so much modified as clarified. For instance, many of the test instructions which were reviewed did not specify starting positions; whether a subject begins with a piece in his/her hand poised over the starting point or with empty hands can affect scores.

The modifications made to the PBWAT, used to evaluate CD glove ensembles, provide an example of the kinds of changes that were made to improve the usefulness of a given test. This test is a large plastic board of ten rows of ten holes each with large wells at either end—one holds nuts and the other holds bolts (Fig. 1). As noted previously, two tests are performed with this equipment: Disassembly and Assembly. The first modification was to eliminate the Disassembly task since it was deemed redundant.

Scoring procedures were examined next. In its original version, the PBWAT was scored by the amount of time required to complete the assemblies: the first two rows were used for practice, and subjects were scored on the remaining eight rows. During pilot testing of this test, it became obvious that it was unnecessarily long. Subjects complained of boredom and muscle fatigue when asked to repeat the test under several conditions. Rather than decrease the number of assemblies required, we established 2 min as the amount of time needed to yield information about the gloves while minimizing subject boredom. The time limit was determined by observing performance in pilot testing and by the minimum amount of time needed to reveal effectively the differences in glove types. In short, if information about the hand wear can be obtained in 2 min, there is no reason to allow the test to drag on for 5 min.

Finally, to further standardize the test, we defined the starting position, which was not specified in the manual. Subjects were instructed to begin with a piece in each hand, their hands resting at the sides of the board. Table 1 is a summary of the tests used in the battery and modifications made.

Experimental Design

The object of standardizing tests is to obtain a "true" measure of the relative performance by controlling for extraneous factors such as fatigue, boredom, learning, and subjects' strategies, all of which can affect scores. While some of these factors are controlled wholly or partially by standardizing individual tests, others are largely controlled through the experimental design.

One factor which can affect performance, perhaps more than any other, is learning. This is controlled by allowing sufficient practice trials and by randomizing the order of the

FIG. 1—*Pennsylvania Bi-Manual Worksample Assembly Test.*

TABLE 1—*Summary of battery tests: objectives, original versions, modifications.*

Tests	Objective	Original Version	Modifications
Minnesota Rate of Manipulation Test-Turning	turn over a series of pegs in a specified order using two hands	scored by number of seconds needed to turn over all pegs	none
Roeder Manipulative Aptitude Test-Rods and Caps	screw rods into a board and screw caps on top, alternating between rods and caps, using the dominant hand	number of rods and caps placed in three minutes (two separate scores); No starting position specified	scored by total number of pieces placed (rods plus caps) in 2 min; starting position specified
Purdue Pegboard-Assembly	complete assemblies of four small parts, using two hands	scored by number of pieces placed in 60 s; no starting position specified	starting position specified
O'Connor Finger Dexterity Test	pick up three pins at a time and place in one hole using the dominant hand	scored by number of seconds used to complete first 50 holes plus number of seconds (multiplied by 1.1) to complete second 50 holes, the sum divided by two; no starting position specified	scored by number of holes filled in 2 min; specified starting position; changed orientation of board so well of pins is farthest from, instead of next to, dominant hand
Pennsylvania Bi-Manual Worksample	assemble and then disassemble nut and bolt combinations using two hands	scored by amount of time needed to assemble/disassemble (two separate tasks) 80 assemblies; no starting position specified	used assembly task only; scored by number of assemblies completed in 2 min; specified starting position.

conditions being evaluated. It is important to note that both of these procedures must be used; while practice allows an individual to reach a point where learning progresses slowly, some learning still occurs. Effects of these remaining small learning increments can be overridden by randomization. If learning effects are not first reduced by sufficient practice, the differences between gloves can become lost even with randomization.

For the number of practice trials required for each test to be determined, the point at which subjects' scores "plateau," or cease to improve, needs to be identified. Figure 2 is a learning curve of the PBWAT; as can be seen, there is no improvement after the fifth trial.

The number of practice trials needed for each test appears to be dependent on two factors: complexity of the tests, and how the practice trials are incorporated into the experimental design. For instance, more practice seems necessary for tasks requiring two hands and assembly of parts than would be needed for tasks requiring one hand and simple placement of parts.

The necessary number of practice trials is also higher if the trials are spread out over several days. This is because subjects must reacquaint themselves with the tests and equipment each day. If the practice trials are allowed one after the other, this reacquaintance is not necessary and the number of practice trials can be reduced. For instance, users of a similar type of battery administer one trial each day for 12 days, using only the data from the last six trials in analysis. In preparing for this battery [1], investigators found no significant differences between the second and third test trials, when administered after allowing six shorter practice trials in the same session.

In addition to learning the test equipment, subjects must also adapt to wearing different types of gloves. To control for this kind of "learning," subjects should practice the tests in the various conditions to be tested. Adaptation to the hand wear seems to work most efficiently if subjects practice a given test in order, from the easiest condition (bare-handed) to the most difficult (wearing the thickest or largest number of layers). Observations made in at least one previous study [2] and by investigators and subjects in this study indicate that people tend to learn faster and more easily by beginning at the simplest level and gradually progressing to the most difficult.

In addition to controlling for remaining learning effects, proper randomization also reduces the possibility that other sources of variance will provide misleading results. There are several methods of randomizing: by randomizing the order of conditions within each test; by performing all tests in one glove type, then the next type and so on; or by randomizing the conditions *and* the order of the tests. A good rule to follow is this: If it is going to be compared, randomize it. Thus, while gloved conditions should always be randomized, the order of tests need not be. Randomizing both conditions and tests is recommended, however, since it relieves subjects' boredom and muscle fatigue.

Suggested Battery

In the testing of CD glove ensembles, five tests have proved to indicate differences in types [3]—the Minnesota Rate of Manipulation-Turning, Purdue Pegboard Assembly, Roeder Manipulative Aptitude Test-Rods and Caps, O'Connor Finger Dexterity Test, and The Pennsylvania Bi-Manual Worksample Assembly Test. Only the first two tests, which were designed to take 1 min or less to complete, were not modified except to specify the starting position of the Purdue Assembly. As noted above, scoring for the other three tests was changed to record the number of pieces placed in 2 min. In these longer tests, practice trials were shortened to 1 min each.

Testing was divided into two sessions, each approximately 2½ to 3 h. Subjects performed three tests in the first session and two tests in the final session; no subject performed all tests in one day. Practice trials for all the tests to be performed in a given session were

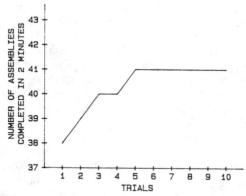

FIG. 2—*A learning curve of the Pennsylvania Bi-Manual Worksample Assembly Test. (Each line point reflects average scores of ten subjects.)*

completed before actual testing began. For scoring purposes, subjects performed three replicates of each test for each condition. The first replicate was to reacquaint the subject with the test and condition, and the last two were averaged and used in the data analysis. By using the average of the final two scores, the effect of a "fluke" or abnormal performance was reduced. Figure 3 is a simple data sheet from the most recent study of aircrew glove ensembles. The numbers to the left of the three trial lines reflect the order of the conditions across all of the tests to be given. This subject, for example, would first perform the Rods and Caps with no gloves, then the Purdue Assembly with the Nomex EB-12.5 ensemble, and so on. Other subjects would perform the tasks in different orders, as specified by different data sheets.

Minnesota (Timed)	Practice		Trials		
No Gloves	___	___	14___	___	___
Nomex B-7	___	___	8___	___	___
Liner EB-12.5 - Nomex	___	___	12___	___	___
Liner B-7 - Nomex	___	___	11___	___	___
Nomex EB-12.5	___	___	3___	___	___
MAT Rods & Caps (2 · in.)					
No Gloves	___	___	1___	___	___
Nomex B-7	___	___	5___	___	___
Liner EB-12.5 - Nomex	___	___	15___	___	___
Liner B-7 - Nomex	___	___	7___	___	___
Nomex EB-12.5	___	___	4___	___	___
Purdue Assembly (1 min.)					
No Gloves	___	___	6___	___	___
Nomex B-7	___	___	13___	___	___
Liner EB-12.5 - Nomex	___	___	10___	___	___
Liner B-7 - Nomex	___	___	9___	___	___
Nomex EB-12.5	___	___	2___	___	___
O'Connor (2 min.)					
No Gloves	___	___	2___	___	___
Nomex B-7	___	___	10___	___	___
Liner EB-12.5 - Nomex	___	___	6___	___	___
Liner B-7 - Nomex	___	___	8___	___	___
Nomex EB-12.5	___	___	5___	___	___
Pennsylvania (2 min.)					
No Gloves	___	___	3___	___	___
Nomex B-7	___	___	9___	___	___
Liner EB-12.5 - Nomex	___	___	7___	___	___
Liner B-7 - Nomex	___	___	1___	___	___
Nomex EB-12.5	___	___	4___	___	___

FIG. 3—*Sample data sheet.*

Summary

The test battery reviewed here was developed from a series of research studies of chemical defense gloves. The tests included in the battery were selected because they proved sensitive to different CD glove types, were easily standardized and administered, required few practice trials, and were easily scored. Whether this battery is applicable to other types of gloves or whether a different group of tests would be required for other types of hand wear remains to be seen. The central point is that it is both possible and desirable to develop standardized dexterity test batteries for given purposes, and that these efforts will pay dividends in the comparative evaluation of existing products and in the development of new ones.

References

[1] Robinette, K. M., Ervin, C., and Zehner, G. F., "Dexterity Testing of Chemical Defense Gloves (U)," Technical Report AAMRL-TR-86-021, AD A173 545, Armstrong Aerospace Medical Research Laboratory, Wright-Patterson Air Force Base, OH, May 1986.
[2] Fisk, A. D., Ackerman, P. L., and Schneider W., "Automatic and Controlled Processing Theory and Its Application to Human Factors Problems," *Human Factors Psychology,* P. A. Hancock, Ed., North Holland Publishing, New York (in press).
[3] Ross, J. L. and Ervin, C., "Chemical Defense Flight Glove Ensemble Evaluation (U)," Technical Report AAMRL-TR-047, Armstrong Aerospace Medical Research Laboratory, Wright-Patterson Air Force Base, OH, June 1987.

Protection From Physical Stressors

Thermal Protection

Martin W. King,[1] *Xiaojiu Li,*[2] *Barbara E. Doupe,*[3]
and Janet A. Mellish[4]

Thermal Protective Performance of Single-Layer and Multiple-Layer Fabrics Exposed to Electrical Flashovers

REFERENCE: King, M. W., Li, X., Doupe, B. E., and Mellish, J. A., **"Thermal Protective Performance of Single-Layer and Multiple-Layer Fabrics Exposed to Electrical Flashovers,"** *Performance of Protective Clothing: Second Symposium, ASTM STP 989,* S. Z. Mansdorf, R. Sager, and A. P. Nielsen, Eds., American Society for Testing and Materials, Philadelphia, 1988, pp. 59–81.

ABSTRACT: Utility workers can be exposed to an acute flammability hazard when they are working close to an arcing fault or flashover from electrical equipment or power lines. In the past, it has not been possible to recommend the type of protective clothing that should be worn in these circumstances because no laboratory test method has been available to measure the thermal protective performance (TPP) of fabrics and other laminae against exposure to a high intensity, infrared dominant, radiant energy source similar to that emitted from electrical flashovers. The study describes the use of a recently developed Flash Tester, which, by generating a controlled electrical discharge across welding electrodes, produces radiant flux densities in the 40 to 600 kW/m² (1 to 15 cals/cm²/s) range. By exposing fabric layers equivalent to various clothing assemblies worn by utility linesmen and a range of 32 flame-retardant (FR) fabrics as single- and multiple-layer specimens, it has been shown that those parts of the body covered by only one fabric layer are exposed to the highest risk of burn injury, whereas for those covered with six or more layers, the risk of burn injury is negligible. The use of single-layer fabrics containing FR fibers or FR finishes provide some additional protection, particularly those with a high mass per unit area (fabric weight), while the use of thicker or lighter colored fabrics does not appear to influence the TPP ratings. More protection is obtained by using two or more fabric layers. However, two layers appear to be effective only when both layers contain fabrics that do not fuse together. Because of strong interactions between fabrics in a multiple-layer assembly, it is not possible to predict the TPP of combined layers worn together from tests of the individual layers.

KEY WORDS: electric arcs, flame-retardant fabrics, flashover, heat measurement, high-temperature test, multiple-layer assemblies, protection time, protective clothing, radiant heat flux, single-layer assemblies, thermal protection

[1] Associate professor, Department of Clothing and Textiles, University of Manitoba, Winnipeg, Man. R3T 2N2, Canada.
[2] Visiting scholar from Tianjin Textile Engineering Institute, Tianjin, Peoples Republic of China; currently at the Department of Clothing and Textiles, University of Manitoba, Winnipeg, Man. R3T 2N2, Canada.
[3] Research assistant, Department of Clothing and Textiles, University of Manitoba, Winnipeg, Man. R3T 2N2, Canada; currently at the Retail Research Foundation of Canada, Markham, Ont., Canada.
[4] Student, Department of Clothing and Textiles, University of Manitoba, Winnipeg, Man. R3T 2N2, Canada; currently at Angel Merchandising Services, Calgary, Alta., Canada.

Employees of utility companies who work close to live lines and electrical equipment are continually exposed to the risk of electric shocks. Various types of protective apparel and special tools, such as rubber gloves, dielectric hard hats and boots, sleeve protectors, conductive Faraday Cage garments, rubber blankets, nonconductive hot sticks, and live line buckets are used to minimize the possibility and extent of injury [1]. At the same time, these workers are also exposed to an acute flammability hazard in the event that an accidental arcing fault or flashover occurs nearby. The explosive intensity of the radiant flux produced can result in the ignition of one or more fabric layers and cause burn injuries through several layers of clothing [2]. This type of hazard is of particular concern to those employees working close to live switching equipment, transformers, and local distribution lines where distances between the live conductors and ground may be about 0.6 m (2 ft) and working clearances for linesmen may be as small as 0.3 m (1 ft) at 20 kV voltages (Fig. 1).

With a view to identifying suitable protective clothing for these workers, we searched for an appropriate test method that would enable us to expose textile fabrics and other laminae to an intense, high-energy radiant source under controlled laboratory conditions. A variety of methods were considered, including the use of radiant panels [3,4], as required by the International Standards Organization (ISO) Clothing for Protection Against Heat and Fire (ISO 6942-1981), and the use of a combination of sources [5,6]. But none of the methods attempt to reproduce a radiant heat flux with a dominant wavelength in the infrared region [7] similar to that emitted from an electrical flashover. We therefore designed a test ap-

FIG. 1—*Linesman wearing winter clothes repairs a local distribution line.*

paratus, the University of Manitoba Flash Tester, and developed a test procedure that has enabled us to measure the Thermal Protective Performance (TPP) of fabrics and fabric assemblies exposed to electrical arcs in the laboratory. The Flash Tester can generate radiant energy flux densities in the range of 40 to 600 kW/m² (1 to 15 cal/cm²/s) for any preset period from 1/60 s to over 2 min. Preliminary test results, which demonstrate the validity and reliability of the method, have been reported previously [8].

Objectives

There were two main objectives for this study. First, it was important to determine the level of protection currently provided by regular workwear worn by linesmen and utility workers in Canada during both the summer and winter seasons. The second objective was to measure the relative TPP of a wide range of flame-retardant (FR) fabrics, both in single and multiple layers, with a view to proposing how the protective performance of current work clothes might be improved by the incorporation of one or more flame-retardant layers. We were particularly interested to learn whether or not the TPP of a multiple assembly could be predicted from the behavior of the individual layers tested separately. It should be noted that our objective was not to establish a specification or pass/fail criteria for utility workers clothing, but rather to undertake an exploratory study to assess the current level of protection and identify whether and how improvements might be achieved [9].

Evaluation of Current Linesmen's Clothing

Since employees of most Canadian utility companies are expected to purchase their own workwear, the clothing worn by any particular linesman will to some extent depend on his or her own preferences and on the season of the year. Even so, we have observed that the type of outdoor clothing worn by most employees does not vary much from the typical summer and winter assemblies described in Table 1 and illustrated in Fig. 2. Note that the number of fabric layers covering the body varies from one to six depending on the part of the body and the season. During spring and fall, it is assumed that components of the summer and winter assemblies are combined to provide intermediate numbers of fabric layers.

Materials

Samples of the clothing items listed in Table 1 were purchased at a retail store in Winnipeg, Man., Canada. Fabric specimens were cut from these garments, the layers were combined to form the four assemblies indicated and the specimens were tested for TPP ratings in the Flash Tester.

Test Apparatus

The Flash Tester consists of four parts: an arc chamber, two calorimeters and a chart recorder to measure and record the temperature profiles in front of and behind the fabric specimen, a welding transformer to supply the power, and an electronic controller to control the duration of the arc (Fig. 3). The two-layer construction of the chamber can be seen in Fig. 4. Dried single- or multiple-layer fabric specimens measuring 100 mm by 100 mm are clamped between two window plates and mounted vertically across the chamber between the two adjustable slots. Two solid carbon, copper-clad a-c welding electrodes measuring 12.7 mm (½ in.) in diameter are mounted with their tips in contact in the adjustable electrode

TABLE 1—*The TPP of typical summer and winter outdoor clothing worn by Canadian linesmen on exposure to radiant energy flux density of 220 kW/m².*

Assembly	Season	Part of Body	No. of Fabric Layers	Items of Clothing	Fiber Content	Type of Fabric	Mass per Unit Area of Fabric, g/m²	Protection Time, s	TPP, MJ/m²
A	summer	legs	1	jeans	cotton	3/1 twill, denim	338	3.3	0.74
B	summer	torso	2	long-sleeved shirt	50/50 polyester/cotton	1/1 plain, chambray	186	5.4	1.38
				jean jacket	cotton	3/1 twill, denim	338		
E	winter	legs	3	long underpants	50/50 polyester/cotton	1/1 rib knit	218	9.1	2.07
				jeans	cotton	3/1 twill, denim	338		
				coveralls	cotton	3/1 twill, drill	282		
G	winter	torso	6	long-sleeved undervest	50/50 polyester/cotton	1/1 rib knit	218	>20.9	>5.19
				long-sleeved shirt	cotton, brushed	1/1 plain, flannel	151		
				parka with hood	shell: nylon	1/1 plain, ripstop	535		
					filling: polyester	non-bonded web			
					lining: nylon	1/1 plain			
				coveralls	cotton	3/1 twill, drill	282		

FIG. 2—*Typical summer* (top) *and winter* (bottom) *outdoor clothing worn by Canadian linesmen showing sequence from inner layer* (left) *to outer layer* (right).

holders and positioned at any distance between 30 mm and 150 mm from the fabric specimen. Each pair of electrodes is prepared by grinding the tips to form a 45° cone and dried over a desiccant for 72 h prior to firing.

The two calorimeters consist of small blackened stainless steel discs measuring 2.2 mm in thickness and 6.4 mm in diameter. They contain chromel-alumel thermocouples and are mounted at the ends of adjustable ceramic tubes above and behind the fabric specimen.

The duration of the arc is preset manually between 1 and 9999 cycles (1/60 s to 166 s) on the controller box. An integral cycle thyristor switch actuates the firing of the arc and controls its duration by switching the primary circuit of the welding transformer that converts the 220-V, 60-Hz supply to 80-V, 225-A on the secondary side for the preset number of cycles. For safety reasons, the arc chamber is located in a vented fume-hood with a shatterproof glass front panel during testing (Fig. 5).

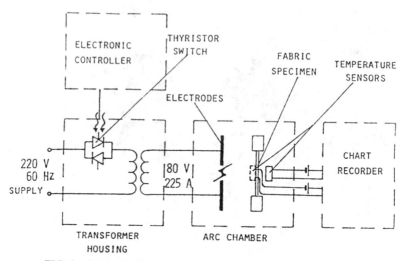

FIG. 3—*Schematic diagram of University of Manitoba Flash Tester.*

FIG. 4—*Internal view of arc chamber with lid raised.*

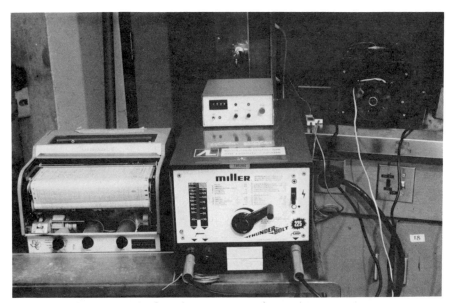

FIG. 5—*View of University of Manitoba Flash Tester in the laboratory.*

Test Method

Five specimens of each assembly were exposed to a 12 s (720 cycle) arc at a preselected distance fo 57 mm from the electrodes. (See next section for rationale.) Under these conditions, the specimens were exposed to an incident energy flux density of 211 ± 14 kW/m² as measured by the front calorimeter. The fabric layers were clamped so as to just touch one another. Also, the back calorimeter was mounted in contact with, but without pressure against, the back of the specimen. The avoidance of additional air spaces ensured that more severe testing conditions were used [10,11]. The curve recording the millivolts of the thermocouple at the back of the specimen during and after each exposure was converted to transmitted flux density, integrated, and compared with Stoll's criteria for second-degree burns in order to calculate the protection time [12]. The TPP ratings were determined by multiplying the protection time by the incident flux density measured by the front calorimeter [5].

Selection of Incident Flux Density

The conditions for the laboratory test were selected so as to reproduce the general level of hazard that might be encountered in the field. This example is not presented to demonstrate the level of protection to be expected under an actual flashover situation.

Given that linesmen are not permitted to work at a distance less than 1.55 m (5 ft) from a live a-c distribution line with a rated phase voltage, E_x, of 230 kV and rated power P_s of 2000 kVA, the maximum arc power condition can be determined as follows [5]. With a three-phase supply, the phase current, I_s, is calculated from

$$I_s = P_s/3E_x$$

Assuming the leakage reactance, X_L, is 5% based on the ratings, the resistance, R_f, voltage drop, E_f, current, I_f, and energy, P_f, can be calculated for the maximum power condition when

$$E_f = E_\infty/\sqrt{2}$$

$$R_f = X_L = 0.05\ (E_\infty/I_s) = 0.05E_\infty/(P_s/3E_\infty)$$

$$= \frac{0.05 \times 3(230)^2 \times 10^6}{2 \times 10^6} = 3.97\ \text{k}\Omega$$

$$E_f = E_\infty/\sqrt{2} = 230/\sqrt{2} = 162.4\ \text{kV}$$

$$I_f = E_f/R_f = \frac{162.4 \times 10^3}{3.97 \times 10^3} = 40.9\ \text{A}$$

$$P_f = E_f{\cdot}I_f = 162.4 \times 40.9 = 6650\ \text{kW}$$

If we assume that the radiant flux of the flashover forms a sphere of radius, D, and that the conversion to radiant energy is 100% efficient (we have confirmed experimentally that the conversion lies between 79 and 97% [8]), then the flux density, Q_o, that reaches the linesman located at a distance of 1.55 m from the center of the arc is given by

$$Q_o = P_f/4\pi D^2 = \frac{6650}{4\pi(1.55)^2} = 220\ \text{kW/m}^2$$

A parallel calculation can now be performed so as to determine the distance, D, between the electrodes and the calorimeters that will generate the same incident flux density in the laboratory Flash Tester. As described earlier, the welding transformer gives a supply voltage, E_∞, of 80 V and a base current, I_s, of 225 A. The resistance, R_a, voltage drop, E_a, current, I_a, and energy, P_a, can be calculated for the maximum power condition using the same assumptions described earlier.

$$R_a = X_L = E_\infty/I_s = 80/225 = 0.36\Omega$$

$$E_a = E_\infty/\sqrt{2} = 80/\sqrt{2} = 56.6\ \text{V}$$

$$I_a = E_a/R_a = 56.6/0.36 = 159\ \text{A}$$

$$P_a = E_a{\cdot}I_a = 56.6 \times 159 = 9.00\ \text{kW}$$

$$D = \sqrt{\frac{P_a}{4\pi Q_o}} = \sqrt{\frac{9.00}{4\pi \times 220}} = 0.057\ \text{m (57 mm)}$$

Results and Discussion

The average TPP curves recording the energy measured by the back calorimeter during the tests are shown in Fig. 6. The protection times and corresponding TPP ratings are listed in Table 1. Both clearly demonstrate that the number of fabric layers is the most important factor in influencing the degree of protection, regardless of the type of fiber and fabric. In

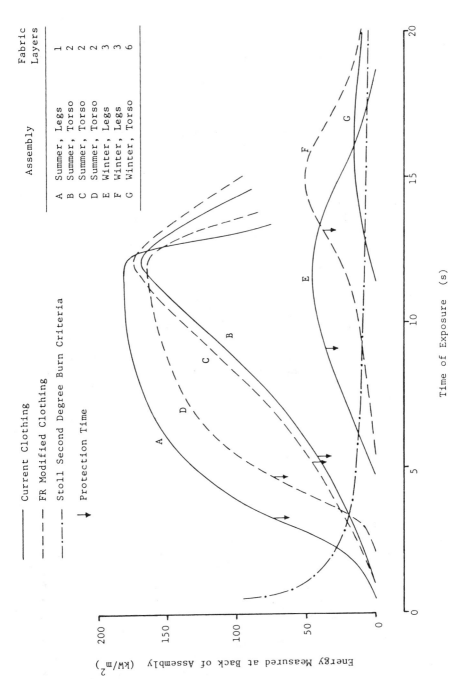

FIG. 6—Average TPP curves of energy transmitted through typical summer and winter clothing assemblies.

other words, the legs, being covered by only one or three fabric layers depending on the season, are more susceptible to burn injury than the torso under two or six layers. Likewise, the linesman is at greater risk during the summer than during the winter when more layers are required for warmth.

At this stage in the study, we were interested to learn whether improvements in the protective performance of these clothing assemblies could be achieved by substituting one or more of the layers with flame-retardant (FR) fabrics. The first step in this direction was to measure the relative TPP ratings of a range of 32 different FR fabrics.

Evaluation of Single-Layer Flame Retardant (FR) Fabrics

Materials

Thirty-two commercial fabrics were chosen to represent a wide range of fiber contents, fabric weights, types of weave, and colors. Only those fabrics that were known to demonstrate some FR behavior, either on account of their inherent chemical composition, or as a result of the application of FR finishes, were included in the sample. Details of the fabrics are given in Tables 2 and 3.

Test Methods

The apparatus and test procedure used to determine the TPP of these fabrics were identical to those just described, except that the duration of the arc was preset to 25 s (1500 cycles) instead of 12 s. In addition, three physical properties of each fabric were measured, namely; thickness, mass per unit area (weight), and surface lightness. These three properties were chosen because it was believed that they might influence the TPP rating. Standard methods: ASTM Method for Measuring Thickness of Textile Materials (D 1777–64) and ASTM Test Methods for Mass per Unit Area (Weight) of Woven Fabric (D 3776–84) were used for the first two tests. For the third, a Hunterlab Color Difference Meter, Model D25, was employed to measure the percent total reflectance of visible light, L, or surface lightness, of each sample. According to this method, a totally absorptive surface (black body) yields an L value of 0%, in contrast to a perfectly reflective surface, which would have a 100% value. Therefore, the lightness value represents how dark or light a shade the fabric appears regardless of its hue.

Results and Discussion

The average thermal properties, thickness, mass per unit area, and lightness values are listed in Tables 2 and 3. They demonstrate a wide variation in thermal behavior. The highest TPP ratings were achieved with a glass fiber fabric laminated to highly reflective aluminum foil (#25) and a heavy woven aramid Norfab® fabric (#22) with values of 2.80 and 1.77 MJ/m², respectively. Other superior performances were observed with a reclaimed aramid felt (#21) and a Proban®-treated cotton 14-oz denim (#8). All four samples gave protection times in excess of 6 s. The lowest TPP ratings were obtained from an SEF® modacrylic (#32), two light-weight woven novoloid fabrics (#27 and #28), and a shirting weight Proban®-treated cotton (#1). All four fabrics provided protection times of less than 2.5 s.

On observing the exposed test specimens, it was evident that different samples experienced different failure mechanisms [13]. The residues were viewed subjectively and classified into

TABLE 2—*Single-layer FR cotton, wool, and blended fabrics tested for TPP ratings.*

Fabric No.	Description				Properties				
	Fabric				Physical			Thermal	
	Fiber(s)	Name	Weave	Finish	Thickness, mm	Mass per Unit Area, g/m^2	Lightness, %	Protection Time, s	TPP Rating, MJ/m^2
1	cotton	shirting	2/1 twill	Proban[a]	0.51	177	34.1	2.4	0.46
2	cotton	drill	3/1 twill	Proban	0.65	288	32.7	4.7	0.92
3	cotton	satin	4/1 satin	THPOH-NH$_3$	0.73	328	52.6	4.7	0.83
4	cotton	denim	3/1 twill	Proban	0.81	348	28.9	4.7	0.89
5	cotton	brushed	4/1 twill	Proban	1.12	414	89.3	5.5	1.26
6	cotton	satin	4/1 satin	P44	0.66	429	87.2	4.4	0.95
7	cotton	denim	3/1 twill	Proban	0.93	491	84.1	5.6	1.12
8	cotton	denim	3/1 twill	Proban	0.89	495	21.3	6.9	1.35
9	85/15 cotton/polyester	denim	2/1 twill	Proban	0.75	361	28.2	4.1	0.84
10	85/15 cotton/polyester	denim	2/1 twill	P44	0.66	453	31.2	5.2	1.29
11	wool	suiting	2/2 twill	FR	0.68	262	14.3	4.6	0.95
12	wool	suiting	2/2 twill	FR	1.11	378	11.6	4.3	0.67
13	60/40 wool/Cordelan[a]	serge	2/2 twill	...	0.61	243	55.2	3.9	0.97
14	60/40 wool/Cordelan	suiting	2/2 twill	...	0.52	259	22.6	3.2	0.76
15	60/40 Cordelan/wool	suiting	2/2 twill	...	0.60	274	23.0	4.4	0.98

[a] Registered trademarks.

TABLE 3—Single-layer FR synthetic fiber fabrics tested for TPP ratings.

Fabric No.	Description			Properties				
				Physical			Thermal	
	Fiber	Trade Name	Fabric Structure	Thickness, mm	Mass per Unit Area, g/m²	Lightness, %	Protection Time, s	TPP Rating, MJ/m²
16	aramid	Nomex[a] III	1/1 plain weave	0.58	178	87.9	4.1	0.72
17	aramid	Nomex III	1/1 plain weave	0.71	234	87.1	4.0	0.79
18	aramid	50/50 Kevlar[a]/Nomex	1/1 plain weave	0.64	250	83.7	4.1	0.77
19	aramid	Nomex III	1/1 plain weave	0.89	293	88.3	4.7	0.90
20	aramid	Nomex	1/1 plain weave	0.81	322	28.8	5.0	1.00
21	aramid	(reclaimed)	Needlepunched felt	3.65	353	26.7	6.1	1.41
22	aramid	Norfab[a]	1/1 plain weave	2.45	747	77.2	9.4	1.77
23	glass		1/1 plain weave	0.28	205	92.1	4.5	0.77
24	glass		1/1 plain weave	1.53	646	68.0	4.9	0.84
25	glass & Al foil		1/1 plain weave	1.42	677	95.9	15.7	2.80
26	novoloid	Kynol[a]	needlepunched felt	2.50	157	49.2	3.8	0.69
27	novoloid	Kynol	1/1 plain weave	0.39	163	46.7	2.4	0.45
28	novoloid/ aramid	Kynol	1/1 plain weave	0.54	201	46.6	2.4	0.44
29	novoloid/glass	Kynol	1/1 plain weave	2.11	561	52.4	5.1	1.07
30	polybanzimid- azole	PBI[a]	2/1 twill weave	0.95	308	27.5	4.1	0.59
31	oxidized acrylic	Pyron[a]	2/1 twill weave	1.28	419	14.3	3.9	0.73
32	modacrylic	SEF[a]	2/1 twill weave	0.52	217	24.7	2.3	0.43

[a] Registered trademarks.

the following four ordinal groups:

(a) discoloration and embrittlement,
(b) embrittlement and shrinkage,
(c) shrinkage and decomposition, and
(d) melting and decomposition.

The results in Table 4 show that the mode of thermal failure is not only dependent on fiber content and FR finish, but also on other fabric properties, since some fiber types appear in more than one or two groups.

In an attempt to ascertain which fabric properties other than fiber content and FR finish

TABLE 4—*Classification of failure mechanisms of 32 single-layer FR fabrics tested for TPP ratings.*

Mode of Failure	Fabric No.	Fiber Content and Finish	Mass per Unit Area, g/m²	TPP Rating, MJ/m²
Discoloration and embrittlement	22	aramid	747	1.77
	23	glass	205	0.77
	24	glass	646	0.84
	25	glass and Al foil	677	2.80
	26	novoloid	157	0.69
	29	novoloid/glass	561	1.07
Means			499	1.32
Embrittlement and shrinkage	2	FR cotton	288	0.92
	3	FR cotton	328	0.83
	4	FR cotton	348	0.89
	5	FR cotton	414	1.26
	7	FR cotton	491	1.12
	8	FR cotton	495	1.35
	28	novoloid	201	0.44
	30	polybenzimidazole	308	0.59
	31	oxidized acrylic	419	0.73
Means			366	0.90
Shrinkage and decomposition	1	FR cotton	177	0.46
	6	FR cotton	429	0.95
	9	FR cotton/polyester	361	0.84
	10	FR cotton/polyester	453	1.29
	15	Cordelan*a*/wool	274	0.98
	16	aramid	178	0.72
	17	aramid	234	0.79
	18	aramid	250	0.77
	19	aramid	293	0.90
	20	aramid	322	1.00
	27	novoloid	163	0.45
Means			258	0.83
Melting and decomposition	11	FR wool	262	0.95
	12	FR wool	378	0.67
	13	wool/Cordelan	243	0.97
	14	wool/Cordelan	259	0.76
	21	aramid	353	1.41
	32	modacrylic	217	0.43
Means			285	0.86

[a] Registered trademark.

might be influencing the TPP ratings, the data from the thickness, mass per unit area, surface lightness, and TPP measurements were plotted in Fig. 7 and analyzed statistically by linear, multiple, and step-wise regression analyses [14]. The closeness of fit to the line, the Pearson's r correlation coefficients (Fig. 7), and the partial F values in Table 5 indicate the significant contribution of mass per unit area compared to the negligible contributions of thickness and lightness in predicting the TPP values. Indeed, the step-wise multiple regression analysis did not proceed beyond the first step (with mass only entered) because the other two variables contributed no improvement to the regression at the 0.01 significance level. The conclusion that fabric mass per unit area (weight) is a strong predictor of TPP values agrees with Krasny's findings on FR cotton fabrics [15] and, assuming the different types of fibers have similar specific heats, supports Schoppee's relationship that the rise in specimen temperature varies inversely with its mass per unit area [7]. These data also corroborate Baitinger's conclusion with cotton duck fabrics that the fabric's lightness to visible light (or color) is not a significant influence on TPP ratings. At the same time the finding that thickness is not a significant variable differs from the reports of other workers who have used mixed radiant/convection sources or much lower radiant flux intensities [10,16].

Notwithstanding, this series of tests has demonstrated that improved protection against flashovers is provided by FR fabrics, and that against an incident radiant flux of 211 kW/m^2, the protection time of 3.3 s can be increased to 4.1 to 5.0 s by replacing a single layer of 10-oz cotton denim (Table 1, Assembly A) by a single layer of a polybenzimidazole (PBI) (#30), FR wool (#11), FR cotton (#4), or aramid fabric (#20) (see Tables 2 and 3). For additional protection using apparel fabrics, that is, avoiding uncomfortably heavy and stiff constructions, one must use more than one layer. Consequently, the next series of tests involved two-layer assemblies.

Evaluation of Two-Layer Flame Retardant (FR) Fabric Assemblies

Materials

With a view to identifying suitable fabrics for two-layer clothing assemblies (for example, a shirt and jacket combination), seven of the 32 FR fabrics previously tested were selected for the next series of tests. Four outer fabrics and three inner fabrics with a variety of fiber contents were chosen, giving a total of 12 different combinations. See Table 6 for details.

Test Method

The average protection time and TPP rating was measured from five specimens of each two-layer assembly. The apparatus and test conditions were identical to those used in the last series: the electrode-to-specimen distance was maintained at 57 mm and an arc of 1500 cycles (25 s) duration was fired. The two fabric layers were clamped so as to just touch one another. Also, the back calorimeter was mounted in contact with, but without pressure against the back of the specimen.

Results and Discussion

The results in Table 6 clearly show that the TPP rating for a two-layer assembly cannot be predicted with certainty from the sum of the TPP's for the individual layers. Some fabric assemblies, such as the FR wool/aramid, aramid/aramid, and aramid/novoloid combinations, give TPP ratings that are less than the sum of the components, whereas others, such as the FR cotton/polyester blend/FR cotton and aramid/FR cotton combinations, give TPPs that are significantly greater than would be expected from summing the individual ratings.

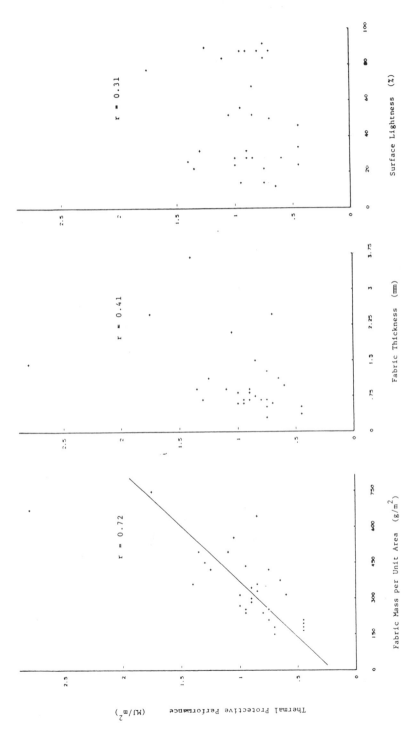

FIG. 7—*Scatterplot showing correlation between TPP of single-layer fabrics and mass per unit area, thickness, and surface lightness as independent variables.*

TABLE 5—*Results of regression analyses to predict the TPP values of single-layer FR fabrics from other physical properties.*

Type of Regression	Independent Variables	Intercept	Regression Coefficient	Pearson's r Correlation Coefficient	Coefficient of Multiple Determination	F Statistic (Partial)
Linear	mass per unit area	4.975	0.051	0.724	0.524	
	thickness	16.556	6.062	0.409	0.168	
	lightness	16.697	0.119	0.314	0.098	
Multiple	mass per unit area		0.044			(19.488)[a]
	thickness	1.714	1.955	0.754	0.568	(2.210)
	lightness		0.072			(0.897)
Step-wise	mass per unit area	4.975	0.051	0.724	0.524	33.063[a]

[a] Significant at $p < 0.01$ confidence interval.

Figure 8 shows the same data graphically and, in addition, suggests that interactions are occurring between certain pairs of fiber types, namely, in wool/FR cotton and aramid/FR cotton combinations. The significance of these interactions can be seen from the results of a two-way analysis of variance in Table 7 [*14*]. They appear to be due in part to the relative softening and melting temperatures of the fibers in the two layers. For example, the aramid/aramid combination fuses together to form one inseparable layer during testing. On the other hand, the two layers in the combinations with PBI or FR cotton maintain their separate identifies (Fig. 9) and provide protection in excess of the sum of the two layers due, no doubt, to some small amount of air space between the two fabrics [*10*].

Because of these interactions, the independent variable, mass per unit area of the combined layers, is no longer a predictor of the TPP rating for two-layer assemblies (Table 8). Furthermore, neither the combined thickness nor the surface lightness of the outer layer appears to serve as a useful predictor. One important observation is that one can take advantage of such interactions by choosing an appropriate fiber combination. For example, it is now possible to achieve protection times against an incident radiant flux of about 220 kW/m² that are in excess of 10 s by using a combination of aramid/FR cotton or PBI/aramid layers (Table 6).

Evaluation of Flame Retardant (FR) Modified Linesmen's Clothing

Since the previous series of tests had demonstrated that FR fabrics afford improved protection, but that in multiple-layer assemblies the amount of improvement depends to a large extent on the types of fibers present in the different layers, it was opportune to assess the impact of including FR fabric layers within the linesmen's clothing assemblies tested earlier.

Materials and Method

Samples of typical summer and winter outdoor clothing (Table 1) were tested again in the Flash Tester, only this time certain layers were replaced by some of the FR fabrics tested previously. Details of the fabrics used and the mean results for each sample are presented in Table 9. The test apparatus and procedure followed were identical to those described for the initial testing of current linesmen's clothing.

TABLE 6—The TPP Ratings for two-layer assemblies.

Outer Fabric				Inner Fabric				Two-Layer Assembly		
Fabric No.	Fiber/Finish	Mass per Unit Area, g/m²	TPP MJ/m²	Fabric No.	Fiber/Finish	Mass per Unit Area, g/m²	TPP MJ/m²	Mass per Unit Area, g/m²	Protection Time, s	TPP MJ/m²
9	85/15 FR cotton/ polyester	361	0.84	1	FR cotton	177	0.46	538	6.7	1.67
9	85/15 FR cotton/ polyester	361	0.84	16	aramid	178	0.72	539	7.3	1.81
9	85/15 FR cotton/ polyester	361	0.84	27	novoloid	163	0.45	524	7.2	1.56
11	FR wool	262	0.95	1	FR cotton	177	0.46	439	7.6	1.55
11	FR wool	262	0.95	16	aramid	178	0.72	440	6.3	1.31
11	FR wool	262	0.95	27	novoloid	163	0.45	425	6.3	1.31
19	aramid	293	0.90	1	FR cotton	177	0.46	470	11.8	2.19
19	aramid	293	0.90	16	aramid	178	0.72	471	6.5	0.97
19	aramid	293	0.90	27	novoloid	163	0.45	456	7.3	0.98
30	PBI	308	0.59	1	FR cotton	177	0.46	485	8.1	1.20
30	PBI	308	0.59	16	aramid	178	0.72	486	10.3	1.47
30	PBI	308	0.59	27	novoloid	163	0.45	471	8.5	1.32

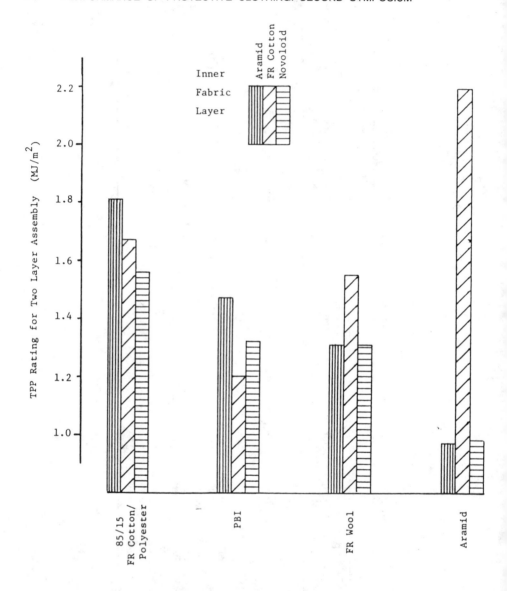

FIG. 8—*The TPP ratings for two-layer assemblies.*

Results and Discussion

In the earlier series, we demonstrated that the use of FR fabrics and finishes usually increased the TPP rating of single-layer fabrics. In contrast, we now find that the incorporation of one FR fabric layer in a two-layer assembly does not necessarily afford significant improvements in thermal protection. For example, compare the TPPs of Assemblies C and D with that for Assembly B in Table 9 and Fig. 6. The lack of a discernible improvement

TABLE 7—*Two-way ANOVA of TPP ratings with different outer and inner fabrics in two-layer assemblies.*

Source	Degrees of Freedom	Sum of Squares	Mean Square	F-Ratio
Outer fabric	3	587.73	195.91	0.61
Inner fabric	2	490.70	245.35	0.76
Interaction	6	1928.11	321.35	16.82[a]
Error	48	917.09	19.11	
Total	59	3923.63		

[a] Significant at $p < 0.01$ confidence interval.

parallels Krasny's finding that there is little difference in the total heat transmitted through two-layer assemblies consisting of an FR outerfabric and various non-FR underwear fabrics [15].

With three fabric layers, however, the incorporation of an outer FR layer does appear to improve the protection of the two non-FR inner layers. Table 9 shows that Assembly F with its FR treated cotton drill coveralls provides additional protection time and TPP over the non-FR layers in Assembly E.

Throughout these series of tests, we have identified three limitations in our use of the Flash Tester to generate TPP ratings. First, the conditions of arcing are idealized in the laboratory test and therefore cannot be used to recreate the exact conditions of any field situation. Second, the calorimeters do not give the same thermal response as human skin [7]. Finally, we recognize that by testing only the fabric layers and not the sewing threads, findings, trim, and closures, we are not testing the TPP of a finished garment. In other words, the results of the Flash Test should not be interpreted to represent any particular field conditions. However, the results can be used to rank the relative TPP of different single- and multiple-layer fabrics against electrical flashovers.

Conclusions

The Flash Tester has proven to be a simple, reliable, and useful instrument to measure the relative TPP of fabrics and other laminae when exposed to severe heat flux densities rich in radiant energy. It has potential as an analytical tool in understanding which properties of fibers, fabrics, and finishes influence TPP ratings. The findings from this study, derived

TABLE 8—*Results of linear regression to predict the TPP values of two-layer FR fabrics from other physical properties.*

Independent Variables	Intercept	Regression Coefficient	Pearson's r Correlation Coefficient	Coefficient of Multiple Determination
Mass per unit area of combined layers	−0.2424	0.0035	0.3887	0.1511
Thickness of combined layers	1.6088	−0.1248	−0.0602	0.0036
Surface lightness of outer layer	1.4835	−0.0009	−0.0841	0.0071

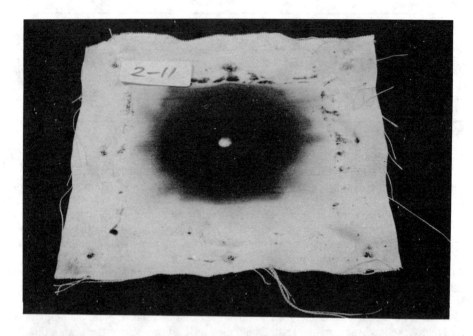

FIG. 9—*The back-side of the aramid inner fabric showing that it maintains a separate identity during TPP testing as a two-layer assembly in combination with the PBI fabric (top) or the 85/15 FR cotton/polyester denim (bottom).*

TABLE 9—*The TPP of FR modified summer and winter outdoor clothing on exposure to radiant energy flux density of 220 kW/m².*

Season	Part of Body Covered	Number of Fabric Layers	Current Clothing				FR Modified Clothing				
			Assembly	Mass per Unit Area of Assembly, g/m²	Protection Time, s	TPP, MJ/m²	Assembly	Modification Made to Current Clothing Assembly	Mass per Unit Area of Assembly, g/m²	Protection Time, s	TPP, MJ/m²
Summer	torso	2	B	524	5.4	1.38	C	cotton denim jacket replaced by FR cotton denim (#4)	534	5.2	1.33
							D	polyester/cotton shirt replaced by FR cotton shirting (#1)	515	4.7	1.28
Winter	legs	3	E	838	9.1	2.07	F	cotton coveralls replaced by FR cotton drill (#2)	844	13.2	3.33

from a wide range of different FR fabrics, point to fiber content and fabric mass per unit area (weight) as the properties that have a significant effect on the TPP ratings.

This study has demonstrated that a single garment or single layer of fabric alone is unlikely to provide adequate protection against second-degree burns when exposed to severe flash-over conditions with high flux densities and long exposure times (that is, several seconds). Two layers appear to be more effective, but only if both layers contain FR fabrics with different softening temperatures. For example, combined layers of aramid and FR cotton have been shown to be particularly effective. The use of assemblies with more than two layers is even more advantageous; each additional layer providing additional protection. Because of strong interactions between fabrics in multiple-layer assemblies, it is not possible at this time to predict the TPP performance of a combined assembly from a knowledge of the properties of the individual layers. Work is continuing in our laboratories to understand more clearly the nature of these multiple-layer interactions.

Disclaimer

The opinions expressed herein are solely those of the authors and do not necessarily represent existing or forthcoming policies of the affiliated or sponsoring institutions. The laboratory results reported in this paper should not be taken as predicting the performance of the same fabrics in the field. Nor is it the intention of the authors to recommend, endorse, exclude, or predict the suitability of any commercial product for a particular end-use.

Acknowledgments

Part of this research has been supported by the Canadian Electrical Association, Montreal, Canada, under Contract No. 114D203. We wish to thank the Association for permission to publish this work. We are indebted to Wallace Behnke, Sheila Brown, Jacques Castonguay, Ken Mount, Myron Musick, Nick Talanow, and Peggy Tyrchniewicz for their continuing support and encouragement. We are sincerely grateful for the technical assistance provided by Brigitta Badour, Cecile Clayton, Angela Dupuis, Keo Nishizeki, Al Symmons, Robert Talbot, Henry Teichroeb, Judy Teerhuis, and Barbara Westcott; and we are indebted to the following companies and institutions who graciously provided the fabric samples: Celanese Fibers Marketing, Charlotte, NC; W. H. Clegg and Associates, Mississauga, Ont., Canada; Cleyn and Tinker, Huntingdon, Que., Canada; Intertex Textiles, Mississauga, Ont., Canada; Kohjin, Osaka, Japan; Lincoln Fabrics, St. Catherines, Ont., Canada; Nippon Kynol, Osaka, Japan; Southern Mills, Atlanta, GA; Stackpole Fibers, Lowell, MA; J. P. Stevens, New York, NY; Testfabrics, Middlesex, NJ; Westex, Chicago, IL; and the Wool Bureau, Woodbury, NY.

References

[1] Watkins, S. M., *Clothing—The Portable Environment,* Iowa State University Press, Ames, IA, 1984, p. 122.
[2] Lee, R. H., "The Other Electrical Hazard—Electric Arc Blast Burns," *Record of 1981 Annual Meeting,* Industrial Applications Society Conference, Institute of Electrical and Electronic Engineers, New York, 1981, pp. 401–406.
[3] Baitinger, W. F., "Product Engineering of Safety Apparel Fabrics: Insulation Characteristics of Fire Retardant Cotton," *Textile Research Journal,* Vol. 49, No. 4, April 1979, pp. 221–225.
[4] Braun, E., Cobble, V. B., Krasny, J. F., and Peacock, R., "Development of a Proposed Flammability Standard for Commercial Transport Flight Attendant Uniforms," Report FAA-RD-75-17, NTIS Report ADA-033 740, National Technical Information Service, Springfield, VA, 1976.

[5] Behnke, W. P., "Thermal Protective Performance Test for Clothing," *Fire Technology,* Vol. 13, No. 1, Feb. 1977, pp. 6–12.

[6] Ernst, E. D., "Laboratory Test Techniques for Evaluating the Thermal Protection of Materials When Exposed to Various Heat Sources," NTIS Report AD-784 923, National Technical Information Service, Springfield, VA, March 1974.

[7] Schoppee, M. M., Welsford, J. M., and Abbott, N. J., "Protection Offered by Lightweight Clothing Materials to the Heat of a Fire," *Performance of Protective Clothing, ASTM STP 900,* R. L. Barker and G. C. Coletta, Eds., American Society for Testing and Materials, Philadelphia, 1986, pp. 340–357.

[8] King, M. W., Menzies, R. W., Doupe, B. E., Mathur, R. M., and Gole, A. M., "An Evaluation of the Protection Provided by Textile Fabrics Against High-Energy Electrical Flashovers: Design of Test Apparatus," *Behavior of Polymeric Materials in Fire, ASTM STP 816,* E. L. Schaffer, Ed., American Society for Testing and Materials, Philadelphia, 1983, pp. 89–106.

[9] Holcombe, B. V. and Hoschke, B. N., "Do Test Methods Yield Meaningful Performance Specifications?" *Performance of Protective Clothing, ASTM STP 900,* R. L. Barker and G. C. Coletta, Eds., American Society for Testing and Materials, Philadelphia, 1986, pp. 327–339.

[10] Baitinger, W. F. and Konopasek, L., "Thermal Insulative Performance of Single-layer and Multi-layer Fabric Assemblies," *Performance of Protective Clothing, ASTM STP 900,* R. L. Barker and G. C. Coletta, Eds., American Society for Testing and Materials, Philadelphia, 1986, pp. 421–437.

[11] Hoschke, B. N., Holcombe, B. V., and Plante, A. M., "A Critical Appraisal of Test Methods for Thermal Protective Fabrics," *Performance of Protective Clothing," ASTM STP 900,* R. L. Barker and G. C. Coletta, Eds., American Society for Testing and Materials, Philadelphia, 1986, pp. 311–326.

[12] Stoll, A. M. and Chianta, M. A., "Method and Rating Systems for Evaluation of Thermal Protection," *Aerospace Medicine,* Vol. 40, No. 11, Nov. 1969, pp. 1232–1238.

[13] Shalev, I. and Barker, R. L., "Analysis of Heat Transfer Characteristics of Fabrics in an Open Flame Exposure," *Textile Research Journal,* Vol. 53, 1983, pp. 475–483.

[14] Imhof, M. and Hewett, S., *Statpro: the Statistics and Graphics Database Workstation,* Wadsworth Professional Software, Boston, MA, 1983.

[15] Krasny, J. F., "Some Characteristics of Fabrics for Heat Protective Garments," *Performance of Protective Clothing, ASTM STP 900,* R. L. Barker and G. C. Coletta, Eds., American Society for Testing and Materials, Philadelphia, 1986, pp. 463–474.

[16] Shalev, I. and Barker, R. L., "Predicting the TPP of Heat Protective Fabrics from Basic Properties," *Performance of Protective Clothing, ASTM STP 900,* R. L. Barker and G. C. Coletta, Eds., American Society for Testing and Materials, Philadelphia, 1986, pp. 358–375.

Ingvar Holmér[1]

Thermal Properties of Protective Clothing and Prediction of Physiological Strain

REFERENCE: Holmér, I., "Thermal Properties of Protective Clothing and Prediction of Physiological Strain," *Performance of Protective Clothing: Second Symposium, ASTM STP 989*, S. Z. Mansdorf, R. Sager, and A. P. Nielsen, Eds., American Society for Testing and Materials, Philadelphia, 1988, pp. 82–86.

ABSTRACT: Heat stress and reduced work performance are often linked to work in protective clothing, especially when the rate of work or temperature or both of the environment are high. The resultant thermal insulation (I_T) and evaporative resistance (R_T) of garments and protective clothing systems have been measured using indirect calorimetry on subjects exercising in a climatic chamber. In different series of wear trials, the physiological responses associated with wearing these garments during 60 min of treadmill walking in hot environments (35 to 45°C, 30 to 40% relative humidity) were evaluated for young, male, healthy subjects. Obtained values for I_T and R_T were used in a physiological model to predict rectal temperature response to the same conditions as for the wear trials in order to compare predicted and measured rectal temperatures. The results showed that a combination of clothing heat exchange measurements on subjects and simple physiological modeling may be a useful technique for assessment of clothing thermal function.

KEY WORDS: heat stress, thermal insulation, evaporative resistance, moisture permeation, methods, predictive models, physiological responses, protective clothing

Personal protective clothing (PPC) is widely used in many areas to prevent workers from being exposed to harmful chemical substances and physical agents. A high level of protection is often achieved only at the expense of significant thermal discomfort and strain. Particularly in warm and hot environments, the thermal impact may, in fact, present a greater health hazard than the toxic substance [1]. In addition, poor designs and functions of PPC sometimes contribute to deterioration in work performance and increased accident risks. The PPC are often made of impermeable fabrics, which interfere with moisture transfer from the skin and, hence, impair body cooling by sweat evaporation [2–5]. Sophisticated methods using thermal manikins are available in a few laboratories for the assessment of the impact of clothing thermal properties on human function and performance [3,4,6]. This paper presents an evaluation of the model by Givoni and Goldman [7] for prediction of physiological strain associated with work in PPC, based on measurements of clothing thermal properties using subjects exercising in a climatic chamber [2].

Determination of Thermal Properties of Clothing

Heat exchange with the environment takes place by radiation, convection, and evaporation. Clothing determines this heat exchange by its thermal insulation (I_T) and evaporative

---

[1] Professor, Climatic Physiology Unit, National Institute of Occupational Health, S-171 84 Solna, Sweden.

resistance (R_T). The following two equations apply and may also serve as definitions of I_T and R_T, respectively [3,4]

$$I_T = \frac{t_{sk} - t_o}{R + C} = \frac{t_{sk} - t_o}{M - W - RES - E - S} \qquad (1)$$

where

I_T = insulation of clothing and boundary air layer, m² °C/W, (or in clo units, where 1 clo = 0.155 m² °C/W);
$R + C$ = radiative and convective heat exchange, W/m²;
t_{sk} = mean skin temperature, °C;
t_o = ambient operative temperature, °C;
M = metabolic energy production, W/m²;
W = external mechanical work rate, W/m²;
RES = respiratory heat exchange, W/m²; and
S = body heat storage rate, W/m².

$$R_T = \frac{p_{sk} - p_a}{E} \qquad (2)$$

where

R_T = evaporative resistance of clothing and boundary air layer, Pam²/W;
E = evaporative heat exchange, W/m²;
p_{sk} = water vapor pressure at the skin surface, Pa; and
p_a = water vapor pressure of ambient air, Pa.

Equations 1 and 2 form the basis of a method for actual measurements of I_T and R_T of clothing, when worn by subjects during exercise in a climatic chamber, that we have used in our laboratory for some years. The technique is reported in detail elsewhere [2]. In brief, it is based on indirect calorimetry of subjects performing exercise in a climatic chamber. Heat transfer and the driving temperature and pressure gradients are determined simultaneously and continuously during the steady-state phase of a 1-h exposure to a standardized environment, normally characterized by a work rate of 50 W, 20°C, and 30% relative humidity. Evaporative heat loss is determined by a continuous weighing technique. Water vapor pressure is measured by continuously sampling air from representative skin surface areas and ambient air. Radiative and convective heat losses are determined by measuring the other relevant factors in the heat balance equation (Eq 1, second part). During the same experiment, physiological and subjective responses are recorded.

The technique measures differences between garments of relevance for their heat transfer properties [2,8]. Similar approaches to the analysis of clothing heat transfer have been made by Goldman and coworkers at Natick [3,7] and Meechels and Umbach in Hohenstein [4]. However, their techniques rely to a great extent on measurements on thermal manikins.

Thermal Properties of Garments

The garment assemblies used in the present analysis and their thermal properties (I_T and R_T) are presented in Table 1. These values were obtained in various investigations carried out in our laboratory [2,8]. Values for each garment represent the mean value of measurements on at least three subjects. The values are averaged for the whole body surface area. It should be emphasized that measurements were made during light bicycle exercise and

TABLE 1—*Total value for thermal insulation and evaporative resistance of selected work clothing and PPC based on measurements on subjects exercising in a climatic chamber [2,8]. All ensembles comprised socks and shoes. Predicted values for rectal temperature (T_{re}) after 60 min of exposure. Values refer to an average man of 1.8 m^2 body surface area with a metabolic rate of 350 W in an environment at 30°C, a relative humidity of 30%, and an air velocity of 0.2 m/s. Mean skin temperature and initial rectal temperature were assumed to be 37.0°C. For further explanations see text.*

Clothing Ensemble	I_T, clo	R_T, Pa m²/W	T_{re}, (60) °C
Shorts	0.8	8	37.7
Overall, underwear	1.2	19	37.9
Coverall, overall, underwear	1.3	35	38.2
Rainwear (Gore-Tex), overall, underwear	1.3	34	38.1
Rainwear (polyurethane-coated), overall, underwear	1.3	77	38.6
Firefighter suit (Nomex), overall, underwear	1.4	30	38.1
Protective suit (Gore-Tex), overall, underwear	1.3	37	38.2
Flame-retardant woolen garment, overall, underwear	1.2	51	38.4
Aluminum-coated suit, overall, underwear	1.1	70	38.5
Chemical protective suit, overall, underwear	0.8	180	38.7

values, accordingly, include some of the dynamic effects caused by moisture absorption and body motion, that is, the pumping effect [4,6,7]. Indeed, one advantage with the method [2] is, that corrections of the I_T and R_T values may not be required to account for these effects, when clothing function in the actual wear situation must be assessed. Such corrections are needed when thermal manikin data are used [7].

It is clear from Table 1 that the garments differed widely with respect to their effect on heat exchange. This was most apparent for evaporative heat exchange. The R_T-values varied between 19 for a standard workwear (cotton-polyester overall plus underwear) and 180 Pa m²/W for a garment assembly comprising the same overall and underwear worn underneath a chemical protective suit. Rainwear, as a rule, is made of impermeable fabric and presents high R_T. New, microporous fabrics (Gore-Tex) have lower R_T-values and, subsequently, make it possible to combine proofness to liquid water with moisture permeation. Thermal insulation, however, was relatively independent of type of clothing and varied between 0.8 and 1.4 clo. First, there is an effect of body motion, that reduces the insulation and makes differences between garments smaller than would have been the case if values were obtained with a static, thermal manikin [4,6]. Second, impermeable clothing causes considerable sweat accumulation, which might further reduce insulation. This effect was most evident for the last four garments in Table 1 and, particularly, for the chemical protective suit.

Physiological Wear Trials

The physiological responses in the garment assemblies were investigated in a different series of experiments comprising 60 min of work (walking on a treadmill) in a climatic chamber at 35 or 45°C operative temperature, 30 to 40% relative humidity, and air velocity <0.5 m/s. Metabolic rate was constant for each 60-min period but varied in the series between 250 to 400 W. Healthy, male subjects volunteered for the experiments and exercised on different days in each of the nine different garment assemblies. Oxygen uptake and skin and rectal temperatures were measured during work, and the values obtained at the end of the 60-min period were used for comparison with predicted values.

Results

It follows from Eqs 1 and 2 that a given temperature and vapor pressure gradient allows more heat to be dissipated when I_T and R_T are low. The physiological strain may be supposed to reflect this variation in clothing thermal properties, but is not so easy to determine. In the model developed by Givoni and Goldman [7], the rectal temperature response to both transient and steady-state conditions can be predicted as function of clothing, climate, and activity level. This model was evaluated by using the measured I_T and R_T values for the garment assemblies in Table 1 and calculating the rectal temperature response after 60 min of work in the same conditions of climate and activity as in the physiological trials. The measured and predicted rectal temperatures were then compared.

In the first analysis, the predicted temperatures significantly overestimated the measured temperatures. Two factors may readily explain most of this difference. Ventilation of clothing is different during walking compared to bicycling [6] and mean skin temperature during the tests averaged 37°C rather than 36°C, as assumed in the model. In the second analysis, the I_T and R_T values were corrected by the formulas given by Givoni and Goldman [7] and 36 was replaced by 37 in the prediction formula. Figure 1 presents the results. There was a significant correlation between measured and predicted rectal temperatures ($r = 0.716$, $p < 0.001$). However, predicted values tended to overestimate the increase in rectal tem-

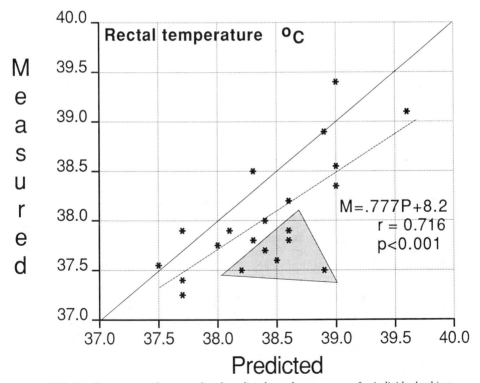

FIG. 1—*Comparison of measured and predicted rectal temperatures for individual subjects during exercise at metabolic rates of 250 to 400 W in ambient conditions of 35 to 45°C, 30 to 40% relative humidity, and air velocity <0.5 m/s. Shaded area denotes six experiments with heavy protective clothing (see text).*

perature. This was particularly true for garments with high absorptive properties and medium resistance to vapor transfer (aluminum-coated garments and woolen garments). The physiological measurements revealed a smaller increase in rectal temperature with these garments. There was also considerable accumulation of sweat in the garments. The effect of moisture absorption on heat exchange is not considered in the predictive model and is difficult to evaluate. Other factors that of course affect the predictions are individual characteristics like physical work capacity and heat acclimatization.

Values plotted in Fig. 1 represent several different combinations of climate and activity level. In order to more clearly illustrate the effect of the different types of PPC, the rectal temperature after 60 min of work was calculated for a metabolic rate of 350 W in an environment at 30°C, 30% relative humidity, and 0.2 m/s. The predicted values are given in Table 1. As expected, the rectal temperature increase was related to the type of clothing worn and their heat transfer properties, especially the evaporative resistance. The increase is smallest with shorts and greatest with heavy protective clothing. It is interesting to note that the rectal temperature increase is much smaller with Gore-Tex rainwear compared to polyurethane-coated rainwear.

In conclusion, it is a well-known and badly solved problem, that personal protective clothing against chemical and physical hazards often interferes with heat exchange and leaves the wearer with discomfort and considerable thermal strain. Moisture permeation of PPC is critical to the development of heat stress and impaired performance. Although more work in this field needs to be done, the present results suggest that a combination of a method to determine clothing heat transfer properties on subjects in climatic chamber tests and physiological modeling can be a useful technique to

1. predict thermal strain associated with work in PPC,
2. classify various types of PPC in terms of their thermal impact on the wearer, and
3. evaluate effects of new fabrics and garment design on clothing thermal function.

This work has been carried out in part with support from the Swedish Work Environment Fund.

References

[1] O'Neill, D. H. and Whyte, T. H., "The Danger of Wearing Impermeable Clothing While Spraying," *Journal of Social and Occupational Medicine,* Vol. 35, 1985, pp. 10–13.
[2] Holmér, I. and Elnäs, S., "Physiological Evaluation of the Resistance to Evaporative Heat Transfer by Clothing," *Ergonomics,* Vol. 24, 1981, pp. 63–74.
[3] Martin, H. de V. and Goldman, R., "Comparison of Physical, Biophysical and Physiological Methods of Evaluating the Thermal Strain Associated with Wearing Protective Clothing," *Ergonomics,* Vol. 15, 1972, pp. 337–342.
[4] Meechels, J. and Umbach, K. H., "Thermophysiologische Elgenschaften von Kleidungssystemen," *Melliand Textilberichte,* Vol. 57, 1976, pp. 1029–1040.
[5] Nishi, Y. and Gagge, A. P., Moisture Permeation of Clothing—A Factor Governing Thermal Equilibrium and Comfort," *Transactions,* American Society of Heating, Refrigeration and Air Conditioning Engineers, Vol. 76, 1970, pp. 137–145.
[6] Olesen, B. W., Silwinska, E., Madsen, T. L., and Fanger, P. O., "Effect of Body Posture and Activity on the Thermal Insulation of Clothing; Measurements by a Moveable Thermal Manikin," *Transactions,* American Society of Heating, Refrigeration and Air Conditioning Engineers, Vol. 88, 1982, p. 2.
[7] Givoni, B. and Goldman, R., "Predicting Rectal Temperature Response to Work, Environment and Clothing," *Journal of Applied Physiology,* Vol. 32, 1972, pp. 812–822.
[8] Holmér, I., "Termofysiologiska egenskaper hos arbets- och skyddskläder, (Thermophysiological Properties of Work Clothing and Protective Clothing)," Arbete och Hälsa 12, in Swedish, National Board of Occupational Safety and Health, Solna, Sweden, 1986, pp. 1–62.

Roger L. Barker,[1] Sandra K. Stamper,[2] and Itzhak Shalev[3]

Measuring the Protective Insulation of Fabrics in Hot Surface Contact

REFERENCE: Barker, R. L., Stamper, S. K., and Shalev, I., **"Measuring the Protective Insulation of Fabrics in Hot Surface Contact,"** *Performance of Protective Clothing: Second Symposium, ASTM STP 989,* S. Z. Mansdorf, R. Sager, and A. P. Nielsen, Eds., American Society for Testing and Materials, Philadelphia, 1988, pp. 87–100.

ABSTRACT: This research demonstrates procedures useful in measuring the ability of fabrics to block heat transfer when in contact with a hot surface. Experiments show how the temperature of the contacting surface, contact pressure, and fabric moisture content affect protective insulation in conductive tests. Single-layer fabrics made with flame-resistant cotton, rayon, and wool; glass fabrics, ceramic fabrics; and fabrics from aramids, novoloids, modacrylic, and stabilized acrylic fibers are evaluated. Statistical methods show the correlation between measured fabric properties and insulative values. An analytical model is developed that predicts a burn protection index from simple measurement of fabric thickness.

KEY WORDS: thermal protective performance, burn injury, protective clothing, conductive exposure, hot surface contact, heat hazards

An important property of fabrics used in safety clothing is the ability to insulate against conductive heat transfer or to provide protection against burn injuries in the case of accidental direct contact with a hot surface. Such hazardous exposures are possible in occupational environments where workers handle hot materials or in fire fighting situations where firemen contact building surfaces that have been highly heated by fire. A hot object picked up by a gloved hand or clothing splashed by hot liquid, both result in direct physical contact between the garment and a solid or liquid heat source. Heat transfer to the fabric is by direct conduction, together with infrared radiation into the pores of the structure. The relative contribution of individual heat transfer mechanisms is largely dependent upon the density of the fabric structure.

Although a considerable amount of research has been conducted to measure the resistance of protective fabrics to flames or to radiant heat, few studies have been made to evaluate thermal insulation in high-intensity contact exposures. Furthermore, a need exists to demonstrate a reliable laboratory procedure for measuring the protective performance of fabrics in hot surface exposures. The purpose of this research, therefore, is to demonstrate the efficacy of such a test procedure by comparing the resistance of a selected group of safety fabrics. It is also the intention of this research to study the influence of fiber composition and fabric construction on the thermal protective performance in high-intensity tests of conductive heat transfer.

[1] Department of Textile Engineering and Science, North Carolina State University, Raleigh, NC 27695-8301.
[2] Department of Textile Engineering and Science, North Carolina State University, Raleigh, NC 27695-8301; presently at General Motors Corporation, Warren, MI 48090.
[3] Department of Textile Engineering and Science, North Carolina State University, Raleigh, NC 27695-8301; presently at Shenkar College of Textile Technology and Fashion, Ramat-Gan 52526, Israel.

Experimental Procedures

Measurement of Thermal Protective Insulation

The experimental apparatus, illustrated in Fig. 1, consists of a laboratory hot plate, with a 16 by 16 cm (6.25 by 6.25 in.) flat heated surface. The exposure surface is a 0.6 cm (0.25 in.) by 10 cm (4 in.) by 15 cm (6 in.) polished aluminum plate. The plate temperature is monitored using an iron-constantan thermocouple, located in a 0.2 cm (3/32 in.) diameter hole drilled from the edge to the center of the aluminum plate.

A blackened copper disk mounted in the center of an insulating board serves as the heat sensor. Four thermocouples, embedded in the back side of the disk, measure temperature of the copper disk.

To make the thermal protective insulation measurements, specimens were attached to the sample holder and the required amount of weight placed on top of the assembly. The start of the exposure was indicated using the event marker on the chart recorder.

The end point of the test was determined using a plot of transmitted thermal energy versus time to second-degree burn [1].

Results and Discussion

Effect of Plate Temperature

Table 1 gives insulation values measured at different plate temperatures. Figure 2 shows that protection time drops off exponentially with surface temperature.

None of the fabric samples were visibly degraded at 200°C. However, several fabrics charred, melted, or embrittled when the exposure was 300°C or higher. The flame resistant (FR) cotton and FR rayon samples scorched at 300°C. Aramids and fabrics made with high-

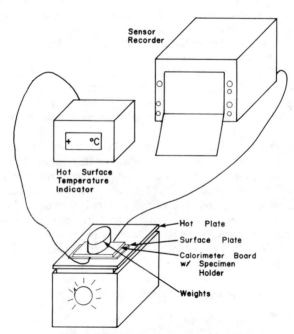

FIG. 1—*Schematic of test apparatus.*

TABLE 1—*Insulation of fabrics in hot surface contact.*

Sample No.	Fabric Composition	Tolerance Time, s,[a] at Surface Temperature, °C				
		100	150	200	300	375
1	Nomex III (6 oz)[b]	23.0	6.6	3.9	1.8	1.2
2	Nomex III (10 oz)	>30.0	12.3	5.8	2.8	2.2
3	Kevlar (8 oz)	23.3	7.8	4.4	2.4	1.8
4	Aluminized Nomex III (10 oz)	12.1	4.1	2.5	1.3	1.3
5	FR cotton (13 oz)	20.2	5.8	3.4	1.8	1.6
6	FR cotton (7.3 oz)	6.6	2.2	1.3	0.0	0.0
7	Aluminized FR cotton (14 oz)	16.2	4.8	2.9	1.7	1.4
8	FR wool (18 oz)	>30.0	>30.0	15.9	5.3	3.0
9	FR rayon (9 oz)	12.0	3.3	2.0	1.2	0.0
10	FR modacrylic (7 oz)	23.9	6.9	3.5	1.6	1.3
11	Novoloid (9 oz)	14.2	4.5	2.5	1.4	1.2
12	Aramid/glass (18 oz)	>30.0	>30.0	19.9	10.0	7.4
13	Silica (35 oz)	>30.0	>30.0	27.1	14.3	11.6
14	PAN (17 oz)	>30	16.3	8.1	4.2	2.7
15	Aluminized PAN (18 oz)	27.3	6.5	3.4	2.0	1.6

[a] Time to second-degree burn injury predicted using the Stoll criterion.
[b] Metric conversion; 1 ounce = 28.35 grams.

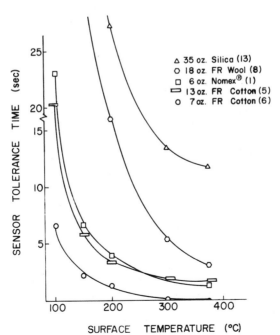

FIG. 2—*Effect of hot plate temperature on protective insulation.*

temperature inorganic fibers showed little visible change in appearance at the highest test temperature, although one Nomex sample discolored at 375°C (Sample 2). A silica sample (Zetex[4] 1100) had a sticky finish that scorched and stuck to the hot surface, especially at high temperatures. The FR wool and FR modacrylic samples were the most visibly damaged by the exposures. The insulative value of wool samples deteriorated rapidly at 300°C and higher exposure temperatures. At these temperatures, wool specimens charred, melted, and stuck to the hot surface. They also shrank and became brittle and stiff. The modacrylic fabric (Sample 10) degraded so severely in the 375°C test that it fell apart when removed from the hot surface.

Effects of Contact Pressure

Applied pressure increased the rate of heat transfer by increasing contact between the fabric and the hot surface and by decreasing the effective thickness of compressible fabrics. In our experiments, pressure was systematically varied to range between 3.45 to 6.89 kPa (0.2 and 1.0 psi). Figure 3 shows that the loss in the protective insulation is much greater between 3.45 to 6.89 kPa (0.2 and 0.5 psi) than between 1.18 to 6.89 kPa (0.5 and 1.0 psi). These results also show that the decrease in insulation is most pronounced in the thickest fabrics, or samples that undergo the greatest loss in thickness in compressive loading (for example, FR wool).

Correlation with Fabric Properties

Figure 4 shows how fabric properties correlate with conductive insulation measured at 200°C. These data indicate that protective insulation is directly correlated with thickness,

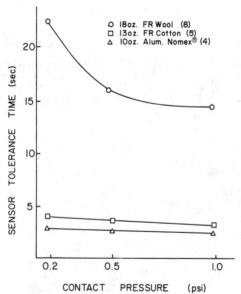

FIG. 3—*Effect of contact pressure on protective insulation in hot surface contact.*

[4] Registered trademark of Newtex Industries, Inc.

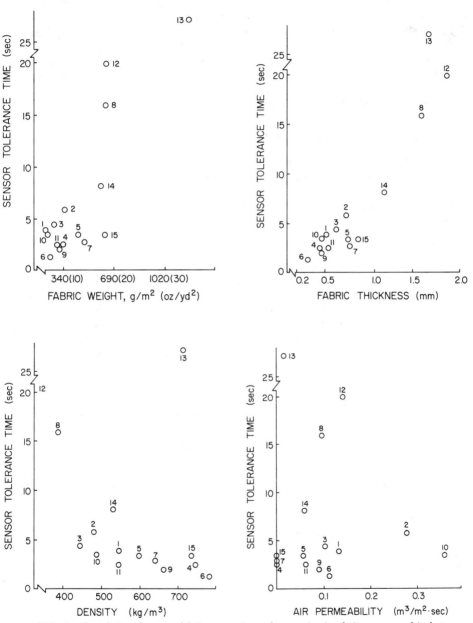

FIG. 4—*Correlations between fabric properties and protective insulation measured in hot plate test.*

although the relationship is not simply linear. There is weaker correlation between insulation and fabric weight. Insulation decreased with increased fabric density, as expected in tests of conductive heat transfer. On the other hand, protective performance is not strongly correlated with moisture regain or fabric air permeability.

Table 2 shows thermal protective insulation adjusted for fabric thickness. This comparison shows that aramid, aramid/glass, and silica samples afford superior insulative values per millimeter thickness. In comparison, FR cotton, FR rayon, novoloid, and aluminized fabrics consistently provide lower insulative values per millimeter fabric thickness. An FR wool sample provided better than average insulator at contact temperatures below 200°C, but at a contact temperature of 300°C, provided only average thickness normalized insulation.

Effect of Moisture on Insulation

Figure 5 shows that protective insulation decreases as the moisture contained by the fabric increases. While differences between dry and conditioned samples are not dramatic, protective insulation is severely reduced in fabrics deliberately soaked with water. Typical calorimetric traces for oven dry, conditioned, and wet specimens are shown in Fig. 6. These graphs show that the fastest rate of heat transfer occurs in wet fabric samples. These data also show that the degree of the moisture effect depends on the capacity of the fabric to hold liquid water. Therefore, the insulation of lightweight fabrics, or fabrics that can hold comparatively little moisture in liquid form (for example, Samples 2 and 6), is affected to a lesser degree (Fig. 6). Samples that hold more liquid (Samples 5 and 9) show substantial differences between fabrics tested in the wet and dry state (Fig. 6). The effect of moisture vaporization is clearly indicated by the leveling of the calorimetric trace at 100°C. Beneficial endothermic effects occur after heat sufficient to produce second-degree burn injury has been transmitted. Apparently, the effect that moisture has on increasing the heat capacity of the fabric is far outweighed by the increased thermal conductivity of wet materials.

TABLE 2—*Thickness normalized protective insulation.*

Sample No.	Fabric Composition	Tolerance Time Per Unit Thickness, s/mm at Surface Temperature, °C	
		200	300
1	Nomex III	7.3	3.4
2	Nomex III	8.2	3.9
3	Kevlar	7.2	3.9
4	Aluminized Nomex III	5.6	2.9
5	FR cotton	4.6	2.4
6	FR cotton	4.1	0.0
7	Aluminized FR cotton	3.9	2.3
8	FR wool	10.0	3.3
9	FR rayon	4.2	2.5
10	FR modacrylic	7.5	3.4
11	Novoloid	4.7	2.6
12	Aramid/glass	10.6	5.4
13	Silica	16.3	8.6
14	PAN	7.3	3.8
15	Aluminized PAN	4.1	2.4

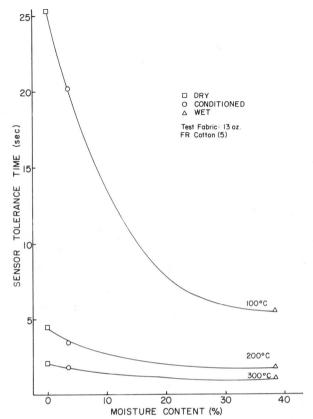

FIG. 5—*Effect of fabric moisture in protective insulation measured in hot plate test.*

Analysis of Correlations with Fabric Properties

Measured fabric properties were used in a statistical analysis to determine which parameters predict thermal insulation [2]. The result of this analysis is a mathematical model that reliably computes insulation time from simple measure of fabric thickness (Fig. 7).

Figure 8 compares experimental data and insulative values predicted using the statistically derived model. Since the model computes insulation based on fabric thickness, it permits a comparison of specific fabrics on the basis of construction and fiber composition. For example, we notice that aramid fabrics (Figs. 8a and b) provide better protected insulation than predicted by the model, based on measured thickness.

We also observe that insulation provided by FR cotton fabrics is determined by the types of flame retardant finish used. One FR cotton sample (Fig. 8c) insulates about the same as predicted on the basis of its thickness. However, another cotton sample (Fig. 8d) provides less thermal insulation per millimeter of thickness. We believe that this is because the finish on this particular sample causes it to be very dense, thereby increasing solid heat conduction.

We found that FR wool is a good insulator, especially at lower contact temperatures (Fig. 8e). However, wool degrades at temperatures above 250°C. This can be beneficial in some exposures, such as to an open flame [3]. In a contact exposure, however, degradation is detrimental to the insulation because wool melts and sticks to the heat source. Wool also

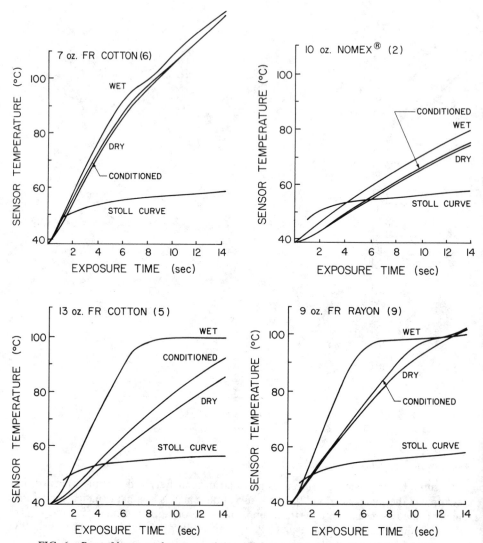

FIG. 6—*Rate of heat transfer measured through dry, conditioned, and wet fabric samples.*

shrinks, reducing the effective thickness. Shrinkage increases the heat transfer by compacting fibers and increasing solid conduction. Figure 8e shows that the effective insulation of wool drops below the predicted insulative value at contact temperatures above 250°C.

The insulation provided by the FR-treated rayon sample is below that predicted on the basis of fabric thickness (Fig. 8f). We believe that this results from the fact that rayon has a high moisture content and density. These factors contribute to increase solid conduction through the fabric.

The modacrylic sample insulates better than predicted at low temperatures, but the insulation provided is less as the material degrades and conductivity increases (Fig. 8g). On the other hand, the insulation afforded by the novoloid fabric is close to that predicted by the empirical model (Fig. 8h).

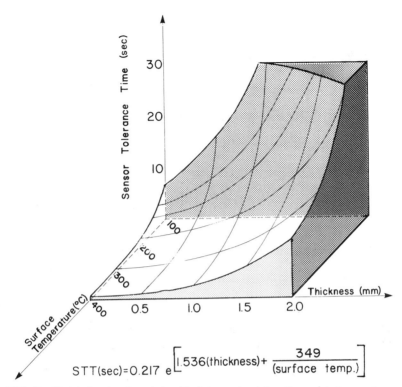

$$STT(sec) = 0.217\ e^{\left[1.536(\text{thickness}) + \dfrac{349}{(\text{surface temp.})}\right]}$$

FIG. 7—*Model showing the relationship between insulation time, plate temperature, and fabric thickness.*

The aramid/glass sample has lower protective insulation than predicted, from its thickness (Fig. 8i). However, the aramid and silica samples are better insulators than predicted (Figs. 8j and 8k). Figure 8i shows that when aramid and glass are combined in a sheath core yarn construction, the protection observed is worse than expected based on this model.

A polyacrylonitrile (PAN) fabric is shown to be a better insulator, especially at lower plate temperatures (Fig. 8l). At higher temperatures, observed and predicted insulation times are in close agreement.

An aluminized layer acts as a good conductor of heat in contact exposures. It is not surprising, therefore, that aluminized samples provide less protective insulation than the substrate materials (Figs. 8m and 8n).

Conclusions

A procedure has been demonstrated for measuring the protective insulation of fabrics in contact with a hot surface. The temperature of the hot surface and the pressure applied during the exposure are shown to be controlling test variables. Resistance to transmission of injurious levels of heat decreases as the test temperature and contact pressure increase. The extent of the temperature and pressure effects is a characteristic of the material sample. Insulative performance of some fabrics deteriorate significantly at surface temperatures above 300°C, as heat degradation produces dramatic changes in physical properties.

This research provides insight into the relationship between fabric properties and conductive insulation of heat-resistant fabrics at high temperatures. Classic correlations between

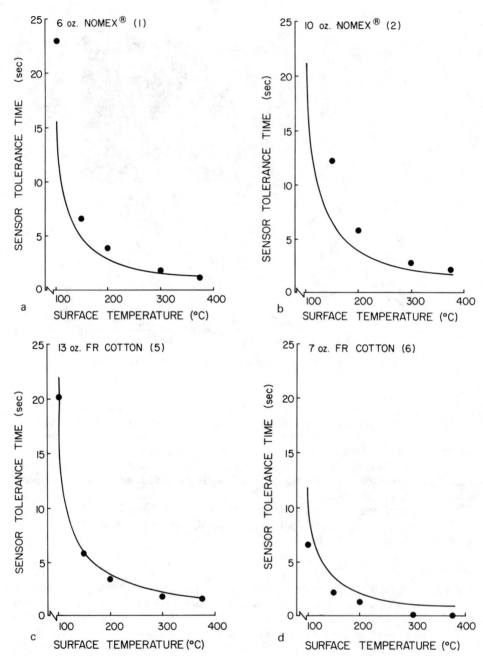

FIG. 8—*Correlation between observed and predicted insulation for selected materials (solid curve indicates predicted insulation).*

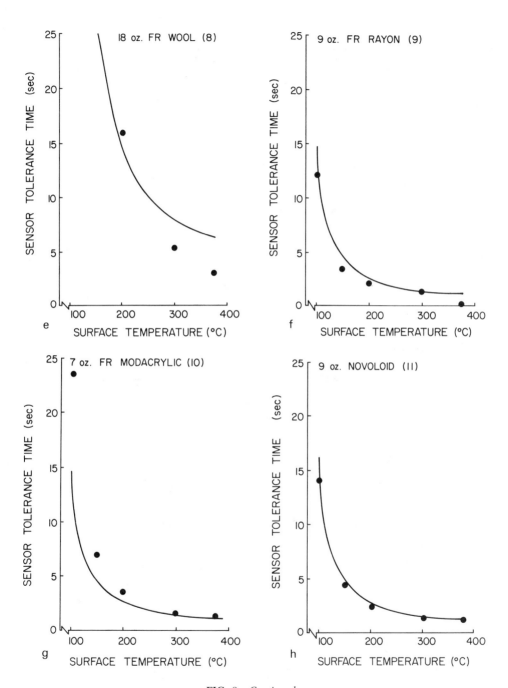

FIG. 8—*Continued.*

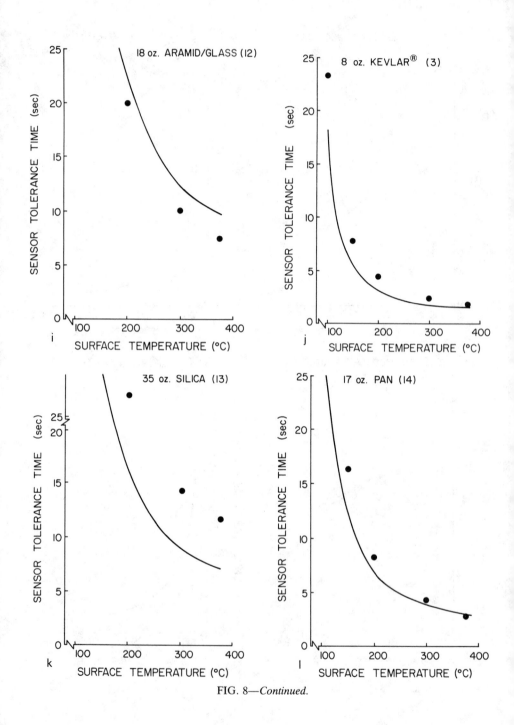

FIG. 8—*Continued.*

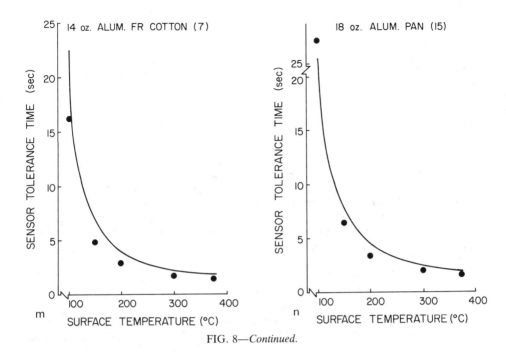

FIG. 8—*Continued.*

fabric thickness, weight, and insulation are observed, although the relationship between fabric structure and insulation is not simple. Confirmation that fabric thickness is an important factor is consistent with research in this field, and predicted by heat transfer theory in fibrous structures, where the air-to-fiber ratio is a deciding factor. In comparison, fabric air permeability has little effect on the rate of heat transfer in conductive exposures.

This research shows that the insulation of single-layer fabrics in hot surface exposures is significantly lowered by soaking the fabric with water. The extent to which moisture affects insulation is determined largely by the capacity of the fabric to hold liquid water. We associate moisture-induced reductions in protective insulation to the fact that water increases the thermal conductivity of the fabric, therefore, increasing the rate of heat transfer to the thermal sensor. This is a significant finding, especially in light of constructions found in protective clothing ensembles and glove constructions, which use membrane barriers to block the transfer of heat by vaporized moisture. Another study [4] has shown that the role of fabric moisture in thermal protective performance is complex and controlled almost entirely by the heat exposure itself.

We have demonstrated an emperical model that uses fabric thickness to predict insulation time with remarkable reliability. This model provided a convenient way of comparing different materials and for engineering fabrics for optimum insulation in hot surface exposures.

Finally, it must be emphasized that the results of this particular study apply only to single-layer fabrics tested in specific laboratory conditions. A useful area for further research would be to conduct experiments using multiple-layer fabric assemblies and in configurations typical of those found in protective apparel, especially glove constructions. Such research would contribute to the confidence with which laboratory findings can be used to predict field performance.

Acknowledgments

We are grateful for technical advice and encouragement provided by W. P. Behnke of the duPont Company. The procedures described in this paper derive from a method developed by Behnke. It is similar to one being considered by the American Society for Testing and Materials.

Author's Note

Care must be exercised in interpreting these test data or in deriving conclusions concerning the safety benefits from these data. These data are the results of a particular laboratory exposure; extrapolations to other types of heat exposures cannot be made. The correlation between this laboratory measurement of thermal properties and field protective performance is unknown, and no claims are made. These data should only be used to measure and describe the properties of materials, products, or assemblies in response to precisely the same controlled laboratory exposure and conditions described herein. They should not be used to predict the thermal hazard or thermal risk of materials, products, or assemblies under actual field conditions.

References

[1] Stoll, A. M., Chianter, M. A., and Piergalline, J. R., "Heat Transfer Measurements of Safety Apparel Fabrics," Technical Report, NADC-78209-60, Naval Air Development Center, Warminster, PA, 1978.
[2] Stamper, S. K., "Measuring the Thermal Protective Performance of Fabrics in Hot Surface Contact," M.S. thesis, School of Textiles, North Carolina State University, Raleigh, NC, 1983.
[3] Shalev, I. and R. L. Barker, "Analysis of Heat Transfer Characteristics of Fabrics in an Open Flame Exposure," *Textile Research Journal,* Vol. 53, No. 8, Aug. 1983, pp. 475–482.
[4] Lee, Y. M. and R. L. Barker, "Effect of Moisture on the Thermal Protective Performance of Heat-Resistant Fabrics," *Journal of Fire Sciences,* Dec. 1986.

Krister Forsberg[1]

Evaluation of Fourteen Fabric Combinations, One Glove Material and Three Face Shield Materials to Molten Steel Impact

REFERENCE: Forsberg, K., **"Evaluation of Fourteen Fabric Combinations, One Glove Material and Three Face Shield Materials to Molten Steel Impact,"** *Performance of Protective Clothing: Second Symposium, ASTM STP 989,* S. Z. Mansdorf, R. Sager, and A. P. Nielsen, Eds., American Society for Testing and Materials, Philadelphia, 1988, pp. 101–107.

ABSTRACT: The heat resistance of fourteen fabric combinations, one glove material, and three face shield materials has been evaluated by controlled splashes of 1 kg of molten steel. The test procedure is described in ASTM Evaluating Heat Transfer through Materials Upon Contact with Molten Substances (F 955-85). Five- and forty-second calorimeter thermal responses were plotted. The heat resistance was determined by measuring the temperature rise and calculating the maximum heat flux. The test materials were also visually examined.

   The test materials were aluminized rayon; jacket/pant materials in wool, cotton, and wool/ Cordelan; underwear materials in wool, cotton/modal, and Cordelan/polyester; Kevlar; and face shield materials in polysulfone polycarbonate and acetate.

   The test results indicated <1 cal/cm^2/s of maximum heat flux in six of the fabric combinations and the three face shields.

   The evaluation of the fabric combinations and face shield materials to molten steel impact was a part of the project named "Development of Protective Clothing Working at Continuous Casting."

   It was concluded that the proper combination of outer clothing and underwear clothing for the "8-h garment" can be a more comfortable substitute for the combination of aluminized clothing and outer clothing used today in front coats and flame-retardent clothing.

KEY WORDS: heat resistance, molten steel impact, heat protective clothing, protective clothing

The technology of continued casting has improved the working environment by reducing the amount of steel splashes and eliminating the radiant heat. This positive development has decreased the risks and the workers are no longer motivated to wear the traditional aluminized protective clothing. However, even if this technology has decreased the risks, steel splashes are not entirely eliminated.

The aim of this project was:

1. To develop an "8-h garment" that has the same protection as traditional protective garments but with better comfort qualities.
2. To develop a glove for the founders.
3. To develop a face shield for the founders.

This paper will describe the evaluation of fourteen fabric combinations, one glove material,

---

[1] Researcher, Department of Work Science, The Royal Institute of Technology, S-100 44 Stockholm, Sweden; presently, AGA, AB, 18181 Lidingö, Sweden.

and three face shield materials to molten steel impact. The splash tests were performed by Southern Research Institute in Birmingham, Alabama under supervision of Charles E. Bates.

Procedure

The standardized conditions for molten steel impact evaluation consist of pouring 1 kg of steel at a temperature of approximately 1580°C and splashing the fabric samples attached to a transite board at an angle of 70° from the horizontal and from a height of 50 cm. The test procedure is described in detail in ASTM "Evaluating Heat Transfer through Materials Upon Contact with Molten Substances" (F 955-85).

Tested Materials

Five of the fabrics were jacket/pant materials. One of the fabrics was aluminized coat material. Another one was a liner to jacket/pant materials. A third fabric was an absorption garment material. A fourth fabric was a woven Kevlar. Two fabrics were underwear materials. One felted Kevlar was a glove material. Finally, three of the plastics were face shield materials.

The test materials were labeled by a code (shown in Table 1).

All fabrics, except the absorption garment material, were flame-retardent treated.

The five outer garment materials (PW1, PW2, PC1, PC2, and PWP) were combined with the different underwear garment materials. The absorption garment material (UA) was placed between PW1 and the underwear materials and PC2, respectively.

Visual Examination

The visual appearance of each of the outer layer fabrics after impact with molten steel was subjectively related in four categories. These were charring, shrinkage, metal adherence, and perforation. A rating system, with 1 representing good behavior and 5 representing poor behavior, was used.

Result and Discussion

The results of the heat transfer tests are presented in Table 2 and the visual ratings of fabrics exposed to molten steel are presented in Table 3. The temperature rise in 5 and 40 s and heat flux curves were plotted for the tested materials. Three of the 40-s curves are shown in Figs. 1 through 3.

The fabric combination PAL/PC2, which is the most common combination in traditional protective clothing, and the Kevlar felt (PU2) indicated the lowest heat flux and temperature rise. However, several combinations of non-aluminized material and underwear materials also included low heat flux (<1 cal/cm^2/s). The maximum temperature rise was less than 12°C for these fabrics. The woven Kevlar performs worst regarding lowest heat flux. The best fabrics in terms of visual appearance after impact and minimum heat transfer were PAL/PC2, PW1/UW, PC1/UA/UM, and PC2/UM.

The face shield materials shed molten metal well, and the maximum heat flux values were less than 0.65 cal/cm^2/s.

Eight replicate tests were also conducted on PW1, and the results are presented in Table 4. Standard deviations for the temperature rise and maximum heat flux were 1.4°C and 0.26 cal/cm^2/s, respectively, to give coefficients of variation of 9.6 and 9.9%, respectively.

The test results showed that UA, the absorption garment material, improves the heat protection without flame retardancy treatment. In addition, UA is a special absorption garment that is more comfortable for the worker.

TABLE 1—*Description of the tested materials.*

Code	Material Description	Fabric Weight, g/m^2
PW1	wool, woven, Zirpro treated	415
PW2	wool 93%/Nylon 7%, woven, Zirpro	375
PC1	cotton, woven, Proban treated	400
PC2	cotton, woven, Pyrovatex treated	335
PWP	wool 50%/Cordelan 50%, Zirpro (wool)	410
PWF	wool, woven, linear, Zirpro	180
PAL	aluminized (2 layers) rayon	550
UW	wool, knit, underwear, Zirpro	300
UM	Cordelan 70%/polyester 30%, knit, underwear	195
UA	cotton 50%/modal 50%, knit, absorption garment	215
PU1	Kevlar, woven	280
PU2	Kevlar, felt	360
PF	polycarbonate, face shield, 1 mm	...
CF	cellulose acetate, face shield, 0.5 mm	...
OF	polysulfone face shield, 0.65 mm	...

There is a tendency for lower heat transfer through heavier fabrics. In this study, however, there are exceptions to that rule (see PW2/UW and PW2/UM). The materials in UM are constructed to build up protection when exposed to heat. However, the thicker outer garment materials insulate better and prevent the UM from building up its protection.

Conclusion

In conclusion, the combination of outer clothing materials and underwear clothing material can be substituted for the combination of aluminized clothing material and clothing used in

TABLE 2—*Results from the steel splash evaluation.*

Code, out/inside	Maximum Rate of Heat Flow, cal/cm^2/s	Maximum Temperature Rise, °C, after 30 s
PW1/UW	0.68	8.6
PW1/UM	1.24	10.4
PW1/UA/UM	0.81	10.1
PW1/PWF	0.97	10.7
PW2/UW	2.00	17.5
PW2/UM	1.04	10.8
PC1/UW	1.12	10.2
PC1/UM	1.47	12.6
PC1/UA/UM	0.72	8.6
PC2/UW	1.05	9.6
PC2/UM	0.95	11.9
PWP/UM	1.43	9.9
PAL/PC2	0.49	7.8
PU1	2.15	29.0
PU2	0.42	11.6
PF	0.55	4.2
CF	0.59	8.2
OF	0.62	8.8
No test material	157.00	917

TABLE 3—*Visual ratings of fabrics exposed to molten steel.*

Material	Rating of Outer (Impacted) Layer			
	Charring	Shrinking	Adherence	Perforation
PW1/UW	moderate	slight	none	slight
PW1/UM	charred	slight	none	moderate
PW1/UA/UM	charred	slight	moderate	moderate
PW1/PWF	moderate	none	none	moderate
PW2/UW	moderate	slight	none	moderate
PW2/UM	moderate	none	none	none
PC1/UW	moderate	none	none	none
PC1/UM	severely charred	moderate	none	slight
PC1/UA/UM	moderate	none	none	none
PC2/UW	charred	none	none	slight
PC2/UM	charred	none	none	none
PWP/UM	charred	significant shrinkage	small	heavy perforation
PAL/PC2	slight	none	none	slight
PU1	slight	none	none	none
PU2	slight	none	none	none
PF	...	...	none	none
CF	...	...	none	none
OF	...	slight	none	none

jacket and pants. Other important factors, such as the design of the garment so that steel splashes can rapidly run off, are not discussed in this paper.

A follow-up study to test the heat resistance of the fabrics after 25 washes and dries is planned. This is necessary to assure the safety for the workers.

Acknowledgments

This study has been supported by the Swedish Work Environment Fund under Contract ASF 85-0754.

TABLE 4—*Repeatability of calorimeter thermal response under PW1 fabric to molten steel impact from 50-cm drop height.*

Test No.	Maximum Heat Flux, cal/cm²/s	Maximum Temperature Rise, °C
1	2.27	14.0
2	2.35	13.0
3	2.75	15.5
4	3.02	17.3
5	2.48	14.7
6	2.48	14.7
7	2.86	16.1
8	2.57	14.4
Mean	2.60	14.9
Standard deviation	0.26	1.4
Coefficients of variation	9.6%	9.9%

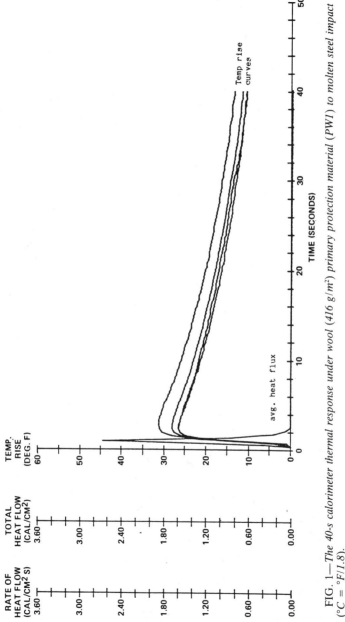

FIG. 1—*The 40-s calorimeter thermal response under wool (416 g/m²) primary protection material (PWI) to molten steel impact* (°C = °F/1.8).

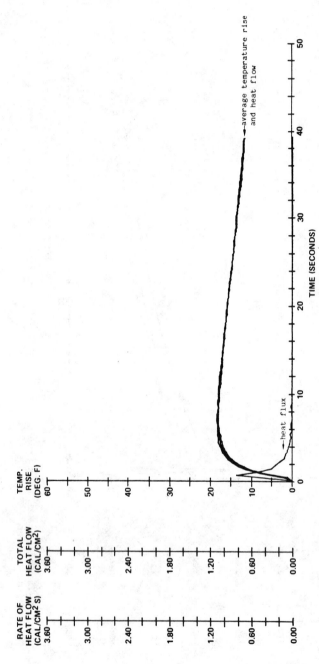

FIG. 2—*The 40-s calorimeter thermal response under 416 g/m² wool over 50% cotton, 50% modal and 70% cordelan, 30% polyester (PW1/UA/UM) to molten steel impact (°C = °F/1.8).*

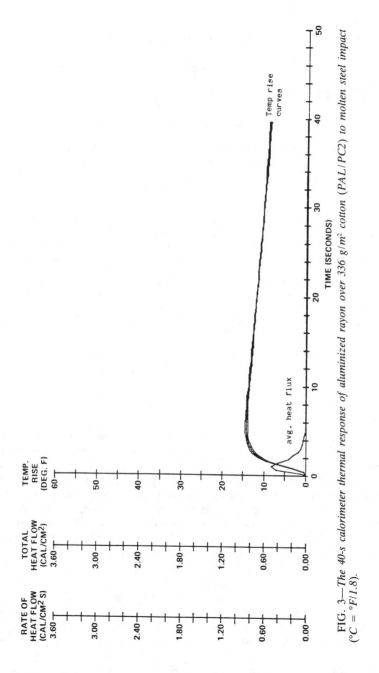

FIG. 3—*The 40-s calorimeter thermal response of aluminized rayon over 336 g/m² cotton (PAL/PC2) to molten steel impact (°C = °F/1.8).*

Michael Day[1]

A Comparative Evaluation of Test Methods and Materials for Thermal Protective Performance

REFERENCE: Day, M., **"A Comparative Evaluation of Test Methods and Materials for Thermal Protective Performance,"** *Performance of Protective Clothing: Second Symposium, ASTM STP 989,* S. Z. Mansdorf, R. Sager, and A. P. Nielsen, Eds., American Society for Testing and Materials, Philadelphia, 1988, pp. 108–120.

ABSTRACT: The American Society for Testing and Materials (ASTM) has developed Test Method for Thermal Protective Performance of Materials for Clothing by Open-Flame Method (D 4108-82). A modified version of this test is utilized in the National Fire Protection Association's NFPA 1971 (1986 Edition) and the International Organization for Standardization's (ISO) Draft Proposal 9151. The influences of these modifications on measured Thermal Protective Performance (TPP) values for composite assemblies have been determined. The sample restraining technique was identified as a critical factor in influencing the results obtained. An alternative system of mounting involving pins, is described, that prevents fabric shrinkage without compression. The behavior of a series of assemblies has been investigated using the technique and the contribution of each component (outer shell, moisture barrier, and thermal liner) to the measured TPP values evaluated.

KEY WORDS: thermal protection, convective heat, radiant heat, heat transfer, thermal injury, test methods, specimen mounting, protective clothing

One of the prime requirements for the fire fighter's protective clothing is to provide protection from flames and heat. The utilization of newer materials and constructions to achieve these objectives has resulted in the identification of a need for suitable test methods to accurately reflect the occupational exposure conditions. Originally, the National Fire Protection Association (NFPA) Standard NFPA 1971 on Protective Clothing for Structural Fire Fighters and Occupational Safety and Health (OSHA) Fire Brigade Standard 29 CFR 1910.156 called for a minimum thickness requirement to provide the fire fighter with effective protection during emergency fire fighting conditions. The most recent edition of NFPA 1971 (1986 Edition), however, calls for a test method to measure the total heat transferred through a protective ensemble exposed to heat flux of 2 cal/cm²/s (83 kW/m²) that would result in burn injury to the wearer. This test method was developed to simulate the exposure condition anticipated in a JP-4 fuel fire such as those encountered in the U.S. Army Fire Pit Facility [1,2], which in turn was chosen to reflect a fire that may surround an aircraft that has crashed, bursting its fuel tanks. Unfortunately, the original test method developed by Behnke [3] has been modified several times with several versions in existence [4–6]. Essentially, however, all three versions [4–6] employ the same principle. A horizontal specimen is exposed to a prescribed flame from a gas burner placed beneath it. The amount of heat

[1] Senior research officer, Division of Chemistry, National Research Council of Canada, Ottawa, Ont. K1A OR6, Canada.

energy passing through the specimen is measured by means of a copper calorimeter placed behind the specimen. The temperature increase of the calorimeter as a function of time is used to determine the exposure time required to cause pain and second degree burn or blister in accordance with specified burn criteria developed by Stoll and Chianta [7]. Based upon the total heat exposure, determined from the time and the exposure level, a direct measure of the protective capabilities fo the specimen can be obtained. This rating is called the Thermal Protective Performance or TPP rating. Thus, a burn time of 17.5 s at a heat flux of 2 cal/cm²s corresponds to a TPP rating of 35 cal/cm² (that is, 17.5 × 2).

In this study, we report on the test results obtained when three standard test garment assemblies were evaluated according to ASTM Test Method for Thermal Protective Performance of Materials for Clothing by Open-Flame Method (D 4108–82) and its various test method modifications, along with some modifications made in this study. The variables investigated and included in the various tests studied were the nature of the heat source and the specimen mounting configuration employed.

In addition to studying the role of test procedure variables on the measured TPP values, the importance of a variety of fabric type on the TPP values, of various garment assemblies has also been investigated. Our study was not concerned with the behavior of individual fabrics in this type of test of which there is a large amount of data [3,8–16], but was concerned with the behavior of fabric assemblies similar to those reported elsewhere [3,9,12–15,17]. This study was restricted to three outer shell fabrics, two moisture barriers, and three thermal barrier fabrics, combined in all garment assemblies.

Experimental

Materials

The fabrics and combinations studied in the evaluation of the test methods are described in Table 1. The fabrics employed in the systematic evaluation of the effect of garment composition on TPP data are described in Table 2. The thickness measurements presented in these tables were determined in accordance with CAN/CGSB-4.2 No. 37M [18] using a standard pressure of 1.72 kPa.

Test Methods Employed

All the test methods used in this evaluation employed essentially the same principle and all claim to be based upon ASTM D 4108–82 [4].

TABLE 1—*Fabric assemblies evaluated.*

	Outer Shell	Moisture Barrier	Thermal Liner	Total Assembly
System A				
Material	Nomex III	Neoprene on Nomex	Nomex needlepunch	
Weight, g/m²	242	261	356	859
Thickness, mm	0.57	0.29	3.23	4.21
System B				
Material	40/60 PBI/Kevlar	Gore-Tex on Nomex	Nomex quilt	
Weight, g/m²	260	145	248	653
Thickness, mm	0.59	0.34	2.40	3.57
System C				
Material	FR cotton	Neoprene on poly/cotton	FR cotton denim	
Weight, g/m²	456	330	386	1172
Thickness, mm	0.76	0.30	0.69	1.76

TABLE 2—*Fabrics employed in garment assembly tests.*

Fabric Code	Type Component	Fabric	Weight, g/m²	Thickness, mm
ON	outer shell	Nomex III	241	0.64
OC	outer shell	FR cotton	435	0.69
OW	outer shell	FR wool	336	0.65
BG	moisture barrier	Gore-Tex on Nomex	128	0.28
BN	moisture barrier	Neoprene on Nomex	224	0.32
TN	thermal liner	Nomex needlepunch	161	2.46
TW	thermal liner	FR wool needlepunch	247	2.62
L	inner liner	Nomex pajama check	112	0.39

ASTM D 4108–82 [4]

Standard Procedure—As written, this method provides several options in the specimen mounting procedure. In this study, we have employed the multiple-layer procedure with the sensor mounted in direct contact with the back surface of the fabric that would be towards the skin. The other layers were then positioned in the order that they would be used in the final garment with the face of the outer shell fabric facing downwards. Figure 1 shows a schematic diagram of the test apparatus. Essentially, the fabric components (100 by 100 mm) are mounted horizontally on a steel mounting plate (150 by 150 mm) that has a 50-mm-square hole in the center. A copper calorimeter consisting of a copper disk containing four 32-gauge chromel/alumel thermocouples mounted in an insulating board is then placed face down on the fabric assembly. The exposed fabric is then subjected to a methane flame

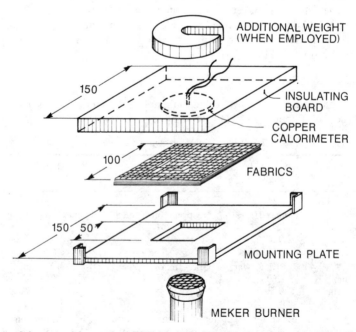

FIG. 1—*Schematic of the standard ASTM D 4108–82 testing apparatus and additional weight when employed (dimensions in mm).*

from a Meker burner adjusted to deliver a heat flux of 84 ± 2 kW/m^2 (2.00 ± 0.05 cal/cm^2/s). The temperature/time response of the calorimeter is then used to determine the exposure time that would be required to cause pain and second-degree burn or blister in accordance with the tolerance curves developed by Stoll and Chianta [7].

1.0-kg Weight Restraining Procedure—The mounting procedure used in the standard procedure uses an unrestrained relaxed mounting system that permits heat shrinkage of the fabrics. The NFPA 1971 procedure [5] (outlined later) uses an additional weight on the sensor block. In this study, a weight was fabricated such that its mass combined with that of the sensor block (shown in Fig. 1) was 1 kg. All the other conditions were identical to those in the standard procedure.

4.0-kg Weight Restraining Procedure—In addition to employing a 1 kg mass, the effect of a 4.0 kg total restraining mass was also evaluated.

Pin Restraining Procedure—The utilization of restraining weights to prevent thermal shrinkage does exert considerable compressive forces on the garment assembly. In order to overcome this drawback, the steel base plate was modified as shown in Fig. 2, to include a series of steel pins. Holes were drilled in the sensor block to accommodate these pins in order to enable the sensor to come into contact with the back surface of the specimen. The fabric assemblies were then mounted on these pins, and the sensor block placed on top. This arrangement prevented appreciable thermal shrinkage of the fabrics while applying minimum compressive forces to the assembly.

TPP Test Method Outlined in NFPA—1971 (1986 Edition) [5]

Although classified as a modified version of ASTM D 4108-82, it does have several important differences. This equipment, instead of employing a single Meker burner to give an approximately 30% radiative, 70% convective heat flux, combines two burners and a bank of nine quartz tubes controlled with a variable voltage supply to provide a 50% radiative and 50% convective flux at a total heat flux of 83 ± 5 kW/m^2 (2.00 ± 0.1 cal/cm^2/s) (see Fig. 3). The specimens (150 by 150 mm) are also slightly larger in this test since a larger mounted frame (200 by 200 mm) is employed with a larger exposure opening (100 by 100 mm). The fabric components are once again laid on the mounting frame in the correct order. Although this method calls for the use of a 1000-g weight to prevent thermal shrinkage, the data presented in this study were obtained using a 4000-g restraining weight, as was required in the original draft before its revision at the December 1985 meeting of NFPA.

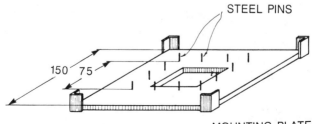

STEEL PINS

150 75

MOUNTING PLATE

FIG. 2—*Modified ASTM D 4108–82 mounting plate incorporating a pin restraining system (dimensions in mm).*

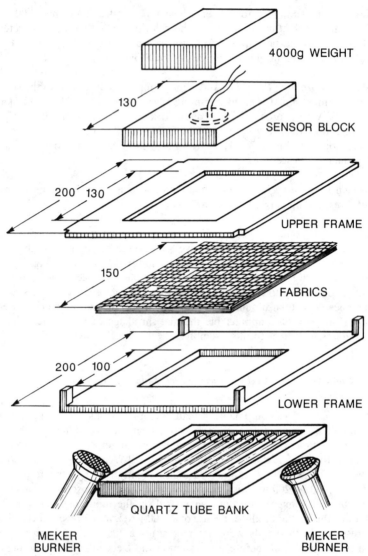

FIG. 3—*Schematic of the NFPA 1971 testing apparatus showing the use of a 4000-g restraining weight (dimensions in mm).*

The exposure time required to cause pain and second-degree burn or blister is determined as described in the ASTM method.

In order to elucidate the role of restraining weight on the measured TPP values, System A was also evaluated employing the following restraining weights: 0, 1000, 2000, and 3000 g in addition to the 4000-g weight.

ISO Draft Proposal 9151 [6]

Standard Procedure—This method, which measures the protection afforded by fabric or garment assemblies, was developed by ISO TC 94/SC 13/WG 2 from ASTM D 4108–82 by introducing a number of refinements. Like the ASTM method, it employs a single Meker burner, but at a calibrated heat flux of 80 kW/m² (slightly less than ASTM). Although it employs specimens 100 by 100 mm and mounts them on a base plate 150 by 150 mm with a central hole 50 by 50 mm as does the ASTM method, copper instead of steel is employed as the construction material. The sample is held in position by an aluminum retaining plate that has a central hole to position the calorimeter in its insulating mounting block (see Fig. 4). Movement of the test specimen due to thermal shrinkage is prevented by squeezing the specimen restraining plate against the base plate with four toggle clamps positioned one at each corner. Unfortunately, the present draft does not specify a standard clamping pressure. This means that it is possible to highly compress the fabric assemblies or lightly compress them depending on the force applied. In this study, we employed a 20% compression ratio obtained by setting the clamping adjustment to give a spacing between the base plate and retaining frame of 80% of the measured fabric assembly thickness. The sensor is positioned 3 mm above the inner fabric surface by the use of an aluminum spacer ring. Although the calorimeter was once again a copper disk, a single copper-constantan thermocouple is used to measure the temperature increase. Once again the temperature/time relationship obtained from the calorimeter is used to determine the exposure time required to cause pain and

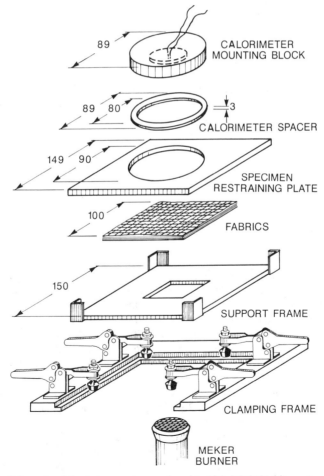

FIG. 4—*Schematic of the test apparatus employed in ISO DP 9151* (*dimensions in mm*).

second-degree burn or blister in accordance with the same specified pain and burn criteria used in the other tests.

Modified Procedure Using No Spacer—In addition to performing tests with the spacer as is required in the ISO test, tests were conducted without the spacer present in order to get a better comparison with the other methods.

Test Protocol

The ASTM and ISO tests, along with the modifications just described were conducted in our laboratories. In all of these tests, the burner was located to the side of the test arrangement prior to exposure of the specimens. To commence the heat exposures, the burner was rapidly positioned as required in the method, triggering the timing mechanism in the process. An Apple II+ microcomputer equipped with an ADALAB interface card and QUICK I/O software (from Interactive Microwave Inc.) was programmed to measure the analog signal from the sensor, converting it to a digital value. These values were then compared with Stoll and Chianta's [7] skin damage criteria that had been primed into the program. When the measured values coincided with the criteria for pain and second-degree burns, the values were recorded.

The test results obtained according to NFPA 1971 were determined during an interlaboratory evaluation involving several independent laboratories [19]. These tests were conducted using commercial equipment (Custom Scientific Instrument Co.) that controls the exposure time to 0.2 s by a pneumatic shutter mechanism controlled by a digital timer.

In all cases, at least five tests were conducted on each garment assembly system using randomly selected specimens. All fabrics were initially preconditioned for not less than 4 h at $50 \pm 2°C$ before being conditioned at $20 \pm 2°C$ and $65 \pm 5\%$ relative humidity for at least 18 h. Samples were then tested no more than 5 min after removal from the standard conditioning atmosphere.

The systematic evaluation of the component fabrics on measured TPP values were performed using the ASTM D 4108–82 test method with the 1-kg sensor weight as outlined in the Experimental Section.

Results and Discussion

Test Method Variables

The results obtained using the various test procedures examined in this study are tabulated in Tables 3, 4, and 5 for the garment assembly Systems A, B, and C, respectively. However, systematic changes due to test method variables may be more easily detected by considering the graphical presentation of the TPP data given in Fig. 5.

Consideration of these data enables some important conclusions to be drawn. The mounting of the specimen clearly has a pronounced effect on the TPP values, probably more so than the actual composition of the thermal heat exposure, that is, 50/50 radiative/convective or 30/70 radiative/convective. The data obtained with the ISO technique clearly indicate that the introduction of an air gap between the sensor and the inside fabric greatly reduces the heat transfer leading to higher TPP values. Based on the three systems studied in this investigation, the 3-mm air gap provided by the spacer in the ISO method is responsible for approximately 40% increase in TPP value over that measured without a spacer present. However, in view of the possible distortion of the fabric from its initial planar configuration, this air gap can change dramatically during exposure, as shown by Holcombe and Hoschke

TABLE 3—*Test results obtained with System A.*

Test	Mean Test Weight g/m²	Mean Test Thickness, mm	Pain Time, s	Burn Time, s	TPP Rating, cal/cm²
ASTM					
No weight	854	ND[a]	16.1 ± 0.3	26.0 ± 0.9	52.1 ± 1.8
1000-g weight	861	ND	15.2 ± 0.5	23.8 ± 0.8	47.6 ± 1.6
4000-g weight	866	4.20	14.1 ± 0.2	22.0 ± 0.7	44.0 ± 1.3
Pins	867	4.29	14.3 ± 0.2	23.2 ± 0.8	46.4 ± 1.5
NFPA					
No weight	NA[b]	4.66[c]	15.4 ± 0.1	25.1 ± 0.7	50.2 ± 1.3
1000-g weight	NA	4.76[c]	15.1 ± 0.1	24.4 ± 2.9	48.9 ± 5.7
2000-g weight	NA	4.58[c]	15.0 ± 0.8	22.9 ± 0.8	46.0 ± 1.8
3000-g weight	NA	4.74[c]	15.0 ± 0.8	23.9 ± 1.6	47.8 ± 3.1
4000-g weight	NA	4.70[c]	13.7 ± 0.4	22.6 ± 0.9	45.2 ± 1.8
ISO					
Spacer	862	4.17	22.1 ± 0.7	36.4 ± 1.6	72.8 ± 3.2
No spacer	854	4.17	17.0 ± 0.5	26.6 ± 1.1	53.2 ± 2.1

[a] ND = not determined.
[b] NA = not available.
[c] Determined using a pressure of 0.34 kPa (0.05 psi).

[20]. The use of an air gap in a test of this type raises several problems especially if the data are to be used in garment performance specifications.

Comparisons of the ISO test data obtained without a spacer to those obtained with the standard ASTM procedure revealed that the ISO method gave consistently higher values. While a slightly lower heat flux is employed in the ISO procedure than is required in the ASTM (that is, 80 kW/m² compared to 84 kW/m²), this difference was not felt to be sufficiently large to be responsible for the observed differences. It should be noted that in the ISO technique, there is a certain amount of variability in the clamping pressure that can be exerted on the fabric to prevent shrinkage. In this study, only a light clamping pressure was applied to avoid undue compression. The application of a larger clamping force would

TABLE 4—*Test results obtained with System B.*

Test	Mean Test Weight, g/m²	Mean Test Thickness, mm	Pain Time, s	Burn Time, s	TPP Rating, cal/cm²
ASTM					
No weight	652	3.64	14.8 ± 0.5	21.1 ± 0.3	42.2 ± 0.7
1000-g weight	652	ND[a]	13.7 ± 0.9	19.4 ± 1.1	38.3 ± 2.1
4000-g weight	652	3.55	11.5 ± 0.4	16.7 ± 0.6	33.4 ± 1.2
Pins	653	3.40	12.4 ± 0.2	17.6 ± 0.5	35.3 ± 0.9
NFPA 4000 g-weight	NA[b]	4.05[c]	11.9 ± 1.3	17.4 ± 1.4	34.8 ± 2.7
ISO					
Spacer	659	3.69	19.4 ± 0.5	29.2 ± 0.8	58.4 ± 1.6
No spacer	653	3.55	15.6 ± 0.5	22.5 ± 0.5	45.0 ± 1.0

[a] ND = not determined.
[b] NA = not available.
[c] Determined using a pressure of 0.34 kPa (0.05 psi).

TABLE 5—*Test results obtained with System C.*

Test	Mean Test Weight, g/m²	Mean Test Thickness, mm	Pain Time, s	Burn Time, s	TPP Rating, cal/cm²
ASTM					
No weight	1180	1.76	12.7 ± 0.3	17.0 ± 0.3	34.1 ± 0.7
1000-g weight	1168	ND[a]	12.2 ± 0.3	16.1 ± 0.4	32.2 ± 0.8
4000-g weight	1180	1.77	11.1 ± 0.3	14.7 ± 0.6	29.4 ± 1.1
Pins	1157	1.76	11.6 ± 0.3	15.4 ± 0.3	30.7 ± 0.7
NFPA 4000-g weight	NA[b]	2.02[c]	13.2 ± 0.8	17.4 ± 1.0	34.8 ± 2.0
ISO					
Spacer	1188	1.74	18.4 ± 0.3	26.3 ± 0.3	52.6 ± 0.7
No spacer	1157	1.75	13.2 ± 0.3	17.6 ± 0.5	35.3 ± 1.0

[a] ND = not determined.
[b] NA = not available.
[c] Determined using a pressure of 0.34 kPa (0.05 psi).

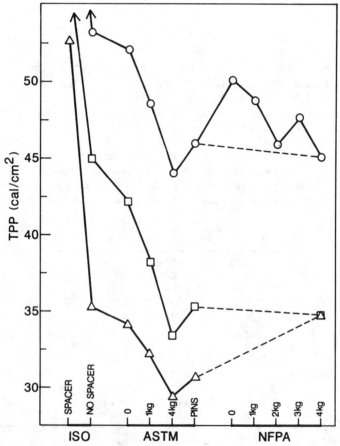

FIG. 5—*Graphical presentation of the TPP data obtained by the various test procedures;*
○ = *System A;* □ = *System B;* △ = *System C.*

no doubt cause some compression and reduce shrinkage resulting in lower TPP values. However, in view of the low contacting force applied to the sensor, the magnitude of the reduction would not be anticipated to be as great as that noted below in the case of the ASTM method.

The standard ASTM test procedure for multiple garment assemblies requires the use of fabric assemblies in the unrestrained mode with the weight of the sensor and its mounting board (260 g) being the only compressive force. Unfortunately, this relaxed mounting system allows heat shrinkage of the fabrics that is capable of resulting in higher TPP values than would be the case if the fabrics were restrained to prevent this thermal shrinkage. The use of weights on the sensor block as well as positioning clamps have therefore been proposed to prevent excessive thermal shrinkage and restrain fabric movement [5]. The data presented in our study clearly indicates that adding weights to the sensor block has an influence on the TPP results. The greater the weight applied to the sensor, the lower the TPP value. This observation is not restricted to the ASTM procedure, but is noted with System A studied using the NFPA 1971 test protocol. Although the addition of weights to the sensor block does play a role in reducing thermal shrinkage, it has also been shown to cause considerable compression of a garment assembly, sufficient to cause considerable reduction in thickness [21]. This thickness compression alone is sufficient to be responsible for a substantial reduction in the measured TPP values.

There are alternative ways to prevent thermal shrinkage without compressing the ensemble and thereby penalizing high loft thermal insulation materials. The use of pins as outlined in this paper appears to offer an excellent method of ensuring control of thermal shrinkage without undue compression. Clearly the results, while in line with the other ASTM data, are correspondingly lower than those obtained with unrestrained samples and yet are higher than those obtained with the use of a 4-kg restraining weight where considerable thickness compression occurs.

Comparison of the TPP values obtained with the ASTM method employing a 4-kg sensor weight with those determined using the NFPA 1971 technique with a 4-kg weight reveal that the results are similar. This observation being in keeping with the data reported by Shalev and Barker [10] who noted that the small variation in incident heat flux from 50/50 to 70/30 convective/radiative had little effect on the TPP values of a whole range of fabrics. Despite the close similarity in the data, the NFPA values were always higher than those obtained with the ASTM techniques. Although it could be speculated that this difference is due to the radiative to convective heat ratio, another reason could be the greater sag of the material into the larger opening of the sample holder in the NFPA method. The extent of this sagging has been demonstrated by Veghte [21] and is clearly a potential source of the variation between the two techniques.

Test Fabric Combinations

Table 6 compares the results obtained for the various test fabric combinations examined in this study. All the data presented in this table were obtained using ASTM D 4108–82 employing unrestrained mounting of the fabric specimens, but with a total sensor block weight of 1 kg.

With regard to the influence of fabric variables on measured TPP values, fabric thickness has been shown to be a very important characteristic parameter in determining protection time [10,13,22,23]. This factor was recognized in the earlier edition of NFPA 1971 where a minimum thickness of the assembled garments of 4.44 mm (0.175 in.) was required (tested with a compressometer with a 3-in.-diameter presser foot set at 0.05 psi). The data presented in Table 6 were therefore analyzed to determine the nature of the relationship between TPP values and total thickness, and the result is presented in Fig. 6. Clearly that there is indeed

TABLE 6—*TPP ratings for various combinations according to ASTM D 4108–82.*

Test Combination			Total Weight, g/m^2	Total Thickness, mm	Pain Time, s	Burn Time, s	TPP Rating, cal/cm^2
Outer Shell	Moisture Barrier	Thermal Liner					
ON	BN	TN	738	4.14	15.2 ± 0.3	21.7 ± 0.9	43.4 ± 1.8
ON	BN	TN	824	4.11	12.4 ± 0.1	18.7 ± 0.3	37.4 ± 0.5
ON	BG	TN	642	3.96	13.6 ± 0.9	18.8 ± 1.3	37.5 ± 2.7
ON	BG	TN	728	4.17	11.3 ± 0.4	17.2 ± 0.3	34.4 ± 0.7
ON	none	TN	514	3.61	10.5 ± 0.7	14.5 ± 1.0	29.0 ± 2.0
ON	none	TW	600	3.71	9.5 ± 0.3	14.1 ± 0.6	28.2 ± 1.2
OC	BN	TN	932	4.09	18.5 ± 0.5	23.8 ± 0.3	47.6 ± 0.5
OC	BN	TW	1018	4.34	17.3 ± 0.3	22.4 ± 0.4	44.7 ± 0.8
OC	BG	TN	836	4.19	15.2 ± 0.9	22.4 ± 1.0	44.8 ± 2.0
OC	BG	TW	922	4.22	13.9 ± 0.6	20.1 ± 0.7	40.2 ± 1.3
OC	none	TN	708	3.76	7.0 ± 0.2	7.7 ± 0.2	15.4 ± 0.3
OC	none	TW	794	3.86	7.6 ± 0.3	8.3 ± 0.4	16.7 ± 0.8
OW	BN	TN	833	4.09	18.8 ± 0.3	25.3 ± 0.3	50.6 ± 0.5
OW	BN	TW	919	4.22	16.7 ± 0.3	23.2 ± 0.3	46.4 ± 0.5
OW	BG	TN	737	4.01	15.9 ± 0.3	23.1 ± 0.4	46.2 ± 0.8
OW	BG	TW	823	4.11	14.1 ± 0.2	20.4 ± 0.3	40.8 ± 0.6
OW	none	TN	609	3.66	8.2 ± 0.5	13.4 ± 0.4	26.8 ± 0.8
OW	none	TW	695	3.81	7.4 ± 0.2	11.0 ± 0.3	22.1 ± 0.5

NOTE—All test configurations had a Nomex inner liner (*L*) between the thermal liner and the sensor.

a reasonable correlation between the measured TPP value and the total thickness of the garment assembly. However, close examination of the data reveals some other interesting observations regarding the effect of the different fabric components of the garment assembly on the measured TPP values.

By comparing the performance of the outer shell fabrics, the TPP ratings can be seen to be greatly dependant upon the presence or absence of a moisture barrier layer. In the absence of a moisture barrier, Nomex is clearly the best followed by wool and then cotton. In the presence of a moisture barrier, wool is clearly the best followed by cotton and then Nomex. This difference is probably due to the moisture present in the hydrophilic fibers of the outer shell of cotton and wool that is capable of being transferred to the sensor in the absence of the moisture barriers. This moisture results in a greater conduction of heat to the sensor causing a reduction in the measured TPP values. Meanwhile, in the presence of the moisture barrier, the effect is no longer of importance and the greater protection of the wool and cotton over Nomex is probably due to the thermal charing that can occur with these fabrics.

The importance of the moisture barrier has been clearly noted for cotton and wool but also for the Nomex sample. The presence of a moisture barrier obviously contributes to the degree of thermal protection, and its complete removal without some substitution in terms of thermal protection would not be wise. Comparing the neoprene moisture barrier to that of Gore-Tex indicate that the former gave slightly higher TPP values than comparable Gore-Tex combinations. However, these differences could be due to the differences in the weight or thickness of these two moisture barriers.

In terms of the behavior of the two thermal liners, the Nomex material consistently gave higher TPP values than the wool despite the greater thickness and weight of the wool. Once again, it appears that the presence of moisture in the hydrophilic wool liner is responsible for the decreased TPP values. This observation clearly identifies the potential problems

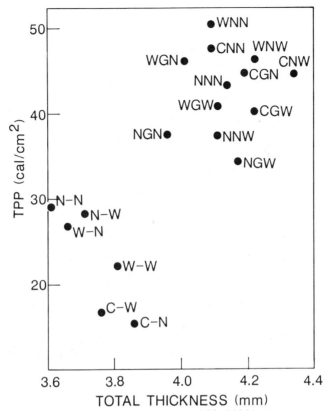

FIG. 6—*TPP data obtained according to the ASTM D 4108–82 test procedure employing the 1-kg restraining weight as a function of assembled garment thickness. Designation XYZ; X = outer shell (that is, N = Nomex, C = cotton, and W = wool), Y = moisture barrier (that is, N = neoprene, G = Gore-Tex, and — = none), Z = thermal liner (that is, N = Nomex and W = wool).*

associated with perspiration on thermal linings in terms of protection offered to the wearer of protective clothing.

Conclusions

The nature of the sources of heat flux, that is, 50/50 or 70/30 convective/radiative, does not appear to be as important as the manner of specimen mounting in determining the heat transfer through garment fabric assemblies. The use of weights on the sensor block to restrain the specimens and prevent thermal shrinkage appears to penalize high loft thermal liners that have been engineered to give the greatest thermal protection with the lowest added weight burden. The use of a pin restraining frame appears to offer the advantage of preventing thermal shrinkage without deforming the assembly. This modification of ASTM D 4108–82 is being seriously considered in the Canadian Standard for Protective Clothing for Fire Fighters.

The data concerning the relative thermal merits of various textile fabrics employed in garment assemblies provides the garment manufacturers with some fundamental information to assist them in the design of clothing to provide satisfactory thermal protection.

Acknowledgments

The help and cooperation of the following individuals is greatly appreciated; without their assistance this work would not have been possible: R. L. Barker, North Carolina State University; W. P. Behnke, E. I. DuPont de Nemours and Company Inc; E. Iannone, Cotton Incorporated; M. T. Stanhope, Celanese Corporation; and J. H. Veghte, Biotherm Inc. The development of the best scientific and technologically practical standard for protective clothing for firefighters is the sole objective of us all. This international cooperation ensures that new ideas and developments are available to all interested parties in this endeavor. The technical assistance of P. Z. Sturgeon is also acknowledged.

This paper has been issued as NRCC No. 28334.

References

[1] Abott, N. J., *Journal of Coated Fabrics,* Vol. 6, July 1976, p. 48.
[2] Abbott, N. J. and Schulman, S., *Fire Technology,* Vol. 12, No. 3, 1976, p. 204.
[3] Behnke, W. P., *Fire Technology,* Vol. 13, No. 1, 1977, p. 6.
[4] ASTM Standard Test Method for Thermal Protective Performance of Materials for Clothing by Open-Flame Method (D 4108–82), *Annual Book of ASTM Standards,* Vol. 07.01, American Society for Testing and Materials, Philadelphia.
[5] NFPA Standard 1971, "Standard on Protective Clothing for Structural Fire Fighters," National Fire Protection Association, Battenymarch Park, Quincy, MA, 1986.
[6] ISO DP 9151, Draft Proposal for an International Standard Protective Clothing: Assessment of the Behavior of Materials for Protection Against Flames (Predominantly Convective Heat), committee document, International Organization for Standards, TC 94/SC 13, 1985.
[7] Stoll, A. M. and Chianta, M., *Aerospace Medicine,* Vol. 40, 1969, p. 1323.
[8] Barker, R. L. and Lee, Y. M., *Tappi Journal,* Vol. 68, No. 12, 1985, p. 62.
[9] Krasny, J. F., Singleton, R. W., and Pettengill, J., *Fire Technology,* Vol. 18, No. 4, 1982, p. 309.
[10] Shalev, I. and Barker, R. L., *Textile Research Journal,* Vol. 54, No. 10, 1984, p. 648.
[11] Perkins, R. M., *Textile Research Journal,* Vol. 49, No. 4, 1979, p. 202.
[12] Behnke, W. P., *Fire and Materials,* Vol. 8, No. 2, 1984, p. 57.
[13] Baitinger, W. F. and Konopasek, L. in *Performance of Protective Clothing, ASTM STP 900,* R. L. Barker and G. C. Coletta, Eds., American Society for Testing Materials, Philadelphia, 1986, p. 421.
[14] Benisek, L., Edmondson, G. K., Mehta, P., and Phillips, W. A. in *Performance of Protective Clothing, ASTM STP 900,* R. L. Barker and G. C. Coletta, Eds., American Society for Testing and Materials, Philadelphia, 1986, p. 405.
[15] Krasny, J. F. in *Performance of Protective Clothing, ASTM STP 900,* R. L. Barker and G. C. Coletta, Eds., American Society for Testing and Materials, Philadelphia, 1986, p. 463.
[16] Shalev, I. and Barker, R. L., *Textile Research Journal,* Vol. 53, No. 8, 1983, p. 475.
[17] Veghte, J. H. in *Performance of Protective Clothing, ASTM STP 900,* R. L. Barker and G. C. Coletta, Eds., American Society for Testing and Materials, Philadelphia, 1986, p. 487.
[18] CAN/CGSB-4.2 No. 37M, Method of Test for Fabric Thickness, National Standard of Canada, Canadian General Standards Board, Ottawa, Ont., Canada.
[19] Behnke, W. P., private communications.
[20] Holcombe, B. V. and Hoschke, B. N. in *Performance of Protective Clothing, ASTM STP 900,* R. L. Barker and G. C. Coletta, Eds., American Society for Testing and Materials, Philadelphia, 1986, p. 327.
[21] Veghte, J. H., private communications.
[22] Holcombe, B. V., *Fire Safety Journal,* Vol. 6, 1983, p. 129.
[23] Baitinger, W. F., *Textile Research Journal,* Vol. 49, 1979, p. 221.

Itzhak Shalev[1]

Effect of Permanent Press Resin Finish on Cotton Fabric Thermal Resistivity

REFERENCE: Shalev, I., **"Effect of Permanent Press Resin Finish on Cotton Fabric Thermal Resistivity,"** *Performance of Protective Clothing: Second Symposium, ASTM STP 989,* S. Z. Mansdorf, R. Sager, and A. P. Nielsen, Eds., American Society for Testing and Materials, Philadelphia, 1988, pp. 121–130.

ABSTRACT: The effect of permanent press resin of the dimethylol ethelene urea (DMEU) type on 100% cotton fabric resistivity, as measured on a guarded hot plate, was studied. Increased add-on reduced average air pore size and increased solid content. A reduction in fabric thickness and an increase in actual contact area with a solid surface were also observed. A maximum of resistivity was obtained for an add-on of 4.4%. Further increase in add-on caused a reduction in resistivity as bulk density increased. These results were compared with theoretical models of heat transfer through randomly oriented fiber assemblies.

KEY WORDS: protective clothing, fabrics, thermal resistivity, resin treatment, heat transfer, comfort

The analysis of heat flow through textiles is complex because textiles are an interdispersed, two-phase system comprised of polymer and air. Assigning a composite conductivity constant to such a system is difficult because radiation effects exist and conductivity may depend on physical dimensions. Representative dimensions such as thickness, fiber orientation, and bulk density are difficult to measure without perturbation of the fabric. This field has received considerable empirical attention. Some of the earlier work [1,2] already quantified the effects of thickness on overall thermal resistance of fabrics. Fink [3] concluded that air and fiber conduction and radiation were the important modes of heat transfer through fibrous insulations. Convective transfer was considered negligible. Recent studies have confirmed this [4].

Textile parameters that affect these modes such as packing fraction and fiber diameter were studied by Pelanne [5]. Plots of packing fraction or fiber diameter versus thermal conductivity showed minima due to the decrease of radiative transfer and concurrent increase in conductive transfer as packing fraction or fiber diameter were increased.

Naka and Kamata [6] measured the conductivity of dense fabrics under different levels of compression in order to decrease thickness and increase bulk density. They found a linear increase in conductivity for filament yarn fabrics but found no clear trend for staple yarn fabrics because of changes in fiber arrangement due to compression.

Several attempts have been made to formulate mathematical models that predict the thermal response of textiles [7–10]. These require input of fabric and fiber properties that affect radiant and conductive transfer through the fabrics.

Fibers conduct heat by point-to-point contact but also resist the penetration of thermal radiation by absorption and scattering, thus shading the fibers in deeper layers. Ideally,

---

[1] Lecturer, Shenkar College of Textile Technology and Fashion, Ramat-Gan, Israel.

they provide a still-air layer and radiation block with a minimized conduction penalty and acceptable textile properties. The extinction length (λ) of radiation in a random fiber assembly was defined by Stuart and Holcombe [11] as

$$\lambda = d/p$$

where

d = fiber diameter, and
p = packing fraction = fiber volume/fabric volume.

Pelanne combined the separate mode "conductivities" to obtain the overall apparent conductivity as a function of p using the following expression

$$k_{app} = ((k_g + k_{rad})/(1 - p)) + k_s p$$

where

k_g = gas conductivity (air),
k_s = solid conductivity (fiber), and
k_{rad} = radiative conductivity (conductivity due to radiation).

Farnworth [4] defines k_{rad} in a random fiber assembly as

$$k_{rad} = 8\sigma T^3 \frac{r}{pe}$$

where

σ = Stefan-Boltzman constant,
T = absolute temperature,
r = fiber radius, and
e = fiber emissivity.

Packing fraction and fiber diameter thus affect the radiative and conductive portions of thermal transmission.

Woven fabrics as opposed to random fiber assemblies, have distinct pores between yarns that are open from side to side. Intrayarn pores also exist. For common apparel fabrics, these average in the range of 35 μm and 5 μm, respectively [12].

The present study varies pore size and packing fraction of fabrics without compression by introduction of a thermoset permanent press resin into a lightweight woven cotton fabric. Gross fiber and yarn geometry is thus largely unaffected during thermal testing, and the effects of resin treatment on pore size and packing fraction are studied.

Experimental

The test fabric was a 100% cotton, plain weave fabric. Warp and fill yarn count was 22/1 Nec Warp, and fill yarn densities were 28 yarns/cm and 15 yarns/cm, respectively.

A commercial resin was chosen for this study. This was a reactant crosslinking resin of the dimethylol ethelene urea (DMEU) type (Fixapret CNF—BASF).

The test fabrics were desized, scoured, and bleached. They were then padded with a solution containing: 80 kg/m^3 Fixapret CNF and 20 kg/m^3 Condensol FB.

After padding, the fabrics were dried at 80°C for 10 s and cured at 170°C for 32 s. The following series of fabrics was obtained:

Fabric No. 1 = no resin.
Fabric No. 2 = 25% pick-up and 4.4% add-on by weight.
Fabric No. 3 = 35% pick-up and 12.0% add-on by weight.
Fabric No. 4 = 45% pick-up and 20.8% add-on by weight.

Textile properties of the test fabrics were measured as follows:

Unit weight: ASTM Tests for Construction Characteristics of Woven Fabrics (D 1910-64) (discontinued).
Thickness: ASTM Method for Measuring Thickness of Textile Materials (D 1777-64) (7.14 kg/m²).
Stiffness: ASTM Test Methods for Stiffness of Fabrics (D 1388-64).
Air permeability: ASTM Test Method for Air Permeability of Textile Fabrics (D 737-75).
Thermal transmission: ASTM Test Method for Thermal Transmittance of Textile Material (D 1518-85) ($\Delta T = 15°C$).

Protruding surface fibers contribute to the effective thermal thickness of fabrics. In order to assess the effect of resin application on surface fibers, the true contact area of the test fabrics with a flat surface under their own weight was measured. This was done by laying the fabrics on a sooted glass surface with a jacking system that prevented sideways motion during laying and removal. The glass plate was then placed on an overhead projector and the image was projected on a calibrated chart. The true contact area relative to the nominal fabric surface area was measured by determining the extent of the bright areas.

Packing fraction was calculated as follows

$$p = V_f/(V_f + V_a) = \rho_{fab}/\rho_{fib}$$

where

V_f = solids volume,
V_a = air volume,
ρ_{fab} = bulk density of fabric, and
ρ_{fib} = density of the solids in the fabric.

The average density of the solids in the test fabrics was determined using a Beckman air comparison pycnometer, Model 930.

The test fabrics were also examined visually under a stereomicroscope.

Results

The test results are given in Table 1. The relationships between resin add-on and the properties measured are plotted in Figs. 1–6. The relationship between thermal resistivity, bulk density, and p are plotted in Figs. 7 and 8.

A decrease in thickness and an increase in contact area with add-on were observed. This is related to the smoothing of surface fibers due to the resin treatment. This was corroborated by microscopic observation. Bulk density increased almost linearly with add-on. Air permeability was significantly reduced on addition of 4.4% resin and changed little on further resin addition.

TABLE 1—*Test results*.

Fabric No.	Add-on, %	Weight, kg/m²	Stiffness, kg/km	Thickness, mm	Air Permeability, m³/m² s	Contact Area, %	Bulk Density, kg/m³	p	R, m²K/W	R/m, mK/W
1	0	145	0.0082	0.84	0.471	0.70	173	0.12	0.065	77.26
2	4.4	151	0.0108	0.81	0.447	0.77	186	0.13	0.078	96.34
3	12.0	162	0.0113	0.79	0.448	0.94	206	0.14	0.071	90.25
4	20.8	175	0.0129	0.79	0.449	2.14	220	0.145	0.059	74.17

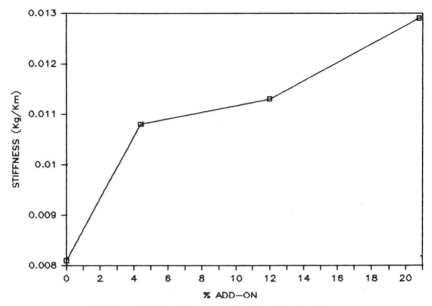

FIG. 1—*Fabric stiffness as a function of resin add-on.*

This is indicative of an initial reduction in the size of interyarn pores that do not then change appreciably at higher add-ons.

The increase in fabric stiffness illustrates the increased adhesion between fibers and yarns. Plots of bulk density or p versus fabric resistivity show a maximum in the region of 5% add-on.

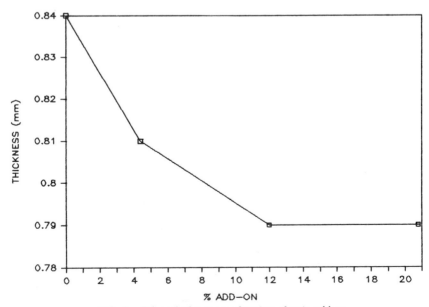

FIG. 2—*Fabric thickness as a function of resin add-on.*

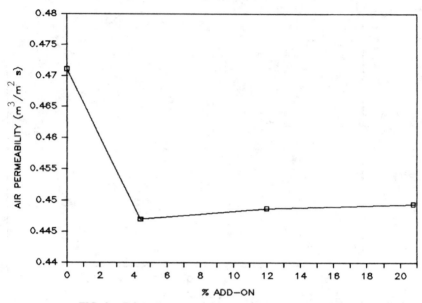

FIG. 3—*Fabric air permeability as a function of resin add-on.*

Discussion

Similar resistivities are observed for an untreated fabric and fabric with nearly 20% add-on. The mechanisms of transfer for these two fabrics are different. The maximum obtained in resistivity versus bulk density demonstrates the interplay of radiant and conductive heat transfer in the fabrics. The decrease in fabric pore size reduces fabric diathermancy and an

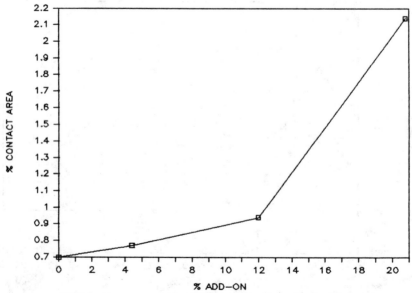

FIG. 4—*True contact area of fabric as a function of resin add-on.*

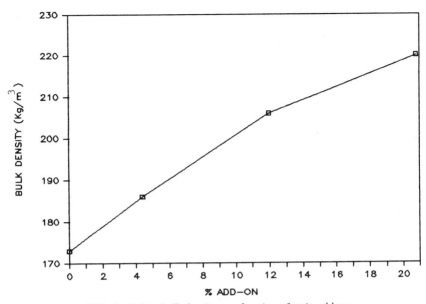

FIG. 5—*Fabric bulk density as a function of resin add-on.*

increase in resistivity is observed. Further increase in solid content seems to have no further effect on diathermancy and a decrease in resistivity ensues as conduction mechanisms become dominant. There is some difficulty in separating the effects of increasing true contact area and decreasing thickness. At steady-state heat flow conditions, the actual contact area of the fabric with the heated plate will have an effect on point-to-point fiber heat transfer from

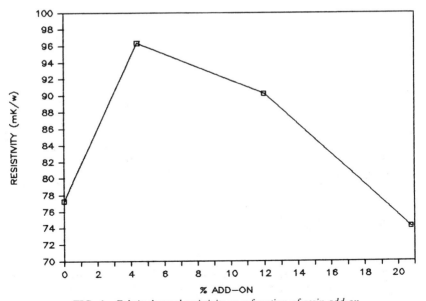

FIG. 6—*Fabric thermal resistivity as a function of resin add-on.*

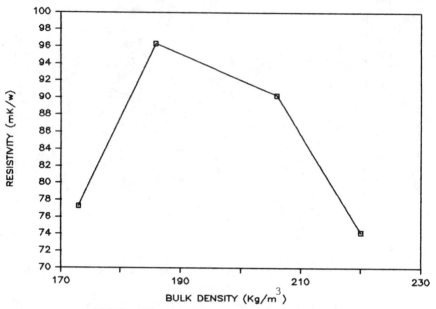

FIG. 7—*Fabric resistivity as a function of bulk density.*

the plate. The reduction in underlying air layer thickness, due to surface fiber smoothing, may also cause decreased resistivity at high resin add-ons.

Note, however, that a large drop in resistivity occurs between Fabrics 3 and 4. The thickness of the fabrics is the same, however, contact area increases sharply. Bulk density increases linearly with add-on. Again, a conductive heat transfer mechanism is indicated.

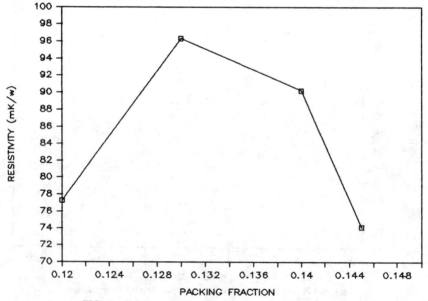

FIG. 8—*Fabric resistivity as a function of packing fraction.*

The increased resistivity at 4.4% add-on occurs despite a small increase in contact area and bulk density and a decrease in thickness. This is strongly indicative of a change in transmissivity affecting radiant heat transfer.

The occurence of a resistivity maximum (conductivity minimum) for random fiber assemblies as a function of packing fraction, predicted and observed by Pelanne and others, occurs then for woven fabrics as well, if bulk yarn and fiber geometry is unaffected by the bulk density change. The maximum for the test fabrics occurs at higher calculated packing fractions than for random fiber beds due to the uneven distribution of open pores in the fabrics. Tight packing of the yarns has little effect on radiative transfer through interyarn pores. Only when these become small enough does diathermancy decrease. An added complication is that woven fabrics present an inhomogeneous density profile across their thickness because of surface fibers.

A numerical computer routine developed by Holcombe and Stuart [13] predicts a minimum in conductivity for a random fiber bed with fiber diameter of 65 μm, at a packing fraction of 0.05. Above this packing fraction, fiber diameter is predicted to have little effect and conductivity increases steadily with packing fraction. This is due to the reduced importance of radiative transfer at high packing fractions so that the surface area available per unit volume of fiber bed to absorb and scatter radiation becomes less significant.

It is not surprising, therefore, that when the measured parameters of Fabrics 2 through 4 are inserted into this model (assuming fiber diameter of 50 μm, solids conductivity of 0.2 W/mK, and emissivities of 0.95), conductivity values similar to the ones obtained experimentally are predicted. The conductivity of the untreated fabric having wide open pores is, of course, not predicted by the random fiber bed model.

Conclusions

Descriptors such as bulk density and packing fraction are obviously inadequate when dealing with heat flow through fabrics, due to significant radiative transfer effects. In this study a resin treated fabric with a nominal packing fraction of 0.13 minimized the effects of radiant and conductive transfer. This can be largely attributed to pore closure due to the resin treatment. The resulting decrease in resistivity at higher add-ons is related to conduction mechanisms due to improved contact between the fabric and the heated plate and increased solids content. Thermal resistivity of woven fabrics is thus affected by resin treatment, even at commercial add-on levels (around 5%). It can be expected that fabrics with an open weave would be most affected in this regard.

Acknowledgments

The author wishes to acknowledge the help of Ms. Dorit Amir in performing the test procedures. This study was funded by a grant from the Bennett and Pauline Rose Foundation.

References

[1] Marsh, M. C., "The Thermal Insulating Properties of Fabrics," *Journal of the Textile Institute*, Vol. 21, 1931, pp. T245–T273.
[2] Peirce, F. T. and Rees, W. H., "The Transmission of Heat through Textile Fabrics—Part 2," *Journal of the Textile Institute*, Vol. 37, 1946, pp. T181–T204.
[3] Finck J. L., "Mechanisms of Heat Flow in Fibrous Materials," *Bureau of Standards Journal of Research*, Vol. 5, 1930, pp. 973–984.
[4] Farnworth, B., "Mechanisms of Heat Flow through Clothing Insulation," *Textile Research Journal*, Vol. 53, 1983, pp. 717–724.

[5] Pelanne, C. M., "Experiments on The Separation of Heat Transfer Mechanisms in Low Density Fibrous Insulations," *Thermal Conductivity,* Proceedings of 8th, Conference, C. R. Ho and R. E. Taylor, Eds., Plenum Press, New York, 1969.

[6] Naka, S. and Kamata, Y., "The Effect of Bulk Density on Thermal Conductivity of Fabrics," *Sen-i Gakkaishi,* Vol. 30, 1974, pp. T43–T52.

[7] Freeston, W. D., Jr., "Flammability and Heat Transfer Characteristics of Cotton, Nomex and PBI Fabric," *Journal of Fire and Flammability,* Vol. 2, 1971, pp. 57–76.

[8] Kalelkar, A. S. and Kung, H. C., "Preignition Heat Transfer in Fabric Skin Systems," *Journal,* American Association of Textile Chemists and Colorists, Vol. 4, 1972, pp. 200–203.

[9] Ross, J. H., "Thermal Conductivity of Fabrics as Related to Skin Burn Damage," *Journal of Applied Polymer Science: Symposia,* Vol. 31, 1977, pp. 292–312.

[10] Morse, H. L. et al., "Analysis of the Thermal Response of Protective Fabrics," *NTIS–AD, 759-525,* National Technical Information Service, Springfield, VA, 1973.

[11] Stuart, I. M. and Holcombe, B. V., "Heat Transfer Through Fiber Beds by Radiation with Shading and Conduction," *Textile Research Journal,* Vol. 54, 1984, pp. 149–157.

[12] Miller, B. and Tyomkin, I., "An Extended Range Liquid Extrusion Method for Determining Pore Size Distributions," *Textile Research Journal,* Vol. 56, 1986, pp. 35–40.

[13] Holcombe, B. V. and Stuart, I. M., "Method of Numerical Solution of Equations Describing Heat Transfer in Fibrous Beds," *Thermal Insulation,* Vol. 7, 1984, pp. 214–227.

Thomas D. Proctor[1] and Harry Thompson[1]

Setting Standards for the Resistance of Clothing to Molten Metal Splashes

REFERENCE: Proctor, T. D. and Thompson, H., **"Setting Standards for the Resistance of Clothing to Molten Metal Splashes,"** *Performance of Protective Clothing: Second Symposium, ASTM STP 989*, S. Z. Mansdorf, R. Sager, and A. P. Nielsen, Eds., American Society for Testing and Materials, Philadelphia, 1988, pp. 131–141.

ABSTRACT: Comparisons are reported between two laboratory methods for assessing the resistance of clothing materials to splashes of molten substances. The methods, ISO DIS 9185:1986 and a modification using calorimeters, were found to rank materials in broadly the same way. The results, including the performance of clothing in industry, have shown that heat transfer occurs due to adhesion of molten substances to clothing and penetration of molten substances through clothing. Information is given on the extent of adhesion of molten iron, copper, brass, aluminum, slag, and the electrolyte used in aluminum smelting, to commonly found clothing materials in order to aid selection for specific uses. The structure of wool melton cloths to provide protection against molten iron, copper, and brass is considered.

KEY WORDS: protective clothing, hazards, molten metals, wool, cotton, leather, aluminized fabrics, laboratory test methods

Several methods have been suggested for assessing the resistance of clothing materials to splashes of molten metal [*1–4*]. All of them use the heat transferred through the material as a measure of performance, and this is measured either by damage to a plastic film or by calorimeters.

In the United Kingdom (UK) the plastic film method has been developed into a British Standard, BS6357 [*5*], and further work in cooperation with the International Standards Organization (ISO) has led to its adoption as a Draft International Standard ISO DIS 9185 [*6*].

The two methods are quite different in the way they respond to heat, and doubt has been expressed as to whether they rank materials in the same order of performance. This paper describes experiments that were designed to clarify this point and goes on to examine the relationship between laboratory tests and real conditions.

Materials and Test Methods

Clothing Materials and Molten Substances

A wide variety of clothing materials have been assessed against a range of molten substances. Results are reported here for some of the more commonly used materials. Their performance has been established against molten iron, copper, 63-37 brass, aluminum, iron slag, steel slag, and the electrolyte used in the smelting of aluminium (a mixture of cryolite

[1] Principal scientific officer and high scientific officer, respectively, Health and Safety Executive, Research and Laboratory Services Division, Safety Engineering Laboratory, Sheffield S3 7HQ, UK.

and fluorspar). The composition of the iron, copper, and aluminum was as specified in ISO DIS 9185, while the slag and electrolyte samples were those obtained from the appropriate smelting processes.

ISO DIS 9185 Method

The laboratory method described in ISO DIS 9185 is based on the apparatus shown in Fig 1. A crucible rotates at constant speed to pour molten metal onto a sloping sample of clothing material pinned on top of, and in contact with, a PVC film. The angle at which the sample is held in relation to the pour can be varied. The weight of molten metal is systematically varied from testpiece to testpiece until a weight is found that results in sufficient transfer of heat to produce a defined degree of damage to the PVC film. This weight is called the molten metal splash index (MMSI) and is characteristic of the material.

As well as assessing the heat insulation provided by the material, the test responds to the penetration of molten droplets into the material and the adhesion of hot metal to the surface.

Calorimeter Method

For this method, the sample holder in the ISO DIS 9185 equipment was fitted with an insulating board supporting the calorimeters. The early experiments were conducted with a 40-mm-diameter circular calorimeter similar to the ASTM design [4]. This, however, proved unsuccessful due to variations in the point of molten metal impact from one experiment to another that resulted in calorimeter temperature rises differing by a factor of two.

A multi-strip calorimeter system was developed that overcame this disadvantage; see Fig. 2a. Four calorimeters, 10 by 40 mm, are placed 10 mm apart down the long axis of the material to cover the main area of molten metal impact. The small spacing between them ensures that the splash covers at least one strip.

The calorimeters have the same heat capacity per unit area as those in the ASTM method [4] and their temperatures are monitored by two thermocouples connected in parallel and positioned as shown in Fig. 2b. The output is taken via amplifiers to a microcomputer that is programmed to record the temperature of all four calorimeters at 0.1-s intervals for 35 s after the initiation of the pouring cycle. The computer then selects the calorimeter showing the largest increase in temperature, displays the results, and records relevant information onto disc for future analysis.

For the purposes of this paper, the largest increase in temperature, T_M, has been taken as a measure of performance following the practice adopted in the United States [7].

Effect of Variations in Experimental Conditions

Reproducibility of Measurements

The repeatability of both methods has been found to depend on the type of material being tested. Where there is adhesion of molten substance to the material or penetration of the material by molten droplets, there is a greater variation from sample to sample than where the material readily sheds the molten substance.

Two materials with good shedding properties were assessed against molten aluminum at two angles of inclination using the ISO method [6]. The coefficients of variation for three repeat determinations of MMSI lay between 1.0 and 5.8%. In interlaboratory trials of this method through ISO, two fabrics were satisfactorily assessed against molten aluminum. The coefficients of variation between the four laboratories were 9.3 and 29%, respectively.

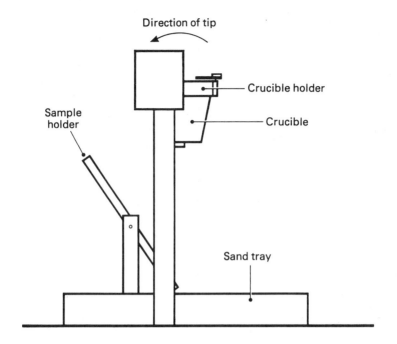

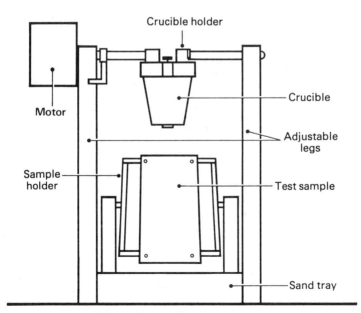

FIG. 1—*Motorized pouring apparatus.*

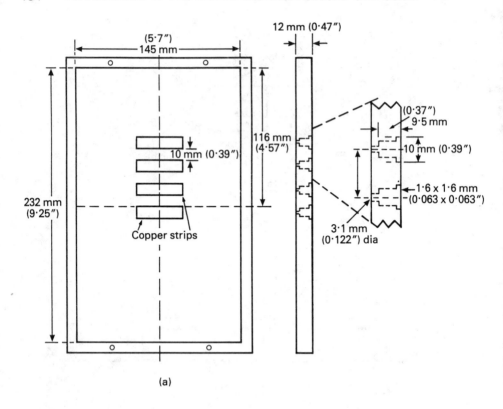

(a)

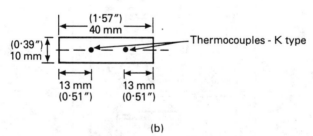

(b)

FIG. 2—*Construction of multi-strip calorimeter:* (a) *layout of multi-strip calorimeter and* (b) *position of thermocouples in each strip.*

For the calorimeter method, the distribution of the results of repeat tests has been found to be positively skewed especially when adhesion or pin-holding of the material occurs. Applying a log transformation allows standard procedures for curve fitting and statistical testing to be applied. Thus, throughout this paper, the mean temperature rise, $(\overline{T}_M)$, is defined by the following equation

$$\log (\overline{T}_M) = \frac{\Sigma \log T_M}{n}$$

where n = number of repeat tests.

When tested against molten aluminum, a fabric with good shedding properties gave a coefficient of variation of 13.9% within our laboratory on ten repeat tests with the calorimeter. Overall, it has been found that the scatter of results increases with $\overline{T}_M$, coefficients of variation averaging 16.5% at a $\overline{T}_M$ of 10°C (18°F) and 76% at 30°C (54°F), the latter figure being due to the unpredictability of molten metal adhesion and penetration.

Although the figures for the calorimeter may seem higher than those for the plastic film, it has to be borne in mind that they represent the scatter of individual tests whereas each MMSI value requires about eight tests to determine. Overall, it would appear that variations are not very different for the two methods and the main source of scatter is due to random differences in the interaction between molten metal and material from test to test.

Effect of Weight of Molten Metal Poured (Calorimeter Method)

Five fabrics were assessed against molten iron at 1400°C (2552°F) using the calorimeter. The weight of molten iron poured in the tests was varied between 100 and 800 g with the samples at 75° to the horizontal. It was found that the fabrics ranked in the same order for different weights of molten iron poured apart from two fabrics with very similar performance for which the ranking reversed at 300 g. This suggested that no particular advantage is to be gained by pouring large weights, and it was decided to standardize on 300 g, a weight that causes considerable damage to the lighter fabrics but is resisted by the heavier fabrics apart from some pin-holing.

Variations in Molten Metal Temperature

The calorimeter was used to examine the effect of variations in the pouring temperature for molten iron, copper, and aluminum. It was found that where fabrics are prone to adhesion of molten metal, for example, Proban-treated cotton, the transfer of heat increases as the pouring temperature approaches the melting point of the metal. On the other hand, where adhesion is not a factor, for example, wool- and Caliban-treated cotton, no significant correlation was found between pouring temperature and $\overline{T}_M$.

For the assessments discussed in this paper, the metals were poured at the temperatures commonly used for casting in the UK foundry industry, that is, iron at 1400°C (2552°F), copper at 1280°C (2336°F), brass at 1100°C (2012°F), aluminum at 800°C (1472°F), and electrolyte at 1000°C (1832°F). Iron and steel slag were poured when molten, since it was found impractical to measure the temperature of this substance due to solidification on the temperature measuring probe.

Variations in the Angle of the Testpiece to the Horizontal

It has been found that for those metals that tend to penetrate fabrics, for example, molten iron and copper, the angle the test specimen makes with the horizontal can have a significant effect on its performance. At shallow angles, the ranking of the more loosely woven fabrics may be reversed with respect to the tightly woven ones. Care needs to be exercised, therefore, in the use of different specimen angles as a means of changing the severity of the test, for example, to assess fabrics by the ISO method that would otherwise be beyond the range of the test. Where metals are not so penetrating, for example, aluminum, slag, and electrolytes, the angle has little effect on the ranking of fabrics and assessments have been made at 60°. A steeper angle of 75° has been used for iron, copper, and brass to simulate the near vertical clothing surfaces during normal wear.

Comparison of Test Methods

The performances of 16 fabrics, including wool, Proban-treated cotton, and Caliban-treated cotton, have been measured by both methods against molten iron (see Table 1). In this table, the fabrics are listed in descending order of their MMSI values and the ranking in Column 6 is on the basis of the better fabrics being given low numbers. Column 7 shows the corresponding rank order of the $\overline{T}_M$ values. The degree of correlation was assessed by calculating Spearman's Rank Correlation Coefficient. Although there is a significant positive correlation ($R = 0.70$), which is significantly different from zero at the 0.005 level, a value of 0.49 for R^2 suggests that the association is not very strong.

To see if there were differences between fabrics, they were split into two groups. Group 1 was wool- and Caliban-treated cotton fabrics that all suffer pin-holing of the material but little adhesion of molten metal. The Proban-treated cotton fabrics, which do show adhesion, made up Group 2.

For Group 1, log ($\overline{T}_M$) was plotted against log (MMSI). A straight line fitted by least squares regression showed a correlation coefficient of -0.796 that is significantly different from zero at the 0.02 level. The relationship between the variables was $\overline{T}_M = 993(\text{MMSI})^{-0.847}$.

Similar calculations for the fabrics in Group 2 showed $R = -0.454$ that is not significantly different from zero. However, the rank correlation for this group gave a value significantly above zero at the 0.02 level.

The $\overline{T}_M$ values for Fabrics 13 and 15, where penetration of molten iron occurred, are higher than those for the more tightly woven Fabrics 14 and 16 of corresponding weight while the MMSI values are similar. The calorimeter method, with its fixed weight of molten metal poured, is thus more favorable towards the tightly woven fabrics than the ISO 9185 method that determines the critical pouring weight for each material. Minor changes in ranking are due to this difference in the principle of the two tests.

Heat Transfer

There are two ways whereby a substantial amount of heat from molten metal splashes may be transferred through clothing materials. The first of these is adhesion of molten metal to the surface of the material that effectively increases the time for which the hot metal is in contact with the clothing. The second involves penetration of the molten metal into the material with subsequent holding and possible break-through of hot metal.

Adhesion of Molten Substances to Clothing Materials

By examining the surface of clothing samples after test, it has been possible to build up a picture of the extent to which various molten substances adhere to the surface of commonly found clothing materials. In Table 2, the materials have been ranked from the top in order of least to most adhesion. Molten substances have been similarly ranked from left to right. Low numbers in the body of the table indicate those combinations of material and molten substance where adhesion is least likely.

Close examination of this table shows that the ranking of the materials is similar for all metals and vice-versa. This suggests that, for these materials and substances, adhesion is a physical phenomenon and not due to chemical interaction between particular substances and materials. In addition, inspection of samples after test has shown that solidified metal is only loosely attached to the material even when adhesion is extensive whereas closer bonding might be expected if chemical interaction were involved. The mechanism seems more likely to be simply a surface cooling process closely related to the thermal properties

TABLE 1—*Comparison of plastic film and calorimeter methods.*

Fabric No.	Fabric Description	Flame Retardant	Molten Metal Splash Index, MMSI, g	$\bar{T}_M$, °C (°F)	Rank Order of MMSI	Rank Order of $\bar{T}_M$
1	cotton mole	Proban	383	13.90 (25.02)	1	7
2	wool melton	Zirpro	276	12.54 (22.57)	2	5
3	cotton mole	Proban	190	10.84 (19.51)	3	1
4	cotton mole	Proban	183	14.21 (25.58)	4	10
5	cotton mole	Proban	182	14.01 (25.22)	5.5	8
6	wool melton	none	182	11.07 (19.93)	5.5	2
7	wool melton	none	152	12.44 (22.39)	7	4
8	cotton beaverteen	Caliban	134	11.81 (20.12)	8	3
9	wool twill	Zirpro	133	14.53 (26.15)	9	11
10	wool melton	Zirpro	126	15.96 (28.73)	10	14
11	cotton denim	Caliban	123	12.63 (22.73)	11	6
12	cotton mole	Proban	104	14.07 (25.33)	12	9
13	cotton denim	Caliban	90	29.68 (53.42)	13	15
14	cotton beaverteen	Proban	82	14.54 (26.17)	14	12
15	wool twill	Zirpro	72	33.52 (60.34)	15	16
16	cotton sateen	Proban	63	15.70 (28.26)	16	13

of the molten substance and its viscosity that influences its ability to spread across the material's surface and hence the area available for heat transfer. Where a substance has low thermal inertia and high ability to spread, the rate of fall of temperature will be quicker than for substances with the reverse properties. Cooling will also be related to the thermal properties and surface construction of the material and, in the case of Proban-treated cotton, by the endothermic action of the flame retardant.

TABLE 2—*Adhesion of molten substances to protective clothing materials.*

Material Type	Molten Substance					
	Brass at 1100°C (2012°F)	Copper at 1280°C (2366°F)	Iron at 1400°C (2552°F)	Aluminum at 800°C (1472°F)	Slag	Electrolyte in Aluminum Smelter at 1000°C (1832°F)
Wool	0[a]	0	0	0	1	2
Zirpro-treated wool	0	0	0	0	1	1 to 3
Aluminized materials	0	0 to 1[b,e]	0	0 to 2[c]	3	1 to 3
Caliban-treated cotton	0	0 to 1	0	1 to 2	3	3
Leather	0	0 to 1	0 to 3[d]	0 to 3	3	3
Proban-treated cotton	0	0 to 1	1 to 3	3	3	3

[a] 0 indicates no more than specks of adhesion.
[b] 1 indicates spots of adhesion not more than 2 mm (0.079 in.) across.
[c] 2 indicates patches of adhesion larger than 2 mm (0.20 in.) and not more than 10 mm (0.39 in.) across.
[d] 3 indicates patches of adhesion more than 10 mm (0.39 in.) across.
[e] Where a range of values is indicated there was variability from test to test.

Holing of Clothing Materials

It was mentioned when discussing the effect of the angle of inclination of the testpiece that some materials are prone to pin-holing due to the open nature of their structure. This occurs particularly with wool melton fabrics against molten metals such as iron and copper. The relationship between the structure of such fabrics and their performance has been investigated for 18 wool melton cloths using the calorimeter. Details of the structure of these materials and their performance are shown in Table 3. Measurements of air permeability were taken in accordance with the method described in *British Standard Handbook 11* [8]. Air is drawn through a circular fabric sample, 22.6 mm in diameter, and the flow rate adjusted by suitable means until the pressure drop across the sample is 10 mm water gage. The air permeability is then taken as the flow rate of air through the sample in 1 min^{-1}.

Fabric sett (the sum of the threads per unit length in the weft and the warp) was measured by scorching off the surface of the felt on a hot plate to reveal the underlying surface structure, on which the thread counts could be made.

The fabrics include both new ones and some obtained from garments that had been worn in the foundry industry (see Table 3). The samples from used garments were taken as far as possible from the least damaged areas. The tests against molten iron were made on the unwashed samples but the fabric weights given are after washing.

Multiple linear correlation coefficients were calculated with various transformations of the variables from which it was found that the highest correlation was obtained using a log transformation of the $\overline{T}_M$ values. Table 4 shows the correlation matrix obtained. Regression of log $(\overline{T}_M)$ against fabric weight, air permeability, and sett gave a multiple correlation coefficient, R, of 0.936. This is not much higher than the R of 0.93 for log $(\overline{T}_M)$ against air permability alone shown in Table 4, and it can be concluded that the effect of weight and sett are contained in variations of air permeability. The linear regression line for log $(\overline{T}_M)$ against air permeability, AP, is given by the following equation

$$\log (\overline{T}_M) = 0.274 \text{ AP} + 0.685$$

TABLE 3—*Effect of fabric construction on performance for wool melton cloths.*

Fabric No.	Source of Fabric	Fabric weight, gm^{-2} (oz yd^{-2})	Sett, threads cm^{-1} (threads in.$^{-1}$)	Air Permeability, 1 min^{-1}	$\overline{T}_M$, °C (°F)
1	used trousers	774 (22.9)	30.2 (76.7)	0.95	10.82 (19.48)
2	used jacket	723 (21.4)	32.3 (82.0)	1.55	10.80 (19.44)
3	used jacket	716 (21.2)	32.3 (82.0)	1.67	11.64 (20.95)
4	new	698 (20.6)	28.0 (71.1)	1.51	11.07 (19.93)
5	used jacket	694 (20.5)	30.7 (78.0)	1.08	12.75 (22.95)
6	new	693 (20.5)	30.3 (77.0)	1.35	11.18 (20.12)
7	new	681 (20.1)	28.0 (71.1)	1.79	11.15 (20.07)
8	used jacket	673 (19.9)	27.0 (68.6)	3.07	40.96 (73.73)
9	new	658 (19.4)	29.6 (75.2)	1.71	12.44 (22.39)
10	used jacket	655 (19.4)	32.3 (82.0)	1.10	8.65 (15.57)
11	used jacket	650 (19.2)	27.3 (69.3)	2.09	15.56 (28.01)
12	used jacket	618 (18.3)	30.3 (77.0)	2.26	20.60 (37.08)
13	new	613 (18.1)	28.0 (71.1)	2.59	20.06 (36.11)
14	used trousers	591 (17.5)	22.0 (55.9)	3.03	56.56 (101.8)
15	used jacket	576 (17.0)	24.2 (61.5)	2.82	29.26 (52.67)
16	new	575 (17.0)	39.7 (100.8)	1.10	12.80 (23.04)
17	used jacket	565 (16.7)	21.8 (55.4)	3.91	45.92 (82.66)
18	used trousers	554 (16.4)	21.1 (53.6)	3.52	46.76 (84.17)

TABLE 4—*Correlation matrix for measurements on wool melton cloths.*

	log, $\overline{T}_M$	Fabric Weight	Air Permeability	Sett
Log, $\overline{T}_M$	1	−0.6904	0.9305	−0.7705
Fabric weight	...	1	0.1244	0.4099
Air permeability	...	...	1	−0.8247
Sett	...	...	...	1

that is shown with the 95% confidence limits in Fig. 3. For this graph, the regression accounts for 86.6% of the variance in log ($\overline{T}_M$). Hence, it can be concluded that a measurement of air permeability could be used as a simple means of checking the performance of this type of fabric.

Comparison Between Laboratory Testing and Real Incidents

Clothing cannot be expected to give adequate protection against sustained contact with molten substances as occur, for example, when persons are exposed directly to a molten metal stream. This type of incident has occurred due to overhead ladles being over-filled and spilling over as a result of erratic movement, molds bursting apart, etc. A particularly serious incident of this type was reported in 1983 [9]. The risks can be reduced by careful attention to plant layout and design, and protective clothing must only be regarded as a secondary line of defense.

However, despite its limitations, some types of clothing have been successful in preventing injury particularly when persons have been showered by molten droplets arising from explosions as a result of contact between molten metal and water. We have examined clothing from eight such incidents involving molten iron, copper, brass, aluminum, and slag. The

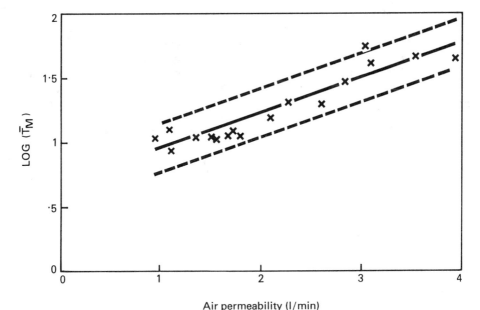

Air permeability (l/min)

FIG. 3—*Relationship between performance and air permeability for wool melton cloths.*

most serious injury, 25% second-degree burns, was caused by molten slag adhering to a 397 gm^{-2} Proban-treated cotton jacket. Table 2 shows that this material is not suitable for protection against this particular hazard. A similar fabric, marginally heavier (470 gm^{-2}), gave adequate protection against molten copper in an incident where the wearer was positioned adjacent to the mold that exploded. Table 2 shows that little adhesion would be expected in this case. These results suggest that it is most important to choose clothing to minimize adhesion of the molten substance against which protection is required.

Wool clothing was being worn in the remaining six incidents. Injuries were caused in three cases, two being due to trapping of metal within the garment and the third to poor quality material (Fabric 17 in Table 3). These injuries were all localized. Where wool garments are used to protect against the high-temperature metals, the material must be heavy (>550 gm^{-2}) and tightly woven to avoid penetration of hot droplets; an air permeability of <1.8 L/min^{-1}, as defined in the section on holing, is necessary. In the UK, the British Steel Corporation have published their own specification for wool melton cloth meeting this criterion, which over the years has prevented many injuries [10]. Protection against molten aluminum that is less penetrating may be afforded by a lighter, more open structure. In one incident, a 285 gm^{-2} Zirpro-treated wool twill gave adequate protection against a cloud of droplets of aluminum alloy. However, because of adhesion, the same material would not be suitable against the electrolyte used in the smelting process (see Table 2). The scale of hazard would be very important there.

Conclusions

In laboratory tests and from examination of clothing after accidents, it has been found that heat is transferred from molten metal splashes in two main ways: adhesion and penetration.

Both methods of measuring heat transfer, that is, plastic film and calorimeter, respond irrespective of the way heat is transferred and hence rank clothing materials in broadly the same order of performance. Differences only occur when the quantity of molten metal poured in the calorimeter method is well beyond the structural capability of the material.

The coefficient of variation of the results from the calorimeter method averages about 16.5% when good quality fabrics are tested. Thus, to attain a standard error of the mean averaging 7.4% would require five repeat tests compared with seven to ten for a single determination of MMSI by the plastic film technique. This is a better level of reproducibility than that obtained in interlaboratory trials of the plastic film method, suggesting that the calorimeter method could offer considerable cost savings by minimizing repeat testing. Although the equipment is more complicated technically, the benefits could outweigh this disadvantage. At the same time, the problems associated with ensuring a source of supply of the particular plastic film would also be avoided.

Assessment of clothing worn in accidents has shown that the materials that offer the least protection were those that performed poorly in laboratory tests against the same molten substances.

Trapping of metal within garments has been found to cause injuries, therefore, it is important to design garments to avoid this possibility as far as possible. A report giving a fuller discussion of these matters has been published [11].

References

[1] Mehta, P. N. and Willerton, K., Textile Institute Industries, Vol. 15, 1977, pp. 334–337.
[2] Cornu, J. C. and Aubertin, G., "Vetements de protection et d'intervention contre les projections de metaux en fusion-Resultats de mesures d'efficacite de la protection offerte par les materiaux

constitutifs," Institute de Recherch et Securite, Nancy, France, Report No 495/RE, 1981: Health and Safety Executive, Library, Sheffield, U.K., Translation No 9872.

[3] Benisek, L. and Edmondson, G. K., *Textile Research Journal*, Vol. 51, 1981, pp. 182–190.

[4] "Evaluating Heat Transfer through Materials upon Impact of Molten Metals and Related Substances," American Society of Testing and Materials, Draft No 1, 1982.

[5] "Method of Assessment of the Resistance of Materials Used in Protective Clothing to Molten Metal Splash," British Standards Institution, BS6357, London, 1983.

[6] "Protective Clothing: Method of Assessment of the Resistance of Materials to Molten Metal Splash," International Standards Organization, DIS 9185, Zurich, Switzerland, 1986.

[7] Jaynes, P. S. in *Performance of Protective Clothing, ASTM STP 900*, R. L. Barker and G. C. Coletta, Eds., American Society for Testing and Materials, Philadelphia, 1986, pp. 475–486.

[8] "Air Permeability of Fabrics," *British Standard Handbook 11*, Section 4, British Standards Institution, London, 1974, pp. 171–173.

[9] Hewerdine, H. I. and Buckley, P. J., *Foundry Trade Journal*, 10 Nov. 1983, pp. 458–460.

[10] British Steel Corporation, Protective Equipment Standards, PES 100-105, London, 1985.

[11] "First Report on Clothing to Protect Against Molten Metal Splash Hazards in Foundries," Her Majesty's Stationary Office, Health and Safety Commission, London, 1985.

Bal Dixit[1]

Development and Testing of Extremely High-Temperature (1093°C and Over) Coatings for Safety Clothing

REFERENCE: Dixit B., "Development and Testing of Extremely High-Temperature (1093°C and Over) Coatings for Safety Clothing," *Performance of Protective Clothing: Second Symposium, ASTM STP 989*, S. Z. Mansdorf, R. Sager, and A. P. Nielsen, Eds., American Society for Testing and Materials, Philadelphia, 1988, pp. 142–158.

ABSTRACT: Various properties critical for satisfactory performance of the fabric at temperatures of 1093°C (2000°F) and over are described. There is a dramatic change in the properties of materials at high temperatures. Various coatings have been developed over the years to enhance the temperature capability of both organic and inorganic materials. These coatings are evaluated. The selection process and the effect of the coatings in different applications are described. The paper also describes the development of an inorganic coating that dramatically changes the high-temperature properties of fiberglass with very little outgassing or smoke. New test methods were developed to evaluate the performance of Zetex Plus made from highly texturized fiberglass yarn and coated with the inorganic treatment. The insulation properties and product life are significantly improved with the application of the Zetex Plus coating.

KEY WORDS: adherence, aramid, asbestos, charring, double-palm mitten, fiberglass, high-temperature coatings, inorganic, organic, perforation, shrinkage, single-palm mitten, thermal insulation, vermiculite

In view of the several Federal and state regulations regarding toxic and hazardous substances, the development of high-temperature coatings involves consideration of several important criteria. The most important is the protection of personnel at high temperatures, but the second most important criterion is to provide this protection without generating excessive smoke or gaseous products which may be hazardous to the individual. Other important properties are flexibility, hand and feel, ability to cut and sew, seam strength, abrasion resistance, and durability. Various coatings have been developed over the years to enhance the temperature capability of both organic and inorganic materials, but the safety of these coatings has not been carefully studied. The problem has become increasingly acute as some of these products are used in the manufacture of safety clothing which are used as a substitute for asbestos and handle temperatures in excess of 1093°C (2000°F).

The maximum temperature handled by a Zetex[2]—a highly texturized form of fiberglass—double-palm mitten has been about 982°C (1800°F). The performance of the mitten suffered considerably when the temperature exceeded 982°C (1800°F). Even though the double-palm mitten could handle a 1.701-kg (3.75-lb) pipe at 1093°C (2000°F) for up to 15 s, the fiberglass fused, causing the pipe to stick to the double-palm mitten. The useful life of the double-palm mitten was considerably shortened because the outside layer of Zetex was destroyed.

[1] President, Newtex Industries, Inc., Victor, NY 14564.
[2] Registered trademark of Newtex Industries, Inc., Victor, NY.

142

During the past eight years, several organic coatings have been developed but most of these coatings have not worked out satisfactorily due to

1. Excessive smoke.
2. Excessive outgassing—particularly carbon monoxide (CO) and hydrogen cyanide (HCN).
3. Once the coating burned off or carbonized, the fiberglass could not handle the excessive heat 1093°C (2000°F).

The primary purpose of developing these coatings was to prevent unravelling during fabrication of gloves, mittens, and other safety products from Zetex. The Zetex fabric was developed to duplicate the performance, appearance, hand, and feel of asbestos; however, due to the slippery nature of fiberglass, the fabric unravelled when the cut patterns were handled or sewn. The secondary purpose was to improve seam strength, which was very weak for untreated fabrics, again due to the slippery nature of fiberglass even though texturization of fiberglass did help. Other advantages gained from the coating were: (1) Some improvement in abrasion resistance of the fabric, (2) By adding some nonhalogenated additive materials to the coating, the flame resistance of the coating was improved. In order to keep the outgassing to a bare minimum level, the treatment was applied to one side of the fabric; this allowed the total organic content of the treatment to be kept at a very low level of 2 to 3%.

The importance of keeping the organic contents to a minimum is demonstrated by the data in Table 1. The smoke density test was conducted in accordance with the ASTM (E 662-79) using a National Bureau of Standards (NBS) smoke chamber. The results clearly show the need for keeping the organic contents as low as possible. The Zetex fabric coated on one side shows a maximum smoke density of 3, whereas the aramid fabric coated with a rubber binder shows a maximum density of 84. Both fabrics are being used as a replacement for asbestos.

Several inorganic coatings have been investigated for Zetex, one of which shows the most promising results. This coating, M729, developed and patented by Imperial Chemical Industries (ICI), is made from chemically exfoliated vermiculite suspension. Vermiculite is a magnesium aluminum silicate having a layered structure. Figure 1 shows vermiculite ore in a test tube and Fig. 2 shows 30-fold expansion of ion-exchanged vermiculite. The aqueous suspension of the high-aspect-ratio lamellae is highly anisotropic, and it has been found that many properties of the film prepared from these aqueous suspensions are determined by the distribution of particle thicknesses and particle diameter. Figure 3 shows the film forming characteristic of the delaminated vermiculite suspension.

According to Ballard and Rideal [1], the coating does an excellent job of penetrating the fiberglass yarn. Figure 4 shows the uncoated fiberglass in the nonwoven mat as seen by scanning electron microscopy (SEM), and Fig. 5 shows the same fiberglass mat after impregnation and drying. The film forming ability of the vermiculite suspension is quite obvious from Fig. 5. The coated and uncoated fiberglass mat, 0.254 to 0.381 mm (10 to 15 mil)

TABLE 1—*Smoke density comparison of Zetex and aramid.*

	Zetex 800 Coated	Aramid with Rubber Coating
Specific optical density at		
1.5 min	1	49
4.0 min	3	73
Maximum density of smoke	3	84
Time to maximum density, min	5.2	7.9

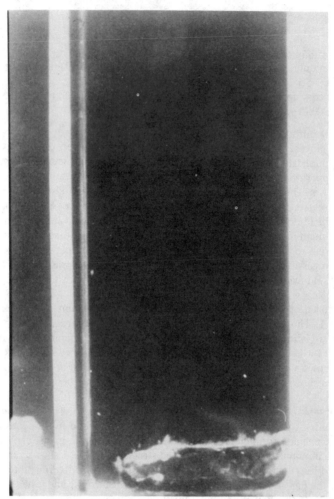

FIG. 1—*Vermiculite ore in a test tube.*

thick, was exposed to 1260°C (2300°F). Figure 6 shows that the uncoated fiberglass mat burns through in seconds, whereas the same mat with vermiculite coating resists flame for 20 min, as shown in Fig. 7. The SEM of coated fiberglass mat, after exposure to the 1260°C (2300°F) flame shown in Fig. 8, demonstrates the physical integrity of the coating. The total encapsulation of fiberglass fiber is shown in Fig. 9, and even though the coating is only 2×10^2 nm thick, it consists of at least 100 lamellae arranged in continuous layers. The multiple layers protect fiberglass from fusion by dissipating heat along the surface rather than though it. This phenomenon is further demonstrated by the photographs shown in Figs. 10 and 11. The uncoated Zetex fabric in Fig. 10 begins to melt in about 25 s, but the same 1085 g/m² (32-oz/yd²) fabric with vermiculite coating withstands the heat continuously and retains the integrity of the fabric structure.

Evaluation of Zetex with Vermiculite

Zetex fabrics of various weights were coated with the modified suspension of delaminated vermiculite. These coated fabrics, called Zetex Plus, were tested for smoke generation,

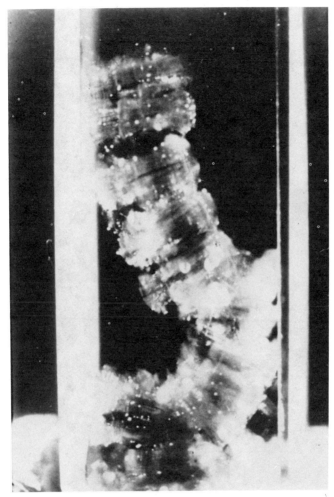

FIG. 2—*Thirty-fold expansion of chemically ion-exchanged vermiculite.*

flame spread, linear shrinkage, breaking strength, abrasion, molten metal splash, and thermal insulation properties. Table 2 gives the smoke density measurement of Zetex Style 800 tested in accordance with ASTM E 662-79. The data from Table 1 have also been included, and the optical density measurements show very clearly that Zetex Plus generates the least amount of smoke compared with regular Zetex and aramid/rubber combinations. This criterion is particularly important because these fabrics are sometimes used in limited environments for welding and stress-relieving applications.

The flame spread, burn length, and afterglow were measured in accordance with Federal Test Method Standard 19, Method 5903. The results given in Table 3 for Zetex Plus Style 1100 show zero flame spread or afterglow and only 2.54 mm (0.1 in.) burn length.

The linear shrinkage of three fabrics was measured by subjecting the fabrics for one hour to a temperature of 816°C (1500°F). The results given in Table 4 clearly demonstrate the protection provided by the vermiculite coating. The shrinkage in the warp direction for Zetex 800 and Zetex 800 heat cleaned was 19.7 and 21.2%, whereas the shrinkage in fill direction was 29.0 and 26.2%, respectively. Zetex Plus 800 exhibited linear shrinkage in

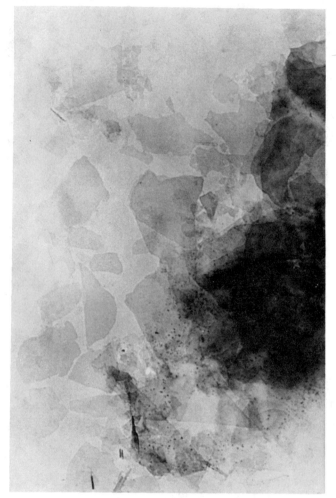

FIG. 3—*Film-forming characteristic of delaminated vermiculite slurry.*

both warp and fill directions of 0.5%. This is an extraordinary benefit of the vermiculite coating and is particularly valuable in gasketing, seals, and thermal insulation applications where shrinkage can cause stresses and failures in standard fiberglass products.

The breaking strength and abrasion resistance properties of Zetex 1100 and Zetex Plus 1100 are presented in Table 5. The average breaking strength in the warp direction for Zetex Plus 1100 is 11.41% higher than regular Zetex, whereas in fill direction the breaking strength for the same fabric is 32.13% greater. The abrasion resistance also shows considerable improvement as shown in Table 5. The average abrasion resistance of Zetex Plus 1100 measured by the Taber Abrasion Tester is 62.65%, better than regular Zetex 1100, and further demonstrates the excellent bonding and coating of each individual yarn.

Molten Metal Testing

The molten metal splash test was described by Dixit [3] and is repeated here. Some of the test results are also reported for comparison.

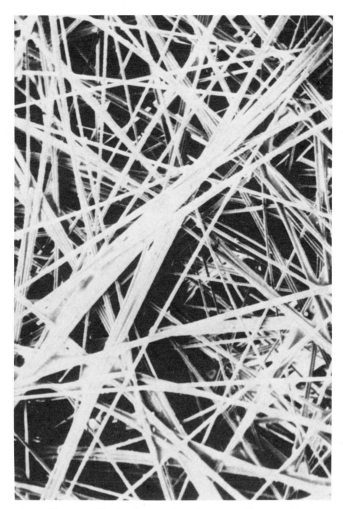

FIG. 4—*SEM of nonwoven fiberglass mat without coating.*

In order to determine the molten metal splash resistance, Zetex, Zetex Plus, and asbestos fabrics were sent to an independent laboratory where these fabrics were tested and compared for their ability to provide thermal protection against controlled splashes of molten iron at 1510°C (2750°F). The test procedure consisted of pouring about 0.907 kg (2 lb) of molten iron at 1510°C (2750°F) onto fabric specimens attached to a board held at an angle of 1.22 rad (70 deg) from the horizontal and from a height of 30.5 cm (12 in.). The test apparatus is shown schematically in Fig. 12. Each fabric was placed on a transite board and held in place with clips along the upper edge. The preheated ladle was filled with molten iron. The weight of molten iron was maintained at 0.907 ± 0.057 kg (2 lb ± 2 oz). The filled ladle was placed in a ladle holder. A fixed delay of 20 s after the start of the furnace pour was used to ensure a consistent temperature. The molten metal was then dumped onto the fabrics, and the results were assessed. The total splash was maintained at 0.5 ± 0.1 s.

Each of the fabrics was examined for both visual appearance and heat transfer through the fabric. The visual appearance of each of the outer layers after molten metal impact was

FIG. 5—*SEM of nonwoven fiberglass mat with delaminated vermiculite coating.*

rated subjectively in four categories:

(*a*) charring,
(*b*) shrinkage,
(*c*) metal adherence, and
(*d*) perforation.

The rating system uses numbers from 1 to 5 in each category, with 1 representing good and 5 representing poor. Table 6 gives the visual ratings of various fabrics exposed to the molten iron test. An outline of the rating system in detail follows.

Grading System Used to Evaluate Fabric Damage

The fabric specimens were evaluated visually for charring, shrinkage, and perforation to provide an indication of the extent of damage to the outer impacted layer. Five grades were

FIG. 6—*Nonwoven fiberglass mat 0.254 to 0.381 mm (10 to 15 mil) thick exposed to 1260°C (2300°F) flame.*

used in evaluating the extent of charring:

1 = slight scorching—the fabric had small brown areas;
2 = slight charring—the fabric was mostly brown in the impacted area;
3 = moderate charring—the fabric was mostly black in the impacted area;
4 = charred—the fabric was black and brittle and cracked when bent; and
5 = severely charred—large holes or cracks appeared, and the fabric was very brittle.

Shrinkage was evaluated by laying the fabric on a flat surface and observing the extent of fabric wrinkling around the splash area. Shrinkage was evaluated using three categories:

1 = no shrinkage,
3 = moderate shrinkage, and
5 = significant shrinkage; and fabric was badly distorted.

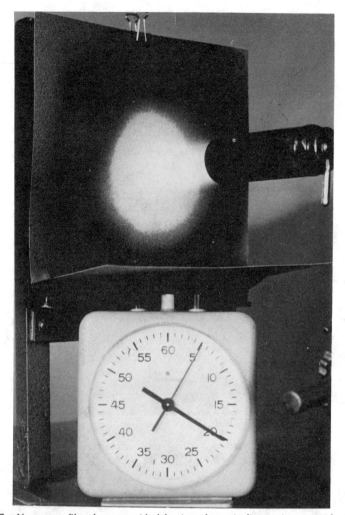

FIG. 7—*Nonwoven fiberglass mat with delaminated vermiculite coating exposed to 1260°C (2300°F) flame after 20 min.*

The adherence rating refers to the amount of metal sticking to the front of the fabric. Adherence of the metal was rated using five categories:

1 = no adherence,
2 = a small amount of metal adherence to the face or back of the fabric,
3 = a moderate amount of metal adherence to the fabric,
4 = substantial adherence of the metal to the fabric, and
5 = a large amount of adherence of metal to the fabric.

Perforation was evaluated by observing the extent of destruction of the fabric, usually by holding it up to a light. Five grades were used in evaluating perforation:

1 = none,
2 = slight, with small holes in the impacted area;

FIG. 8—*SEM of vermiculite-coated nonwoven fiberglass mat after exposure to 1260°C (2300°F) flame.*

3 = moderate, with holes in the fabric;
4 = metal penetration through the fabric, with some metal retained on the fabric; and
5 = heavy perforation; the fabric exhibited gaping holes or large cracks or substantial metal penetration to the back side.

The heat-transfer data for various fabrics attached to the transite board are obtained by a 25.4-mm-diameter (1-in.) and 1.6-mm-thick (1/16 in.) copper calorimeter located under the point of molten metal impact. Table 7 gives the average calorimeter temperature rise and the maximum rate of heat flow in molten iron splash evaluation. After comparing the results tabulated in Tables 6 and 7, the following conclusions can be derived.

1. Zetex Plus 1100, which is about 8.5% lighter than Zetex 1200 and 20% lighter than asbestos, shows no penetration of the molten metal through the fabric, whereas both asbestos and Zetex allowed the metal to penetrate the fabric.

FIG. 9—*Single fiberglass filament coated with delaminated vermiculite.*

2. The heat-transfer data offer the most convincing proof of the superior thermal insulating properties of Zetex Plus coating. The maximum temperature rise for Zetex Plus is almost half that of regular Zetex, and the maximum rate of heat flow was about 70% of Zetex for Zetex Plus. The differences in these values is even more dramatic when compared with asbestos. The maximum rise in temperature for Zetex Plus is only ⅓ to ⅙ of asbestos, whereas the maximum rate of heat flow was only 8% of the maximum rate of heat flow for asbestos.

Thermal Insulation

The test procedure for measuring the thermal insulation properties of protective gloves and mittens, as well as the superiority of Zetex over asbestos gloves and mittens, was described by Dixit [3] in a previous paper. The same procedure was followed for determining the thermal insulation properties of Zetex Plus except that the temperature of the 1.70-kg

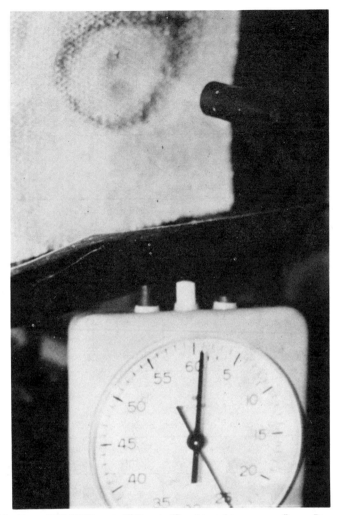

FIG. 10—*Zetex 1100 (1100 gm/m² [32 oz/yd²]) subjected to propane flame. Burns through in 25 s.*

(3.75-lb) pipe and 76-mm-diameter (3-in.) pipe was 1093°C (2000°F). Figure 13 shows the schematic of the testing of double-palm mitten. The same double-palm mitten was used to determine the effect of temperature on the performance of the mitten. The results of testing are presented in Table 8.

It is quite apparent from the test results that the regular double-palm mitten cannot be used more than twice, since the outer layer of Zetex fused and the pipe stuck to the fused fabric. Even though there is another layer of Zetex underneath, the testing was discontinued due to the risk involved. The Zetex Plus double-palm mitten, however, performed remarkably well. Results in Table 8 show that

1. The 1093°C (2000°F) pipe could be held for 20 to 24 s, which is more than double the length of time for regular Zetex.

TABLE 2—*Smoke density comparison of Zetex Plus.*

	Zetex Plus	Zetex 800 (Control)	Aramid
Specific optical density at			
1.5 min	0	1	49
4.0 min	1	3	73
Maximum density of smoke	1	3	84
Time to maximum density, min	12.2	5.2	7.9

TABLE 3—*Flame spread of Zetex Plus Style 1100.*

	After Flame, s	Burn Length, in./mm	Afterglow, s
Warp	0.0	0.1/2.54	0.0
	0.0	0.1/2.54	0.0
	0.0	0.1/2.54	0.0
	0.0	0.1/2.54	0.0
	0.0	0.1/2.54	0.0
Filling	0.0	0.1/2.54	0.0
	0.0	0.1/2.54	0.0
	0.0	0.1/2.54	0.0
	0.0	0.1/2.54	0.0
	0.0	0.1/2.54	0.0

TABLE 4—*Linear shrinkage of Zetex and Zetex Plus at 816°C (1500°F) for 1 h.*

	Shrinkage Warp, %	Shrinkage Fill, %
Zetex 800	19.7	29.0
Zetex 800HC (heat cleaned)	21.2	26.2
Zetex Plus 800	0.5	0.5
High silica fabric	12% areal shrinkage[2]	

TABLE 5—*Physical test data for Zetex Style 1100 and Zetex Plus 1100.*

	Zetex 1100		Zetex Plus 1100	
Average Breaking Strength	Warp	Fill	Warp	Fill
kg, (lb)	153.7 (338.2)	126.2 (277.6)	171.3 (376.8)	167.7 (366.8)
Abrasion resistance Taber—cycles to failure (H-18 wheel, 500 gm/wheel, 100% vacuum level)	1783		2900	

TABLE 6—*Visual ratings of fabrics used in a single layer over a tee-shirt fabric lay-up exposed to molten iron.*

Material Number	Material Designation	Ratings of Outer (Impacted) Layer			
		Charring	Shrinkage	Perforation	Adherence
1	asbestos	2	1	4	4
2	Zetex 1200	2	1	4	3
3	Zetex Plus 1100	3	1	1	3

TABLE 7—*Average calorimeter temperature rise and maximum rate of heat flow in molten iron splash evaluations for a single layer over tee-shirt fabric lay-up.*

Material Number	Material Designation	Maximum Temperature Rise, °C (°F)		Maximum Rate of Heat Flow, cal/cm²/s
		5 s	40 s	
1	asbestos	48.22 (86.8)	64.00 (115.2)	3.74
2	Zetex 1200	13.44 (24.2)	38.28 (68.9)	0.43
3	Zetex Plus 1100	7.44 (13.4)	20.56 (37.0)	0.30

TABLE 8—*Zetex and Zetex Plus double-palm mitten handling 1092°C (2000°F) pipe.*

Cycle No.	Average Time in Seconds	
	Zetex	Zetex Plus
1	10.5	23.5
2	9.0	22.0
3	fused	20.5
4		20.0
5		20.0
6		20.0

TABLE 9—*Testing Zetex, Zetex Plus, and aramid mittens at 593°C (1100°F) and 982°C (1800°F).*

Product/Temperature	Average Time in Seconds		
	Zetex	Zetex Plus	Aramid
Single-palm mittens/593°C (1100°F)	11	11	7
Double-palm mittens/981°C (1800°F)	21	25	0
	17	23	(instant
	17	24	combustion)
	17	21	
	17	21	
	17	22	

FIG. 11—*Zetex Plus 1100 (1100 gm/m² [32 oz/yd²]) coated with delaminated vermiculite after 1 h 10 min exposure to propane flame.*

2. The outer layer of the Zetex Plus double-palm mitten remained soft and flexible.
3. After the third cycle, the time to handle the 1093°C (2000°F) pipe stabilized at 20 s.
4. There was very little smoke, if any, with either of the mittens.

In order to see the effect of lower temperatures on the performance of mittens, we conducted another series of tests using Zetex, Zetex Plus single- and double-palm mittens, as well as single- and double-palm aramid mittens. The 76-mm (3 in.) pipe weighing 1.70 kg (3.75 lb) was used, and two series of tests were conducted at 593°C (1100°F) and 982°C (1800°F). The test results presented in Table 9 show that

1. At 593°C (1100°F), both Zetex and Zetex Plus single-palm mittens could handle the hot pipe for 11 s, whereas the aramid single-palm mitten handled the pipe for 7 s. There was considerable smoke during testing of the aramid mitten.

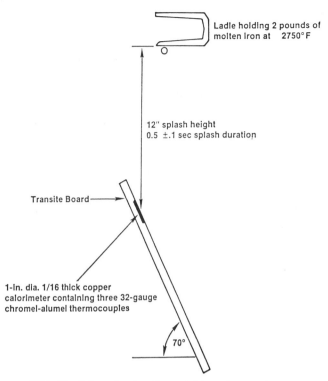

FIG. 12—*Schematic of molten metal impact apparatus.*

2. At 982°C (1800°F), the differences in the performance of the three double-palm mittens are quite significant. The Zetex double-palm mitten could handle the hot pipe repeatedly, and the length of time stabilized at 17 s. However, the Zetex Plus double-palm mitten could handle the same pipe repeatedly for 21 to 22 s, which is about 24% better than the Zetex double-palm mitten.

3. The aramid mitten caught fire upon contact with the hot pipe and also smoked heavily. Further testing of the aramid mitten was discontinued.

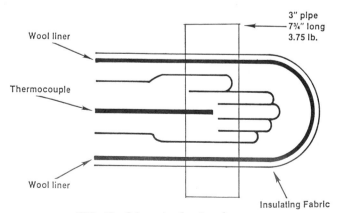

FIG. 13—*Schematic of testing glove.*

Summary and Conclusions

The delaminated vermiculite coating completely encapsulates fiberglass filaments and prevents the melting of the filaments by conducting the heat along the surface of the yarns. The fabric shrinkage of Zetex Plus is also reduced very dramatically compared with regular fabric. The net result has been a significant improvement in the thermal insulating properties of vermiculite-coated fabric at temperatures up to 1093°C (2000°F). The Zetex Plus fabric also showed considerable improvement over regular Zetex and asbestos fabrics in molten metal splash tests in terms of metal penetration and maximum rate of heat flow.

References

[1] Ballard, D. G. H. and Rideal, G. R., *Journal of Materials Science,* Vol. 18, 1983, pp. 545–561.
[2] Ametek Technical Bulletin, HS-116, Ametek, HAVEG Division, 900 Greenbank Road, Wilmington, DE 19808.
[3] Dixit, B. in *Performance of Protective Clothing, ASTM STP 900,* R. L. Barker and G. C. Coletta, Eds., American Society for Testing and Materials, Philadelphia, 1986, pp. 446–460.

Protection From Physical Stressors

Cut Protection

Jean Arteau[1] and Denis Turcot[1]

Performance Evaluation of Chain Saw Protective Leggings

REFERENCE: Arteau, J. and Turcot, D., **"Performance Evaluation of Chain Saw Protective Leggings,"** *Performance of Protective Clothing: Second Symposium, ASTM STP 989,* S. Z. Mansdorf, R. Sager, and A. P. Nielsen, Eds., American Society for Testing and Materials, Philadelphia, 1988, pp. 159–173.

ABSTRACT: The Standardization Office of Quebec mandated us to develop a test that could rate protective pads or leggings in terms of performance.

In the test, a chain saw was mounted on a horizontal axis about which it was allowed to rotate freely. A test consisted of letting the blade of this chain saw drop perpendicularly to the longitudinal axis of the legging that was mounted on a wood cylinder to simulate the leg of the worker. The legging was fixed on the specimen support in such a way that the textile threads could be pulled out by the chain so as to cause minimum interference with the protective mechanism of the textile. Wires on the legging and underneath the pad and an appropriate data acquisition system permitted measurement of the cut-through time and the chain saw speed.

The number of plies, the material thickness, and the chain saw speed were shown to change the cut-through time and stopping speed, which measures the maximum speed that the chain saw can run without actually cutting completely through the leggings.

For the first time, both performance criteria were measured with the same apparatus on the same type of leggings to confirm that these two different performance criteria were required to rate legging performance when no correction is made to the cut-through time for the free-fall time. The threshold stopping speed seems to be a preferable criterion.

The cut-through time was shown to be sensitive to the thickness of the material; the threshold stopping speed, to the strength of the assembly of fabric layers. Observations with a high-speed video camera (2000 frames/s) demonstrated the slowing of the chain by the fabrics, the stretching of the yarns up to the stitching on the perimeter of the pad, and the absence of bouncing during testing.

KEY WORDS: protective clothing, leg protection, chain saws, performance, laboratory tests

Early in the 1960s, the Quebec Pulp and Paper Association introduced a protective pad sewn on trousers to protect workers from leg injuries. These protective pads are more or less an assembly of layers of fabrics made from high-strength fiber such as nylon or aramid. The layers are sewn around their edges to the trouser legs. Protective pads can be used with two types of clothing: chaps or open-back, and trousers or closed-back.

In spite of various developments since then, standards to evaluate pad performance have not been set by any standardization organization, such as the American Society for Testing and Materials, and, test methods to check compliance are not available. The testing methods have been based on two types of performance criteria: the threshold stopping speed of the

[1] Senior project engineer and project engineer, respectively, Safety and Engineering, Laboratory Division, Institut de recherche en santé et en sécurité du travail (IRSST), Montreal, Quebec H3A 3C2, Canada.

chain saw [1] and the cut-through time [2–4]. Each of the test methods present differences with respect to the means of driving the chain (gas or electrical motor), to the force applied by the guide bar on the specimen (9 to 50 N), to the means of attaching the specimen to the simulated leg (tightly to loosely), to whether the simulated leg rotates or not, and to whether the guide bar bounces or not [5]. Test results are available for three of the four methods [1,2,4], but only two of these assess the type of specimens that are used in North America [1,2].

However, the rating or classification of protective pad performance gives different results depending on the performance criteria and the test method [1,2,5]. We were therefore mandated by the Standardization Office of Quebec to develop a test that could rate protective pads or leggings in terms of performance.

To do this, a test method that uses the best elements of the other methods was perfected. Five types of protective pads underwent exhaustive tests that validate the test method. In addition, interesting observations provide a preliminary hypothesis for the understanding of the protective mechanisms.

Procedure

Equipment

Three main components make up the test method: a specimen support, a gas-powered chain saw, and a measurement and data acquisition system. (Fig. 1)

FIG. 1—*Testing equipment: the chain saw, the specimen support and the data acquisition system.*

Specimen Support—The specimen support is a wooden cylinder with a 152 mm (6 in.) diameter and a 356 mm (14 in.) length. It is held immobile at both ends. It is partially covered with a resilient material having a thickness of 20 mm (¾ in.); this material, a vinyl nitrile foam (Ensolite Type M), avoids bouncing of the guide bar. Two attachment points for the protective pad are located on an axis parallel to the axis of the wooden cylinder and 65 mm from the vertical on the side away from the saw; they are located 45 mm from each end of the specimen support. Each one is made up of two metal plates (76 by 25 mm) attached by two screws. They are located 210 mm from each end of the specimen that is described later.

Chain Saw—A Husqvarna Model 266 gas-powered chain saw is used. It has a maximum power of 3 kW (4 hp) at 8500 rpm, and a maximum torque of 4.0 N · m (35 lb · in) at 6100 rpm. The Oregon guide bar is 457 mm (18 in.) long. The Oregon 72 LP chain has a pitch of 10 mm (⅜ in.). The saw is attached to a metal plate that rotates freely around an axle whose axis is parallel to that of the specimen support. The horizontal distance between the pivot of the saw and the central axis of the specimen support is 380 mm.

Thus, the guide bar falls perpendicularly on the support axis, and it exerts a static force of 15 N at the point of contact. The vertical dropping distance to the point of contact is 50 mm (2 in.). The longitudinal axis of the guide bar is horizontal at the moment of contact. A trigger-release system allows free fall without initial acceleration.

The motor's speed of revolution is maintained constant before the test by controlling the position of the gas throttle lever.

Data Acquisition System—The cut-through time of the protective pad and the motor's speed of revolution are the two parameters measured electronically. The cut-through time is measured by a timer (± 0.001 s) that measures the time between the cutting of a wire on the top of the specimen being tested and the cutting of a wire under the specimen, the wires being connected to the appropriate electronic circuitry of the controller.

The speed of revolution of the motor ($\pm 3\%$) is evaluated by counting the number of times that the voltage is induced at the spark plug. The two parameters are recorded on an oscillograph.

In addition, a few tests were analyzed by means of a video camera at high speed (2000 frames/s).

Specimens

The test specimens are described in Table 1. Those from Groups A and B are rectangular (711 by 203 mm; 28 by 8 in.), with stitching 25.4 mm (1 in.) from the edge. These dimensions are similar to those of protectors sewn inside of trousers. Those of Group C are chaps available commercially, their dimensions (663 by 290 mm) are close to those of Groups A and B. The specimens are conditioned and tested at 21 $\pm$ 1°C (70 $\pm$ 2°F) and at a relative humidity of 50 $\pm$ 5%. They are new.

Test Procedure

The chain saw is refueled and oiled after every ten trials in order to avoid variation in mass.

Using a template, six holes are made in the specimen to allow the passage of the four attachment screws and the upper wire. The under wire is positioned and connected to the circuit terminals. The specimen is attached to the specimen support by means of metal

TABLE 1—*Description of specimens.*

Identification	Combinations of Layers
A2 A3 A4	Exterior cotton shell. 2, 3, or 4 strata made from one ply (two layers) of 2108 nylon[a] and one layer of taffeta.
B[b]	Polyester shell. 1 layer polypropylene knit RVY 021 7 layers knit wrap Nylon RVY 022 1 layer Nylon lining.
C	Exterior Cordura nylon shell. 1 layer woven Kevlar[c] 1 layer felgg Kevlar[d] 1 layer woven Kevlar 1 layer felt Kevlar interior Cordura nylon shell.

[a] 2108 Nylon: one ply is made from two layers interwoven with parallel strips at 17.8 cm (7 in.) Thread count: 19.7 by 17.3/cm (50 by 44/in.) 840 denier by 840 denier Nylon 66; Canadian Patent No. 1088701. U.S. Patent No. 427956.

[b] Brand name: Chainguard I by Safeco.

[c] Woven Kevlar: Type 352 Kevlar 49 continuous filament 1140 denier yarns per inch: warp 16 fill 16 yarns per 10 cm: warp 63 fill 63. U.S. Department of Agriculture, Forest Service: Specification for Chaps Chain Saw M-1980. Spec. 6170-4C, Feb. 1981.

[d] Felt Kevlar: style 970 Kevlar 29 3.5 ounces per square yard needle punched; U.S. Department of Agriculture, Forest Service: Specification for Chaps Chain Saw M-1980. Spec. 6170-4C, Feb. 1981.

plates. The upper wire is positioned and passed through the holes to reach its terminals. The continuity of the circuits is verified by the controller during zero adjustment of the timer.

The motor of the chain saw is turned on and the speed adjusted. The oscillograph paper drive is turned on in order to record, with precision, the speed of rotation of the motor before the test. The trigger release frees the saw that rotates around the axle. The chain cuts the upper wire that starts the timer. From this moment, there are two possibilities.

First, the speed is sufficiently high and the specimen is cut through. In this case, the under wire is cut, which stops the timer, the oscillograph, and the motor of the chain saw. The result is the cut-through time as a function of a motor speed for a given specimen.

Second, the speed is low and there is jamming. Jamming is the stopping of the rotating chain by the specimen (protective pad) before it is cut through. In certain tests, the saw motor can continue to turn even though the chain is stopped, because the protective pad slowed the motor sufficiently for the centrifugal clutch to stop driving the chain. The stopping speed is the speed of the motor before the chain comes into contact with the protection where the jamming occurs. In order for jamming of the saw to be considered acceptable, the specimen's lower shell must not be cut at all. In this case, the controller automatically stops the saw's motor and the oscillograph after several seconds. The result is the stopping speed and, therefore, the cut-through time is infinite.

After each test, the entire driving mechanism of the chain is cleaned.

The procedure just described was used to prepare the draft standard NQ 1923–095 "Protective component for chaps and trousers for chain saw users. Determination of the resistance upon contact with this tool" for the Bureau de normalisation du Québec (BNQ).

A variation in this procedure allows the effect of extensive protection to be checked. It consists of placing up to 10 plies (20 layers) of Type A specimens by combining the appropriate number of 2, 3, and 4 plies (A2, A3, and A4). The rotation speed of the motor before contact is maintained at successive 1000 rpm increments between 5000 and 12 000 rpm. The chain falls on the pile of specimens, cuts a certain number, and jams. A relationship is then obtained between the stopping speed and the number of plies.

Finally, the free-fall time is measured. It is the time that the bar travels from the upper wire to the under wire with no pad. Both wires are maintained at the appropriate vertical distance by placing one pad on each side of the contact point where the bar touches the specimen support. The corrected cut-through time is then the cut-through time minus the free-fall time.

Results

General Aspects

Each of the five groups of specimens underwent tests for the entire range of chain saw speeds, namely, between 4000 and 11 000 rpm. For each type, specimen cut-throughs were observed as well as was jamming of the chain. The general form of the relationship between cut-through time and motor speed is a curve of the type $y = 1/x$, where the x asymptote is the threshold stopping speed (TSS), and the y asymptote is the average cut-through time (ACT) for each type of specimen. Figure 2 shows the curves for each type of protective specimen. Actually, the curves are represented by two stright lines due to the scale of the revolutions per minute axis.

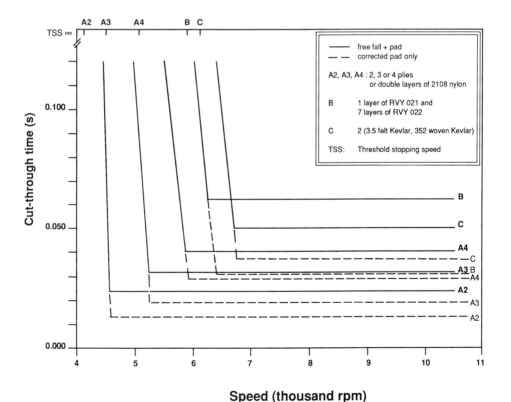

FIG. 2—*Speed and average cut-through time.*

The same trend is observed with the corrected cut-through time; the corrected cut-through time (CACT) being the horizontal asymptote (Fig. 2). The free-fall times are A2 = 0.012 s, A3 = 0.011 s, A4 = 0.010 s, B = 0.032 s, and C = 0.011 s. Because none of the other testing methods [2–4] corrects the cut-through time for the free-fall time, the cut-through time results are presented in the first instance. In other respects, the corrected cut-through times will be briefly presented and discussed.

Average Cut-Through Time (ACT)

The average cut-through time for each type of protective pad is presented in Table 2a. It represents the horizontal part (at high speed) of the curves in Fig. 2. The absence of any relationship between the cut-through time and the speed is statistically shown, and the zero slope hypothesis is retained. Therefore, prior control of the speed of the chain saw is not critical for this test at high speed. If the test takes place at a speed where there is maximum motor power (Table 2b) or at a speed where there is maximum motor torque (Table 2c), no significant difference is noted for the cut-thorugh time. The ACT results from Table 2b (7400 to 8600 rpm) will be used in the following text, unless otherwise stated.

Type A2, A3, and A4 specimens have the same construction, only the number of plies (that is, the thickness) varies from two to four. The differences between their average cut-through times are statistically significant and are not due to chance. The test used is a Student's *t* comparison of means (independent samples, populations with common variance) with a confidence level of 95%.

In increasing order of average cut-through time, the types of protection pads are classed: A2 < A3 < A4 < C < B. In increasing order of corrected average cut-thorugh time, the pads are classed: A2 < A3 < A4 < B < C.

TABLE 2—*Mean cut-through time.*

Type of Protective Pad	Number Tested	Speed Range, rpm	Mean Cut-Through Time, s	Standard Deviation, s	Minimum, s	Maximum, s
			(A) ALL RESULTS			
A2	14	4440 to 9230	0.027	0.006	0.015	0.036
A3	19	5040 to 9300	0.032	0.005	0.024	0.042
A4	17	5480 to 9230	0.040	0.004	0.034	0.047
B	8	6350 to 10260	0.062	0.006	0.050	0.070
C	4	7000 to 8450	0.047	0.008	0.041	0.059
		(B) RESULTS AT MAXIMUM POWER OF THE CHAIN SAW: 7400 TO 8600 RPM				
A2	4	8220 to 8570	0.025	0.005	0.019	0.030
A3	4	8000 to 8280	0.034	0.006	0.028	0.042
A4	5	7450 to 8300	0.041	0.010	0.034	0.057
B	2	7750 to 8400	0.065	0.008	0.059	0.070
C	3	7550 to 8450	0.049	0.009	0.041	0.059
		(C) RESULTS AT MAXIMUM TORQUE OF THE CHAIN SAW: 5500 TO 6700 RPM				
A2	3	6520 to 6560	0.028	0.011	0.015	0.036
A3	5	5630 to 6700	0.035	0.006	0.026	0.042
A4	4	5580 to 6600	0.039	0.001	0.038	0.041
B	3	6350 to 6630	0.065	0.001	0.064	0.066
C	0	The threshold stopping speed is reached.				

Threshold Stopping Speed (TSS)

The threshold stopping speed is the maximum stopping speed. For the same test conditions and for identical protective components, jamming will occur for saw motor speeds equal to or less than the threshold stopping speed. This speed represents the vertical asymptote of the curves in Fig. 2. Table 3 presents the values obtained for each type of protector. In order to obtain the threshold stopping speed, several tests were carried out by progressively reducing the speed where jamming does or does not occur, and the maximum stopping speed, which is then a single value, is retained.

In increasing order of threshold stopping speed, the types of protective pads are classed: A2 < A3 < A4 < B < C.

Thickness and Surface Density

The thickness of each specimen type is measured according to Method 37 of Standard CAN 2 − 4.2 − M1977 using a 0.5 cm^2 presser foot and an applied mass of 170 g. The average cut-through time (8000 to 9000 rpm, Table 2*b*) and the threshold stopping speed with respect to the thickness are shown in Figs. 3 and 4. For Type A protective pad, we observe a linear relationship between the average cut-through time and the thickness (ACT = 0.0110 T + 0.005; R^2 = 0.990) and a linear relationship between the threshold stopping speed and the thickness (TSS = 692 T + 2870; R^2 = 0.980). In the same way, the average cut-through time and the threshold stopping speed with respect to the surface density are shown in Figs. 5 and 6.

Large Number of Protective Pads

Figure 7 shows the results of the stopping speed and the number of plies cut for the Type A protective pad. The points are obtained using the alternate test procedure. A linear relationship is observed between the stopping speed and the number of plies (or the thickness), and the plateau observed at 11 000 rpm is the maximum speed of the saw's motor. This shows the relationship between the amount of protection and the measured values which reach the limit of the measuring instrument when a large number of layers are tested.

Conversely, the lack of protection (namely, ordinary work pants) gives a cut-through time of 0.004 s.

TABLE 3—*Threshold stopping speed according to IRSST and Putnam et al [1].*

Type of Protective Pad	IRSST			Putnam et al. [1]		Δ, fpm	Δ, %
	rpm	fpm	m/s	fpm	m/s		
A2	4200	2100	10.7	1900	9.7	−200	−9.5
A3	4550	2300	11.7	2200	11.2	−100	−4.3
A4	5200	2640	13.5	2550	13.0	−90	−3.4
B	6060	3070	15.6	3050	15.5(7)	−20	−0.7
C	6190	3140	16.0	3350	17.1	+210	+6.7
Measurement error	±180	±90	...	±100	...	±190	...

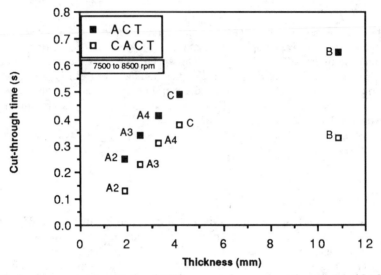

FIG. 3—*Average cut-through time (ACT), corrected average cut-through time (CACT), and thickness.*

Discussion

Three aspects will be discussed: (1) the validation of the test method, (2) the existence of two performance criteria, and (3) the understanding of the protection mechanism.

Validation of the Test Method

The observations using the video camera showed the absence of bounce, in contrast to other test methods [1]. Thus, the cut-through times and the stopping speeds are measured during continuous contact and without the random bounce effects.

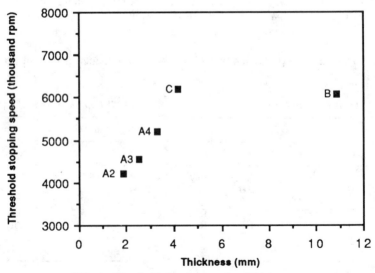

FIG. 4—*Threshold stopping speed and thickness.*

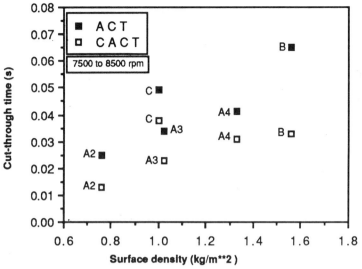

FIG. 5—*Average cut-through time (ACT), corrected average cut-through time (CACT), and surface density.*

Reproducibility—The measurements are reproducible. The sensitivity of the cut-through time measuring method is ±0.003 s. This is due to the fact that one tooth can come into contact with the protective pad while the preceding tooth has not yet cut the upper wire. This fact was observed and quantified by means of the high-speed video camera. The average coefficient of variation (standard deviation divided by the average) is 15% for all of the average cut-through times (Table 2a). The standard deviation is slightly greater than the sensitivity due to the wire. In recent commercial tests [6], this standard deviation is reduced to 0.003 s (coefficient of variation of 9%).

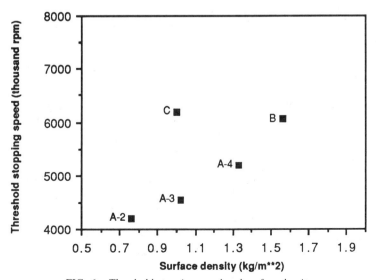

FIG. 6—*Threshold stopping speed and surface density.*

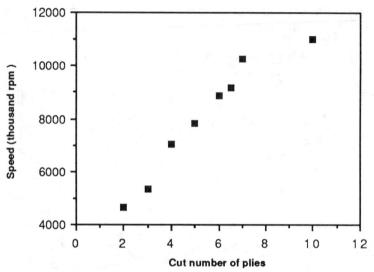

FIG. 7—*Speed and cut number of plies, Type A (two plies = A2, three plies = A3, etc. in Table 1).*

Linear Response—The test apparatus and the procedure are a good measuring instrument. In fact, the response of the instrument is proportional to the amount of protection. The cut-through time for ordinary work pants (no protection) is 0.004 s. The cut-through time and the threshold stopping speed increase with the thickness or the number of plies for Type A protective pad (Figs. 3, 4, and 7). At this stage, the increase seems linear. Therefore, more data for the Type A pad, such as A5, A6, etc., and data for other types are required in order to confirm the linear trend. For infinite protection, the threshold stopping speed becomes infinite, namely, the maximum speed of the chain saw (Fig. 7). Finally, the differences between the average cut-through times of the A2 and A3, and the A3 and A4 protection are significant at a 95% confidence level, and the instrument is sufficiently discriminating.

Comparison with Other Methods

The measured threshold stopping speeds are identical to those measured by Putnam [1] (Table 3). The zero difference hypothesis is retained at a 95% confidence level. Taking into account the similarities and differences between our procedure and that of Putnam (Table 4), it seems that the force of contact and the dropping distance have no significant influence on the threshold stopping speed. Maybe the similarity, namely, the chain, is the important factor. For cut-through times, no comparison is possible. We observed that the means of specimen attachment had a great effect on cut-through time. Thus for Type C protection, the average cut-through time is 0.049 s with the two-point attachment described earlier. The cut-through time becomes 0.191 s when the specimen is attached as it would be to a leg, with straps. Since each method attaches the specimens differently, no comparison is possible. Perhaps this explains the great variation in requirements in terms of cut-through time from 0.15 to 2 s [5] for the different test methods [2–4].

TABLE 4—*Comparison between Putnam et al. [1] and IRSST test apparatus.*

	Putnam et al.	IRSST
	SIMILARITIES	
Chain	Oregon 72LP chisel	Oregon 72LP chisel
Specimen dimensions	711 by 203 mm	711 by 203 mm
Specimen support diameter	160 mm	152 mm
Simulated flesh	16 mm vinyl nitrile foam	19 mm vinyl nitrile foam
	(Ensolite Type M)	(Ensolite Type M)
	DIFFERENCES	
Chain saw	Homelite 350	Husqvarna 266
Cubic capacity of motor	57.2 cm^3	66.7 cm^3
Power (maximum)	2.6 kW at 8000 rpm	3.1 kW at 8500 rpm
Torque (maximum)	...	4.0 N·m at 6100 rpm
Weight of the motor without bar, chain, gasoline, and oil	5.7 kg	6.2 kg
Length of the bar	61 cm	45.7 cm
Gasoline tank capacity	...	750 mL
Oil tank capacity	...	450 mL
Contact force	50 N	15 N
Drop height	6 mm	50 mm

Existence of Two Performance Criteria

A significant difference is observed in the classification of the Type B protective pad. The type B protective pad has the longest average cut-through time (Fig. 3), whereas it has only the fourth best stopping speed (Fig. 4). On that basis, it seems that there are two performance criteria: the cut-through time and the threshold stopping speed. This was previously observed [5] in a comparison between the average cut-through time [2] and the threshold stopping speed [1] of several types of protective pads. On the other hand, the corrected average cut-through time (CACT) and the threshold stopping speed (TSS) give the same rating as shown in Figs. 2, 3, and 4. The absence of correction for the free-fall time in Ref 2 could be the source of the discrepancy previously observed.

The cut-through time is a performance criterion that is linked to a person's reaction time. We shall bear in mind that the laboratory cut-through time differs from one testing facility to another and that laboratory cut-through time differs from real field cut-through time. The threshold stopping speed measures the capacity of the chap to slow the saw.

For testing purposes, the two performance criteria have pro and con aspects. The cut-through time is measured at a constant power (3.0 to 3.1 kW) of the chain saw motor by means of the controlled speed (8000 ± 500 rpm). Interpretation is trivial because the cut-through time is shown on a digital display and the speed is easily checked on the paper chart. Few trials are needed to evaluate the average cut-through time or to control the quality on a negative result basis. The negative side is that the cut-through time varies from one to two orders of magnitude according to the testing facilities. In addition, the subtraction of the free-fall time to the cut-through time increases the relative error of the corrected cut-through time. This correction is required in order to measure the effect of the protective textile only, though this correction was not reported [2–4]. The threshold stopping speed seems to be a more universal measurement of the performance, despite some differences in testing procedure (Tables 3 and 4). It will require more trials to determine the performance of one type of pad.

Protection Mechanism

Even though a protection mechanism model does not exist, certain interesting observations can be shown in Figs. 3 through 6. Figure 3 shows that the average cut-through time is mainly affected by thickness, because Type B has the greatest ACT and because Type A shows an increase in ACT or CACT with an increase of the thickness. But, one can conclude that corrected average cut-through time is not solely affected by thickness, because Type B has the fourth best CACT. This latter statement is the correct one. It is obvious that the thickness affects the performance when layers of the same textile are added (for example, A2, A3, and A4 in Figs. 3 and 4). But the rating of Type B shows clearly that extra thickness does not imply an equivalent improvement of performance in CACT or in TSS (Figs. 3 and 4). The type of textile is then an explanatory factor. In addition, when surface densities are equal, Type C has a significantly higher threshold stopping speed (Fig. 6). This demonstrates well the effect of the superior mechanical resistance of aramid fibers.

Ideal protection for users must maximize performance, minimize weight (surface density) and bulk (thickness), while maintaining an affordable price. The analysis is based on the first three factors. With the threshold stopping speed and the corrected average cut-through time as performance criteria, Type C offers the best performance when the thickness and the weight are taken into account (Figs. 4 and 6). When using the cut-through time as a performance criterion, the choice varies. When the thickness is minimized, an improved Type C or a Type A7 would have the same performance as a Type B (Fig. 3). When the weight is minimized, Type B seems to be the best choice, even though it would be interesting to measure the performance of Type C protection with more layers (Fig. 5).

Conclusions and Recommendations

The method presented is satisfactory for carrying out a comparative evaluation of leg protection for chain saw users, and it is reproducible and discriminating. The TSSs determined by this method are comparable with the TSSs determined by one other testing method. For the first time, ACT, CACT, and TSS were measured with the same apparatus on the same type of leggings. The ratings with ACT and TSS as performance criteria are different, so when using ACT, as done by other methods, both performance criteria are required to rate legging performance. The ratings with CACT and TSS as performance criteria are the same. The TSS seems to be a preferable criterion in spite of some differences between testing procedures. The fabric structure is also an important factor in the performance of the protective pad.

However, questions remain unanswered. Which of the criteria gives a better simulation of the true protective behavior in real working situations? The effects of washing and drying cycles, of gasoline, of oil, of dry cleaning, and of other conditions on the performance of protective equipment are also unknown.

Acknowledgments

We wish to thank J.-G. Martel, coordinator of Safety and Engineering, and A. Lajoie, director of laboratories, IRSST, for permission to publish this account. We also acknowledge the cooperation of the members of the BNQ committee, of G. Perrault for his stimulating commentaries, and of R. Daigle for the carefully done tests.

References

[1] Putnam, T., Jackson, G., and Davis, J., "Chain Saw Chaps Redesign," EDT Report 9102, U.S. Department of Agriculture, Forest Service, Missoula, MT, Aug. 1982.

[2] Monroe, G., "Test Device and Procedure for Testing Protective Pads for Chain Saw Operators," Agricultural Engineering Department, Virginia Polytechnic Institute, Blacksburg, VA, Dec. 1980, unpublished report written for American Pulpwood Association.

[3] "Leg Protective Devices for Workers Operating Power Chain Saws," British Columbia Workers' Compensation Board, Standard P.P.E. 14.1, 1975.

[4] Turtiainen, K., "Comparison of Test Methods used for Durability Testing of Leg Shields for Chain Saw Users," VAKOLA, Finnish Research Institute of Engineering in Agriculture and Forestry, Helsinki, 1979.

[5] Arteau, J. and Turcot, D., "Faisabilité d'un critère de performance pour les jambiéres de protection portées par les opérateurs de scies à chaîne. Étude E-015. (Feasibility of a Performance Criterion for Protective Leggings Worn by Chain Saw Operators)," Research Report, Institut de recherche en santé et en sécurité du travail du Québec, Montreal, 1985.

[6] Arteau, J., "Performance de l'élément de protection après cinq cycles de lavage et de séchage (Performance of Protective Pad after Five Washing and Drying Cycles)," Report presented to the BNQ Standardization Committee on Protective Equipment for Loggers, Bureau de normalisation du Quebec, Nov. 1986.

[7] G. Mark, personnal communication, Safety Supply Ltd, Montreal, 1985.

Protection From Industrial Chemical Stressors

Dermal Toxicology

Larry L. Hall,[1] Henry L. Fisher,[1] Martha R. Sumler,[2]
Robert J. Monroe,[3] Neil Chernoff,[1] and P. V. Shah[2]

Dose Response of Skin Absorption in Young and Adult Rats

REFERENCE: Hall, L. L., Fisher, H. L., Sumler, M. R., Monroe, R. J., Chernoff, N., and Shah, P. V., **"Dose Response of Skin Absorption in Young and Adult Rats,"** *Performance of Protective Clothing: Second Symposium, ASTM STP 989,* S. Z. Mansdorf, R. Sager, and A. P. Nielsen, Eds., American Society for Testing and Materials, Philadelphia, 1988, pp. 177–194.

ABSTRACT: The effect of skin dosage on percutaneous absorption was determined in young and adult rats using 14 pesticidal chemicals. Young (33-day old) and adult (82-day old) female Fischer 344 rats, with previously clipped mid-dorsal skin, were treated with three dosages of labeled pesticides in acetone and sacrificed 72 h following application. The treated area was 2.3% of the body surface area and was protected by a perforated plastic blister. Skin penetration was determined by dividing the radioactivity in the body plus excreta by the total radioactivity recovered. The dosages used were 0.02 to 0.35, 0.54, and 2.68 μmol/cm^2 for the low, medium, and high dosage levels, respectively.

Skin penetration of the pesticides in young animals ranged from 2.9 to 81.5%, 2.0 to 84.4%, and 0.9 to 90.1% at the low, medium, and high dosage levels, respectively. In the adult, it ranged from 7.7 to 86.4%, 2.7 to 90.5%, and 1.0 to 93.3% at the low, medium, and high dosage levels, respectively.

Chlorpyrifos, dinoseb, mono- and disodium methanearsonate did not show a significant effect of dosage on dermal penetration. Dinoseb, mono- and disodium methanearsonate displayed constant fractional penetration in both young and adult animals over the dosage range examined. The other compounds showed dose-dependent dermal absorption although the total amount absorbed usually increased with dose. Dose-response curves for young and adult animals were not parallel in 8 of the 14 pesticides studied.

Eleven of the 14 pesticides showed significant age-dependent differences in skin penetration. Four of these eleven compounds had greater absorption in the young than adult at some dose. The highest and lowest significant young/adult penetration ratios were 1.53 and 0.19, respectively.

In summary, dermal absorption of the majority of the pesticides studied was found to be dependent on dose and age of the animal. With this heterogeneous group of chemicals, the correlation between skin penetration and octanol/water partition coefficients was marginal. Stratum corneum thickness did not appear to be a factor in the penetration differences between young and adult.

KEY WORDS: protective clothing, pesticides, skin, dose, absorption, rats, age

Skin is the major barrier between man and his environment and is the second largest organ in the body. It represents 18% of the body weight and has a surface area of approximately 1.7 m^2 [1]. Composed of epidermis and dermis containing hair follicles, sebaceous

---

[1] Toxicologist, biophysicist, and toxicologist, respectively, Perinatal Toxicology Branch, Health Effects Research Laboratory, U.S. Environmental Protection Agency, Research Triangle Park, NC 27711.

[2] Scientist and toxicologist, respectively, Northrop Services Inc., Environmental Sciences, Research Triangle Park, NC 27709.

[3] Emeritus professor, North Carolina State University, Department of Statistics, Raleigh, NC 27650.

glands as well as eccrine and apocrine sweat glands, it is a very heterogeneous organ that provides mechanical protection in addition to isolating the organism from the hostile environment. Skin contact, rather than the pulmonary or oral routes, has been shown to be more important during occupational exposure to pesticides [2]. Many factors, such as dose, age, species, vehicle, etc., that affect skin permeation and subsequent dermal and systemic toxicity have been identified and evaluated to some degree. Dosage and age, however, have had only limited assessment.

The effect of the dose response on percutaneous absorption has been recently reviewed by Wester and Maibach [3]. Published studies (Holland et al. [4]; Maibach and Feldman [5]; Sanders et al. [6]; Scheuplein and Ross [7]; and Wester and Maibach [8]) indicate increasing the applied dose results in an increase in total absorbed dose. However, the form of the dose response was not constant.

Few studies have been conducted to compare the ability of a chemical to penetrate the skin in the young and adult animal, as well as any physiological or morphological differences. Singer et al. [9] studied the development of the stratum corneum in the rat and guinea pig and found no difference in ultrastructure between the neonate and the adult. Rasmussen [10] suggested that the normal full-term infant probably has a fully developed stratum corneum and intact barrier function.

Nachman and Esterly [11] suggested that there may be increased skin permeability in the premature infant. Blood levels of hexachlorophene were higher in premature infants than in fully matured infants after bathing (Greaves et al. [12]). Wester et al. [13], however, reported similar extents of percutaneous penetration of testosterone in the newborn and adult rhesus monkey. McCormack et al. [14] compared in vitro penetration of series of alcohols and fatty acids in premature and full-term human infant skin with that of the adult. They have suggested no significant difference in the penetration of alcohols between premature, full-term infant, and adult skin. However, they report differences in the penetration of fatty acids that could be due to differences in the solubility of fatty acids in corneum lipids.

Knaak et al. [15] studied the percutaneous penetration of triadimefon on young and adult male and female Sprague-Dawley rats and reported that, based on the ^{14}C counts, triadimefon was lost most rapidly from the skin of young animals than from the skin of adult animals.

Since very limited information is available on the effect of dose and age on percutaneous penetration of chemicals, this study was conducted to determine the effect of dose on the percutaneous penetration of pesticides of various classes in young and adult rats.

Materials and Methods

Chemicals

The chemicals studied, abbreviations used in this report, their specific activity, radiolabeling, sources, and doses are listed in Tables 1 and 2. The radiochemical purity of all the chemicals was greater than 98% and was reconfirmed by appropriate thin-layer chromatography (TLC) systems. Unlabeled chemicals were supplied by the Pesticide and Industrial Chemicals Repository (U.S. Environmental Protection Agency (EPA)).

Animals

Time-pregnant Fischer 344 female rats were purchased from the Charles River Breeding Farm, Kingston, New York. The date of birth of the offspring was recorded. On day one,

TABLE 1—*List of the pesticides tested in the present study, specific activity, radiolabeling position, and source.*

Common Name (abbr)	Specific Activity, mCi/mM	Radiolabeling Position	Source
Atrazine (ATR)	3.71	^{14}C-ring(U)	Ciba-Geigy, Greensboro, NC
Captan (CAP)	10.00	^{14}C-ring(U)	Midwest Research Institute Kansas City, MO
Carbaryl (CAB)	5.8	naphthyl-1-^{14}C	Amersham, Arlington Heights, IL
Carbofuran (CAF)	39.4	^{14}C(U)benzofuranyl ring	FMC Corporation, Princeton, NJ
Chlordecone (KEP)	3.2	^{14}C-ring(U)	Midwest Research Institute Kansas City, MO
Chlorpyrifos (CHL)	1.99	^{14}C-2,6 pyridyl ring	Dow Chemical, Midland, MI
Dinoseb (DNS)	4.29	^{14}C-ring(U)	Pathfinder Laboratories Inc. St. Louis, MO
DSMA[a] (DSM)	10.00	^{14}C-methyl-carbon	ICN Radiochemicals, Irvine, CA
Folpet (FOL)	10.6	trichloromethyl-^{14}C	Amersham, Arlington Heights, IL
MSMA[b] (MSM)	10.00	^{14}C-methyl-carbon	ICN Radiochemicals, Irvine, CA
Nicotine (NIC)	60.0	pyrrolidine-2-^{14}C	New England Nuclear, Boston, MA
Parathion (PAR)	20.00	^{14}C-ring(U)	NC State University Raleigh, NC
PCB[c] (PCB)	8.2	^{14}C-ring(U)	New England Nuclear Boston, MA
Permethrin (PER)	57.01	^{14}C-methylene group of the *m*-phenoxybenzyl alcohol moiety	FMC Corporation, Princeton, NJ

[a] Disodium methanearsonate.
[b] Monosodium methanearsonate.
[c] 2,4,5-2,4,5-hexachlorobiphenyl.

the male pups were discarded and a dam and eight female pups were assigned to a cage. At 28 days of age, all animals were weight ranked and put into two groups such that the mean and standard deviation for the groups were approximately equal. Animals in Group 1 were treated with a test chemical at 33 days of age (young) while those in Group 2 were treated at 82 days of age (adults). Thirty-three day old animals were used for the young prepuberal group because the restraint used to protect the skin application site did not retard growth significantly at this age as it did in younger animals, while vaginal opening, a hormone dependent event, occurs at approximately 40 days of age in this strain of rat. Eighty-two day old mature adult female animals were chosen for logistical reasons. All animals were housed in a Bioclean room with controlled lighting (6:00 am to 6:00 pm) and constant temperature (20 to 24°C). They were provided with a regular Purina Rat Chow (Ralston Purina Company, St. Louis, Missouri) and tap water *ad libitum*.

TABLE 2—*Dosages and molecular weights of compounds studied.*

Compound	GMW	Low Dose[a] $\mu mol/cm^2$	Low Dose[a] $\mu g/cm^2$	Medium,[b] $\mu g/cm^2$	High,[c] $\mu g/cm^2$
ATR	215.7	A 0.25 Y 0.29	53.9 61.6	115.6	577.9
CAB	201.2	0.15 0.19	30.9 37.4	107.8	539.0
CAP	300.6	0.09 0.11	26.8 32.2	161.1	805.3
CAF	221.3	0.02 0.03	5.1 6.3	118.6	592.9
KEP	491.0	0.29 0.34	140.3 164.8	263.2	1315.4
CHL	350.6	... ...	... ...	187.9	939.3
DNS	240.2	0.21 0.25	51.5 60.1	128.7	643.5
DSM	183.9	0.09 0.11	16.4 19.7	98.6	492.7
FOL	296.5	0.08 0.10	24.9 29.7	158.9	794.3
MSM	162.0	0.09 0.11	14.5 17.4	86.8	434.0
NIC	162.2	0.01 0.02	2.3 2.9	86.9	434.5
PAR	291.3	0.05 0.05	15.6 15.6	52.1[d]	... ...
PCB	360.9	0.11 0.13	39.3 47.7	193.4	966.9
PER	391.3	0.02 0.02	6.3 7.0	209.7	1048.3

[a] Adult (A) values first, young (Y) values second.
[b] $0.54\ \mu mol/cm^2$.
[c] $2.68\ \mu mol/cm^2$.
[d] $0.18\ \mu mol/cm^2$.

Experimental Procedure

The backs of the animals were clipped with an electric animal hair clipper (Type 40 blade, Oster, Milwaukee, Wisconsin) 24 h prior to treatment. The clipped area was swabbed with reagent-grade acetone to remove dirt and sebaceous gland secretions.

Animals were anesthetized with ether before treatment and their weights were recorded. The treatment area (young, 2.8 cm^2; adult, 5.6 cm^2), which is approximately 2.3% of the body surface area, was marked. Acetone solution, usually 100 μL (young) and 200 μL (adult), that contained ^{14}C labeled compounds, 3 μCi (young) and 6 μCi (adult), was slowly dropped over the marked area. Care was taken to prevent run-off and damage to the treatment site during application. The area and site of application were kept constant for all experiments. For the chemical captan, 150 μl (young) and 300 μl (adult) of acetone solution was applied due to the lack of sufficient captan solubility in acetone. Water was used as the vehicle for DSMA and MSMA due to lack of sufficient solubility in acetone.

The young and adult animals treated with DSMA and MSMA received 60 and 120 μL of the aqueous solutions, respectively. Heat-sealed polyethylene tubing applicators were used to spread the water evenly over the treatment area.

The treated area was protected by gluing a 5-mL perforated plastic disposable beaker (Fisher Scientific Co., Raleigh, North Carolina) on the young animals and a perforated plastic blister from a Cathavex[4] single-use filter on the adult animals with cyanoacrylate adhesive. The treated area and blister were protected by placement of a collar of rubber tubing (1.12 cm diameter), tied with 20-gage wire, behind the forelegs as described by Bartek et al. [16]. The treated animals were housed in Nalgene[5] rat metabolism cages equipped with urine and feces collection devices. All treated animals were provided with dustless precision pellets (45 mg rodent chow formula, Bio Serv. Inc., Frenchtown, New Jersey) and tap water *ad libitum*.

At 72 h following dermal application, treated animals were killed by cardiac exsanquination under ether anesthesia. Blisters, treated skin, urine and feces, and carcasses (remains of the body) were analyzed to determine absorption and radioactive recovery.

Three doses with three animals per dose level were tested. The low dose consisted only of radioactive material in both the young and adult animals. In the young animal, 1.5 μmol (medium) and 7.5 μmol (high) dose levels were tested. In the adult animal, 3.0 μmol (medium) and 15.0 μmol (high) dose levels were tested.

Radioactivity Determination

Treated skin was cut into 8 to 10 pieces and combusted in a Packard Tricarb Sample Oxidizer (Model 306B, Packard Instruments Co., Downers Grove, Illinois). Air-dried fecal samples were ground in a Mini Coffee Grinder (Moulinex Products, Inc., Virginia Beach, Virginia) and aliquots were combusted similarly. Aliquots of the urine samples (100 μL) were pipetted directly into scintillation vials and 15 mL of Insta-Gel[6] scintillation fluid were added. Radioactivity in the blisters was determined by adding 15 mL of Insta-Gel scintillation fluid into the scintillation vials containing the blisters. Carcasses were cut into 10 to 12 pieces, frozen in liquid nitrogen and ground in liquid nitrogen with a Waring blender. The resultant powder was resuspended in water and the final volume recorded. Aliquots (in triplicate) of the homogenates were combusted.

All samples were counted in a Packard Tri-Carb 2660 (Packard Instrument Company, Downers Grove, Illinois) scintillation counter equipped with a quench correction and disintegration per minute (dpm) conversion.

Statistical Analysis

Fractional absorption was calculated by dividing the radioactivity in the body plus excreta by the total radioactivity recovered. Multivariant analysis using polynomial regression of fractional absorption on age, log (dose), $\log^2$ (dose), age × log (dose) interaction, and age × $\log^2$ (dose) interaction was done using SAS (Helwig and Council [17]). Age was chosen to be a class variable. The t tests computed by the General Linear Model Program (SAS) were used to compare the fractional absorption between young and adult animals at individual dose levels, when necessitated by age-dose interactions. Comparison at the low dose was accomplished by interpolation using the regression equation to estimate the fractional absorption of the adult at the young low-dose value. The interpolation was necessary due to differences in low dose in young and adult rats.

---

[4] Registered trademark of Millipore Corporation.
[5] Registered trademark of Nalge Company.
[6] Registered trademark of the Packard Instrument Company.

Results

The mean skin penetration values of the 14 compounds studied are given in Table 3, and the individual absorption values are plotted in Fig. 1. At the low dose, penetration in adult rats ranged from 7.7% (ATR) to 86.4% (DNS) with a rank median of 38.2% (CAP), while in young rats, penetration ranged from 2.9% (MSM) to 81.5% (PAR) with a rank median of 26.7% (CAP). Skin absorption for the medium dose (0.54 μmol/cm^2) in adults ranged from 2.7% (FOL) to 90.5% (DNS) with two rank medians of 15.3% (DSM) and 19.8% (CAB), while in young penetration ranged from 2.0% (DSM) to 84.4% (NIC) with two ranked medians of 9.2% (CAF) and 12.2% (CAB). Penetration at the high dose (2.68 μmol/cm^2) in adults ranged from 1.0% (KEP) to 93.3% (DNS) with a rank median of 6.0% (CAF), and in young ranged from 0.9% (FOL) to 90.1% (CHL) with a rank median of 3.7% (CAF).

Radioactive recoveries are shown in Table 3. Recovery from the 14 compounds averaged 88.75% with a range of 57 to 100% in adult and 68 to 109% in young. The low radioactivity recovery may be due to volatility.

The results of the statistical analysis of skin absorption of the 14 compounds in young and adult female rats are shown in Table 4, and the young/adult ratios are shown in Table 5. Eight chemicals showed a statistically significant age main effect, and 11 of the 14 compounds showed age-related effects in the main (age) or interaction [age $\times$ log (recovered dose), age $\times$ log^2 (recovered dose)] parameters.

The compounds DNS, DSM, and MSM showed significant age differences in skin penetration with nonsignificant interactions. The average 72-h skin absorption in young (33 days) was much less than adult (82 days). The mean young/adult ratio for DNS, DSM, and MSM was calculated as 0.90, 0.25, and 0.19, respectively. These ratios for DSM and MSM are the smallest of any of the compounds tested that were significant statistically.

There was a significant age effect in all model parameters for CAF, CAP, and NIC. Since the interaction terms were significant, indicating that the curves are not parallel, individual young/adult comparisons were performed at each dose level. Significant differences as a function of age were seen at only the low dose for these three compounds. The young/adult 72-h skin penetration ratios at the low dose are 0.32 for CAF, 0.65 for NIC, and 0.81 for CAP.

Atrazine (ATR) showed significant age and age $\times$ dose interaction effects. Individual contrasts at each dose level revealed significant young/adult differences at the low and medium dose groups. The 72-h dermal penetration for the young was greater than that of the adult by 38% at the low dose level and by 49% at the medium dose. These are the second and third largest young/adult differences found for any compound at any dose for which significance was obtained (Table 5).

Chlorpyrifos (CHL) showed significant age and age $\times$ dose interaction between young and adult rats. Data for only the medium and high-dose levels were collected because of low specific activity. Both doses showed differences as a function of age, and average absorption for CHL was 23% greater in young than adult at the medium dose and 53% greater at the high dose.

Significance between young and adult was observed at all dose levels for PCB. This was due to the lack of parallelism between the dose-absorption curves only, as no significant age-only effect occurred. Young/adult 72-h dermal penetration ratios were 0.88, 1.29, and 0.47 at the low, medium, and high dose, respectively.

Carbaryl (CAB) had a nonsignificant age main effect, while all other model parameters were significant. Age comparisons were significant at the low and medium-dose levels and the young/adult ratios were 1.29 at the low dose, 0.62 at the medium dose, and a nonsignificant 1.22 at the high dose.

Dose levels of PAR were reduced because of acute toxicity. It showed no significant age main effect but a significant age × log dose interaction. Age comparison at the low dose was not significant. Comparison of the adult interpolated value to the young medium dose was significant. The young/adult ratio for the low dose was 0.99, and interpolated ratio for the medium dose was 0.76.

Three compounds (FOL, KEP, PER) produced no age-related effects (age, age × dose). The young/adult ratios ranged between 0.76 to 0.97 for FOL, 1.21 to 1.87 for KEP, and 0.87 to 1.06 for PER.

Six compounds (DNS, MSM, DSM, FOL, KEP, and PER) showed parallel dose-absorption curves in young and adult rats. Four compounds (DNS, DSM, MSM and CHL) showed a lack of significant ($P = 0.05$) dose effect while 10 of the 14 compounds have a significant changing fractional penetration with dose. The compounds CAP, CAF, and NIC showed higher penetration at the low dose with equal absorption in the medium and high-dose groups. The compounds ATR, CAB, FOL, KEP, PCB, and PER showed decreasing absorption with increasing dose. Nicotine (NIC) had the least absorption at the low dose and no difference in absorption between the medium and high dose.

The magnitude of the effect of dose on skin penetration was evaluated by comparing the ratio of low dose/high dose for each compound and age. The compounds CAB, CAP, FOL, and PCB had ratios greater than ten. The compounds CAF, CAB, CAP, FOL, KEP, and PCB had ratios greater than seven. Toxicity in the PAR experiment precludes any substantive analysis of dose-skin penetration other than the existence of a dose effect.

Discussion

This paper presents the results from the preliminary percutaneous penetration screen to assess the effects of dose and age on skin penetration of pesticides and other toxic substances. The percutaneous absorption of 14 compounds, including two organophosphates, two carbamates, two organometals, two chlorinated hydrocarbons, two chloroalkyl thio heterocyclics, two biological insecticides, one nitrophenol, and one triazine, was determined in young (33 days) and adult (82 days) female Fischer 344 rats at 72 h postdosing. Three dose levels of each compound were studied in each age group. This approach requires only 33 animals per compound compared to 90 (2 ages, 3 doses, 5 times, 3 animals/group) animals for a serial sacrifice study. It also provides information at a single sacrifice time on dose response, age differences, organs or tissues that may store the chemical or its metabolites, and the experimental duration necessary to characterize adequately the disposition of the substance in a serial sacrifice study.

Examination of the rank order of skin absorption of the 14 compounds shows that as dose is increased, the median decreases in both young and adult rats. The median is always higher in the adult than in the young. The maximum and minimum value is always higher in the adult at each dose. No other generalization of skin penetration characteristics were noted among the diverse compounds examined. Critical evaluation must rest with individual compounds and specific doses. Advancement in understanding of penetration mechanism(s) will require some type of structure analysis of the compounds.

Ideally one would like to know the population distribution curve for skin penetration. Therefore, the variability of the observed penetration values about the mean were examined. The sample standard errors given in Table 3 were multiplied by square root of three to give the observed population standard deviation. The ratio of this value to the mean is the coefficient of variation for the population (CVP). The individual animal variability can be observed in Fig. 1. The two median CVP were 8.8 and 9.2% for adults and 11 and 13% for the young. Of the 40 CVP values at each age, five values for adult and six values for young

TABLE 3—Skin penetration of pesticides in young and adult female rats at 72 h.

Compound	Low Dose μmol/cm²	Fractional Penetration of the Recovered Dose, mean (SEM)			Percent Radioactivity Recovered, mean (SEM)		
		Low Dose	Medium Dose[d]	High Dose[d]	Low Dose	Medium Dose	High Dose
Atrazine	0.25	0.0765(0.0039)[b]	0.0455(0.0041)	0.0278(0.0049)	79.6(0.7)	83.2(0.5)	78.9(1.2)
	0.29	0.0963(0.0059)	0.0676(0.0017)	0.0322(0.0020)	88.7(0.2)	90.2(2.3)	88.9(0.7)
Carbaryl	0.15	0.3013(0.0027)	0.1975(0.0156)	0.0396(0.0053)	90.5(1.3)	92.4(1.8)	93.1(0.3)
	0.19	0.3669(0.0232)	0.1221(0.0127)	0.0485(0.0038)	95.5(1.8)	95.4(0.8)	95.8(0.4)
Carbofuran	0.02	0.8341(0.0078)	0.0827(0.0018)	0.0597(0.0013)	97.2(0.7)	92.3(1.1)	93.2(0.2)
	0.03	0.2453(0.0850)	0.0923(0.0089)	0.0369(0.0024)	94.6(0.5)	97.8(0.6)	95.3(0.3)
Captan	0.09	0.3824(0.0057)	0.0372(0.0004)	0.0365(0.0097)	70.5(0.7)	87.7(0.8)	88.2(0.7)
	0.11	0.2674(0.0022)	0.0378(0.0029)	0.0264(0.0026)	69.8(1.1)	86.8(0.1)	81.2(1.3)
Chlorpyrifos	...	...	0.6633(0.0353)	0.5870(0.0225)	...	97.4(0.2)	99.8(0.3)
	...	...	0.8153(0.136)	0.9005(0.0159)	...	95.1(0.2)	85.8(4.3)
Dinoseb	0.21	0.8639(0.0107)	0.9051(0.0114)	0.9325(0.0060)	87.9(1.8)	91.5(0.6)	90.4(0.7)
	0.25	0.7770(0.0589)	0.8152(0.0278)	0.8288(0.0096)	94.9(0.7)	96.2(1.4)	105.9(3.0)
DSMA[c]	0.09	0.0901(0.0286)	0.1530(0.0222)	0.1188(0.0791)	85.9(0.5)	88.6(0.5)	90.2(0.6)
	0.11	0.0600(0.0361)	0.0200(0.0063)	0.0120(0.0025)	95.3(0.4)	90.0(3.7)	94.5(1.3)

Compound							
Folpet	0.08	0.1477(0.0241)	0.0271(0.0048)	0.0112(0.0008)	71.8(1.4)	82.4(0.1)	75.7(0.7)
	0.10	0.1226(0.0063)	0.0262(0.0028)	0.0085(0.0011)	83.5(0.7)	90.0(0.9)	87.9(0.7)
Chlordecone	0.29	0.0920(0.0137)	0.0596(0.0092)	0.0103(0.0005)	86.5(1.3)	95.1(0.2)	86.8(0.6)
	0.34	0.1017(0.0064)	0.0723(0.0008)	0.0193(0.0015)	89.1(1.5)	90.9(0.9)	89.9(0.8)
MSMA[d]	0.09	0.2204(0.0216)	0.1378(0.0431)	0.1888(0.0788)	89.9(0.7)	87.4(1.5)	87.9(0.8)
	0.11	0.0294(0.0126)	0.0206(0.0039)	0.0516(0.0127)	89.7(4.5)	92.2(2.6)	90.1(1.9)
Nicotine	0.01	0.7503(0.0148)	0.8296(0.0040)	0.8588(0.0055)	89.0(4.6)	57.4(1.2)	63.8(1.6)
	0.02	0.4887(0.0138)	0.8440(0.0275)	0.8820(0.0086)	90.9(2.5)	77.8(1.3)	86.1(4.8)
Parathion[e]	0.05	0.8200(0.0386)	0.7066(0.0209)	…	100.0(0.2)	97.5(1.0)	…
	0.05	0.8145(0.0356)	0.5785(0.0579)	…	95.3(2.9)	94.4(0.8)	…
PCB[f]	0.11	0.4070(0.0082)	0.2080(0.0171)	0.0582(0.0114)	88.3(1.0)	85.5(0.5)	85.4(1.1)
	0.13	0.3345(0.0059)	0.2673(0.0070)	0.0272(0.0037)	83.2(0.7)	79.3(1.6)	79.6(3.3)
Permethrin	0.02	0.5677(0.0065)	0.2692(0.0066)	0.1576(0.0243)	88.2(0.3)	87.9(1.4)	85.9(0.8)
	0.02	0.4850(0.0058)	0.2729(0.0635)	0.1671(0.0023)	92.8(1.0)	92.0(1.4)	93.9(0.9)

[a] Medium and high dosage were 0.54 and 2.68 $\mu mol/cm^2$ in young and adult rats, respectively.
[b] Adult values first, young values second. The treatment area was 2.8 and 5.6 cm^2 in young and adult rats, respectively.
[c] DSMA = disodium methanearsonate.
[d] MSMA = monosodium methanearsonate.
[e] Medium parathion dose in young was 0.18 $\mu mol/cm^2$.
[f] PCB = 2,4,5-2',4',5'-hexachlorobiphenyl.

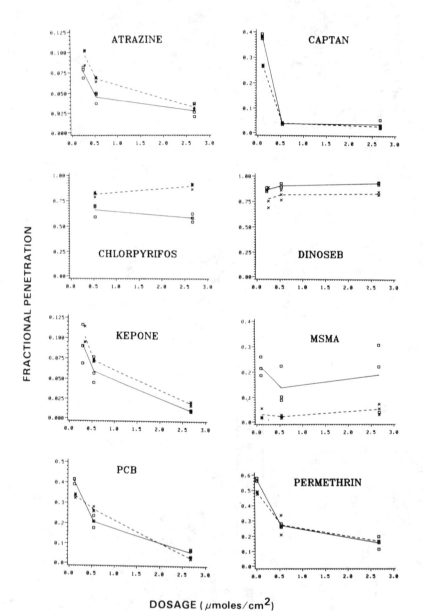

DOSAGE (μmoles/cm^2)

FIG. 1—*Mean fractional skin penetration in young (--) and adult (—) female rats as a function of dosage. Individual values are* X = *young,* □ = *adult.*

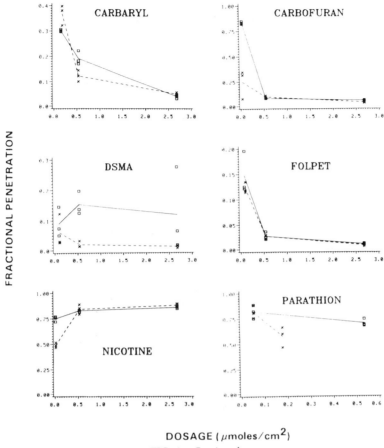

DOSAGE (μmoles/cm^2)

FIG. 1—*Continued.*

were above 36%. The compounds with the largest values were DSM and MSM for both adult and young. The largest CVP values were 115% for adult and 91% for young for DSM and 72% for adult and 74% for young for MSM. For the adult, both values occurred in the high-dose group while for young both values occurred in the low-dose group.

There were three compounds (CAF, DNS, and NIC) with all three CVPs below 4% in the adult. The two compounds with the lowest CVPs in the young were CHL and NIC. The values for CHL, for which only two doses were tested, were below 4% in the young, and the three CVP values for NIC ranged to 5.6% in the young. There was no obvious categorization as to which dose group the high and low CVP values were found. Variability between animals and for different locations on an animal needs to be determined versus the age of the animal for a proper hazards analysis to be made. With a paucity of data and with analytical variability confounding with natural biological variability, we tentatively find that the young exhibit more variability than adults.

Eight of the 14 compounds showed statistically significant age main effects (see Table 4) while 11 of the 14 compounds showed significant age differences in either age main effect or age × dose interaction. Only FOL, KEP, and PER failed to show any effect of age on skin absorption. Four compounds (CHL, CAB, ATR, and PCB) showed greater skin penetration in the young animal at some dose. Chlorpyrifos (CHL) had greater penetration in

TABLE 4—Statistical significance (%P) of variables in the model[a] and young versus adult comparisons when indicated.

Compound	Age	Log(RD)	Log²(RD)	Age × Log(RD)	Age × Log²(RD)	Young versus Adult Comparison		
						Low Dose	Medium Dose	High Dose
Atrazine	0.06	0.01	0.60	1.40	77.00	0.06	0.23	46.00
Carbaryl	88.00	0.01	0.12	1.20	0.01	0.07	0.25	63.00
Carbofuran	0.01	0.01	0.03	0.01	0.49	0.001	85.00	66.00
Captan	0.01	0.01	0.01	0.01	0.06	0.001	94.00	17.00
Chlorpyrifos	0.01	85.00	...	0.88	...	...	0.18	0.001
Dinoseb	0.14	6.10	44.00	86.00	86.00	...	...	...
DSMA[b]	1.30	84.00	53.00	35.00	36.00	...	...	...
Folpet	28.00	0.01	0.04	74.00	78.00	...	...	...
Chlordecone	10.00	0.01	18.00	52.00	75.00	...	...	...
MSMA[c]	0.05	87.00	19.00	58.00	55.00	...	...	...
Nicotine	0.01	0.01	3.70	0.01	0.57	0.001	50.00	28.00
Parathion	14.00	0.64	...	2.60	...	93.00	1.50	...
Permethrin	91.00	0.01	63.00	5.70	61.00	...	...	...
PCB[d]	9.30	0.01	3.30	26.00	0.01	0.67	0.11	4.70

[a] Model; (fractional penetration) = age, log(RD), log²(RD), age × log(RD), age × log²(RD), where RD is the recovered dose.
[b] DSMA = disodium methanearsonate.
[c] MSMA = monosodium methanearsonate.
[d] PCB = 2,4,5-2',4',5'-hexachlorobiphenyl.

TABLE 5—*Ratio of fractional penetration in young and adult female Fischer 344 rats at each dosage.*

Compound	Young/Adult Ratio		
	Low Dose[a]	Medium Dose	High Dose
Atrazine	1.38	1.49	1.16
Carbaryl	1.29	0.62	1.22
Carbofuran	0.32	1.12	0.62
Captan	0.81	1.02	0.72
Chlorpyrifos	...	1.23	1.53
Dinoseb[b]	0.89	0.90	0.89
DSMA[b,c]	0.59	0.13	0.10
Folpet	0.93	0.97	0.76
Chlordecone	1.23	1.21	1.87
MSMA[b,d]	0.14	0.15	0.27
Nicotine	0.65	1.02	1.03
Parathion	0.99	0.76	...
PCB[e]	0.88	1.29	0.47
Permethrin	0.87	1.01	1.06

[a] Ratio computed with interpolated values.
[b] Mean young/adult ratios for DNS, DSM, and MSM are 0.90, 0.25, and 0.19, respectively.
[c] DSMA = disodium methanearsonate.
[d] MSMA = monosodium methanearsonate.
[e] PCB = 2,4,5-2′,4′,5′-hexachlorobiphenyl.

the young at both doses and showed the highest significant young/adult ratio of 1.53. Eight compounds (DNS, DSM, MSM, PCB, CAP, NIC, CAB and CAF) showed greater penetration in adults at some dose. Absorption of DNS, MSM, and DSM was greater in adults at all doses. The largest difference between young and adult was seen with MSM, which has a ratio of 0.19. The other arsenical, DSM, has the second largest young/adult ratio (0.25), which is a four-fold greater penetration in the adult.

A paucity of information on the effect of age on skin penetration of chemicals is found in the open literature. Using *in vitro* techniques with hairless mouse skin, Behl et al. [*18*] observed that skin permeability to hydrocortisone increased and then decreased from birth and was coincident with the development of the hair growth cycle. Solomon et al. [*19*] reported no significant differences between newborn and adult guinea pigs, in blood or brain concentrations following dermal application of gamma-benzene hexachloride. Wester et al. [*13*] also noted no difference in skin penetration of testosterone between newborn and adult rhesus monkeys. The very limited information of the effect of age, species, and chemical structure should provoke caution in the extrapolation of health effects data to the human agricultural situation.

The dose-absorption response was determined in young and adult rats using three dosages (mol/cm²) and scaled for increasing body size by increasing the area of exposure. Several different dose-absorption curves were noted. The compounds DNS, MSM and DSM displayed constant fractional penetration, the classical diffusional behavior. The compounds CAP and CAF showed greater penetration at the low dose while the middle and high doses were quantitatively less but equal. The compounds ATR, CAB, FOL, KEP, PCB and PER showed a decrease in absorption in each successive dose. Nicotine (NIC) showed equal penetration at the middle and high dose and lower absorption at the low dose. Insufficient data were available to evaluate PAR and CHL although CHL was the only compound where absorption in the young was greater than adult at both doses, which spanned the dosage range of the other compounds.

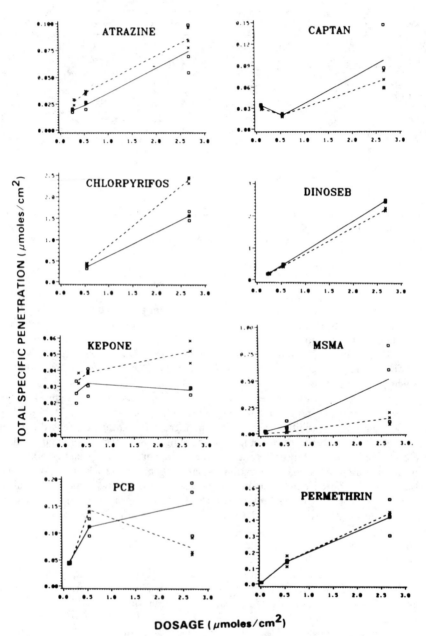

FIG. 2—*Mean total specific penetration (μmol/cm²) in young (--) and adult (—) female rats as a function of dosage. Individual values are* X = *young,* □ = *adult.*

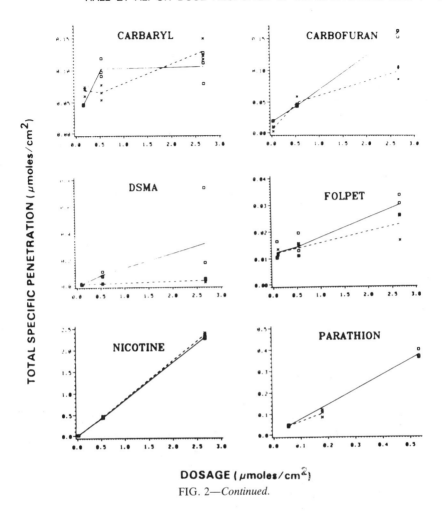

DOSAGE (μmoles/cm²)

FIG. 2—*Continued.*

Little information is available in the literature on the effects of dose on percutaneous absorption. Scheuplein and Ross [7] using *in vitro* techniques with human skin found that the skin absorption of cortisone was dose-dependent. Wester and Maibach [8] reported dose-dependent skin absorption of testosterone, hydrocortisone, and benzoic acid in the rhesus monkey and man. Maibach and Feldmann [5] found that the dermal absorption of parathion was independent of dose while lindane showed dose-dependent skin absorption in humans.

Although the number of compounds for which information is available on the nature of the dose-skin absorption curve is not large, the percentage showing nonclassical membrane behavior is somewhat surprising. Also the occurrence of "regions of classical" skin absorption at the higher doses and disappearance at the lower complicate the picture. Dosage, therefore, must be included in any extrapolation in risk assessment.

Total specific penetration is shown in Fig. 2. Multiplying the fractional 72-h penetration by the applied dose in micromoles yields the total 72-h penetration in micromoles. For a constant fractional penetration, the resulting total penetration plotted against dose is a straight line passing through the origin. Excluding CHL and PAR for which only two dose levels were tested, the only compounds for which a linear function passing through the origin is obtained in both young and adult are DNS and NIC. The variability of the data for DSM and MSM is such that it is possible that these compounds satisfy the criteria also. A second type of curve observed was linear in all three points with the extended line having a nonzero intercept. This type of curve was observed in young and adult for ATR and FOL. This type of function was also obtained for CAF in the adult and PER in the young. A third type of curve observed was a concave downward whose intercept could very well be the origin. This type of curve was observed for CAF in the young and PER in the adult. The PCB in the adult has this form also, while in the young the total absorption at the high-dose level is about half of that at the medium-dose level, a unique characteristic in any of the 14 compounds examined. Chlordecone (KEP) also displayed a concave downward dose-penetration curve in young and adult. However, the data are such that more information at lower doses is needed to characterize the curve as it approaches zero dose. The only compound having a concave-up curve was CAP. This occurred for both young and adult. Carbaryl (CAB) penetration in young and adult showed different types of behavior. In the adult, the curve was concave downward and the intercept could very well be the origin. In the young, the curve was concave upward with the low and medium-dose groups having about the same total absorption. Total specific penetration is an additional way of examining absorption data. Characteristics not readily apparent in the fractional absorption are displayed in the total absorption data and lend an extra dimension to the analysis of dermal absorption.

Skin thickness differences between young and adult animals could affect the rate of penetration. Therefore, skin morphometry in young and adult female rats was performed as described by Bronaugh et al. [20]. Preliminary results for the thickness of the stratum corneum, epidermis, and dermis (mean $\pm$ SEM) was 3.6 $\pm$ 0.2, 13.4 $\pm$ 0.5, and 532.5 $\pm$ 12.0 μm in young, respectively, and 3.3 $\pm$ 0.2, 14.0 $\pm$ 0.5, and 641.4 $\pm$ 19.9 μm, in the adult, respectively, suggesting that thickness is not a factor in the observed age differences in penetration. Histopathologically, the skin of the young rat was less organized, particularly the adnexal structures.

Bronaugh et al. [20] reported the thickness of the stratum corneum (18.4 $\pm$ 0.5 μm) and epithelium (32.1 $\pm$ 1.3 μm) in female Osborne-Mendel rat. Significant differences in skin morphometry between rodent strains may exist in addition to differences between species, thus the effect of skin structure on penetration remains unclear.

The correlation of chemical structure with skin absorption was examined using the octanol/water partition coefficient (P) and water solubility (WS) as chemical descriptors and the logarithm of the percent absorbed for both young and adult rats. The r^2 in young and adult using log P and log WS was 0.47 and 0.35, respectively. This low correlation is due possibly to the diversity of the chemicals studied as well as the manner of defining the rate of penetration [21]. Additional research is necessary to identify a better rate of penetration descriptor in finite dose experiments.

In conclusion, skin ingress of the majority of the 14 pesticides studied in young and adult female rats was found to be dose and age dependent. Constant fractional penetration was seen in only three compounds, and no common pattern to the other dose-absorption curves was noted. Skin penetration was only marginally correlated with the octanol/water partition coefficients of this diverse group of compounds.

Acknowledgments

The authors would like to thank Dr. Robert A. Robinson of FMC, Princeton, New Jersey for providing radiolabeled carbofuran and permethrin. We also would like to thank Dr.

Homer M. LeBaron of Ciba-Geigy, Greensboro, North Carolina, for providing ^{14}C-labeled atrazine.

The research described in this paper has been reviewed by the Health Effects Research Laboratory, U. S. Environmental Protection Agency and approved for publication. Approval does not signify that the contents necessarily reflect the views and policies of the Agency nor does mention of trade names or commercial products constitute endorsement or recommendation for use.

References

[1] Skelton, H., "The Storage of Water by Various Tissues of the Body," *Archives of International Medicine*, Vol. 40, 1927, pp. 140–152.

[2] Durham, W. F. and Wolfe, H. R., "Measurement of Exposure of Workers to Pesticides," *Bulletin of World Health Organization*, Vol. 26, 1962, pp. 75–91.

[3] Wester, R. C. and Maibach, H. I., "Interrelationships in the Dose Response of Percutaneous Absorption," *Percutaneous Absorption*, R. L. Bronaugh and H. I. Mailbach, Eds., Marcel Dekker, NY, 1985, pp. 347–357.

[4] Holland, J. M., Kao, J. Y., and Whitaker. M. J., "A Multisample Apparatus for Kinetic Evaluation of Skin Penetration in vitro: The Influence of Viability and Metabolic Status of the Skin," *Toxicology and Applied Pharmacology*, Vol. 72, 1984, p. 272–280.

[5] Maibach, H. I. and Feldmann, R. J., "Systemic Absorption of Pesticides Through the Skin of Man," *Occupational Exposure to Pesticides Report to the Federal Working Group on Pest Management from the Task Group on Occupational Exposure to Pesticides*, Washington, DC, 1974, pp. 120–127.

[6] Sanders, C. L., Skinner C., and Gelman, R. A., "Percutaneous Absorption of [7,10-^{14}C]benzo [a]pyrene and [7,12-^{14}C]dimethylbenz[a]anthracene in Mice," *Environmental Research*, Vol. 33, 1984, pp. 353–360.

[7] Scheuplein, R. J. and Ross, L. W., "Mechanism of Percutaneous Absorption V.," *Journal of Investigative Dermatology*, Vol. 62, 1974, pp. 353–360.

[8] Wester, R. C. and Maibach, H. I., "Relationship of Topical Dose and Percutaneous Absorption in Rhesus Monkey and Man," *Journal of Investigative Dermatology*, Vol. 67, 1976, pp. 518–520.

[9] Singer, E. J., Wegmann, P. C., Lehman, M. D., Christensen, M. S., and Vinson, L. J., "Barrier Development, Ultrastructure, and Sulfhydryl Content of the Fetal Epidermis," *Journal of the Society of Cosmetic Chemists*, Vol. 22, 1971, pp. 119–137.

[10] Rasmussen, J. E., "Percutaneous Absorption in Children" in *Year Book of Dermatology*, R. L. Dobson, Ed., Year Book Medical Publications, Inc., Chicago, 1979, pp. 25–38.

[11] Nachman, R. L. and Esterly, N. B., "Increased Skin Permeability in Preterm Infants," *Journal of Pediatrics*, Vol. 79, 1971, pp. 628–632.

[12] Greaves, S. J., Ferry, D. G., McQueen, E. G., Malcolm, D. S., and Buckfield, P. M., "Serial Hexachlorophene Blood Levels in the Premature Infant," *New Zealand Medical Journal*, Vol. 81, 1975, pp. 334–336.

[13] Wester, R. C., Noonan, P. K., Cole, M. P., and Maibach, H. I., "Percutaneous Absorption of Testosterone in the Newborn Rhesus Monkey: Comparison to the Adult," *Pediatric Research*, Vol. 11, 1977, pp. 737–739.

[14] McCormack, J. J., Boisits, E. K., and Fisher, L. B., "An in vitro Comparison of the Permeability of Adult versus Neonatal Skin," *Neonatal Skin, Structure and Function*, H. I. Maibach and E. K. Boisits, Ed., Marcel Dekker, Inc., New York and Basel, 1982, pp. 149–164.

[15] Knaak, J. B., Yee, K., Ackerman, C. R., Zweig, G., and Wilson, B. W., "Percutaneous Absorption of Triadimefon in the Adult and Young Male and Female Rat," *Toxicology and Applied Pharmacology*, Vol. 72, 1984, pp. 406–416.

[16] Bartek, M. J., LaBudde, J. A., and Maibach, H. I., "Skin Permeability in vivo. Comparison in Rat, Rabbit, Pig and Man," *Journal of Investigative Dermatology*, Vol. 58, No. 3, 1972, pp. 114–123.

[17] Helwig, J. and Council, K. A., "*SAS User's Guide*," Statistical Analysis System (SAS) Institute, Cary, NC, 1979.

[18] Behl, C. R., Flynn, G. L., Linn, E. E., and Smith, W. M., "Percutaneous Absorption of Corticosteroids: Age, Site, Skin-Sectioning Influences on Rate of Permeation of Hairless Mouse Skin by Hydrocortisone," *Journal of Pharmaceutical Sciences*, Vol. 73, 1984, pp. 1287–1290.

[*19*] Solomon, L. M., West, D. P., Fitzcoff, J. F., and Becker, A. M., "Gamma Benzene Hexachloride in Guinea Pig Brain after Topical Application," *Journal of Investigative Dermatology,* Vol. 68, 1977, pp. 310–312.
[*20*] Bronaugh, R. L., Stewart, R. F., and Congdon, E. R., "Methods for in vitro Percutaneous Absorption Studies. II. Animal Models for Human Skin," *Toxicology and Applied Pharmacology,* Vol. 62, 1982, pp. 481–488.
[*21*] C. Hansch, personal communication, 1985.

Protection From Industrial Chemical Stressors

Chemical Breakthrough Parameters

Charles M. Hansen[1] and Kristian M. Hansen[2]

Solubility Parameter Prediction of the Barrier Properties of Chemical Protective Clothing

REFERENCE: Hansen, C. M. and Hansen, K. M., **"Solubility Parameter Prediction of the Barrier Properties of Chemical Protective Clothing,"** *Performance of Protective Clothing: Second Symposium, ASTM STP 989*, S. Z. Mansdorf, R. Sager, and A. P. Nielsen, Eds., American Society for Testing and Materials, Philadelphia, 1988, pp. 197–208.

ABSTRACT: Solubility parameter correlations based on breakthrough times are presented. Solubility parameters have been used previously to estimate whether or not a given permeant will swell or dissolve in a given membrane. Increased solubility is predicted when the solubility parameters for the polymer and penetrant approach each other. Some experimental work is required to characterize either the permeant or the membrane if no data are already available. Since the diffusion coefficients for organic permeants in organic membranes increase exponentially with permeant concentration, there is a very strong dependence of permeation on solubility and therefore on solubility parameters. All other things being equal, penetrants with bulky structures and side groups will permeate more slowly than those with more linear and less voluminous structures. For this reason elimination of permeants with either very small or very large molecular size significantly improves most of the correlations presented, but limits their range of usefulness.

KEY WORDS: solubility parameters, diffusion coefficients, breakthrough time, permeation, protective clothing

The three parameter approach to the solubility parameter enables characterization of a polymer membrane by all the solvents that dissolve or swell it [*1–3*]. These solvents will be located as points in a three-dimensional system in a manner such that they define a symmetrical volume of interaction for the solute. This enables ready choice of solvents or swelling agents in a given situation, and represents solubility relations so well that even nonsolvents can be predictably mixed to give a dissolving or swelling mixture. The nonsolvents in these cases need only lie on opposite sides of the volume of interaction from each other.

The basis of the system is the assumption that the energy of evaporation, that is, the total cohesive energy which holds a liquid together, ΔE, can be divided into contributions from dispersion (London) forces, ΔE_d, polar forces, ΔE_p, and hydrogen bonding forces, ΔE_h. Thus

$$\Delta E = \Delta E_d + \Delta E_p + \Delta E_h \tag{1}$$

Dividing this equation by the molar volume of a solvent, V_m gives

$$\frac{\Delta E}{V_m} = \frac{\Delta E_d}{V_m} + \frac{\Delta E_p}{V_m} + \frac{\Delta E_h}{V_m} \tag{2}$$

[1] Senior scientist, Danish Isotope Center, 2 Skelbaekgade, DK-1717 Copenhagen, Denmark.
[2] Consultant, Jens Bornosvej 16, 2970 Horsholm, Denmark.

or

$$\delta^2 = \delta_d^2 + \delta_p^2 + \delta_h^2 \tag{3}$$

where

$$\delta = (\Delta E/V_m)^{1/2} \tag{4}$$

is the usual equation for the solubility parameter [4,5]

$$\delta_d = (\Delta E_d/V_m)^{1/2} \tag{5}$$

is the dispersion component of the solubility parameter

$$\delta_p = (\Delta E_p/V_m)^{1/2} \tag{6}$$

is the polar component of the solubility parameter, and

$$\delta_h = (\Delta E_h/V_m)^{1/2} \tag{7}$$

is the hydrogen bonding component of the solubility parameter.

The units for each of these parameters are $(cal/cm^3)^{1/2}$ or more recently $(J/cm^3)^{1/2}$ that is 2.046 times as large.

The principle on which the solubility parameter system is based is that materials having similar solubility parameters are physically similar to each other and mix easily with small heat effects [4,5]. Even many surfaces can be characterized by their individual interactions with a series of solvents, in which case they attract and adsorb those that are physically similar [6]. This then gives a characterization of the surface in terms of solvent properties.

Knowing δ_d, δ_p, and δ_h for a series of solvents enables characterization of membranes by solubility or swelling measurements. The material tested can be characterized by an essentially spherical volume in the system when the distance for a unit along the δ_d axis is twice that along the δ_p and δ_h axes as demonstrated in Fig. 1.

The Dispersion Parameter

The initial approach to the division of the solubility parameter, δ, into the components representing dispersion, polar, and hydrogen bonding forces was based on the homomorph comparison to estimate dispersion forces, and trial and error placements of solvents as points in a three-dimensional system. The homomorph of a solvent is the hydrocarbon solvent that most nearly resembles it in size and shape [7]. The energy of evaporation of the homomorph is taken as ΔE_d in Eq 5. The homomorph approach failed in the case of solvents containing chlorine or sulfur atoms. It was also obvious that homomorphs were difficult to choose properly for cyclic solvents [8].

The Hydrogen Bonding Parameter

Infrared spectroscopy and other measurements indicate a reasonable value for the OH...O bonds is 5000 cal/mol [9]. If one quite simply ascribes 5000 cal of the energy of evaporation of a solvent to the presence of each alcohol group it contains, one can estimate the hydrogen-bonding solubility parameter for the solvent.

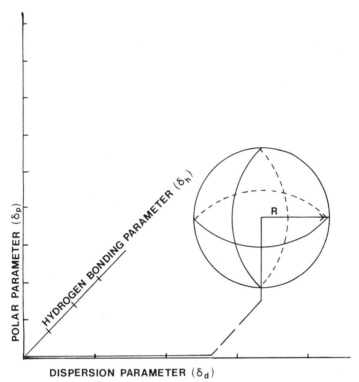

FIG. 1—*Sketch of the Hansen three-dimensional solubility parameter system.*

Thus,

$$\delta_H = \sqrt{\frac{5000 \ N}{V_m}} \tag{8}$$

where N is the number of alcohol groups in the molecule and V_m is the solvent's molar volume.

The Polar Bonding Parameter

Böttcher [10] has developed a relation for calculating the contribution of the permanent dipoles to the cohesion energy of a liquid or a gas. This energy is given as W in Eq 9, and has been divided by V_m to fit the definition of the polar solubility parameter as given in Eq 10.

$$W = \frac{-4\pi}{3} \frac{d N_A^2}{M} \cdot \frac{\epsilon - 1}{2\epsilon + n_D^2} \cdot \frac{n_D^2 + 2}{3} \mu^2 \left[\frac{\text{erg}}{\text{mole}}\right] \tag{9}$$

$$\frac{W}{V_m} = \delta_P^2 = \frac{12108}{V_m^2} \frac{\epsilon - 1}{2\epsilon + n_D^2} \cdot (n_D^2 + 2)\mu^2 \left[\frac{\text{cal}}{\text{cm}^3}\right] \tag{10}$$

where for the condensed phase

d　= density, g/cm^3;
M　= molecular weight;
N_A　= Avogadros number;
ϵ　= dielectric constant, static value;
n_D　= index of refraction for sodium − D line;
μ　= dipole moment, debyes (E_p 10); and
V_m　= molar volume, cm^3.

The Present State—Solubility and Swelling

The values for the dispersion, polar, and hydrogen bonding contributions to the solubility parameter found from trial and error placements based on solubility data were revised. This revision was based on independent calculations of the polar and hydrogen bonding contributions as suggested by Klemen Skaarup. These are the parameters listed in Refs 1, 2, and 3.

Many correlations and the general usefulness of the system should emphasize that the three-dimensional solubility parameter concept is not just a strictly empirical correlation, but rather reflects physical interactions in a predictable manner by use of independently measured physical quantities. Data are given in the literature for well over 200 liquids and many polymers. These data can be used to predict solubility and swelling by mixtures as well as by pure liquids.

The data given below are actually characterizations of the membrane materials using "shorter than" breakthrough times for permeation as the criterion for higher affinity, that is, greater similarity of energy. This criterion is thought to function well because of concentration-dependent diffusion coefficients.

Concentration-Dependent Diffusion—The Exponential Case [1,11–14]

Concentration-dependent diffusion coefficients with an exponential variation are of considerable interest, since the diffusion coefficients of organic solvents in polymeric material vary in this manner at low and moderate concentrations, see Fig. 2. These diffusion coefficients are reported for data interpreted with solutions to the diffusion equation with an exponential diffusion coefficient.

These solutions to the diffusion equation and procedures for interpreting diffusion data are given elsewhere [1,11–14]. Let it be noted here that interpretation of data using solutions to the diffusion equation for a constant diffusion coefficient will cause errors.

In addition to concentration dependence, surface effects can be also very significant, particularly at higher flux rates and solvent concentrations [13–15]. Such conditions prevail where penetration/permeation rates are high, that is, for situations of short breakthrough times. It has been shown that the diffusion coefficient for penetrant chlorobenzene in polyvinylacetate varies nine decades over the whole concentration range, when data are interpreted including the preceding effects. The concentration dependence at low solvent concentrations is particularly strong for this pair as it also is for other typical penetrant/organic membrane combinations. In the same membrane, if the local concentration varies between zero and 0.20 volume fraction chlorobenzene, the diffusion coefficient varies from 10^{-14} cm^2/s up to 10^{-8} cm^2/s, that is, six decades. This same difference is found when comparing the speed of a snail crawling in the woods with that of a jet plane.

While the preceding discussion has been particularly directed toward the measurement of diffusion coefficients, the same effects are also found for permeation cell testing and for

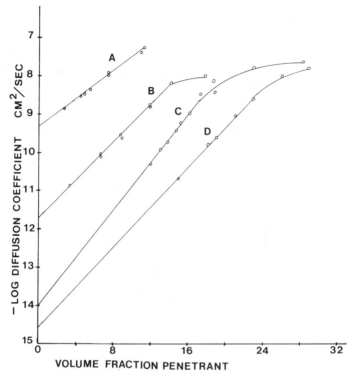

FIG. 2—*Diffusion coefficients in poly(vinyl acetate) at 25° for: (A) methanol, (B) ethylene glycol monomethyl ether, (C) chlorobenzene, and (D) cyclohexanone.*

practical experience. Exposure to liquid penetrant means the local concentration at that surface will be very high (if solubility parameters match well), and the local diffusion coefficient will be very high. At the other side of the membrane, the local concentration will be lower, diffusion coefficients will be lower and the major resistance to mass transport in the membrane will be found here. All the intermediate local concentrations with their respective local diffusion coefficients will be found within the film as well. Permeation through gloves is a very complex phenomena.

It follows that one preferably uses a glove where the solubility on exposure to liquid is very low. While this is not a unique finding, the consequences of even slight increases in solubility are important and have been given more attention in the next section.

Permeation of Organic Penetrants in Polymer Membranes

The permeation of penetrants through a membrane, P, can be described as the product of the diffusion coefficient, D, and the solubility, S, $P = DS$. If we hypothetically assume the solubility of a penetrant increases from 1 to 4% (perhaps due to a new additive in a glove), it is clear that for a constant diffusion coefficient the permeation rate would be four times as great. For the exponential concentration dependence expected in diffusion in polymers, the diffusion coefficient at the higher concentration will in fact be about ten times as great. The permeation rate is far more dependent on the amount of absorbed penetrant than would normally be expected. This strong concentration dependence is because the penetrant plasticizes the membrane by a high free volume contribution. One can thus expect the solubility parameter to correlate with permeation data, since it correlates with both the solubility and the solubility-dependent diffusion coefficient. Where solubilities are only a few percent, that is, where the solubility parameters of the penetrant and glove material are quite different, the size and shape of the penetrant molecules will be more important since these most strongly affect differences in diffusion coefficients, that is, permeation and penetration rates.

The solubility parameter has been demonstrated as a reasonably reliable, if imperfect, correlating variable for the amount of penetrant taken up by a membrane [16]. Large differences in solubility parameters between penetrant and membrane give low degrees of swelling. Beerbower used a criterion of greater than 25% uptake to define penetrants that were too good for use with various types of packing material. These excessive swellers were used to define a volume of highest affinity, just like a breakthrough criterion has been used to define highest affinity in this paper. Extensions of this type of correlation to the area of protective clothing and permeation rates were first presented by Henriksen [17]. Perkins and Tippit [18] using weight gain, as well as Christensen and Banke [19] who used permeation rates, have also applied solubility parameters to predicting glove performance. A number of pitfalls in the use of solubility parameters are reviewed by Barton [20]. These include methanol self-association to yield a hydrophobic entity that swells fluorocarbon elastomer Viton A, and other examples of association including water. The solubility parameter correlates physical phenomena; where reactions occur a look at the resulting system can be a help, but this then becomes much more complicated. It would thus appear relevant that the design and function of protective clothing can be improved by including the solubility parameter among the various other factors that determine the development of protective clothing. The lower the solubility, the better the protective potential of the membrane. As just stated, lowered solubility is predicted for greater differences in the solubility parameters of penetrant and membrane material, respectively.

Solvent Quality and Solubility Parameters

Solubility parameters have been traditionally used to determine which of a group of solvents or solvent blends is best for a given polymer. The best will have parameters such that R_A for it in Eq 11 is less than R_A for the others.

$$R_A^2 = 4(\delta_{d_s} - \delta_{d_p})^2 + (\delta_{p_s} - \delta_{p_p})^2 + (\delta_{h_s} - \delta_{h_p})^2 \qquad (11)$$

The subscripts, s and p, are for the solvent and polymer, respectively.

The correlating procedure for finding the polymer parameters is to choose a level of affinity to arbitrarily define whether a solvent is good or bad. Such criteria can be soluble or not at a given concentration, or whether swelling occurs to above a given percentage of interest for crosslinked polymers, for example. In the present case, breakthrough times were used to arbitrarily define good or bad solvents. A breakthrough time of 2 h for a given glove/solvent combination would mean the solvent is (too good) a sweller/penetrant for a 4-h breakthrough requirement. If the breakthrough time requirement being considered is 20 min or 1 h, then this same system becomes acceptable in practice under these conditions, and the solvent is here considered sufficiently bad to be called bad when correlations are made.

Correlations are done using Eq 11 for a large number of solvents that can be divided into good and bad groups. The object of the correlation is to find the polymer solubility parameters that cause all the good solvents to have their respective R_A values less than that of the R for the polymer. Similarly, R_A for bad solvents should be greater than R. The four polymer parameters are systematically varied until the data fit cannot be improved. At this point, one accepts these four parameters as those describing the polymer for the case being studied.

The data fit criterion is the desirability function, which is a useful statistical tool for just this type of human judgment situation where several factors are involved as discussed by Harrington [21]. The fit to be maximized, hopefully to 1.00, is

$$\text{FIT} = \sqrt[N]{D_1 D_2 \ldots D_N} \tag{12}$$

Here, N is the number of solvents considered; D is 1 when a solvent has R_A less than R and is good; D is also 1 when R_A is greater than R and the solvent is bad; D is equal to $e^{-(R_A - R)}$ when good solvents are outside the sphere $(R_A > R)$; and D is similarly equal to $e^{-(R - R_A)}$ for bad solvents located within the sphere $(R_A < R)$. When such noncorrelating systems or "outliers" are involved, the data fit is obviously less than 1.0. The data fit is severely penalized by good solvents well outside the sphere describing the polymer. Experience has shown that data fits are typically above 0.90 for polymer solubility data. Only rarely are they perfect.

For the preceding reasons, the data presented are in no way a substitute for experiment. They are useful to predict trends when systems are altered, as in development work. They are also useful to quickly reject hopelessly poor, untested systems, and to point up which glove(s) are most likely to be satisfactory in given cases where exact data are not immediately available.

A more complete analysis of the outlier problem for the results in Table 1 is given in Table 2 and in the comments to Table 2.

Results and Discussion

Breakthrough times of less than 20 min, 1 h, or 4 h, respectively, were selected as criteria for determining useful lifetimes of the chemical protective membranes. Breakthrough and permeation rate data for many systems have been collected by Forsberg and Olsson [22]. This reference was the only data source used for this study. The data are primarily those reported by the manufacturers of protective clothing.

Data are often found that indicate a given membrane had breakthrough times for a given chemical both longer than and shorter than the lifetime being considered. Such systems were omitted from consideration. As an example, breakthrough times of 0, 28, 30, and 90 min as well as greater than 8 h were reported for the system ethanol/natural rubber. One could unethically choose those data that fit the correlation in such cases, but as just stated, the whole system was treated as if no data existed at all. In each case reported, a satisfactorily large number of penetrants could be included in the correlations as can be seen in Table 1. "Degradation" with no data on breakthrough times was considered to be a failure in all cases.

A home computer (Commodore 64) was used to correlate the data. Printing results in both alphabetical and molecular volume order maximizes their usefulness. Poor correlations were generally found when all penetrants were considered regardless of molecular size (and shape). The errors in the correlations were most frequently for penetrants having smaller molecules that penetrated faster than expected, and those that have larger molecules that did not penetrate as fast as expected. There are many ways to approach correcting this situation; we choose to simply limit the correlations to molecules of reasonably similar size. Table 1 confirms that this still allows inclusion of many penetrants. It improves the correlations markedly. When using these modified correlations, the penetrants involved must have molecular volumes within the limits indicated (lower limit/upper limit) in Table 1.

TABLE 1—*Solubility parameter correlations of penetration breakthrough times.*

Time	δ_d	δ_p	δ_h	R	FIT	V_M LOW/UP	No.	Outliers
				NITRILE RUBBER				
20 min	17.2	7.4	7.6	4.9	0.739	all	57	12
1 h	16.7	9.2	5.5	9.2	0.665	all	71	14
4 h	22.6	12.5	9.0	16.6	0.642	all	72	13
20 min	17.5	7.3	6.5	5.1	0.870	58/177.5	48	7
1 h	16.6	9.1	4.4	10.0	0.911	61/177.5	58	8
4 h	19.0	12.6	3.8	13.3	0.946	57.5/177.5	61	5
				BUTYL RUBBER				
20 min	16.2	4.7	0.8	5.5	0.730	all	58	14
1 h	17.3	1.8	1.3	6.8	0.768	all	58	7
4 h	11.7	−9.0	3.7	20.0	0.468	all	57	17
20 min	16.5	1.0	5.1	5.0	0.953	71/110	40	4
1 h	15.8	−2.1	4.0	8.2	0.953	71/126	43	3
4 h	17.8	−7.7	−3.0	17.5	0.666	71/−	48	9
				NATURAL RUBBER				
20 min	15.0	7.8	4.2	10.4	0.680	all	64	12
1 h	17.0	5.6	10.9	13.7	0.822	all	66	5
4 h	20.0	12.6	8.7	18.4	0.880	all	66	2
20 min	14.5	7.3	4.5	11.0	0.905	61/267	53	5
1 h	15.6	3.4	9.1	14.0	1.000	55.6/325	61	0
4 h	19.4	13.2	7.7	19.0	1.000	−/325	65	0
				POLYVINYL CHLORIDE				
20 min	17.3	6.9	6.3	8.5	0.774	all	56	7
1 h	16.1	12.9	3.4	13.7	0.681	all	65	8
4 h	17.3	7.1	10.4	13.7	0.750	all	65	3
20 min	16.1	7.1	5.9	9.3	0.920	61/267	46	3
1 h	14.9	11.0	3.8	13.2	0.970	61/267	53	2
4 h	24.4	4.9	9.9	22.7	0.980	−/148.9	58	2
				POLYVINYL ALCOHOL				
20 min	11.2	12.4	13.0	12.1	0.861	all	64	7
1 h	15.3	13.2	13.5	8.8	0.940	all	60	4
4 h	17.2	13.6	15.4	10.9	0.845	all	59	9
				POLYETHYLENE				
20 min	15.7	2.3	5.3	9.3	0.740	all	33	4
1 h	15.7	3.7	6.6	8.4	0.729	all	33	6
4 h	24.5	14.7	2.6	24.9	0.853	all	33	2
20 min	16.9	3.3	4.1	8.1	0.906	40/−	32	2
1 h	17.1	3.1	5.2	8.2	0.880	40/−	32	4
4 h	24.1	14.9	0.3	24.3	0.943	55.6/−	31	2
				VITON				
20 min	10.9	14.5	3.1	14.1	0.896	all	50	5
1 h	16.0	9.2	8.4	7.7	0.840	all	45	7
4 h	13.3	13.9	9.5	13.4	0.836	all	45	5
1 h	16.5	8.1	8.3	6.6	0.921	56/177.5	42	5
4 h	13.6	15.4	8.6	14.4	0.880	−/177.5	44	6
				NEOPRENE RUBBER				
20 min	17.7	1.6	6.8	7.0	0.710	all	54	11
1 h	16.0	8.8	4.0	10.1	0.574	all	66	11
4 h	15.1	10.3	4.5	12.3	0.642	all	66	10
20 min	17.6	2.5	5.9	6.2	0.877	75/177.5	39	4
1 h	19.0	8.0	0	13.2	1.00	69/177.5	50	0
4 h	14.6	13.9	2.3	15.9	0.970	61/266	53	2

TABLE 2—Range of outliers for solubility parameter correlations of the performance of protective clothing.[a]

Membrane	R'_Amax/R'_Amin, 20 min	R'_Amax/R'_Amin, 1 h	R'_Amax/R'_Amin, 4 h	Remarks, R'_A
Nitrile rubber (NR)	1.21/0.87	1.20/0.89	1.08/0.95	20-benzylalcohol (1.47) 20-styrene (1.39)
Butyl rubber (BR)	1.0/0.91	1.14/0.98	1.24/0.81	20-ethyl acrylate (1.29)
Natural rubber (NAT)	1.02/0.88	1.0/1.0	1.0/1.9	20-ethylene dibromide (1.15)
Polyvinyl chloride (PVC)	1.11/0.92	1.01/0.97	1.04/0.98	20-1-octanol (0.78) 1 h-1-octanol (0.90)
Polyvinyl alcohol (PVA)	1.25/0.91	1.17/0.91	1.18/0.82	
Polyethylene (PE)	1.0/0.88	1.13/0.84	1.07/0.99	20-ethylene dibromide (1.26)
Viton (VIT)	1.10/0.89	1.10/0.94	1.08/0.96	1 h-dioxane (1.23) 4 h-dioxane (1.21)
Neoprene rubber (NEO)	1.05/0.99	1.0/1.0	1.0/0.94	20-ethylene dibromide (1.38) 20-cyclohexanone (0.63)

NOTE—The ranges are expressed relative to the radius, $R'_A = R_A/R$ for the sake of easier comparison. Thus, a range of 1.21/0.87 for NR means the maximum R_A for a penetrant with breakthrough time less than that specified is 1.21 times 5.1 or 6.17. Correspondingly, the minimum R_A for a penetrant that did not breakthrough at a time less than that specified is 0.87 times 5.1 or 4.44. These ranges of outliers are not unusual when working with solubility parameters. Special situations where some special problem is suspected are given under the remarks. It appears that the solubility parameters for ethylene dibromide should be reviewed for possible reassignment, for example. With this type correlation, even the worst data fits (BR = 4 h) allow the reasonable estimation of performance for many untested penetrants.

The data here are for the correlations where limits have been placed on the molecular volumes being considered.

The smaller molecules excluded from these correlations typically include methanol, ace-tonitrile, acetic acid, and carbon disulfide. The larger molecules excluded from these cor-relations typically include dioctyl phthalate, oleic acid, tricresyl phosphate, the dibutyl phthalate. These two groups of molecules can not directly compared with each other in this context. There are those who will question this apparently arbitrary nature of limiting the molecular size ranges of the correlations. It can be only said that considering solubility parameters as a single correlating variable for the behavior of all molecular sizes is simply not the best approach. One can not on the basis of solubility parameters alone compare the permeation of a penetrant of molecular volume say 80 cm^3/mol with that of a penetrant of say 300 cm^3/mol. With the same solubility parameters, their permeation behavior will be vastly different as confirmed both by experimental data and the improved data fits for the correlations.

There were no thickness data for a large number of penetrant/membrane systems in the data collection. One could have estimated breakthrough times at a common film thickness, but this would also introduce another source of variation, and was not done.

There is considerably more variation in breakthrough data at the shorter times than at the longer times. This is presumed due to experimental problems. This significantly reduced the number of systems included for polyvinyl chloride, for example.

The correlations can be expected to be improved if the data were limited to single mem-brane sources. The groupings here included data from many manufacturers, and the additives they use differ. This can be the reason for lower data fits in some cases, for example, for butyl rubber.

Some specific comments to the individual membrane types are discussed in the following paragraphs.

Nitrile Rubber

The nonpenetrating larger molecules in particular are the cause of poor data fits when all data points are considered. No penetrants having molar volumes above 123 have broken through after 20 min. This limit is 177.5 at 1 h. Those penetrants with larger molecular volumes cause a considerable reduction in the data fit in this case.

Butyl Rubber

The correlations for 20 min and 1 h are good, but that for 4 h is quite poor. The errors in the 4-h correlation are not systematic, and no reason for this is obvious, except as just noted.

Natural Rubber

Practically all penetrants had breakthrough times less than 4 h. The only two that did not were ethylene glycol and dioctyl phthalate. Elimination of the latter because of high mo-lecular size yields a perfect correlation.

Polyvinyl Chloride

General durability for polyvinyl chloride is not good. Correlations for 1 h and 4 h are high because only a few liquids are included that do not penetrate within these times, respectively. These liquids have high solubility parameters.

Polyvinyl Alcohol

The correlations are useful and can not be improved by selectively limiting the data sets as in other cases. One can perhaps interpret this as a more chemically resistant barrier rather than a primarily physical or diffusion resistant barrier. Variations in the content of water in the systems can lead to deviating results because of its severe softening effect. Some systems had widely different breakthrough times reported from different sources.

No molecules having molecular volumes larger than 104 penetrated at 20 min or 1 h although this did not contribute significantly to lower the data correlation coefficient.

Polyethylene

There was considerable variation in some of the reported data. At 4 h, only three non-penetrating liquids were included, which makes the high correlation coefficient less meaningful.

Neoprene

Penetrants having either very large or very small molecules cause poor data fits when all data points are considered. Surprisingly good correlations for 1 h and 4 h are found when these extremes are not considered. The preceding discussions about diffusion, solubility, and molecular size are strongly supported by these results.

Conclusion

Solubility parameters for a number of typical glove materials have been determined from their breakthrough times for a large number of penetrants. The correlations used to establish these parameters can be used for estimating the performance (breakthrough times) of the polymers for penetrants where experimental permeation data are not yet available. Greater differences in solubility parameters for the glove and penetrant imply lower penetrant uptakes, lower diffusion/permeation rates, and better protection. Such correlations can be also used to locate questionable experimental data (data that do not seem to fit) as experience in other areas has confirmed. A computer printout in order of increasing molecular volume and that includes a large data file (about 240 permeants) gives an overview that maximizes the potential usefulness of these correlations.

Acknowledgment

Funding for this study was generously provided by the Dr. and Mrs. A. N. Neergaard Foundation (Denmark). The help of Susan M. Hansen for assistance with the computer is also acknowledged.

References

[1] Hansen, C. M., doctoral thesis, Denmarks Tekniske Højskole, Danish Technical Press, Copenhagen, 1967.

[2] Hansen, C. M. and Beerbower, A., "Solubility Parameters" *Encyclopedia of Chemical Technology*, Supplementary Volume, 2nd ed., Wiley, New York, 1971, pp. 889–910.

[3] Barton, A. F. M., *Handbook of Solubility Parameters and Other Cohesion Parameters*, CRC Press Inc., Boca Raton, FL, 1983.

[4] Hildebrand, J. and Scott, R., *The Solubility of Nonelectrolytes*, 3rd ed., Reinhold Publishing Corp., New York, 1949.

[5] Hildebrand, J. and Scott, R., *Regular Solutions*, Prentice-Hall Inc., Englewood Cliffs, NJ, 1962.
[6] Hansen, C. M. and Wallström, E., *Journal of Adhesion*, Vol. 15, 1983, pp. 275–286.
[7] Blanks, R. F., and Prausnitz, J. M., *Industrial and Engineering Chemistry, Fundamentals*, Vol. 3, No. 1, 1964, pp. 1–8.
[8] Hansen, C. M., *Journal of Paint Technology*, Vol. 39, No. 505, 1967, pp. 104–117.
[9] Pimentel, G. C. and McClellan, A. L., The Hydrogen Bond, W. H. Freeman and Co., San Francisco, Chapter 7, 1960.
[10] Böttcher, C. J. F., *Theory of Electric Polarization*, Elsevier, New York, Chapter 5.
[11] Crank, J., *The Mathematics of Diffusion*, Oxford, 1957.
[12] Hansen, C. M., *Industrial Engineering Chemistry, Fundamentals*, Vol. 6, No. 4, 1967, pp. 609–614.
[13] Hansen, C. M., *Polymer Engineering Science*, Vol. 20, No. 4, 1980, pp. 252–258.
[14] Skaarup, K. and Hansen, C. M., *Polymer Engineering Science*, Vol. 6, No. 4, 1980, pp. 259–263.
[15] Hansen, C. M., *Journal of Applied Polymer Science*, Vol. 26, 1981, pp. 3311–3315.
[16] Beerbower, A. and Dickey, J. R. in *Transactions*, American Society of Lubrication Engineers Transactions Vol. 12, 1969, pp. 1–20.
[17] Henriksen, H. R., "Materials for Protective Gloves" (in Danish), Report No. 8, Direktoratet for Arbejdstilsynet, Copenhagen, 1982; exists in English translation.
[18] Perkins, J. L. and Tippit, A. D., *Journal, American Industrial Hygiene Association*, Vol. 46, No. 8, 1985, pp. 455–459.
[19] Christensen, U. L. and Banke, O., *Dansk Kemi*, Vol. 3, 1982, pp. 70–76; in Danish.
[20] Barton, A. F. M., *Pure and Applied Chemistry*, Vol. 57, No. 7, 1985, pp. 905–912.
[21] Harrington, E. C., Jr., *Industrial Quality Control*, Vol. 10, No. 10, April 1965, pp. 494–498.
[22] Forsberg, K. and Olsson, K. G., "Riktlinjer för val av kemiskyddshandskar," (Guidelines for Selecting Chemical Protective Gloves), Föreningen Teknisk Företagshälsovård (FTF), Stockholm, 1985; in Swedish.

Alan P. Bentz[1] and Clare B. Billing, Jr.[1]

Determination of Solubility Parameters of New Suit Materials

REFERENCE: Bentz, A. P. and Billing, C. B., Jr., **"Determination of Solubility Parameters of New Suit Materials,"** *Performance of Protective Clothing: Second Symposium, ASTM STP 989,* S. Z. Mansdorf, R. Sager, and A. P. Nielsen, Eds., American Society for Testing and Materials, Philadelphia, 1988, pp. 209–218.

ABSTRACT: The U.S. Coast Guard set out to determine three-dimensional solubility parameters for several new materials. The final approach employed three measures of responses to exposure of material to each of 56 solvents of diverse solubility parameters. The intent was to identify the best measure for this purpose in terms of ease of conducting the test and its statistical reliability. The responses measured were: visual observation of changes, volume increase, and weight gain. Weight gain proved to be the best measure of the three, providing the greatest amount of information with the least experimental error. Tests were conducted on: butyl rubber, chlorobutyl rubber, "three-ply" chlorinated polyethylene, Viton A (copolymer), Viton B (terpolymer), and Teflon. Tests conducted for 24 h gave essentially the same results as those conducted for one week. No three-dimensional solubility parameter could be obtained for Teflon, because of its minimal interaction throughout the solubility parameter space.

KEY WORDS: protective clothing, three-dimensional solubility parameter, butyl rubber, chlorobutyl rubber, chlorinated polyethylene, Viton A, Viton B, Teflon

The purpose of this paper is three-fold: first, to report solubility parameter data on some new materials; second, to evaluate three measures of solvent effects; and third, to provide a reasonable approach as a guide to those about to embark in the measurement of solubility parameters.

In the development of the new U.S. Coast Guard Chemical Response Suit, a testing program was instituted for the purpose of evaluating candidate suit materials for permeation, penetration, and degradation. These tests, including the solubility parameter studies, were designed to enhance our basic understanding of the behavior of various polymeric materials when exposed to hazardous chemicals. The goal was to predict behavior of untested materials from a knowledge of their chemical structures and solubility properties, as well as to specify changes in existing materials to improve their properties.

One outstanding success in this area is the work by Henriksen [1] in which he used the concept of solubility parameter to modify disposable polyethylene gloves so that they could be used by industrial workers for protection against epoxy resins.

Background

Permeation is envisioned as diffusion through a membrane at the molecular level. The energy holding the molecules in the liquid phase directly affects their ability to permeate,

---

[1] Project manager and senior statistician, respectively, U.S. Coast Guard Research and Development Center, Groton, CT 06340-6096.

since their cohesive energies must be overcome for the molecules to enter the matrix and permeate individually. Materials will dissolve when the solvation energies exceed the forces holding the polymer matrix molecules together, unless the polymer molecules are too large because of chain lengths or cross-linking. In those cases, the polymer will swell and exhibit a weight gain.

Hildebrand and Scott [2,3] advanced the concept of solubility parameter for correlating the solubilities of various polymers. There were numerous developments leading to Hansen's postulate [4] that the energy of evaporation of the solvent, ΔE, is the sum of the energies arising from the dispersion forces, ΔE_d, the polar forces, ΔE_p, and the hydrogen bonding forces ΔE_h. Dividing ΔE by molar volume gives the cohesive energy density. The solubility parameter, δ, is defined as the square root of the cohesive energy density for each solvent

$$\delta = \left(\frac{\Delta E}{V_m}\right)^{1/2}$$

where $\Delta E = \Delta E_d + \Delta E_p + \Delta E_h$.

The three-dimensional solubility parameter is defined as a vector of magnitude, δ, with components δ_p, δ_h, and δ_d, derived from the energies arising from the three types of molecular forces. The solubility parameter of a given substance, therefore, can be visualized as a fixed point in three-dimensional space. Hansen [4] conjectured that the closer the solubility parameters of two substances lie within the three-dimensional system, the greater their affinity and similarity of response to other substances. (Note that substances with similar magnitudes of δ may, or may not, be close in three-dimensional space, since the magnitude represents only the distance from the origin.)

Hansen found that a sphere could be defined in solubility parameter space for each polymer such that when exposed to solvents with solubility parameters lying within the sphere, the polymer would interact (that is, dissolve, swell, etc.), whereas those solvents lying outside had relatively little effect. Figure 1 shows this schematically with two-dimensional projections on Cartesian planes.

In the work reported here, interaction results for each polymer tested by exposure to 56 chemicals were plotted in two-dimensional projections. Circles were constructed to incorporate as many points that show interaction as possible. These experimentally determined circles (shown as dotted-line projections in Fig. 1) were then used to define a sphere, the center of which is the three-dimensional solubility parameter of the polymer being tested.

The effects of solvent were noted visually, by volume change, and by weight gain. A preliminary evaluation of the relative merits of each type of data was presented by Derrickson at the 1986 Pittsburgh Conference [5], and further details were given by Billing at the Second Scandinavian Symposium [6].

Experimental

The technique used to study the solubility parameter incorporates that of ASTM Standard Test for Solubility Range of Resins and Polymers (D 3132-84) and that of Holcomb [7]. The degree of material solubility in each solvent was determined visually. Weight gain and volume increase (swelling) were determined by simple measurements. The yes/no criteria were selected from examination of the data for positive interaction by weight gain and volume increase. Visually observed changes signified a positive solvent interaction, regardless of degree of change.

Each candidate material was tested in triplicate, by exposing 4 by 1 cm pieces of the material to 15 mL of liquid solvent in a test tube. The test tubes were placed in a rack and

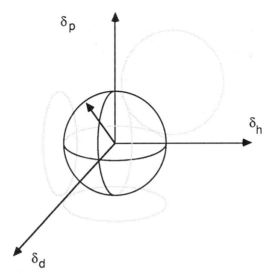

FIG. 1—*Conceptual diagram of three-dimensional solubility parameter showing sphere and its radius for a given material.*

tumbled end-over-end at 6 rpm (to give thorough mixing) for 24 h. A few tests were also conducted for seven days to determine whether extended times showed an increase of interaction in borderline cases.

Six materials were tested: butyl rubber (one of the standard Coast Guard suits), chlorinated polyethylene (CPE), chlorobutyl rubber, Viton A[2] (copolymer), Viton B[2] (terpolymer), and CHEMFAB Teflon[3] (used in making Challenge 5100 material).

The 55 chemicals used by Holcomb [7] and water were tested against each material (all tests were conducted at room temperature). Chemicals of 17 functional group classes were represented as shown in Table 1. The chemicals were selected by Holcomb to give as wide

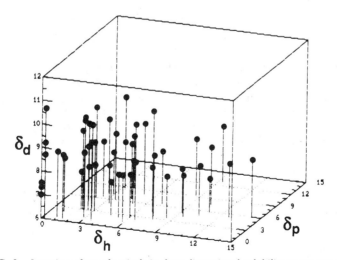

FIG. 2—*Location of test chemicals in three-dimensional solubility parameter space.*

[2] Registered trademark of E. I. duPont de Memorus and Company, Inc.
[3] Registered trademark of Chemical Fabrics Corporation.

TABLE 1—*Chemical classes tested (number in each class).*

Alcohols (8)	Ethers (4)
Aldehydes (2)	Halocarbons (5)
Alkanes (3)	Heterocyclic Compounds (1)
Amides (4)	
Amines (2)	Ketones (6)
Anhydrides (1)	Nitriles (2)
Aromatic Amines (1)	Nitro Compounds (3)
Aromatic Hydrocarbons (3)	Polyols (4)
Esters (4)	Sulfur Compounds (2)

a range as possible of solubility parameter in each of the three dimensions (polarity, hydrogen bonding, and dispersion). The three-dimensional distribution of the 55 liquids is shown in Fig. 2. (The value of the hydrogen bonding parameter for water is off the scale of the figure, so is not plotted.)

The three-dimensional schematic plot of Fig. 1 shows the projections of all points from the solubility parameter sphere onto two-dimensional planes. Hanson found that, by doubling the values of the dispersion component, the region of interaction was made more nearly spherical. We achieved the same result by expanding the δ_d axis. For each tested material, the 55 points were plotted for each of the three measures of interaction, showing whether or not interaction occurred.

The yes/no criteria for the weight gain and volume increase measures of interaction were determined for each material from the statistical characteristics of the measurements. This included the variation of the three replicates and the distribution of measurements for the 56 chemicals. For each material, except Teflon, the weight or volume increase determined to be a "significant" interaction was approximately 10% of the maximum value encountered by each measure for that material. Teflon had such small weight gain and volume increases overall, that the criteria for a significant interaction represented a much higher percentage of the total range.

Results

Typical results are shown in Fig. 3(*a* through *c*), where the filled diamonds indicate interaction as measured by a significant weight gain, and the open diamonds indicate no significant weight gain. This figure shows the three planar projections for butyl rubber from the weight increases during 24-h exposure. These projections, as previously discussed, also show circles constructed manually to incorporate as many filled points as possible. (Note that some open diamonds appear within the circular projections, but may not lie within the sphere. They may actually lie in front of or behind the sphere.) The centers of the circles give the coordinates for the center of the sphere in three-dimensional space, and the radius (size of sphere) for butyl rubber.

TABLE 2—*Estimate of solubility parameter for test materials (in $(J/cm^3)^{1/2}$).*

Material	δ	δ_p	δ_h	δ_d	Radius
Butyl rubber	18	0.8	0.8	18	7.6
Chlorobutyl rubber	18	1.8	0.6	18	7.2
CPE	20	8.2	4.3	18	9.6
Viton A	22	11.5	6.8	18	9.0
Viton B	22	9.8	8.6	18	7.4
CHEMFAB Teflon	...	...	...	...	...

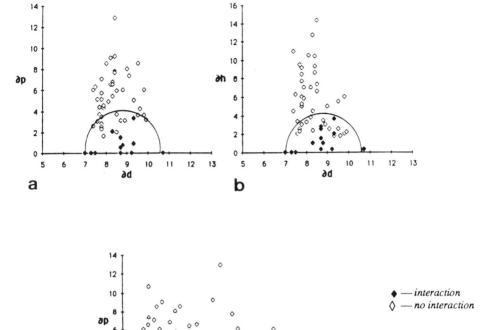

FIG. 3—*Two-dimensional projections of solubility parameters based on weight gain measurements of butyl rubber:* (a) *polarity versus dispersion,* (b) *hydrogen bonding versus dispersion, and* (c) *polarity versus hydrogen bonding.*

Of the three projections, the most useful is that showing the polarity and hydrogen bonding components. The dispersion parameter gives less information, since its range of values is relatively small, and there is scatter throughout the range of interaction and noninteraction. Figures 3c, 4a, and 4b show the polarity versus hydrogen bonding parameters for butyl rubber as determined by weight gain, volume increase, and visually observed changes, respectively. It is apparent that the results are similar, but the visually observed changes show significantly more outliers than the other changes. The outliers were found to include observations of curling that may or may not be solubility related effects. The figures also show the radius of the sphere in which solvents interact with the polymer. Weight gain shows the most well-defined interaction region and is used in the remaining illustrations.

Figure 5 shows corresponding plots for weight gain of Viton A for the polarity parameter versus the hydrogen bonding, Fig. 5a, and dispersion, Fig. 5b, parameters. It can be readily seen that the sphere is shifted from that for butyl rubber. Figure 5b shows the interacting chemicals spanning the entire range of the dispersion parameter.

A special Teflon prepared by CHEMFAB gave results as shown in Figure 6 (a and b). Both figures show a scatter throughout the three-dimensional space of points that indicate positive and negative interactions interspersed so that no circle could be drawn to delineate them. The implications of this could be: (1) the tests were not conducted long enough for

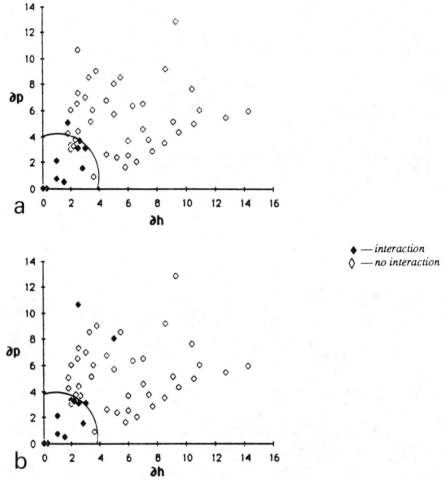

FIG. 4—*Two-dimensional projections for butyl rubber showing plarity versus hydrogen bonding parameters for* (a) *volume change and* (b) *visually observed changes.*

the Teflon to gain sufficient weight; (2) the crystallinity in the Teflon material introduces a new factor that is not accounted for by the three-dimensional solubility parameter; and (3) permeation through Teflon by most chemicals appears not to be dependent upon solubility-type interactions, but rather molecular size and shape limited diffusion.

The solubility parameters estimated for each material using weight gain data are shown in Table 2. They were originally computed in $(cal/cm^3)^{1/2}$ from the raw data shown in the figures, then multiplied by 2.046 to tabulate the final results in SI units $(J/cm^3)^{1/2}$ (Tables 2 and 3). The cited values are the center and radius of the sphere that contain solubility parameters for most of the test chemicals with positive interactions. Table 2 shows that we selected the mid-range for the dispersion parameter. Our present method could not estimate this parameter accurately, because of the small range of dispersion parameter values of the test chemicals and the scatter of interaction/noninteraction throughout the range. Thus, our results effectively are reduced to two-dimensional solubility parameters (that is, circles rather than spheres). Another point is that, although both Vitons have the same overall solubility

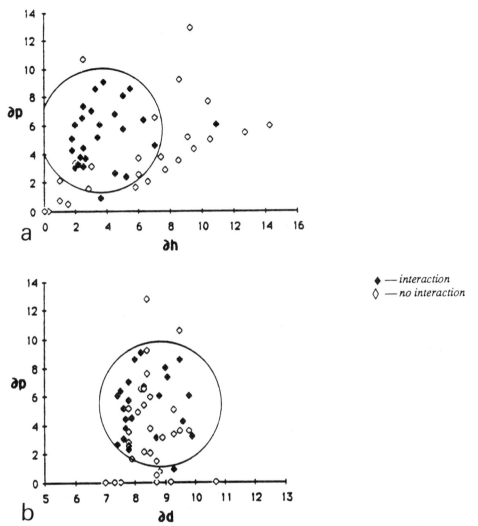

♦ — *interaction*
◊ — *no interaction*

FIG. 5—*Two-dimensional projections of solubility parameters from weight gain of Viton A:* (a) *polarity versus hydrogen bonding and* (b) *polarity versus dispersion.*

parameter (distance from origin), they are in different spatial locations, and the smaller radius of Viton B indicates that it would be unaffected by more chemicals than Viton A.

Volume increase measurements always showed significantly higher variations than the other methods. Volume changes also were not consistent. For instance, Viton A in ketones, nitriles, or acetates may swell lengthwise in one or two tests and increase thickness in the third. There was no way of predicting which dimension would show change. Another problem was the loss of solvent after exposure. In some instances as the solvent evaporated, the length of the test piece would decrease; in others it did not. The thickness measurements for CPE posed another problem. In halocarbons, solvents pockets would form, the material would tend to split into layers, or pieces would etch off, all causing a variation in thickness. As a result of these problems, the volume measurements would have to be reviewed along

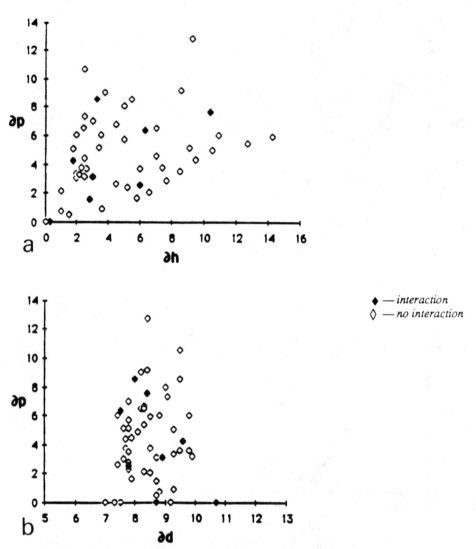

FIG. 6—*Two-dimensional projections of solubility parameters from weight gain measurements of CHEMFAB Teflon:* (a) *polarity versus hydrogen bonding and* (b) *polarity versus dispersion.*

with the notations of physical change in order to be considered a valuable or even useful tool for determining the solubility parameter.

In Table 3, we have compared our solubility parameter values, expressed in $(J/cm^3)^{1/2}$ for butyl rubber and Viton with selected literature values [8–10]. The comparison was to reassure ourselves that the results we obtained for unreported materials were reasonable. The most recent value, obtained by Perkins [11] for butyl rubber is virtually identical to ours, since his value of zero for the polarity and hydrogen bonding components "is really 0.0 to 1.0." Perkins' results for Viton [10] are also very close to ours. This similarity was not surprising, since we were using Perkins' approach. The Beerbower data [8] are much older (1969), and both his and Henriksen's [9] were determined with different, though generically the same, materials.

TABLE 3—*Comparison of values (in $(J/cm^3)^{1/2}$).*

Data Source	δ_p	δ_h	δ_d	Radius
Butyl rubber				
Coast Guard	0.8	0.8	18.0	7.6
Perkins [11]	0	0	18.0	...
Beerbower and Dickey [8]	2.3	3.3	16.0	...
Henriksen et al. [9]	5.9	6.5	19.1	6.7
Viton A				
Coast Guard	11.5	6.8	18.0	9.0
Perkins and Tippitt [10]	10.6	6.1	17.0	8.6

The weight gain and volume increase measurements were compared statistically as shown in Table 4. Confidence intervals for the percent increases were calculated from their variation in the triplicate tests. The weight gain measurements exhibit greater precision, as shown by their smaller confidence intervals. The only exception is CHEMFAB Teflon, which showed a low degree of interaction as measured by either method.

A measure of the amount of information available from each method in determining interaction regions in solubility parameter space is also shown in Table 4. The listed ratios of the variation in responses among the different chemicals to the error variation, represents how well each method can differentiate each material's interaction with the 56 chemicals. They were calculated by dividing the pooled variances of each measurement among all chemicals by the pooled variances within triplicates of the same chemical. The results in Table 4 show that more information is derived from weight gain measurement than from volume increase. Little information is provided for CHEMFAB Teflon by either measurement.

The seven-day experiments conducted for Viton A and butyl rubber provided no obvious advantage over the one-day experiments. The measured values were in most cases slightly higher, but the interaction regions derived were similar. Therefore, subsequent experiments were conducted for only 24 h.

Acknowledgments

The authors are indebted to Jennelle Derrickson who conducted the exposure tests and did the visual observations, weight gain, and volume measurements, and reported the initial results at the 1986 Pittsburgh Conference [5]. The authors are also indebted to Vincent A. Bastecki for assistance in computer calculations of volume changes, and in graphics for the two-dimensional projections.

TABLE 4—*Variation of weight and volume gain measurements.*

Material	Confidence Interval, $\alpha = 0.01$		σ_b^2/σ_w^2	
	Weight	Volume	Weight	Volume
Butyl rubber	7.0	31.3	104.2	2.9
Chlorobutyl rubber	37.7	128.2	29.5	16.0
CPE	30.7	43.5	57.6	20.2
Viton A	13.0	80.0	130.6	20.1
Viton B	13.5	51.4	70.8	27.4
CHEMFAB Teflon	2.7	1.5	5.2	3.6

References

[1] Henriksen, H. R., "Development of Protective Gloves Against Epoxies and Solvents," *Proceedings*, Scandinavian Symposium on Protective Clothing Against Chemicals, Copenhagen, Denmark, 26–28 Nov. 1984.

[2] Hildebrand, J. and Scott, R., *Solubility of Non-electrolytes*, 3rd ed., Reinhold Publishing Corp., New York, 1949.

[3] Hildebrand, J. and Scott, R., *Regular Solutions*, Prentice-Hall, Inc., Englewood Cliffs, NJ, 1962.

[4] Hansen, C. M., "The Three Dimensional Solubility Parameter and Solvent Diffusion Coefficient," Danish Technical Press, Copenhagen, Denmark, 1967.

[5] Derrickson, J. L., Man, V., Billing, C. B., Jr., and Bentz, A. P., "Permeation of Protective Clothing Materials 3. Comparisons of Methods to Determine Solubility Parameter and Examination of Variables Affecting Breakthrough Time," Abstracts of the 1986 Pittsburgh Conference and Exposition on Analytical chemistry and Applied Spectroscopy, No. 079, Atlantic City, NJ, March 1986.

[6] Billing, C. B., Jr., Derrickson, J. L., and Bentz, A. P., "Evaluations of Methods for Determining Solubility Parameter of Protective Clothing Materials," *Proceedings*, Second Scandinavian Symposium on Protective Clothing Against Chemicals and Other Health Risks, Stockholm, Sweden, Nov. 1986.

[7] Holcomb, A. B., "Use of Solubility Parameters to Predict Glove Permeation by Industrial Chemicals," M.Sc. thesis, University of Alabama in Birmingham, 1983.

[8] Beerbower, A. and Dickey, J. R., "Advanced Methods for Predicting Elastomer/Fluids Interaction," *Transactions*, American Society of Lubrication Engineers, Vol. 12, 1969, pp. 1–20.

[9] Henriksen, H. R., Styhr Petersen, H. J., and Madsen, J. O., "Solubility Parameters and Permeation Resistance (Tightness)," Report from the Institute for Chemical Industries, The Technical University of Denmark on Project 1982–39, 1986.

[10] Perkins, J. M. and Tippit, A. D., "Use of Three-Dimensional Solubility Parameter to Predict Glove Permeation," *Journal*, American Industrial Hygiene Association, Vol. 46, Aug. 1985, pp. 455–458.

[11] J. L. Perkins, private communication, University of Alabama in Birmingham, 1987.

Nader Vahdat[1]

Permeation of Polymeric Materials by Chemicals: A Comparison of 25-mm and 51-mm ASTM Cells

REFERENCE: Vahdat, N., **"Permeation of Polymeric Materials by Chemicals: A Comparison of 25-mm and 51-mm ASTM Cells,"** *Performance of Protective Clothing: Second Symposium, ASTM STP 989,* S. Z. Mansdorf, R. Sager, and A. P. Nielsen, Eds., American Society for Testing and Materials, Philadelphia, 1988, pp. 219–225.

ABSTRACT: Chemical permeation measurements were made for six chemical/polymer systems using the 51-mm (2-in.) ASTM cell and the smaller version of the cell (25 mm [1 in.]), and the results were compared. An open loop system with gas flow rates of 1000 mL/min (33.8 fl oz/min) (ASTM cell) and 250 mL/min (8.45 fl oz/min) [25 mm (1 in.) cell] was used to determine the breakthrough times and steady-state permeation rates. For each material/ chemical pair, at least four experiments were conducted in each cell. The 25-mm (1-in.) cell, compared with the ASTM cell, showed a longer breakthrough time and a lower permeation rate. However, statistical treatment of the results demonstrated that the variation of breakthrough times and permeation rates between the two cells was insignificant and that the two cells were equivalent.

KEY WORDS: chemical protective clothing, permeation rate, breakthrough time, ASTM permeation cell, equivalency

Workers involved in the production, use, and transportation of hazardous chemicals can be exposed to numerous compounds capable of causing harm upon contact with the human body. In order to evaluate dermal exposure to potentially hazardous chemicals, permeation measurements are made. Two important parameters are obtained: breakthrough time and steady-state permeation rate. Breakthrough time serves as an estimation of the protection provided by protective clothing. The steady-state permeation rate permits quantitation of the amount of hazardous chemical to which the skin is exposed through clothing. The ASTM Test Method for Resistance of Protective Clothing Materials to Permeation by Liquids or Gases (F 739-85) uses a commercially available test cell. The present study compared the ASTM cell with a 25-mm (1-in.) cell which is essentially a smaller version of the ASTM cell. Six polymer/chemical systems were studied. Breakthrough times and steady-state permeation rates were determined for both cells, and the results were compared.

Experimental

The protective clothing materials, manufacturers, and nominal thicknesses (measured thickness average rounded to two significant figures) are given in Table 1. Specimens were cut from the palm of a glove, except the butyl specimen, which were cut from a sheet of the material. The chemicals from Fisher Scientific were used without additional purification.

[1] Professor, Chemical Engineering Department, Tuskegee University, AL 36088.

TABLE 1—*Materials tested.*

Material	Average Thickness, mm	Supplier
Neoprene	0.53	Edmont (No. 29-870)
PVC	1.68	Edmont (No. 3-318)
NBR[a] Solvex	0.58	Edmond (No. 37-165)
Butyl	0.53	NASA

1 mm = 0.039 in.
[a] NBR = nitrile butadiene rubber.

The ASTM test cell consists of a two-chambered cell for contacting the specimen with a hazardous liquid on the specimen's normal outside surface. The 51-mm (2-in.) cell is constructed of two sections of straight glass pipe, each nominally sized to 51 mm (2 in.) diameter. The section that is designed to contain the collecting medium is 35 mm (1.37 in.). The volume of challenge chamber is 45 mL (1.52 fl oz), and the volume of collecting medium chamber is 100 mL (3.38 fl oz). The 51-mm (2-in.) cell is constructed of two sections of straight pipe, each normally sized to 25 mm (1 in.) diameter. The challenge side is 22 mm (0.86 in.) in length and has a volume of 15 mL (0.5 fl oz). The collecting medium chamber is 27 mm (1.06 in.) in length and has a volume of 15 mL (0.5 fl oz).

A procedure similar to that previously reported was employed [1–3]. Breakthrough values are dependent on limit of detection, limit of quantitation, specimen area, and rate of flow through cell. The flow rate (Q) through the ASTM cell should be set constant for all comparison, and Q in the alternate cell should be set as

$$Q_{ALT} = Q_{ASTM} \times A_{ALT}/A_{ASTM} \qquad (1)$$

where

A = specimen surface area exposed,
ALT = alternate cell, and
ASTM = ASTM cell.

The specimen surface area exposed in the ASTM cell is four times the area in the 25-mm (1-in.) cell; therefore, the flow rate in the ASTM cell should be also four times the flow rate in the 25-mm (1-in.) cell. Nitrogen, with a flow rate of 1000 mL/min (33.8 fl oz/min) [for 51-mm (2-in.) cell] and 250 mL/min (8.45 fl oz/min) [for 25-mm (1-in.) cell], was used as the collection medium. Discrete specimens [0.5 mL (0.0169 fl oz)] were taken from the sampling chamber. Analysis was performed using a Perkin-Elmer Sigma 3B gas chromatograph with Apezion column. The sensitivity of the gas chromatograph for toluene, trichloroethane, and perchloroethylene in nitrogen was 0.1 to 0.2 µg/mL (1×10^{-7} to 2×10^{-7} oz/fl oz), and for methylene chloride in nitrogen was 2.0 µg/mL (2×10^{-7} oz/fl oz). For each material/chemical pair, at least four experiments were conducted in each cell. The experiments were continued until a steady-state permeation was achieved. The steady-state permeation rates were determined by [3–5]

$$J = \frac{FC_s}{A} \qquad (2)$$

where

F = flow rate of nitrogen, mL/min,
C_s = steady-state concentration of hazardous chemical in the collection medium, μg/mL,
J = steady-state permeation rate, μg/cm^2/min, and
A = surface area of protective material exposed to the chemical, cm^2.

Statistical Treatments

For each material/chemical and test cell combination, the following quantities were computed [6,7]
1. Mean

$$X = \frac{\sum X_i}{n} \tag{3}$$

2. Standard deviation

$$S = \sqrt{\frac{\sum (X_i - X)^2}{n - 1}} \tag{4}$$

3. Coefficient of variation

$$CV(\%) = 100(S/X) \tag{5}$$

where X_i ($i = 1, 2, 3 \ldots n$) is the measured value for the ith specimen (breakthrough time or permeation rate) and n is the number of specimens tested in each permeation cell.
4. Pooled standard deviations for each cell

$$S_p = \sqrt{\frac{\sum S_j^2}{K}} \tag{6}$$

where S_j ($j = 1, 2, \ldots K$) is the standard deviation of the jth material and K is the number of materials.
5. Absolute percent difference of means of breakthrough time and permeation rate

$$d(\%) = (100) \left[\frac{X_{1'j} - X_{2'j}}{X_{2'j}} \right] \tag{7}$$

For comparison of means, a t-test will be used. The acceptable difference for means is $\pm 12.5\%$. If one assumes a CV for both cells of 6%, and uses a specimen size of 4 for each cell test level, this difference can be detected with an α-error of 0.01 and β-error of 0.1.

For comparison of variances, an F-ratio will be used. Since CV_{ASTM} values are thought to be small, increasing these in the alternate cell by a factor of 1.7 (variance ratio would be about 3) would not seem significant provided that the means t-test was not significant. Therefore the ratio of variances ($S_{P\text{-ALT}}^2/S_{P\text{-ASTM}}^2$) should be less than 3 for the two cells to have equivalent precision.

TABLE 2—*Chemical permeation data, 25-mm (1 in.) cell.*

System	Category	Run				
		1	2	3	4	5
Toluene-neoprene	breakthrough time, min	12	10	10	10	10
	permeation rate, $\mu g/cm^2/min$	2050	2100	2200	2150	2000
	nominal thickness, mm	0.48	0.51	0.48	0.58	0.51
Methylene Chloride-PVC	breakthrough time, min	12	10	12	10	...
	permeation rate, $\mu g/cm^2/min$	7150	6700	6500	6000	...
	nominal thickness, mm	1.78	1.75	1.65	1.60	...
Trichloroethane-neoprene	breakthrough time, min	24	26	26	24	24
	permeation rate, $\mu g/cm^2/min$	2200	2250	2100	2400	2250
	nominal thickness, mm	0.56	0.53	0.53	0.56	0.53
Perchloroethylene-PVC	breakthrough time, min	44	40	40	40	...
	permeation rate, $\mu g/cm^2/min$	550	610	500	600	...
	nominal thickness, mm	1.68	1.62	1.60	1.68	...
Trichloroethane-NBR[a]	breakthrough time, min	116	104	117	126	...
	permeation rate, $\mu g/cm^2/min$	525	480	525	500	...
	nominal thickness, mm	0.58	0.56	0.63	0.58	...
Trichloroethane-butyl	breakthrough time, min	104	106	106	104	...
	permeation rate, $\mu g/cm^2/min$	120	145	120	128	...
	nominal thickness, mm	0.51	0.53	0.53	0.51	...

1 mm = 0.039 in.
[a] NBR = nitrile butadiene rubber.

TABLE 3—*Chemical permeation data, 51-mm (2 in.) cell.*

System	Category	Run				
		1	2	3	4	5
Toluene-neoprene	breakthrough time, min	12	10	10	10	10
	permeation rate, $\mu g/cm^2/min$	2500	2550	2540	2350	2000
	nominal thickness, mm	0.48	0.48	0.53	0.51	0.56
Methylene chloride-PVC	breakthrough time, min	8	10	8	10	...
	permeation rate, $\mu g/cm^2/min$	8200	8160	8150	7200	...
	nominal thickness, mm	1.78	1.83	1.85	1.80	...
Trichloroethane-neoprene	breakthrough time, min	24	20	20	24	22
	permeation rate, $\mu g/cm^2/min$	2500	2450	2650	2500	2550
	nominal thickness, mm	0.56	0.53	0.48	0.53	0.46
Perchloroethylene-PVC	breakthrough time, min	36	36	36	40	...
	permeation rate, $\mu g/cm^2/min$	720	640	680	800	...
	nominal thickness, mm	1.62	1.60	1.57	1.65	...
Trichloroethane-NBR[a]	breakthrough time, min	114	100	115	105	...
	permeation rate, $\mu g/cm^2/min$	480	550	600	570	...
	nominal thickness, mm	0.56	0.58	0.56	0.58	...
Trichloroethane-butyl	breakthrough time, min	103	104	104	104	...
	permeation rate, $\mu g/cm^2/min$	154	130	140	130	...
	nominal thickness, mm	0.51	0.51	0.53	0.53	...

1 mm. = 0.039 in.
[a] NBR = nitrile butadiene rubber.

TABLE 4—*Mean* (X), *standard deviation* (S), *and percent coefficient of variation* (PCV) *of permeation parameters.*

System	Cell Size, mm	Breakthrough Time, min			Permeation Rate, $\mu g/cm^2/min$		
		X	S	PCV	X	S	PCV
Toluene-neoprene	25	10	0.9	8.6	2100	79	3.8
	51	10	0.9	8.6	2388	231	9.7
Methylene chloride-	25	11	1.1	10.0	6587	520	7.9
PVC	51	9	1.5	18.0	7927	218	2.7
Trichloroethane-	25	25	2.2	8.8	2240	108	4.8
neoprene	51	22	2.0	9.0	2530	150	5.9
Perchloroethylene-	25	41	1.9	4.6	565	88	15.0
PVC	51	37	2.0	5.4	727	50	6.9
Trichloroethane-	25	116	9.0	7.8	507	22	4.3
NBR[a]	51	108	7.2	6.7	550	51	9.3
Trichloroethane-	25	105	1.1	1.0	128	12	9.2
butyl	51	104	0.8	0.8	138	11	8.2

1 mm = 0.039 in.
[a] NBR = nitrile butadiene rubber.

Results and Discussion

Tables 2 and 3 give breakthrough time, permeation rate, and nominal thickness of specimens for all the runs of the material/chemical systems in the 25 and 51-mm (1 and 2-in.) permeation cells. For toluene-neoprene and trichloroethane-neoprene pairs, five test specimens were tested in each cell. For the other four material/chemical pairs, four permeation tests were carried out in each cell. Table 4 gives the mean, standard deviation, and percent coefficient of variation (PCV) of breakthrough time and permeation rate in both cells. The coefficients of variation are low in both cells; in fact, all of them are less than 10% with the exception of methylene chloride-polyvinyl chloride (PVC) [with coefficient of variation of breakthrough time equal to 10 and 18% in the 25 and 51-mm (1 and 2-in.) cells, respectively], and perchloroethylene-PVC [with coefficient of variation of permeation rate equal to 15% in the 25-mm (1-in.) cell]. This indicates that the PVC material used for the permeation test is not sufficiently homogeneous or uniform.

The pooled standard deviations for the cells are given in Table 5. Because of the non-uniformity in PVC material, the pooled standard deviations were calculated twice, once for all the six polymer/chemical pairs and once for only four pairs (not considering the two pairs containing PVC material). The pooled standard deviations for breakthrough time are

TABLE 5—*Comparison of pooled standard deviations.*

Parameter	S_p, 25-mm (1-in.) Cell	S_p, 51-mm (2-in.) Cell	$\dfrac{S_p \ 25\text{-mm (1-in.)}}{S_p \ 51\text{-mm (2-in.)}}$
Breakthrough time	3.9[a](4.7)[b]	3.2[a](3.7)[b]	1.2 (1.3)
Permeation rate	222[a](68)[b]	146[a](140)[b]	1.5 (.49)

[a] Based on all material/chemical pairs.
[b] Based on four material/chemical pairs.

TABLE 6—*Comparison of means.*

System	Breakthrough Time, Absolute Percent Difference	Permeation Rate, Absolute Percent Difference
Toluene-neoprene	0.0	−12.0
Methylene chloride-PVC	22.0	−16.9
Trichloroethane-neoprene	12.7	−11.4
Perchloroethylene-PVC	12.0	−22.0
Trichloroethane-NBR[a]	6.6	−8.0
Trichloroethane-butyl	2.9	−7.2

[a] NBR = nitrile butadiene rubber.

very close in the two cells. The pooled standard deviation for permeation rate is much smaller in the 25-mm (1-in.) cell (not considering PVC pairs). It may be concluded that the 25-mm (1-in.) cell has an equal or greater precision when compared with the 51-mm (2-in.) cell.

Equation 7 was used to compute the percent absolute difference of means of permeation rate and breakthrough time for each chemical/material system. The percent of differences is less than 12.5% in all cases except for the two PVC/chemical pairs. This was anticipated because of the nonuniformity of PVC material. Therefore, it may be concluded that the two cells are equivalent. Table 6 shows that breakthrough time percent differences are all positive and that permeation rate percent differences are all negative. In other words, the 25-mm (1-in.) cell always gives a longer breakthrough time, and a smaller permeation rate, but the differences are well within the acceptable limits and the 25 and 51-mm (1 and 2-in.) ASTM cells are equivalent. Bentz et al. [8] have compared breakthrough times for 25 and 51-mm (1 and 2-in.) cells in a closed-loop system. They have concluded that the 25-mm (1-in.) cell shows an overall greater precision with a generally higher breakthrough time. This is consistent with the results of the present investigation. The advantage of the smaller cell is that it requires the handling and disposal of one order of magnitude less-hazardous chemicals. On the other hand, its disadvantage is that the specimen surface area exposed to chemical is so small [5 cm² (0.775 in.²) compared with 20 cm² (3.10 in.²) for the 51-mm (2-in.) cell] that it is difficult to get a representative specimen from the material.

Acknowledgment

This work was supported by the National Aeronautics and Space Administration, Kennedy Space Center, under Contract No. NAG-10-0019.

References

[1] Dillon, I. G. and Obasyui, E., "Permeation of Hexane through Butyl Nomex," *American Industrial Hygiene Association Journal,* Vol. 46, No. 5, May 1985, pp. 233–235.
[2] Vahdat, N., "Permeation of Polymeric Materials by Toluene," *American Industrial Hygiene Association Journal,* Vol. 48, No. 2, Feb. 1987, pp. 155–159.
[3] Vahdat, N., "Permeation of Protective Clothing Materials by Methylene Chloride and Perchloroethylene," *American Industrial Hygiene Association Journal,* Vol. 48, No. 7, July 1987, pp. 646–651.
[4] Stampfer, J. F., McLeod, M. J., Betts, M. R., Martinex, A. M., and Berardinelli, S. P., "Permeation of Eleven Protective Garment Materials by Four Organic Solvents," *American Industrial Hygiene Association Journal,* Vol. 45, No. 9, Sept. 1984, pp. 642–654.

[5] Berardinelli, S. P., Mickelson, R. L., and Roder, M. M., "Chemical Protective Clothing: A Comparison of Chemical Permeation Test Cells and Direct Reading Instruments," *American Industrial Hygiene Association Journal,* Vol. 44, No. 12, Dec. 1983, pp. 886–889.
[6] Hoel, P., *Elementary Statistics,* Wiley, New York, 1960.
[7] Sterling, T. D. and Pollack, S. V., *Introduction to Statistical Data Processing,* Prentice-Hall, Englewood Cliffs, NJ, 1968.
[8] Derrickson, J. L., Billing, C. B., Man, V. L., Bastecki, V. A., and Bentz, A. P., "Permeation of Protective Clothing Materials Comparison of Methods to Determine Solubility Parameter: Examination of Variables Affecting Breakthrough Time," presented at the 1986 Pittsburgh Conference on Analytical Chemistry and Applied Spectroscopy, Paper 079, Atlantic City, NJ, 1986.

Clare B. Billing, Jr.,[1] *and Alan P. Bentz*[1]

Effect of Temperature, Material Thickness, and Experimental Apparatus on Permeation Measurement

REFERENCE: Billing, C. B., Jr., and Bentz, A. P., **"Effect of Temperature, Material Thickness, and Experimental Apparatus on Permeation Measurement,"** *Performance of Protective Clothing: Second Symposium, ASTM STP 989,* S. Z. Mansdorf, R. Sager, and A. P. Nielsen, Eds., American Society for Testing and Materials, Philadelphia, 1988, pp. 226–235.

ABSTRACT: Despite the use of a standard test method, variations in experimental conditions make comparison of permeation results difficult. Studies which quantify the effect of several variables on chemical permeation and its measurement were conducted in order to improve the interpretation of permeation test results. The variables studied included temperature, material thickness, experimental setup (cell size, open versus closed loop), and detector sensitivity. Both theoretical and experimental approaches were taken. Fickian diffusion principles were used to characterize permeation mathematically and model the effect of the above variables. Permeation experiments were conducted for several material/chemical combinations using different-sized permeation cells and several cell temperatures. Also, permeation through several thicknesses of fluorinated ethylene-propylene [FEP (Teflon)] was measured. The 2.5-cm-diameter (1-in.) ASTM-type cell provided greater precision than the standard 5-cm (2-in.) cell, but with longer breakthrough times. Breakthrough times were seen to be profoundly affected by temperature, material thickness, and detector sensitivity.

KEY WORDS: protective clothing, chemical permeation, permeation testing, breakthrough time, Viton, Teflon, chlorobutyl rubber

The objective of conducting standard permeation tests is to obtain measurements which allow direct comparison of different protective clothing materials and challenge chemicals and which are representative of the properties of the materials in actual use. Clearly, this objective is compromised by variation in both test procedures and conditions of protective clothing use. Experimental variables such as test cell temperatures, experimental setup (closed versus open loop), sampling frequency, cell type, and detector sensitivity make comparison of results difficult unless conditions are tightly controlled. Other variables such as ambient temperature, material thickness and uniformity, degree of contamination, and presence of mixtures make questionable whether the standard test results are truly representative of real-world conditions.

With the above objective in mind and cognizant that wide variations in permeation results are a consequence of varying test techniques, ASTM Committee F23 on Protective Clothing established the Test Method for Resistance of Protective Clothing Materials to Permeation by Liquids or Gases (F 739-85) to quantify the permeation of liquids through protective clothing materials under conditions of continuous contact [1]. Consequently, this method

---

[1] Senior statistician and project manager, respectively, Chemistry Branch, U.S. Coast Guard Research and Development Center, Groton, CT 06340-6096.

has been widely used for obtaining chemical resistance data for materials used under a wide range of field conditions. Despite the use of a standard test method, however, difficulties in comparing test results and applying them to the field remain. This has been particularly true with measurement of breakthrough times which, although commonly used, is extremely sensitive to variations in analytical detection limits of the test method.

In order for permeation test results to be better interpreted, given the above limitations, a series of studies is reported which quantify the effect of several of these variables on permeation and its measurement. In particular, the variables examined are temperature, material thickness, cell size, and detector sensitivity. Differences in interpreting results from open- versus closed-loop experimental setups are also considered. Both theoretical (based on Fickian diffusion principles) and experimental approaches (using actual permeation test results) are presented.

Permeation Theory

Permeation is the molecular process by which chemicals move through protective clothing materials. The mechanism of permeation is usually envisioned as including absorption of individual molecules of the chemical into the exposed surface of the material, molecular diffusion through the material matrix along a concentration gradient, and desorption of the chemical from the inside surface. The mathematical formulation of molecular diffusion through a plane sheet of material has previously been well described [2] and applied to chemical permeation of polymeric materials [3–6]. The mathematical theory of diffusion is based on Fick's first law, which states that the permeation rate of a chemical through a unit area of material is proportional to its concentration gradient through the material. The proportionality constant D is the diffusion coefficient, which may be a function of chemical concentration and therefore time. In many cases, however, D is treated as a constant. If D is constant, the one-dimensional differential equation of diffusion (Fick's second law) can be expressed as

$$\frac{\partial c}{\partial t} = D \frac{\partial^2 c}{\partial x^2}$$

where c is the concentration of the diffusing chemical. This equation is derived from Fick's first law by considering diffusion through a volume element.

If it is assumed that chemical permeation through protective clothing material obeys Fick's laws, the diffusion equation can be solved to obtain the permeation rate and chemical mass permeated as a function of time. The following assumptions and conditions are applied:

1. The diffusion coefficient D is constant (independent of solvent concentration in the material).
2. Swelling of the material and degradation by the permeating chemical are negligible.
3. At time zero, one surface is exposed to the chemical and immediately reaches a constant concentration.
4. The concentration of chemical on the unexposed side of the material is kept negligible compared to that of the exposed side.

Solving the above differential equation by Laplace transformation techniques, we find the total amount of chemical which has permeated a unit area of the material at time t

$$m_t = \frac{\overline{D}c_1 t}{a} + \frac{2c_1 a}{\pi^2} \sum_{n=1}^{\infty} \frac{(-1)^n}{n^2} [1 - \exp(-\overline{D}n^2\pi^2 t/a^2)]$$

where

c_1 = concentration of chemical on the exposed side,
a = material thickness, and
$\overline{D}$ = assumed constant diffusion coefficient (integral value of the diffusion coefficient over the range of chemical concentrations encountered).

At large values of t (after long exposure), the exponential terms in the above equation become small, such that amount of permeated chemical versus time can be expressed by

$$m_t \approx \frac{\overline{D}c_1 t}{a} - \frac{c_1 a}{6}$$

This is the permeation equation at steady state where the total amount of chemical permeating increases linearly with time (constant permeation rate). From the equations for m_t, the concentration of challenge chemical in the collection side of a closed-loop permeation test cell can be expressed as a function of time into the test by

$$c_t = \frac{m_t A}{V}$$

where A is the cross-sectional area of the exposed material and V is the volume of the collection side of the cell.

The preceding equations model the permeation curve obtained from testing a material against a challenge chemical. Figure 1 is a plot obtained from the equations of the concentration versus time. The curve has a transient portion of increasing permeation rate, followed by steady state. If the limit of detection for the permeating chemical is c_d, then the time the chemical is first "detected" on the collection side is shown on the curve as time t_b. This is defined as the breakthrough time of the chemical through the material. The lag time τ is defined as the extrapolation of the steady-state portion of the permeation curve to zero concentration. The value for this can be found by setting $m_t = 0$ in the above steady-state

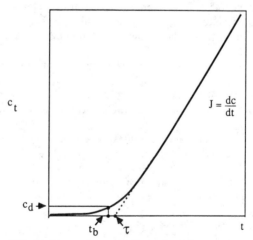

FIG. 1—Permeation curve for a closed-loop system.

equation and solving for time

$$\tau = \frac{a^2}{6\overline{D}}$$

An equation for the permeation rate of the chemical, J_t, as a function of time, per unit area of material may be also obtained from the diffusion equation

$$J_t = \frac{\overline{D}c_1}{a} \left[1 + 2 \sum_{n=1}^{\infty} (-1)^n \exp(-\overline{D}n^2\pi^2t/a^2) \right]$$

At large t, the exponential terms vanish, leaving a constant (steady-state) permeation rate.

The concentration of chemical on the collection side of an open-loop permeation test cell is expressed by

$$c_t = \frac{J_t A}{F}$$

where F is the flow rate on the collection side of the cell. The chemical concentration on the collection side of an open-loop system at steady state is obtained by substituting the steady-state expression for J_t, giving

$$c_t \approx \frac{\overline{D}cA}{aF}$$

The breakthrough time of a chemical through a material, as measured by the standard permeation test method, can be modeled by the above permeation equations. It is defined as the time t at which the concentration c_t is detected in the collection side of the cell. If detection occurs prior to the beginning of steady-state permeation, breakthrough time occurs both in the linear term and in the exponential term within the infinite series of the permeation equation. Thus, no simple function can be derived relating breakthrough time of a given chemical/material pair to the variables of material thickness, detector sensitivity, or cell size.

Cell size in the closed-loop configuration is expressed as the ratio of the surface area of the exposed material to the volume of the collection side of the permeation cell. For an open-loop setup, volume is replaced in the expression by collection medium flow rate.

Since no simple analytical expression is derived to predict the effect on breakthrough time of variation in the above variables, simulations of the permeation equation must be conducted for the desired set of experimental conditions. Results of such simulations have been used by us and others [4] to relate breakthrough times under different conditions. One result of such work is that the exact relationship between breakthrough time and any variable depends on the value of the other variables. For example, the curve relating breakthrough time to material thickness would vary with the cell size or detector sensitivity. Obviously, this can complicate the interpretation and application of standard permeation test results.

Although breakthrough time is commonly used, steady-state permeation rate is perhaps a better measurement for comparing test results. The permeation rate J_t at steady state is a simple inverse function of material thickness and, if measureable, is independent of cell size and analytical detection limit. However, limitations to the practical use of steady-state permeation rate are the possible long experimental times required to reach steady state and the large chemical concentrations which sometimes must be measured. Also, information important in protective clothing use, such as the time required for accumulation of an appreciable amount of permeated chemical, is not always provided. Dangerous amounts of chemical may have permeated well before steady-state permeation has been reached. In such cases, breakthrough time or mass permeated by a given time is a better measure of permeation.

Permeation Experiments

A series of permeation tests was conducted using ASTM F 739-85 to quantify the effect of several variables experimentally on measured breakthrough time. Results of breakthrough time measurements were obtained for several material/chemical combinations under conditions of different cell sizes, cell-bath temperatures, and material thicknesses.

Cell Size and Temperature

The ASTM standard permeation test method requires the use of a standard commercially available cell with specified dimensions. Smaller cells than this 5-cm-diameter (2-in.) ASTM cell are more desirable, however, because they require handling and disposing of significantly smaller amounts of toxic material (about 0.1). Before shifting to a smaller cell, however, it is important to evaluate its performance relative to the standard size. Several studies have been reported comparing the ASTM cell with different designs of smaller cells [7,8]. The smaller (2.5-cm-diameter [1-in.]) cells evaluated are marketed by the manufacturer of the standard ASTM cell.

Our experience from previous permeation testing and that of others [9] is that temperature has a profound effect on chemical permeation. It is critical to know the magnitude of temperature effects in order to assess the impact of temperature variations on permeation measurements and to relate results from standard tests to field conditions which may exhibit a wide variety of temperatures. It is important to know, for example, whether or not breakthrough might occur within a given time at an elevated temperature, when it did not at lower temperatures.

Changes in temperature may have an influence on permeation by several mechanisms. Increased temperatures may increase the concentration of the challenge chemical adsorbed onto the material surface by increasing solubility of matrix by the chemical or by increasing the vapor pressure of the chemical. The rate of the diffusion step may also increase with temperature following an Arrhenius equation type of relationship [9–11]. Temperature, therefore, will exhibit its effect on breakthrough time and permeation rate through the diffusion coefficient (D) and challenge surface concentration (c_i) terms of the permeation equations. The expected effect would be a complex logarithmic-like relationship between breakthrough time and cell temperature.

To measure the effects of cell size and temperature on breakthrough time, we conducted permeation experiments using two types of Viton[2]/chlorobutyl rubber laminate material. Closed-loop tests were conducted at a temperature of 25°C for three challenge chemicals using both the 5 and 2.5-cm (2 and 1-in.) ASTM cells and at three temperatures for methylene chloride using both types of cells. Table 1 gives the mean and standard deviation of breakthrough times for each combination of conditions. An analysis of variance (ANOVA) on results from the methylene chloride tests with Viton/chlorobutyl rubber on polyester shows both cell size and temperature to have significant effects on breakthrough times. The interaction term was not significant, suggesting independence of the two effects.

---

[2] Registered trademark of E. I. duPont de Nemours and Company, Inc.

TABLE 1—*Comparison of breakthrough times obtained using 5- and 2.5-cm (2 and 1 in.) ASTM-type permeation cells.*

Chemical	Cell Temperature °C	5 cm (2-in.) (min)			2.5 cm (1-in.) (min)		
		n	t_b	s_t	n	t_b	s_t
Tetrahydrofuran[a]	25	2	17.9	2.7	2	18.4	7.6
Methylene chloride[a]	25	4	21.8	1.4	4	18.5	1.4
Ethyl acetate[a]	25	5	11.2	9.2	4	25.0	0.4
Methylene chloride[b]	25	6	17.5	6.6	5	23.8	2.0
Methylene chloride[b]	15	2	58.5	0.7	3	64.7	0.3
Methylene chloride[b]	35	4	7.4	5.5	2	13.3	0.4

[a] Material tested was Viton/chlorobutyl with cotton.
[b] Material tested was Viton/chlorobutyl with polyester.

Inspection of Table 1 shows breakthrough times in general to be slightly longer with the small cell. ANOVA using the entire data set shows this cell effect to be significant.

Table 2 summarizes the comparison of the two cell types. Overall mean breakthrough times and pooled standard deviations are given from the above set of permeation experiments. Although the standard deviations given in Table 1 for the individual test combinations vary greatly, no systematic trend is indicated, and the number of data points for each is small. Pooling of the standard deviations for each cell type is therefore justified. The results show the 2.5-cm (1-in.) cell to have greater precision than the ASTM standard 5-cm (2-in.) cell. An F-test on the two variances with the appropriate degrees of freedom shows this to be significant at the $p > 0.99$ level.

The lengthening of breakthrough time by the smaller cell, although significant by the analysis of variance, is small and not easily understood. An evaluation of the equations in the previous section indicates that a cell with a greater ratio of exposed surface area to collection medium volume should exhibit higher concentrations at a given time and, therefore, lower breakthrough times. The 2.5-cm (1-in.) cell in our experiments gives longer times despite having an area to volume ratio about 38% greater than the larger cell. This and the lower precision are perhaps due to a greater probability of leakage with the large cell, resulting in outliers of short breakthrough times. When such outliers were tested for and removed from the data set, the difference in mean breakthrough times between cells decreased and the precision (pooled variances) became similar. The effect of replacing specimens with "clean" solution may also have an effect of diluting the smaller collection volume of the small cell to a greater degree and thereby delaying breakthrough detection. The collection volume of the large and small cells were 85 and 15 mL, respectively. The specimen volumes varied between 5 and 100 μL.

Figure 2 shows the effect of temperature on methylene chloride breakthrough times using each cell. The confidence intervals were left off the curve for the 2.5-cm (1-in.) cell for clarity. An analysis of the functional relationship between breakthrough time and temper-

TABLE 2—*Summary comparison of 5 and 2.5-cm (2 and 1 in.) ASTM-type permeation cells.*

	n	Mean t_b, (min)	Pooled σ_t, (min)
5-cm (2-in.) cell	23	20.4	6.2
2.5-cm (1-in.) cell	20	25.8	2.4

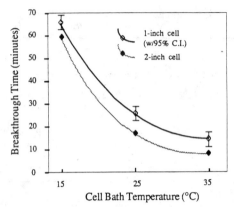

FIG. 2—*Effect of permeation cell size and bath temperature on permeation of methylene chloride through Viton/Chlorobutyl laminate.*

ature yields no definitive simple relationship. However, it is clearly a logorithmic-like relationship, as expected. Because of the complex relationship between breakthrough time and the surface concentration and diffusion coefficient, breakthrough time cannot be expected to vary with temperature in a simple Arrhenius manner. Steady-state permeation rate measurements likely would yield a better defined Arrhenius relationship predictable from theory. However, since the main objective of our testing was to determine breakthrough times, steady-state permeation rate was not measured for most tests.

Figures 3 and 4 show the effect of temperature on permeation measurements for two more material/chemical combinations. Both the curves for methyl ethyl ketone versus Viton/ chlorobutyl rubber and those for methylene chloride against Challenge 5100,[3] a tetrafluoroethylene (Teflon)[2]/Nomex[2] composite (the Coast Guard's new suit material) exhibit a similar relationship to that shown in Fig. 2. A wider range of temperatures was tested using the 2.5-cm (1-in.) cell in a closed-loop configuration. The error bars on the curves represent the range of breakthrough times obtained on replicated tests.

Examples of permeation curves obtained from the tests for methylene chloride versus Challenge 5100 material for three cell temperatures are shown in Fig. 5. The portions of the permeation curve shown are early stages, prior to reaching steady state. At the levels shown, the detector response was linear with specimen concentration. The figure shows an

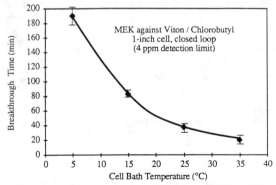

FIG. 3—*Effect of temperature on permeation of MEK through Viton/Chlorobutyl laminate.*

[3] Registered trademark of Chemical Fabrics Corporation.

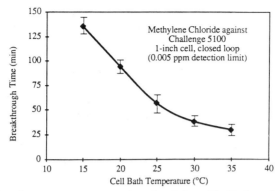

FIG. 4—*Effect of temperature on permeation of methylene chloride through Challenge 5100.*

effect of temperature not only on breakthrough time, but also on the change in permeation rate with time. The curve for the permeation test at 15°C is also a good example of the potentially large effect which might be imposed by the minimum detection limit of the analytical procedure on the measured breakthrough time. If the detection limit of that test were to change by a factor of two, it can be seen that the measured breakthrough time would change by 25 min or more, because the slope of the permeation curve is so low.

Test Material Thickness

The literature has a number of reports on the effect of material thickness on permeation and on methods for normalizing breakthrough measurements to compensate for thickness variations [12–14]. Linear, logarithmic, square, and quadratic relationships have been found or proposed or both.

Simulation of the permeation equation can allow prediction of the effect of material thickness on breakthrough time for material/chemical combinations which have Fickian permeation behavior. In general, the specific function of the effect will be complex and depend on the other permeation variables such as the concentration of the chemical at the exposed surface (which depends on temperature and solubility), test temperature, and cell

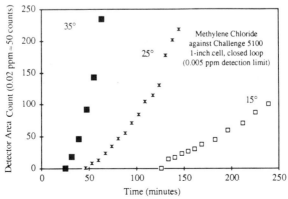

FIG. 5—*Permeation of methylene chloride through Challenge 5100 at several cell bath temperatures.*

size and configuration, as well as detector sensitivity. Predicted effects of thickness on breakthrough time from such simulations will therefore be specific to the particular measurement conditions.

Permeation experiments were conducted for various thicknesses of the fluorinated ethylene-proplyene [(FEP) Teflon] material used as faceplate material for the new Coast Guard suit. Methylene chloride was tested against four thicknesses of the material (1, 5, 10, and 14 mil) and carbon disulfide against all but the 1-mil film. All experiments were conducted using 2.5-cm (1-in.) cells in an open-loop configuration at 25°C (77°F). Each test was conducted at least in triplicate. Figure 6 shows the results of the tests. The functional relationship between breakthrough time and thickness appears to be quadratic. It is similar to that predicted in general from simulations of the permeation equations [4].

Conclusions

These results show the importance of considering, controlling, and reporting the experimental conditions under which permeation tests are conducted. This is critical not only in accurately interpreting and comparing results, but in applying test results to real-world situations. Compound/material combinations that show no detected breakthrough in standard 3-h tests, may in fact permeate within that time under elevated temperatures or permeate if the material used is actually thinner than the test specimens.

We found the smaller ASTM-type cell offers greater precision, in addition to requiring handling and disposal of one-order-of-magnitude less hazardous chemicals. In addition, we found from permeation theory and experimental permeation curves that the analytical detection limit can profoundly affect measured breakthrough time. Cell size and experimental configuration (open versus closed loop), also play an extremely large role in determining breakthrough detection.

Temperature has a quasi-logarithmic effect on breakthrough time, indicating the need for close temperature control and reporting of temperature for all permeation tests. In addition, permeation data should be obtained, or results predicted, for the full range of temperatures that a material might encounter in use.

Both theory and experiments indicate an apparent quadratic increase of breakthrough time with material thickness. The exact functional relationship, however, does not have a simple closed form and varies as a function of other variables such as temperature and analytical sensitivity.

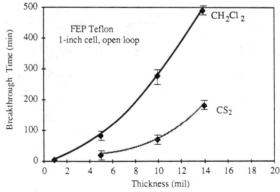

FIG. 6—*Effect of material thickness on permeation of methylene chloride and carbon disulfide through FEP.*

Acknowledgments

The authors express their sincere appreciation to Victoria Man, Vincent A. Bastecki, Larisa Bekman, and Kenneth Williams for their efforts in conducting the permeation experiments.

References

[1] Henry, N. W., III, and Schlatter, C. N., *American Industrial Hygiene Association Journal,* Vol. 42, March 1981, pp. 202–207.

[2] Crank, J., *The Mathematics of Diffusion,* 2nd ed., Clarendon Press, Oxford, U.K., 1975.

[3] Bomberger, D. C., Brauman, S. K., and Podoll, R. T., "Studies to Support PMN Review: Effectiveness of Protective Gloves," report to Environmental Protection Agency under Contract 68-01-6016, SRI International, Menlo Park, CA, Sept. 1984.

[4] Schwope, A. D., Goydan, R., and Reid, R. C., "Breakthrough Time—What Is It?" paper presented at the 2nd Scandinavian Symposium on Protective Clothing Against Chemicals and Other Health Risks, Stockholm, 5–7 Nov. 1986.

[5] Colletta, G. C. et al., "Development of Performance Criteria for Protective Clothing Used Against Carcinogenic Liquids," report to National Institute for Occupational Safety and Health under Contract 210-76-0130, Arthur D. Little, Cambridge, MA, Oct. 1978.

[6] Bhown, A. S. et al., "Predicting the Effectiveness of Chemical-Protective Clothing: Model and Test Model Development," report to Environmental Protection Agency under Contract 68-03-3133, SRI, Birmingham, AL, 1986.

[7] Berardinelli, S. P., Mickelson, R. L., and Roder, M. M., *American Industrial Hygiene Association Journal,* Vol. 44, Dec. 1983, pp. 886–889.

[8] Forsberg, K., "Permeation Testing. Round Robin Test Results from a NORDTEST Project," paper presented at the Scandinavian Symposium on Protective Clothing Against Chemicals, Copenhagen, 26–28 Nov. 1984.

[9] Spence, M. W., "Chemical Permeation Through Protective Clothing Material: An Evaluation of Several Critical Variables," paper presented at the American Industrial Hygiene Conference, Portland, OR, 25–29 May 1981.

[10] Nelson, G. O. et al., *American Industrial Hygiene Association Journal,* Vol. 42, March 1981, pp. 217–225.

[11] Huang, R. Y. M. and Lin, V. J. C., *Journal of Applied Polymer Science,* Vol. 12, 1968, p. 2615.

[12] Holcomb, A. B., "Use of Solubility Parameters to Predict Glove Permeation by Industrial Chemicals," M.Sc. thesis, J. L. Perkins, Advisor, University of Alabama, Birmingham, AL, 1983.

[13] Schlatter, C. N. and Miller, D. J., "Influence of Film Thickness on Permeation Resistance Properties of Unsupported Glove Films," *Performance of Protective Clothing, ASTM STP 900,* R. L. Barker and G. C. Coletta, Eds., American Society for Testing and Materials, Philadelphia, 1986, pp. 22–31.

[14] Sansone, E. B. and Jonas, L. A., *Environmental Research,* Vol. 26, 1981, pp. 340–346.

Norman W. Henry III[1]

Comparative Evaluation of a Smaller Version of the ASTM Permeation Test Cell Versus the Standard Cell

REFERENCE: Henry, N. W., III, "**Comparative Evaluation of a Smaller Version of the ASTM Permeation Test Cell Versus the Standard Cell**," *Performance of Protective Clothing: Second Symposium, ASTM STP 989*, S. Z. Mansdorf, R. Sager, and A. P. Nielsen, Eds., American Society for Testing and Materials, Philadelphia, 1988, pp. 236–242.

ABSTRACT: An evaluation of a smaller version of the ASTM permeation test cell used in ASTM Test Method For Resistance of Protective Clothing Materials to Permeation by Liquids and Gases (F 739-85) was conducted. The smaller test cell was tested using acetone as the challenge liquid chemical and milled neoprene as the test material. The tests were performed using both air (in open-loop mode) and water (in closed-loop mode) as collection media. Results of measurements of both breakthrough time and permeation rate were compared to those obtained using the larger ASTM permeation test cell. Based on statistical evaluation by two-way analysis of variance, the small test cell gave results that were not significantly different at the 5% level from the large ASTM cell.

KEY WORDS: permeation test cell, equivalency, breakthrough time, permeation rate, open-loop and closed-loop systems, analysis of variance, protective clothing

Since the adoption of ASTM Test Method for Resistance of Protective Clothing Materials to Permeation by Hazardous Liquid Chemicals (F 739-81), concern has been raised about the use of alternative permeation test cells rather than the one specified in the method. The major concern has been the preferred use of smaller test cells, so that less hazardous liquid chemical is used for testing. While cell size was originally considered in the development of ASTM F 739-81, the consensus was that a large permeation test cell would allow for a more representative test sample including the possibility of testing seamed areas of protective clothing materials. Consequently, the 5.08 cm (2-in.) diameter glass permeation test cell was specified as a compromise in the permeation test method. This test cell has been used routinely over the past few years for generating much of the reported permeation data on protective clothing materials. However, several laboratories doing permeation tests have developed their own smaller test cells and would like to demonstrate equivalency with the large ASTM cell. One laboratory has already reported data indicating that their small test cell gives comparable results to the large ASTM cell[2]. Efforts to draft a permeation test cell equivalency method have been initiated within Subcommittee F23.30 on chemical resistance, but no method has been adopted to date. The purpose of this study was to evaluate and

---

[1] Research chemist, E. I. duPont de Nemours and Company, Inc., Haskell Laboratory for Toxicology and Industrial Medicine, Newark, DE 19714.

[2] Personal communication of permeation data sent by S. P. Berardinelli of NIOSH, Division of Safety and Research, Morgantown, WV, 25 Feb. 1986.

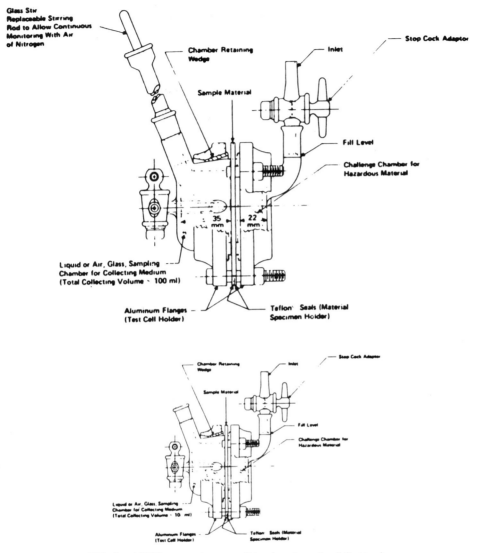

FIG. 1—*ASTM permeation test cell* (top) *and small cell* (bottom).

determine whether a small 2.54 cm (1-in.) diameter test cell is equivalent to the large 5.08 cm (2-in.) ASTM permeation test cell using a standard statistical test method.

Materials and Methods

Three 5.8 cm (2-in.) diameter ASTM permeation test cells (Fig. 1) and three 2.54 cm (1-in.) diameter scaled-down versions of the ASTM cell were used for the evaluation (Peace Labs., Kennett Square, Pennsylvania). Both test cells were evaluated using a sample of 0.40-mm-thick (16 mil) milled neoprene (Fairprene Industrial Products Company Inc. of Fairfield, Connecticut). This is the same stock of neoprene that was used in the original round-robin evaluation of the permeation test method. Reagent-grade acetone was used as the challenge hazardous liquid chemical. Both water and air were used as collection media with both cells to demonstrate equivalency with the closed-loop and open-loop sampling and analytical systems, respectively. Critical orifices were used to maintain an air flow rate of approximately 50 mL/min through the collection media side of both the large and small test cells in the open-loop sampling system. In the closed-loop system, acetone was measured in aliquots of water samples removed at selected time intervals using a gas chromatograph (HP-5880) equipped with a 3 m by 3.2 mm stainless steel column packed with 10% FFAP coated on 80/100 mesh AW Chromosorb W. Acetone was identified by its retention time and quantitatively measured by comparison of peak areas to a standard calibration curve prepared by injecting known concentrations of acetone into the gas chromatograph and measuring peak area.

TABLE 1—*Summary of breakthrough times and permeation rates for the large and small permeation test cells.*

Cell Type	Collection Medium	Breakthrough Time, min	Permeation Rate, mg/min/m^2
Large ASTM	water	20	3057
	(closed-loop)	20	2445
		20	3301
		10	2690
		20	2443
		20	1791
		19	3175
	mean =	18	2700
	standard deviation =	3.7	528
Large ASTM	air	20	2150
	(open-loop)	19	2660
		20	2330
		16	2445
		17	2364
		16	2445
		12	2338
	mean =	17	2390
	standard deviation =	2.9	155
Small cell	water	16	2016
	(closed-loop)	18	2016
		18	2304
		16	1836
		16	2282
		20	2936
	mean =	17	2232
	standard deviation =	1.6	388
Small cell	air	8	2300
	(open-loop)	18	2592
		18	2568
		14	3648
		14	2520
		14	2088
		14	1992
		14	2520
	mean =	14	2529
	standard deviation =	3.1	506

For the open-loop system, with air as the collection medium, acetone standards were prepared in Tedlar[3] bags and a calibration curve prepared by plotting peak area versus concentration of acetone in the air sampling system. The minimum detection limit for acetone in water was 1 µg/mL and in air 1 ppm (volume/volume).

Both test cells were evaluated for equivalency by following the procedure specified in ASTM F 739-85. Breakthrough times and permeation rates were determined for both the large ASTM and small test cell and the results reported. The mean values were then tested for equivalency by using a two-way analysis of variance, where statistical significance was judged at the 5% level.[4]

Results

A summary of the results of breakthrough times and permeation rates for both the large ASTM and small permeation test cells using both air and water as collection media are reported in Table 1. A minimum of six tests were run per permeation test cell and collection medium. The average breakthrough time using the large cell with water as the collection medium was 18 min. The average breakthrough time for the large cell using air as the collection medium was 17 min. Breakthrough times for the small cell were 17 min using water as the collection medium and 14 min using air as the collection medium.

The average permeation rates using water and air as collection media with the large ASTM cell were 2700 and 2390 mg/min/m², respectively. The average permeation rates found with the small permeation test cell were 2232 mg/min/m² using water as a collection medium and 2529 mg/min/m² using air as the collection medium. Plots showing the average concentration of acetone versus time for the large and small permeation test cells using both water and air as collection media are shown in Figs. 2 and 3.

A summary of the statistical results comparing the large ASTM cell versus the small cell using two-way analysis of variance is shown in Table 2. The F and p values for both breakthrough time and permeation rate are reported for cell and collection media differences. The overall cell and collection media differences are reported as the interaction. All the p values are greater than 0.05. Therefore, there is no significant difference between test cells or collection media. Graphs showing a comparison of the large ASTM cell versus the small cell for both breakthrough times and permeation rates are shown in Fig. 4.

Discussion

The results of the evaluation show that there is no statistically significant difference in breakthrough times and permeation rates between the two permeation test cells. The smaller cell, like the larger one, had a maximum coefficient of variation of 22% for breakthrough time and 20% for permeation rate using either water or air as a collection medium. This is comparable to the 25% variation observed and reported in the original interlaboratory testing of the ASTM permeation test, F 739-81, using the 5.08 cm (2-in.) ASTM cell, 0.40-mm (16-mil) neoprene and acetone as the challenge liquid chemical. Although the 2.54 cm (1-in.) cell with air as a collection medium does appear to have a shorter average breakthrough time, the mean value was not statistically significantly different from the mean value of the 5.08 cm (2-in.) permeation cell.

---

[3] Registered trademark of E. I. duPont de Nemours and Company, Inc.
[4] Youden, W. J., "Use of Variance to Compare Averages," *Statistical Methods for Chemists*, Wiley, New York, 1951, pp. 50–58.

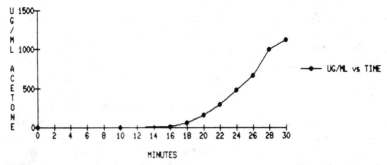

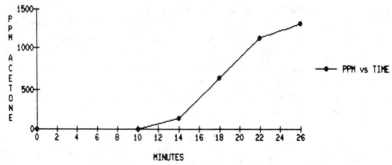

FIG. 2—*Acetone versus neoprene:* (top) *ASTM large cell* (water) *and* (bottom) *ASTM small cell* (water).

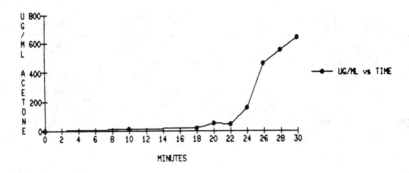

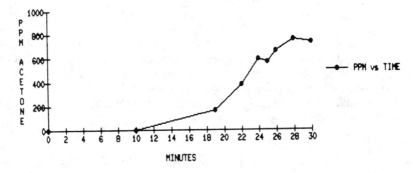

FIG. 3—*Acetone versus neoprene:* (top) *ASTM large cell* (air) *and* (bottom) *ASTM small cell* (air).

TABLE 2—*Summary of statistical results comparing the large ASTM cell versus the small cell using two-way analysis of variance.*

	F Value	P Value[a]
	BREAKTHROUGH TIME	
Cell	3.10	0.0911
Collection medium	3.72	0.0657
Interaction	0.63	0.4353
	PERMEATION RATE	
Cell	1.04	0.3179
Collection medium	0.00	0.9679
Interaction	3.51	0.0733

[a] If $p < 0.05$, then difference is significant.

Despite the four-fold difference in surface area between the two test cells, similar results were obtained. From a practical point of view, the small cell was much easier to work with because a smaller volume (10 mL) of acetone was needed for continuous liquid contact with the neoprene sample compared to 50 mL of acetone needed in the large ASTM permeation test cell. This volume difference could cause both safety and waste disposal problems when working with more hazardous liquid chemicals. One would also expect that the volume difference on the collection medium side to effect sensitivity, since a smaller liquid or air volume would contain a higher concentration of acetone. Apparently, the surface area-to-

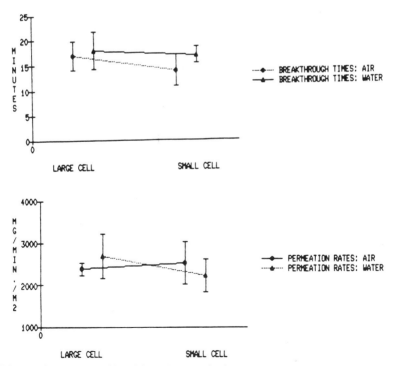

FIG. 4—*Comparison of breakthrough times* (top) *and comparison of permeation rates* (bottom).

volume ratio is not large enough to make a difference, although the smaller cell tends to have shorter breakthrough times, probably because of better mixing in the smaller volume. Obviously, the larger ASTM test cell allows for the evaluation of a more representative sample of clothing material including the possibility of testing seamed areas. However, if the main concern is permeation, the smaller cell gives comparable results as the results of this evaluation have shown.

Gary L. Patton,[1] *Meredith Conoley,*[1] *and Laurence H. Keith*[1]

Problems in Determining Permeation Cell Equivalency

REFERENCE: Patton, G. L., Conoley, M., and Keith, L. H., **"Problems in Determining Permeation Cell Equivalency,"** *Performance of Protective Clothing: Second Symposium, ASTM STP 989,* S. Z. Mansdorf, R. Sager, and A. P. Nielsen, Eds., American Society for Testing and Materials, 1988, pp. 243–251.

ABSTRACT: Permeation tests were conducted to determine if an alternate 1-in. stainless steel cell was equivalent to the ASTM standard 2-in. glass cell. Permeation tests were conducted according to ASTM Test Method for Resistance of Protective Clothing Materials to Permeation by Liquids or Gases (F 739–85). Sheet stock neoprene (0.41 and 0.81 mm) and sheet stock nitrile (0.41 mm) were challenged with *n*-hexane (99%). Seven replicates of each chemical/material were tested for each cell.

Data sets of breakthrough time and permeation rate were compared using a proposed ASTM Cell Equivalency Method, Standard Practice for Determining Equivalency of Optional Chemical Permeation Test Cells to that of the ASTM Cell—Draft, 10 April 1985. This method first determines the equivalence of cells in terms of precision, via an *F* statistic. Equivalence in accuracy (bias relative to the ASTM cell), is then determined via a comparison of absolute percent difference and a calculated *t* statistic.

Analysis of ASTM and alternate cell data demonstrated standard deviations were dependent on mean breakthrough times and mean permeation rates. Therefore, the proposed ASTM procedure utilizing pooled standard deviations could not be used to determine cell equivalency. Further test results to characterize both cells are presented along with alternative statistical methods to evaluate data in cell validation tests.

KEY WORDS: protective clothing, alternate cell, breakthrough time, cell validation, equivalency, permeation, permeation cell, permeation rate

Radian Corporation has patented a permeation cell to be used as an alternative to the ASTM standard permeation cell in systems using a gas collection medium. This study was conducted to validate the Radian Microcell (RMC) as equivalent to the ASTM standard cell.

The Radian Microcell was designed to address three attributes associated with the ASTM cell. These attributes and the RMC changes are listed in Table 1.

Procedure

The ASTM draft cell validation procedure, Standard Practice for Determining Equivalency of Optional Chemical Permeation Test Cells to that of the ASTM Cell—Draft, 10 April 1985, was used as a guide for our experimental design. Specifically, three materials—0.41-mm neoprene, 0.81-mm neoprene, and 0.41-mm nitrile—were challenged with *n*-hexane. Table 2 contains a list of the major sources of variation (error) associated with this study and the solutions used to hold known variation relatively constant. A total of nine replicates

---

[1] Radian Corporation, Austin, TX 78720-1088.

TABLE 1—*Radian microcell versus ASTM cell.*

Attribute	ASTM Cell	Radian Microcell
1. Challenge side volume	approximately 75 to 100 mL	approximately 2 mL
2. Cell material	glass easily broken, service life approximately six months	stainless steel (or other), service life approximately 2 to 5 years
3. Collection side mixing	gas collection stream directed towards material but in diffuse manner	gas collection stream directed across face of material creating a "vortex" directed out of the cell

were generated for each chemical/material/cell type. Two replicates were discarded as outliers when visibly lower values were attributed to incorrect collection flows. Therefore, seven replicates of each chemical/material for each cell type were used, as directed in the draft procedure, for statistical comparison. Tests were performed in groups of three replicates plus one blank for each chemical/material/cell type. One triplicate test per day was performed with tests randomly assigned to working days. Materials were obtained from Fairprene Industrial Products, Fairfield, CT, as sheet stock. Hexane was obtained from Burdick and Jackson.

Each permeation test was conducted according to ASTM Test Method for Resistance of Protective Clothing Materials to Permeation by Liquids or Gases (F 739–85). Data were tabulated and reduced according to the ASTM draft cell validation procedure.

TABLE 2—*Major sources of variation.*

Four major sources of variation (error) were addressed while forming the experimental design. They were:

1. MATERIALS
   Problem—Sheet stock materials are not consistent in thickness or density.
   Solution—(1) A population of swatches was prepared that were of nominal thickness ±0.025 mm. Swatches then were randomly assigned to tests.
   (2) Density problems could not be limited; therefore variation was recognized but accepted.

2. OPERATOR
   Problem—Different operators performing the same tasks on the same or consecutive days would introduce variation to the test results.
   Solution—One operator would perform all tests. All tests were performed according to ASTM F739-85 to minimize operator variation.

3. AUTOMATED TEST SYSTEM
   Problem—(1) Inter-system variation is difficult to quantify.
   (2) Collection flows are different for both cell types.
   Solution—(1) The operator used a single automated test system for all tests to eliminate inter-system variation.
   (2) Collection flows were set to 35 mL/min for the alternate cell and 50 mL/min for the ASTM cell.

4. CELL TYPE
   Problem—Variation in torque load used to tighten cells will create variation in permeation results that is not due to cell type.
   Solution—The operator used torque loads of 40 kg · cm (35 lb in.), but this may have to be investigated as an untested source of error.

TABLE 3—*Statistics for performance evaluation—breakthrough time.*

	ASTM Cell			RMC Cell		
Material	Mean, min	Standard Deviation, min	Coefficient of Variation, %	Mean, min	Standard Deviation, min	Coefficient of Variation, %
0.41-mm neoprene	11	2.1	18.9	14	0.8	5.8
0.81-mm neoprene	37	5.3	14.3	42	2.7	6.3
Nitrile	245	19.5	8	282	25	8.9

Results

Permeation test results are summarized in Table 3 for breakthrough time and in Table 4 for permeation rate. The statistical values presented in these tables were to be used in the validation procedure. Figures 1 and 2 show ASTM and RMC standard deviations plotted against ASTM means for each corresponding material for breakthrough time and permeation rate, respectively. Figures 3 and 4 show ASTM and RMC coefficients of variation versus ASTM cell means for breakthrough time and permeation rate, respectively.

The first validation criterion, that the coefficient of variation for each cell/material combination be less than 20%, was met. A coefficient of variation greater than 20% would have indicated the test material may not have been sufficiently uniform or the test procedure may not have been properly performed. The second validation criterion, demonstration that standard deviations were not dependent on means for both cell types and each material, was not met. Standard deviations dependent on means indicates variation in test results is not consistent between materials and thus data sets and variances should not be pooled. Figures 1 and 2 show increasing standard deviation with increasing ASTM mean value for breakthrough time and permeation rate, respectively, for both cell types. Accordingly, the draft procedure could not be used as written to test for equivalence. However, the draft procedure does allow each material to be evaluated separately if the sample size is increased to retain statistical power and the data is proved to be normally distributed. Coefficients of variation were plotted against ASTM cell means to determine if the relative variation for each material was similar. A line connecting the three data points with a slope not significantly different from zero would indicate the relative variation for each material was similar. However, the percent variation for each material was not similar, as can be seen in Figs. 3 and 4 for breakthrough time and permeation rate, respectively. A comparison of trends for each cell type indicated that the amount of relative variation was not consistent between

TABLE 4—*Statistics for performance evaluation—permeation rate.*

	ASTM Cell			RMC Cell		
Material	Mean, mg/m² · s	Standard Deviation, mg/m² · s	Coefficient of Variation, %	Mean mg/m² · s	Standard Deviation, mg/m² · s	Coefficient of Variation, %
0.41-mm neoprene	36.14	3.34	9.24	25.86	1.35	5.20
0.81-mm neoprene	20.71	1.89	9.12	16.57	1.27	7.68
Nitrile	0.01	0.00	0.00	0.01	0.00	0.00

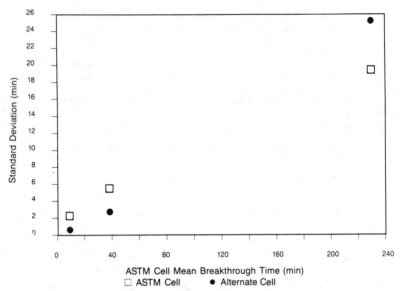

FIG. 1—*RMC and ASTM cell breakthrough time standard deviations versus ASTM means.*

materials. For example, in Fig. 3, the ASTM cell breakthrough time results indicated decreasing relative variation with longer mean breakthrough time, whereas the RMC breakthrough time results indicated increasing variation with longer mean breakthrough time. According to the draft equivalency procedure, the only option with these results is to evaluate the results for each material independently and proclaim equivalency for "ranges" of break-

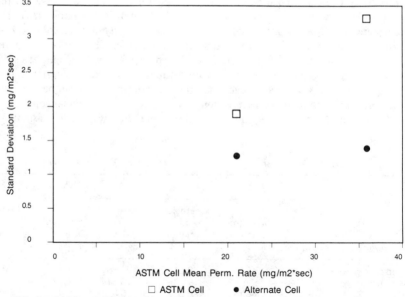

FIG. 2—*RMC and ASTM cell permeation rate standard deviations versus cell means.*

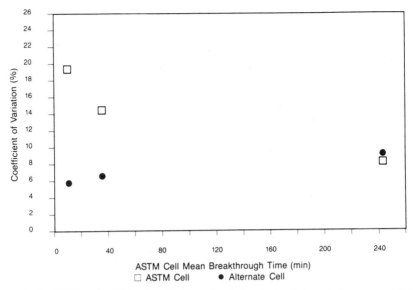

FIG. 3—*RMC and ASTM cell coefficients of variation for breakthrough time versus ASTM cell means.*

through times and permeation rates. In order to use this option, however, we would need to conduct more tests to increase the number of replicates available for each material/cell combination. The larger number of replicates would allow us to retain the statistical power we would have had if we had been able to combine results for all materials and calculate pooled statistics.

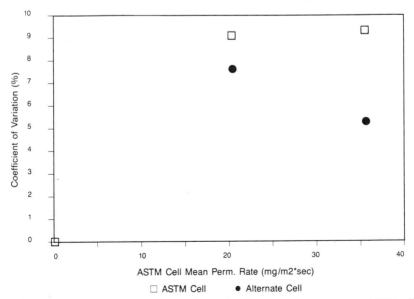

FIG. 4—*RMC and ASTM coefficients of variation for permeation rate versus ASTM cell means.*

Discussion

Since laboratories conducting ASTM permeation tests may utilize different cells, the validation procedure was developed to ensure that differences in inter-laboratory results would not be attributable to different cell types. To determine if this validation procedure was statistically appropriate for permeation data, we reviewed historical data for 0.81-mm neoprene versus acetone. This data represents permeation testing for quality control conducted over the last two years. Analysis of this data set would determine if statistical assumptions for populations could be met for small data sets. Using a pooled standard deviation for statistical testing in the draft procedure requires that variances be similar within test groups. Our results showed that variation (and hence variance and standard deviation) was dependent on means. Our inability to pass this criterion of the draft equivalency procedure for both the ASTM cell and the alternate cell made us question the population distribution from which we were drawing samples (that is, the statistical tests are based on the assumption that the samples come from a normally distributed population). We chose a Radian Microcell data set (490 observations) rather than ASTM cell data because we had evidence that this RMC permeation data was consistent within itself, and more data points existed than for the ASTM cell. We conducted a Shapiro-Wilks "W" test for normality using SAS,[2] a statistical software package, and found that breakthrough time and permeation rate data were not normally distributed at the 95% (α = 0.05) significance level. Frequency distributions of breakthrough time and permeation rate data appeared to represent a pattern such that a transformation could be used to attain normality (see Figs. 5 and 6, respectively). The following data transformations were attempted:

1. Breakthrough time divided by material thickness.
2. Breakthrough time divided by the material thickness squared.

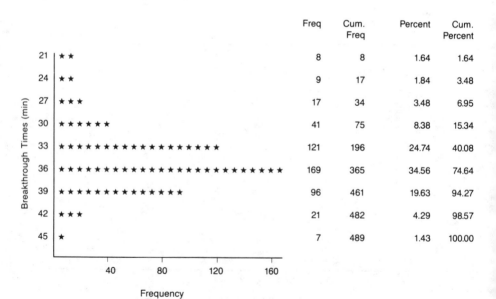

FIG. 5—*Frequency distribution of breakthrough times.*

[2] *SAS® Procedure Guide for Personal Computers,* 6th ed., Cary, NC, SAS Institute Inc., 1985, pp. 341–357.

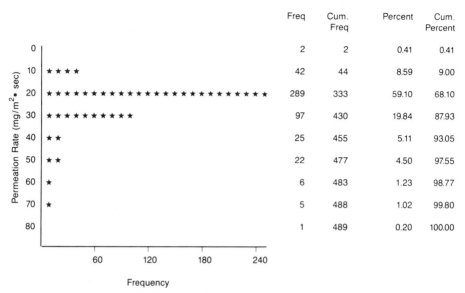

FIG. 6—*Frequency distribution for permeation rate.*

3. Permeation rate divided by material thickness.
4. Permeation rate divided by the material thickness squared.
5. Natural logarithm of permeation rate.
6. Natural logarithm of (permeation rate divided by material thickness).
7. Natural logarithm of (permeation rate divided by material thickness squared).

None of these transformations resulted in an normally distributed data set. Other transformations may be appropriate; however, they should be chosen based on theoretical and empirical permeation data. We then tested material thickness results for normality. Material thickness was not normally distributed even though swatches for these tests were chosen at random from 0.81-mm sheet stock. Figure 7 shows a frequency distribution for material thickness. This may not be a good evaluation of thickness, however. The instruments used to determine thickness measure a relatively large (or macro) section of the material. Since permeation takes place on a micro or molecular level, our thickness measurements may be too gross to be indicative of thickness at permeation sites. Because variations in manufacturing techniques also may lead to sheet stock with irregular thickness variation, truly random sampling of swatches may not be possible without using a much larger sample population.

We conclude that permeation results for any cell, and especially the ASTM cell, need to be fully characterized. Otherwise, cell equivalency tests using parametric statistics, which assume the sample comes from a normally distributed population, as in the draft validation procedure, may be inappropriate due to non-normally distributed data. We suggest the following for future cell equivalency validations:

1. As with the current draft procedure, perform permeation tests with three or more materials that will give results covering a broad range of breakthrough times and permeation rates.
2. Calculate necessary sample size using accepted alpha (α), beta (β), and coefficient of variation levels to standardize accepted error levels for both parametric and non-

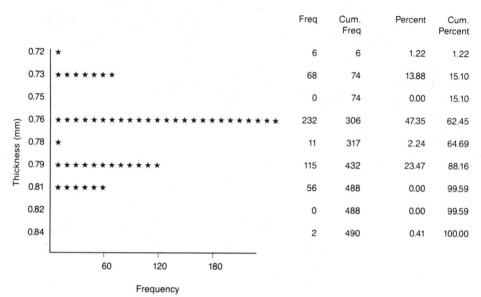

FIG. 7—*Frequency distribution of material thickness.*

parametric methods of analyzing data. An example is:

(a) For parametric statistical analysis: $\alpha = 0.10$, $\beta = 0.10$, known coefficient of variation = 10%, acceptable coefficient of variation = 20%; number of replicates necessary = six replicates.[3]

(b) For non-parametric statistical analysis, the required sample size is difficult to calculate because non-parametric analyses depend on the sample size to determine critical values rather than predetermined α and β levels. We suggest a minimum sample size of six, because the power of a non-parametric test may be critically effected with sample sizes less than six.

This will provide enough replicates to be used to retain statistical power in both types of statistical analyses.

3. Plan to treat materials independently. Although this will result in a large sample size for each material, it will preclude having to perform another round of tests. An additional set of tests under different experimental conditions could result in increased experimental error.

4. Choose a rounding criteria for data that leaves a realistic data value but does not round out all variation in results. This is especially critical for low permeation rates. For example, if all results are rounded up to 0.1 mg/m^2 s for data less than 0.1 mg/m^2 s, data will artificially have no variation between replicates. This diminishes the power of a statistical test to detect differences in populations.

5. Test each material/cell combination's data for normality. The Shapiro-Wilk "W" test has been observed to be excellent for detecting departure from normalty;[4] however, it is cumbersome when analyzing large data sets without a computer. Other tests for normality are the Chi Square and the Kolmogorow-Smirnov goodness-of-fit tests.

---

[3]Davies, O. L., *Design and Analysis of Industrial Experiments,* Hafner Publishing Co., New York, 1956, Table N.

[4] Zar, J. H., *Biostatistical Analysis,* Prentice-Hall, Inc., Englewood Cliffs, NJ, 1974, p. 82.

6. If raw data sets are not normally distributed, examine realistic data transformations such as a logarithmic transformation. If the data can be normalized, the generally more powerful parametric statistics such as the Student's t and the F test can be used in comparisons.

7. If realistic transformations do not normalize data, proceed with non-parametric statistics for comparison of cell types. Non-parametric statistics only assume that (1) samples are independent, and (2) the population is continuous.[5] A possible procedure to follow would be:

   (a) Test for equivalency in location (that is, medians of sample sets are similar) using the Mann-Whitney U or the Wilcoxin test.

   (b) If the cell results for each material are equivalent in location, test for equivalence in scale (that is, dispersion or variability in the sample set is similar). The Mood test has been shown to be a good test of scale with an efficiency of 0.76 relative to the parametric F test (see Footnote 5).

A more complete characterization of permeation results obtained with the ASTM standard permeation test cell can contribute valuable information to a cell equivalency test procedure that evaluates both the alternate cell under consideration and the permeation test procedure used by the researcher. However, cell variation, system variation, and test method variation may be integrated such that a cell equivalency test is valid only for the system and test parameters used in that particular equivalency evaluation.

Conclusions

The current draft cell equivalency test, Standard Practice for Determining Equivalency of Optional Chemical Permeation Test Cells to that of the ASTM Cell—Draft, 10 April 1985, could not be used to evaluate the Radian Microcell relative to the ASTM cell. The variation in test results was shown to be dependent with respect to mean breakthrough time and permeation rate, thus precluding the use of this proposed procedure. Analysis of a data set for a single material/chemical (0.81-mm neoprene versus acetone) indicated that it may be more appropriate to use non-parametric statistics to evaluate equivalency of permeation test cells.

Acknowledgments

The authors would like to thank the following individuals for their contributions: Damon Doss, for conducting the laboratory tests, and Jana Steinmetz, for guidance with the statistical analyses.

[5] Gibbons, J. D., *Non-Parametric Statistical Inference,* McGraw-Hill, New York, 1971, pp. 140, 164, 176.

Karen L. Verschoor,[1] Larry N. Britton,[1] and Ed D. Golla[1]

Innovative Method for Determining Minimum Detectable Limits in Permeation Testing

REFERENCE: Verschoor, K. L., Britton, L. N., and Golla, E. D., "**Innovative Method for Determining Minimum Detectable Limits in Permeation Testing,**" *Performance of Protective Clothing: Second Symposium, ASTM STP 989*, S. Z. Mansdorf, R. Sager, and A. P. Nielsen, Eds., American Society for Testing and Materials, Philadelphia, 1988, pp. 252–256.

ABSTRACT: A method has been developed to determine minimum detectable limits for permeation testing of protective clothing. A syringe pump is used to deliver the chemical of interest into the ASTM standard permeation cell at a very slow rate. A constant low-level concentration is sent to the detector via the same pathway the permeant would travel. Efforts have been made to optimize the method. Validation experiments designed to check the linearity of response have been performed. The syringe pump method for determining minimum detectable limits is applicable to other permeation testing systems.

KEY WORDS: permeation testing, minimum detectable limits, protective clothing, syringe pump, photoionization detection

The protection of workers from exposure to hazardous chemicals is of major concern to industry and government agencies. Protective clothing is currently being evaluated by measuring its resistance to permeation by potentially hazardous chemicals following the American Society for Testing and Materials (ASTM) Test Method for Resistance of Protective Clothing Materials to Permeation by Hazardous Liquid Chemicals (F 739-81). This method is designed to measure both breakthrough times and steady-state permeation rates. The breakthrough time is defined as the elapsed time between contact of the challenge chemical with the test material and the time when the chemical is first detected in the collection media. Breakthrough times are inherently dependent on the sensitivity of the analytical technique used. A more sensitive method will detect the chemical sooner than a less sensitive technique and would therefore give a shorter breakthrough time for the given material and chemical.

An assessment of the duration of protection provided by the protective clothing is given by the measurement of breakthrough times. To give meaning to determined breakthrough times, it is important to report the minimum detectable limits (MDLs) for the permeation test system used. A method using a syringe pump was designed to efficiently and accurately measure MDLs within the apparatus used to perform the permeation testing. This paper outlines the method used, the efforts made to optimize the method, and the results from a series of validation experiments.

[1] Program manager, director of Environmental Sciences Department, and laboratory director, respectively, Texas Research Institute, Inc., Austin, TX 78733.

Experimental Procedure

Equipment

The apparatus used to perform the permeation testing consisted of standard 2-in. or 1-in. glass permeation cells with polytetrafluoroethylene (PTFE) gaskets and a photoionization detector. Stainless steel tubing and short pieces of flexible PTFE tubing allowed a flow of nitrogen to continually sweep through the collection side of the cell to the detector. The photoionization detector was an HNU Model PI-52-02 outfitted with either an 11.7 or 10.2 eV lamp. The response from the detector was recorded on a Houston Instruments strip chart recorder.

A Sage Instruments syringe pump Model 341 was used with an SGE, gas tight, removable-needle, 5-μL glass syringe to pump the chemical of interest. The syringe was outfitted with needles cut from small-diameter vitreous silica tubing. The syringe was modified by the addition of a metal stop near the end of the syringe barrel to facilitate the pump's ability to tightly hold the syringe.

Methods

The permeation testing apparatus was operated by methods consistent with ASTM Test Method for Resistance of Protective Clothing Materials to Permeation by Hazardous Liquid Chemicals (F 739-81). Standard permeation cells were used in which the sample was sandwiched between the challenge side and the collection side of the cell. Nitrogen flowed at 100 cm^3/min into the collection side of the cell, across the sample surface, and out to the photoionization detector in an open loop system. The collection side of the cell was continually monitored for the presence of the challenge chemical over the specified time period.

Minimum detectable limits (MDLs) were determined by pumping the chemical of interest into the collection side of a standard permeation cell at a very slow rate using a syringe pump. The 2-in. cell was modified by the placement of an impermeable aluminum sheet between the challenge and collection sections of the cell. The chemical was filtered prior to filling the syringe using a 0.2 μm disposable filter assembly. The tip of the needle was placed into a specially fabricated glass joint adapter that fits into the glass stoppered joint of the permeation cell (see Fig. 1).

A constant low-level concentration was sent to the detector via the same pathway a permeant would travel. The pump rate could be adjusted from 0.116 μL/h to many higher

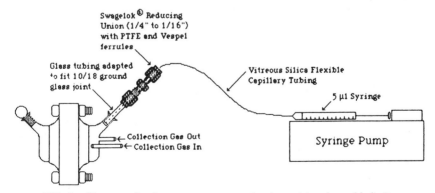

FIG. 1—*Diagram of syringe pump apparatus for determining detectable limits.*

settings. The concentration of the chemical of interest being delivered to the detector was calculated using the following equation:

$$ppm = \frac{d \times MV \times PR}{MW \times F}$$

where d is the density, MV is the molar volume (24 450 (μL/mol)), PR is the syringe pump rate (μL/h), MW is the molecular weight, and F is the nitrogen flow rate (L/h).

The response generated from the determined concentration was used to calculate the MDL. The MDL was subjectively defined as the concentration corresponding to the response that is twice the noise level. The noise level was determined as the long-term fluctuation from the average baseline in a control cell.

Results and Discussion

The syringe pump provides an effective method of delivering a low concentration of chemical to the detector enabling the calculation of MDLs. To determine the accuracy and linearity of the syringe pump method for MDLs, known concentrations of standard toluene gas were introduced into the detection system and compared with the detector response from toluene introduced via the syringe pump (Fig. 2). As expected, the responses generated from standard toluene gas were linear with respect to concentration (square symbols in Fig. 2). Neat toluene delivered into the system by the syringe pump is shown with the triangular symbols in Fig. 2. The lowest concentration, 4.45 ppm, was calculated from the slowest pump rate (0.116 μL/h) and a flow rate of 100 mL/min. Lower levels of toluene (circular symbols) were achieved by diluting the toluene in acetonitrile, a volatile solvent that is not detected by the photoionization detector. Figure 2 illustrates that the calculated concentrations of toluene delivered by the syringe pump are the same as known levels of standard

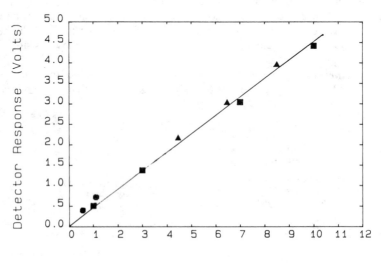

FIG. 2—*Comparison of the responses of toluene introduced via the syringe pump with known concentrations of toluene gas: standard toluene gas concentrations (■); calculated toluene levels from neat toluene introduced with the syringe pump (▲); and calculated toluene levels from toluene diluted in acetonitrile and introduced with the syringe pump (●).*

toluene gas and that the syringe pump method can reproducibly introduce toluene vapors into the permeation test system in a linear fashion.

As shown with the solution of toluene in acetonitrile, the dilution of the chemical of interest in a volatile solvent that is not detected by the method of analysis is an efficient technique to achieve very low concentrations for MDL determinations. For example, for systems using electron capture detectors, 2,2,4-trimethylpentane or other appropriate alkanes would be useful as a dilution solvent. In addition to dilution, the solvents provide an effective method for introducing less volatile and viscous compounds into the system for MDL determination. The highly volatile solvents would act in vaporizing the chemicals that would tend to remain at the tip of the needle in the neat, liquid state.

Chemicals of varying detector sensitivities and volatilities were analyzed using the syringe pump method and were shown to produce linear detector responses. Table 1 outlines the response of these chemicals. In general, relative sensitivities compared to those reported in the literature for the photoionization detector.[2]

Optimization of the Method

The response generated by the slow introduction of the chemical into the permeation cell was not initially a smooth recorder tracing. First attempts were made by placing the tip of the needle directly into the stream of nitrogen entering the collection side of the permeation cell. This resulted in a wildly pulsing response that centered around the expected response. One possible explanation for this varied response is that microdroplets were formed at the tip of the needle. An increased response was produced when the droplets were dispersed

TABLE 1—*Detector response of various chemicals tested by the syringe pump method.*

Chemical	Boiling Point, °C	Response, mV/ppm
Toluene	110.6	590
Styrene	145 to 146	500
Nitrobenzene	210 to 211	450
Benzene	80.1	588
m-Cresol	202	620
Ethyl acetate	77	37
Ethyl ether	34.6	120
Methyl acrylate	70	40
Butyraldehyde	74.8	210
Propionic acid	141.1	14
Propyl alcohol	97.2	28
Carbon disulfide	46.5	91
Dimethylacetamide	164	590
Dimethyl sulfate	188	14
Acetone cyanohydrin	95	15
Vinylidene chloride	31.7	200
Propylene oxide	34.2	48
Allyl chloride	44	26
Vinyl acetate	72 to 73	230
Tetrachloroethylene	121	340
Trichloroethylene	86.7	480
Furfural	161.8	320
Acrolein	52.5	31

[2] Langhorst, M. L., *Journal of Chromatographic Science*, Vol. 19, Feb. 1981, pp. 98–103.

by the force of the nitrogen stream and evaporated. This was followed by a period of lower response while the microdroplet was reforming.

The needle was then placed into the adapter at the glass joint of the permeation cell. This removed the tip of the needle from the turbulent nitrogen stream. It also forced the chemical to diffuse down the glass adapter before entering the outlet stream. The process of diffusion helped to average out minor concentration variations. It was found that the placement of the tip of the needle closer to the nitrogen stream caused a more varied response.

Different diameter capillary tubes were used as needles in the syringe pump system. It was expected that a smaller diameter tubing would decrease the size of the microdroplet formed and help decrease the amplitude of the pulsing response. Tubes ranging from 0.075 to 0.025 mm inner diameter were tried with no apparent effect on response.

The concentration delivered to the detector is strongly dependent on the flow rate of nitrogen to the detector and the flow rate of the chemical of interest into the cell. Any unintentional leaks in the system had a substantial effect on the response.

The syringe itself was a potential source of error. It was essential that a complete seal be formed around the needle. Another problem area was the syringe plunger. If the PTFE tip on the plunger did not fit tightly in the barrel of the syringe, less chemical was delivered to the cell. It was believed that at slow pump rates a portion of the liquid escaped around the tip of the plunger.

Temperature had a strong effect on the response generated. Minor increases in temperature, such as those produced by touching the needle, created a spiked response. The signal would then fall below the expected value before resuming the initial response. This effect could be explained by thermal expansion of the liquid within the barrel of the syringe and the needle itself. Efforts were made to insulate the needle, although this had little effect on improving the response.

The cells and syringe pump were placed in an incubator in an effort to insulate the system. This effectively smoothed the signal. Some pulsing was observed that could be attributed to the turning on and off of the heating element to produce slight fluctuations in temperature. The liquid within the needle and the barrel of the syringe emulated a very sensitive thermometer. The expansion and contraction due to temperature changes altered the rate of delivery. Because of this "thermometer effect," it was critical that temperature be precisely and smoothly maintained.

Conclusion

It is agreed that reported breakthrough times in permeation testing are strongly dependent on the sensitivity of the analytical method used. The standard method, ASTM F 739-81, gives guidelines for performing permeation testing but does not specify the analytical methods or the complete test apparatus. There is a need to compare and correlate results from different systems with different detectors. A universal technique should be employed to meet this objective. The syringe pump method is an effective technique that delivers a known concentration through the same pathway that the permeant would travel. This is a significant advantage. It is possible that the permeant concentration changes in the plumbing between the cell and the detection system due to sorption of the chemical on tubing surfaces and valving. The syringe pump method eliminates the tedious preparation of static gas standard solutions. It also eradicates the error due to surface effects involved in the preparation of standard solutions by series dilutions. The method described in this paper can be employed without modification in any permeation system. The method was developed to work in open loop systems in which the collection medium is a gas.

Rosemary Goydan,[1] Arthur D. Schwope,[1] Robert C. Reid,[2]
S. Krishnamurthy,[3] and Kin Wong[4]

Approaches to Predicting the Cumulative Permeation of Chemicals Through Protective Clothing Polymers

REFERENCE: Goydan, R., Schwope, A. D., Reid, R. C., Krishnamurthy, S., and Wong, K., "**Approaches to Predicting the Cumulative Permeation of Chemicals Through Protective Clothing Polymers,**" *Performance of Protective Clothing: Second Symposium, ASTM STP 989,* S. Z. Mansdorf, R. Sager, and A. P. Nielsen, Eds., American Society for Testing and Materials, Philadelphia, 1988, pp. 257–268.

ABSTRACT: Approaches for predicting the permeation resistance of chemical protective clothing polymers were investigated and assessed for accuracy and applicability to the Premanufacture Notification (PMN) review process of the U.S. Environmental Protection Agency (EPA) Office of Toxic Substances (OTS). The approaches emphasize the prediction of cumulative permeation and include: (1) a predictive model based on Fickian diffusion theory and (2) test methods for directly assessing permeation resistance.

The predictive model is based on refinements of existing theoretical approaches for estimating diffusion coefficients and solubilities from the physical properties of the solute and polymer. Permeation model predictions were compared to well-documented permeation data from the literature. Examples of the model's range of applicability and its limitations are provided. Although the results of the permeation modeling approach are promising, direct testing is the most accurate approach to assess clothing performance for the variety and complexity of chemicals and chemical mixtures that are typical of PMN substances. Consequently, chemical resistance test methods were reviewed for accuracy and applicability. Test methods reviewed include permeation tests, degradation tests, liquid immersion weight change tests, and sorption/desorption methods. Based on this review, a hierarchy is proposed that ranks chemical resistance tests according to their ability to generate data needed to assess PMN clothing requirements.

KEY WORDS: permeation, protective clothing polymers, predictive model, Fick's law of diffusion, solubility, diffusion coefficient, chemical resistance test methods, protective clothing

Section 5 of the Toxic Substances Control Act (Public Law 94-469) requires prospective manufacturers to submit Premanufacture Notifications (PMNs), which are reviewed by the U.S. Environmental Protection Agency (EPA) Office of Toxic Substances (OTS) prior to the manufacture or import of new chemicals. The OTS is permitted only 90 days to review each of the approximately 1800 PMNs submitted annually. While many substances are not subjected to all aspects of the review process, those that are judged potentially toxic require detailed assessments of the potential for their environmental release and human exposure

[1] Chemical engineer and unit leader, respectively, Arthur D. Little, Inc., Cambridge, MA 02140.
[2] Professor Emeritus, Massachusetts Institute of Technology, Cambridge, MA 02139.
[3] Physical scientist, U.S. Environmental Protection Agency, Edison, NJ 08837.
[4] Chemical engineer, U.S. Environmental Protection Agency, Washington, DC 20460.

during manufacture, processing, and end use. If concerns are raised to warrant regulation, engineering controls, work practice restrictions, or protective clothing and equipment are investigated as a means to reduce exposure risks.

The submitter of the PMN often recommends protective clothing as the means to limit dermal exposures. Occasionally, the type of clothing is specified; more often it is not. In either case, OTS must have a means of assessing the exposure reduction provided by protective clothing. Two options are available to OTS for estimating the adequacy of protective clothing: (1) OTS can request that PMN submitters test the clothing materials and submit the resultant data, or (2) OTS can estimate clothing performance using available information and predictive models. Because of the volume of PMNs and the limited time permitted for each review, the development of a reliable model is desirable. If a model is to be used, it must estimate exposure protection using only the limited chemical property data available for PMN substances. If testing is to be specified, OTS must have an awareness of relevant test methods and the limitations of the data that are produced.

The overall assessment of protective clothing requirements must consider the potential health effects of the PMN substance, the probable exposure conditions, and the effectiveness of clothing in limiting exposures. This study focused on predicting the barrier effectiveness of clothing materials. Two general approaches for assessing clothing performance were pursued—one to develop Fick's law predictive models, and a second to identify appropriate chemical resistance test methods and analyze existing performance data generated by these methods.

The criteria used to guide development of a methodology for OTS use are that the approach should (1) be simple for the reviewer to use; (2) be applicable to a wide range of solute/polymer combinations; (3) be truly predictive, that is, apply to new chemicals; and (4) not require data other than those readily available or measured by simple reproducible tests. The methodology should predict the cumulative mass permeated as a function of time and enable estimation of breakthrough times. At a minimum, the methodology should allow classification of the protective clothing performance: not acceptable for limited use, acceptable for limited use, or requires more data to evaluate.

Predictive Models

Permeation theory was used to estimate chemical permeation through clothing materials under continuous exposure to pure substances. The model development effort focused on predicting the cumulative mass-permeated as a function of time and not solely a breakthrough time. The OTS needs more than a single breakthrough time to judge whether clothing performance will be acceptable. The OTS requires estimates of the time at which the cumulative amount permeated reaches a limit of unacceptable human risk. The time when this toxicity limit is reached may be well beyond the time when breakthrough is detected.

Fickian diffusion behavior was assumed and the classic mathematical relationships were used for the estimation of the rate, J, and cumulative amount, Q, of chemical that permeates a polymer film at any time following the initiation of the exposure. These relationships require two key parameters: the diffusion coefficient, D, and the solubility, S. For a planar film of thickness, ℓ, the appropriate solutions to Fick's second law of diffusion with standard initial conditions and simple boundary conditions (no external phase resistances) are

$$J = (DS/\ell) \left[1 + 2 \sum_{n=1}^{\infty} (-1)^n \exp(-\pi^2 n^2 \psi) \right] \qquad (1)$$

and

$$Q = (S\ell) \left\{ \psi + 2 \sum_{n=1}^{\infty} [(-1)^n/(\pi n)^2] [1 - \exp(-\pi^2 n^2 \psi)] \right\} \qquad (2)$$

where

J = permeation flux, $g/cm^2 - s$;
Q = cumulative amount permeated, g/cm^2;
ψ = Dt/ℓ^2, dimensionless;
D = diffusion coefficient, cm^2/s;
S = solubility, g/cm^3;
ℓ = membrane thickness, cm; and
t = time, s.

The detailed derivation of Eqs 1 and 2 is presented elsewhere [1]. Using these equations, one can calculate values for the breakthrough time, t_b, and the steady-state permeation rate, J_∞, for a given chemical/polymer system. As discussed later, the breakthrough time must be defined either on the basis of a minimum detectable permeation rate, J_b, or a minimum detectable cumulative amount permeated, Q_b, for the existing experimental conditions.

Five generic polymers were emphasized: butyl rubber, natural rubber, Neoprene, nitrile rubber, and low-density polyethylene (LDPE). This paper summarizes the progress to date; refinements to the permeation estimation model are continuing.

Solubility Estimation

There are no well-tested, purely theoretical approaches that provide accurate predictions of S for the systems of interest. A theoretical equation of state (EOS) approach shows promise although the technique requires further refinement. Three group contribution approaches were also investigated and the Oishi and Prausnitz approach was determined the most accurate and broadly applicable.

Theoretical Approach—Kumar Equation of State—A statistical mechanics-based lattice model EOS was recently used with success by Kumar et al. to model the phase behavior of supercritical fluids containing polymer molecules [2,3]. A simplified form of the Kumar EOS was used here to estimate chemical/polymer solubilities at room temperature and pressure. The S was predicted using this approach for natural rubber, butyl rubber, and Neoprene. In general, the predictions are accurate to an order of magnitude but tend to underestimate values reported in the literature. An important advantage is that this approach requires only three pure component properties of the solute to estimate S: the molecular weight, vapor pressure, and liquid density at room temperature.

Group Contribution Approaches—Group contribution approaches require only that the generic structure of the solute and polymer be known. Most methods proposed in the literature stem from an activity coefficient correlation method called UNIQUAC (universal quasi-chemical activity coefficients) [4]. In this approach, γ, the activity coefficient, is expressed as a sum of combinatorial, residual, and free volume contributions [5].

Solubilities were predicted using three group contribution approaches, the Oishi and Prausnitz UNIFAP (UNIQUAC for polymers) [6], the Holten-Andersen [7], and the Ilyas and Doherty [8] techniques, and compared with measured values for butyl rubber, LDPE,

TABLE 1—*Solubility predictions for natural rubber/solute systems using group contribution approaches.*

Chemical Name	Average[a] Experimental S, g/cm^3	Estimated S, g/cm^{3b}		
		UNIFAP	Holt-And[c]	Ily-Doh[d]
Acetic acid	0.078	0.062	na	0.002
Acetic anhydride	0.039	0.012	0.009	0.009
Acetone	0.095	0.10	0.13	0.090
Benzene	3.2	1.6	1.7	6.5
Benzyl alcohol	0.15	0.028	0.019	na
Butylamine	1.4	0.76	na	0.33
Carbon tetrachloride	8.3	7.3	2.4	na
Cyclohexane	2.8	4.1	0.93	5.8
Cyclohexanone	2.4	2.4	0.62	9.8
Dimethylaminopropylamine	1.1	0.74	na	na
Dimethylethanolamine	0.17	0.020	na	0.010
Dimethylformamide	0.039	0.052	na	na
Ethanol	0.007	0.019	0.019	6.2
Ethyl acetate	0.43	0.34	0.75	0.52
2-ethyl-1-butanol	0.42	0.11	0.068	0.007
Ethylene dichloride	2.1	0.21	2.9	na
Ethylenediamine	0.087	0.038	na	6.5
n-heptane	1.6	3.5	2.0	4.5
n-hexane	1.3	3.4	1.8	4.5
Isopropyl alcohol	0.040	0.048	0.035	0.014
Methanol	0.002	0.008	0.008	5.7
Methyl acrylate	0.52	0.15	0.14	na
Methyl chloroform	4.5	4.4	5.0	na
Methyl ethyl ketone	0.46	0.18	0.24	0.14
Methyl isobutyl ketone	1.2	1.1	0.55	0.26
Methyl methacrylate	1.1	0.28	0.37	0.27
n-pentanol	0.12	0.088	0.085	0.17
t-pentanol	0.39	0.082	0.085	0.028
n-propanol	0.088	0.047	0.036	0.015
n-propyl acetate	1.3	0.40	0.22	1.4
Tetrachloroethylene	7.0	4.2	na	na
Tetralin	4.4	4.6	1.6	3.8
Toluene	3.5	3.7	1.2	1.4
Trichloroethylene	7.5	4.8	na	na
o-xylene	3.8	5.0	0.79	1.5

[a] Experimental solubility data are average values for that chemical in natural rubber in the temperature range from 20 to 30°C.
[b] (na) indicates that the required group interaction parameters are not available for this solute/polymer system.
[c] Holten-Andersen approach.
[d] Ilyas and Doherty approach.

natural butyl rubber, Neoprene, and nitrile rubber, when applicable. Table 1 presents results for natural rubber in which predictions by the three techniques are compared with solubility data. The UNIFAP and Holten-Andersen approaches are both very accurate for a wide range of chemicals; the predictions are accurate within an order of magnitude and many are within a factor of two (that is, two times higher or half the measured value). Predictions are less accurate using the Ilyas and Doherty method and can only be made for a more limited set of chemicals. The results for the UNIFAP technique are graphed in Fig. 1 with the measured S values on the *x*-axis versus the predicted values plotted on the *y*-axis. The

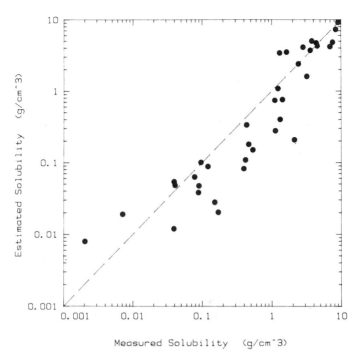

FIG. 1—*Comparison of solubility predictions using UNIFAP technique with experimental data for natural rubber.*

data generally fall along the 45° line and are accurate within an order of magnitude. This accuracy is quite good considering the range and complexity of chemicals involved. The results for butyl rubber, LDPE, neoprene, and nitrile rubber were comparable to the natural rubber results; solubility prediction accuracy was generally within an order of magnitude.

The Oishi and Prausnitz method is the recommended approach. It can be used to estimate S for the five polymers emphasized here. In some cases, the Holten-Andersen approach is more accurate but its applicability is more limited. As present, the Holten-Andersen approach can treat natural rubber, butyl rubber, LDPE, and solutes from chemical classes including ketones, alcohols, esters, ethers, simple chlorinated compounds, and most hydrocarbons. The Ilyas and Doherty approach is more limited and demonstrates only marginal accuracy. For cases in which the UNIFAP technique cannot be applied, the Kumar EOS approach is recommended although the approach is generally less accurate. Chemical functional groups that cannot be treated by the UNIFAP method include nitriles, tertiary amines, phosphorous containing compounds, and compounds containing fluorine.

Diffusion Coefficient Estimation

Models to estimate D for solute/polymer matrix systems are much less advanced than solubility estimation procedures. There are no broadly applicable theoretical models to estimate D in concentrated polymer solutions. At present, a useful approach is an empirical correlation of experimental diffusion coefficients with physical properties of the solute molecule.

Theoretical Approach—Theoretical approaches to diffusion coefficient prediction generally involve application of free volume theory [9,10]. Although these models provide a good qualitative representation of variations in D with temperature and solute concentration, they are difficult to apply and require physical property data that are not generally available. Consequently, these approaches are not suitable for PMN review evaluations where simplicity and broad applicability are essential requirements.

Empirical Approach—Because of the limitations of theoretical approaches, the project focused on correlating diffusion coefficient data with properties of the solute. The goal was a simple, broadly applicable technique that requires only the physical property data typically available in a PMN. Published diffusion coefficients were obtained by literature review. Additional values were calculated from permeation versus time data reported in the literature.

Correlations of measured D values with properties representative of solute size and shape were investigated. These properties include molecular weight, molecular connectivity, surface/volume ratios, and the acentric factor. The best correlations were with molecular weight, although, for the polymers with less extensive data sets, the correlations are not well defined.

In Fig. 2, the D values are plotted for organic liquids and gases in natural rubber versus the molecular weight of the solute. In general, the data show a consistent trend with molecular weight. For the straight-chain hydrocarbons and other approximately linear molecules, the D values decrease approximately linearly by two orders of magnitude as the molecular weight increases from 10 to 1000 on the log-log scale. However, the values for branched or cyclic molecules lie above the general trend.

The development of correlations requires a sufficiently large set of D values. Few diffusion coefficient values were found for butyl rubber, Neoprene, and nitrile rubber, (less than 10

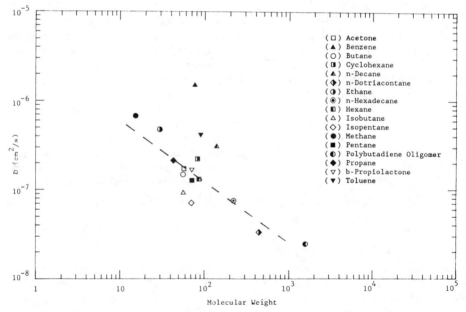

FIG. 2—*Correlation of diffusion coefficients with solute molecular weight for solute/natural rubber systems.*

in each case), and the degree of confidence in the correlation equations is less for these polymers. In general, the approach is not accurate for cyclic molecules. Analysis of additional permeation data sets continues and should enable significant improvements to the preliminary correlations presented here.

Prediction of Permeation Behavior

To judge the accuracy and applicability of the model, permeation predictions were compared to experimentally derived permeation curves reported in the literature. A permeation estimation model prototype was developed based on the preceding techniques to estimate S and D. The model was coded in FORTRAN and designed to run on personal computers at OTS. The overall accuracy of the model is determined by the accuracy to which D and S can be predicted and the applicability of simple Fick's law equations to describe the permeation behavior. As noted earlier, the model uses Fick's law with the assumption of a constant diffusion coefficient.

Permeation data from the literature were used to judge the accuracy of the model in predicting the full permeation curve. Four sources of good quality, well-documented data were used [11–14]. Each of the studies involve permeation testing of common protective clothing polymers and the permeation data, generated according to ASTM Test Method for Resistance of Protective Clothing Materials to Permeation by Liquids or Gases (F 739-85), are provided as a function of time. In total, the data set for model validation included data for approximately two to seven solutes for each of the five polymers of interest. Predictions were generated using the computer model; two approaches were used when applicable— one using UNIFAP for S and the molecular weight correlation for D (UNIFAP S/Corr D) and the second using the Kumar EOS for S and the molecular weight correlation for D (Kumar S/Corr D).

The permeation data and the model predictions are summarized here in terms of breakthrough time and steady-state permeation rate estimates. Graphs comparing the model predictions with the data over the entire permeation curve were also generated for each of the 28 polymer/solute combinations evaluated but are not included here. Breakthrough time values were predicted using the minimum detected flux or minimum detected amount permeated reported for each data set.

The results are summarized in Table 2 and, as an example, Fig. 3 compares the model predictions with experimental data for natural rubber/acetone. In general, predictions using the UNIFAP S/Corr D are accurate to within a factor of two to three for breakthrough times and to within a factor of five to ten for steady-state permeation rates. Predictions made using the Kumur S/Corr D are less accurate and tend to underestimate the permeation behavior by predicting lower permeation rates and longer breakthrough times. In most cases, the permeation predictions using a constant D are accurate within factors of two to five and are often well within the range of experimental values reported in the literature.

Test Methods and Test Data Interpretation

For instances when the predictive models do not apply or clothing performance data for the chemical of interest do not exist, OTS may need to request protective clothing performance data to facilitate a thorough review. This is often the case for chemical mixtures. Thus, it is very important to have well-defined test methods and data reporting requirements. Consequently, chemical resistance test methods were reviewed for accuracy and applicability relative to OTS needs. The test methods were compared on the basis of the types of results obtained, their approximate cost, the relative skill level required to perform them, and their

TABLE 2—*Comparison of breakthrough time and steady-state permeation rate data with permeation model predictions.*

	Thickness, cm	Breakthrough Time, min		Steady-State Permeation Rate, $\mu g/cm^2$-min		Ref
		Measured	Predicted[a]	Measured	Predicted[a]	
NATURAL RUBBER						
Acetone	0.06	17 (0.76)[b]	8 / 20	34	34 / 2	11
Cyclohexane	0.06	19 (1.8)	8 / 13	525	959 / 59	11
Isopropanol	0.06	5340 (1.0)	11 / 115	1.0	16 / 1	11
β-propiolactone	0.03	25 (19.7c)	-- / 7	3.6	-- / 22	13
Toluene	0.06	10 (109.7)	21 / 77	782	792 / 126	11
BUTYL RUBBER						
Acetone	0.06	ND (0.47)	38 / 39	--	3 / 3	11
Cyclohexane	0.06	56 (6.4)	29 / ND[c]	437	328 / 5	11
Dimethylhydrazine	0.08	1650 (542c)	-- / 395	1.5	-- / 2	13
Epichlorohydrin	0.08	4740 (9.7c)	131 / 65	0.2	1 / 23	13
Ethylenimene	0.08	1380 (2426c)	197 / 737	3.8	28 / 4	13
Hexamethylphos-phoramide	0.08	60 (19.7c)	-- / 99	1.2	-- / 38	13
Toluene	0.06	24 (7.4)	48 / 139	396	66 / 10	11
	0.04	18 (7.3)	19 / 41	144	99 / 15	14
LOW-DENSITY POLYETHYLENE						
Hexamethylphos-phoramide	0.006	25 (19.7c)	-- / 7	3.4	-- / 87	13
β-propiolactone	0.006	30 (39.5c)	-- / 4		-- / 29	13
NEOPRENE RUBBER						
Acetone	0.05	22 (3.8)	12 / 14	151	100 / 56	11
Cyclohexane	0.05	722 (8.0)	13 / 33	26	370 / 21	11
Dimethylhydrazine	0.08	40 (78.9c)	-- / 51	445	-- / 33	13
Epichlorohydrin	0.08	83 (157.8c)	-- / 57	111	-- / 77	13
Ethylenimene	0.02	5 (2209c)	-- / 104	5595	-- / 22	13
Toluene	0.05	12 (93.3)	-- / ND	642	-- / 58	11
	0.05	6 (16.3)	-- / 29	308	-- / 58	14
Trichloroethylene	0.08	23 (414.2c)	60 / 84	55	369 / 81	13

	Thickness, cm	Breakthrough Time, min Measured	Predicted[a]	Steady-State Permeation Rate, $\mu g/cm^2$-min Measured	Predicted[a]	Ref
		NITRILE RUBBER				
Acetone	0.03	5 (64)	1	4445	4135	12
			ND		6	
	0.06	10 (5.6)	3	962	2068	11
			ND		3	
Hexamethylphos-phoramide	0.09	105 (39.5c)	--	13	--	13
			44		147	
Hexane	0.03	ND (1.0)	1		109	12
			2		51	
Methanol	0.04	33 (0.12)	0.4	53	382	12
			1		2	
2-nitropropane	0.09	50 (750c)	--	296	--	13
			50		80	
Toluene	0.03	11 (33.6)	2	815	4070	12
			3		192	
	0.06	52 (0.22)	4	232	2035	11
			5		96	
Trichloroethylene	0.09	20 (375c)	36	53	651	13
			51		122	

[a] All breakthrough time and permeation rate predictions used the molecular weight correlation for D. The results using the UNIFAP approach for S are reported on the first line followed by predictions using the Kumar EOS S on the second line. The (--) indicates that predictions were not possible.

[b] The values in parenthesis are either the permeation rate, J_b, in units of $\mu g/cm^2$-min or the cumulative mass permeated, Q_b, in units of g/cm^2 at breakthrough as reported in the noted reference. The letter c denotes Q_b values. These values were used as the basis for prediction of breakthrough time.

[c] ND = "none detected" or no breakthrough detected.

inherent limitations. Based on this comparison, a testing hierarchy was developed, as discussed later.

Permeation Tests

In a permeation test, the chemical of interest is placed on one side of the clothing material and the other side is monitored for the appearance of the chemical. From the results, the total amount (mass) of chemical permeating a known surface area of the clothing at any given time can be calculated. The cumulative permeation (mass/area) or the corresponding permeation rate (mass/area/time) can be used along with the estimated frequency, duration, and exposed body surface area to estimate dermal exposures for specific workplace activities.

Although the test is straightforward and a standard method (ASTM F 739-85) exists for its performance, variable results can be obtained under different testing conditions for the same chemical/clothing material. Consequently, in describing or interpreting the results of a permeation test, there is a certain, minimum amount of information required. This information includes the breakthrough time, the steady-state permeation rate, the clothing material thickness and surface area, the analytical sensitivity, the collection medium flow rate (open-loop systems) or volume (closed-loop systems), and temperature [15].

Immersion Tests

In an immersion test, the clothing material is exposed on one or both sides to the chemical of interest for some period of time. The change in weight or in other physical characteristics is measured. In tests in which the weight is accurately monitored as a function of time, D

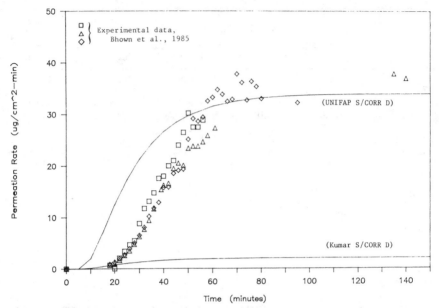

FIG. 3—*Comparison of model predictions for permeation flux as a function of time with experimental data for acetone/natural rubber.*

and S of the chemical in the material can be calculated. Such tests are referred to as sorption/ desorption tests and may be performed using either liquid or vapor exposures.

In addition to weight, other physical characteristics of the clothing material can be used to assess the overall resistance to a chemical or solution of chemicals. These characteristics include dimensions, puncture resistance, tear resistance, elongation resistance, strength, and so forth. Such tests are referred to as degradation tests since significant detrimental changes to the clothing due to the chemical are of interest.

Testing Hierarchy

Both the chemical resistance and the physical properties of the clothing material must be considered in judging its acceptability. While the physical property requirements are specific to the particular application, chemical resistance is a more general concept. In Table 3, a test hierarchy is proposed for assessing chemical resistance. The testing hierarchy ranks chemical resistance tests according to their ability to generate data that can be used directly to estimate the effectiveness of the clothing material in reducing exposure risks.

Conclusions and Recommendations

A system for estimating clothing performance should consider the cumulative amount of chemicals that crosses the protective clothing boundary. The predictive model described here was successful in estimating the cumulative permeation to within factors of five to ten. If greater accuracy is required in PMN review, additional refinement of the model is necessary. Alternatively, OTS can estimate protective clothing performance on the basis of data for "existing" chemical analogues. Where there is no such chemical, the PMN chemical/ clothing combination must be tested.

The recommended approaches to estimating permeation behavior as a function of time

TABLE 3—*Test data priority for estimating the chemical resistance of protective clothing polymers.*

Chemicals	Test Method(s)
Single-Component Liquids/Gases	1. ASTM F 739-85 permeation test
	2. Permeation cup (if chemical has sufficient vapor pressure)
	3. Weight change from liquid immersion
	4. *D* and *S* from vapor or liquid sorption/desorption test
Multicomponent Solutions	1. ASTM F739-85 permeation test

are those based on Fick's law. These require the estimation of the diffusion coefficient and the solubility of the chemical in the polymer. Techniques for estimating *D* and *S* were developed and used in a permeation model prototype based on Fick's law. In most cases, the permeation predictions were well within the range of experimental values reported for a specific permeant/polymer system. In other cases, the predictions were less consistent with the experimental data and this result may be because certain assumptions of the methodology were not valid. For example, the assumption of a constant diffusion coefficient is likely to be invalid for polymers that are swollen to a high degree by the permeant.

For PMN substances that are mixtures and those with no chemical analogues for which there are permeation data, testing is required. Permeation testing is preferred and should be performed according to ASTM F 739-85 or similar procedure under well-controlled and documented experimental conditions. Degradation and sorption/desorption test methods provide either fundamental parameters for use with predictive models or general indications of clothing performance. However, the results of these tests require interpretation by persons with some training in polymer permeation science.

To achieve the goal of developing an accurate and efficient method for judging the suitability of PMN protective clothing recommendations, OTS should continue its efforts to develop an integrated system. The system should incorporate (1) predictive models for estimating chemical resistance, (2) existing chemical resistance and physical property databases, and (3) the identification and specification of appropriate test methods.

Acknowledgment

The research described in this paper was funded by the U. S. Environmental Protection Agency (Contract No. 68-03-3293). The document, however, does not necessarily represent the views of the Agency, and no official endorsements should be inferred.

Mention of trade names or commercial products does not constitute endorsement or recommendation for use.

References

[1] Crank, J., *The Mathematics of Diffusion,* Oxford University Press, New York, 1956.
[2] Kumar, S. K., Reid, R. C., and Suter, U. W., "A Statistical-Mechanics Based Lattice Model Equation of State: Applications to Mixtures with Supercritical Fluids," American Chemical Society, Division of Fuel Chemistry Preprints, Vol. 30, No. 3, 1985, pp. 66–77.
[3] Kumar, S. K., Suter, U. W., and Reid, R. C., "A Statistical-Mechanics Based Lattice Model Equation of State," *Journal of Chemistry and Physics,* to be published.
[4] Abrams, D. S. and Prausnitz, J. M., "Statistical Thermodynamics of Liquid Mixtures: A New Expression for the Excess Gibbs Energy of Partly or Completely Miscible Systems," *Journal, American Institute of Chemical Engineers,* Vol. 21, 1975, p. 116.
[5] Fredenslund, A., Jones, R. L., and Prausnitz, J. M., "Group-Contribution Estimation of Activity Coefficients in Non-Ideal Liquid Mixtures," *Journal, American Institute of Chemical Engineers,* Vol. 21, 1975, p. 1086.

[6] Oishi, T. and Prausnitz, J. M., "Estimation of Solvent Activities in Polymer Solutions Using a Group Contribution Method," *Industrial Engineering and Chemistry, Process Design and Development,* Vol. 17, 1978, p. 333.

[7] Andersen, J. H., Rasmussen, P., Fredenslund, A., and Carvoli, G., "Phase Equilibria of Polymer Solutions by Group-Contribution," SEP 8610, Institute for Chemical Industries, The Technical University of Denmark, Lyngby, Denmark, April 1986.

[8] Ilyas, S., "Group Contributions for Fluid Phase Equilibria," Ph.D. thesis, University of Massachusetts, Amherst, MA, 1982.

[9] Fujita, H., *Diffusion,* in *Encyclopedia of Polymer Science and Technology,* Vol. 5, Wiley, New York, 1966, pp. 65–81.

[10] Vrentas, J. S. and Duda, J. L., "Diffusion in Polymer-Solvent Systems. II. A Predictive Theory for the Dependence of Diffusion Coefficients on Temperature, Concentration and Molecular Weight," *Journal of Polymer Science, Polymer Physics Edition,* Vol. 15, 1977, pp. 417–439.

[11] Bhown, A. S., Philpot, E. F., Segers, D. P., Sides, G. D., and Spafford, R. B., "Predicting the Effectiveness of Chemical Protective Clothing: Model and Test Method Development," EPA/600/S2-86/055, U.S. Environmental Protection Agency, Cincinnati, OH, 1985.

[12] Schwope, A. D., Carroll, T. R., Huang, R., and Royer, M. D., "Test Kit for Field Evaluation of the Chemical Resistance of Protective Clothing," this publication, pp. 314–325.

[13] Coletta, G. C., Schwope, A. D., Arons, I., King, J., and Sivak, A., "Development of Performance Criteria for Protective Clothing Used Against Carcinogenic Liquids," DHEW (NIOSH) Publication No. 79-106, National Institute for Occupational Safety and Health, Oct. 1978.

[14] Vahdat, N., "Permeation of Polymeric Materials by Toluene," *Journal, American Industrial Hygiene Association,* Vol. 48, No. 2, 1987, pp. 155–159.

[15] Goydan, R., Schwope, A. D., and Reid, R. C., "Methodologies for Estimating Protective Clothing Performance—Development, Validation and Assessment," Interim Draft Report on EPA Contract 68-03-3293, Work Assignment 08, to Arthur D. Little, Inc., Dec. 1986.

Participants in Chemical Breakthrough Parameters Session

Roundtable Discussion on Solubility Parameter

The five papers in Session 7 were on the Solubility Parameter. All speakers were invited to participate in a Roundtable Discussion following the session. The Roundtable Discussion participants included the Session Chair, Alan P. Bentz (*AB*); the Keynote Speaker, Charles M. Hansen (*CH*); Henning R. Henriksen (*HH*); Jimmy Perkins (*JP*); and C. Nelson Schlatter (*NS*). Identified from the audience were: Rosemary Goyden (*RG*), Norman W. Henry III (*NH*), and Zack Mansdorf (*ZM*). Unidentified questioners (or commenters) from the floor are designated (*Q*).

Discussion

[The Session Chair opened the panel discussion by calling on Dr. Hansen to comment on the treatment of the solubility parameter concept in the session on that subject.]

CH: Speakers have all used the concept intelligently and the limitations have been pointed out. There may be a few differences of opinion as to where the limitations are, but future research will clear all of that up.

HH: [Gave discussion with viewgraphs. In it, he introduced a new achronym, MAXI-PARDIF, for MAXimum Interaction PARameter DIFference, which he coined on the premise that the word "interaction" is more consistent and informative than the word "solubility." See the Appendix for details.] Referring to the presentations, I would like to emphasize some aspects. Regarding the individual polymer of each type, that is, neoprene or nitrile rubber, the intra-type variation may be quite large. Conversely, the inter-type variations may be very small, for example, the parameters for neoprene and butyl rubber overlap almost perfectly. Thus, the barrier properties of a polymer cannot be determined merely by the generic name. This may be severely misleading. Instead, it is necessary to know the permeation data for the specific brand of glove/polymer.

[Comment was made from the audience about different additives being responsible for different behavior of the same type of polymer; some behavior enhanced and some degraded, depending on the additive.]

CH: That should be used as a challenge to the manufacturers not to ruin an inherently good polymer by doing something wrong with it.

[General discussion followed of economics driving the development of the polymers, especially if they are not good.]

HH: In the Danish Technical University, the technique used is similar to Hansen's original method, using a test tube or a small can and dropping a polymer in it along with solvent, and covering it. These polymeric materials for protective clothing rarely dissolve, so we measure the swelling. After one week, take it out and dry it and weigh to measure the weight gain and determine crudely the degree of swelling.

Now we are starting a new vapor phase method, using special equipment (See Fig. 2 in the Appendix) to put saturated vapor in contact with the material for a period of one month.

The results on the swelling are the same as for direct liquid contact with the material.

JP: If results are comparable, what is the advantage?

RG: I have a comment on your vapor phase exposure—your statement is true as long as you have good mixing and have a vapor activity of 1 for the duration of the exposure. With no mixing, I don't know that it *will* be 1.

CH: I fully agree. I tried experiments like that and have seen liquid collecting at the bottom of the upper tube. I've seen so much collected that it oversaturated . . . my experiments were not in the glove area.

HH: [Discussed condensation problems including "raining" in the tubes, but indicated that at constant temperature this posed no problem.] In the first experiments conducted, some condensation was observed, but was attributed to poor thermostatting. Just now, we are investigating if good temperature control solves the "rain" problem completely.

JP: Have you compared your results using this technique versus the liquid?

HH: We have done a few preliminary experiments and they were so promising that we are continuing the experiments.

JP: If they are equivalent, why do the experiments this way?

HH: With the liquid ("old" method), there will be many cases where parameters cannot be determined at all. In other cases, they have little validity.

JP: I agree with what you are saying, I've given this a lot of thought. We used your technique to try to determine the three-dimensional solubility parameter for skin. But the problem I had with this technique is that this is not the way we do permeation tests with liquids where we leach out plasticizers and so forth, and that affects the permeation rate. The other point I wanted to make is that we gain a lot of information by looking at the solution when we actually put the polymer into the liquid, because if it changes color, we know something has happened.

Regarding the point that Alan raised earlier about 24 h being the sufficient time to run the test, I have found that if you run the test for seven days, or even 14 days, you will see some things that you won't see in 24 h. For example, I gave you information earlier about ethanolamine dissolving Viton, which takes three days. In other cases amines harden Viton in two or three days. You won't see that, I don't think, if you use a vapor . . . .

HH: Well, I think that is academic. What we need here is to get a good estimate of the parameter before taking the next step, namely, examination of MAXIPARDIF and finally using this to predict permeation rate, etc.

JP: It's not academic at all. We are using this information to predict permeation properties, and permeation properties are obtained by exposing a polymer to a liquid, not to a vapor.

HH: Regarding permeation, the most important problem is permeation resistance to liquid chemicals. Diluted vapors may also be present, but their chemical potential (driving force) for permeation will be substantially smaller, since the driving force across the membrane is a function of the concentration of the permeant at the absorption and the desorption surfaces of the polymer membrane.

What we are talking about is the MAXIPARDIF concept. That is, we are trying to find out the true physical chemistry . . . can we correlate this coincidence with the laws of thermodynamics? The next step is to apply it to the workplace. In the workplace, there is a difference between vapor contact and the actual exposure.

JP: In four to five of the 55 exposures to different chemicals [in determining solubility parameter], you are going to get some color changes and other changes in the polymer, and its going to occur with the liquid phase [and not with the vapor phase].

HH: I would like to emphasize that, in my opinion, the most urgent task in this field is to scrutinize MAXIPARDIF scientifically. I think the concept is valid, but we have serious reliability problems and problems with the mathematical modeling as well. To complete the necessary comparisons and correlation analyses, we require well-defined and consistent data which are presently lacking.

Q1: In doing these tests, you are getting the solubility parameter at equilibrium. We don't carry on that test [permeation] at equilibrium. How do we know that the rate we obtain correlates with the equilibrium value? We have to account for some of the differences in correlation.

JP: I don't think there is a problem here. Once we get past the surface phenomenon effect that Dr. Hansen talked about this morning—essentially what is happening is what occurs at the molecular level across the polymer, and the only difference between what happens at steady-state permeation rate and what happens at breakthrough time is that we have established whatever gradient is going to be established in the polymer.

Q1: There is an interaction between solvent and polymer which might not reach equilibrium. You change the polymer with such interactions.

JP: You may be right if you are talking about swelling. With swelling, you have created areas within the polymer which are more solvent than polymer.

NS: How long is long enough [for Henriksen's technique]?

HH: Liquid contact for one week is long enough to reach equilibrium. I feel confident that parameters can be determined this way for many polymers, but definitely not for all. According to our preliminary measurements, one month is long enough for the vapor phase method, but much more development work has to be done.

It doesn't matter that the polymer membranes of protective gloves, etc. are not fully saturated at steady-state permeation or breakthrough time. The validity of MAXIPARDIF has been verified in those cases where proper parameters and permeation data were available. Qualitatively, the interactions are approximately the same at low and high concentrations in the polymer.

Q2: I'd like to raise the question of measurement of breakthrough time. It seems to me that you are at the tip of the iceberg in measuring a very complicated process. There is a good reason for us to look at the permeation rate at steady state. As we develop more and more sensitive equipment and better models, we should actually be able to predict how long it takes for the first molecule to cross a given barrier. My question is, after you do that, what are you going to do with that information? It seems that what seems to be lacking is toxicity, dermal toxicity, and acute toxicity versus concentration, because as your data show, some cases will show a relatively fast breakthrough time and a relatively slow permeation rate . . . and you are probably never going to get a good correlation between breakthrough time and permeation, especially in your cases where you are measuring polymers which are filled, like a Viton that has cross-linkers in it, has a filler in it, etc. A molecule making its way through a filled polymer may get adsorbed and desorbed on top of every pigment in its breakthrough to that steady state. A lot of those effects take place as it comes to equilibrium, so I don't understand how better and better measurements of breakthrough time are going to give you anything unless you correlate it with dermal toxicity.

AB: I'd like to discuss that briefly and dismiss it—not out of hand, but because we are going to have a session this afternoon to discuss the variables that affect breakthrough time. This is as complex as the three-dimensional solubility parameter. As you know, ASTM has a standard method on the books for permeation which is undergoing revision right now with a number of things being considered. One is the detectability limit and its influence. Our statistician has indicated that we should get the first molecules through instantaneously— just on a probability basis, not a chemical one. I wasn't quite willing to buy that, but I found out when we started to do carbon-14 tracer work (subject of another paper to be given later by Mr. Ursin in another session), that indeed when you get sensitive enough, it's almost literally true that breakthrough occurs immediately. This generates a profound philosophical question. If you are sending a man into a life-threatening situation, whether you are in the Coast Guard or in industry, and you know that there is breakthrough time early on, but the rate is very low, who is going to decide "I know you have benzene in your suit, but it's not enough to be concerned about"?

Q3: In industry we have up to 15 threshold limits published. . . .

AB: We have a number of people here that can address that. What the Coast Guard is doing is tabulating the known TLVs to make sure that the detection limits used are sufficiently sensitive to see a given chemical before it reaches its TLV. One of the problems other people can answer better than I is regarding the dermal toxicity on exposure to vapors and/or in the presence of perspiration. There isn't that much in the literature.

JP: Alan, I'm going to give a paper tomorrow afternoon that addresses all these things from a philosophical point of view. I think the questioner is absolutely right—I think we have been putting the cart before the horse for the last ten years. We know a lot of things about polymers, but not very much about skin permeation.

RG: Also this afternoon, I'll be presenting a paper on the general topic that the focus has always been on breakthrough time, rather than the amount coming through the polymer, and how that amount changes with time. It focuses on one of the topics Dr. Hansen discussed this morning—using Fick's Law equations to predict culmulative permeation. You don't have to settle for just the breakthrough time, but can compute the total amount permeating over time.

AB: Norm Henry is an industrial hygienist and can address this as well.

NH: I work in a toxicology laboratory where determining a toxic dose is the main endpoint in evaluating chemicals. In evaluating permeation resistance of protective clothing, we determine both the breakthrough time and permeation rate. NIOSH has taken these three parameters and derived an equation for selecting clothing based on a permeation dose. It is as follows:

$$PD = (MD - BT)(PR)(BSA)$$

where PD = permeation dose, MD = mission duration, BT = breakthrough time, PR = permeation rate, and BSA = suit surface area (body surface area).

If the PD is greater than, or equal to, the toxic dose for a particular chemical and item of protective clothing, then a more resistant clothing material should be selected. If the PD is less than the toxic dose, then that particular clothing material ought to provide adequate protection for the duration of the mission or job.

This decision logic or algorithm for selecting protective clothing is based on a toxic dose which may be defined as a dose which causes harmful effects in the body. There is a lot published on toxic doses for animals from which extrapolations are made to set limits for human inhalation exposure (TLVs). In the permeation test method, an "acceptable contact level" is suggested as a possible limit for dermal exposure by permeation through clothing. To date, no acceptable levels have been defined, but may be set later on to help in protective clothing decision logic.

AB: Thank you Norm. We could go on with this at great length. I think that because of time limitations, we will proceed with a statement from Dr. Perkins, if he has anything to add.

JP: Nothing.

NS: I would like to add one comment. Jimmy, you said thickness had no statistically significant effect on breakthrough time—with the materials you were working with. This intrigues me because I have a report published in *STP 900* [ASTM, 1986] on the influence of film thickness on the permeation resistance properties of unsupported glove films [p. 75] in which the most important effect was that on breakthrough time. I was wondering why we got such different results.

JP: It's probably [in the definition of] "breakthrough time"—it's an elusive variable. I've been preaching for two years that we should throw it out the window. Unfortunately, we have thousands of data points now that we want to use and we don't have the corresponding lag times to go with them.

AB: You see no difference in breakthrough time with thickness?

JP: In my detection system, I don't have the same detection limits for nitrobenzene as I do for methyl alcohol.

Q4: On a single chemical?

JP: On a single chemical with the same detection system, yes. It's a function of the curve, and for every detector for a given chemical in the same system, the detection limits should be exactly the same from test to test.

CH: I'm glad you all agree. I'd like to throw, as we say, something into the bicycle wheel. If the major resistance is at the surface, and this happens with thin membranes, then you don't see the effect of film thickness—because the resistance is at the surface. This is another facet that deserves research.

AB: Before we close the session, our symposium chairman has a few comments.

ZM: My technical comment is one addressed earlier. I believe that a lot of the polymers that have fillers and so forth are where you will find development of a reservoir. And in some cases, after one molecule breaks through, you will have continuous release over a period, perhaps up to years . . . somewhat predicted by solubility, based on other work.

AB: We could go on and on. I would like to comment on two points. Speaking about brand-specific polymers, there is a publication out by NIOSH in which generic gloves from different manufacturers were examined. It was no surprise that differences were found when comparing gloves from different sources. It was surprising however, to find that some much thicker gloves were poorer barriers to the same chemical than gloves of the same generic typed from another manufacturer. This is an important point.

I had promised earlier that I was going to comment on Dr. Hansen's use of psoriasis scales as a good example for skin. He mentioned keratin, and probably it is through my own ignorance that I don't understand, but I thought keratin was a component of hair and nails. However, the point I wanted to bring out here is that the natural skin is a pretty good barrier. It wasn't original with me, but I've heard that when you have auto body paint shops, for example, when workers clean their hands with solvent, they defat the skin. Once it is defatted, it is no longer the barrier it once was. Here we have a very complex system. When you are talking about psoriasis scales, I'm not sure there is any fat present. I'm not contradicting what he said, because he defined what his model was. Looking at the little finger on his three-dimensional model, I'm wondering whether that is just keratin, or is natural skin with all the oils present.

CH: Well, I can't argue the point. I can only comment that there is also a correlation for fat solubility. So you have them both. But, again, you get the same correlations with the skin warning labels of skin data in the TLV book and the solubility parameter. The correlations are there—which again suggests to me that the people who print the labels should consider some of the solvents within this correlation as to whether or not they should be labeled for skin absorption. I have nothing more to say than that. But when you see psoriasis scales swell up, and I've smelled these same chemicals on my hands long after and so forth, I'm convinced there is a correlation between the solubility parameters for keratin and the real world.

AB: [Closed the roundtable discussion and thanked the participants.]

Protection from Industrial Chemical Stressors

New Laboratory Test Methods

Mark W. Spence[1]

An Analytical Technique for Permeation Testing of Compounds with Low Volatility and Water Solubility

REFERENCE: Spence, M. W., **"An Analytical Technique for Permeation Testing of Compounds with Low Volatility and Water Solubility,"** *Performance of Protective Clothing: Second Symposium, ASTM STP 989,* S. Z. Mansdorf, R. Sager, and A. P. Nielsen, Eds., American Society for Testing and Materials, Philadelphia, 1988, pp. 277–285.

ABSTRACT: One of the most challenging problems currently facing researchers in the field of chemical protective clothing (CPC) permeation testing is how to perform tests for compounds that have low volatility and low water solubility. Permeation testing with these compounds is difficult because an inert gas or water are currently the only permeant collection media that one can use with the assurance that the CPC material will be unaffected by the collection medium.

An approach to solving this problem using an automated sample concentrating technique in conjunction with gas chromatography is described. Such a technique allows determinations of breakthrough times and permeation rates while maintaining the permeant concentration in the collection medium at very low levels through the use of relatively high collection medium flow rates. By keeping the permeant concentration low enough to remain within the volatility limit of the compound being tested, the range of compounds that can be tested using air or nitrogen as the collection medium is greatly expanded.

KEY WORDS: chemical protective clothing, permeation, permeation testing, permeation resistance, permeation test methods, low-volatility compounds, low-solubility compounds, protective clothing

As permeation testing has become a more common technique for assessing the effectiveness of chemical protective clothing (CPC), the variety of chemicals tested has increasingly included compounds with low volatility and low water solubility. Many of the chemicals of greatest interest from an industrial hygiene standpoint are of this type; for example, polycyclic aromatic hydrocarbons (PAHs), poly chlorinated biphenyls (PCBs), and most pesticides and herbicides fall into this category. Such chemicals present a challenging problem for those involved in permeation testing. To understand why, we must consider the permeation process and conventional permeation testing techniques.

According to the ASTM Test Method for Resistance of Protective Clothing Materials to Permeation by Hazardous Liquid Chemicals (F 739-81), the permeation process consists of three steps. First, the chemical is sorbed onto the surface of the CPC material. Next the chemical moves through the CPC material by a vapor diffusion process. Finally, the chemical is desorbed from the inner surface of the material. In actual CPC use, the permeated chemical present on the inner surface of the material may be desorbed by any of three actions:

[1] Senior research chemist, Health and Environmental Sciences Research, The Dow Chemical Company, Midland, MI 48674.

evaporation into the air present, dissolution into the water present on the skin surface, or mechanical transfer by the contact of the skin with the inner surface of the CPC material. The physical properties of the chemical (principally vapor pressure and water solubility) determine the relative importance of each of these actions in facilitating movement of the permeated chemical from the inner surface of the CPC material to the skin surface. For compounds having both a low vapor pressure and a low water solubility, mechanical transfer is probably the most important action.

In permeation testing, the environment inside the CPC material is simulated by a water or inert gas permeant collection medium. Thus, two of the three actions discussed earlier are represented but the mechanical transfer action, which is of primary importance for low volatility and low water solubility compounds, is not. Ideally, for such compounds, testing should be conducted using a mechanical transfer collection technique that is both quantitative and reproducible. Since this appears at present to be an impossible task, alternative approaches must be considered.

One such approach is modification of the collection medium by the addition of solvents, surfactants, emulsifiers, etc., to increase the solubility of the permeant. While this may work in certain cases, it is likely that the additive(s) used would have to be tailored to each compound tested and that such additives could affect permeation through different CPC materials in different ways. This approach, then, would be costly and time consuming since the modified collection medium would have to be validated for each set of test chemicals and CPC materials.

Another approach that avoids these complications is to use the conventional collection media, but apply some analytical technique sensitive enough to allow monitoring permeation at permeant concentrations in the collection medium that are low enough to avoid solubility or volatility limitations. This paper will describe our use of a "sample trapping" concentration technique to accomplish this goal.

Sample trapping is a technique that has found wide application in trace analysis, particularly in the environmental area [1–4]. Figure 1 shows schematically the basic concept of sample trapping. The inert gas or water collection medium from a permeation test cell is passed through a "sample trap," that is, some device that will reversibly strip the permeant molecules from the collection medium. For analysis, the sample trap is subjected to a different set of conditions that reverses the stripping process, releasing the accumulated permeant molecules for quantitation by an appropriate analyzer. For ease in automation, the sample trap may be connected to an automatic stream sampling valve in the same configuration as a conventional volumetric sample loop. The key difference between a conventional volumetric sampling system and a sample trapping system is that the sample trap works essentially as a "variable volume" sample loop, allowing increased analytical sensitivity for the permeant by increasing the concentration factor through the use of comparatively long trapping times.

Experimental

Sample Trapping System

The sample trapping system we have used in permeation testing employs a piece of thick film, fused-silica-capillary gas chromatography (GC) column as the sample trap. In the trapping mode, the collection medium is passed through the trap at a relatively low temperature and the permeant molecules are stripped from the collection medium and absorbed in the stationary phase film that lines the column. For analysis, the trap is heated and backflushed with a carrier gas to desorb the permeant molecules and transport them to a GC for quantitation.

A diagram of the sample trapping permeation testing system used is depicted in Fig. 2.

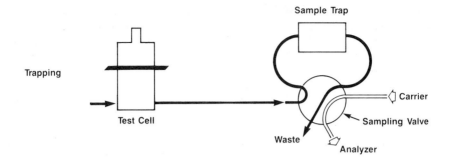

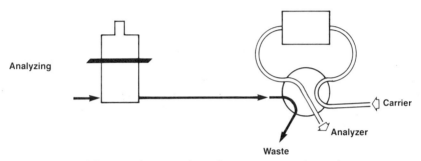

FIG. 1—*Schematic of automated sample trapping permeation testing system.*

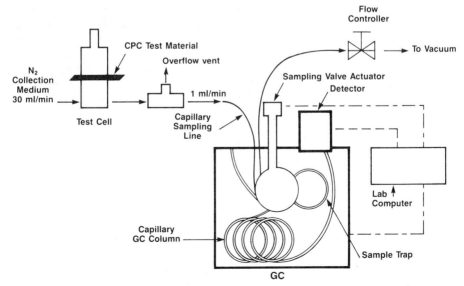

FIG. 2—*Details of automated sample trapping permeation testing system used.*

The sample trap and stream sampling valve are located inside the oven of the GC used for analysis. A regulated vacuum supply is used to draw a 1 mL/min portion of the permeant collection medium through the sample trap. To make an injection into the GC, the sampling valve is rotated to put the trap in line with the GC column and the GC oven temperature is raised. Following an analysis, the GC oven is cooled back to the trapping temperature and the sample valve is rotated back to place the sample trap back in line with the test cell collection medium to collect another sample. A laboratory computer system is used to control the operation of the stream sampling valve and the GC as well as to collect and process the analysis results.

The cycle time required for this trapping and analyzing cycle depends on the trapping time required to obtain the desired sensitivity, the time required to analyze the compounds of interest, and the GC oven cooldown time. For the examples described in the present work, cycle times of 3 to 5 min were achieved.

The specific equipment used in the present work included the following items.

Gas chromatograph	Hewlett Packard 5890 with split/splitless injector and flame ionization detector
Computer system	Hewlett Packard 3357 Laboratory Automation System
Sampling valve	Valco 6N6WTFSR capillary six-port injection valve
Sampling line	1.5 m by 0.2 mm (inside diameter) deactivated vitreous silica capillary tubing
Sample trap	1 m × 0.32 mm (inside diameter) J&W DB-1701 (1 μm film) capillary column
Analytical column	10 m × 0.32 mm (inside diameter) Alltech RSL-200 capillary column
GC carrier gas	nitrogen at 95 cm/s

Test Chemical Properties and Testing Parameters

To demonstrate the utility of this capillary GC sample trap system, some permeation testing we have conducted with a series of halogenated pyridines will be discussed. Selected physical properties and testing parameters for each of the halogenated pyridines tested are given in Table 1. The vapor pressures of the compounds ranged over three orders of magnitude, the lowest being 0.02 mm Hg for Compound E. This represents a very low vapor pressure for the use of a gaseous collection medium; however, the concentrating effect of the trapping system affords a detection limit of 0.01 ppm, which is less than 0.1% of the

TABLE 1—*Properties of halogenated pyridines tested.*

Compound	Vapor Pressure, mm Hg	Saturation Concentration, ppm (v/v)	Trapping[a] Temperature, °C	Time, min	Detection[b] Limit, ppm (v/v)
A	30	40 000	30	0.5	0.04
B	5	6 600	30	0.5	0.05
C	2	2 700	30	0.5	0.05
D	2 to 3	3 000	30	0.5	0.04
E	0.02	13.2	100	2.0	0.01

[a] Sample stream flow through trap = 1 mL/min.
[b] At trapping time indicated, signal to noise ratio = 3.

theoretical saturation concentration of 13.2 ppm indicating that the breakthrough of the compound through the CPC material being tested would be detected long before the compound's volatility became limiting.

The trapping temperatures and times are also given. It is interesting to note that even Compound A with a vapor pressure of 30 mm Hg could be effectively trapped for a trapping time of 0.5 min at 30°C, a temperature that can be maintained in most modern GCs without any auxiliary means of cooling. For Compound E, the least volatile compound, the trap was found to work effectively at even higher temperatures. This illustrates a useful feature of this sample trapping system: the lower the volatility of the permeant, the higher the trapping capacity at a given temperature. This effect is consistent with what one would expect from a study of gas chromatographic theory [5].

Results and Discussion

Linearity with Trapping Time

To investigate the linearity of the sample trapping system with varying trapping time, a test atmosphere containing 2.8 ppm of Compound D was generated using a CPC material that was at conditions of steady-state permeation with Compound D. This test atmosphere was then sampled using the sample trapping system at trapping times ranging from 0.05 to 5.0 min and the resulting GC/FID response (in area units) was recorded. Table 2 shows the average area and percent relative standard deviation (%RSD) for each trapping time. The %RSD figures indicate that for trapping times above 0.3 min, very good sample-to-sample reproducibility is found. At sample times less than 0.3 min, variations on the order of 5 to 10% can be expected.

The actual linearity comparisons were made by dividing the average area found by the corrected trapping time (the nominal trapping time corrected for an experimentally determined valve switching time of 0.017 min per valve cycle). This yields values in terms of area per minute of trapping time that should be constant for all trapping times. As can be seen from Table 2, good linearity is indeed obtained since the relative standard deviation (RSD) of the mean of the area per minute values over two orders of magnitude of trapping time is 3%.

Effect of Trapping Time on Peak Shape

Figure 3 shows chromatograms from the trapping time variation experiments discussed in the previous section. At a trapping time of 1 min, the peak for Compound D still is quite

TABLE 2—*Linearity with trapping time (test atmosphere: 2.8 ppm Compound D).*

Trapping Time, min		Mean Peak Area	RSD, %	Area/min
Nominal	Corrected[a]			
0.05	0.033	5 769	1.6	174 800
0.10	0.083	44 500	6.0	174 700
0.30	0.233	43 461	9.6	186 530
0.50	0.483	86 150	1.6	178 400
1.00	0.983	179 615	1.6	182 720
5.00	4.98	926 923	1.1	186 000
			Mean =	180 524
			RSD =	3%

[a] Corrected for 0.017 min valve switching time.

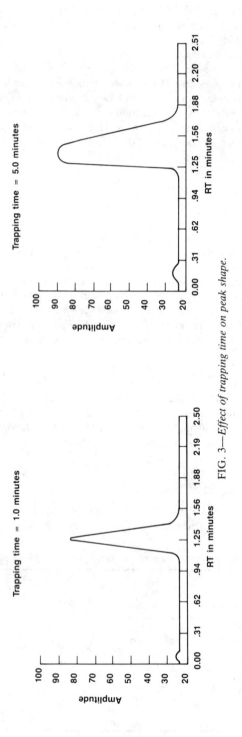

FIG. 3—*Effect of trapping time on peak shape.*

sharp. For a 5-min trapping time, however, the peak has a broad, plateau-like top even though there are still no signs of fronting or tailing. Interestingly, recalling the results shown in Table 2, the area found for the 5-min trapping time is entirely consistent with results for shorter sampling times. This indicates that a longer portion of the film in the capillary sample trap was loaded with Compound D during the 5-min trapping time but the trap was not overloaded. Under overload conditions, the same peak shape as that obtained for the 5-min time would be expected, but the area units per minute of trapping time would be low.

Effect of Permeant Volatility on Trap Capacity

Chromatograms from trapping experiments at three different trapping times using halogenated pyridine Compounds A, B, and C are shown in Fig. 4. Referring to Table 1, these compounds have different vapor pressures (A = 30 mm Hg, B = 5 mm Hg, and C = 2 mm Hg) and these chromatograms show the effect this has on trap capacity. The peak for Compound A has essentially the same amplitude and width (and therefore the same area) for all three trapping times indicating that the entire length of the trap is at its capacity for this compound. For Compound B, the peak amplitude and width increase in going from the 0.5-min trapping time to the 1.0-min time, and the peak width increases at constant amplitude on going from 1.0 to 5.0 min. These results are indicative of a case similar to that discussed for Fig. 3; the initial part of the trap is at capacity for the compound so the sample is trapped using a longer portion of the trap but the trap as a whole is not overloaded. Finally, the peak for Compound C grows in both amplitude and width with each increase in sample time indicating that the trap is even farther away from overload conditions than for Compound B. This series of compounds clearly demonstrates that the trapping system is more effective and has a higher capacity as the volatility of the permeant decreases.

Sensitivity Control

An additional benefit of the variable trapping time feature of the sample trapping system is the possibility of sensitivity control. Permeation testing demands different (and often conflicting) analytical requirements during different portions of the test run. At the point of breakthrough, high sensitivity (low detection limit) is critical but as permeation increases, higher and higher concentrations must be measured so a wide linear dynamic range is critical. The conflict between these two requirements can be lessened through the use of the sample trapping system at relatively long trapping times to obtain high sensitivity until breakthrough occurs, then reducing sampling time to reduce the dynamic range required to measure permeation rate. For example, in testing with the halogenated pyridines A through D described earlier, an initial trapping time of 0.5 min was used until the peak area reached a preset value; at this point the computer controlling the valve automatically reduced the trapping time to 0.05 minutes thus reducing by one order of magnitude the linear dynamic range required of the GC. This feature becomes especially important when using analytical detection techniques with inherently narrow linear dynamic range (for example, GC with electron capture detection).

Future Directions

Two areas of probable future directions in the technique of sample trapping for permeation testing that would broaden its applicability include improvement of gas-phase traps and extension of the technique to water collection medium.

Capillary column manufacturers are continually improving their columns to make them

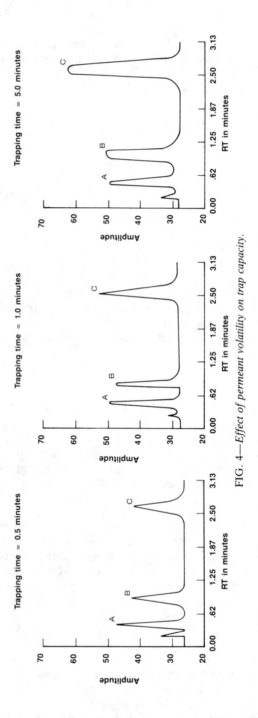

FIG. 4—*Effect of permeant volatility on trap capacity.*

more retentive for low-boiling-point compounds. This is generally done through the use of new and thicker films on the inside of the columns. These same features make them better sample traps, expanding the application of the sample trapping technique described here to higher vapor pressure compounds and increasing the capacity of a given length of sample trap.

While the gas-phase technique described will greatly expand the range of compounds that can be tested using a gaseous collection medium, there still will be some whose volatility is so low that water will be the preferred collection medium. For such compounds, whose water solubility still may be very low, the development of an analogous trapping system for water collection medium would offer the same advantages as the system described. Such a system could be based on liquid chromatographic analysis or, through the use of an automated extraction technique, gas chromatographic analysis.

Summary

In summary, an automated sample trapping technique has been described that facilitates permeation testing of low volatility compounds. The technique works better as the test compound volatility gets lower and offers the additional benefit of easy variable sensitivity thus allowing one to tailor the system to the sensitivity and dynamic range requirements for the particular test compound.

References

[1] McLenny, W. and Pleil, J., *Analytical Chemistry*, Vol. 56, 1984, p. 2947.
[2] Lonneman, W. A., Kopczynski, S. L., Darley, P. E., and Sutterfield, F. D., *Environmental Science and Technology*, Vol. 8, 1974, p. 229.
[3] Seila, R. L., EPA Report No. 600/3-79-010; U.S. Environmental Protection Agency, Research Triangle Park, NC, 1979.
[4] Pellizzari, E. D., Bunch, J. E., Burkley, R. E., and McRae, *Journal of Analytical Chemistry*, Vol. 48, 1976, p. 803.
[5] Jennings, W., *Gas Chromatography with Glass Capillary Columns,* 2nd ed., Academic Press, New York, 1980.

Christian Ursin[1] and Iver Drabaek[1]

Carbon-14 Tracers in Permeation Studies: Feasibility Demonstration

REFERENCE: Ursin, C. and Drabaek, I., **"Carbon-14 Tracers in Permeation Studies: Feasibility Demonstration,"** *Performance of Protective Clothing: Second Symposium, ASTM STP 989,* S. Z. Mansdorf, R. Sager, and A. R. Nielsen, Eds., American Society for Testing and Materials, Philadelphia, 1988, pp. 286–294.

ABSTRACT: This paper deals with results from a feasibility study initiated by the U. S. Coast Guard under Contract DTCG39-85-R-80272 on Radioactive Tracers in Permeation Studies.

Permeation tests using ^{14}C-labeled permeant chemicals have been performed according to ASTM Test Method for Resistance of Protective Clothing Materials to Permeation by Hazardous Liquid Chemicals (F 739-81). The method has included water as the collecting medium, discrete sampling with replenishment, and counting of the samples with liquid scintillation counting. Tests with Viton/chlorobutyl laminate on polyester support and acetone as permeant plus neoprene and nitrobenzene are reported. Due to a high sensitivity, breakthrough is seen in both cases in the first samples, that is, after 2 min for Viton/chlorobutyl laminate/acetone and after 2 s for neoprene/nitrobenzene. Steady-state permeation rate is achieved after 3.5 h at 0.49 mg/m²/s[1], and 1 and 0.3 h at 5.93 mg/m²/s[1] for the Viton/chlorobutyl laminate/acetone system and the neoprene/nitrobenzene system, respectively.

For the Viton/chlorobutyl laminate/acetone system, the propagation versus time of the permeant front inside the membrane has been visualized by autoradiography using a freezing technique and a cryomicrotome.

An autoradiographic calibration technique has been developed and applied giving the absolute concentration (mg/g) versus depth in microns from the exposed side of the membranes.

The tracer technique in combination with dry combustion has been applied to well-known decontamination procedures with the neoprene/nitrobenzene system. After thermal treatment, a residual amount of 0.08% nitrobenzene was found, implying a decontamination efficiency of 99.9%.

KEY WORDS: carbon-14, radioactive tracer, autoradiography, protective clothing, chemical permeation, decontamination, Viton/chlorobutyl laminate, neoprene, nitrobenzene, acetone

The use of carbon-14 (^{14}C) labelled organic compounds provides several advantages in the field of permeation studies. The ^{14}C isotope can be detected with very high efficiency, and permeation rates below 0.0001 mg/m²/s can be measured.

The combined use of radioactive labeled organic permeants and autoradiography makes it possible to visualize concentration gradients within a protective clothing membrane for one given chemical exposure time with a spacial resolution in the order of 5 μm. By doing sequential experiments, the propagation of the permeant "front" during the process of permeation can be revealed.

Autoradiography is a well-established technique within biological science [1], but despite its potential it has hitherto only been used rarely in combination with polymer membranes [2,3]. One of the main reasons for this might be due to the inherent problems caused by

---

[1] Danish Isotope Centre, Copenhagen, Denmark.

incompatibilities between the materials used in normal autoradiography and the polymers. Furthermore, the volatile nature of most organic solvents force the individual autoradiographic steps to be made at very low temperatures in order to fix the organic permeant within the protective clothing material [4,5].

Exploration of the feasibility of the ^{14}C tracer technique in combination with autoradiographic visualization has been performed by the Danish Isotope Centre under contract with the U. S. Coast Guard.

Work has been limited to three materials, that is, Viton[2]/chlorobutyl laminate on polyester support, Teflon coated Nomex and neoprene, and a limited number of test chemicals (for example, acetone and nitrobenzene).

Besides looking at permeant concentration gradients within these materials after normal exposure tests, special attention has been given to the efficacy of decontamination.

The work is still in progress. This paper describes the technical approach and the results obtained so far.

Experimental

Chemical Exposure

Chemical exposures are performed using an ASTM 1-in. cell.

An aqueous collecting medium has been used with discrete sampling and replenishment. The samples were counted in a liquid scintillation counter. Both Viton-chlorobutyl laminate and neoprene have been tested in this way. The tests were performed according to ASTM Test Method for Resistance of Protective Clothing Materials to Permeation by Hazardous Liquid Chemicals (F 739-81), using the ASTM 1-in. cell (slightly modified), water as the collecting medium (15.8 mL), and discrete sampling (2 mL) with replenishment. The specific activities of both the ^{14}C-labeled acetone and the nitrobenzene were 0.5 MBq/g. Blind and background corrections were made using samples collected before addition of the ^{14}C-labeled compounds.

Autoradiographic Techniques

The autoradiographic technique includes an ordinary chemical exposure test using a ^{14}C labeled test chemical. By the end of the exposure, the test material is removed from the test cell, and in order to keep the test chemical at a fixed position, frozen at −120°C above liquid nitrogen. A cross-section of the test material is obtained by cutting with a cryomicrotome. A nuclear emulsion is put on top of the sectioned material and after a suitable autoradiographic exposure time the emulsion-coated section is processed by conventional autoradiographic developing techniques. Silver grains induced by the ^{14}C-radioactivity now reveal both the position and a concentration equivalence of the test chemical within the cross-section of the test material. The final visualization and quantification is done through a series of steps including microscopical grain counting, high temperature dry combustion, liquid scintillation counting, and different calibration procedures.

Autoradiographic techniques are normally used for biological materials. The application to protective clothing polymer materials has called for a number of technical modifications within almost all individual steps. So far, no universal freezing and cutting technique for the different test materials were found, and a lot of efforts have been directed towards developing the technical approach.

---

[2] Registered trademark of E. I. duPont de Nemours and Company.

The technique used at present is based on Kodak Fine-Grain Autoradiographic Stripping Plate, AR 10. Using this stripping film, contact without change of position between specimen and film during all steps was shown feasible. The technique was originally developed for autoradiography of freeze-dried sections [6] and has been modified for the current application. After the cryosectioning at −30°C, the sections are collected on Scotch (3723) low temperature tape. The collected sections are transferred at −50°C to the darkroom, and are then glued to the stripping film with ethyl cyanoacrylate.

The resulting "sandwich" is illustrated in Fig. 1.

The autoradiographic exposure period is typically carried out somewhere between 2 to 21 days. The exposure is performed at −120 to −150°C.

After exposure, the autoradiograms are processed according to Kodak's recommendations.

Evaluation and Calibration

The autoradiograms are evaluated in a Leitz orthoplan or a Leitz Metallux 3 microscope using a counting grid. Either *normal* transmitted bright field or incident dark field illumination is used, the latter especially in the case of the opaque neoprene.

Conversion of the grain density to concentration units is carried out in two ways. Either using a calibrated certified polymer reference standard kit in pre-cut strips containing a selected range of known specific activities (Amersham), or by cutting thin sections parallel to the exposed surface of the test materials (diskotomy) followed by extraction and counting with liquid scintillation counter (LSC). (In the autoradiographic procedure, perpendicular sections are cut.) The diskotomy is further used to verify and calibrate the concentration gradients found by autoradiography. In general, it has been found that especially for permeant chemicals with a high vapor pressure there is an inevitable evaporative loss during the autoradiographic steps. Curved shapes, however, do not seem to change significantly compared to those obtained by diskotomy. In terms of spatial resolution and the concurrent relationship between concentration and structural position, autoradiography is far more advantageous than diskotomy.

With the Viton/chlorobutyl laminate, parallel slices of 15 μm thickness extending the whole way through the material are cut at −30°C. The radioactivity and thus the concentration is determined in the liquid scintillation counter.

Chemographic effects [1] are evaluated by parallel experiments using both nonexposed membranes and membranes exposed to unlabeled compounds.

Decontamination

Two decontamination procedures have been tried in the project: immersion in Freon 113 for 10 min; and hot-air oven for 24 h at 100°C.

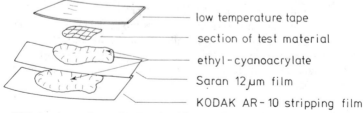

— low temperature tape
— section of test material
— ethyl-cyanoacrylate
— Saran 12 μm film
— KODAK AR-10 stripping film

FIG. 1—*Autoradiographic "sandwich" using the Kodak AR 10 stripping film.*

After a normal chemical exposure using a ^{14}C labeled test chemical, the test material is subjected to the decontamination procedure. The chemical exposure times were varied from 2 s to a time equal to previously reported breakthrough times plus 1 h.

A small piece of the decontaminated material is burned in a LECO combustion device, and the evolved ^{14}C-CO$_2$ is trapped and counted using LSC.

Another piece is subjected to autoradiography.

Results and Discussion

It should be emphasized that, at present, results are not available for all the techniques described in the experimental section.

Viton/Chlorobutyl Laminate

For the Viton/chlorobutyl laminate, chemical exposure tests have been performed using ^{14}C-labeled acetone (0.47 MBq/g). The tests have been run for predetermined total exposure times of 20 min, 20 h, and 72 h. A permeation curve including discrete samples up to 72 h is presented in Fig. 2.

From the permeation curve, a steady-state permeation rate, reached after 3.5 h, is calculated to be 0.49 mg/m^2/s. Breakthrough is significantly seen in the first sample collected after 2 min (breakthrough was found in 54 min using ordinary gas chromatography technique and unlabeled acetone).

Figure 3 shows an example of an autoradiogram of the yellow Viton/chlorobutyl laminate where the exterior side has been exposed to ^{14}C-acetone for 72 h.

It is clear from Fig. 3, that silver grains are seen on top of the exposed Viton/chlorobutyl laminate as black specks, and that there is a diminishing gradient from the exposed exterior side of the material towards the inside.

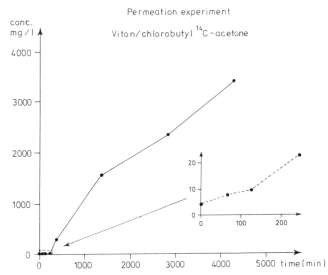

FIG. 2—*The concentration C_n (mg/L) versus time (min) for Viton-chlorobutyl laminate tested with ^{14}C-labeled acetone (0.5 MBq/g) as the permeant: ASTM 1-in. cell, water as collecting medium, discrete sampling (2 mL) with replenishment, counting with LSC. Breakthrough is significant in the first sample after 2 min.*

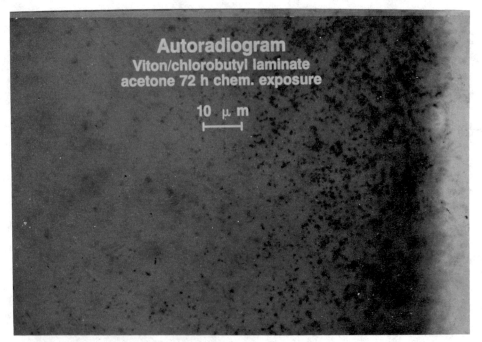

FIG. 3—*Autoradiogram (Kodak AR 10 stripping film—"sandwich" technique) of Viton/ chlorobutyl laminate exposed to ^{14}C-labeled acetone, 0.5 MBq/g, for 72 h. Autoradiographic exposure time is three days, normal light microscopic optics (magnification X1000). Exposed side to the right.*

Figure 4 shows the calibrated autoradiographic results. Through diskotomy and high temperature dry combustion, the autoradiographic results (grains per 100 μm^2) have been converted to chemical concentrations (mg/g) of acetone inside the Viton/chlorobutyl laminate. The eyefitted curves represent autoradiograms from the chemical exposure times 20 min, 20 h, and 72 h.

Comparison of the three curves in Fig. 4 shows that the descendent gradient of the activity versus penetration depth is very steep for the 20-min chemical exposure time, while it tends to flatten out at the 20-h chemical exposure time. The curve for the 72-h chemical exposure time is not significantly different from that for 20 h. This is in accordance with the permeation characteristics of the acetone for the Viton/chlorobutyl laminate, where a steady-state permeation rate is found after 3.4 h.

The drop in concentration towards the exposed side is probably due to inevitable evaporation loss.

Neoprene Elastomer

For the neoprene elastomer, chemical exposure tests have been performed using ^{14}C-labeled nitrobenzene. A permeation curve including discrete samples up to 75 min is presented in Fig. 5.

A steady-state permeation rate is reached after 30 min. The steady-state rate is calculated to 5.9 mg/m²/s. Significant breakthrough is seen in the first sample after 2 s in the splash test.

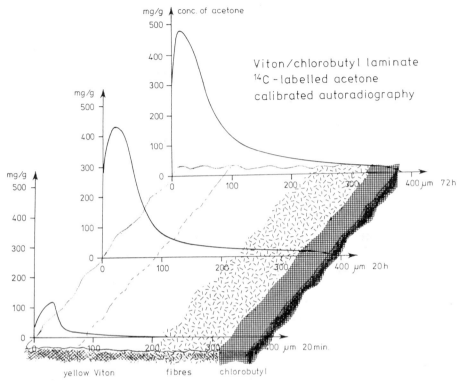

FIG. 4—*Autoradiography in combination with diskotomy and high-temperature dry combustion.*

Figure 6 shows the calibrated autoradiographic results after 2 s and after 75 min of chemical exposure. The autoradiogram after the 2-s splash test showed slight decendent grain density from the exposed exterior side to the unexposed side. In the outermost 15 μm, the concentration was 142 mg/g, but already after 20 to 30 μm, the concentration had decreased to 90 mg/g. Near the unexposed side, the concentration has decreased to less than 30 mg/g.

After 75 min of chemical exposure, the grain densities were much higher, and the concentration at the outermost 15 μm was 545 mg/g with a single peak value on 767 mg/g at 30 μm. (The peak is not visible on the smooth curve.) Near the unexposed side, the concentration was 450 mg/g. Compared to the Viton/chlorobutyl laminate/acetone system (for example, after 20 h), the neoprene/nitrobenzene system has an almost uniform concentration the whole way through the elastomer membrane.

Decontamination Experiments

Decontamination experiments have been carried out with the neoprene elastomer exposed to ^{14}C-labeled nitrobenzene for 75 min at 20°C.

Table 1 shows the results of the decontamination procedures with "Freon" 113 and thermal exposure for 24 h at 100°C. The decontamination efficiency is measured using dry combustion and liquid scintillation counting of the trapped $^{14}CO_2$. Evaluation was further done by autoradiography. Freon 113 for 10 min at 20°C gave a decontamination/efficiency of 87%,

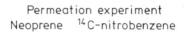

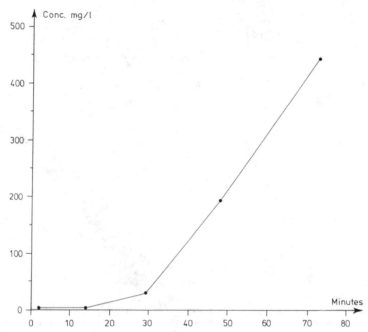

FIG. 5—*The concentration C_n (mg/L) versus time (min) for neoprene tested with ^{14}C-nitrobenzene (0.5 MBq/g): ASTM 1-in. cell, water as collecting medium, discrete sampling (2 mL) with replenishment, counting with LSC. Breakthrough is significant in the first sample after 2 s.*

while thermal treatment (100°C for 24 h) resulted in a decontamination efficiency of 99.9%. A significant residue of 0.08% nitrobenzene was found.

The corresponding autoradiograms of the neoprene decontaminated with Freon 113 showed very little difference between the exposed side, the unexposed side, and the intermediate region. The residual nitrobenzene was considered homogeneously distributed with a mean value of 4.3 grains per 100 μm^2. With the applied specific activity, thermal decontamination does not leave autoradiographically detectable amounts of ^{14}C-nitrobenzene within the neoprene.

Conclusion

Radioactive tracers in permeation studies seem to be very well suited to study permeation behavior, residual contamination, and efficiency of decontamination due to higher sensitivity than most other analytical procedures. In the Viton/chlorobutyl laminate/^{14}C-acetone permeation test, breakthrough was seen in the first sample collected after 2 min, while breakthrough time using gas chromatography with FID was found to be 54 min. This and other results suggest that the present concept of breakthrough might be related to the detector sensitivity, and if so, the concept of breakthrough should be reconsidered. The results show that the autoradiographic technique enables visualization of the chemical permeant inside the polymer matrix with a resolution in the order of 5 μm. Furthermore, an accurate

Neoprene
^{14}C-labelled nitrobenzene
calibrated autoradiography

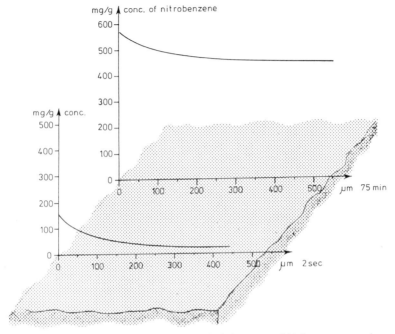

FIG. 6—*Autoradiography in combination with diskotomy and high-temperature dry combustion.*

TABLE 1—*Decontamination neoprene/nitrobenzene (after chemical exposure for 75 min at 20°C).*

Treatment after Chemical Exposure	Disintegrations per Minute Unit (exposed) Membrane (dry combustion)[a]	Nitrobenzene Residue, %	Decontamination Efficiency, %[b]	Autoradiography, g/100 μm²
None	119 600 ±1 555	(100)	...	...
Freon 113 for 10' min at 20°C (1 wash)	15 000 ±1 050	12.5 ±0.9	87 ±1	4.25
Thermal hot air oven 100°C for 24 h	90 ±32	0.08 ±0.03	99.9 ±0.1	below background

[a] One unit membrane is a circular disk of 5 mm diameter.
[b] % decontamination = % decontamination efficiency = $(Ca - Cb)/Ca \times 100$ where Ca is the contaminant concentration in the unwashed, exposed disk and Cb is the contaminant concentration in the washed exposed disk.

calibration method to convert the grain density to a chemical concentration appears feasible. The technique has proven very useful for evaluation of decontamination procedures, and with the high resolution autoradiography combined with dry combustion and liquid scintillation, counting both the position and the concentration of chemical residues within the polymer matrix can be established.

Acknowledgment

This work has been carried out under Contract DTCG39-85-R-80272 on Radioactive Tracers in Permeation Studies for the U. S. Coast Guard. The work was originally inspired by Dr. Alan Bentz, U. S. Coast Guard R&D Center.

References

[1] Rodgers, A. W., *Techniques of Autoradiography,* Elsevier Publishing Company, 3rd ed., New York, 1979.
[2] Joks, Z., Krejci, M., Menclova, B., Freyer, K., Treutler, M. C., and Birkholz, W., "Application of Autoradiography for Penetration Studies of Chemicals in Plastics," *Nukleonika,* Vol. 25, No. 10, 1980, pp. 1277–1284.
[3] Flessner, M. F., Fenstermacher, J. D., Blasbjerg, R. G., and Dedrick, R. L., "Peritoneal Absorption of Macromolecules Studies by Quantitative Autoradiography," *American Journal of Physiology,* Vol. 248, No. 1, 1985, pp. H26–H32.
[4] Kobayashi, T. and Bakay, L., "Autoradiography for Diffusible Substances and Its Application to Central Nervous Tissue," *Journal of Medical (Basel),* Vol. 2, No. 1, 1971, pp. 35–44.
[5] Cohen, E. N. and Hood, N., "Application of Low-Temperature Autoradiography to Studies of the Uptake and Metabolism of Volatile Anesthetics in the Mouse, I, Chloroform," *Anesthesiology,* Vol. 30, No. 3, 1969, pp. 306–314.
[6] Wedeen, R. P., "Autoradiography of Freeze-Dried Sections: Studies of Concentrative Transport in the Kidney," in *Autoradiography of Diffusible Substances,* L. J. Roth and W. E. Stumpf, Eds., Academic Press, New York, 1969, pp. 147–160.

Petr Salz[1]

Testing the Quality of Breathable Textiles

REFERENCE: Salz, P., "Testing the Quality of Breathable Textiles," *Performance of Protective Clothing: Second Symposium, ASTM STP 989,* S. Z. Mansdorf, R. Sager, and A. P. Nielsen, Eds., American Society for Testing and Materials, Philadelphia, 1988, pp. 295–304.

ABSTRACT: New laboratory methods have been developed for the evaluation of the water-proofness and water vapor permeability of breathable rainwear. The waterproofness was measured using a dynamic shower tester that combines mechanical action and wicking effects, thus closely simulating actual wear conditions. It is shown that some microporous fabrics that have a high water entry pressure allow leakage when they are periodically compressed. Multiple-layer assemblies of water-repellent fabrics were also evaluated. The water vapor transmission of breathable fabrics in rain was determined using a heated cup method in combination with an artificial rain installation. The results indicate that vapor transport characteristics of fabrics are subject to change depending on the weather.

The new methods provide a means for selecting fabrics for high-performance rainwear.

KEY WORDS: protective clothing, water-repellent fabrics, breathable fabrics, testing, multiple-layer fabrics

In the last few years, the diversity of waterproof, water vapor permeable fabrics has grown with the refinement of coating and laminating techniques. All major rainwear fabric manufacturers now offer breathable products that are meant to increase comfort in sportswear or relieve thermal stress in protective clothing. Different ways of achieving a combination of water repellency and vapor transport are used.

The oldest principle can be found in densely woven cotton fabrics. When wet, the swelling cotton fibers seal the pores in the fabric rendering it impermeable [1]. Several layers are necessary to make this type of fabric really waterproof. A giant step forward was made with the introduction of polytetrafluoroethylene (PTFE)-laminates in the late seventies. Miniature pores (size ~0.2 μm) in the highly hydrophobic PTFE-membrane prohibit the passage of liquid water but allow diffusion of water vapor [2]. Since the seventies, many microporous coatings, most of them based on polyurethane (PU), have been developed [3]. Also some hydrophobic microfiber fabrics have been constructed by combining 1.3 denier fibers in the warp- and 0.4 denier fibers in the fill direction, the latter having a diameter of only 6 μm [4]. The latest breathable fabrics are covered with a continuous hydrophilic layer. The waterproofness of these fabrics is inherent in the non-porous structure of the hydrophilic film. Transmission of vapor through the layer takes place by a complex molecular diffusion process [5].

In the face of the ever growing number of breathable rainwear fabrics, the users are confronted with the problem of choosing one that will meet their particular demands. A pre-selection based on durability can be made, but when it comes to waterproofing and comfort, an objective comparison is almost impossible: many of the fabrics withstand a hydrostatic head of 1 m (1 cm of water pressure = 98.0665 Pa) that makes them seem more

---

[1] Physicist, Fibre Research Institute TNO, Delft, The Netherlands.

than adequate for all but most extreme conditions. But are they? The vapor transmission rates are often difficult to compare because of the variety of the test methods, but also because the conditions during the tests do not even remotely reflect the conditions of rainy weather for which the fabrics are designed. Therefore, the practical value of the results is doubtful. New test methods have been developed that allow objective evaluation of breathable rainwear fabrics on grounds of waterproofing and comfort as a part of a research program on design of functional clothing.

Water Vapor Permeability

Test Principles

At present, the vapor transmission characteristics of fabrics are usually evaluated using one of the standardized methods similar to ASTM Test Methods for Water Vapor Transmission of Materials (E 96–80) and the Canadian Method of Test for Resistance of Materials to Water Vapor Diffusion (Can2-4.2-M77 Method 49, Control-Dish Method). In these methods, a specimen covers a dish containing either water or a desiccant that is placed in a well-defined low- or high-humidity environment. The change of weight of the dish with time gives the measure of vapor transport through the specimen. Other methods using more sophisticated techniques have been developed in an effort to simulate the microclimate in clothing [6–8].

In all these methods, the outer side of the specimen remains dry. In the apparatus described here, water vapor transmission is determined from the weight loss from a heated water-filled cup. Contrary to the previous methods, however, the specimen is continuously wetted by artificial rain during the test. Moreover, the rain cools the fabric and this can cause condensation of the escaping vapor on the reverse side or within the specimen in a similar way as in rainwear.

Apparatus and Methods

The cup and the rain installation are shown in Fig. 1. The rain installation produces 4 L/ h of water in drops of ~65 mg. The raindrops fall from a height of 27 cm over an area of 50 cm². The cup is installed at an angle of 20° to allow water to run off from the surface of the specimen. The temperature of the rain water is 20°C in the standard test described here, but it can be varied in order to simulate different conditions. The specimen is sealed to the aluminum cup, which is partly filled with water, with paraffin wax. Figure 2 shows a close up of the cup during a test.

The cup holder contains a hot-plate by which the temperature in the cup is regulated. The installation further consists of electronic control for the hot-plate temperature, recording devices for water and air temperature in the cup, and a water reservoir with a pump to refill the funnel of the rain installation.

During a test, the air temperature in the cup is kept at 36°C. Each test lasts 1 h, during which the specimen is continuously exposed to the artificial rain. The water vapor transmission is established in terms of weight loss from the cup.

The cup is weighed with the sample in place. This means that absorption of rain water on the outer side of the fabric causes an effective increase in weight. In order to be able to calculate the real loss of water vapor from the cup, the absorption of each specimen is measured in a separate test and added to the measured weight loss.

Fabrics that are not showerproof cannot be evaluated with this method because leakage will cause a weight gain in the cup. However, this is not an important restriction because such fabrics are not suitable for protective rainwear.

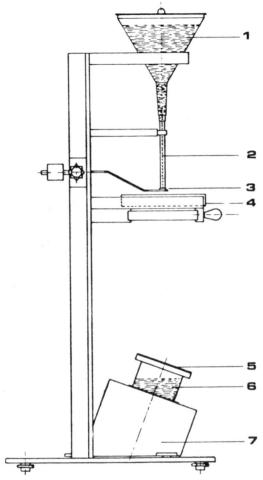

FIG. 1—*Test apparatus for determination of water vapor permeability in rain conditions: (1) funnel filled with water, (2) flow control tube, (3) rubber valve, (4) perforated PTFE plate, (5) specimen, (6) metal cup, and (7) cup holder with hot-plate.*

The heated cup can be also used for determining the water vapor transmission in dry conditions, simply by switching the rain installation off. This option (dry heated cup test) allows a comparison of fabric performance in dry and rainy weather.

Results

To facilitate the comparison of the results of the dry heated cup test with other more frequently used methods, the vapor resistance of a number of samples was also determined using the control-dish method.

Figure 3 shows that in spite of the differences in test conditions and techniques between the two methods, there is good correlation. The theoretical relationship between the tests, assuming an average air layer of 13 mm in the dry heated cup test, is also plotted in Fig. 3.

FIG. 2—*Close-up of the cup with specimen in place during a test.*

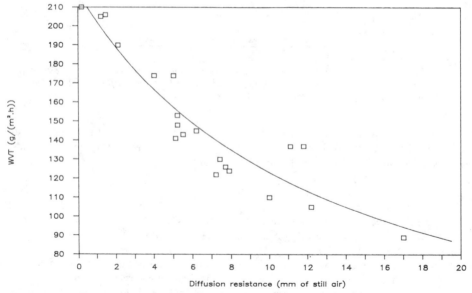

FIG. 3—*Relationship between the dry heated cup test and the control-dish method for breathable rainwear fabrics. The solid line shows the theoretical relationship.*

The water vapor permeability of different types of rainwear fabrics has been evaluated in dry and in wet conditions. The results are shown in Table 1. The data indicate, that rain has a major influence on vapor transport. In most cases, the vapor transmission rate for a dry specimen is higher than for a specimen wetted by rain.

It was not the goal of this study to analyze the complex mechanisms that determine the vapor transport through fabrics in wet conditions But it can be assumed that the vapor transport in rain is influenced by several factors:

1. Microporous fabrics can become virtually impermeable in rain due to the blocking of the micropores by water.
2. If the pores in the fabric are very small and highly hydrophobic, the blocking will not occur.
3. Saturation of hydrophilic membranes with rain water can prevent the adsorption of water vapor from the heated cup.

The two-layer PTFE-laminate shows an increase in vapor transport when it is exposed to rain. This increase is caused by the condensation of escaping water vapor on the reverse side of the specimen due to the cooling of the face side by the artificial rain. Thus, the air layer, which forms a vapor resistance in the dry test, is bridged in rain conditions. This increases the water vapor gradient over the specimen resulting in a higher rate of vapor transmission.

Dynamic Waterproofness

Test Principles

Microporous fabrics sometimes allow leakage that can not be predicted using hydrostatic head or shower tests. The water entry pressure, determined according to the hydrostatic pressure test (ISO 811-1981(E)), is often used to describe the waterproofness because shower tests like Bundesmann (DIN 53888-1979) and WIRA (British Standard 5066:1974) do not differentiate sufficiently between high-performance fabrics. The suitability of this parameter for evaluation of clothing has often been doubted because there is no impact of the raindrops on the specimen [9]. The other major disadvantage of the test is the absence of mechanical action and wicking effects to which clothing is subjected in normal use.

A new dynamic shower test has been devised in which clothing is simulated in a simple, straightforward way. A sleeve-like cylinder is made from the fabric, which is then subjected

TABLE 1—*Water vapor transport through rainwear fabrics in dry and rain conditions.*

	Conditions	
Material	Dry, $g/h/m^2$	Rain, $g/h/m^2$
Microporous PU-coated Fabric A	142	34
Microporous PU-coated Fabric B	206	72
Two-layer PTFE laminate	205	269
Three-layer PTFE laminate	174	141
Hydrophilic PU laminate	119	23
Microporous AC-coated fabric	143	17
Microfiber fabric	190	50
PU-coated fabric	18	4

to periodic bending in artificial rain. Thus, a pumping effect is created. Water-absorbent paper is placed at the inner side of the specimen to induce wicking.

Apparatus

The dynamic shower test, which has been named Vidybel, is shown in Fig. 4. It consists of the same artificial rain installation as the vapor permeability tester just described, a polypropylene arm, a motor to move the arm, and a programmable electronic counter.

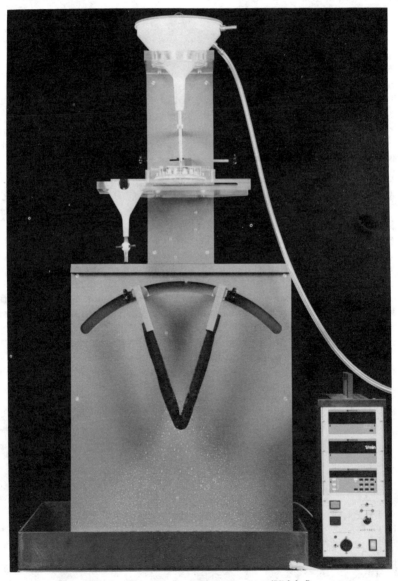

FIG. 4—*Dynamic shower test apparatus (Vidybel).*

The polypropylene arm has one joint, which is located below the rain source. The sleeve, which is made from the sample fabric, is fitted over the arm and a folded piece of filter paper is placed inside the sleeve over the joint area. Stripes of water-soluble ink are drawn on the filter paper. Any water penetration through the specimen is clearly visible as one or more stains on the filter paper.

During the test, the arm is bent 22 times each minute. The angle of the joint changes from 85 to 20° during each cycle. The standard length of each test is 20 min. This period is sufficient to differentiate between most fabrics. For fabrics that should offer a very high degree of protection over a long period of time, a longer test period may be necessary. For fabrics that are only water repellent, the test period is restricted to 5 or 10 min.

The water penetration is evaluated using a scale from 5 (no penetration) to 1 (very poor resistance to water penetration) shown in Fig. 5.

Results

One of the factors that influences the penetration is the way in which the specimen is folded in the joint area. Whereas in some thin fabrics sharp creases develop during the test, stiff fabrics develop blunt folds over a larger area. This is because the angle over which the

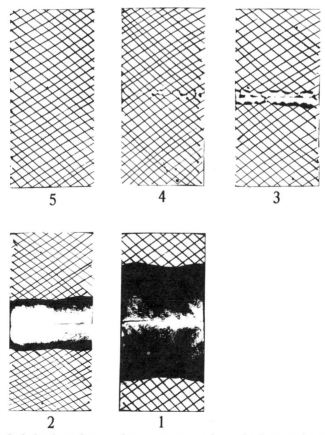

FIG. 5—*Scale for the evaluation of water penetration during the dynamic shower test.*

arm moves is fixed, rather than the deformation of the specimen. The sleeve fits snugly over the arm but is not fixed to it, to allow natural movement. The wrinkling of the fabric in the joint area is therefore very similar to wrinkling of rainwear in use.

The Vidybel test is more severe than conventional shower tests. Some fabrics, which allow no penetration on Bundesmann or Wira shower tests, are penetrated by water in the first few minutes of the Vidybel test. The cause of the leakage is found in the pumping effect, which is created by movement of the arm, combined with wicking.

Using the Vidybel test, water penetration was found through some microporous fabrics with a water entry pressure of 100 cm and more. Table 2 shows the results of an evaluation of a number of current rainwear fabrics.

Two or more water-repellent layers are sometimes used in rainwear in an effort to avoid the use of relatively costly microporous fabrics. None of the layers is truly waterproof, but the total assembly offers sufficient protection, as was shown in comparative wear trials by the Dutch Automobile Association and Dutch Mountaineering Association [10]. Several assemblies in cotton and polyester/cotton were evaluated using the Vidybel test, the hydrostatic head test, and the Wira shower test. The results are shown in Table 3.

Not all the fabrics tested are equally suitable for use in multiple-layer assemblies. Two of the fabrics allow considerable leakage within 5 min on the Vidybel test even when three layers are used. On the other hand, the KG 245 seems very suitable. Although the water entry pressure is only 23.6 cm, no penetration was found for a three-layer assembly after 30 min on the Vidybel tester.

Discussion

Water Vapor Transport

The change in vapor transport, which occurs when breathable fabrics are wetted, has a significant influence on the comfort of the wearer. To achieve thermal balance, the evaporative heat loss must usually be between 20 and 300 W/m^2 depending on the metabolism, ambient temperature, and clothing. The best fabrics described in this article allow water vapor transport of approximately 200 $g/m^2/h$ in dry conditions, which is approximately equivalent to an evaporative heat loss of 135 W/m^2. In rain, the transport through most of the fabrics is restricted to 50 $g/m^2/h$ ($\sim$35 W/m^2) or less. Particularly for persons in strenuous working conditions, this may prove to be insufficient.

The cooling effect of rain on rainwear fabrics can cause condensation of the escaping vapor. Although in some fabrics this leads to an increase of vapor transport, it also means that the wearer can become wet due to backward wicking. However, the total amount of

TABLE 2—Comparison of water entry pressure and dynamic waterproofness of rainwear fabrics.

Material	Hydrostatic Pressure Test, cm water	Vidybel Shower Test, rating[a]
Microporous PU-coated Fabric A	>100	2
Microporous PU-coated Fabric C	>100	3 to 4
Two-layer PTFE laminate	>100	4 to 5
Hydrophilic PU laminate	>100	5
Microporous AC-coated fabric	91	3 to 4
PU-coated fabric	>100	5
Ventile fabric	80	3 to 4

[a] Rating from 5 (no penetration) to 1 (very poor resistance to water penetration).

TABLE 3—*Waterproofness of water-repellent fabrics in multiple-layer assemblies.*

Material	No. of Layers	Wira Shower Test, mL passed	Hydrostatic Pressure Test, cm water	Vidybel Shower Test, Time, min	Vidybel Shower Test, Rating
KG 245	1	2.5	23.6	10	2
(PES/CO,65/35)	2	0	31.6	10	4 to 5
	3	0	35.2	30	5
KL 107 (CO)	1	28	21.3	5	2 to 3
	2	3	29.2	5	3
	3	0	31.5	15	1 to 2
KL 275 (CO)	1	26	28.1	5	2 to 3
	2	1.5	34.8	5	2 to 3
	3	0	39.1	5	3
KL 249 (CO)	1	2	29.1	10	2
	2	0	34.4	10	2
	3	0	39.7	10	2 to 3
KL 266 (CO)	1	52	23.6	5	1
	2	23	30.4	5	2
	3	0	32.9	5	5

moisture inside the clothing will be significantly lower if the breathability of the rainwear in rain conditions is high.

Dynamic Waterproofness

The Vidybel shower test has proved to be a useful tool for predicting the waterproofness of protective clothing. It is suitable for evaluating workwear and sportswear fabrics that might be worn in rain for long periods of time. By combining artificial rain, pumping, and wicking, the Vidybel test simulates the conditions to which rainwear is exposed. However, for applications where the kinetic energy of the rain drops on impact is very high, other test methods may be necessary.

Conclusions

The variation in water vapor permeability in different atmospheric conditions means that the choice of the most suitable fabric will depend on the demands of the user. For some, vapor permeability under dry conditions will be sufficient. However, people who work outdoors for longer periods will only remain dry and comfortable in clothing that allows water vapor to escape even when it is raining.

Rainwear fabrics that show good results on the hydrostatic head test should be checked for leakage due to pumping and wicking to ensure high performance under adverse conditions. It can be expected that in the future, dynamic shower tests will replace the hydrostatic head test, because the simulation of real conditions allows a broader range of fabrics to be evaluated.

Acknowledgments

The author wishes to thank J. Verboekend for his contribution in designing the apparatus and conducting the experiments.

References

[1] Taylor, H. M., "Ventile Fabrics: How and Why They were Developed," *Textile Institute and Industry*, Nov. 1975, pp. 359–360.

[2] Tanner, J. C., "Breathability, Comfort and Gore-tex Laminates," *Journal of Coated Fabrics*, Vol. 8, 1979, pp. 312–322.

[3] Lomax, R., "Recent Developments in Coated Apparel," *Journal of Coated Fabrics*, Vol. 14, 1984, pp. 91–99.

[4] "Active Sportswear Materials of Japanese Synthetic Fiber Producers," *Japan Textile News*, Nov. 1984, pp. 21–44.

[5] Lomax, G. R., "Coated Fabrics: Part 1-Lightweight Breathable Fabrics," *Journal of Coated Fabrics*, Vol. 15, 1985, pp. 115–126.

[6] Day, M. and Sturgeon, P. Z., "Water Vapor Transmission Rates through Textile Materials as Measured by Differential Scanning Calorimetry," *Textile Research Journal*, Vol. 56, 1986, pp. 157–161.

[7] Farnworth, B. and Dolhan, P., "Apparatus to Measure the Water-Vapour Resistance of Textiles," *Journal of the Textile Institute*, Vol. 75, 1984, pp. 142–145.

[8] Mecheels, J. and Umbach, K. H., "Thermofysiologische Eigenschaften von Kleidungsystemen," *Melliand Textilberichte*, Vol. 58, 1977, pp. 73–81.

[9] Norris, C. A., in *Waterproofing and Water-Repellency*, J. L. Moillet, Ed., American Elsevier Publishing Company, Inc., New York, 1963, Chapter 9, pp. 265–296.

[10] "Ventilating Rainwear," (in Dutch), *Op Pad*, No. 1, 1983, pp. 12–17.

Protection from Industrial Chemical Stressors

Field Test Methods and Application of Laboratory Data

Gao Fang,[1] Hsin Dingmao,[1] and Wang Huaimin[1]

Using Dipropyl Sulfide to Test the Leakage of Protective Clothing

REFERENCE: Fang, G., Dingmao, H., and Huaimin, W., **"Using Dipropyl Sulfide to Test the Leakage of Protective Clothing,"** *Performance of Protective Clothing: Second Symposium, ASTM STP 989,* S. Z. Mansdorf, R. Sager, and A. P. Nielsen, Eds., American Society for Testing and Materials, Philadelphia, 1988, pp. 307–313.

ABSTRACT: Dipropyl sulfide is used as a leakage testing gas for either a carbon-containing permeable chemical protective suit or a permeation-resistant suit. It has the advantages of low toxicity, easiness of adsorption by activated carbon, relatively high vapor pressure, and commercial availability.

A sensitive Chloramide-Ester-Congo Red (CEC) indicator paper (cloth) was developed that can detect 0.014 mg/dm³ of dipropyl sulfide vapor by color change within 2 min.

Testing procedures are demonstrated that can evaluate the protective suit's airtightness.

KEY WORDS: protective clothing, dipropyl sulfide, airtight suit

Protective clothing must isolate man from toxic environments; this is one of the basic principles in protective clothing design work. The general method of testing the leakage of protective clothing, especially the leakage of closures, is to let a man wear an indicator suit under the protective suit to be tested, and to do some scheduled exercises in a testing gas chamber. The location of leaking point will be clearly indicated by the color change on the inner suit. If the concentration of testing gas in the chamber is C_0, the sensitivity of the indicator suit to the testing gas is C_1, and the indicator suit does not change its color at all, the protective factor (*PF*) of this protective suit will be

$$PF \geq \frac{C_0}{C_1}$$

In our laboratory, chlorine or ammonia have been used as testing gases and fluorescein or pH indicator impregnated suit have been used as an indicator suit. These methods have the advantages of simplicity and distinctness; nevertheless, neither of these two methods meets our requirements. The chlorine method can not measure high *PF* value because the sensitivity of fluorescein is not high enough. The ammonia method cannot be used in the recently developed carbon-containing permeable protective suit, since the adsorption equilibrium value of ammonia on activated charcoal is too low.

In the development of the Chinese carbon-containing M-82 permeable protective suit, we

---

[1] Professor, professor, and engineer, respectively, Research Institute of Chemical Defence, Beijing, China.

have used dipropyl sulfide as a leakage testing gas,[2] developed a sensitive Chloramide-Ester-Congo Red (CEC) indicator, and found this method is very convenient to test the leakage of protective clothing, either carbon-containing permeable or permeation resistant.

Search for a Testing Gas

For safety reasons, it is advantageous to use nontoxic or lower toxic chemicals (to respiratory organs and to skin) to test the leakage of protective suits, moreover, the testing gas must have following properties:

1. relatively high vapor pressure, easy to establish desired concentration in testing chamber;
2. easy to be adsorbed by activated charcoal, so the man can be protected by respirator and the testing gas does not penetrate the carbon-containing clothing directly;
3. can be sensitive to detection by a color change indicator; and
4. easy to obtain commercially.

In our laboratory, n-tolyl,n-chlorobenzoylamide-Congo Red test paper (TCBA-CR paper) has been widely used for a long time to detect mustard gas vapor.[3] Although this paper is very sensitive to mustard gas (4 μg/cm^2), it was found to be insensitive to dipropyl sulfide under dynamic conditions.

Cadogan et al.[4] reported that many phosphorus-sulfur or carbon-sulfur bond containing

TABLE 1—*Reaction of TCBA-CR paper.*

Compound	Boiling Point, K	With TCBA-CR Paper		With Congo Red Paper	
		Liquid	Vapor	Liquid	Vapor
n-C$_3$H$_7$\,S / n-C$_3$H$_7$	416.0	+	+	−	−
n-C$_5$H$_{11}$\,S / n-C$_5$H$_{11}$	500.6	+	+	−	−
CH$_3$S\,P\,CH$_3$ / C$_2$H$_5$O / O		+	+	−	−
C$_2$H$_5$S\,P\,CH$_3$ / C$_2$H$_5$O / O	339.2 to 339.7 (133.3Pa)	+	+	−	−
C$_4$H$_9$S\,P\,CH$_3$ / C$_4$H$_9$O / O	543.2	+	+	−	−

[2] Fang, G., in *Proceedings,* Second International Symposium on Protection Against Chemical Warfare Agents", FOA Report C-40228-C2,C3, National Defence Research Institute, Stockholm, Sweden, 1986, pp. 51–58.

[3] Chaowen, Y., private communication.

[4] Cadogan, J. I. G., et al., *Journal of Chemical Society,* The Chemical Society, London, 1961, p. 5524.

TABLE 2—*Physical properties of dipropyl sulfide (293.2 K).*

Density, g/cm^3	Surface Tension, N/cm	Vapor Pressure, Pa	Maximum Concentration in Air, mg/dm^3	Adsorption Equilibrium Value,[a] mg/g
0.8377	2.634×10^{-4}	629.28	30.47	277.5

[a]Adsorption equilibrium value is calculated from Dubinin's equation based on the Chinese Type 111 charcoal's structural constants ($W_0 = 0.367 \text{ cm}^3/\text{g}$, $B = 1.3 \times 10^{-6}$) when the concentration of dipropyl sulfide is 1 mg/dm^3.

compounds react with chloramide, breaking down these bonds and liberating hydrogen chloride (HCl), the liberated HCl reacts with Congo Red, changing the color of TCBA-CR paper from red to blue sharply.

Preliminary testing shows that thioethers and esters of phosphonothiolic acid react with TCBA-CR paper very sensitively (Table 1).

With regard to the toxicity, esters of phosphonothiolic acid are somewhat toxic and their toxicity is similar to that of 0,0-dimethyl-2,2-dichloro-vinyl phosphate (DDVP). Thioethers are practically nontoxic compounds, the LCt_{50} of dipropyl sulfide is 17 mg/dm^3 (2 h for mice).[5] It has no irritating effect on human skin and is relatively volatile. Moreover, its vapor can easily be adsorbed by activated charcoal (Table 2) and is commercially available.

Sensitivities of TCBA-CR paper (cloth) to dipropyl sulfide under different conditions are shown in Table 3. It shows that although the TCBA-CR paper (cloth) has enough sensitivity under static conditions, it appears insufficient under dynamic conditions. Therefore, the remaining problem is how to improve the sensitivity of TCBA-CR paper (cloth) under dynamic conditions.

Improvement of TCBA-CR Paper (Cloth)

The low sensitivity of TCBA-CR paper (cloth) under dynamic conditions is possibly due to the low absorption rate of dipropyl sulfide by detection reagents. A method is suggested that some absorbents such as dibutyl phthalate (DBP) can be added to TCBA-CR paper (cloth). The new, more sensitive Chloramide-Ester-Congo Red paper (cloth) is called CEC paper (cloth), and its sensitivity under dynamic conditions is shown in Table 4.

The addition of dibutyl phthalate really increases the dynamic sensitivity rapidly. The best TCBA:DBP composition to be chosen is 1:1, because the sensitivity is not enough when the composition is lower and the paper (cloth) will become tacky when the ester content is higher. The sensitivity of this composition (1:1) is about 0.014 mg/dm^3 (within 2 min).

Experiments also show that the static sensitivity of CEC paper (cloth) is increased to 3 to 4 $\mu g/cm^2$ (293.2 K).

TABLE 3—*Sensitivity of TCBA-CR paper (cloth) to dipropyl sulfide (293.2 K).*

Condition	Sensitivity
Static	6 $\mu g/cm^2$
Dynamic (air velocity = 40 cm/min, exposure time = 2 min)	2 mg/dm^3

[5] Lazarev, N. V., et al., *Toxic Substances in Industry*, 7th ed., Khimiya, Leningrad, USSR, 1976, Vol. 2, p. 387, in Russian.

TABLE 4—*Sensitivity of CEC paper (cloth) (299.2 K, air velocity = 40 cm/min, exposure time = 2 min).*

Weight Concentration (TCBA:DBP)	Intensity of Color Change of CEC Paper (Cloth) at Different Dipropyl Sulfide Concentration		
	0.010 mg/dm^3	0.014 mg/dm^3	0.023 mg/dm^3
1:0	−	−	−
1:0.01	−	−	−
1:0.05	−	−	−
1:0.1	−	−	−
1:0.5	−	+	+ +
1:1.0	−	+ +	+ + +
1:2.0	+	+ +	+ + +
1:5.0	+	+ +	+ + +

Therefore, if the concentration of dipropyl sulfide in the testing chamber is 1.3 mg/dm^3, and if there is no color change on the indicator suit (worn under protective suit), the *PF* of the testing protective suit will be greater than 90 (= 1.3/0.014) and satisfies the requirements for the Chinese M-82 permeable suit.

Before and after testing photographs are shown in Figs. 1 and 2.

Testing Procedures

The leakage testing procedures using dipropyl sulfide are as follows:

1. Preparation of indicator suit
   (1) Material:
   Indicator suit is made of thin white cotton cloth.
   (2) Procedures:
   (*a*) Degreasing: wash with soda water.
   (*b*) First impregnating: soak in 0.1% Congo Red alcohol water solution (1:1) for 5 min and dry at room temperature.
   (*c*) Second impregnating: soak in chloramide-dibutyl phthalate-chloroform solution (1:1:35) for another 5 min and dry again at room temperature.
2. Method of testing
   (1) Testing chamber:
   The testing chamber can be any size preferably 10 to 20 m^3. Its walls are made of less adsorptive materials such as plastics.
   (2) Wearing equipment:
   Man performing test must wear an indicator suit (with hat pad) under the testing protective suit and wear respirator, protective gloves, as well as protective over-boots.
   (3) Vapor concentration:
   Spray dipropyl sulfide in the testing chamber, make its concentration to 1.3 mg/dm^3 (by calculation).
   (4) Schedule:
   Enter testing chamber and do scheduled exercises for a definite time. The scheduled exercises in our laboratory are described in Table 5.
3. Examination of leakage:
   After testing, take off the outer protective suit outdoors and examine the inner suit.

FIG. 1—*The testing protective suit (gray) and the indicator suit (dark gray).*

The blue spots on the indicator suit will show the location of leakage of the protective suit. If no color change takes place anywhere, the *PF* value of the testing protective suit is well above 90.

Conclusions

We have used this method in the development of the M-82 Chinese permeable protective suit to improve its construction and to evaluate its tightness. We have found that this method is practically useful to permeable suits as well as permeation-resistant suits.

TABLE 5—*Schedule of exercises.*

Item	Shaking Head	Bending Down and Squating	Running	Working	Walking
Duration, min	2	2	2	3	6

FIG. 2—*After testing, the gray color spot on the indicator suit shows the location of the leakage.*

The following conclusions can be drawn.

1. Dipropyl sulfide can be utilized to test the leakage of protective suits. It has the advantages of safety and ease of operation.
2. A Chloramide-Ester-Congo Red (CEC) indicator has been developed. This indicator reacts very sensitively to dipropyl sulfide.

Acknowledgments

The authors wish to thank Fang Jiagu, Yu Chaowen, and Jiang Lintai for careful performance of the experiments.

APPENDIX

Synthesis of *n*-tolyl,*n*-chlorobenzolamide

$$0\text{-}CH_3C_6H_4NH_2 + C_6H_5COCl + NaOH \quad 0\text{-}CH_3C_6H_4NHCOC_6H_5 + NaCl + H_2O$$

$$0\text{-}CH_3C_6H_4NHCOC_6H_5 \xrightarrow{[Cl]} 0\text{-}CH_3C_6H_4NClCOC_6H_5$$
$$\text{(TCBA)}$$

Ten millilitres 0-toluidine and 40 mL ethyl alcohol are placed in a beaker, the mixture is cooled in an ice bath, and 15 mL 20% NaOH is added. In another beaker, 8.5 mL benzoyl chloride and 14 mL ethyl ether are mixed. These two solutions are added through two dropping funnels separately to a cooled beaker. The dropping speed is adjusted so that the two solutions will exhaust nearly at the same time.

The reaction mixture is filtered and the precipitate is washed with distilled water. If the product remains a pink color, it will be washed with ethyl alcohol or ether. After drying at 50 to 60°C, the product obtained is a white needle crystalline with melting point 142 to 144°C. The yield is about 95%.

Six grams of the product and 20 mL ethyl alcohol are placed in a flask. The mixture is stirred to make a suspension. A 250 mL $3Ca(OCl)_2:2Ca(OH)_2$ solution with 6% available chlorine is prepared, and 150 mL of its clear solution is added to flask. The flask is placed for more than 4 h at room temperature (it is best to cool the flask in an ice bath when the room temperature is above 30°C).

The mixture is filtered and the precipitate is washed with distilled water. The precipitate and 50 mL distilled water are added to a beaker. 10% HCl is added to neutralize the excess $3Ca(OCl)_2:2Ca(OH)_2$ until no longer rising in bubbles.

The mixture is filtered and the precipitate is washed with distilled water again until the filtrate is neutral. The precipitate is dried in a desiccator. The product (TCBA), melting point 71 to 72°C, is a white powder. Available chlorine of TCBA is above 25% (theoretically, it is 28.8%).

The TCBA must be preserved in a dark brown bottle, and its available chlorine content must be analyzed before use.

Arthur D. Schwope,[1] *Todd R. Carroll,*[1] *Robert Huang,*[1]
and Michael D. Royer[2]

Test Kit for Field Evaluation of the Chemical Resistance of Protective Clothing

---

REFERENCE: Schwope, A. D., Carroll, T. R., Huang, R., and Royer, M. D., "**Test Kit for Field Evaluation of the Chemical Resistance of Protective Clothing,**" *Performance of Protective Clothing: Second Symposium, ASTM STP 989,* S. Z. Mansdorf, R. Sager, and A. P. Nielsen, Eds., American Society for Testing and Materials, Philadelphia, 1988, pp. 314–325.

ABSTRACT: Personnel involved in emergency response and hazardous waste site activities often have the need to reach on-scene decisions regarding the effectiveness and limitations of chemical protective clothing. Three gravimetric techniques were evaluated as means for providing essential information for aiding such decisions. For four neat chemicals, three two-component solutions, and two clothing materials, permeation cup tests yielded breakthrough time and permeation rate data comparable to that from testing performed according to ASTM Test Method for Resistance of Protective Clothing Materials to Permeation by Liquids and Gases (F 739-85). Three prototype kits based on the cup test are presently undergoing field trials. The permeation cup test was selected over a simple immersion test and a proposed degradation test. Recommendations for further development of the cup test and the degradation test are given.

KEY WORDS: protective clothing, permeation, permeation cup, standards, field kit, chemical resistance, gravimetric measurements, nitrile rubber, butyl-coated nylon, degradation tests, break through time

The materials of construction of chemical protective clothing should be resistant to permeation and physical degradation by chemicals. The ASTM Test Method for Resistance of Protective Clothing Materials to Permeation by Liquids and Gases (F 739-85) and analogous procedures[3] were developed, and are widely applied for measuring the permeation properties of clothing materials. Although no standard yet exists specifically for protective clothing, resistance to physical degradation can be measured by application of one or more standard physical property tests to the materials before and after exposure to the chemical, namely, ASTM Test Method for Rubber Property—Effect of Liquids (D 471-79) and ASTM Test Method for Resistance of Plastics to Chemical Reagents (D 543-84).

In general, the preceding tests were designed for and are performed in the laboratory under controlled conditions. For example, the detection of permeation is typically achieved by means of gas chromatography, infrared spectroscopy, atomic absorption, and other meth-

---

[1] Unit leader, consultant, and research assistant, respectively, Arthur D. Little, Inc., Cambridge, MA 02140.

[2] Project officer, U. S. Environmental Protection Agency, Edison, NJ 08837-3679.

[3] Berardinelli, S. P., Mickelsen, R. L., and Roder, M. M., "Chemical Protective Clothing: A Comparison of Chemical Permeation Test Cells and Direct-Reading Instruments," *American Industrial Hygiene Association Journal,* Vol. 44, No. 12, 1983, pp. 886–889.

ods requiring expensive instrumentation and skilled personnel. Physical property measurements often require mechanized tension testing equipment. These tests are fairly reproducible and the results have allowed significant advances in clothing development and selection. One source of the permeation data through November 1986 is the third edition of the *Guidelines for the Selection of Chemical Protective Clothing*.[4]

In some applications, the accuracy and precision of laboratory testing may not be necessary nor is there sufficient time to initiate and await the results of such tests. Examples include emergency spill response and, in many cases, hazardous waste cleanup. These applications would benefit from a field test that could rapidly provide information on the limitations of the available clothing, or provide immediate direction to the selection and acquisition of clothing or work practices trade-offs. Such a test should:

1. Provide some indication of "breakthrough time" and permeation rate (But, it is not necessary that the specific permeant of a mixture be identified. In other words, the permeation of any chemical, regardless of its toxicity, would be interpreted as a breach of the barrier. Also, it is not necessary that the test actually measure breakthrough time or permeation rate in cases where good correlations exist between the results of other tests and these parameters.).
2. Be durable, portable, and self-contained and require no external power source (The number of parts should be small.).
3. Be simple and easy to learn and to perform (Minimal calibration should be required.).
4. Be of immediate benefit, requiring minimal development time and cost.
5. At a minimum, be applicable to a wide variety of liquid organic chemicals.

A decision matrix approach was used in a recent investigation of methods that would satisfy the preceding criteria.[5] Approximately 50 techniques were considered, including those based on ultraviolet and infrared absorption, ionization detectors, conductivity, colorimetry (for example, chemical length of stain tubes), and gravimetry. Of these, three gravimetric methods were identified as being the most expedient, cost-effective, and practical approaches to providing benefit to field personnel. These three methods are an immersion test, a degradation test, and a permeation cup test, and are the subject of this paper.

Experimental

Chemicals

Each of the above methods was performed with seven, chemical challenges: acetone, hexane, methanol, toluene, and 25/75, 50/50, and 75/25 volume/volume mixtures of acetone and hexane.

Materials

Nitrile rubber and butyl rubber-coated nylon were the test materials; the applicability of the test to both free-standing and coated materials was of interest. The nitrile was 0.04 cm

---

[4] Schwope, A. D., Costas, P. P., Jackson, J. O., Stull, J. O., and Weitzman, D. J., *Guidelines for the Selection of Chemical Protective Clothing,* 3rd ed., American Conference of Governmental Industrial Hygientists, Inc., 6500 Glenway Ave., Cincinnatti, OH 45211.
[5] Stampfer, J. F. and Schwope, A. D., "Feasibility Study of a Field Test Kit for Chemical Protective Clothing," Los Alamos Report LA-UR-85-3717, Los Alamos National Laboratory, Los Alamos, NM, Sept. 1985.

in thickness and taken from a glove (Pioneer StanSolv Nitrile (A-15)). The butyl-coated nylon was 0.03 cm in thickness and of the type that is used in full-body encapsulating ensembles.

Immersion Test

In the immersion test, a specimen of the clothing material is weighed, then completely immersed in the chemical or chemical mixture, and at a specified time removed, patted dry, and reweighed. In some cases, the specimen is removed, reweighed, and reimmersed several times during the course of the test. This was the case in our experiments since one objective was to establish an exposure duration that would provide insight into the weight change characteristics of the clothing specimen as a function of time. Percent weight change is calculated by subtracting the initial weight from the final weight, dividing the result by the initial weight, and multiplying by 100%. The initial weight of the specimens in this study ranged from 0.414 to 0.481 g. Weights were measured using a four decimal place, analytical balance (Mettler Model AE163). Each chemical/material pair was tested in triplicate.

Degradation Test

The degradation test was performed according to a draft of a proposed standard test.[6] A 5.08 cm (2-in.) diameter specimen of the clothing material was exposed on one side to the challenge chemical(s) for 1 h. The weight, length, and elongation of the specimen were measured before and after the exposure period. Similar to the immersion test, percentage changes were then calculated. Weights were measured with the Mettler AE163 balance and thickness with a digital, linear micrometer (Ono Sokki EG-307). In the thickness measurements, the foot was 0.52 cm in diameter and the load was 200 g. Elongation was measured using a linear scale having 0.25 cm (0.10-in.) demarcations and read to the nearest 0.13 cm (0.05-in.). Each chemical/material pair was tested in triplicate.

The test cell used in this study was a modification of that in the ASTM/NIOSH draft method. A standard Telon gasket (Corning Style 1-2, solid TFE, Type T, Corning Glass Inc.) was used in place of the specified gasket. A 0.63-cm thick stainless steel base plate was used in place of the specified polyethylene base plate. A standard, metallic flange (Corning Style 1, aluminum, Corning Glass Inc.) and associated bolts were used in place of the specified upper polyethylene ring/wing nut fastener system. These changes resulted in no perceptible leakage from the cell and, more importantly, a better defined and controlled area exposed to the chemical. With the soft gasket specified in the draft method, lateral diffusion of the chemical beyond the gasket was observed. The weight gained due to this lateral diffusion was found to confound the weight change calculation.

Permeation Cup Test

The permeation cup was 1.50 cm deep and 7.67 cm in diameter (46.2 cm^2) and was fabricated from 6061 aluminum, Fig. 1. The challenge chemical was placed in the cup and

---

[6] "Test Method for Evaluating Protective Clothing Materials for Resistance to Degradation by Liquid Chemicals," NIOSH 200-84-2702 Deg. (Rev. 4), National Institute for Occupational Safety and Health, Morgantown, WV, 1985.

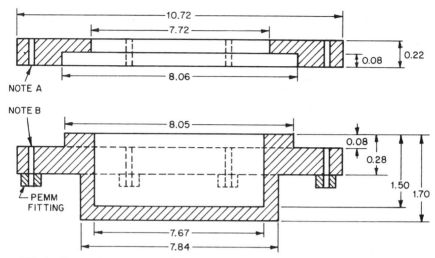

FIG. 1—*Permeation cup: material is 6061 aluminum. All dimensions are in centimetres (not to scale). Note A—top ring drill #16 drill (0.45) six equally spaced on a 9.88 B.C. Note B— bottom retainer drill #3 drill (0.45) six equally spaced on a 9.88 B.C.*

then the cup covered with the clothing material, sealed, weighed, and inverted so that the chemical was in direct contact with the clothing material. The normally outside surface of the clothing was placed toward the chemical; exposure was single-sided. The cup was weighed several additional times during the exposure period. The weight change per unit area was plotted versus time in order to obtain the typical cumulative permeation curve. Breakthrough time can be estimated and the permeation rate calculated from the slope of the cumulative permeation curve. One hour is suggested as the minimum test duration although useful results were obtained in less time with certain clothing/chemical combinations.

The test is identical in concept to ASTM Test Methods for Water Vapor Transmission of Materials (E 96-80) for measuring the moisture permeability of polymeric films. However, the ASTM E 96-80 cell was redesigned to obviate the need for rubber gaskets and wax seals and to meet weight constraints imposed by presently available battery-powered, top loading balances. A seal was obtained by means of a lip on the inner circumference of the cup. The capacity of the most sensitive battery-powered balance suitable for field use is presently 100 g with a two decimal place scale and a sensitivity of ±0.01 g (Whatman Model 100, Portable Electronic Balance, Fisher Scientific). Permeation data for the cup reported herein were obtained with the four decimal place balance identified earlier and at temperatures ranging from 18 to 24°C. The tests were performed in triplicate.

Standard Permeation Tests

Baseline data for comparison of the results of the preceding tests were obtained according to ASTM F 739-85. In all cases, the tests were performed in duplicate and a Miran 80A infrared spectrophotometer was the detector (Foxboro Corp). Each permeation test was begun in closed-loop mode in order to maximize sensitivity for detection of breakthrough, and completed in open-loop mode in order to maintain the permeant concentration within the range of the calibration curve. The volume of the closed-loop system was approximately 6 L and the flowrate of air in open-loop mode was approximately 10 L/min. The permeation rates at breakthrough detection are summarized in Table 1. In cases where mixtures of the two chemicals constituted the challenge, the permeation of each chemical was monitored by measuring the absorbances at both characteristic wavelengths and then reducing the data by a matrix technique described in the instrument instruction manual.

Results

All results are summarized in Table 2 for the nitrile rubber and Table 3 for the butyl-coated nylon material.

Immersion and Degradation Tests

In general, the relative orders of the precent weight changes are similar for the immersion and degradation tests for each material. However, the weight change values of the degradation tests were always less than those of the immersion tests. This finding was not unexpected in view of the differences between the test procedures. In the immersion test, the specimen was engulfed in chemical; while in the degradation test, the exposure was single-sided *and,* more importantly, the unexposed side was open to the atmosphere. The consequence of this second fact is that chemical could evaporate from the unexposed surface thereby forming a concentration gradient across the specimen. The specimen of the degradation test could not become saturated as would the immersion specimen in the times indicated in the third column of Tables 2 and 3, except if there were no evaporation. The degree of air movement over the unexposed surface will determine the shape of the concentration gradient—the steeper the gradient, the less chemical in the specimen; the more shallow the gradient, the more chemical in the specimen. Thus, with the proposed degradation test, different weight change values can be obtained for the same chemical/material pair, depending on the air flowrate over the clothing material surface. For this reason, we would propose that the degradation test apparatus should be modified so that no evaporation

TABLE 1—*Permeation rate at breakthrough detection by ASTM F 739-85.*

| | Infrared Conditions | | | | |
| | | Pathlength, m | | Permeation Rate[a] | |
	Wavelength,[c] nm	Nitrile	Butyl	Nitrile	Butyl
Acetone	7.3,[N] 8.5[B]	0.75	20.25	73.1	ND
Hexane	3.35	20.25	0.75	ND[b]	1.6
Methanol	9.74	0.75	14.25	3.2	ND
Toluene	13.8	6.25	2.25	62	175
Acetone/hexane (25/75, v/v)	7.3[A]/3.35[H]	2.25	3.75	14.6	28.9
Acetone/hexane (50/50, v/v)	7.3[A]/3.35[H]	2.25	3.75	63.5	18.7
Acetone/hexane (75/25, v/v)	7.3[A]/3.35[H]	0.75	0.75	152	11.2

[a] Permeation rate in units of $\mu g/cm^2$ min.
[b] No breakthrough detected in 6 h.
[c] Reference wavelength 4 nm.

TABLE 2—Test results (means) for nitrile rubber.

	Immersion Test		Degradation Test			Permeation Cup		ASTM F 739-85	
	Weight Change, %	Time, h	Weight Change, %	Thickness Change, %	Elongation Change, %	BT^a	PR^b	BT	PR
Acetone	178	<1	79	16	554	4	2500	4.5	89.50
Hexane	5	>1	0.2	−0.1	25	ND^c	ND	ND	ND
Methanol	35	1	2	6	241	40	49	39	86
Toluene	132	<1	103	19	270	13	720	12	1370
Acetone/hexane (25/75, v/v)	52	≪1	4	11	258	9	550	11	650
Acetone/hexane (50/50, v/v)	100	≪1	54	13	307	6	1200	6	1900
Acetone/hexane (75/25, v/v)	166	≪1	2	17	361	4	1800	5	3300

[a] Breakthrough time, min.
[b] Steady-state permeation rate, μg/cm² min.
[c] No breakthrough detected in 1 h for the cup test or 6 h for ASTM F 739-85.

TABLE 3—Test results (means) for butyl-coated nylon.

	Immersion Test		Degradation Test			Permeation Cup		ASTM F 739-85	
	Weight Change, %	Time, h	Weight Change, %	Thickness Change, %	Elongation Change, %	BT[a]	PR[b]	BT[a]	PR[b]
Acetone	8	<1	6	5	NM[c]	ND[d]	ND	ND	ND
Hexane	72	<1	59	60	NM	30	55	13	83
Methanol	11	1	4	2	NM	ND	ND	ND	ND
Toluene	103	<1	31	31	NM	7	290	7	528
Acetone/hexane (25/75, v/v)	59	<1	40	46	NM	7	250	4	465
Acetone/hexane (50/50, v/v)	22	≪1	14	15	NM	6	220	6	330
Acetone/hexane (75/25, v/v)	13	≪1	10	10	NM	13	80	16	92

[a] Breakthrough time, min.
[b] Steady-state permeation rate, $\mu g/cm^2$ min.
[c] None measured.
[d] No breakthrough detected in 1 h for the cup test or 6 h for ASTM F 739-85.

can occur from the unexposed surface; that is, the backing plate should be a solid flat plate with no hole.

For nitrile rubber, there appears to be little or no correlation of the elongation results with the weight change results, nor as will be seen later with the results of the permeation tests. No results were obtained from the elongation test of the butyl-coated nylon material since the nylon reinforcement in this material prevented any elongation of the material either before or after the exposure to chemicals. For these reasons, there would appear to be no merit in including the measurement of elongation change in a field test procedure.

The results of the thickness measurements mirrored fairly well the weight change results trend in that greater thickness changes generally occurred where there were greater weight changes. An observation that remains unexplained is that the ranges of thickness changes for the two materials are significantly different—for nitrile 0 to 20%, and for butyl-coated nylon 2 to 60%. This observation is interesting because the ranges of results for the other parameters are similar to each other. One could speculate that the variance in thickness ranges was due to the somewhat subjective nature of the thickness measurement procedure. Although the pressure applied to both clothing materials was the same, it may be that the same pressure applied to one material will produce results that differ from those of another material (for example, a coated versus a free-standing material). If this were so, then it would not be possible to generalize results of thickness change measurements. Separate scales correlating thickness change to permeation properties would be required for each material. While the preceding discussion is speculative, it would seem imperative that it be refuted or validated before thickness change is adopted as a key parameter for estimating the chemical resistance of clothing materials.

Permeation Cup and ASTM F 739-85 Tests

The agreement between the breakthrough times measured gravimetrically with a four decimal place balance and those measured by infrared spectroscopy were remarkably similar. Only in the case of hexane permeation through the butyl material did the breakthrough times differ by more than 3 min. This consistency of the breakthrough time results applies to both the neat chemicals and the acetone/hexane mixtures. For the mixtures, the permeation of both components was monitored independently in the F 739-85 tests, and the results reported in Tables 2 and 3 are the composite of the results—first breakthrough regardless of chemical identity and total permeation rate. An example of the permeation curves obtained for the mixtures by ASTM F 739-85 is shown as Fig. 2. Figures 3 and 4 exemplify the similarities of the results from the permeation cup and ASTM F 739-85.

There was less agreement between the permeation rates obtained by the two permeation methods. In general and especially at higher permeation rates, higher values were obtained with the ASTM F 739-85. We speculate that the controlling factor in producing this result was the air velocity over the surface from which the chemical evaporated following its diffusion through the material. The velocity in the ASTM apparatus was high, whereas that over the cups was no more than that resulting from their placement face down in a fume hood. It is possible that airflow in the immediate location of the cups was insufficient to prevent the formation of boundary layer resistances to permeation. Another possibility is that the amount of chemical added to the cup did not adequately or evenly cover the clothing specimen when the cup was inverted; thus, permeation did not occur at the maximum possible rate over the entire surface area of the clothing material. Neither of these possibilities was investigated in this study, but are recommended for future investigation to better judge this apparently promising approach to a field test.

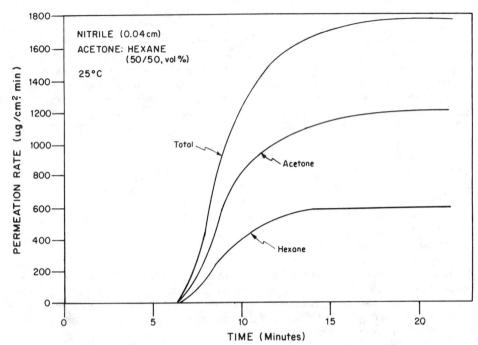

FIG. 2—*Permeation of acetone and hexane from a 50/50% by volume solution through nitrile rubber according to ASTM F 739-85.*

Discussion

Persons involved in emergency response and waste site cleanup are often required to reach rapid, safe, or cost-effective decisions regarding the chemical resistance of the materials of construction of chemical protective clothing. In some cases, such decisions can be guided by past experience, vendor literature, or compilations of chemical resistance data. However, it may be more likely than not that the scope of these sources does not include the situation at hand. Most of the literature addresses only neat chemicals, and at that only about 600 of the thousands of chemicals of commerce. Chemical mixtures represent a particularly difficult area for clothing selection decisions. A field method for testing clothing is necessary.

Three approaches to a field test were investigated in the laboratory for a small number of chemicals and two clothing materials. This preliminary effort indicates that tests based on weight change provide considerable insight to the permeation resistance of the materials. Indeed, one of the methods, that based on a permeation cup, yields directly breakthrough time and permeation rate information. Consequently, the uncertainties of correlations of weight change (or any other secondary parameter) with breakthrough time or permeation rate are avoided. Simply stated, if one is going to measure weight change, he may as well conduct a permeation cup test.

The sensitivity of the cup test is a function of the balance and the available surface area of the cup. Laboratory tests using a four decimal place balance yielded results comparable to those attained using the fairly sensitive instrumental technique. Present battery powered (that is, portable) balances have scales that are limited to only two decimal places. As the sensitivity of the balance is decreased, the detection of breakthrough in a gravimetric method

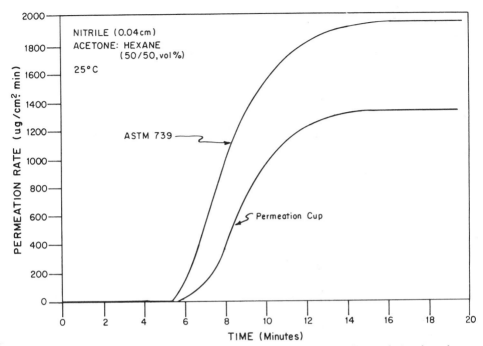

FIG. 3—*Permeation of acetone plus hexane from a 50/50% by volume solution through nitrile rubber.*

is delayed. The length of the delay will be dependent on the permeation rate. For chemicals permeating at high rates, the delay (or offset) will be small. For example, breakthrough times with two and four decimal place balances were, respectively, 10 and 8 min for toluene/butyl-coated nylon. For more slowly permeating chemicals, the offset is greater. For example, breakthrough times with the two and four decimal place balances were, respectively, 65 and 53 min (approximately) for methanol/nitrile. (A different nitrile was used than in the earlier tests.) Thus, if the portability of a battery-powered balance is essential, a trade-off with sensitivity must be accepted, at least at the present time.

Alternatively, if alternating current is available, sensitivity that matches that of traditional analytical chemistry instrumentation can be obtained.

The permeation cup can also serve as the exposure cell for investigation of the degradation of the physical properties of the clothing material. Although the concerns just expressed for an open-backed degradation cell must be borne in mind, the specimen used in the permeation test can be evaluated for physical properties following the permeation test. Alternatively, the permeation cup can be modified to accommodate a solid plate that would back the surface of the specimen not exposed to the chemical.

To this point in the discussion, the permeation cup test has been viewed from the perspective of its utility as the basis for a field test. However, it should be readily apparent that the cup offers significant promise as a means for rapidly and inexpensively assessing in the laboratory the barrier properties of clothing materials. In this mode, the test could greatly speed up the screening of candidate materials for specific applications and the development of new clothing materials.

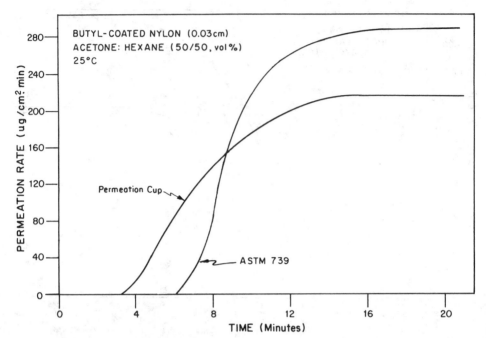

FIG. 4—*Permeation of acetone plus hexane from a 50/50% by volume solution through butyl-coated nylon fabric.*

While all the foregoing bodes well for the permeation cup test, it must be remembered that the preceding investigation was preliminary in nature. Considerably more effort is necessary to troubleshoot, validate, and better understand the nuances of the method. In the laboratory, the effects of temperature, air velocity, and amount of chemical in the cell must be documented. In the field, the benefit, ease of use, and durability of the method must be demonstrated. To this end, a field trial involving three kits has been initiated. The kits comprise three permeation cups, a battery-powered, two decimal place balance and associated parts. Also included are detailed instructions for the performance of the tests and the calculation of the results.

Conclusions

Gravimetric methods are an expedient means for providing field personnel with considerably more information than is presently available to them on the likely chemical resistance of protective clothing materials. Of the methods evaluated, that based on a permeation cup would seem to optimize the information available with minimal materials, training, and performance costs. The permeation cup procedure can yield breakthrough time and permeation rate data comparable to those from tests involving traditional instrumentation such as gas chromatography, dependent on the sensitivity of the balance. However, the applicability of the gravimetric test in its present form is limited to chemicals that evaporate from surface of the clothing. Many chemicals (for example, polychlorinated biphenyls and many pesticides) have low vapor pressures and the detection of their permeation by gravimetry is unlikely.

Field trials and further laboratory development and evaluation are warranted and required. In addition to its utility as a field test, the permeation cup procedure holds promise as an expedient laboratory procedure for the screening of clothing materials for chemical resistance for specific applications and in new product development.

Acknowledgments

The information in this document has been funded wholly by the U. S. Environmental Protection Agency under Contract 68-03-3293. Mention of trade names or commercial products or materials does not constitute endorsement or recommendation for use.

Richard B. Gammage,[1] William G. Dreibelbis,[2] D. Allen White,[1] Tuan Vo-Dinh,[1] and Judy D. Huguenard[3]

Evaluation of Protective Clothing Materials Challenged by Petroleum and Synfuel Fluids

REFERENCE: Gammage, R. B., Dreibelbis, W. G., White, D. A., Vo-Dinh, T., and Huguenard, J. D., "**Evaluation of Protective Clothing Materials Challenged by Petroleum and Synfuel Fluids,**" *Performance of Protective Clothing: Second Symposium, ASTM STP 989,* S. Z. Mansdorf, R. Sager, and A. P. Nielsen, Eds., American Society for Testing and Materials, Philadelphia, 1988, pp. 326–338.

ABSTRACT: The permeation characteristics of eleven petroleum, coal, and shale oil hydrocarbon liquids through eight different types of glove were measured over 24 h. Two analytical techniques, photoionization and room-temperature phosphorescence of polynuclear aromatic compounds, were used to measure breakthrough times by volatile and low-volatility constituents, respectively. Serious drawbacks to the general use of these techniques for measuring steady-state rates of permeation were noted.

The lighter, smaller molecular-size constituents permeated faster than the larger, multi-ringed aromatic constituents. For the light hydrocarbon fuels, especially gasoline, there was preferential permeation by benzene and toluene. Nitrile was severely corroded after extended exposure to hydroxybenzene-containing coal-derived liquids. A general ranking, from worst to best, of the protection afforded by the different gloves was latex $\ll$ neoprene $<$ butyl rubber, polyvinyl chloride (PVC) $<$ nitrile $<$ Viton, Tyvek/Saranex 23, and polyvinyl alcohol (PVA). No breakthroughs within 24 h were observed with the latter three glove materials.

KEY WORDS: liquid hydrocarbon fuels, glove materials, breakthrough time, photoionization detection, room-temperature phosphorescence, volatile organic compounds, polynuclear aromatics, protective clothing

To place the current work in perspective, we note that many of the permeation studies to date have focused on single-component liquid chemicals. Examples are solvents such as hexane [1], benzene [2], toluene and *N,N*-dimethylformamide [3], chlorobenzenes and chlorotoluenes [4], epichlorohydrin, 1,2-dibromoethane, perchloroethylene, trichloroethylene [5], polychlorinated biphenyls [6], and common laboratory solvents [7]. Recently, permeation studies have been reported for binary [8] and multicomponent solvents [9]. Further, some work involving selected hazardous non-solvents, including polychlorinated biphenyls (PCBs) in trichlorobenzene or paraffin oil [6], two pentachlorophenol formulations [10], and epichlorohydrin in water [11], has been conducted.

Workers in the petroleum industry often require protective clothing to prevent skin contact by liquid organics more complex than those just reported, including those found in gasoline

[1] Manager, Occupational Health Research Programs; technical associate, Environmental Compliance and Health Protection; and group leader, Advanced Monitoring Development, respectively, Health and Safety Research Division, Oak Ridge National Laboratory, Oak Ridge, TN 37831.
[2] Industrial hygienist, Pennsylvania State University, University Park, PA 16802.
[3] School teacher, Jefferson County High School, Talbott, TN 37877.

and refined oils as well as their parent petroleum crudes or synthetic fossil-derived crudes. Limited permeation studies have been conducted on such complex mixtures. One such study of the permeation through glove materials by a coal-derived, heavy fuel oil was found in the literature. This study was accomplished by using radioactively-tagged ^{14}C phenol as an additive [12]. Whether the permeation by phenol was indicative of the behavior of other toxicologically significant compounds individually or the whole mixture of compounds is unclear. Another study of glove permeation by shale oil and coal tar extract limited itself to non-compound specific measurement of penetrating vapors without reference to any higher-boiling and nonvolatile constituents that could also be penetrating [13].

Our goal in this study was to evaluate the permeation characteristics of several polymeric materials commonly used in protective clothing (eight in number) by both volatile and low-volatility constituents of complex petroleum, oil shale, and coal-derived products (11 in number). The low-volatility constituents were principally polynuclear aromatic (PNA) hydrocarbons. A practical objective was to identify the protective clothing affording the worker nearly complete skin protection from the mixture of permeating organic compounds contained within a particular hazardous material. Another goal was to glean information on select permeants of toxicological significance. We did not presume that the original mixtures of compounds would permeate intact, so we measured a few individual compounds (benzene, phenanthrene, and pyrene) to look for evidence of preferred penetration by individual chemicals.

Our measurements (total photoionization or total phosphorescence) proved to be quite effective for determining breakthrough times representative of photoionizable volatile organic compounds (VOC) or phosphorescing PNA. With these same measurements, it proved more difficult to obtain a handle on the steady-state permeation rates of whole complex mixtures. We actually measured steady-state permeation rates of the photoionizable VOC or the PNA components of whole complex mixtures relative to either toluene or phenanthrene, respectively. In defense of emphasizing breakthrough, or the first detectable amount permeated, this is a more important quantity than the steady-state permeation rate when the permeant carries either a carcinogenic risk or is absorbed rapidly through the skin [10]. Such is the case with some of the gasoline and heavy oil products that we were studying.

Experimental Procedures

Materials and Challenge Chemicals

Protective clothing, manufacturers, nominal thicknesses, and densities are shown in Table 1. For our experiments, gloves were used as sources of sample specimens with the samples cut from the palm or guantlet area. The exception was Tyvek,[4] which was cut from a sheet of the material. The polyvinyl chloride (PVC) and polyvinyl alcohol (PVA) glove materials were supported with cotton fabric liners. The other materials were unsupported. For PVC and PVA, thickness measurements included the liner. The materials were stored and used in the laboratory at a temperature of 21 to 24°C and a relative humidity of 40 to 60%.

The challenge fluids and their sources are listed in Table 2. The fluids were stored at room temperature inside a ventilated compartment. Three of these, slurry recycle oil, crude shale oil, and hydrotreated shale oil, were highly viscous at room temperature. To study these viscous fluids, both the permeation cell and the storage bottles were initially warmed with hot tap water to improve flow during pouring and to ensure intimate contact between the glove material and the challenge fluid. The filled permeation cell was then cooled rapidly by running cold tap water over its outer surface for a few minutes.

---

[4] Registered trademark of E. I. duPont de Nemours & Company.

TABLE 1—*Protective garment fabrics.*

Material	Source	Average Thickness, mm	Average Density, g/cm³
Latex	LRC Safety Products Co., Surety Glove Division	0.41	1.02
Neoprene	Stanzoil Whitecaps by Pioneer	0.41	1.36
Butyl rubber	North Hand Protection, Industrial Gloves	0.44	1.13
PVC (I) (cotton liner)	JOMAC Products		
PVC (II) (cotton liner)	Best Co.	1.10	1.04
Nitrile	Stansolv by Pioneer	0.38	1.04
Viton[a] (polyvinylidene fluoride)	Latex Glove Co.	0.34	2.30
Tyvek[a]/Saranex[a] 23 (spunbonded olefin)	Kappler Co.	0.15	0.67
PVA (thick cotton liner)	Edmont Co.	0.75	0.82

[a] Registered trademarks.

Permeation Test Protocols

The permeation testing for VOCs was conducted using the American Society for Testing and Materials (ASTM) Test Method for Resistance of Protective Clothing Materials to Permeation by Hazardous Liquid Chemicals (F 739-81) developed by Committee F-23 and commercially available test cells [15] operated in the open-loop mode [6]. The permeating organic vapors were measured non-compound selectively with an HNU photoionization detector using either a 10.2 or an 11.7 eV lamp.

Breakthrough time, which is defined in ASTM F 739-81 as time at which the permeant is first detected by means of the chosen analytical technique [3], is a function of the sensitivity of the analytical method. The definitions and quantities for first detection of permeant (lower limits of detection) are given in Table 3.

Volatile Organic Compounds (Toluene-Equivalent Units)—Measurements of toluene were made in the gas phase in an open-flow system to determine limits of detection (LOD) (see Table 3).

TABLE 2—*Challenge fluids (arranged in order of increasing viscosity).*

Fluid	Source
Unleaded gasoline	local commercial distributor (from ORNL gas pump)
Kerosene "AVJET A"	local commercial distributor
Diesel fuel	Phillips reference DF2, Lot C-345 provided by API
H-Coal light fuel oil (naptha)	#1326 [14]
Fuel Oil #6	local commercial distributor
Used motor oil	oil change on newly rebuilt V-8 engine after 5000 miles
Aromatic extracts	1H 32/33 provided by API
Crude petroleum A	CRM-3 [14]
Slurry recycle oil (catalytic cracked bottoms)	4547, 4-DHL-100 provided by API
Hydrotreated Parahoe/SOHIO shale oil	4602 [14]
Crude Parahoe shale oil	CRM-2 [14]

TABLE 3—*Definitions and quantities for first detection of permeant.*

| Lamp Energy | LOD | | Flux, $mg \cdot m^{-2} \cdot min^{-1}$ |
| | Concentration | | |
	ppm	$mg \cdot m^{-3}$	
10.2 eV	0.023[a]	0.088[b]	0.0090[c]
11.7 eV	0.033[a]	0.126[b]	0.0128[c]

[a] LOD is defined as 0.1 units on the most sensitive 0 to 20 scale (0.5% of full scale deflection and the smallest incremental division that can be discerned visually) of the HNU reader with zero span pot settings for maximum response; single point calibration with Matheson standard of 1.143 ppm toluene.
[b] 1 ppm = 3.82 mg·m^{-3} at 20°C and 760 mm Hg
[c] Flux $J = FC/A$ where F = rate of gas flow of 200 cm^3 · min^{-1}, C = concentration of toluene in mg · m^{-3}, and A = challenged glove area of 19.6 cm^{-2}.

Nonvolatile Organic Compounds (*Phenanthrene-Equivalent Units*)—The filter paper sorbent acts as a closed collection medium. The LOD is defined as that quantity of phenanthrene-equivalent phosphorescing permeant that, with excitation at 295 nm, produces a room-temperature phosphorescence (RTP) signal that is 15% greater than the corresponding signal intensity from clean filter paper:

$$\frac{\text{RTP signal}}{\text{295 nm excitation}} \quad \frac{\text{LOD } (mg \cdot m^{-2})}{0.14}$$

The actual size of the disk of filter paper onto which phenanthrene solution was spotted, or onto which other permeants were collected, was 0.32 cm². For phenanthrene, a calibration curve of molar concentration versus net RTP response was determined. The range of linearity extended from 0.14 mg · m^{-2}, the limit of detection, to about 2 mg · m^{-2}.

A sketch of the cell for measuring permeation of less-volatile constituents of the hydrocarbon fuels is shown in Fig. 1. The body of the cell was a glass vial partially filled with the challenge fluid. The protective clothing material with filter paper on the outside was mounted over the mouth of the vial. Challenge was initiated by inverting the vial. Any phosphorescing materials that permeated were contained within the sorbent filter paper. The phosphoresce was measured directly from a filter paper at room temperature after inserting the filter paper into a spectrometer. A separate vial, clothing material, and filter paper were used for each challenge time to measure the amount accumulated over that particular period of time. Three randomly selected specimens of each material were tested at challenge times of 4, 8, 16, and 24 h. More complete details concerning the cell and the spectroscopic method of analyses are described elsewhere [16,17].

Select Compound Measuring Techniques

Portable Gas Chromatograph—A Photovac 10A10 gas chromatograph (GC) with a photoionization detector was used to measure individual VOC. Small samples of up to 1 cm³ of reference compounds taken from an air bag, or permeant-containing air collected after exiting the open-loop permeation cell, were injected directly from a syringe into the GC.

Room Temperature Phosphorescence—The RTP spectra of original and permeated hydrocarbon challenge fluids were measured and compared with the RTP spectra of standard

HOLDING PAPER DISC   ANALYTICAL PAPER DISC

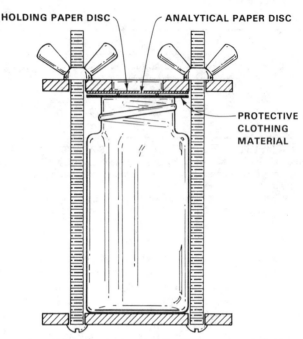

PROTECTIVE
CLOTHING
MATERIAL

FIG. 1—*Permeation cell for collecting low-volatility PNA compounds on filter paper pressed against the inside surface of glove material.*

PNA compounds for the purposes of identifying specific PNA in the challenge and permeant fluids. The RTP technique is based on the detection of organic compounds adsorbed on solid substrates such as filter paper treated with a heavy-atom agent to enhance the phosphorescence emission [18]. The disks of filter paper, after impregnation with the challenging materials, were removed to a spectrofluorimeter with a phosphoroscopic attachment and analyzed at excitation wavelengths of 295, 315, and 343 nm, which are the optimal wavelengths for maximizing the responses to phenanthrene, quinoline, and pyrene, respectively [16,17].

Results

Breakthrough Times

The matrices in Figs. 2 and 3 show measured (not normalized for unit thickness) breakthrough times for volatile and nonvolatile component mixtures, respectively.

For volatiles, the breakthrough times are in minutes (Fig. 2). Times greater than 100 and 1000 min are rounded to the nearest 5 and 100 min, respectively. Three and occasionally more replicates were usually tested for breakthrough times of less than about 100 min. Precision ranged from ±9% to ±32%. The butyl rubber/H-Coal light fuel oil pairing produced results (three replicates) with the worst precision of ±32%. For breakthrough times well in excess of 100 min, two and sometimes only one specimen of each protective fabric was tested.

The breakthrough times for phosphorescing aromatic compounds (Fig. 3) are, by comparison, on a much coarser scale of resolution. The numbers in Table 2 indicate whether breakthrough was observed at one of the four standard challenge times of 4, 8, 16, and 24

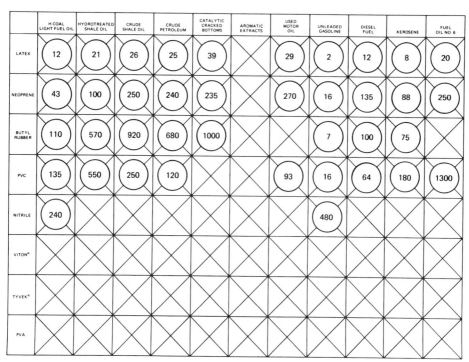

FIG. 2—*Breakthrough times (min) of photoionizable VOC; crossed-out squares indicate no breakthrough in 24 h.*

h. For example, a number of eight means that breakthrough was observed if the glove material was challenged for 8 h, but not if the challenge time was 4 h. The breakthrough would thus be in the time range of 4 to 8 h. Crosses indicate that no breakthrough had occurred after 24 h, the maximum time of challenge.

Permeation Rates of Mixtures

Volatile Organic Compounds—Steady-state permeation was measured for only a limited number of combinations of glove material/challenge fluid. One such combination was neoprene/gasoline where steady-state flux could be consistently achieved (Fig. 4) in repeat experiments. For several other pairings, however, the photoionization detector recorded abrupt discontinuities, unsteadiness, or reductions in the apparent permeation rate. An example of sudden reversal in permeation rate is shown in Fig. 4 for butyl rubber/gasoline.

There are a number of explanations for the high proportion of abnormal permeation curves. In many instances, the glove material became severely swollen, distorted, or corroded. The concurrently changing physical properties might be causing the permeation rate to change. The different types of permeation behavior we observed were the same as those reported by Nelson et al. [7] where the garment materials were being structurally modified by challenging solvents.

The performance of the HNU photoionization detector deteriorated significantly during the course of some of these experiments. This was especially the case with the heavier hydrocarbon fluids where dark deposits built up on the electrode and optics, thus diminishing the responsiveness of the detector. It soon became standard practice to dismantle and wash

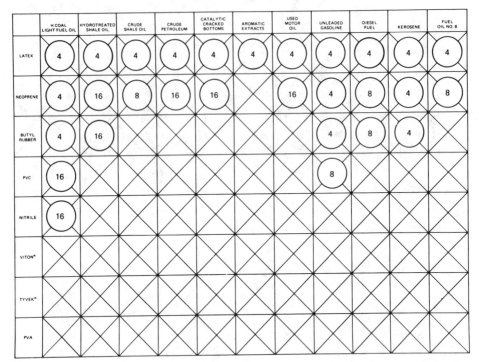

FIG. 3—*Breakthrough times (h) of phosphorescing aromatic compounds on a resolution scale of 0 to 4, 4 to 8, 8 to 16, and 16 to 24 h; crossed-out squares indicate no breakthrough in 24 h.*

the detector head with toluene and then acetone after each permeation test. This practice restored the prior instrumental sensitivity to the toluene standard.

Nitrile was especially prone to corrosion by coal-derived liquids including the H-Coal light fuel oil. The culprit was suspected to be hydroxybenzenes; H-Coal liquids typically contain $mg \cdot g^{-1}$ quantitites of hydroxybenzenes [19], which petroleum analogues do not. A derivative ultraviolet absorption spectrometer [20] was used to measure the headspace vapors above the H-Coal light fuel oil; a strong signal characteristic of phenol was produced that verified that hydroxybenzene was present in this light fuel oil. To test the likelihood of nitrile being corroded by hydroxybenzenes, pieces of nitrile were immersed in dilute phenol for several hours. The nitrile was corroded and readily disintegrated after the phenol treatment.

The excessive corrosion wrought by overnight exposure to a coal-derived liquid (comparative research material called Coal Oil A with a repository number CRM-1 [14]) is shown in Fig. 5. Nitrile is judged, therefore, to be unsuitable in providing protection against coal-derived liquids.

Polynuclear Aromatic Hydrocarbons—For coarse measurement of breakthrough time, readings were taken at 4-h intervals. By reducing the time interval between consecutive measurements during a permeation test, the filter paper/RTP technique can be used to construct permeation curves. Three such permeation curves are shown in Fig. 6 where H-Coal light fuel oil was the challenge fluid. Each point represents a single separate measurement on a single sample of clothing material.

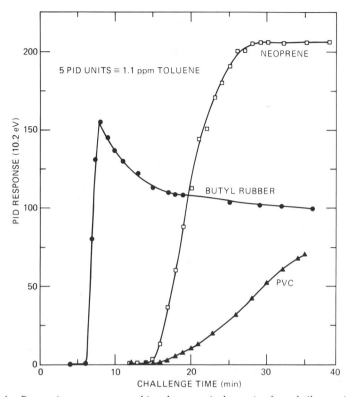

FIG. 4—*Permeation rates, expressed in toluene-equivalent units, for volatile constituents of gasoline.*

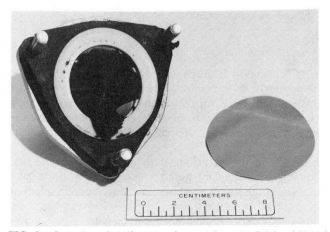

FIG. 5—*Corrosion of nitrile exposed overnight to Coal Oil A (CRM1).*

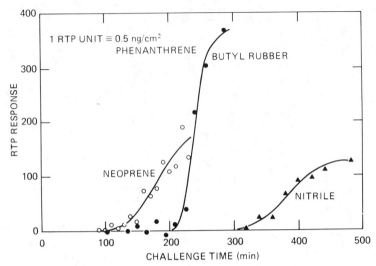

FIG. 6—*Permeation rates, expressed in phenanthrene-equivalent units, for low-volatility PNA compounds in H-Coal light fuel oil.*

The slopes of the linear portions of the curves of Fig. 6 should denote the steady-state rates of permeation. The ranges of linearity, however, are quite restricted. This is brought about by early signs of saturation in the RTP response as phosphorescing material accumulates. In fact, it is likely that saturation becomes significant in magnitude before steady-state permeation is reached, as evidenced by the s-shaped curves in Fig. 6 for neoprene and nitrile. The start of nonlinear response for pure phenanthrene of 2 mg/m^2 corresponds to the RTP response level of 100 in Fig. 6. Thus for neoprene and nitrile, the phosphorescence characteristics of permeant quite closely mimic those of pure phenanthrene. In the case of butyl rubber, however, a different permeant composition is indicated because the permeant phosphorescence saturates less readily then it does for pure phenanthrene.

As a result of these difficulties, the measurement of permeation curves were only attempted for gasoline and H-Coal light fuel oil on neoprene, butyl rubber, PVC, and nitrile. Measurements were also attempted using latex, but rapid breakthrough and very high permeation rates produced inaccurate results.

Individual Compound Permeation

Volatile Organic Compounds—Headspaces above challenge fluid were sampled and analyzed with the Photovac GC. For compound identification, the chromatograms (retention times) were compared qualitatively with chromatograms produced by standard compounds of toxicological significant such as benzene, toluene, and several mercaptans and sulfides. Benzene and toluene were prevalent, especially in the vapors of the lighter hydrocarbon fuels.

The permeating VOC in the air exiting the permeation cell were also collected and analyzed. The ensuing chromatograms were then compared with the gas chromatograms obtained for the vapors of the parent challenge fluid; differences in relative peak intensities were indicative of enhanced or restricted permeation by select compounds. The most obvious and striking results were obtained for light-hydrocarbon fuels; Figs. 7 and 8 show that there is preferential permeation by the benzene and toluene in the gasoline and kerosene challenge fluids.

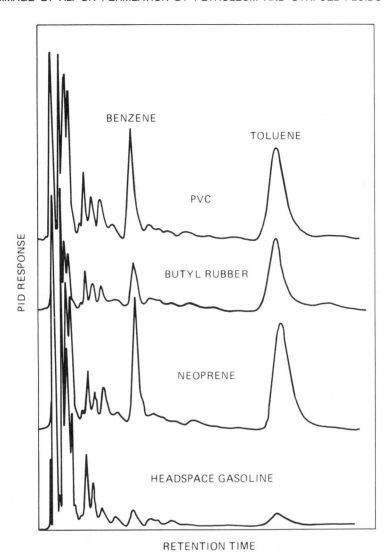

FIG. 7—*Gas chromatograms for permeated gasoline vapors compared with the chromatogram for direct-headspace vapor.*

Polynuclear Aromatic Hydrocarbons—The emission spectra show broad-band features to be expected of complex mixtures of phosphorescing compounds that compose petroleum, coal, and shale-oil products; the excited PNA compounds range from simpler three-ring to more complex higher-ring compounds. At the optimum excitation wavelengths, resolution in the emission spectra becomes sufficient to identify phenanthrene and pyrene [16,17]. In comparison to challenge fluids spotted directly onto filter paper, permeated PNA compounds are skewed in favor of a mixture of lower molecular weight (smaller ring size) compounds. The RTP spectra for the original Fuel Oil #6 and the oil permeated through neoprene are reproduced in Fig. 9; relative to phenanthrene, pyrene has been largely excluded from the permeant. The RTP spectra obtained by excitation at 343 nm are relatively weaker than those produced by 295 and 315 nm excitations [21].

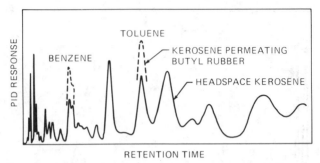

FIG. 8—*Comparison of gas chromatograms for kerosene vapors in direct headspace and after permeating butyl rubber.*

Discussion

Breakthrough times by lower and higher (aromatic) molecular weight fractions of complex mixtures of hydrocarbons were measured successfully. The two essentially non-compound selective techniques of photoionization of permeated vapors and RTP of permeated low-volatility PNA compounds proved to be sensitive analytical tools for detecting breakthrough.

Neither the photoionization or RTP technique proved universally adequate for making measurements of steady-state flux. The permeation curves measured with the photoioni-

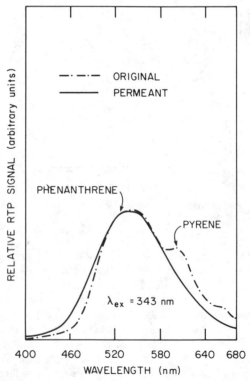

FIG. 9—*Comparison of RTP spectra of original Fuel Oil #6 and after permeating neoprene.*

zation detector often produced permeation curves of abnormal appearance. The problem was sometimes due, at least in part, to fouling of the surfaces of the detector head. Swelling and distortion of glove materials was also a complicating factor. The difficulty posed by the RTP technique is a generally quite limited range of linearity of response; suppression of the RTP response often occurred before a steady-state flux was achieved.

Breakthrough, if it occurred within 24-h, was clearly faster for VOC than for multi-ringed PNA compounds. Leaving aside for the moment the latex-aromatic extracts combination, there were no obvious exceptions to this rule for any of the pairings listed in Figs. 2 and 3. Common sense would dictate that the larger, less volatile molecular constituents of these complex hydrocarbon liquids should permeate more slowly than the smaller and more volatile molecules. Within the class of PNA compounds, the molecular size has a dominant effect on permeation; on a relative scale there was little penetration by pyrene or larger ring size PNA compared to three-ring PNA compounds such as phenanthrene.

With the aromatic extracts, even direct headspace vapors produced only a small response equivalent to 1 ppm toluene on the HNU photoionization meter. The aromatic extracts are nearly devoid of volatile constituents; this then is the reason for the apparent faster breakthrough by PNA than VOC for the combination of latex-aromatic extracts.

The preferential permeation of neoprene, butyl rubber, and polyvinyl chloride by benzene and toluene is also probably related, in part, to molecular size and geometry; the planar aromatic benzene nucleus projects a smaller minimum edge-on size than nonplanar aliphatic compounds of similar overall size. Thus, benzene and toluene are more easily able to penetrate the polymeric structures of the glove fabrics. Chemical-polymer interactions involving parameters other than just size are undoubtedly influencial as well.

The failure of nitrile when challenged by coal liquids presents a special situation. It is the chemical attack of the nitrile, presumably by hydroxybenzenes, that causes corrosion and outright failure of the fabric.

The hydrocarbon fuels and oil residues we have studied each contain constituents that are rapidly absorbed by the skin or are carcinogenic. In such situations, minimum or no contact with the skin should be the aim when selecting a glove material [10]. Breakthrough time rather than steady-state permeation rate is therefore the better means of assessing the protection afforded a wearer. On this basis, a general ranking of the usefulness of the eight different types of protective garment fabric can be made:

Poor protection	1. latex (very poor)
	2. neoprene
Moderate protection	1. butyl rubber
(except gasoline)	2. PVC
Good protection	nitrile (except phenolic synfuel)
Excellent protection	1. Viton
(no breakthrough)	2. Tyvek/Saranex 23[5]
	3. PVA

Acknowledgments

This research was sponsored jointly by the American Petroleum Institute (DOE Contract No. ERD-83-322) and the Office of Health and Environmental Research, U. S. Department of Energy, under Contract No. DE-AC05-84OR21400 with the Martin Marietta Energy Systems, Inc. Use of a company name or product does not constitute endorsement by the American Petroleum Institute.

[5] Registered trademark of Dow Chemical Company.

References

[1] Dillen, I. G. and Obasuyi, E., *Journal,* American Industrial Hygiene Association, Vol. 46, No. 5, May 1985, pp. 233–235.
[2] Weeks, R. A., Jr. and McLeod, M. J., *Journal,* American Industrial Hygiene Association, Vol. 43, No. 3, March 1982, pp. 201–211.
[3] Henry, N. W., III and Schlatter, C. N., *Journal,* American Industrial Hygiene Association, Vol. 42, No. 3, March 1981, pp. 202–207.
[4] Mikatavage, M., Que Hee, S. S., and Ayer, H. E., *Journal,* American Industrial Hygiene Association, Vol. 45, No. 9, Sept. 1984, pp. 617–621.
[5] Stampfer, J. F., McLeod, M. J., Betts, M. R., Martinez, A. M., and Berardinelli, S. P., *Journal,* American Industrial Hygiene Association, Vol. 45, No. 9, Sept. 1984, pp. 642–654.
[6] Stampfer, J. F., McLeod, M. J., Betts, M. R., Martinez, A. M., and Berardinelli, S. P., *Journal,* American Industrial Hygiene Association, Vol. 45, No. 9, Sept. 1984, pp. 634–641.
[7] Nelson, G. O., Lum, B. Y., Carlson, G. J., Wong, C. M., and Johnson, J. S., *Journal,* American Industrial Hygiene Association, Vol. 42, No. 3, March 1981, pp. 217–225.
[8] Mickelsen, R. L., Roder, M. M., and Berardinelli, S. P., *Journal,* American Industrial Hygiene Association, Vol. 47, No. 4, April 1986, pp. 236–240.
[9] Forsberg, K. and Faniadis, S., *Journal,* American Industrial Hygiene Association, Vol. 47, No. 3, March 1986, pp. 189–193.
[10] Silkowski, J. B., Horstman, S. W., and Morgan, M. S., *Journal,* American Industrial Hygiene Association, Vol. 45, No. 8, Aug. 1984, pp. 501–504.
[11] Stampfer, J. F., McLeod, M. J., Betts, M. R., and Martinez, A. M., "The Permeation of Eleven Protective Garment Materials by Organic Solvents," Technical Report LA-UR 83-1926, National Institute for Occupational Safety and Health, Morgantown, WV, 1983.
[12] Bennett, R. D., Feigley, C. E., Oswald, E. O., and Hill, R. H., *Journal,* American Industrial Hygiene Association, Vol. 44, No. 6, June 1983, pp. 447–452.
[13] Nelson, G. O., Carlson, G. J., and Buerer, A. L., "Glove Permeation by Shale Oil and Coal Tar Extract," Technical Report UCRL-52893, Lawrence Livermore Laboratory, Livermore, CA, 14 Feb. 1980.
[14] Griest, W. H., Coffin, D. L., and Guerin, M. R., "Fossil Fuels Research Matrix Program," Technical Laboratory Report ORNL/TM-7346, Oak Ridge National Laboratory, TN, June 1980.
[15] Pesce Laboratory Sales, Kennett Square, Pa.
[16] Gammage, R. B., Vo-Dinh, T., and White, D. A., "Measurement by Room Temperature Phosphorescence of Polynuclear Aromatic Containing Hydrocarbon Fuels that Permeate Glove Materials," Radiation Protection Dosimetry, Vol. 17, 1986, pp. 263–265.
[17] Vo-Dinh, T. and White, D. A., "Development of Luminescence Procedures to Evaluate Permeation of Multi-Ring Polyaromatic Compounds through Protective Materials," *Journal,* American Industrial Hygiene Association, Vol. 48, 1987, pp. 400–405.
[18] Vo-Dinh, T., *Room Temperature Phosphorimetry for Chemical Analysis,* Wiley, New York, 1984.
[19] Yeatts, L. B., Jr., Hurst, G. B., and Caton, J. E., Analytica Chimica Acta, Vol. 151, 1983, pp. 349–358.
[20] Hawthorne, A. R., *Journal,* American Industrial Hygiene Association, Vol. 41, No. 12, Dec. 1980, pp. 915–921.
[21] Vo-Dinh, T. and White, D. A., *Journal,* American Industrial Hygiene Association, Vol. 48, 1987, pp. 400–405.

Protection from Industrial Chemical Stressors

Field Experiences

Daniel N. Eiser[1]

Problems in Personal Protective Equipment Selection

REFERENCE: Eiser, D. N., "**Problems in Personal Protective Equipment Selection,**" *Performance of Protective Clothing: Second Symposium, ASTM STP 989*, S. Z. Mansdorf, R. Sager, and A. P. Nielsen, Eds., American Society for Testing and Materials, Philadelphia, 1988, pp. 341–346.

ABSTRACT: The development and validation of protective clothing test methods are very important to industrial hygiene practitioners. These methods will eventually help to ensure that protective clothing performance data are comparable, meaningful, and available to field personnel. However, field problems are typically very complex, with multiple challenges placed on protective clothing, as well as the users. The "perfect" protective clothing item is rarely, if ever, found in practice.

Protective clothing challenges can often include combinations of heat, tear, and cut hazards, rough surfaces, puncture hazards, and complex chemical mixtures. Fitting problems, differences in performance of similarly-described clothing from different manufacturers, and occasional poor quality control can cause additional complications for users. Training in the proper wearing of clothing, whether or not the equipment is considered disposable, worker acceptance, comfort, flexibility, compatibility with the work being done, and other more subtle issues also influence whether or not clothing is actually used. Dermatitis aggravated by personal protective equipment use, skin cleaners, and hand creams can also present problems to clothing users. In addition, practical guidelines for assessing toxic risks from dermal exposures to chemical agents are critical if we are to meaningfully apply protective clothing performance data.

This paper presents some of the factors that must be considered during personal protective equipment selection, and gives specific examples of typical selection problems. Engineering, job layout, and skills' training solutions are needed to minimize the need for personal protective equipment and to ensure effective use when it must be specified for protection.

KEY WORDS: personal protective equipment, selection, field experience, protective clothing, skin protection, dermal hazards, industrial hygiene, safety, dermatitis, chemical hazards, gloves, occupational medicine, electronics industry, allergens, sensitization

There has been considerable progress in the development of test methods and standards for evaluating protective clothing performance over the last several years. However, there are a number of other factors that must enter into the selection process for personal protective equipment. The inherent chemical or physical resistance of the materials of construction are critical factors that must be considered early in the selection process, thus information generated by permeation, penetration, and degradation testing can help to eliminate candidate materials that are clearly unsuitable for a particular application. However, there are a number of issues that can complicate the selection process in "real world" applications. The purpose of this paper is to explore a few of these issues, support them with some specific examples, and offer some practical management techniques and engineering solutions that have been used to resolve or control them.

[1] Senior industrial hygienist, AT&T Network Systems Equipment Division, Winston-Salem, NC 27102.

Chemical Mixtures

The problems imposed by chemical mixtures include considerable variation in their response versus neat substances through protective clothing [1]. Also, most tests have been restricted to simple binary or tertiary mixtures. Actual permeation results may be quite different for mixtures than for single component chemicals. Consider the following actual mixture employed in a commercial ink and paint thinner. The actual formation is often variable, but all components represent greater than one percent by volume of the mixture.

toluene
acetone
xylene
acetone
methyl isobutyl ketone
methyl ethyl ketone
2-ethoxyethyl acetate
glycol ether (unspecified)
n-butyl acetate
methyl alcohol
ethyl acetate
ethyl alcohol
isopropyl acetate
butyl alcohol

Another good example of a complex protective clothing challenge is the following commercial mixture coating for some types of printed circuit boards:

toluene
methyl methacrylate
n-butyl acetate
perchloroethylene
2-ethoxyethyl acetate
2-methoxyethanol
methylene chloride
butyl methacrylate
acrylic ester resin

The range of properties required to confidently protect against such diverse mixtures is not easily found in one barrier material.

Of course, this pre-supposes that accurate information on the composition of the chemical mixture is available. Many suppliers of chemicals will not provide exact chemical compositions for their products. They will often give information on the general composition by chemical class, or omit materials not included in the Occupational Safety and Health Administration (OSHA) standards [2], or in the Threshold Limit Values (TLVs) developed by the American Conference of Governmental Industrial Hygienists [3]. In addition, toxicological data or analytical methods for the components in the mixture may not be readily available.

Chemical Mixture/Puncture Resistance

One solvent mixture that has been used in the cleaning of printed circuit boards is perchloroethylene and a glycol ether. Perchloroethylene has been implicated as an animal carcinogen [4], and several glycol ethers have been implicated as animal teratogens that can be absorbed through the intact skin [5]. This combination is complicated by the fact that sometimes the mixture is heated to over 120°C (as in a vapor degreaser). Another factor that must be overcome in normal use is that the protective clothing must provide puncture resistance to "solder icicles" that are often present on printed wiring boards.

Permeation resistance data is used when available to select protective clothing. However, compromises must often be made to achieve the best overall combination of chemical resistance and protection against physical damage. A glove exhibiting superior permeation resistance will do no good if it is easily punctured or torn during use.

In the case of a perchloroethylene/glycol ether mixture, Buna-N nitrile gloves (15 mil thickness, 28 cm (11 in.) length) can be used for normal operations where incidental contact is possible. However, permeation tests run on this solvent blend versus these gloves have indicated a drastic reduction in barrier protection against the mixture versus a neat perchloroethylene or glycol ether challenge [6,7]. Similar effects were demonstrated with "Norfoil" film as a barrier material [8–10]. It has been postulated that the glycol ether may act as an accelerant for perchloroethylene permeation through the nitrile elastomer [6]. Another possibility is that solvent mixture components may selectively attack the laminate layers in the "Norfoil" film, allowing the other solvent to follow behind, but a clear mechanism has not yet been defined [11].

When an unusual operation requiring immersion is anticipated, (specialized maintenance operations or during emergency response) gloves made from "Norfoil" film have been specified to be worn as liners under Buna-N nitrile gloves (22 mil thickness, 46 cm (18 in.) length). Field trials with this glove combination are still being conducted, but preliminary results indicate the gloves can be effectively used in this combination providing that the work being conducted does not require considerable dexterity or "feel."

Electrostatic Characteristics

A consideration that may not be present in other industries is the electrostatic properties of the protective clothing material. Electrostatic discharges can damage sensitive components, and grounding of personnel is important while handling electronic products.

Engineering Controls to Minimize Protective Clothing Requirements

Engineering controls such as air knives (to dry parts as thoroughly as possible prior to handling) and proper operating techniques (drainage, use of drying booths, tools, hoists, etc. with vapor degreasers) are employed to minimize the chance of skin contact.

When the boards can be adequately dried prior to leaving process equipment, vinyl-impregnated stretch fabric gloves or nitrile-dipped fabric gloves provide reasonable hand protection, comfort, and dexterity. Robotics are also being employed in a number of operations to minimize the chance of hand contact with solvents. In addition, new flux formulations and cleaners are being investigated to minimize the need for cleaning solvents as a part of an overall waste minimization plan.

Heat, Combustion, and Chemical Resistance

Another interesting and somewhat unique problem is presented during Tantalum Capacitor manufacturing. Manganese nitrate is used in one part of the manufacturing process. Parts are dipped in various concentrations of this chemical and processed through pyrolyzer ovens.

Manganese nitrate presents a mild acid exposure (pH of 1.5), and can cause irritation and a burning sensation if entrapped against the skin (especially if there are any open cuts in the skin). In addition, the chemical is an oxidizer and, although thermal protection is required while handling the parts' racks as they come out of the ovens, the glove material must be chosen carefully due to the increased risk of gloves catching on fire. The gloves must provide thermal protection for handling racks with a surface temperature exceeding 200°C. The gloves must also provide enough dexterity to allow the physical rotation of the bars containing the capacitors (approximately 1.27 cm (0.5 in.) square bar stock), and protect against thermal burns due to contact with the hot racks.

Although hot mill gloves could provide adequate dexterity and thermal resistance, if they contact the manganese nitrate solution and are not changed, the gloves can ignite when they contact the hot rack surfaces. Asbestos gloves would be a potential answer to the ignition problem, but are excluded from consideration due to the obvious toxicity problems that could be presented as they wear and abrade [12].

To address this problem, administrative controls and personal protective clothing specifications should be established. For example, manganese nitrate solutions can be mixed, delivered, and poured into the dip pans by an employee other than an oven tender (using adequate solvent and acid-resistant gloves). This minimizes the chance of thermal gloves contacting the solution. Leather work gloves (sometimes with flannel liners) can be used to provide adequate thermal protection and allow flexibility to manipulate the parts. Finally, operators should be instructed to change the gloves if they accidentally contact the manganese nitrate solution. When these procedures are followed, ignition of heat-resistant gloves should not occur, and skin irritation should not be a problem.

Dexterity Problems, Chemical Resistance, Puncture Hazards, and Cut Resistance

A manufacturing example where dexterity problems, chemical resistance, puncture hazards, and cut resistance are important would include fine assembly work using epoxy adhesives. Since the printed circuit boards have rough edges, components can present puncture hazards, and epoxies can lead to dermatitis when improperly handled [13]. When you consider the extreme sensitivity and dexterity required to handle and manipulate small electronic components, the situation is obviously complicated.

Solutions to these problems include designing fixtures to hold and manipulate the parts to help minimize skin contact, using automatic insertion and dispensing machines, supplying tools such as fine paint brushes or specialized syringes to apply epoxy adhesives, and training all operators in "no-touch" techniques when handling these materials.

Disposable protective clothing (sleeves, gloves, and aprons) is typically specified to avoid skin contamination upon reuse. Special attention must be paid to bulk mixing, dispensing equipment, clean-up, and maintenance operations, since these tasks typically present the most likely chances for chemical contact with the skin.

Fitting Problems

There is a wide range of glove and garment sizes encountered in the industrial work population. This is becoming particularly evident as more women enter the industrial environment. There is a continuing need to design protective equipment for all workers to ensure proper fit and comfort. Safety problems such as poor grip, tripping and falling hazards, and loose fitting clothing around machinery can be caused by ill-fitted clothing. In addition, an understandable resistance to using personal protective equipment increases when properly fitting clothing is not available. Solutions must incldue ergonomic considerations in clothing design, and a wider range of sizes made available to a changing industrial work population.

Relative Dermal Toxicity Guideline Needs

Permeation, penetration, and degradation data on personal protective equipment and elastomer performance are becoming more readily available. However, there is still an important need to develop relative dermal toxicity guidelines and to be able to apply this data meaningfully in a field context. A clear method for differentiating significant and insignificant dermal risks, and defining dermal doses that could cause injury or illness would be of great help to field personnel.

Training Needs

Those charged with specifying personal protective equipment must be made aware of the advances in protective clothing research, and the problems in selection. More importantly, they must be made aware of the limitations of protective clothing.

Management and protective clothing users must be made aware that the skin can present a significant route of entry of chemicals into the body [14]. Workers must have at least a basic understanding of the limitations of protective clothing, and develop a commitment to ensure that they follow work practices to minimize all skin contact with chemicals. This should include not only production personnel, but especially maintenance crews that may operate at unusual hours, under difficult conditions, or with minimal supervision.

Training needs include proper donning and doffing of equipment, inspection techniques to determine defects prior to use, operating techniques to minimize exposure to the chemical or physical stressor, proper storage and care of the equipment, when to change gloves, and the specific hazards of the materials being used.

Hazard data covered in Material Safety Data Sheets (MSDSs) should include more specific information on proper protective clothing selection to aid in training and to be used as a reference for end-users. In light of "Hazard Communication" and "Right-to-Know" legislation, those responsible for preparing Material Safety Data Sheets must realize that a statement like "Use Impervious Gloves" is no longer acceptable to more sophisticated users.

Labeling of Personal Protective Equipment

One serious problem for users of protective equipment is that many articles of protective clothing (in particular, gloves) are not adequately labeled to indicate critical information such as the elastomer used in construction, thickness, size, etc. This can be confusing and may even lead to injury when a large number of different styles and manufacturers' products are used, and an inappropriate article of clothing is selected for protection. A consistent labeling standard should be developed and adopted by manufacturers to assist users in proper selection.

Protective Creams, Hand Cleaners, and Dermatitis

It is important to minimize the need for extended personal protective equipment use whenever possible since this can sometimes promote skin irritation and interfere with worker efficiency and comfort. One related source of dermatitis occasionally encountered is the extended use of skin cleaners and hand creams in the workplace. Some workers become sensitized to components in some skin cleaners, creams, and ointments, and this may complicate and confound treatment of dermatitis cases.

Summary and Control Tactics

In summary, there are a number of difficult problems to overcome in the successful selection and use of personal protective equipment. Engineering controls, good operating techniques, and thoughtful workplace design and layout are the first important steps to help minimize employee exposures. The selection process must include a consideration of the chemical, physical, and psychological aspect of use, in addition to compatibility with the work being done. Employee training, clear protective clothing specifications, and awareness of clothing limitations are all critical parts of any effective program. Additional needs include the development of useful guidelines for relative dermal risks presented by different chemical and physical exposures, a uniform labeling standard for protective clothing, and aggressive work in the development of validated consensus standards and test methods to make the information accessible and useful to users of protective clothing.

References

[1] Davis, S. L., Feigley, C. E., and Dwiggins, G. A., "Comparison of Two Methods Used to Measure Permeation of Glove Materials by a Complex Organic Mixture" *Performance of Protective clothing, ASTM STP 900*, R. L. Barker and G. C. Coletta, Eds., American Society for Testing and Materials, Philadelphia, 1986, pp. 7–21.

[2] "Occupational Safety and Health Standards for General Industry," 29 CFR Part 1910, Subpart Z, Occupational Health and Environmental Control, 1910.1000 Air Contaminants, 1984, Washington, DC.

[3] TLVs Threshold Limit Values and Biological Exposure Indices for 1986–1987, American Conference of Governmental Industrial Hygienists, Cincinnati, OH.

[4] *Documentation of the Threshold Limit Values and Biological Exposure Indices*, 5th ed., American Conference of Governmental Industrial Hygienists. 1986, Cincinnati, OH.

[5] *Current Intelligence Bulletin 39, Glycol Ethers*, National Institute for Occupational Safety and Health. 1983, Cincinnati, OH.

[6] Schlatter, C. N., personal communication, Edmond Corporation, 1986.

[7] *Edmont Chemical Resistance Guide*, product literature from the Edmont Divison of Becton, Dickinson and Co., Coshocton, OH, 1983.

[8] Seebode, W. F., personal communication, Siebe North, Inc., 1986.

[9] *Siebe Norton Permeation Resistance Guide*, product literature from Siebe North, Inc., Charleston, SC, 1983.

[10] *Silver Shield Gloves IH-244*, product literature from Siebe North, Inc. Charleston, SC, 1986.

[11] Perkins, J. L., personal communication, Department of Environmental Health Sciences, School of Public Health, University of Alabama at Birmingham, 1986.

[12] Dixit, B., "Performance of Protective Clothing: Development and Testing of Asbestos Substitutes," *Performance of Protective Clothing, ASTM STP 900*, R. L. Barker and G. C. Coletta, Eds., American Society for Testing and Materials, Philadelphia, 1986, pp. 446–460.

[13] Adams, M. A., *Occupational Skin Diseases*, Grune and Stratton, Inc., 1983, New York, NY.

[14] Schwope, A. D., "Permeation of Chemicals Through the Skin," *Performance of Protective Clothing, ASTM STP 900*, R. L. Barker and G. C. Coletta, Eds., American Society for Testing and Materials, Philadelphia, 1986, pp. 221–234.

Donald H. Gittelman[1]

Selected Protective Clothing for Semiconductor Manufacture

REFERENCE: Gittelman, D. H., "**Selected Protective Clothing for Semiconductor Manufacture,**" *Performance of Protective Clothing: Second Symposium, ASTM STP 989,* S. Z. Mansdorf, R. Sager, and A. P. Nielsen, Eds., American Society for Testing and Materials, Philadelphia, 1988, pp. 347–355.

ABSTRACT: State-of-the-art technologies in the semiconductor industry require exotic processing techniques using unusual, hazardous chemicals. The very nature of some of these chamicals makes it necessary to very carefully select protective clothing to assure the protection of the worker.

Selection techniques are described for safety glasses, side shields, goggles, visitors' goggles, face shields; ear protection; respirators, in-line air breathing equipment, self-contained breathing apparatus; gloves—acid, solvent and heat-resistant as well as disposable types; oversleeves; aprons; head protection; footwear; and protective suits. Properties of materials—strength, permeability, durability, and impact resistance as well as chemical resistance—must be considered. Finally, recommendations for personal protective equipment to be used in the semiconductor industry are offered. Included will be an overall list of suggested equipment and possible suppliers.

KEY WORDS: hazardous chemicals, personal protective equipment (PPE), electronic component operations, III-V chemistry, single-crystal ingots, photo-resist, vapor prime, diffusion, in-process gases, head protection, hearing protection, eye/face protection, respiratory protecton, hand protection, foot protection, body protection, layering, training, self-contained breathing apparatus (SCBA)

The semiconductor industry is a "high-tech" industry requiring the use of many hazardous chemicals during the processing operations. As a result, the selection of personal protective equipment (PPE) is becoming high-tech in its own right. Long gone is the procedure of, for example, picking up a pair of gloves and putting them on without any thought regarding what they are made of or how they are made. This is true for all forms of protective clothing. The bottom line is that considerable study precedes the selection of proper PPE to avoid such hazards as absorption of chemicals, impact of flying particles, splashing, or eruption of solutions.

An overview of the processing of semiconductor devices should give the reader a sense of the various problems that must be handled in the selection of PPE. Because clean rooms are associated with electronic component operations, a casual observer might assume that there would be few problems associated with processing because of the clean air required in the operations. Nothing could be further from the truth: some of the chemicals are insidious in nature and must be dealt with very carefully.

After review of the following section on semiconductor processing, it can be seen that the semiconductor industry has a tremendous complexity of processes and materials used

---

[1] Senior engineer, environmental safety and health consultant, AT&T, Reading, PA 19612-3396.

in the production of electronic components. Industrial hygiene and safety personnel, therefore, must have a tremendous depth of knowledge and storehouse of information on all
processes and chemicals involved so that workers may be protected from any unusual or
unsuspected eventualities in production.

It is also important to point out that frequently engineering controls are used to reduce
or even eliminate protective clothing that had to be used in the past. Scaling down hazardous
conditions is the procedure that should always be followed so that the least amount of
protective apparel need be worn. This may be accomplished by improving equipment, using
less hazardous chemicals, or using disposable parts that need not be cleaned but may be
thrown away so that there is no contact with the hazardous material.

Semiconductor Processing

In the recent past, the major substrate used for device manufacture had been silicon, but
newer technology involves the use of III-V chemistry; that is, Group III elements such as
aluminum, gallium, and indium and Group V elements such as phosphorus and arsenic.
Source materials for the Group III elements are the metals themselves; whereas, the Group
V elements may use the metals, but for some applications, materials such as arsine, arsenic
trichloride, phosphine, or phosphorus trichloride may be used in the diffusion procedure
for doping the substrate with metal ions to provide different characteristics to the device.

After growth of the single-crystal ingot in a crystal grower, the ingot is sawed into wafers.
During this sawing process, residual amounts of dopant are released from the interstices.
This material along with the sludge requires special handling as well, including peripheral
exhausting around the saw blade to keep evolutes away from the operators. The wafer is
then polished with such chemicals as acetic acid and silica slurries in surfactant solutions.
Occasionally it is necessary to use trichloroethylene for cleaning the mounting substrates or
removing mounting material from the wafers.

After this operation is accomplished, the wafers progress to the clean rooms where they
undergo such procedures as photo-resist application—expose and develop, diffusion techniques, ion implantation, continuous vapor deposition, reactive ion etch, plasma etch, etc.
To the uninitiated, these may sound like space-age terms, but the real concern lies in the
chemicals that are used.

For photo-resist procedures, photo-sensitive polymers are used containing solvents such
as n-butyl acetate, cellosolve acetate, and xylene. This is done after a vapor prime with
hexamethyl disilazane (HMDS) in xylene. The developers for these photo-resists can range
from dilute potassium hydroxide to acetone. Diffusion techniques and ion implantation use
materials such as boron tribromide, phosphorus oxychloride, phosphorus trichloride, arsine,
phosphine, and diborane. Diffusion furnace tubes are cleaned with aqua regia and hydro-
fluoric acid. In-process wafers can be cleaned with ammonium hydroxide, hydrogen per-
oxide, hydrochloric acid, acetic acid, etc., depending upon what must be removed from the
wafer by the specific cleaning process. In-process gases include not only the previously
mentioned toxic hydrides of arsenic, phosphorus, and boron but also, for etching processes,
gases such as several different Freons to provide reactive fluorides, such as chlorine trifluoride, along with oxygen in plasma etchers. Nitrogen trifluoride and hydrogen chloride
gases are also used in these processing procedures. In all cases, exhaust systems are provided,
calculated for total capture of evolutes. This is not to say, however, that the worker cannot
come in contact with hazardous materials, and personal protective equipment provides the
final link in safeguarding employees.

This capsule summary gives some hint of the wide range of problems that must be dealt
with in the environment of the semiconductor industry. With this background, methods used

to select PPE for the broad range of problems and the rationale behind the choices made will become clear.

Personal Protective Equipment

This review will begin with head protection, followed by protection for hearing, eye/face, respiratory system, hands, feet, and finally, the all-important body protection.

Head Protection

Within the operating structure of a semiconductor facility, equipment operators have little need for hard hats. Maintenance and operating engineers, however, need hard hats when working in the equipment yards, boiler room, electrical services, crane and hoist operations, etc. Because of the concern about blows to the side of the head, in-plant studies indicated that the best fit with the most protection and comfort in wearing for a long period of time was an absolute necessity.

The in-plant fire brigade uses high-impact-resistant firemen's helmets for protection. In addition, all fireman are issued a thermal protective hood. The hood of choice is made from polybenzimidazole (PBI). These items are of proven service and were selected by the fire brigade itself, many of whom have had considerable experience as volunteer firemen with local fire companies.

Hearing Protection

The electronic components industry is not a noisy industry, but there are some areas that require hearing protection—such areas where there are trim-and-form tools (trimming and shaping leads on devices), where impact noise is a problem, and areas with ultrasonic cleaners or large air-compressors, which can result in continuous noise problems. Ear plugs are not recommended because workers have a tendency to lay them on work surfaces between uses, where they pick up materials which can cause ear infections. Earmuffs, on the other hand, because of their design, do not cause this problem. Two styles have worked well, which allows the worker to choose the one most comfortable for him. The first is a heavy-duty, impact-resistant earmuff type that attenuates sound pressure efficiently. The other is an insert type, but because the inserts are on an overhead band, they cannot roll around working surfaces or on the floor, as ear plugs are likely to do.

Eye/Face Protection

In the evaluation of eye and face protection, concern has centered in the area of performance rather than on design or dimension of the lenses. Design restrictions are being reconsidered and removed from the American National Standards Institute (ANSI) Z87.1-1979 standard while an overall performance test is being introduced. This new standard is expected to allow greater flexibility in selection to please more workers and make it easier to make people accept eye/face protection.

In an effort to make workers more comfortable, two choices were given to those workers with prescription glasses: standard safety glasses using prescription lenses or the polycarbonate visitors' goggles which can be worn over cosmetic frames or can be used as a plano with built-in side-shields for both workers and visitors as well. The visitors' goggles is the preferred eye protection because it can be used for all applications. In addition, side-shields must be worn with the regular plano safety glasses to protect the eyes from the side; thus, it is easier to use the all-in-one visitors' goggles. A universal fit clip-on cellulose acetate side shield is used on the prescription safety glasses to provide side protection. It affords good protection and fits all frames regardless of manufacturer. Safety goggles, often referred to as monogoggles, are used in areas of heavy chemical use such as bottling rooms. Recent work has shown that newer types of goggles provide better vision, especially peripheral, and have more fog-free capabilities. These are also of polycarbonate construction.

When welding operations need to be performed or when reactions at extremely high temperatures are viewed through sight-glasses on reactors or crystal growers, welding goggles are required, of which there are many satisfactory types available.

Face shields are of polycarbonate construction and are capable of providing good impact resistance. Insofar as chemical usage is concerned, face shields are confined to large-scale handling areas. It is important to be very careful in using shielding with chemicals because vapors may get in back of the shield and irritate facial areas by concentrating the vapors and condensing with the moistness of the skin. This is particularly true where operators are working over tanks and wells in, for example, plating facilities, where they are likely to bend down and expose their faces to the vapors below the exhaust flow. In addition, certain chemicals will react with the polycarbonate face piece, causing deterioration of the resin and thus reducing the effectiveness of the protection.

This is true not only of face shields and goggles, but also of the visors on protective hoods and fully encapsulated suits. The suit material may be coated with a Saran or Viton material which will withstand most chemicals, but it is no better than the shield material if it deteriorates. Efforts are underway by many suppliers to find an answer to this problem. Perhaps the answer might lie in the use of a coating, such as a good chemical-resistant urethane, or else an as-yet-undeveloped laminated material. Saranex-coated Tyvek material with a Mylar polyester film facepiece seems to offer satisfactory protection in extremely hazardous conditions, such as handling and cleaning up spills of toxic chemicals, especially those used in III-V device technology, including compounds of gallium, indium, arsenic, and phosphorus. Some of these chemicals are hazardous by absorption through the skin; therefore, protective gear must assure complete skin protection. Because it is relatively impermeable to most chemicals, though very expensive, Viton, also, is an excellent choice as a protective material, with a Mylar polyester film visor used only when extremely heavy-duty chemical operations are being conducted. Still, the eye-shield problem needs to be solved.

Respiratory Protection

Similar problems can occur with the shielding material on respirators or self-contained breathing apparatus (SCBA). No such problems have been seen with the equipment used by workers or the fire brigade. The preferred respirator is a silicone rubber facemask model selected because of its flexibility in fitting various facial configurations. The combination cartridge used is an all-purpose unit acceptable for dusts, fumes, mists, radionuclides, organic vapors, and acid gases.

The pressure-demand SCBA used by the fire brigade is also provided during highly toxic operations such as changing arsine, diborane, or phosphine cylinders in ion-implantation equipment. This equipment is available from several suppliers, and has proven to be the best approach for these particular conditions.

Hand Protection

Perhaps the most complicated concern of PPE is hand protection. Chemical resistance, degradation, and permeation are under continuous study by researchers in attempts to come up with better or perhaps universal-use gloves for acid, alkalis, and solvents. A substantial amount of this testing review is in the capable hands of the ASTM F-23 Chemical Resistance Committee.

It is, however, not the intent of this paper to discuss testing but to discuss what is most satisfactory for use in the semiconductor industry. An aspect which probably has not been considered by developers as a criterion for glove selection, especially with chemically resistant gloves, is that of "feel." Many of the gloves tested in the past were rejected by users for this reason. Either the thickness prevented workers from "feeling" the work they were handling or the gloves were too slippery when wet to handle the work without dropping or breaking it. These constraints greatly narrowed the choice of gloves that could be used, and it was necessary to test all gloves that could be found for these parameters. Glove selection would be much easier if one glove could be used to protect the hands against all chemicals, with good fit and "feel," and to provide good grip without being slippery, especially when wet.

Research on chemical- and abrasion-resistant gloves differs as to whether or not a universal glove can ever be developed. Significant glove research is continuing in this direction, but many doubts remain. Until this work comes to fruition, the preferred gloves are an acid- and alkali-resistant type based on sheer natural latex, which gives the glove good workability along with good chemical resistance. Nitrile rubber gloves are good, pliable gloves for use in aromatic, petroleum, and chlorinated solvents, outperforming natural rubber and neoprene gloves in these applications. Even with these gloves, problems remain. There is some crossover from solvent to acid- and alkali-resistance gloves, since the solvent-resistant glove is especially good with chlorinated solvents but only fair in acetone use. This may be a function of the solubility parameters of the solvents, since chlorinated solvents are weakly hydrogen-bonded and ketones are moderately hydrogen-bonded. Nonetheless, if both these gloves are available, they will, by and large, cover the gamut of chemicals used. The acid- and alkali-resistant glove has the wider range of use because it is also satisfactory with many solvents.

Viton gloves are particularly good for a wide range of chemical exposures, but their usage is narrowed by the economics involved—they are very expensive. Because they are very good where serious problems could result from highly toxic materials, they find application on those facilities where arsenic or chemicals that are capable of being absorbed through the skin are used.

Specialty gloves are very important as well for specific applications. Listed below are some examples that may be useful for special applications.

1. Nomex knit gloves with leather palms: used on applications where equipment is at elevated temperatures yet "feel" is of great importance when attempting to do intricate work. An example of this is working with the hardware on sputtering equipment.

2. Fire-resistant, Kynol gloves: used where highly elevated temperatures or flames are involved because Kynol will not ignite at temperatures up to 2500°C, but will char slowly at 260°C. These characteristics make this glove satisfactory for use at furnace operations.

3. Heavy-duty neoprene gloves: required by maintenance and construction personnel for certain manufacturing jobs where not only resistance to a wider range of chemicals but also protection from punctures and abrasions are required.

4. Cotton gloves: especially important for use where there is possible contact with heaters and where the gripping motion is particularly essential. This is required in molding operations where epoxy preforms are picked up manually from preheaters and placed into the ram cylinder of the molding presses. With workers coming in such close contact with the heaters in the press, burns would be possible in that location as well as in the preheaters.

5. Firemen's gloves: selected for use by the in-house fire brigade, upon recommendation by the Environment, Safety and Health Engineering Department. After review of several different gloves, the one selected was a heavy-duty abrasion- and puncture-resistant glove. Kevlar construction is inherently flame resistant and will not support combustion; it also surpasses leather for chemical resistance as well as resistance to puncture. It has a Gore-Tex glove insert and a reinforced lining that is aluminized. If chemical exposure is possible after a conflagration, Saranex-coated Tyvek is used by the firemen. These gloves have given the brigade satisfactory performance for a long period of time.

Foot Protection

In the semiconductor industry, the trend is toward the use of electrostatic dissipating materials and the more chemical-resistant polyurethane soling to provide better, more comfortable footwear for shop personnel. Because of the problems in electronic components, footwear's ability to control electrostatic discharge is extremely important. Many styles are provided for men and women for general use; however, where chemicals are in heavy use or are being repackaged, a specific chemical-resistant model was selected for men and a similar model is available for women. All of these models are electrostatic dissipating, while the two special models provide good chemical resistance as well.

Because of the minimum permeation through chemical-resistant shoes, there is an added problem for people required to wear these shoes for extended periods of time. This involves the excessive perspiration which occurs in shoes that do not "breathe." For this reason those employees are provided with tube socks, which absorb moisture.

When personnel are being outfitted for spill-control operations, the preferred procedure is for them to don foot protection that is relatively fast to put on. The boot that is provided for this purpose is a natural rubber over-the-shoe boot. The one that is used has good chemical resistance, is easy to put on, and provides the necessary protection from chemicals.

The fire brigade requires foot protection as well, and depending on the need, members will wear long or short firemen's boots.

Body Protection

Arm protection is very important in day-to-day operations in a semiconductor facility in two areas of work in particular: at acid and alkali facilities and at high-temperature operations where there is the possibility of thermal burns. At acid and alkali operations, the vinyl oversleeves are used. This seems to be satisfactory because, at the wrist, this sleeve provides a snug fit over the glove so that if chemicals are spilled on the sleeve, they will not run down into the glove.

A typical use for the sleeves employed in high-temperature operations is on molding presses where the operator must reach between the heaters to place the epoxy preforms into the cylinder. A 50/50 cotton/polyester sleeve prevents thermal burns. When this type product is used along with the cotton glove, it provides a close fit over the wrist area, it has been very successful in the prevention of burns.

Two types of aprons, another important facet in personal protective apparel, are commonly used:

1. Flame-resistant aprons are required where there is a possibility of flashes or fires, such as with phosphorus or with sputterer hardware coated with titanium-platinum. A 280-g (10 oz) duck apron provides adequate flame resistance for these applications.

2. Chemical-resistant aprons are used in all light chemical operations where body suits are not required. The material used is polyethylene-coated Tyvek, which is capable of handling splashes and small spills.

There are four grades of garments for whole-body protection, which are listed here from lowest to highest protection level:

1. Tyvek suit used in general application, especially by maintenance and construction personnel to protect against oils and greases. These suits are not applicable for chemical service.

2. Polyethylene-coated Tyvek suits are used by spill-control personnel for light chemical spills under contained conditions. Because these suits have limited impermeability, they are used for short durations and if they become impregnated with chemical spills materials, they may be disposed.

3. Saran-coated Tyvek Suit (Saranex) is fully encapsulated with a Saran-coated surface for good impermeability. Used where there are problems with highly toxic chemicals, this will protect the operator and his SCBA from overhead leaks, spills, etc. The SCBA is also protected from corrosion. This suit should not be used where there is the possibility of puncture or of severe abrasion. It is, however, perfectly satisfactory for open-area use.

4. The CPE (chlorinated polyethylene) Suit (Chem Proof) made from Cloropel is fully encapsulated. It has all the impermeability characteristics of the Saran-coated suit, but is more durable. This suit is to be used in confined areas where abrasion and puncturing are strong possibilities. It is also capable of use for longer periods of time.

A coverall made of flame-retardant Nomex III is used by maintenance personnel in servicing or cleaning large-scale equipment where flash fires from reactive/pyrophoric residues are possible. This type of coverall, for example, would also be required while working on ion-implantation equipment where pyrophoric gases such as phosphine or silane are in service.

Layering

Layering is used to provide optional protection from certain chemicals and materials when a single means of protection is inadequate to shield the person from the hazard. Layering may also be done to protect expensive protective equipment. Examples of this type "extra" protection would be:

1. A fully-encapsulated saran-coated (Saranex) Tyvek suit to provide first-line protection over Tyvek or poly-coated Tyvek inner suits. The outer suit will exclude most chemicals and provide considerable abrasion resistance. These layers are worn by spill- and damage-control teams.

2. For most hazardous spills, spill-control teams will don the very expensive, CPE (Cloropel) Chem Proof suit. Even though this suit is more durable, is capable of use over a longer period with a high degree of abrasion resistance, and is, itself, fully encapsulated, the fully encapsulated saran-coated (Saranex) Tyvek suit is worn as an outer garment. This provides not only first-line protection for the individual, but also protection for the CPE Chem Proof suit itself because it is so expensive.

3. Firemen's apparel is also very expensive, so in order for turnout coats and other firefighting gear to be protected from hazardous chemical conditions resulting from spills or breakage after fires, cleanup is handled by workers donning fully encapsulated Saran-

coated (Saranex) Tyvek suits as outer garments. After the cleanup, the Saranex suits can be disposed of at considerably less expense than can contaminated firefighting gear.

It is very important to point out that when layering is used, because of the impermeability of the materials, heat becomes a special problem, especially where several layers of material are used. Personnel involved in such operations must be trained not to overexert nor to work for too long a period of time without relaxing. Heat exhaustion is a real possibility under these conditions, and these people must be watched closely during their work periods to ensure that no health problems result from these hot environments. If SCBA's are being used under hot, exerting conditions, air is used up faster than normal. It is necessary not only to allow time for this eventuality, but also to provide time for decontamination of equipment and apparel before removing the air-breathing equipment. Dual-purpose SCBA's may be used when in-line breathing air is available so that operators can hook into this continuous source while decontamination is taking place.

There are other examples of layering that could be mentioned, but these are the principal procedures used in high-hazard areas as found in the semiconductor industry.

Training

The possibility of heat exhaustion in the situation mentioned above demonstrates the need for training in the use of personal protective equipment. Educating personnel in the proper use of this equipment not only requires care to avoid heat exhaustion but, even more important, awareness of what equipment must be used to protect the workers from the various hazards in their work environment.

As can be seen from the wide selection of equipment available for use, each item must be reviewed and its specific use must be ingrained in the minds of the operators so that they do not select the wrong equipment for a particular manufacturing function. Training may involve slide-and-tape or videocassette presentations with periodic repetition for emphasis. This procedure has been shown to be useful, but it cannot carry the total burden of education: other aids must be used in addition to this training.

It is necessary to specify in the written manufacturing procedures exactly what protective apparel is to be used for each function. End use is a critical matter of concern when selecting protective apparel. In addition, at that facility used to carry out that function, pictorial labels descriptively showing the type of equipment to be used at that work station are placed conspiciously on the equipment so that workers cannot mistake what they are to don. Each chemical or hazardous procedure in the plant follows this training and reinforcement technique and markedly aids in keeping injuries at a low level in the plant. The importance of this cannot be over-stressed. The inspection organization carries out annual safety audits of all facilities to insure conformance with established practice by the Environment, Safety and Health Engineering Organization.

Conclusions

As a result of the complexity and diversity of operations in the semiconductor industry, the selection of personal protective equipment is not a simple matter. A wide diversity of specialized equipment must be used to protect the worker from all the hazardous chemicals and conditions involved in such operations.

Despite all the work that has been put into providing the very best in protective apparel, industrial hygiene and safety engineers continue to work toward the ultimate protection of the worker. Newer materials are being produced and newer designs are being developed to make the work environment a safer place. Reduction in lost-time injuries does save corporations much more than the actual costs of development and use of personal protective equipment, so all industries should be taking a profound interest in these developments to reduce their costs as well. It must be remembered that safety is everybody's concern. It costs very little to work safely, but it can be very expensive not to take all safety precautions and, especially, not to wear the personal protective equipment recommended.

Acknowledgment

The author wishes to thank the members of the Environment, Safety and Health Engineering Department for their encouragement and suggestions in the preparation of this paper. Special thanks are given to S. R. Brady, fire safety engineer; A. C. Hilbert, safety advisor; D. M. Reigle, industrial hygiene engineer; and W. S. Simon, senior industrial hygiene engineer for their input of source material.

Protection from Industrial Chemical Stressors

Decontamination

Stephen P. Berardinelli[1] and Rotha Hall[1]

Decontamination of Chemical Protective Clothing Exhibiting Matrix Release

REFERENCE: Berardinelli, S. P. and Hall, R., **"Decontamination of Chemical Protective Clothing Exhibiting Matrix Release,"** *Performance of Protective Clothing: Second Symposium, ASTM STP 989,* S. Z. Mansdorf, R. Sager, and A. P. Nielsen, Eds., American Society for Testing and Materials, Philadelphia, 1988, pp. 359–367.

ABSTRACT: This study was conducted to evaluate decontamination methods for a glove material/liquid chemical combination that, together, demonstrated matrix release. A glove material/liquid chemical combination was assumed to exhibit matrix release if the following two conditions were met: (1) the glove material, which was exposed to the challenge liquid chemical until equilibrium saturation was attained, slowly released the entrapped challenge chemical for at least 72 h after exposure; and (2) upon re-exposure to the challenge chemical, the glove material demonstrated a significantly different breakthrough time. A neoprene/ *n*-butyl acetate (glove/liquid chemical) combination, meeting these conditions, was investigated in detail. Neoprene/*p*-xylene and nitrile/*n*-butyl acetate combinations were reviewed, but did not exhibit matrix release.

Three decontamination methods were tested: (1) air drying at 25°C, (2) 15-min manual detergent wash, and (3) elevated temperature washing-drying. For the neoprene/*n*-butyl acetate combination, both air drying and a single 15-min manual detergent wash were ineffective in cleaning the saturated matrix. Breakthrough time changes and measurement of the concentration of *n*-butyl acetate in the material demonstrated that the matrix was still saturated. However, elevated temperature washing and drying (45°C wash and 95°C drying) did remove the residual *n*-butyl acetate from the neoprene.

KEY WORDS: decontamination, protective clothing, chemical protective clothing, matrix release, chemical permeators

Decontamination of chemical protective clothing has not been researched in-depth. Several recent investigations have demonstrated that some decontamination methods are ineffective. This study was conducted to evaluate decontamination methods for a glove material/liquid chemical combination that, together, demonstrated matrix release. A glove material/liquid chemical combination was assumed to exhibit matrix release if the following two conditions were met: (1) the glove material, which was exposed to the challenge liquid chemical until equilibrium saturation was attained, slowly released the entrapped challenge chemical for at least 72 h after exposure; and (2) upon re-exposure to the challenge chemical, the glove material demonstrated a significantly different breakthrough time. A neoprene/*n*-butyl acetate (glove/liquid chemical) combination, meeting these conditions was investigated in detail. Neoprene/*p*-xylene and nitrile/*n*-butyl acetate combinations were reviewed, but did not exhibit matrix release.

[1] Industrial hygiene chemist and engineering technician, respectively, National Institute for Occupational Safety and Health, Division of Safety Research, Injury Prevention Research Branch, Protective Equipment Section, Morgantown, WV 26505-2888.

Three decontamination methods were tested: (1) air drying at 25°C, (2) 15-min manual detergent wash, and (3) elevated temperature washing-drying. For the neoprene/n-butyl acetate combination, both air drying and a single 15-min manual detergent wash were ineffective in cleaning the saturated matrix. Breakthrough time changes and measurement of the concentration of n-butyl acetate in the material demonstrated that the matrix was still saturated. However, elevated temperature washing and drying (45°C wash and 95°C drying) did remove the residual n-butyl acetate from the neoprene.

Dr. S. Z. Mansdorf was first to postulate matrix release as an explanation for persistent permeation of chemical from some chemical protective clothing materials after exposure to the chemical has ceased [1]. Matrix release, well known in the pharmaceutical and marine coatings industries, occurs when a small amount of a chemical is retained by a polymetric matrix and then slowly permeates from that matrix over an extended period of time. Permeation is the process by which a chemical moves through protective clothing on a molecular level. Permeation involves: (1) sorption of chemical into the protective clothing (outer) exposed surface, (2) diffusion of molecules through the protective clothing, and (3) desorption of the molecules from the unexposed (inner) protective clothing surface. Matrix release effects, coupled with ineffective decontamination methods, can cause increased worker exposure to liquid and gaseous chemicals.

Two presentations at the 1986 American Industrial Hygiene Conference in Dallas, TX, have demonstrated that many traditional decontamination methods are not effective in decontaminating chemical protective clothing (CPC) [2,3]. The purpose of this study by the National Institute for Occupational Safety and Health (NIOSH) was two fold. First, a glove material/liquid chemical combination that would exhibit matrix release was to be identified. Secondly, this combination would then be used to evauate the effectiveness of three selected decontamination methods for materials exhibiting this trait.

Apparatus

The chemical permeation test system used was an AMK permeation cell in an open loop mode (Fig. 1). Clean dry air at 500 mL/min flowrate was used at the collection medium. The detector, an H-Nu PI-101 photoionization detector, had a lower detection limit of 6.4 mg/m³ (1.3 ppm) for n-butyl acetate and 0.8 mg/m³ (0.2 ppm) for p-xylene. This system has been discussed in detail elsewhere [4].

Materials

The liquid challenge chemicals, n-butyl acetate and p-xylene, were American Chemical Society (ACS) reagent grade, as was the extraction solvent, n-hexane. All were purchased from Fisher Scientific Company of Fairlawn, New Jersey.[2] The decontamination wash reagents were Alcojet[3] made by Alconox, Inc., New York, New York, or a 5% mixture of sodium phosphate tribasic and sodium carbonate from Fisher Scientific Company.

Chemical protective gloves used for this study were purchased from two suppliers. The gloves selected were the Edmont unlined neoprene, No. 29-870, and the Ansel unlined nitrile, No. 632.

Weights of the gloves before and after exposure were determined using a Mettler balance (±0.01 mg). Specimens were weighed in a tared polyethylene bag. In this way, the effects

---

[2] Mention of a company name or product does not constitute endorsement by the National Institute for Occupational Safety and Health (NIOSH).
[3] Registered trademark of Alconox, Inc.

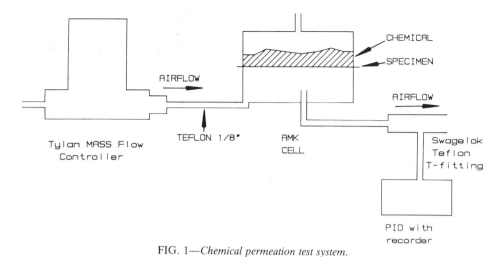

FIG. 1—*Chemical permeation test system.*

of weight loss due to challenge chemical evaporation were minimized and an accurate specimen weight could be determined.

Procedures

Specimens

Specimens were cut from the glove palm or back. Specimen thickness was measured using an Ames Micrometer (± 0.002 mm) available from Ames Instrument, Waltham, Massachusetts in accordance with Federal Test Method Standard No. 191A, Method 5030, Determination of Thickness of Textile Materials. A material specimen from a glove was mounted in the AMK chemical permeation test cell. The airflow was regulated by a calibrated Tylan Mass Flow Controller at 500 mL/min. The challenge liquid chemical was added to the upper chamber of the AMK permeation cell and a timer (± 0.01 min) was started. The breakthrough time was determined and exposure continued for 30 min after steady-state permeation was observed to ensure a saturated specimen matrix. After testing, the specimen was patted dry, decontaminated by one of the methods described later, and the entire procedure (weighing, thickness determination, and chemical permeation) repeated. Repeated exposure to challenge chemical was used to simulate reuse of the glove material. Each specimen was exposed three times. The first exposure established a baseline for a virgin specimen then two reuses.

Decontamination

Three decontamination methods were evaluated. Air drying at 25°C, hand detergent washing at 25°C, and automatic dishwashing (elevated wash temperature and drying cycle).

Except for air drying and hand detergent washing, exposed material specimens were extracted with *n*-hexane by single Soxhlet extraction of 12 h as a final step to determine residual liquid chemical in the CPC matrix. Extracts were analyzed by gas chromatography—flame ionization detection with separation by a 1.83 m (6 ft) Porapak Q Column at 210°C. The injector temperature was 240°C. Soxhlet extractions of nitrile and neoprene samples at 6, 9, 12, and 18 h demonstrated that >95% of *n*-butyl acetate was extracted after 12 h.

For air drying, the amount of sorbent lost was reflected simply by the weight change. For the hand detergent washing method, results were biased by the weight of the wash water.

Here, the hand detergent wash solution was extracted twice with n-hexane in a separatory funnel. These two hexane extractions removed >90% of n-butyl acetate in water.

Hand detergent washing decontamination methods consisted of a 15-min manual agitation at 25°C with 5% (w/w) trisodium phosphate–sodium carbonate in distilled water. The automatic dishwashing decontamination washing method consisted of two 5-min washes with Alcojet at 45°C and two 10-min rinses at 45°C followed by 25-min drying at 90°C.

Decontamination Calculation

The percentage decontamination for each decontamination method was calculated by

$$\frac{\text{weight loss}}{\text{weight gain}} \times 100$$

The difference between the weight of the virgin specimen and weight of the exposed specimen (before decontamination) was recorded as the weight gain. The difference between the weight of the exposed material specimen before and after decontamination was recorded as the weight loss. For reuse, the weight and thickness after drying were used as the baselines for the next run.

Data Analysis

Statistical analysis of the data was performed at the U. S. Public Health Service Parklawn Computer Facility using SAS (Statistical Analysis System), GLM (General Linear Model) at a 95% confidence limit (alpha = 0.05). Breakthrough times were normalized by dividing the experimentally observed breakthrough time by the square of the specimen thickness [5].

Results

The experimental results for nitrile versus p-xylene and n-butyl acetate appear in Tables 1 and 2, respectively. Only air drying of the nitrile gloves was evaluated since matrix release was not prevalent (that is, breakthrough times were equivalent). A glove material/challenge liquid chemical was assumed to exhibit matrix release if the following two conditions were met: (1) the glove material, which was exposed to the challenge liquid until equilibrium/saturation was attained for at least 72 h after exposure; and (2) upon re-exposure of the challenge chemical, the glove material demonstrated a significantly different breakthrough time.

Experimental results for neoprene versus n-butyl acetate that exhibited matrix release appear in Tables 3 through 5. Neoprene was not tested versus p-xylene. Rather, decontamination methods for neoprene versus n-butyl acetate were investigated. Table 3 contains the data for the air drying method. Table 4 has the data for the hand detergent wash method. Table 5 contains the data for the automatic dishwasher method. Note that in Table 3, that the drying time was extended as long as five days.

Analysis of Results

Air Drying Decontamination

The statistical analysis for the air drying decontamination method for nitrile versus p-xylene (Table 1) and nitrile versus n-butyl acetate (Table 2) demonstrated that at a 95% confidence limit there is no statistical difference between the breakthrough times, normalized

TABLE 1—*Air drying decontamination method: nitrile versus* p-*xylene.*

Specimen	Run	Breakthrough Time, min	Thickness, mm	% Decontamination Efficiency
1	0[a]	5	0.373	99.5
1	1[b]	5	0.378	99.8
1	2[c]	4	0.371	100.2
2	0	4	0.351	99.5
2	1	3	0.353	100.4
2	2	3	0.350	99.3
3	0	11	0.389	100.6
3	1	7	0.388	99.8
3	2	7	0.394	100.1
4	0	8	0.358	100.3
4	1	4	0.358	98.9
4	2	7	0.353	100.3
5	0	7	0.350	100.2
5	1	6	0.353	99.6
5	2	7	0.347	100.1
6	0	6	0.363	100.6
6	1	5	0.364	99.4
6	2	5	0.353	100.3
7	0	9	0.408	98.4
7	1	8	0.401	101.2
7	2	8	0.394	100.3
8	0	8	0.374	98.5
8	1	7	0.363	101.3
8	2	7	0.356	100.3
9	0	6	0.354	98.4
9	1	5	0.362	99.6
9	2	5	0.353	101.6

[a] Original or virgin specimen.
[b] First reuse.
[c] Second reuse.

breakthrough times, percentage decontamination, and thickness. The inference is that matrix release was not predominate hence the decontamination methods employed were effective. Since breakthrough times were measured, permeation did occur; however, the nitrile did not exhibit matrix release (that is, act as a reservoir to hold or trap the challenge liquid chemical).

Air Drying

Table 3, air drying of neoprene versus *n*-butyl acetate, demonstrated that breakthrough times, normalized breakthrough times, and percentage decontamination were statistically different at the 95% confidence limit. Further, Duncan's multiple range test showed that the breakthrough times of the virgin specimens were statistically different when compared to the reused specimens. The breakthrough times of the virgin specimens were longer than the reuse (air dried) specimens even at protracted drying times. The breakthrough times and normalized breakthrough times for all virgin specimens were not significantly different. Additionally, no significant statistical differences were found between all first and second reuse breakthrough times or normalized breakthrough times.

The percentage decontamination was different for the original and each reuse. Although statistically significant, the percentage decontamination should be decreasing, not increasing,

TABLE 2—*Air drying decontamination method: nitrile versus* n-*butyl acetate.*

Specimen	Run	Breakthrough Time, min	Thickness, mm	% Decontamination Efficiency
1	0[a]	6	0.328	98.6
1	1[b]	6	0.332	103.1
1	2[c]	6	0.329	100.3
2	0	12	0.377	100.1
2	1	12	0.378	102.6
2	2	12	0.369	100.1
3	0	13	0.402	99.6
3	1	10	0.377	102.9
3	2	10	0.398	99.8
4	0	6	0.373	101.6
4	1	9	0.371	99.3
4	2	5	0.373	99.4
5	0	9	0.398	101.7
5	1	9	0.389	99.6
5	2	7	0.387	99.6
6	0	8	0.357	102.2
6	1	10	0.349	99.8
6	2	9	0.348	99.5
7	0	7	0.371	99.7
7	1	7	0.368	101.8
7	2	6	0.306	99.6
8	0	11	0.367	100.0
8	1	9	0.372	101.3
8	2	10	0.368	99.6
9	0	12	0.371	100.3
9	1	10	0.364	100.9
9	2	9	0.363	99.6

[a] Original or virgin specimen.
[b] First reuse.
[c] Second reuse.

as breakthrough time decreases. This is the result of using a "new baseline" weight for each reuse. The matrix was saturated during the initial exposure and still contained a certain amount of chemical after decontamination. Before beginning the first reuse, the sample was weighed and this amount used as "new baseline" in calculating decontamination for that reuse. The same procedure was used for second reuse.

Additionally, the observed increasing percentage decontamination is due, we believe, to extraction of a specimen component, for example, plasticizer, filler, etc., when exposed to the *n*-butyl acetate. Thus the numerator, weight loss, will be greater and the percentage too large. However, thickness of the virgin and reused specimens were not different.

Hand Detergent Washing Decontamination

Table 4, hand detergent washing decontamination method for neoprene versus *n*-butyl acetate, clearly shows that breakthrough times for the initial runs (virgin material) are statistically different than reuse runs. Thickness was measured before the initial run only. No further thickness determination was made because of the inaccuracy introduced from measuring wet specimens. *N*-butyl acetate concentration was measured in the detergent wash solution (after the initial run) and in the specimen matrix (after the 1st reuse). After the first reuse, the specimen was not detergent washed. Instead it was placed directly into

TABLE 3—*Air drying decontamination method: neoprene versus* n-*butyl acetate.*

Specimen	Run	Breakthrough Time, min	Thickness, mm	% Decontamination Efficiency
1	0[a]	23	0.468	97.5
1	1[b]	6	0.470	99.0
1	2[c]	6	0.470	101.6
2	0	22	0.480	98.7
2	1	2	0.483	99.5
2	2	2	0.479	101.0
3	0	20[d]	0.456	97.9
3	1	8[d]	0.462	98.7
3	2	2	0.462	101.3
4	0	19	0.443	97.8
4	1	10	0.448	97.8
4	2	8	0.450	101.4
5	0	19	0.458	97.6
5	1	11	0.465	99.7
5	2	2	0.465	100.3
6	0	23[e]	0.476	96.9
6	1	3[e]	0.478	97.5
6	2	2	0.480	102.4
7	0	20	0.461	96.0
7	1	2	0.461	99.5
7	2	2	0.464	102.5
8	0	18	0.450	96.6
8	1	2	0.457	96.5
8	2	2	0.459	103.1
9	0	17	0.445	98.5
9	1	2	0.443	99.7
9	2	2	0.449	102.1
10	0	19	0.469	97.1
10	1	2	0.472	98.6
10	2	2	0.476	103.2
11	0	17	0.444	99.2
11	1	2	0.448	99.5
11	2	2	0.452	101.4

[a] Original or virgin specimen.
[b] First reuse.
[c] Second reuse.
[d] Specimen was air dried for six days.
[e] Specimen was air dried for seven days.

the Soxhlet extractor. These data revealed that not much was extracted with the hand detergent wash and a great deal of *n*-butyl acetate remained in the glove matrix. The percentage decontamination was determined only for the initial run; these values are all quite low. If the specimen matrix was saturated with *n*-butyl acetate, instantaneous breakthrough on reuse would be observed—this appears to be the case.

Automatic Dishwasher Decontamination

The automatic dishwasher decontamination data (neoprene/*n*-butyl acetate), Table 5, demonstrated that the breakthrough times of virgin as well as reused specimens were equivalent. Thickness was measured only before the initial run. No further thickness determination was made because of the inaccuracy introduced from measuring wet specimens. After the

TABLE 4—*Hand detergent wash decontamination method: neoprene versus* n-*butyl acetate.*

Specimen	Run	Breakthrough Time, min	Thickness, mm	Concentration, mg Wash[c]	Sox[d]	% Decontamination Efficiency
1	0[a]	20	0.453	3.4	ND	8.9
1	1[b]	0	ND	ND	28.3	ND
2	0	20	0.456	3.9	ND	4.9
2	1	0	ND	ND	28.7	ND
3	0	20	0.467	3.5	ND	4.2
3	1	0	ND	ND	20.6	ND

[a] Initial run—virgin material.
[b] First reuse.
[c] Concentration in milligrams of *n*-butyl acetate in wash solution.
[d] Concentration in milligrams of *n*-butyl acetate extracted from specimens using soxhlet extractor.
ND = not determined.

second reuse, residual *n*-butyl acetate in the neoprene matrix was removed by Soxhlet extraction. Only trace quantities below the limit of quantitation were observed. The percentage decontamination efficiency was determined only for the initial use; the values were high (82 to 97%). However, the percentage decontamination efficiency of 82.2 (Specimen 1, Run 0) is an anomaly that may be due to a weighing error.

Summary and Conclusions

Matrix release explains why breakthrough time varies with reuse. Further, for those glove material/liquid chemical combinations that exhibit matrix release, persistent permeation cannot be alleviated by drying at room temperature even for protracted periods. Hand detergent washing is ineffective as well.

Breakthrough time is the most sensitive measure of decontamination efficiency as demonstrated by the data. Decontamination efficiency as determined by weight change varies

TABLE 5—*Automatic dishwasher decontamination method: neoprene versus* n-*butyl acetate.*

Specimen	Run	Breakthrough Time, min	Thickness, mm	Concentration, mg, Soxhlet	% Decontamination Efficiency
1	0[a]	22	0.457	ND	82.2
1	1[b]	21	ND	ND	ND
1	2[c]	20	ND	TR	ND
2	0	20	0.462	ND	95.5
2	1	20	ND	ND	ND
2	2	19	ND	TR	ND
3	0	21	0.449	ND	96.8
3	1	20	ND	ND	ND
3	2	21	ND	TR	ND

[a] Initial run—virgin material.
[b] First reuse.
[c] Second reuse.
ND = not determined.
TR = trace.

only a few percentage points with a drastic change in the observed breakthrough time. Thus, measuring decontamination efficiency by weight change is imprecise. Thickness, which remained constant before and after exposure, is not an effective means to monitor decontamination efficiency.

Before chemical protective clothing that exhibits matrix release can be effectively decontaminated and safely reused, a specific laboratory experiment is required documenting the effectiveness of decontamination along with a minimum effect on breakthrough time and permeation rate.

References

[1] Mansdorf, S. Z. in *Performance of Protective Clothing, ASTM 900*, R. L. Baker and G. C. Coletta, Eds., American Society for Testing and Materials, Philadelphia, 1986, pp. 207–213.

[2] Garland, C. E., "Chemical Contamination and Decontamination of Protective Clothing," Abstract #28, presented at the 1986 American Industrial Hygiene Conference, Dallas, 19 May 1986.

[3] Perkin, J. L., Johnson, J. S., Swearengen, P. M., and Weaver, S. C., "Residual Splashed Solvents in Butyl Protective Clothing and Usefulness of Decontamination Procedures," Abstract #29, presented at the 1986 American Industrial Hygiene Conference, Dallas, 19 May 1986.

[4] Berardinelli, S. P., Mickelsen, R. L., and Roder, M. M., "Chemical Protective Clothing: A Comparison of Chemical Permeation Test Cells and Direct—Reading Instruments," *Journal,* American Industrial Hygiene Association, Vol. 44, No. 12, Dec. 1983, pp. 886–889.

[5] Davis, S. L., Feigley, C. E., and Dwiggins, G. A. in *Performance of Protective Clothing, ASTM STP 900*, R. L. Barker and G. C. Coletta, Eds., American Society for Testing and Materials, Philadelphia, 1986, pp. 7–21.

Charles E. Garland[1] and Ana M. Torrence[2]

Protective Clothing Materials: Chemical Contamination and Decontamination Concerns and Possible Solutions

REFERENCE: Garland, C. E. and Torrence, A. M., "**Protective Clothing Materials: Chemical Contamination and Decontamination Concerns and Possible Solutions,**" *Performance of Protective Clothing: Second Symposium, ASTM STP 989*, S. Z. Mansdorf, R. Sager, and A. P. Nielsen, Eds., American Society for Testing and Materials, Philadelphia, 1988, pp. 368–375.

ABSTRACT: The theory that supports permeation testing indicates that contaminants will saturate a protective material given sufficient exposure. Once the challenge is removed, chemicals within the matrix diffuse out in both directions. The purpose of this study was to develop a method of measuring contamination of protective clothing materials and to investigate solutions to the problem of matrix contamination and release. Experimental procedures used provided the basis for the ASTM standard development in this area.

Experiments consisted of challenging neoprene, Viton, and butyl with nirobenzene. Percent contamination was measured as a percent by weight of the virgin material at several time intervals up to saturation. Up to 46% contamination was detected.

Three possible solutions to the problem of contamination were investigated: decontamination between wearings, materials with very long breakthrough times, and disposable protective clothing. Disposable protective clothing appears to be the most practical solution at the present time.

Decontamination using solvent flushing (Freon 113) was not effective even after two washes. The data generated also indicated that chemical exposure can alter the barrier properties of the material. Thermal decontamination proved effective with nitrobenzine and neoprene, but more research is needed to determine the extent to which it can be applied. Teflon-coated Nomex, which has a long breakthrough time, exhibited no detectable matrix contamination. As expected, contamination was found to be inversely related to breakthrough time. Where exposure times are very small fractions of breakthrough time, it may be possible to decontaminate by surface cleaning. Disposable protective clothing of laminated Tyvek offers excellent barrier properties to many chemicals. With disposable clothing worn by itself or over a reusable clothing, the contamination problem is discarded with the contaminated garment.

KEY WORDS: protective clothing, chemical protective clothing, neoprene, nitril, polyvinyl chloride, butyl, Viton, Tyvek/Saranex-23, Teflon-coated Nomex, permeation, breakthrough time, contamination, decontamination, decontamination efficiency, solvent flushing, thermal decontamination, contamination profile, disposable protective clothing

Historically, most barrier materials used in chemical protective clothing are made of elastomers such as neoprene, nitrile, polyvinyl chloride (PVC), butyl, Viton,[3] and combinations of these materials. In general, they are too expensive to be considered disposable and multiple uses are expected. Repeated use implies that some service is performed on

[1] President, C. P. C. Genwood, Inc., 314 Brockton, Rd., Wilmington, DE 19803.
[2] Marketing representative, E. I. duPont de Nemours & Co., Inc., Textile Fibers Department, Fibers Marketing Center, Wilmington, DE 19898.
[3] Registered trademark of E. I. duPont de Nemours and Company.

TABLE 1—*Effect of heating neoprene prior to permeation testing.*

Heat Applied	Physical Change	Mcg of Nitrobenzene, 15-min exposure
None	...	2.2
100°C	none	4.9
125°C	none	2.5
150°C	yellow with stiffening	3182.5

the protective clothing between wearings, such as chemical decontamination or hygienic cleaning or both. Often, these services are overlooked and, even when conscientiously applied, are limited to the outside surfaces only.

Background

Surprisingly, the theory that supports permeation testing has been ignored in considering contamination and subsequent decontamination. The ASTM Test Method for Resistance of Protective Clothing Materials to Permeation by Liquids and Gases (F 739-85) offers proof that chemical contaminants diffuse through a cross section of barrier material. Breakthrough times vary from minutes to days, depending upon the particular chemical and protective clothing material used. Therefore, given sufficient exposure, there is a time at which an elastomer matrix becomes saturated with chemical. Once the challenge is removed, chemicals within a matrix diffuse out in both directions. This theory also supports an inverse relationship between breakthrough time and the amount of chemical contamination for any given exposure time.

Anyone performing permeation tests with sulfuric acid and polyvinyl chloride (PVC) has probably observed matrix contamination and release. This occurs with saturated PVC after breakthrough and subsequent water wash during drying in ambient air, or in an oven. Sulfuric acid exudes from the fabric and is observed as droplets on the surface.

The matrix release phenomenon is also known in medicine where it is used for transdermal therapeutic systems. Nitroglycerine for control of anginal attacks and scopolomine for the control of motion sickness are often administered by controlled release from the matrix of a semipermeable membrane.

Purpose and Scope

The purpose of this study was to develop a method for measuring contamination and to investigate possible solutions to the problem of matrix contamination and release. Three

TABLE 2—*Properties of barrier fabrics.*

Type of Fabric	Vendor	Thickness, mm	Weight, kg/m^2	Breakthrough Time, min[a]
Butyl	RICO	0.32	0.427	900
Neoprene	RICO	0.45	0.642	15
Viton	Andover	0.20	0.340	90
Nomex/Teflon	Chemfab	0.39	0.373	>1440
Tyvek/Saranex-23	Durafab	0.18	0.110	240

[a] Measured using ASTM F 739-85.

TABLE 3—*Percent contamination after a 2-s splash.*

Barrier Fabric	Exposure Time	Percent Contamination[a]
Neoprene	2 s	3.3
Viton	2 s	0.3
Butyl	2 s	0.2
Tyvek/Saranex-23	2 s	0
Nomex/Teflon	2 s	0

[a] Percent by weight of virgin material. The analytical detection limit is 0.1 mcg/mL methanol extract.

possible solutions were investigated in this study:

1. decontamination between wearings,
2. materials with very long breakthrough times, and
3. disposable protective clothing.

Of these solutions, disposable protective clothing appears to be the most practical at the present time.

Experimental Design

Matrix Contamination

Experimental techniques developed in this study to measure contamination and decontamination efficiency have formed the basis for an ASTM draft "Standard Test Method for Measuring Matrix Contamination and the Efficiency of Decontaminating Barrier Fabrics Used in Protective Clothing" now under development.

Contamination was controlled by timing the exposure in a permeation test cell to two extremes:

1. a 2-s exposure representing a splash that might occur in the field; and
2. breakthrough time plus 1 h to reach steady-state permeation and saturate the elastomer matrix.

Immediately following exposure, the matrix was removed from the cell and blotted dry with paper towels. Disks, ~19 mm (3/4 in.) in diameter, were punched from the exposed area of fabric, Soxhlet extracted with ethanol, and the extract analyzed by gas chromatography for the challenge chemical.

TABLE 4—*Percent contamination of a saturated matrix.*

Barrier Fabric	Exposure Time	Percent Contamination[a]
Neoprene	75 min	46.4
Viton	150 min	3.9
Butyl	15 h	3.3
Tyvek/Saranex-23	5 h	1.5
Nomex/Teflon	24 h	0

[a] Percent by weight of virgin material. The analytical detection limit is 0.1 mcg/mL methanol extract.

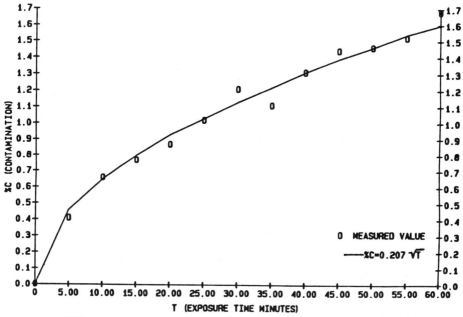

FIG. 1—*Sixty-minute contamination profile for nitrobenzene and butyl.*

Percent contamination as a percent by weight of the virgin fabric was determined for a 2-s splash and for a saturated matrix.

An experiment was also conducted to determine the relationship between exposure time and level of contamination. By measuring contamination ($\%C$), at several time (T) intervals up to 1 h, we obtained data for plotting 60-min contamination profiles.

Decontamination Between Wearings

Two decontamination procedures were evaluated for effectiveness: solvent flushing and heating.

Freon 113, an innocuous solvent that is commonly used for decontaminating equipment, was selected as our flushing solvent. Following controlled contamination to saturate the matrix, three 19 mm (¾ in.) disks were punched from the exposed area of the material. One was analyzed directly to determine the level of contamination; the second was washed for 10 min with Freon 113 and analyzed; and the third disk was given two 10-min washes

TABLE 5—*Contamination coefficients—nitrobenzene versus three elastomers.*

Barrier Fabric	A^a	Multiple R^{2b}
Butyl	0.21	0.998
Viton	0.58	0.985
Neoprene	6.86	0.996

[a] Contamination coefficient from $\%C = A(T)^{1/2}$.
[b] Multiple R^2 = Regression SS/Total SS: where SS = sum of squares.

TABLE 6—*Decontamination efficiency with Freon 113 washing.*

	Percent Decontamination Efficiency[a]	
Barrier Fabric	First Wash	Second Wash
Butyl	44.7	89.5
Neoprene	66.7	92.4
Viton	82.9	99.1
Nomex/Teflon	100.0	100.0

[a] Nitrobenzene was not detected in the unwashed, exposed disk, or in either of the washed, exposed disks.

with Freon 113 prior to analysis. Decontamination efficiencies were obtained per the following

$$\% \text{ decontamination efficiency} = (C_a - C_b) \times 100$$

where C_a is the contaminant concentration in the unwashed, exposed disk, and C_b is the contaminant concentration in the washed, exposed disk.

Experiments involving thermal decontamination were also conducted. For this study, it was necessary to determine the temperature at which neoprene could be heated without degrading. Samples of neoprene were heated in an air oven for 16 h at temperatures of 100, 125, and 150°C. Following heating, the samples were examined, placed in a permeation cell, and exposed for the normal breakthrough time of 15 min. Results are summarized in Table 1. There was no apparent change in physical or permeation properties after heating to 125°C. At 150°C, physical and permeation properties both changed significantly.

With this information, we controlled the contamination of a neoprene sample with a 2-s splash. Three disks were punched from the exposed area. One was extracted and analyzed as is; a second disk was heated to 70°C in an air oven for 24 h, extracted and then analyzed; and the third disk was heated in an air oven at 100°C for 24 h, extracted and analyzed. Percent decontamination efficiencies were calculated.

Impact of Long Breakthrough Times

Impact of long breakthrough time on contamination was investigated. A material for which chemical breakthrough was not detected during the exposure period (24 h) was analyzed for percent contamination.

TABLE 7—*Effect of matrix flushing with Freon 113 on permeation.*

	Nitrobenzene Breakthrough Time, min	
Barrier Fabric	Normal	Flushed[a]
Butyl	900	60
Neoprene	15	<15
Viton	90	<15
Nomex/Teflon	>1440	>1400

[a] Each material sample was exposed to Freon 113 for 4 h in a permeation cell. The Freon 113 was then replaced with nitrobenzene.

TABLE 8—*Thermal decontamination: efficiency of thermal contamination following a 2-s nitrobenzene splash on neoprene.*

Heat Applied	Percent Contamination	Decontamination Efficiency
None	3.62	...
70°C	0	100%
100°C	0	100%

Disposable Protective Clothing

Disposable protective clothing was investigated as a solution to the contamination/decontamination problem. No decontamination experiments were run since the clothing would be discarded after each run. Breakthrough times for many chemicals were established.

Materials and Chemicals

The following barrier fabrics were selected for the study: butyl, neoprene, Viton, Tyvek[3]/Saranex[4]-23 and Teflon[3]-coated Nomex.[3] Pertinent properties of these fabrics are shown in Table 2. Nitrobenzene was chosen as the contaminating chemical because it is used in large quantities at Du Pont's Chambers Works, where the experiments were conducted.

Results and Discussion

Percent contamination of nitrobenzene as a percent by weight of the virgin material for a 2-s splash and for a saturated matrix are shown in Tables 3 and 4, respectively. Significant

TABLE 9—*Permeation data for Nomex/Teflon fabric.*[a]

Challenge Chemical	Breakthrough Time, h
Acetic acid	>8
Acetonitrile	>8
Bromine	>20
Carbon disulfide	>8
Dimethylamine	>8
Diethylether	>8
Dimethylformamide	>8
Ethyl acrylate	>8
Fluorosulfonic acid	>20
Freon TF	>8
Hexane	>8
Methanol	>8
Methyl ethyl ketone	>8
Nitric acid, 90%	>20
Nitrobenzene	>8
Oleum, 65%	>20
Tetrahydrofuran	>8
Toluene	>8

[a] All 8-h permeation tests were performed by Radian Corp. The 20-h tests were performed at Du Pont's Chambers Works Industrial Hygiene Laboratory.

[4] Registered trademark of Dow Chemical Company.

TABLE 10—*Permeation data for Tyvek/Saranex-23.*

Challenge Chemical	Breakthrough Time, min
Acetic acid	>480
Acetone	33
Acetonitrile	97
Chlorine gas	>480
n,n-dimethylacetamide	64
2,4-dimethylacetamide	>2880
Hydrochloric acid, 37%	>2880
Methanol	>480
Methyl chloroacetate	>480
Methyl ethyl ketone	29
Nitric acid, 90%	107
Nitrobenzene	165
50% PCB, 50% mineral oil	>480
Sodium hydroxide, 40%	>480
Sulfuric acid	>480
Triethylamine	>480
Toluene	<5
Toluene diisocyanate	>480

levels of contamination were found in the exposed elastomers, even for the 2-s splash. As predicted, contamination levels are inversely related to breakthrough times. Since breakthrough time for Teflon-coated Nomex was in excess of 24 h, it was considered impractical to reach steady-state permeation. However, there was no measurable contamination after the 24-h exposure time.

Contamination profiles were obtained by measuring contamination at several time intervals. Figure 1 represents a contamination profile for nitrobenzene and butyl. After an induction period of about 5 min, in which there is a rather rapid adsorption of contaminant on the surface, the contamination process moderates to a steady-state mass transfer through the elastomer matrix.

Since the curve is parabolic, we used the general equation

$$\%C = A(T)^{1/2}$$

to compute a fit to the measured data. In this example, the best fit was obtained with $A = 0.21$. The constant, A, called the contamination coefficient, is invariant for a specific contaminant/fabric combination and proportional to contamination level with different contaminant/fabric combinations.

Contamination coefficients for the other elastomers in the study are shown in Table 5. With these A values, one can determine the contamination level for any exposure period. Multiple R^2 is a measure of curve fit when appropriate A values are substituted into the basic equation. A perfect fit of the equation to the measured values would have a Multiple R^2 equal to 1.00.

Information provided by these relationships may help in the future to:

1. develop new decontamination protocols, and
2. quantify hazard potential to the wearer.

However, since this approach requires extensive laboratory work, it may be feasible only in those operations where a few chemicals are handled in large volume.

Decontamination using solvent flushing (Freon 113) was evaluated. Results of these tests in Table 6 show that even after two washes, the contaminant is not completely removed by this technique.

Furthermore, extraneous peaks observed during gas chromatography (GC) analysis indicated that the rubber additives (curing agents, etc.) leached from the elastomers by either nitrobenzene, freon, ethanol, or a combination of these solvents. This phenomenon raised another question: if rubber additives are extracted from elastomers by exposure to a chemical, what effect will this have on permeation properties?

Experiments were run to determine if chemical exposure would alter permeation properties. A comparison of normal breakthrough times with those obtained following exposure to Freon 113 are shown in Table 7. There was a considerable reduction in breakthrough times for the elastomeric fabrics.

Test results for thermal decontamination are shown in Table 8. The results indicate that thermal decontamination is effective with nitrobenzene following exposures of short duration. We have not yet determined a minimum time and temperature at which thermal decontamination is effective following a 2-s splash, nor do we know its effectiveness at higher contamination levels.

It should be emphasized that observations made here pertain only to nitrobenzene and neoprene. Each combination of challenge chemical and barrier are expected to behave differently; therefore, extensive research would be needed to determine thermal decontamination effectiveness and establish a protocol for each fabric/chemical combination. In addition, thermal decontamination does not eliminate the problem of the effect that repeated chemical exposure may have on the barrier properties of the fabric.

Another possible solution to the problem of matrix contamination is to wear protective clothing that offers very long breakthrough times. Where exposure times are small fractions of breakthrough times, it may be possible to decontaminate by surface cleaning. Teflon-coated Nomex provides long breakthrough times for nitrobenzene (Table 9). Teflon-coated Nomex, however, is expensive, and it is still considered experimental in nature. More field experience will help determine its overall effectiveness.

A third possible solution to the contamination/decontamination problem is to discard the clothing after each chemical exposure. While this is not always economically feasible, a variety of protective garments are currently available at prices sufficiently low to be considered disposable. Table 10 shows some breakthrough times for Tyvek laminated to Saranex-23. The data indicates that it is an excellent barrier to many chemicals.

It should be noted that the breakthrough time of disposable materials need only exceed the wear time since they are discarded after each use. Layering with limited-use protective clothing can also be used to reduce contamination in situations where reusable clothing is needed. For example, at Du Pont's Chambers Works, Tyvek garments are often worn over butyl encapsulating suits.

Conclusions

In this study, we have seen experimental evidence of matrix contamination and release. Solvent flushing was not an effective decontaminating procedure for the chemical/barrier combinations used in this study. The data generated during solvent flushing also indicates that chemical exposure can alter the barrier properties of a material. Thermal decontamination shows promise, although extensive laboratory work is necessary to determine the extent to which it can be applied. The problems of matrix contamination can be avoided by using either:

1. clothing that provides very long breakthrough times, or
2. disposable clothing where the problems are discarded with the contaminated garments.

C. Nelson Schlatter[1]

Effects of Water Rinsing on Subsequent Permeation of Rubber Chemical-Protective Gloves

REFERENCE: Schlatter, C. N., **"Effects of Water Rinsing on Subsequent Permeation of Rubber Chemical-Protective Gloves,"** *Performance of Protective Clothing: Second Symposium, ASTM STP 989,* S. Z. Mansdorf, R. Sager, and A. P. Nielsen, Eds., American Society for Testing and Materials, Philadelphia, 1988, pp. 376–385.

ABSTRACT: The author investigated four solvents of varying volatility and water solubility in order to determine if workers can use gloves longer in chemical-resistant applications if they rinse their gloves periodically. The solvents were tested against glove materials selected to give moderate resistance to permeation, on the assumption that these would be most sensitive to changes in the way that gloves are handled. The results of the test were unexpected: water washing improved performance of the glove only for the solvent DMF. The author's conclusion is that water rinsing can help protect workers only when the solvent in use is toxic on skin contact, very water soluble, and not very volatile; it must also permeate relatively slowly.

KEY WORDS: dimethyl formamide (DMF), ethyl acetate, gloves, natural rubber, neoprene, nitrile, perchloroethylene, permeation, resistance, solvent, toluene, volatility, water rinsing, water solubility

As an employee of one of the world's principal manufacturers of industrial gloves, the author is a chemist who responds to customer inquiries on how to use gloves in chemical-resistant applications. In this position, he occasionally hears ideas from end users on how they hope to increase the service life of gloves.

Frequent water rinsing of the gloves to remove chemicals from their surfaces is one idea sometimes proposed. Intuitively, this notion makes some sense. But since surface washing cannot remove chemicals that have already entered the glove matrix and begun to diffuse through, there is also a chance that a water rinse might have little effect. The true benefit of water rinsing cannot be defined until it is tested in a controlled experiment.

Four glove material-solvent combinations were selected for such an experiment: dimethyl formamide (DMF) versus natural rubber, ethyl acetate versus neoprene, perchloroethylene versus lightweight nitrile rubber, and toluene versus heavyweight nitrile. These combinations were chosen because:

1. All of the solvents are moderately volatile. A highly volatile chemical that evaporates before it can be washed off would lead to confusing results. On the other hand, a totally non-volatile solvent would not be typical of those commonly found in industry.

2. The solvents represent a range of water solubilities. Perchloroethylene and toluene are only slightly soluble, ethyl acetate is soluble at about 10%, and DMF is miscible with water in any proportion.

[1] Chemist, Edmont, Division of Becton Dickinson & Co., 1300 Walnut Street, Coshocton, OH 43812.

TABLE 1—*Gloves used in this experiment.*[a]

Style No.	Trade Name	Type of Rubber	Nominal Thickness mm	(in.)
37-145	Sol-Vex	nitrile	0.28	(0.011)
37-165	Sol-Vex	nitrile	0.56	(0.022)
29-840	Neoprene	neoprene	0.38	(0.015)
36-124	Long Service	natural rubber	0.46	(0.018)

[a] All gloves are unsupported; that is, they consist only of hand-shaped rubber films, with no cotton lining or other support layer. All gloves are manufactured by the Edmont Division of Becton Dickinson, Coshocton, Ohio.

3. All four solvents are commonly used in industry.

4. The glove materials have moderate resistance to the solvents. Totally resistant materials are unsuitable for this experiment because they would never fail, whether washed or not; poor performing gloves would never pass. Gloves with moderate resistance are most likely to show some improvement from this cleaning technique.

5. The glove materials are all rubber, not plastic. Plastics such as polyvinyl chloride (PVC) must have plasticizers added to make them flexible enough for use in gloves. If the plasticizer dissolves in a solvent, properties of the base plastic will be affected. This is a source of complications that is better avoided in a first experiment.

TABLE 2—*Effects of water rinsing on subsequent permeation, perchloroethylene versus 37-145 nitrile.*[a]

Time, min	First Test Water Rinse	No Rinse	Second Test Water Rinse	No Rinse	Third Test Water Rinse	No Rinse
5	0	0	0	0	0	0
10	0	0	0	0	0	0
15	0	0	0	0	0	0
20	0	0	0	0	0	0
25	0	0	0	0	0	0
30	0	0	0	0	0	0
40	0	0	0.01	0	0	0
50	0.01	0.4	0.19	0.17	0.01	0
60	0.01	0.67	0.4	0.2	3.13	0
70	0.01	...	2.24	...	1.72	0
80	0.01	...	4.63	...	...	...
90	0.01	2.99	4.53	5.46	...	...
100	...	...	...	...	...	...
110	...	...	...	...	1.7	0.01
120	0.23	4.49	3.31	6.96	1.42	0.01
140	...	...	...	...	...	...
150	...	...	...	...	1.15	0.37
180	1.41	4.23	1.92	6.7	1.02	0.01
210	...	...	...	...	...	...
240	1.42	3.4	1.19	5.7	...	...
300	1.19	1.8	0.87	4.92	...	...
360	0.83	...	0.62	...	...	...

[a] All rates are in milligrams per square metre per second.

TABLE 3—*Effects of water rinsing on subsequent permeation, toluene versus 37-165 nitrile.*[a]

Time, min	First Test		Second Test		Third Test	
	Water Rinse	No Rinse	Water Rinse	No Rinse	Water Rinse	No Rinse
5	0	0.75	0	0	0	0
10	0	13.9	0	0	0	1.84
15	0	14.4	0	0	0.12	0.53
20	0.24	13.6	7.81	0	8.22	0.69
25	4.77	13.45	5.76	...	5.19	...
30	9.3	13.3	4.82	0.01	5.37	0.85
40	10.4	11.6	5.04	...	4.91	...
50	9.8	8.95	4.36	...	...	...
60	9.2	6.3	4.31	0.31	3.95	0.33
70	...	...	3.48	...	3.6	...
80	8.5	4.1	3.21	...	...	...
90	11.1	6.1	3.35	1.66	3.13	1.74
100	8	4.3	2.86	...	...	...
110	...	...	...	...	2.69	...
120	...	...	2.28	1.37	2.36	1.42
140	5.8	2.2	...	...	...	...
150	...	...	...	...	...	...
180	...	...	...	0.68	...	0.68
210	...	...	...	...	...	...
240	...	...	...	0.4	...	0.4
300	...	...	...	0.22	...	0.23
360	...	...	...	0.18	...	0.17

[a] All rates are in milligrams per square metre per second.

Specimens of each glove were exposed to the appropriate solvent for long enough to initiate permeation: 10 min for most combinations, 1 h for perchloroethylene/light nitrile. The specimens were then rinsed with water and left in a continuous stream of nitrogen to simulate the normal air circulation in a workplace. Control specimens were handled in the same way, except that the water rinse was omitted. Permeation rates were measured over the next several hours.

Experimental

The permeation test method used was a modification of the ASTM Test Method for Resistance of Protective Clothing Materials to Permeation by Liquids or Gases (F 739). Seventy-five-millimetre (3 in.) circles were cut from Edmont gloves made from unsupported rubber only, with no flocking or lining fabric. The styles used are listed in Table 1. Each test cell was assembled using two collecting halves of the standard 5-cm (2 in.) ASTM cell. A stream of nitrogen at about 100 mL/min flowed past the hand-contact side of the cut specimen and carried any permeated chemical through a 2-mL loop on a sampling valve on a Gow-Mac 750 gas chromatograph. The valve was thrown at intervals to permit detection and quantitation of permeant by the flame ionization detector (FID) on the gas chromatograph.

The other side of the test cell was filled with the appropriate solvent, and timing was started. After 10 min (60 min for perchloroethylene), the solvent was dumped out and replaced with water; then the cell was emptied again. The water rinse was omitted for control specimens. A nitrogen stream flowed through this side of the cell at 50 mL/min for the

TABLE 4—*Effects of water rinsing on subsequent permeation, ethyl acetate versus 29-840 neoprene.*[a]

Time, min	First Test		Second Test		Third Test	
	Water Rinse	No Rinse	Water Rinse	No Rinse	Water Rinse	No Rinse
5	0	0	0	0	25.9	0
10	0.7	39.6	1.7	0	37.4	14
15	26.3	38.6	53	19.7	33.1	38.3
20	24.9	30.2	49.7	23.3	...	...
25	24.5	25.835	...	...	...	...
30	22.5	21.47	...	...	14.6	28.3
40	16.8	12.74	28.6	14.5	...	...
50	13.6	10.27	18.4	...	6.7	11.3
60	11.3	7.8	12.7	9.9	5.1	8.7
70	...	...	...	...	4	6.6
80	...	...	...	...	...	...
90	7.1	...	6.2	...	2.6	4.1
100	...	...	...	...	...	...
110	...	...	...	...	1.8	2.6
120	4.7	3.2	3.8	4.1	0.4	2.1
140	...	...	...	...	...	...
150	...	...	...	...	...	...
180	2.6	1.8	1.6	2.3	...	...
210	...	...	...	...	...	...
240	1.5	1.1	0.6	1.3	...	...
300	0.9	0.5	0.5	0.8	...	...
360	...	0.01	...	0.7	...	...

[a] All rates are in milligrams per square metre per second.

remainder of the test. All data are reported as permeation rate in milligrams per square metre per second versus time in minutes from the first contact of the chemical with the material specimen.

Results and Discussion

The data from these experiments are somewhat erratic. Other researchers have found that this is sometimes the case when running permeation tests [1,2]. The most complete study of this problem is by Mickelsen [3] who ran a preliminary experiment, had the data analyzed by a statistician, and found that he needed seven replicate tests to be certain that his results were statistically significant. The work currently being reported included only three replicates of each test, in conformance with ASTM Method F 739. Small differences in performance might easily be lost in the background noise.

The test solvents with the lowest water solubility are perchloroethylene and toluene. Complete data on these two solvents are given in Tables 2 and 3. To show the ranges in these data, we plotted the results of the best and worst individual tests, both washed and unwashed, as Figs. 1 and 2. The dotted vertical lines on these figures indicate when the solvents were drained from the test cells.

Figure 1 shows that the best and worst runs for perchloroethylene versus 0.28-mm (11 mil) nitrile were both unrinsed. The trials with rinsed specimens all gave intermediate results.

Figure 2 shows the same pattern out to 40 min from the start of the test. At 50 min two of the data plots cross, so that at longer times the plots of washed and unwashed data

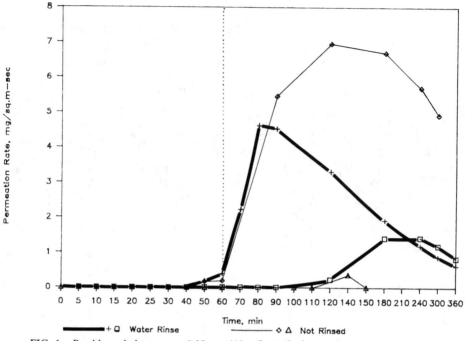

FIG. 1—*Perchloroethylene versus 0.28-mm (11 mil) nitrile, best and worst test results.*

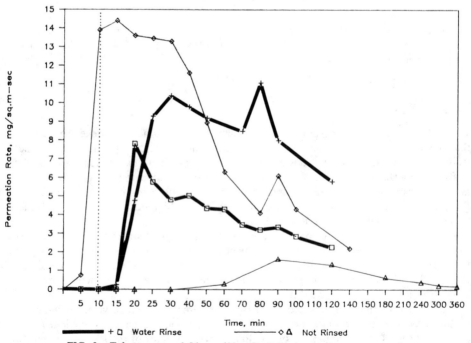

FIG. 2—*Toluene versus 0.56-mm (22 mil) nitrile, best and worst test results.*

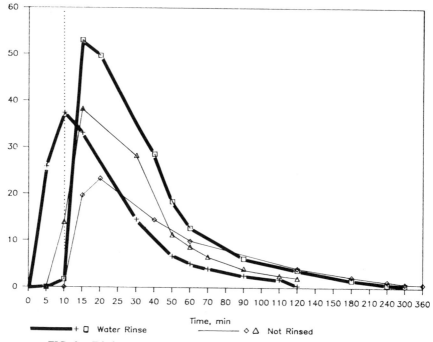

FIG. 3—*Ethyl acetate versus neoprene, best and worst test results.*

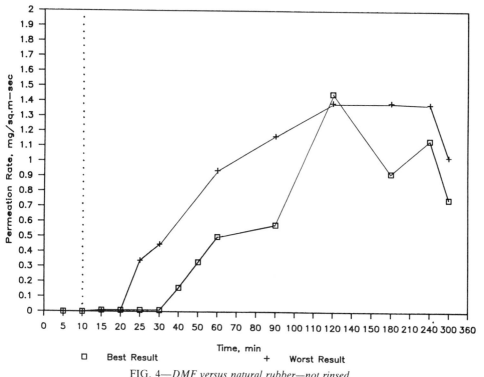

FIG. 4—*DMF versus natural rubber—not rinsed.*

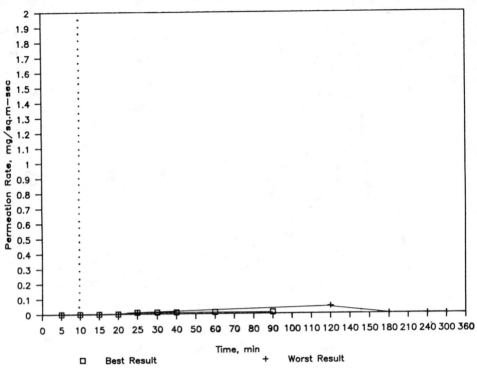

FIG. 5—*DMF versus natural rubber—rinsed.*

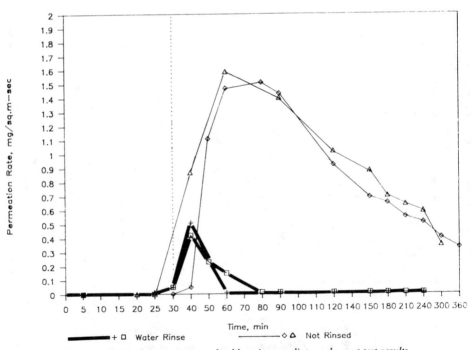

FIG. 6—*Cellosolve versus natural rubber, intermediate and worst test results.*

TABLE 5—*Effects of water rinsing on subsequent permeation, dimethyl formamide versus 36-124 natural rubber.*[a]

Time, min	First Test		Second Test		Third Test	
	Water Rinse	No Rinse	Water Rinse	No Rinse	Water Rinse	No Rinse
5	0	0	0	0	0	0
10	0	0	0	0	0	0
15	0	0.01	0	0.01	0	0.01
20	0	0.01	0	0.01	0	0.01
25	0.01	0.01	0.01	0.34	0.01	...
30	0.01	0.01	...	0.45	0.01	0.28
40	0.01	0.16	...	...	0.01	...
50	0.01	0.33	...	...	...	...
60	0.01	0.5	...	0.94	0.01	0.43
70	0.01	...	...	...	...	...
80	0.01	...	...	...	...	...
90	0.01	0.58	...	1.17	0.01	0.71
100	0.01	...	...	...	...	...
110	...	...	...	...	...	...
120	0.01	1.45	0.05	1.39	...	1.32
140	...	...	...	...	...	...
150	...	...	...	...	...	1.12
180	0	0.92	0	1.39	...	0.91
210	...	...	...	...	...	0.86
240	0	1.14	0	1.38	...	0.94
300	0	0.75	0	1.03	...	0.69
360	...	...	...	...	...	...

[a] All rates are in milligrams per square metre per second.

alternate on the graph. There is no evident influence of rinsing on subsequent permeation data.

Table 4 lists full data on the next combination, ethyl acetate versus the neoprene glove material. Figure 3 is the corresponding graph of best and worst trials. Ethyl acetate is more soluble in water than either perchloroethylene or toluene, so a greater effect from water rinsing might have been predicted. It is certainly not observed.

Data for DMF if given in Table 5. This time there is a strong effect from rinsing, which is emphasized in this presentation by separating the plots of rinsed and unrinsed data as Figs. 4 and 5. These two graphs both have the same vertical scale, and it is necessary to look twice at Fig. 5 to see that any data are plotted on it at all. Water rinsing while the gloves are in service can clearly reduce worker exposure to DMF.

Why DMF, and not the others? There are several factors which might contribute: Of these chemicals, DMF has the highest water solubility, so it is the easiest to remove by water washing. It has the lowest vapor pressure, so it is the hardest to remove by evaporation into a stream of nitrogen gas. And it has the lowest permeation rate through the glove that it was tested against, a fact that may be obscured by the differing vertical scales on the various graphs. A low permeation rate implies a small reservoir of permeating chemical in the rubber film where a surface wash cannot reach it, so washing can remove almost all of the DMF left on and in the film.

It is also worth remembering that DMF is widely used and toxic enough that the result of this experiment is not just of academic interest.

After data on the first four chemicals in this experiment were evaluated, an additional

TABLE 6—*Effect of water rinsing on subsequent permeation, Cellosolve 2-ethoxyethanol versus 36-124 natural rubber.*[a]

Time, min	First Test		Second Test		Third Test	
	Water Rinse	No Rinse	Water Rinse	No Rinse	Water Rinse	No Rinse
5	0	0	0	...	0	0
10	...	...	...	...	...	...
15	...	...	...	...	...	...
20	...	...	0	...	0	0
25	0	...	0.01	0	0.01	0.01
30	0.05	0	0	0.05	0.05	...
40	0.42	0.05	0.01	0.01	0.51	0.87
50	0.23	1.11	...	...	...	...
60	0.15	1.47	0.01	0.01	0.01	1.59
70	...	...	...	...	...	...
80	0.01	1.51	0.01	...	...	...
90	0.01	1.43	0.01	0.21	0.01	1.4
100	...	...	...	...	...	...
110	...	...	...	...	...	...
120	0.01	0.92	0.01	0.2	0.01	1.02
140	...	...	...	...	...	...
150	0.01	0.69	0.01	...	0.01	0.88
180	0.01	0.65	0	0.16	0.01	0.7
210	0.01	0.55	0	...	...	0.64
240	0.01	0.51	0	0.01	0.01	0.59
300	...	0.4	...	...	...	0.35
360	...	0.33	...	...	...	...

[a] All rates are in milligrams per square metre per second.

test was carried out on Cellosolve,[2] Union Carbide 2-ethoxyethanol, against natural rubber. The natural rubber was exposed to the Cellosolve for 30 min to initiate permeation. This combination has the high water solubility, low volatility, and low permeation rate of DMF on neoprene, and it is also toxic enough to be of interest. The results are given in Table 6 and Fig. 6. Cellosolve permeation was also greatly reduced by water rinsing.

Conclusion

If a toxic chemical is highly water soluble, is not very volatile, and gives a low permeation rate through the gloves being used, then exposure to that chemical can be further reduced by water rinsing of the gloves after each contact with the chemical.

It is also important to note that if the solvent in use does not meet these requirements, then water rinsing probably will not help. Other approaches to improving worker protection should be tried.

This conclusion is still qualitative, because we have tested only five chemicals in all. Must a solvent mix in all proportions with water to be "highly soluble"? Precisely what evaporation rate defines "not very volatile"? What is the highest permeation rate that can be tolerated? It would be necessary to test many more solvents to obtain quantitative answers to these questions.

[2] Registered trademark of Union Carbide Corporation.

References

[1] Bennett, R. D., Feighley, C. E., Oswald, E. O., and Hill, R. H., "The Permeation by Liquified Coal of Gloves Used in Coal Liquefaction Pilot Plants," *American Industrial Hygiene Association Journal*, Vol. 44, June 1983, pp. 447–452.

[2] Laidlow, J. T., Connor, T. H., Theiss, J. C., Anderson, R. W., and Matney, T. S., "Permeability of Latex and Polyvinyl Chloride Gloves to 20 Antineoprene Drugs," *American Journal of Hospital Pharmacy*, Vol. 41, Dec. 1984, pp. 2618–2622.

[3] Mickelsen, R. L., "Intermanufacturer's Performance Variability of Unsupported Nitrile and Neoprene Glove Materials," paper presented at the American Industrial Hygiene Conference, Dallas, TX, May 1986.

Protection from Industrial Chemical Stressors

Selection and Use

Riitta Jolanki,[1] *Tuula Estlander,*[1] *Maj-Len Henriks-Eckerman,*[2]
Lasse Kanerva,[1] *and Tuula Stjernvall*[1]

Skin Protection and Sensitizing Epoxy Compounds in Electron Microscopy Laboratories

REFERENCE: Jolanki, R., Estlander, T., Henriks-Eckerman, M.-L., Kanerva, L., and Stjernvall, T., "**Skin Protection and Sensitizing Epoxy Compounds in Electron Microscopy Laboratories,**" *Performance of Protective Clothing: Second Symposium, ASTM STP 989,* S. Z. Mansdorf, R. Sager, and A. P. Nielsen, Eds., American Society for Testing and Materials, Philadelphia, 1988, pp. 389–395.

ABSTRACT: A questionnaire survey was carried out among electron microscopy (EM) laboratory technicians in Finland. The questionnaire was returned by 18 out of the total of 45 to 50 EM workers, covering 13 laboratories. Epoxy resins, especially glycidyl ethers of glycerol, anhydride hardeners, amine accelerators, and propylene oxide, were used for embedding. On average, the technicians had been exposed to unpolymerized epoxy resins for 3 h a week over a 10-year period. Ten workers reported skin symptoms. Seventeen used protective gloves when working with unpolymerized resins.

In gas chromatography, the epoxy resins were found to contain many skin sensitizing compounds, that is, diglycidyl ether of bisphenol A (molecular weight 340) and epoxy reactive diluents. This potential allergenic capacity of the epoxy resins emphasizes the importance of preventing skin contact.

Selection of glove material for the handling of tissue specimens in electron microscopy laboratories was based on the results of permeation tests with propylene oxide, because the chemical construction of propylene oxide is similar to that of sensitizing epoxy compounds. Because it is a low viscosity compound, however, propylene oxide penetrates glove material more rapidly than the resin compounds.

Thin polyethylene gloves provide better protection against propylene oxide, and thus also against embedding resins, than disposable polyvinyl chloride (PVC) or natural rubber gloves.

KEY WORDS: electron microscopy, embedding, chemical exposure, gas chromatographic method, epoxy resins, glycidyl ethers, dermatoses, skin symptoms, contact allergy, sensitizing compounds, protective clothing, gloves

The handling of tissue specimens for electron microscopy (EM) requires working with a number of sensitizing, irritant, and toxic chemicals. Resins are widely used as embedding materials prior to microscopy. The resins used for microscopy fall into three broad categories: epoxy, polyester, and acrylic resins [1].

Many, including researchers at our Institute, have paid much attention to epoxy resins and other epoxy compounds as causes of occupational skin diseases [2–4]. Epoxy resins were found to be one of the most frequent causes of occupational allergic contact dermatitis.

---

[1] Chemist, dermatologist, chief, and chief laboratory technician, respectively, Section of Dermatology, Institute of Occupational Health, Helsinki, Finland.
[2] Chemist, Turku Regional Institute of Occupational Health, Turku, Finland.

The only agents causing occupational allergic dermatitis more often are chromium and rubber chemicals [4]. Epoxy hardeners, especially aliphatic amines, and reactive diluents contained in epoxy products are moderate to strong sensitizers [5,6]. Not only do epoxy compounds present the risk of skin dermatitis; they are also toxic and even carcinogenic [7–9].

A questionnaire survey was carried out in order to evaluate the skin problems now being induced by epoxy compounds in EM laboratories. The use of epoxy resins and other chemicals and current practices for protecting the skin against epoxy resins were clarified in the questionnaire. The sensitizing compounds contained in the epoxy resins were then determined by gas chromatography [10,11], and proper skin protection was selected.

Materials and Methods

Questionnaires

The questionnaires were distributed to laboratory technicians working in EM laboratories at an annual meeting of EM technicians. The questionnaire form requested information about the following: the types of resins, hardeners, and accelerators used for embedding; exposure to embedding resins; skin symptoms in the handling of unpolymerized resins; and the use, type, and construction of protective gloves.

To our knowledge 45 to 50 workers regularly handle tissue specimens in EM laboratories in Finland.

Sensitizing Epoxy Compounds in Embedding Resins

The amounts of diglycidyl ether of bisphenol A and of nine reactive diluents in the embedding resins were determined.

The epoxy resins now in use were ultrasonically dissolved in acetone (2 to 30 mg/cm³). Extracted glycidyl ethers and esters were determined by gas chromatography [10,11]. A 25-m fused-silica capillary column (crosslinked 5% phenylmethylsilicone) coupled to a flame-ionization detector (at 573 K) was used with temperature programming at 10 K/min (0.167 K/s) from 323 to 533 K. The carrier gas (helium) flow-rate was 2 cm³/min (0.033 cm³/s). The injector was used in the split mode (ratio 1:15) at 533 K. One epoxy resin oligomer—that is, the diglycidyl ether of bisphenol A (molecular weight 340)—and nine reactive diluents (Table 1) were used as external standards in gas chromatography.

TABLE 1—*Reactive diluents used as external standards in gas chromatography and a list of reports on contact allergy caused by the diluents.*

Reactive Diluents	Reports on Contact Allergy (reference numbers)
allyl glycidyl ether (AGE)	*3,12*
1,4-butanediol diglycidyl ether (BDDGE)	*3,6,13*
n-butyl glycidyl ether (BGE)	*3,12,14*
glycidyl esters of synthetic fatty acids (GEFA, Cardura E 10)	*6,13,15,16*
cresyl glycidyl ether (CGE)	*3,12,14*
glycidyl ethers of aliphatic alcohols (Epoxide 8)	*6,14,17*
1,6-hexanediol diglycidyl ether (HDDGE)	*3*
neopentyl glycol diglycidyl ether (NPGDGE)	*3,6*
phenyl glycidyl ether (PGE)	*3,12,18*

Diglycidyl ether of bisphenol A and nine reactive diluents in one polymerized block of Araldit M were also analyzed in the same manner.

The detection limits were 0.04 to 1 μg/mg; the limit depended on the amount of resin analyzed.

Results

Questionnaire Survey

The questionnaire was returned by 13 out of 31 laboratories and by 18 out of the total of 45 to 50 EM workers. The worker response rate was thus approximately 40%.

Chemicals Used in EM Laboratories for Embedding

Resins—The chemical structures of the epoxy resins now in use and those previously used are shown in Table 2. Only epoxy-type resins were used for embedding.

The resin now used by most of the workers was LX-112 (16 out of the total of 18 workers who responded, 89%). It is a mixture of diglycidyl and triglycidyl ethers of glycerol. One of those sixteen workers used Araldit M (diglycidyl ethers of bisphenol A); another one used Em-Bed 812 (diglycidyl ethers of bisphenol A), ERL 4206 (vinyl cyclohexane dioxide), and DER 736 (diglycidyl ether of tetrapropyleneglycol). Three workers had used Quetol 651 (diglycidyl ether of ethyleneglycol).

Epon 812 (ten workers, 56%), Epicote 812 (four workers, 22%), and Glycidether 100 (one worker, 6%) had previously been used, but were no longer in use.

Hardeners and accelerators—Epoxy resins were hardened using anhydrides accelerated by amines. Three anhydride hardeners—DDSA (dodecenyl succinic anhydride), NMA (nadic methyl anhydride), and NSA (nonenylsuccinic anhydride)—and two amine accelerators—DMP-30 (2,4,6-tri(dimethylaminomethyl)phenol) and S-1 (dimethyl aminoethanol)—were used.

Propylene oxide—Propylene oxide was commonly used after dehydration, to remove ethanol, and for infiltration.

Exposure Period—The laboratory technicians had been exposed to unpolymerized epoxy resins for about ten years (range: 1 to 21 years) for 3 h (range: 0 to 6 h) a week. Polymerized blocks were handled more often than unpolymerized resins, on average over 20 h (range: 0 to 25 h) a week.

Symptoms—Ten workers reported skin symptoms. In addition, three of them had watery rhinitis, and one reported irritation of the respiratory tract. The skin symptoms included itching (nine workers, 50%), redness (two workers, 11%), swelling of the skin (two workers, 11%), and burning sensation (two workers, 11%). The symptoms were localized on the hands in three cases, on the face in one case, and on the hands and face in three cases. Three workers did not specify any localization of their skin symtpoms. Two of the workers with skin symptoms had previously been examined for occupational dermatosis. One of them had occupational allergic contact dermatitis caused by epoxy resin. There was no relationship between the duration of exposure and the occurrence of symptoms on the basis of the questionnaire responses.

TABLE 2—*Trade names and types of epoxy resins used for embedding in electron microscopy laboratories, and the number of workers using the resins.*

Trade Name	Chemical Structure	Number of Workers
Epicote 812[a]		4
Epon 812[a]		10
Glycidether 100[a]		1
LX-112		16
Araldit M		1
Em-Bed 812		1
Quetol 651		3
ERL 4206		1
DER 736		1

[a] No longer in use.

Protective Gloves—One of the workers did not use protective gloves when working with resins. Five others had previously worked without gloves for 1 to 9 years. Fourteen of the workers wore disposable natural rubber (NR) gloves. Two of them had used disposable polyvinyl chloride (PVC) gloves temporarily, and eight regularly wore thin polyethylene (PE) gloves under the disposable NR gloves. Three of the workers had used only PVC gloves. Both the PVC and the NR gloves were about 0.08 mm thick, and the PE gloves were 0.02 to 0.03 mm. There was no relationship between the glove material used and the occurrence of skin symptoms on the basis of the questionnaire responses.

Sensitizing Epoxy Compounds in Embedding Resins

Diglycidyl Ether of Bisphenol A—Em-Bed 812 and Araldit M contained the sensitizing epoxy resin oligomer, that is, diglycidyl ether of bisphenol A (molecular weight 340), 400 and 650 μg/mg, respectively (Table 3). Four other resins did not contain diglycidyl ether of bisphenol A.

The oligomer content in the Araldit M block was below the detection limit (0.04 μg/mg).

Epoxy Reactive Diluents—The type and amounts of the nine reactive diluents are shown in Table 4. Four of the six resins analyzed contained 0.2 to 430 μg/mL reactive diluents, whereas two resins had none. The amounts of four reactive diluents—AGE, GEFA, CGE, and NPGDGE—were below the detection limits (0.05 to 1 μg/mg) in every product.

The amounts of the reactive diluents in the polymerized block made of Araldit M were below the detection limit (0.04 μg/mg).

Discussion

Only two (LX-112 and ERL 4206) of the six epoxy resins used for embedding did not contain either diglycidyl ether of bisphenol A or epoxy-reactive diluents, both known to be sensitizing epoxy compounds [3,6,7,12–18]. LX-112 produced an allergic patch test reaction in one of the patients with contact allergy to BDDGE [3], indicating that LX-112 can crossreact with BDDGE. Thus, not only the epoxy resins known to be sensitizers but also chemically related epoxy resins, that is, LX-112, Quetol 651, and DER 736, should be considered potential sensitizers. To our knowledge, there are no reports on contact allergy to vinyl cyclohexane dioxide (ERL 4206), but it is a documented animal carcinogen [8].

More than half of the workers had skin symptoms, and one out of three had symptoms on the hands. It is possible that EM laboratory workers with symptoms returned the questionnaire, whereas those without symptoms did not. Prevalence of hand eczema among

TABLE 3—*The quantity of diglycidyl ether of bisphenol A (molecular weight 340) in epoxy embedding resins, determined by gas chromatography.*

Epoxy Resin	Diglycidyl Ether of Bisphenol A (Molecular Weight 340), μg/mg
LX 112	<1[a]
Araldit M	650
Em-Bed 812	400
Quetol 651	<1[a]
ERL 4206	<1[a]
DER 736	<0.05[a]

[a] Detection limit.

TABLE 4—*The quality and quantity of nine reactive diluents* (*Table 1*) *in epoxy embedding resins, determined by gas chromatography.*

Epoxy Resin	Reactive Diluent	Amount, μg/mg
LX 112	all nine diluents	<detection limit[b]
Araldit M	BDDGE,[a] PGE, glycidylether	
	of an aliphatic alcohol	total 5[b]
Em-Bed 812	BDDGE	430[b]
Quetol 651	glycidylether of	
	an aliphatic alcohol	57[b]
	BGE, PGE	total 5[b]
ERL 4206	all nine diluents	<detection limit[b]
DER 736	HDDGE, glycidylether	
	of an aliphatic alcohol	total 0.2[c]

[a] Abbreviations are found in Table 1.
[b] Detection limit 1 to 10 μg/mg.
[c] Detection limit 0.05 to 0.5 μg/mg.

laboratory technicians handling embedding resins is, however, suspected of being higher than among the general population (7.1%) [19].

The potential allergenic capacity of the chemicals used in EM laboratories stresses the importance of preventing skin contact with sensitizing chemicals.

There are no governmental guidelines [20,21] for the selection of clothing for protection against epoxy resins in EM laboratories. Hence, the selection of glove material for the handling of tissue specimens was based on the results of permeation tests with propylene oxide, as the chemical construction of propylene oxide is similar to that of the sensitizing epoxy compounds. Propylene oxide, however, is a low-viscosity compound, and thus penetrates glove material more rapidly than the resin compounds.

Propylene oxide completely degrades PVC material, immediately penetrating through thin NR material [21]. Our experiments showed that gloves of thin PE material give much better protection than those of natural rubber or PVC material. The experiments were performed using fingers of thin PVC, NR, and PE gloves. Five cubic centimetres of propylene oxide was pipetted into the fingers. The fingers made of PVC were completely destroyed immediately. Those made of NR or PE material were not destroyed, but the entire amount of the test liquid penetrated through NR in a couple of minutes, whereas it did not penetrate through PE glove material in 10 min, nor were any visible changes detected in the PE material. Exact breakthrough times were not determined. New gloves of thin laminated membrane, that is, an ethylene vinyl alcohol copolymer film laminated on both sides with linear-low density polyethylene, have been developed for epoxy products [22]. Gloves made of this new material, however, are rather stiff and are suitable only for rough work, for example, the weighing and mixing of resin components.

Practical experience has shown that it is better to wear PE gloves under NR gloves, because PE alone is slightly slippery. If the hands sweat profusely, cotton gloves should be worn as inside liners.

Disposable gloves contaminated with hazardous resins must be removed immediately. Other protective clothing, for example, aprons and separate cuffs, should be worn when embedding resins are handled. In order to prevent exposure to hazardous vapors and dusts, the handling should take place only under a laboratory hood.

Conclusion

This was a preliminary study on skin problems induced by epoxy compounds among EM laboratory workers. Future plans include a detailed questionnaire survey on skin symptoms among all laboratory technicians in EM laboratories in Finland and a dermatological examination to investigate the specific causes of the symptoms in each case.

References

[1] Causton, B. E., Ashhurst, D. E., Butcher, R. G., Chapman, S. K., Thomson, D. J., and Webb, M. J. W., "Resins: Toxicity, Hazards and Safe Handling," *Proceedings RMS*, Vol. 16/4, June 1981, pp. 265–269.
[2] Jolanki, R., Estlander, T., and Kanerva, L., "Occupational Skin Diseases caused by Epoxy Products. Cases Diagnosed during 1974–83 at the Institute of Occupational Health in Finland," *Työterveyslaitoksen tutkimuksia*, Vol. 4, No. 1, 1986, pp. 47–54, 70 (in Finnish, with English and Swedish summaries).
[3] Jolanki, R., Estlander, T., and Kanerva, L., "Contact Allergy to an Epoxy Reactive Diluent: 1,4-Butanediol Diglycidyl Ether," *Contact Dermatitis*, Vol. 16, No. 2, 1987, pp. 87–92.
[4] Kanerva, L., Estlander, T., and Jolanki, R., "Occupational Skin Disease in Finland. Ten Years' Material from an Occupational Dermatology Clinic," *International Archives of Occupational and Enviornmental Health* (in press).
[5] Thorgeirsson, A., "Sensitization Capacity of Epoxy Resin Hardeners in the Guinea Pig," *Acta Dermatovenerologica*, Vol. 58, No. 4, 1978, pp. 332–336.
[6] Thorgeirsson, A., "Sensitization Capacity of Epoxy Reactive Diluents in the Guinea Pig," *Acta Dermatovenerologica*, Vol. 58, No. 4, 1978, pp. 329–331.
[7] NIOSH Criteria for a Recommended Standard. Occupational Exposure of Glycidyl Ether, 1978. *DHEW (NIOSH) Publication* No. 78–166.
[8] *International Agency for Research on Cancer*, 1-Epoxyethyl-3,4-epoxycyclohexane. IARC (Lyon) Monographs on the Evaluation of the Carcinogenic Risk of Chemicals to Humans (Her Majesty Stationery Office Books), Vol. 11, 1976, pp. 141–145.
[9] Afzelius, B., "Occupational Hazards," *Electron Microscopy in Human Medicine*, Vol. 1, J. V. Johannessen, Ed., McGraw-Hill, 1978, pp. 328–339.
[10] Henriks-Eckerman, M.-L., "Gas Chromatographic Determination of the Oligomer of Molecular Weight 340 in Epoxy Resins," *Journal of Chromatography*, Vol. 244, 1982, pp. 378–380.
[11] Henriks-Eckerman, M.-L. and Laijoki, T., "Glycidyl Ethers in Epoxy Resin Products," *Työterveyslaitoksen tutkimuksia*, Vol. 4, No. 1, 1986, pp. 41–46, 70 (in Finnish, with English and Swedish summaries).
[12] Fregert, S. and Rorsman, B., "Allergens in Epoxy Resins," *Acta Allergologica*, Vol. 19, 1964, pp. 296–299.
[13] Thorgeirsson, A., Fregert, S., and Magnusson, B., "Allergenicity of Epoxy-Reactive Diluent in the Guinea Pig," *Berufsdermatosen*, Vol. 23, No. 5, 1975, pp. 178–183.
[14] Dahlquist, I. and Fregert, S., "Allergic Contact Dermatitis from Volatile Epoxy Hardeners and Reactive Diluents," *Contact Dermatitis*, Vol. 5, No. 6, 1979, pp. 406–407.
[15] Dahlquist, I. and Fregert, S., "Contact Allergy to Cardura E, an Epoxy Reactive Diluent of the Ester Type," *Contact Dermatitis*, Vol. 5, No. 2, 1979, pp. 121–122.
[16] Lovell, C. R., Rycroft, R. J. G., and Matood, J., "Isolated Cardura E 10 Sensitivity in an Epoxy Chemical Process," *Contact Dermatitis*, Vol. 11, No. 3, 1984, pp. 190–191.
[17] Björkner, B., Dahlquist, I., Fregert, S., and Magnusson, B., "Contact Allergy to Epoxide 8, an Epoxy Reactive Diluent," *Contact Dermatitis*, Vol. 6, No. 2, 1980, p. 156.
[18] Rudzki, E. and Krajewska, D., "Contact Sensitivity to Phenyl Glycidyl Ether," *Dermatosen in Beruf und Umwelt*, Vol. 27, No. 2, 1979, pp. 42–44.
[19] Lantinga, H., Nater, J. P., and Coenraads, P. J., "Prevalence, Incidence and Course of Eczema on the Hands and Forearms in a Sample of the General Population," *Contact Dermatitis*, Vol. 10, No. 3, 1984, pp. 135–139.
[20] Forsberg, K. and Olsson, K. G., "*Riktlinjer för val av kemikalieskyddshandskar,*" Föreningen Teknisk Företagshälsovård, Stockholm, 1985, p. 188 (in Swedish).
[21] Schwope, A. D., Costas, P. P., Jackson, J. O., and Weitzman, D. J., "*Guidelines for the Selection of Chemical Protective Clothing,*" 2nd ed., American Conference of Governmental Industrial Hygienists, Inc., Vol I, Cincinnati, 1985.
[22] Henriksen, H. R. and Petersen, H. J. S., "*Protective Clothing Against Chemicals: Better Gloves Against Epoxy Products and Other Chemicals,*" Arbejdsmiljöfondets forskningsrapporter, Arbejdsmiljöfondet, Copenhagen, 1986 (in Danish, with an English summary).

S. Z. Mansdorf[1]

Development of a Comprehensive Approach to Chemical Protective Clothing Use

REFERENCE: Mansdorf, S. Z., "**Development of a Comprehensive Approach to Chemical Protective Clothing Use,**" *Performance of Protective Clothing: Second Symposium, ASTM STP 989*, S. Z. Mansdorf, R. Sager, and A. P. Nielsen, Eds., American Society for Testing and Materials, Philadelphia, 1988, pp. 396–402.

ABSTRACT: Chemical protective clothing (CPC) is a "last line of defense." As such, proper selection and use are essential to the safety and health of the wearer.

A comprehensive approach to CPC selection and use should include at least eight key elements. They are: (1) assessing the need; (2) determination of the level and type of chemical and physical protection required through a hazard assessment of the proposed task; (3) proper selection based on the hazard assessment; (4) lab and field validation of the effectiveness of the selection; (5) establishment of reuse and decontamination guidelines; (6) worker and supervisory training in the use and limitations of the selected equipment; (7) establishment of a routine inspection, maintenance, and repair program; and (8) development of a management audit scheme to assure all elements of the CPC selection process are being properly followed.

Chemical protective clothing can be an effective hazard control measure; however, the protection provided will only be as efficacious as the quality of the methodology or program utilized for its proper selection, use, and maintenance.

KEY WORDS: chemical protective clothing, hazard assessment, selection, laboratory validation, field validation, decontamination, training, inspection program, audit scheme, performance criteria, protective clothing

Occupational skin exposures to toxic materials in the work environment are a major health problem [1–6]. This is demonstrated by the 1980 Bureau of Labor Statistics (BLS) report that skin diseases accounted for 43% of all occupational illnesses in the private sector [7]. In their most recent report of illnesses among workers in private industry, the BLS noted 125 000 new cases of occupational illness with over two-thirds of these cases involving skin diseases or disorders associated with repeated trauma [8]. As a consequence, the National Institute for Occupational Safety and Health (NIOSH) has included dermatologic disorders in its list of the ten leading occupational health problems [9].

Historically, most protective clothing was considered to be "impermeable" and an absolute safeguard for the worker. Application restrictions were generally based on the salesperson's recommendations or simple immersion tests published in promotional brochures by the manufacturers. This approach has been shown to be wholly inadequate for evaluating the level of protection provided [10]. Development of standard test methods that measure degradation, penetration, and permeation resistance of chemical protective clothing has resulted in publication by the manufacturers, users, and researchers of specific recommendations for selection of chemical protective clothing keyed to the materials of construction such as the widely used *Guidelines for the Selection of Protective Clothing* [11].

[1] President, S. Z. Mansdorf & Associates, Inc., Cuyahoga Falls, OH 44223-1323.

Presently, the Occupational Safety and Health Administration (OSHA) requires that ". . . all personal protective equipment shall be of a safe design and construction for the work to be performed" [12]. Unlike respiratory protection products that are regulated by the government through OSHA and the Mine Safety and Health Administration (MSHA), chemical protective clothing is not regulated by performance certifications or warnings on use limitations.

Manufacturers have been slow to provide similar information to the users in terms of warnings or proper use and maintenance instructions. Hence, the burden of proper selection, use, and maintenance is placed on the employers and users. This can easily result in misuse by the uninformed.

Chemical protective clothing (CPC) is intended to provide protection for personnel against hazardous chemicals when other more effective methods of protection such as engineering controls are either inappropriate or infeasible. Use configurations of CPC can range from simple latex gloves to totally encapsulating suits, depending on the body zone potentially affected and the level of protection desired. Some configurations such as fully encapsulating suits can provide significant overall protection [13]. However, it must be recognized that the actual protection afforded the user of CPC will depend on the adequacy and effectiveness of the management system instituted by the provider to assure proper selection, use, and maintenance of the protective clothing.

The relative complexity of a CPC program will be determined by the frequency of CPC use, the risk to the users of CPC failures, and available staff and financial resources. This paper has been written to address the higher risk situations where a formalized comprehensive program is required. However, most of the described program aspects are equally applicable to less potentially threatening selection and use situations.

Key Program Elements

The following sections describe the key elements of a systematic approach to the selection and use of CPC. They are:

1. Assessing the need for CPC.
2. Determining the protection level and performance required from the CPC.
3. Proper selection of CPC based on the hazard assessment.
4. Laboratory and field validation of the level of protection provided by the CPC.
5. Establishment of effective decontamination procedures;
6. Training of personnel in the proper use and limitations of the selected CPC.
7. Routine inspection, maintenance, and repair of CPC.
8. Establishment of a management audit scheme to assure the effectiveness of the program.

Assessing the Need

The first step in determining the need for CPC is to characterize the nature and extent of the potential hazards. This requires an initial health hazard evaluation of the process or job for which CPC is being considered. This task should be conducted by a person competent in health and safety evaluations. Normally, such evaluations require a physical inspection of the job or process to identify the potential stressors, their physical state and properties, routes of exposure and body zones of potential chemical contact, type of contact reasonably expected (for example, splash), maximum expected concentrations for airborne contaminants, physical demands of the job, toxicity of the chemical or chemicals being used, and the level of hazard associated with routine, intermittent, and potential emergency exposure to physical or chemical agents. This hazard assessment and subsequent selection of protective equipment should be documented. A form, such as that shown in Fig. 1, may be used for this purpose.

The second step of the need assessment is the evaluation of alternatives to the initial use of CPC. Alternative control approaches would include the following:

1. Substitution with a less hazardous chemical.
2. Redesign of the process to reduce or eliminate worker contact with the hazardous chemical.

CHEMICAL PROTECTIVE CLOTHING SELECTION WORKSHEET

Job Classification or Task:	Process or Task Summary:	
Potential or Actual Hazards Chemical: Physical:		Contact Period:
Type of Potential Contact:	Body Zones of Potential Contact:	
Toxicology for Chemical Exposures:		
Potential Effects of SKIN Exposure:		
Permeation, Penetration, Degradation Data & Source:		
Recommended Base Materials of Construction for Each or All Component CPC:		
Specifications for Each CPC Component Required:		
Respiratory Protection:		
Type/Level of CPC Required:		
Training Required:		
Decontamination Procedures Required:		
Worksheet Completed By:	Date:	Checked By:	Date:

FIG. 1—*The form used for hazard assessment and chemical protective clothing selection.*

3. Engineering controls (for example, local exhaust ventilation) to reduce process emissions.
4. Use of robotics or other mechanical materials handling techniques.

Certain high risk tasks and special personnel assignments such as hazardous waste site work, spill response, chemical fire control, and other such situations where alternative control techniques are not feasible generally require a high level of personal protection by CPC. This is especially true when the type and extent of chemical hazards are not known. In these situations, the use of a worst-case scenario for selection of CPC may be required. The Environmental Protection Agency (EPA) has developed guidelines for personal protective equipment selection for site entry to hazardous waste sites [14]. However, this scheme generally requires a knowledge of possible air concentrations of potential contaminants or the use of the highest level of protection. In many cases such information is simply not available, hence the use of Level A protection usually resulting in significant work restrictions. These work restrictions are a result of a number of factors but principally those of heat stress, breathing air (Level A) limitations, and ergonomic problems. Alternately, NIOSH has published a selection guide that also generally results in Level A protection when the contaminants are unknown [15].

Development of Performance Criteria

The performance criteria for selection of CPC should be based on an evaluation of the physical resistance property (for example, tear and puncture resistance) requirements imposed by the job in addition to the level of chemical protection required. For example, will the garment or accessories be required to be flame retardant or thermally protective? Will the garment or accessories need to be especially abrasion or tear resistant? Job characteristics such as these should be profiled to assure that these factors are taken into consideration during the selection process. Chemical resistance requirements should be determined based on the potential adverse effects of skin contact with the hazardous chemical. Acute effects may be simplified into four general classes as follows:

1. Minimal irritation or other non-permanent effect.
2. Moderate irritation or other non-permanent effect.
3. Severe toxicity, burns, or other permanent effect.
4. Immediately dangerous to life.

Chemical protective clothing selected for a Class 1 hazard such as skin contact with isopropyl alcohol could be simply a latex glove while Class 3 and 4 hazards such as chromic acid exposure require more complete protection. Chemical carcinogens are not included in this rating scheme because their effects are generally chronic rather than acute. However, they would probably fall within the range of Class 2 through 4 depending on the relative potency of the carcinogen.

A key criteria should be a thoughtful consideration of the severity of chemical contact coupled with the anticipated frequency of contact. The severity of contact (consequences) should be the overriding factor in this analysis.

Laboratory and Field Validation

Once selection of candidate materials of construction has been completed based on the required performance criteria, laboratory testing should be conducted for confirmation of the vendors specifications [1]. In many cases, this information may be extracted from data already published using standard methods such as the American Society for Testing and Materials (ASTM) methods for permeation and penetration namely, ASTM Test Method for Resistance of Protective Clothing Materials to Permeation by Hazardous Liquid Chemicals (F 739-81) and ASTM Test Method for Resistance of Protective Clothing Materials to Penetration by Liquids (F 903-84). After completion of the literature review or laboratory studies, field evaluations should be conducted. This would include trials under actual field conditions of use for routine long-term tasks or assignments. No published standard methods to determine potential permeation or penetration in the field presently exist. However, Berardinelli and Roder have presented a number of proposed techniques [16].

Decontamination Procedures

The establishment of effective decontamination procedures is important for both single-use and reusable CPC. Once single-use protective clothing has become contaminated, it may be transferred to the wearer's clothing if care is not taken in removing the CPC. In this situation, single-use items should be decontaminated prior to disrobing for high hazard materials that have the potential for reentrainment in the air or such factors as spontaneous combustion when in contact with other materials that may be in the disposal container. Multiple-use CPC items will normally require some form of decontamination. Therefore, both the inside and outside surfaces may need to be routinely checked for residual contamination before reuse.

Decontamination techniques will vary depending on the reactivity and solubility of the chemical agent to be removed as well as the base material of construction of the CPC to be decontaminated. Water-soluble chemicals may be removed with a detergent and water wash while some chemicals may require complexing or another method of inactivation before removal [17]. Volatile chemicals may be removed by simply drying the CPC at elevated temperatures in some cases [18]. Proper procedures, facilities, and wash solution disposal methods for decontamination should be predetermined and their use explained to the wearer of the protective clothing before entry into the work area.

Training Program

All workers utilizing CPC should be trained in the use and limitations of assigned equipment. This training should include, as a minimum, the following information:

1. The nature, extent, and effects of chemical hazards posed by the job.
2. The proper use and limitations of the CPC assigned.
3. Decontamination procedures, as appropriate.
4. Inspection, maintenance, and repair of the chemical protective equipment.
5. First aid and emergency procedures.

Training should be conducted prior to the actual use of the CPC and should be conducted in a manner that allows actual "hands-on" experience. The trainer should be familiar with the specific chemical protective equipment assigned and be able to answer questions posed by the user. A mechanism, such as a written or simulated performance test, which measures the understanding of the training by the CPC user should be included in the training plan. A roster of participants and outline of the information presented should be maintained by the CPC provider. Retraining of CPC users should be conducted if there are process or situation changes that affect the use of CPC or a refresher course on an annual basis.

Inspection, Maintenance, and Repair

A routine inspection program should be established. This program should be conducted on three levels. The first level should be routine worker inspection of the CPC before and after each use. The second level should include scheduled supervisory inspections of the CPC. The third level should include an audit of these functions by the management person designated as being responsible for the CPC program. Maintenance evaluations of CPC should also be conducted on user, supervisory, and audit levels. Workers should be responsible for turning in defective or worn equipment to supervision. Supervision should also inspect the protective clothing for any additional defects or wear before then turning it in for repair.

Major repair work should be done by a fully competent person who is able to assure that the repaired gear is equal to, or better than, the original manufacturers' performance specifications. This may require repair training by the original manufacturer or return of their product to them for repair.

Auditing the Program

Management should establish a mechanism to assure that all elements of the CPC program are carried out effectively. Typically, this consists of an audit procedure where competent personnel, who are independent of the daily operation of the CPC program regularly review documentation and procedures of that program (on a yearly or other basis,) providing recommendations for improvement. This peer review mechanism is an integral part of maintaining an effective program.

References

[1] "Report of the Advisory Committee on Cutaneous Hazards," report to the Assistant Secretary of Labor, Occupational Safety and Health Administration, Washington, DC, 19 Dec. 1978.
[2] "Summary of the NIOSH Open Meeting on Chemical Protective Clothing," Rockville, MD, 3 June 1981, National Institute for Occupational Safety and Health, Cincinnati, OH, 1981.
[3] Mansdorf, S. Z. and Miles, B., "A Protective Dermal Film System," American Industrial Hygiene Association Conference, Los Angeles, CA, May 1978.
[4] The Prevention of Occupational Skin Diseases, 3rd printing, Soap and Detergent Association, New York, 1981.
[5] Hogan, D. J., "Skin Disorders are High on the List of Occupational Health Hazards," Occupational Health and Safety, Vol. 55, 1986, pp. 42–45.
[6] Banning, M., "Rethinking Chemical Protective Clothing—It's Not That Simple," Ohio Monitor, Vol. 59, 1986.
[7] "Occupational Injuries and Illness in the United States by Industry, 1980," Bulletin 2130, Bureau of Labor Statistics, Washington, DC, 1982.
[8] Occupational Safety and Health Reporter, Vol. 16, Bureau of National Affairs, Washington, DC, Nov. 1986, p. 628.
[9] Occupational Safety and Health Reporter, Vol. 16, Bureau of National Affairs, Washington, DC, Nov. 1986, p. 630.
[10] Mansdorf, S. Z., "Risk Assessment of Chemical Exposure Hazards in the Use of Chemical Protective Clothing—An Overview," Performance of Protective Clothing, ASTM STP 900, R. L. Barker and G. C. Coletta, Eds., American Society for Testing and Materials, Philadelphia, 1986, pp. 207–213.
[11] Schwope, A. D., Costas, P. P., Jackson, J. O., and Weitzman, D. J., Guidelines for the Selection of Chemical Protective Clothing, American Conference of Governmental Industrial Hygienists, Cincinnati, OH, 1983.
[12] General Industry Standards, 29CFR1910.132(c), Occupational Safety and Health Act of 1970 (84 Stat. 1593), Government Printing Office. Washington, DC, Revised 1978.
[13] Schwope, A. D. and Hoyle, R. E., "Tame Hazardous Waste Hazards with Personal Protective Equipment," Hazardous Materials and Waste Management, Vol. 3, 1985, pp. 14–22.

[14] "Interim Standard Operating Safety Procedures," U. S. Environmental Protection Agency, Office of Emergency and Remedial Response, Hazardous Response Division, Washington, DC, 1982.

[15] Ronk, R., White, M. K., and Linn, H., *Personal Protective Equipment for Hazardous Materials Incidents: A Selection Guide,* National Institute for Occupational Safety and Health, Washington, DC, 1984.

[16] Berardinelli, S. P. and Roder, M. M., "Chemical Protective Clothing Field Evaluation Methods," *Performance of Protective Clothing, ASTM STP 900,* R. L. Barker and G. C. Coletta, Eds., American Society for Testing and Materials, Philadelphia, 1986, pp. 250–260.

[17] Ashley, K. A., "Polychlorinated Biphenyl Decontamination of Fire Fighter Turnout Gear," *Performance of Protective Clothing, ASTM STP 900,* R. L. Barker and G. C. Coletta, Eds., American Society for Testing and Materials, Philadelphia, 1986, pp. 298–307.

[18] Perkins, J. L., "Residual Splashed Solvents in Butyl Protective Clothing and Usefulness of Decontamination Procedures," American Industrial Hygiene Conference, Dallas, TX, 18–23 May 1986.

Rosemary Goydan,[1] Arthur D. Schwope,[1] Sean H. Lloyd,[1] and Lucinda M. Huhn[1]

Cpcbase, a Chemical Protective-Clothing Data Base for the Personal Computer

REFERENCE: Goydan, R., Schwope, A. D., Lloyd, S. H., and Huhn, L. M., **"Cpcbase, a Chemical Protective-Clothing Data Base for the Personal Computer,"** *Performance of Protective Clothing: Second Symposium, ASTM STP 989,* S. Z. Mansdorf, R. Sager, and A. P. Nielsen, Eds., American Society for Testing and Materials, Philadelphia, 1988, pp. 403–408.

ABSTRACT: In recent years, interest in and testing of chemical protective clothing has increased significantly. Permeation and immersion testing have produced large amounts of breakthrough time, permeation rate, and weight-change information on clothing materials. This information has been compiled into a relational data base system which operates on a personal computer and is available as a commercial product, CPCbase. Also included in the system are descriptions of the clothing materials tested, their sources, and the literature references for the data. The system is menu-driven with fixed search routines and produces formatted reports. In addition, the user can add new or proprietary information and can tailor searches and reports to meet specific needs.

KEY WORDS: software, personal computer, protective clothing data base, permeation, degradation, weight change, volume change, diffusion coefficient, protective clothing vendors, bibliography

Along with engineering controls and safe work practices, chemical protective clothing (CPC) is fundamental to helping reduce worker exposure to chemicals. CPC is available from hundreds of vendors in the United States and is fabricated in a range of thicknesses from a wide variety of plastic and rubber materials. The effectiveness of CPC is, in part, determined by the resistance of the clothing materials to chemicals. Since no single material is resistant to all chemicals or chemical mixtures, proper selection of CPC is critical.

The task of selecting protective clothing is generally performed by seeking guidance from qualitative and quantitative listings of chemical resistance information for clothing materials. Technical reports and journal articles reporting CPC breakthrough time and permeation rate data have increased in recent years. Many clothing vendors also publish such data, or at least qualitative recommendations, on the chemical resistance of their specific clothing products. Until the publication of the *Guidelines for the Selection of Chemical Protective Clothing [1]*, this information was not available in a standard format.

While this and other published compilations of the available data are major advances, the printed format is not the most useful for certain applications. With a computerized data base, the data can be rapidly searched, customized reports can be generated, and data base expansion and information updates can be simplified. Personal computers and data base software offer the additional benefits of low cost (individuals can own one) and portability.

---

[1] Chemical engineer, unit leader, senior programmer, and technical writer, respectively, Arthur D. Little, Inc., Cambridge, MA 02140.

Several computer data bases for mainframe and personal computers are currently under development [2–5]. This paper describes CPCbase,[2] a commercially available chemical protective clothing data base for the personal computer.

CPCbase Design and Development

The primary goal of the design of CPCbase was a data base system that was easy to use, did not require access to sophisticated computer hardware and could be used immediately and productively by persons having little experience with computer data bases.

In light of this goal, CPCbase was designed as a personal-computer-based system and developed using a commercial software data base management system development package called Dataflex (Data Access Corp., Miami, FL). CPCbase was designed to provide simple searching and reporting routines as well as the capability to input additional resistance or product data to the existing data base. The system is completely menu-driven and provides help screens with brief definitions and system instructions. CPCbase provides three main functions: Data Review/Reporting, Data Input/Editing, and Data Base Querying; which are described later in this paper.

CPCbase runs on IBM and IBM-compatible personal computers and requires a minimum of 384k memory. The present system can be used in both floppy and fixed-disk modes, although the latter is significantly faster. These minimal computer requirements enable a broader access to the data base since many potential users may not have access to larger mainframe systems.

CPCbase is a strict data base (that is, a compilation of data organized for rapid search and retrieval) and not a clothing recommendation system. CPCbase is an unabridged compilation of the published chemical resistance data and their sources and is intended to guide selection of the most appropriate clothing. Such a data base can serve as the foundation of an expert system for clothing recommendations at such time when the factors affecting clothing performance and selection are more fully understood. Recently, a preliminary attempt toward development of an expert system for glove selection has been described [6].

At present, four types of test data compose the CPCbase system:

1. Breakthrough times and permeation rates.
2. Immersion weight changes.
3. Immersion volume changes.
4. Diffusion coefficients.

In addition, the system includes detailed descriptions of the clothing item tested and the clothing vendor name and address. A literature reference is also included for each data record. Further description of the information provided in CPCbase and the data base structure is provided in the following section.

Organization of Data Records

The data base structure and data record format used in CPCbase originate from that developed by Arthur D. Little, Inc., for the EPA *Guidelines for the Selection of Chemical Protective Clothing*. Many of the coding systems developed in that work and used in CPCbase, such as the codes for the clothing chemical-resistant materials and chemical classifications, are currently under consideration as part of an ASTM standard data reporting format guide.[3]

---

[2] CPCbase is a registered trademark of Arthur D. Little, Inc., Cambridge, MA.
[3] Proposed ASTM Standard Guide for Reporting Information from a Permeation Test to be Submitted for Inclusion in a Data Base, Draft 6, ASTM Subcommittee F-23.40, American Society for Testing and Materials, 1916 Race Street, Philadelphia, PA 19103.

CPCbase uses a relational data base structure as opposed to a flat file system in which the various fields would simply be joined into a single data record. The data base is structured using a master data base file which then points to smaller related data bases containing the actual data. This structure is depicted in Fig. 1. This structure allows for more efficient data storage and enables rapid data base searching and information retrieval.

Each data record identifies the chemical tested and its concentration, the specific clothing product, the vendor for the product, and reports the test data and a literature reference for the data. Table 1 provides a list of the various fields in CPCbase and is organized according to the individual data base files.

The chemical is identified using five fields and includes information such as the Chemical Abstract Service Registry Number (CAS Number), the chemical concentration (if in water), alternative chemical names, and the chemical classification. The clothing material is specified using both a resistant material and a product description field. The clothing vendor is defined by two fields, one for vendor name and a second for vendor address, including telephone number.

The test data fields for the four test types are all organized similarly to one another. In addition to reporting the specific resistance data for each test type, each record includes fields for the clothing material thickness, the test temperature, and the bibliographic reference. CPCbase also allows test data to be reported as a range by including high and low fields (Table 1).

As the standard chemical resistance test methods are refined and the data reporting formats are standardized (see footnote 3), additional fields may be required and can be added to the CPCbase system.

CPCbase Operation

As stated earlier, CPCbase was designed to provide simple searching and reporting routines as well as the capability to input additional data to the existing data base. The system operation is completely menu-driven and includes help screens with brief definitions and instructions for every field in the data base. CPCbase provides three main functions:

1. Data Review/Reporting.
2. Data Input/Editing.
3. Data Base Querying.

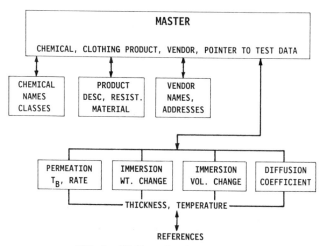

FIG. 1—*CPCbase data base structure.*

TABLE 1—*List of data base fields.*

Chemical	Chemical Abstract Service (CAS) Registry Number
	Chemical concentration (in water)
	Chemical name
	Chemical name sort specification
	Synonym
	Primary chemical class
	Alternate chemical class
Chemical classification	Chemical super-class
	Chemical class type
Clothing material	Resistant material name
	Product description
Clothing vendor	Vendor name
	Vendor address
Bibliographic reference	Bibliographic reference abstract
	Bibliographic reference
Permeation test data	Breakthrough time (low)
	Breakthrough time (high)
	Permeation rate (low)
	Permeation rate (high)
	Thickness
	Temperature
	Bibliographic reference
Immersion weight change test data	Percent weight change (low)
	Percent weight change (high)
	Immersion time
	Thickness
	Temperature
	Bibliographic reference
Immersion volume change test data	Percent volume change (low)
	Percent volume change (high)
	Immersion time
	Thickness
	Temperature
	Bibliographic reference
Diffusion coefficients	Mantissa (low)
	Mantissa (high)
	Exponent
	Thickness
	Temperature
	Bibliographic reference

which are selected from the main menu. Each of these is described in the following paragraphs.

The Data Review/Reporting function enables the retrieval and reporting of the data by chemical and by resistant material type. The output is first displayed on the video display terminal for on-screen analysis and comparison of the data for the various CPC items tested. An example of the on-screen display is shown in Fig. 2. The results of the search also may be printed in predefined report formats.

The Data Input/Editing function is a full-screen data input and edit feature which allows the user to add data that may be edited or deleted later. The original CPCbase data base is protected against accidental or intentional alteration. Data for new or proprietary chemicals and clothing materials may be added as well as additional information on CPC vendors and bibliographic references.

```
                            CPCbase
                    Permeation Test Data

Chemical Abstract Service Registry Number:   79016  Concentration: N
    Chemical Name: Trichloroethylene
```

Resistant Material	Prd Dsc #	Ven #	Thick -ness (cm)	Breakthrough Time (hours) low	high	Permeation Rate ($\mu g/cm^2$-min) low	high	Temp (°C)	Bib Ref #
Butyl	14		0.04	0.08		2044.08		23	75
Butyl	14	118	0.06	0.23		3306.60		23	266
CPE	70		0.05	0.20				23	3
Natural Rubber	1	100	0.03	0.03		9418.80		25	177
Natural Rubber	17		0.02	< 0.02		> 656.31		23	234
Neoprene	18	118	0.08	0.38		1302.60		25	177
Neoprene	18	120	0.05	0.14		2304.60		25	177
Neoprene	31		0.08	0.17	0.25	53.11		22	59
PVA	4	100		0.05		0.08	0.90	25	79
Viton	9	118	0.03	7.35		1.44		23	266

FIG. 2—*Example of on-screen display of output from CPCbase search/report routine. Detailed product descriptions, vendor's names and addresses, and full bibliographic references are obtained by moving cursor to any line and pressing the ⟨ENTER⟩ key.*

Help screens and selection panels are provided in both of the above functions and are accessed by means of a single key. Help screens provide brief definitions and instructions at the field level. Selection panels, provided for many of the entry fields in CPCbase, give an alphabetical listing of all possible entries for the field. Selection panels are included for the selection of chemicals, resistant materials, product descriptions, and vendor names and eliminate the need for typing in either the code or name of the specific entry of interest. Using a predefined function key, the user may access a selection panel so that the list of all possible entries appears on the computer screen. A selection is made by moving the cursor to the line of interest and pressing the ⟨Enter⟩ key. Thus, a user can search and select from a list instead of directly typing in the information for each search.

The Data Base Querying function is a more advanced feature and requires some knowledge of the overall data base structure to be used successfully. This function utilizes a basic feature provided by the Dataflex data base development software package which has not been customized to operate as efficiently as the other CPCbase functions. The Data Base Querying feature enables access to the data at the field level of the individual data bases and allows the user to define specific criteria for the searching and retrieval of information. The output from this search can be directed on-screen, to a printer, or to a disk file.

Conclusions

Personal computers and data base software facilitate the wide dissemination and rapid updating of data that describe the chemical resistance of protective clothing. The ready availability of such data enables industrial hygienists and safety specialists to reach informed decisions regarding clothing selection and specification. To be useful to these groups as well as persons involved in clothing research and product development, a data base should be comprehensive, be easy to use while allowing sophisticated searches, contain information on the clothing materials and data sources, and be fast. CPCbase meets these requirements and also can be readily expanded as additional information becomes available.

References

[1] Schwope, A. D., Costas, P. P., Jackson, J. O., and Weitzman, D. J., *Guidelines for the Selection of Chemical Protective Clothing,* 3rd ed., American Conference of Governmental Industrial Hygienists, Cincinnati, OH, 1987.
[2] Roder, M. M., paper presented at the Second International Symposium on the Performance of Protective Clothing, Tampa, FL, 18–22 Jan. 1987.
[3] Forsberg, K., "Chemical Degradation and Permeation Database and Selection Guide for Resistant Protective Materials," L. Keith, Ed., Instant Reference Sources, Austin, TX, 1986.
[4] Mellstrom, G., paper presented at the Second International Symposium on the Performance of Protective Clothing, Tampa FL, 18–22 Jan. 1987.
[5] Blank, T., paper presented at the Second International Symposium on the Performance of Protective Clothing, Tampa, FL, 18–22 Jan. 1987.
[6] Keith, L. H. and Stuart, J. D., "A Rule-Induction Program for Quality Assurance-Quality Control and Selection of Protective Materials," ACS Symposium Series 306, Artificial Intelligence Applications in Chemistry, American Chemical Society, Washington, DC, 1986.

Alan P. Bentz,[1] Clare B. Billing, Jr.,[1] and Martha S. Hendrick[2]

Use of a Relational Data Base for Protective Clothing Research

REFERENCE: Bentz, A. P., Billing, C. B., Jr., and Hendrick, M. S., "Use of a Relational Data Base for Protective Clothing Research," *Performance of Protective Clothing: Second Symposium, ASTM STP 989,* S. Z. Mansdorf, R. Sager, and A. P. Nielsen, Eds., American Society for Testing and Materials, Philadelphia, 1988, pp. 409–412.

ABSTRACT: This paper describes the use of a relational data base for the selection of priority hazardous chemicals to give a reasonable starting point and a logical sequence for conducting development programs involving chemical pollutants. Chemicals considered were those most likely to be encountered in or near a marine environment. The strategy was to gather available information on spill incidence, hazard factors, and need for clothing protection associated with each chemical. The information was stored in a computer with commercially available data base management software. Selection criteria were established for toxicity hazard, spill history, and requirement for an encapsulated suit. Chemicals for testing were sorted using these criteria.

KEY WORDS: hazardous chemicals, priority list, permeation testing, protective clothing, data base

The value of this paper lies not in any novelty of using a data base for a decision-making process, but rather as an example of a practical application and "how to" use a data base for protective clothing research in the selection of highest priority hazardous compounds.

Chemical analysis and analytical development for environmental spill response is a formidable task because of the infinite number of possible scenarios. Researchers developing chemical analytical schemes for environmental pollution, whether it be in air, soil extracts, or water, generally all start with the selection of test compounds. These may be compounds of particular interest such as polychlorinated biphenyls (PCBs) in sediments, solvent vapors in workspaces, chemicals transported, etc., or they may be extensive lists of chemicals. Depending on the criteria used, a number of chemical lists have been generated for various purposes.

The U.S. Coast Guard has published a list of over 1000 chemicals that it has included in its Chemical Hazard Response Information System (CHRIS) [1]. These are hazardous chemicals commonly transported and likely to be found in spill situations. It is illogical to start at the beginning of the list and develop methods or conduct tests in alphabetical order. It is far more useful to divide up the chemicals into solids, liquids, and gases and then subdivide each of these groups into chemical classes and by their relative hazards.

The Coast Guard required a list of about 100 chemicals to start testing for permeation resistance of the components of materials to be used in a new Chemical Response Suit. Test results were incorporated into a suit user's manual as a guide for response personnel in selection of appropriate suits for specific chemical environments.

[1] Project manager and senior statistician, respectively, Chemistry Branch, U.S. Coast Guard R&D Center, Groton, CT 06340-6096.
[2] Senior chemist, U.S. Coast Guard Central Oil Identification Laboratory, Groton, CT 06340-6096.

This task of selecting priority chemicals requires a data base to handle the large number of chemicals, numerous sources of information regarding spill frequency, and health hazards, and the ultimate sorting and organizing based on criteria established. In the present work, the choice of computer and software were predicated on their availability. The specifications of the data base far exceeded the original needs of the project. However, the memory and power of the relational data base program permitted our data base to expand and evolve in ways that we did not anticipate at the beginning of the project. The flexibility of the data base, which permits modification and expansion after data have been stored, proved to be essential to the continued usefulness of the data base.

For entry and reporting of data, three separate files were established: one for chemical protective clothing (CPC) product information, one for information on the analytical method, and one for the challenge chemical and permeation results. By using the feature of the data base to generate customized reports, the results for each chemical can be printed in standard ASTM format. The product, method, and results files can be linked internally in the program that generates final reports, so that the product and method information does not have to be reentered for each test. The results can be transmitted to another data base via an ASCII file sent by electronic mail. National Oceanic and Atmospheric Administration's (NOAA's) Hazardous Material Spills Response System provides one such electronic network.

Another application is to include an inventory file that contains additional information on amounts of material, properties and analytical conditions. This is particularly important for the current regulations governing the handling, tracking, storage, and ultimately the disposal of hazardous chemicals. Files also have been set up for penetration and degradation results linking them to the product file.

To facilitate the selection, the chemicals were grouped into four main classes, shown as quadrants in Fig. 1, depending on whether or not they had a spill history or were deemed

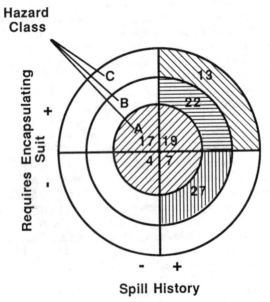

FIG. 1—*Selection criteria used for 109 highest priority hazardous chemicals for permeation testing. Numbers in shaded areas are numbers of compounds in each category.*

TABLE 1—*U.S. Coast Guard R&D Center classification scheme.*

1. Pesticides	9. Halogenated organics	16. Organic sulfur compounds
2. Monomers	10. Alcohols	17. Aromatics
3. Inorganic acid	11. Aldehydes and ketones	18. Organometallics/
4. Inorganic clusters	12. Glycols and epoxides	organosilicons
5. Inorganic halogen	13. Carboxylic acids/	19. Phenol
compounds	derivatives	20. Nitro compounds
6. Other inorganic cations	14. Nitriles and isocyanates	21. Heterocyclics
and anions	15. Amines and imines	22. Phosphorus compounds
7. Saturated hydrocarbons		23. Ether
8. Unsaturated hydrocarbons		24. Peroxides
		25. Oils

to require a totally encapsulating suit [2]. The spill history was obtained from seven sources, including the Coast Guard's Pollution Incident Reporting System (PIRS) [3,4] and data base (CHEMREPS) of NOAA on chemicals encountered in real spills as described by Ernst [5].

The chemicals were further divided into hazard classes (Levels A through C). Level A includes compounds that have been assigned: a carcinogen, Class 1, highly toxic, Class 2, or toxic through skin absorption, Class S, in the Hazard Assessment Index [6]; a Class 4 by NFPA [7] is its highest health hazard rating. Level B includes a Hazard Assessment Index of 3 or an NFPA rating of 3. Level C includes all other chemicals not in A or B.

All chemicals of Level A were included, whether or not they had been spilled yet, or required a suit, in order to provide an additional margin of safety. This group included 47 liquid chemicals, only eleven of which had not been deemed to require a suit. Level B compounds included a total of 49 liquid chemicals with a history of spills. Thirteen compounds, which were included with Level C toxicity, had been spilled and required a suit. These liquids totaled 109 chemicals and include at least one member of each of the 25 chemical classes in the Coast Guard scheme (Table 1). Classes represented by several chemicals may give reasonable estimates of behavior of untested members of the same class.

Computer programs generated reports that divided the chemicals into groups. These were written according to simple high level commands that are spelled out in a document by Hendrick and Billing [8].

The data base contained files on chemical classification, properties, spill information, test conditions, and results. Typical files are shown in Table 2.

The first file in Table 2 contains chemical classification information in 19 separate fields as shown in Table 3. These fields stored three general types of information: (1) identification and classification, (2) chemical hazard information, and (3) spill history—all from various sources.

TABLE 2—*Typical files in data base.*

Chemical classification information
Solubility parameter
National response center list
NOAA CHEMREP information
Gas chromatography conditions
Protective clothing material data
Permeation test method information
Permeation Test Results

TABLE 3—*Fields in CHEMCLASS file.*

Chemical name	1
CHRIS code	1
Chemical state (S, L, G)	1
Chemical classification	3
Hazard information	5
Encapsulated suit requirement	1
Spill history	7
Total	19

The first two fields included the chemical name and CHRIS code, which is a unique three-letter identifier for each compound. The CHRIS code is used to cross-check to ensure against two records of information being independently established for the same chemical (because of synonyms for the chemical name).

The physical state of the chemical at room temperature was entered in one field and the next three fields contained chemical group classifications (Table 1) and two other systems. The next six fields were used for hazard information from different sources. Seven fields were used for spill information.

The initial use of the data base (and the subject of this paper) was to select the 100+ most hazardous chemicals of highest priority for testing. The data base was queried, for example, on what compounds have the highest toxicity rating, have a history of being spilled, and require an encapsulating suit; what compounds have the second highest toxicity rating, etc. From this information, those compounds were identified that were in most urgent need of testing.

The user of test results is the operational Coast Guard. The relational data base concept permits continued expansion as new data are received, and efficient transfer to field units (by modem).

The data base, set up as just described, has been a helpful tool for information storage and retrieval, in formulating lists of priority hazardous chemicals for further research, and in providing test results immediately to the user in the field.

A more important ultimate use as a relational data base will be to investigate interaction of variables and their effects on permeation.

References

[1] Chemical Hazards Response Information System, Commandant Instruction M.16465.12A, U.S. Coast Guard, Nov. 1984.
[2] Friel, J. V., McGoff, M. J., and Rodgers, S. J., "Material Development Study for Hazardous Chemical Protective Clothing Outfit," MSA Research Corp., Evans City, PA, CG-D-58-80, NTIS No. ADA 095-993, Aug. 1980.
[3] Stull, J. O., "Considerations for the Development of a Hazardous Chemical Personnel Protection System," Second Annual Technical Seminar on Chemical Spills, Toronto, Canada, 5–7 Feb. 1985.
[4] Fang, P. C. I. et al., "Analysis of Hazardous Chemical Spills Along the Coasts and Major Waterways of the United States," CG-123-1, U.S. Coast Guard, 1981.
[5] Ernst, W. D., "NOAA's Chemical Advisory Report (CHEMREP) System for Spill Response," *Proceedings,* 1984 Hazardous Materials Spills Conference, Nashville, TN, 5–9 April 1984.
[6] Marine Hazardous Substance Data System, SWRI Project 06-7223, Oct. 1984.
[7] "Fire Protective Guide on Hazardous Materials," Code 704M, 4th ed., National Fire Protection Association, Boston, 1972.
[8] Hendrick, M. S. and Billing, C. B., Jr., "Selection of Priority Chemicals for Permeation Testing and Hazardous Chemical Spill Detection and Analysis," Final Report No. CG-D-22-86, NTIS No. ADA 172370, U.S. Coast Guard, July 1986.

Protection from Industrial Chemical Stressors

Emergency Response and Military Applications

Helena Mäkinen,[1] Juhani Smolander,[1] and Hille Vuorinen[1]

Simulation of the Effect of Moisture Content in Underwear and on the Skin Surface on Steam Burns of Fire Fighters

REFERENCE: Mäkinen, H., Smolander, J., and Vuorinen, H., **"Simulation of the Effect of Moisture Content in Underwear and on the Skin Surface on Steam Burns of Fire Fighters,"** _Performance of Protective Clothing: Second Symposium, ASTM STP 989_, S. Z. Mansdorf, R. Sager, and A. P. Nielsen, Eds., American Society for Testing and Materials, Philadelphia, 1988, pp. 415–421.

ABSTRACT: About 9% of the injuries to fire fighters in Finland are burns. The majority of the burns on the shoulders and upper arms were caused by a high condensation rate of steam on the skin. The moisture transport from the skin of the subjects to the outer layers of clothing, as well as the heat transfer through the same dry and wet garment layers were measured by fabric tests. The differences between types of underwear, that is, cotton knit, aramid knit, and wool knit fabrics, were sought by the simulation method. Humidity measurements were performed with Vaisala Humicap sensors. The sensor mechanic and computer control systems of the sensor were developed at the Institute of Occupational Health.

Four male firemen served as subjects at 30°C ambient temperature and at 30% humidity. Each exposure lasted for 70 min (20-min rest, 25-min walk, and 25-min rest). The humidity between the skin and the underwear of the back and on the shoulders and between two-garment layers were recorded. The skin temperature was recorded at nine sites. Core temperature, heart rate, and oxygen consumption were also measured. The same layers of clothing were moistened with different amounts of water, and their thermal behavior was evaluated by exposing the garments to a source of radiant heat, 20 kW/m², according to the second draft of ISO 6942. The lowest moisture content between skin and underwear on the shoulders was measured when woolen underwear was worn, and the highest content when cotton underwear was worn. The moisture content of the garments was lowest in the aramid underwear. The differences were not significant, however, the moisture in clothing decreased the transmission factor, TF_{20}, the time-to-pain, and second-degree burn time. The decrease was drastic when the moisture content rose from 20% to 30 to 40% in underwear.

KEY WORDS: protective clothing, moisture, radiant heat protection, burn injury, underwear

In Finland, the number of fire fighter injuries causing three days or more of absence from work in 1980 to 1982 averaged 34 per 1000 fulltime fire fighters, or 5.8 per 1000 fire fighting incidents. About 9% of the accidents were burns. The number of small burns without lost working days is even higher. Most of the burns were located on the upper arms and shoulders. The majority are steam burns resulting from the heat released by the condensation of steam. The condensing moisture comes from sweat and water. It is important to prevent this condensation from taking place on the skin or in the clothing [1].

The moisture in the clothing decreases the protection against both radiant and convective heat [2,3]. The vapor barrier also decreases the effect, particularly when the innermost layer is wet from perspiration [3].

[1] Research engineer, researcher, and laboratory engineer, respectively, Institute of Occupational Health, Laajaniityntie 1, 01620 Vantaa, Finland.

From among the natural fibers, cotton and wool are the most suitable materials for the underwear of fire fighters because of their charring properties. However, a wet cotton knit often feels cold on the skin, and a woolen knit does not have good laundering properties. In addition, some people may get skin irritation and allergic reactions when wearing a woolen knit next to the skin.

Most synthetic fibers are not suitable as underwear in fire fighting circumstances because they melt easily. One synthetic fiber suitable for underwear in fire fighting is aramid, with a higher humidity absorption quality than polypropylen and polyester, but much lower than that of cotton and wool fibers [4].

The purpose of this study was to examine the effect of three different underwear materials (cotton, aramid, and wool) on the humidity of the layers of air between the skin and fabric, and between the layers of clothing in a warm atmosphere.

Another aim was to determine the underwear fabrics, when wet and when worn under outer garments to protect against radiant heat.

Material and Methods

Wear Trials

The underwear fabrics used in the trials were cotton, aramid, and wool knits and the outer garments were made of cotton twill fabric and Karvin satin (65/30/5 fr.viscose/aramid (Nomex)/aramid (Kevlar)). The fabrics are described in Table 1.

Four male students from the State Fire Institute volunteered as subjects. All were physically trained and healthy (age 23 ± 4 years, weight 68 ± 3 kg, and height 1.75 ± 0.04 m).

The wear trials consisted of a 20-min rest period, a 25-min exercise period, and a 25-min recovery period. During the exercise period, the subjects walked on a treadmill at 50 to 60% of maximum oxygen consumption (VO_2 max). All tests were carried out in a climatic chamber at a temperature of 30°C; relative humidity was 30%. The air velocity was 0.3 m/s and there was no radiant heat.

During the wear trials, the physiological responses and subjective evaluations were monitored by the following procedures. The total amount of sweat was calculated by weighing the subjects nude before and after the tests (± 10 g). The amount of absorbed sweat was obtained by weighing the different items of clothing before and after the tests (± 0.1 g). Sensation of warmth and thermal comfort [5], wetness of skin [6], and perceived exertion [7] were estimated with rating scales. The relative humidity and temperature (water vapor pressure) were measured on the shoulders and on the back in three layers: (1) between skin and underwear; (2) between underwear and middle clothing layer; and (3) between middle layer and outer garment layer. The method of measuring humidity has been developed at

TABLE 1—Specifications of the tested materials.

	Structure	Weight, g/m²	Thickness, mm
Innermost Layer			
100% cotton	1 by 1 rib	186	0.90
100% aramid	1 by 1 rib	191	1.05
100% wool	1 by 1 rib	225	1.25
Middle Layer			
(100% cotton)	twill	270	0.70
Outer Layer			
"Karvin"ᵃ	satin	300	0.60

ᵃ (65/30/5 fr. viscose/aramid Nomex/aramid Kevlar).

TABLE 2—*Proportion of water as a percentage of dry weight, %.*

Outer layer	10	10	15
Middle layer	20	30	20
Cotton knit underwear	40	...	...
Aramid knit underwear	...	30	...
Wool knit underwear	...	...	40

the Institute of Occupational Health. The method is based on Vaisala Humicap sensors, and is controlled by a desk-top computer [8]. The core temperature was measured by an esophageal catheter (YSI 511) introduced at heart level. Skin temperatures were measured at nine points with YSI 427 thermistors. The electrocardiogram was continuously monitored and heart rate was recorded every minute. During each period, oxygen consumption was measured by the Morgan Exercise Test System.

Fabric Tests

Radiant heat transmission was measured according to the second draft of ISO (International Organization for Standardization) 6942. In this method, the desired heat flux is produced with silicon carbide rods with a mean temperature of 1100°C. The temperature rise is measured as a function of time on the other side of the specimen.

The measured parameter is the transmission factor, TF_{20}, of the test specimen at a particular incident heat flux as the ratio of the transmitted and incident heat flux measured by an aluminum block calorimeter. In these measurements, the incident heat flux was 20 kW/m^2. The other measured parameters were the time-to-pain and second-degree burn time.

The moisture content in the various textile layers in the test series was 0, 20, 60, and 80% water from the dry weight of the specimen. In one of the test series, the wettness of the different layers varied according to the wear test results shown in Table 2.

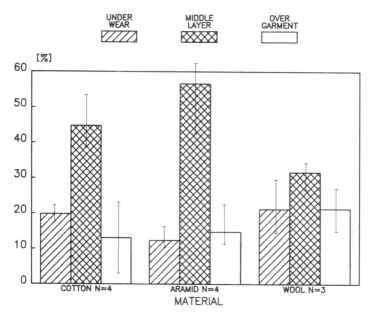

FIG. 1—*The relative amounts of gathered moisture in the different garments.*

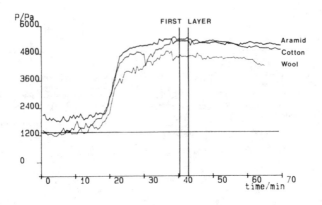

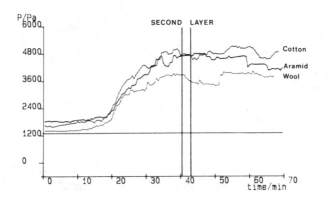

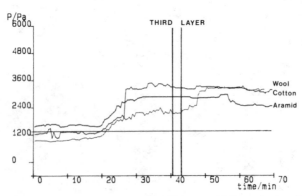

FIG. 2—*The water vapor pressure at the shoulder between different layers of garments.*

Results

Wear Trials

In the wear trials, the sweat amounted to 780 ± 166 g. No differences in the amount of sweat between the three types of underwear were found. Figure 1 shows the relative amounts of gathered moisture in the different garments. The aramid knit underwear absorbed the least moisture, but in this case the middle layer of clothing was the wettest. With woolen underwear, the moisture was evenly distributed in all layers of clothing.

The highest water vapor pressure was measured between the skin and underwear when the subject wore cotton and aramid knit underwear. The water vapor pressure was lowest with woolen underwear, in which case the humidity rose at the end of the trial between the middle and outer layers. Figure 2 shows the water vapor pressures in the different air layers.

The type of underwear did not significantly affect body temperatures or heart rate. At the end of exercise, the average heart rate was 168 ± 10 beats/min. The mean skin temperature was highest when wearing the woolen underwear. During the exercise period, the oxygen consumption did not differ between the three tests.

At the end of the whole test, the sensation of skin wettness was the highest when the subjects wore cotton underwear. The subjective sensations did not differ significantly.

Fabric Tests

The humidity decreased the overall protective value of the garments to radiant heat. The estimated time-to-pain was 12 s in cotton underwear, and 15 s when using woolen and aramid knit underwear testing with dry specimens. Second-degree burn time was 18.2 s in cotton, and 24 s when using woolen and aramid knit underwear.

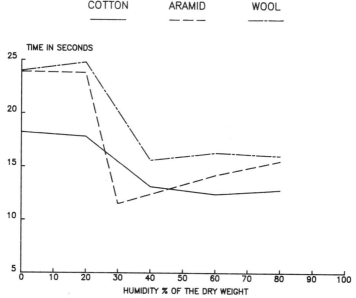

FIG. 3—*The second-degree burn times (20 kW/m²) at different moisture levels.*

In all fabrics, the estimated times were shortest when the moisture content was 30 to 40% of the dry weight. With higher moisture contents, the times were constant when cotton and woolen underwear was used and increased when aramid knit underwear was used. Figure 3 gives the second-degree burn times at different moisture levels.

Table 3 shows the correlation of the thickness of the clothing layers to transmission factor, TF_{20}, to estimated time-to-pain, and to second-degree burn time. The correlation was moderately high with all measured parameters at a moisture level of 60% and with TF_{20} also at the level of 30 to 40%. On this level, however, the correlation was poorest with time-to-pain and second-degree burn.

Discussion

The wear trials indicated that the different underwear materials have different moisture absorption and transfer properties, but the differences were not significant. Morris has studied the relationship between subjective evaluations of wetness and moisture transport properties of undergarments and found that when there was significantly less moisture in the undergarments the wearers felt dryer in the warm environment [9].

The present study also confirms this. With synthetic aramid knit underwear, the moisture transfer to the second layer was the highest, but at the same time, the relative humidity between the skin and underwear was highest because this fiber type absorbs least moisture. When woolen undergarments were used, the moisture was most evenly distributed in the various clothing layers.

The moisture in the clothing layers clearly decreased protection against radiant heat. The decrease was significant when the moisture level rose from 20% to 30 to 40%, and it was the highest with aramid underwear. The correlation to the thickness of the layers was not clear at all moisture levels. These results show that the differences in moisture absorption and the drying speed of fibers also have an effect on the protection capacity of clothes against radiant heat.

TABLE 3—*Correlation of the thickness of the layers to measured parameters.*[a]

Material	Thickness, mm	Moisture Level, %				
		dry	20	30 to 40	60	80
		TRANSMISSION FACTOR TF_{20}				
Cotton	2.20	0.46	0.46	0.47	0.47	0.52
Aramid	2.35	0.36	0.35	0.43	0.45	0.47
Wool	2.55	0.35	0.33	0.39	0.41	0.46
Correlation	$r =$	-0.866	-0.895	-0.997	-0.994	-0.901
		TIME TO PAIN				
Cotton	2.20	12.1	10.7	8.9	8.8	9.6
Aramid	2.35	15.3	12.3	8.0	10.6	12.0
Wool	2.55	15.3	13.1	10.8	12.5	11.9
Correlation	$r =$	0.822	0.963	0.724	0.998	0.800
		SECOND-DEGREE BURN TIME				
Cotton	2.20	18.2	17.8	13.1	12.4	12.8
Aramid	2.35	23.9	23.8	11.5	14.2	15.5
Wool	2.55	24.0	24.8	15.6	16.3	16.0
Correlation	$r =$	0.830	0.890	0.668	0.999	0.896

[a] All test specimens were taken from materials washed three times. The presented values are averages of three experiments.

The advantages of aramid knit and woolen knit fabrics were shown over cotton knit fabric as fire fighters' underwear materials. Furthermore, in most cases, the wool underwear is preferable if the fire fighter can wear it.

Acknowledgments

The authors wish to thank the Finnish Academy for financial support and senior researcher Brita Lisa Irjala from the Technical Research Centre of Finland for guidance in the measurements of radiant heat.

References

[1] Watkins, S. M., *Clothing—The Portable Environment*, IOWA State University Press, Ames, IA, 1984.

[2] Benisek, L., Edmondson, G. K., Mehta, P., and Phillips, W. A. in *Performance of Protective Clothing, ASTM STP 900*, R. L. Barker and G. C. Coletta, Eds., American Society for Testing and Materials, Philadelphia, 1986, pp. 405–420.

[3] Veghte, J. H. in *Performance of Protective Clothing, ASTM STP 900*, R. L. Barker and G. C. Coletta, Eds., American Society for Testing and Materials, Philadelphia, 1986, pp. 487–496.

[4] Haudek, H. W. and Viti, E., *Textilfasern*, Verlag Johann L. Boundid Sohn, Wien-Perch Toldsdorf, Melliand Textilberichte Heidelberg, 1980.

[5] Fanger, P. O., *Thermal Comfort, Analysis and Applications in Environmental Engineering*, Mc-Graw–Hill Book, Co., New York, 1972.

[6] Vokac, Z., Köpke, V., and Keul P., "Assessment and Analysis of the Bellows Ventilation of Clothing," *Textile Research Journal*, Vol. 43, 1973, pp. 474–482.

[7] Borg, G., Perceived Exertion: a Note on "History and Methods," *Medical Science Sports*, Vol. 5, 1973, pp. 90–93.

[8] Vuorinen, H., "Mikroilmaston kosteuden mittaus (The Measurement of Humidity between the Skin and Clothing Layers)," Masters thesis, The Technical University of Tampere, 1986 (in Finnish).

[9] Morris, M. A., Prato, H. H., Chadwick,.S. L., and Bernauer, E. M., "Comfort of Warm-Up Suits During Exercise As Related To Moisture Transport Properties Of Fabrics," *Home Economics Research Journal*, Vol. 14, 1985, pp. 163–170.

C. J. Abraham,[1] Malcolm Newman,[1] and Jesse H. Bidanset[1]

Analysis of Standards and Testing for Firemen's Protective Clothing

REFERENCE: Abraham, C. J., Newman, M., and Bidanset, J. H., **"Analysis of Standards and Testing for Firemen's Protective Clothing,"** *Performance of Protective Clothing: Second Symposium, ASTM STP 989,* S. Z. Mansdorf, R. Sager, and A. P. Nielsen, Eds., American Society for Testing and Materials, Philadelphia, 1988, pp. 422–438.

ABSTRACT: A study was made of standards published since 1970 for firemen's protective clothing. In addition, the state of the art in material science from the early 1960's through 1986, relating to protective clothing, was examined, including testing techniques and development of fire-resistant materials. Studies were performed in conjunction with the remains of protective clothing and equipment exposed to a flashover during residence fire.

The authors conclude that the standards set forth by the National Fire Protection Association, the American National Standards Institute, the American Society for Testing and Materials, and the Occupational Safety and Health Administration do not correlate with the foreseeable exposure temperatures in a burning structure. The materials currently used by fire fighters cannot withstand the extreme heat conditions generated by a flashover. Based upon the findings of the authors, it is necessary to continue to upgrade the preexisting minimum standards to incorporate the state of the art available for protective clothing and equipment.

Upgrading existing standards to reflect the level of available technology would lead to significant improvements in the design and construction of protective gear. In turn, this would result in greater safety for firemen caught in catastrophic fires by reducing foreseeable thermal exposure injury.

KEY WORDS: firemen's protective clothing, testing, ablation, standards, warnings, research, state of the art of technology, flashover

The necessity of firemen to enter a burning building in order to save human lives and property is inescapable. This paper discusses the current performance of and possible improvements in protective clothing and respiratory equipment. Statistically, accident records relating firemen's injuries to the number of fires show that fatalities and severe thermal injuries occur relatively infrequently. Considering the limitations of thermal protective clothing and respiratory equipment, it is truly remarkable that of the 39 900 injuries reported in 1982, only 11% of those injuries accounted for combined burns, cold, and heat exhaustion [1].

Over the course of the past 16 years, the authors have performed extensive studies of individuals injured from exposure to a variety of fires and explosions. The various types of apparel worn were examined and correlated with the thermal injuries. In general, the nature of the thermal exposure was not anticipated by the injured party, in contrast to the case of a fireman entering a burning building.

---

[1] Associate director, director, and director of forensic sciences, respectively, Inter-City Testing & Consulting Corp., 167 Willis Avenue, Mineola, NY 11501. Dr. Bidanset is also professor of Toxicology and Pharmacology, St. John's University, Queens, NY.

There is an expectation that a fireman, prior to entering a burning building, has had sufficient training, background, and experience toward protecting himself, to the extent possible, from the foreseeable thermal exposure which could result in irreversible injuries or death. After examining typical protective clothing and respiratory equipment used by volunteer firemen entering a burning residence, the authors were surprised to find that, although firemen were aware of the inherent dangers involved in entering a burning building, they were not aware of the protective limitations of their garments and respiratory systems.

Case Study

On December 18, 1982, a fire broke out in a frame addition to a residential home. A volunteer fireman who had been with the fire department for approximately ten years arrived with his company at the scene. A bystander ran up to them and stated that some people were still inside.

The ten-year veteran, along with two other veteran fire fighters, entered the structure crawling on their hands and knees with a firehose between their legs. During that time the fire had vented itself through the roof of the addition. Other members of the department were spraying water into the vented hole at the time the three men entered the building at the opposite end.

The subject fireman was wearing a helmet, turnout coat, gloves, boots, and a self-contained breathing apparatus (SCBA). He was familiar with the structure and led the two other men through the darkened area. Shortly after they entered, they saw an "orange glow" in the corner of a room near the ceiling. The men attempted to get out of the building. However, they observed an instantaneous explosion, and the flashover traveled down the hallway leading from the frame addition to the residence. The subject fireman was facing that area directly. As a result, he experienced the full force of the flashover.

Realizing that he was on fire, he had the presence of mind to roll on the floor in an attempt to smother the flames. The next thing he knew, he was being pulled out of the house, which was now fully engulfed in flame. Prior to the flashover, the three men had been in the house for approximately five to ten minutes.

The volunteer fireman received extensive and permanent scarring to his eyes, nose, mouth, side of his face (Fig. 1), right arm (Fig. 2), and hand (Fig. 3), since his right side was facing the flashover on impact. He had eversion of the lower lids, bilaterally, which interfered with his eye movement and function. He had extensive scarring and retraction of the nasal alae, especially on the right side, and perinasal areas were also notably burned. Multiple surgical procedures involved the restructuring of his lips and closure of the tympanic membrane perforation. There was a large percentage of right-sided hearing loss. There are also known risks of repeated infections to the ear, resulting complications, and possible spread to areas such as the brain and the neck tissues.

It is clear that the intense heat, possibly in the range of 815°C (1500°F), was sufficient to melt the plastic visor of the helmet (Fig. 4), melt parts of the SCBA around the mouth, shrink the right glove (Fig. 5), and melt portions of the fireman's clothing (Fig. 6). The duration of the flashover was extremely short, as evidenced by the fact that the clothing did not continue to burn once the fireman's exposure to the flash fire ended.

The SCBA (Figs. 7 to 9) clearly showed evidence of burning, charring, and excessive heat exposure. The approximate melting range of typical aluminum alloys is in the range of 501 to 676°C (935 to 1250°F). Therefore, from the melted material obtained and examined, it is evident that the fireman was exposed to temperatures above this range.

The right glove had shrunk considerably in the finger area, and the resulting thermal exposure caused permanent deformity of the right hand.

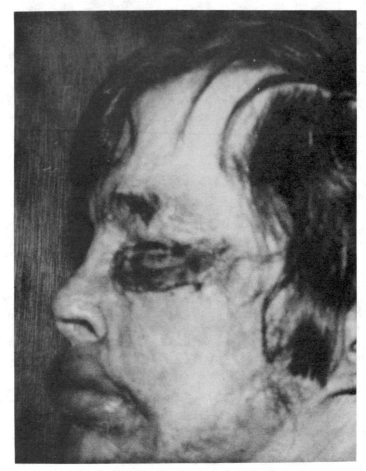

FIG. 1—*Side of subject fireman's face after flashover.*

The helmet and shield (Figs. 10 and 11) were of "65% fiberglass construction for strength, heat- and impact-resistance." The helmet manufacturer stated in the specifications that the helmet and protective system met "any one or more of the currently recognized specifications (OSHA,[2] ANSI,[3] NFPCA,[4] military or state municipal specifications), and that fire departments could effectively accomplish protection against the hazards to which their fire fighters were exposed." It was claimed that the safety face shield provided protection for the face and eyes and that the helmet could be adjusted satisfactorily to prevent interference with the breathing mask or apparatus. The helmet also contained a sweat-liner with adjustable Nomex earlaps. However, upon exposure to the flashover, the face shield melted, and the front part of the helmet melted, charred, and ablated (Fig. 4).

The right side of the turnout jacket, which was directly exposed to the flashover, was completely destroyed. The boots (Fig. 12) showed evidence of decomposition but, since the

[2] Occupational Safety and Health Administration.
[3] American National Standards Institute.
[4] National Fire Prevention and Control Association.

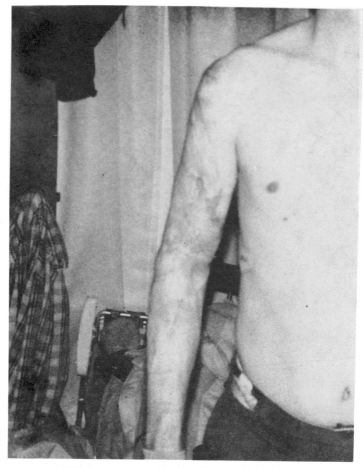

FIG. 2—*Right arm of subject fireman after flashover.*

highest intensity of the heat during flashover occurs in the uppermost part of a room, the boots were not exposed to the extreme temperatures. Therefore, the least number of thermal injuries were observed in the boot area.

Firespread and Flashover

Firespread beyond the initial burning item occurs as a result of involvement of adjacent materials in addition to the finish on the surrounding walls and ceiling. The rate of fire growth and firespread and the severity of the fire depend on the thermophysical and flammability properties of the materials in the fire environment.

Flashover occurs as a result of radiative heat transfer from the hot smoke layer and surfaces in the upper part of a room to the combustibles in the lower part of the room. This usually occurs when the temperatures of the burning products exceed approximately 600°C (1112°F). Flashover does not occur before the concentration of gases is past the point of physiological limits established for the incipient incapacitation of occupants.

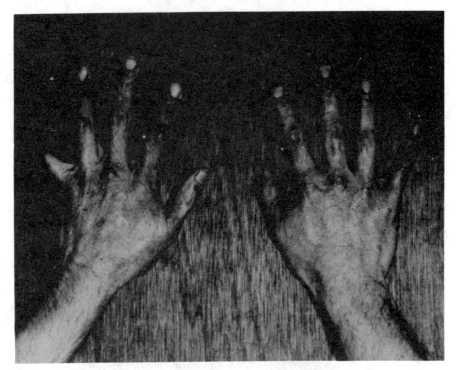

FIG. 3—*Hands of subject fireman after flashover.*

Since there is always a danger of flashover in a burning building, a fireman entering such a structure should be provided with protective gear with optimum protection against the associated severe thermal exposure.

Standards and Test Requirements

The Standard on Protective Clothing for Structural Fire Fighting (NFPA 1971–75) sets forth the minimum recommended standards for protective clothing for firemen. Section 1-2, "Scope," states

This standard applies to protective clothing for structural fire fighting worn for protection against extremes of temperature, hot water, hot particles and other hazards encountered during fires and related life-saving.

Paragraph 3-4.1.2 states that

The manufacturer shall provide the purchaser with the following information prior to purchase:
(*a*) A statement of the wearability of the garment.
(*b*) A statement of the stability of the fabric at high temperature, showing the effects of temperature exposures up to 500°F (260°C), in a forced-air laboratory oven for a period of 10 minutes, shall also include temperature/shrinkage curve, tensile strength retention/temperature curve, and an indication of the temperature at which the fabric will char, separate, or melt and drip.

It is incumbent upon the manufacturers of all materials and equipment used by firemen to document and report the above test results. The inference is that this standard should be applied to all protective equipment.

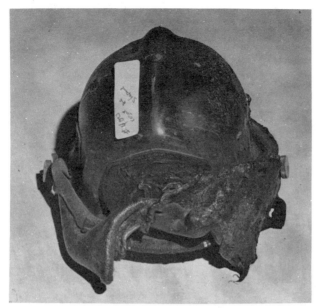

FIG. 4—*Remains of the helmet's plastic visor.*

Paragraph 2-2.2, titled "Labels," states

Each outer shell have sewn to the inside, in a location not covered by the lining (e.g., inside front or fly flap), one or more permanent labels stating the following:
(*a*) Fiber content of the outer shell fabric to conform with the rules and regulations under the Textile Fiber Products Identification Act.
(*b*) The size of the garment.
(*c*) Care instructions, including minimum instruction for washing or cleaning. These instructions shall include instructions for home machine laundering and a cautionary statement that this washing can remove the water-repellent treatment.
(*d*) A warning that the garment is not a proximity or entry suit and should not be kept in direct contact with flames.

In the National Fire Protection Association (NFPA) Standard 1972 (1985), Structural Fire Fighters' Helmets, Paragraph 1-1.1 states

This Standard provides minimum performance criteria and test methods for structural fire fighters' helmets designed to mitigate adverse environmental effects to the fire fighter's head.

Paragraph 2-1.1 states,

Structural fire fighters' helmets shall essentially consist of a shell, an energy absorbing system, a retention system, retroreflective markings, air covers, and face shields.

Face shields are covered in Paragraph 3-9. Paragraph 3-9.1 requires that the helmet shall be equipped with a face shield with a minimum height of 7.6 cm (3 in.) below the reference plane. Paragraph 3-9.2 states that face shields shall be tested for impact protection in accordance with Paragraph 4-2.8.1. Paragraph 3-9.3 states

Face shields shall meet the flame resistance characteristics of Paragraph 6-2.4, ANSI Z87.1, 'Practice for Occupational and Educational Eye and Face Protection.'

In meeting the requirements of Paragraph 3-9.2, the face shield in the case study had to

FIG. 5—*Remains of fireman's gloves.*

resist impact at a temperature of 50°C (122°F). The latter temperature was far below the level at which the face shield melted, and far below the foreseeable temperatures to which such a face shield will be exposed.

In meeting the requirements of Paragraph 3-9.3, the face shield in the case study did not burn at a rate greater than 7.6 cm (3 in.) per minute—it melted instead. In fact, there was scarcely any evidence of burning at all. Thus, although a helmet can meet all the minimum requirements set forth in the NFPA 1972 and ANSI Z-87.1 standards, these requirements are not suitable for the foreseeable use in burning buildings, much less exposure to a flashover.

Polycarbonate, from which the face shield in this case study was manufactured, has a melting temperature of 150°C (302°F). A temperature exceeding 121°C (250°F) therefore would be sufficient to soften the plastic so that it would deform greatly. In view of exposure in a flash fire to temperatures as high as 815°C (1500°F), a refractory material such as tempered glass would probably be the only suitable choice for use in a fireman's face shield.

Given these deficiencies of the polycarbonate face shield in the case study, it was easy to relate the eye injuries and facial burns to the melting of the plastic.

FIG. 6—*Remains of fireman's clothing.*

The NFPA 1973 Standard covers "Gloves for Structural Fire Fighters." Section 1-1.1 states

This standard specifies minimum performance criteria and test methods for gloves for structural fire fighters designed to mitigate adverse environmental effects to the fire fighters' hands and wrists.

The gloves are required to meet four heat performance requirements. Paragraph 2-4.1, which covers heat resistance, states

Sample gloves shall be tested in accordance with 3-2.1, Heat Resistance Testing, and may char or discolor, but shall not separate, melt, or drip. Materials used in glove construction shall not shrink more than five percent in length and width. The glove shall be measured from the tip of the middle finger to the bottom of the glove body.

The test for this is specified in Paragraph 3-2.1.2:

The sample glove shall be suspended in the center of an insulated circulating air oven, with a minimum 4 KVA capacity, with air circulation of 0.18 $^{+}$0.06f cubic meters per minute, which is preheated to 485 $^{+}$10°F (250 $^{+}$5°C).

FIG. 7—*Remains of fireman's self-contained breathing apparatus.*

Paragraph 3-2.1.3 then states

After 5 minutes $^+10$ seconds, the glove shall be removed and shall be examined to ascertain any adverse effects of the heat exposure.

The actual gloves, which presumably met this requirement, shrank considerably more than that, since the temperature to which they were exposed was more than three times higher than the 251°C (485°F) specified in the standard.

NFPA 1973 specifies a second flame resistance test for gloves, covered by Paragraph 2-4.2, which states

Sample specimens shall be tested in accordance with 3-2.2, Flame Resistance Testing, and shall not exceed the following average values:
(a) After flame—2 seconds;
(b) After glow—4 seconds;
(c) Char length—1.0 inches.

The test is conducted by exposing a 5-by-10-cm (2 by 4 in.) specimen of the actual glove material to a Bunsen burner flame inside a closed cabinet for a period of 12 s.

The gloves used in our case study (Fig. 4) were removed from the injured fireman by cutting them off after his exposure to the flashover. The left glove was largely intact, whereas the right glove, which was directed to the flashover, was missing the thumb and the upper part of the glove, leaving only the four glove fingers intact. There was little evidence of charring. The only molten material appeared to be electrical wire insulation that had adhered to the glove material along with a length of the wire.

FIG. 8—*Disassembled remains of fireman's self-contained breathing apparatus.*

The third heat test specified in NFPA 1973 is the conductive heat resistance test described in Paragraph 2-4.3. In this test, the palm of the glove is to be exposed to a conductive heat load of 500°C (932°F) for a period of 4 s with a pressure loading of 27.5 kPa (4 psi). As specified in Paragraph 2-4.3.1

> The temperature of the inner surface of the glove at the palm and at the gripping surface of the fingers shall not exceed 111°F.

To achieve the thermal injuries shown on the injured fireman's hands, assuming the outside temperature remained at 500°C (932°F) and that the glove continued to operate as an insulator, would have required a period of 14 to 16 s; the inside temperature would then reach between 71 and 82°C (160 and 180°F), which is sufficient to produce third-degree burns. To produce the same condition within 12 s, with the same insulating characteristics, would have required an outside temperature of nearly 759°C (1400°F). If the gloves were exposed to a 815°C (1500°F) flame, the time required to reach the same condition inside the glove would be approximately 7 s.

The fourth heat test in NFPA 1973 is the Radiant Heat Resistance Test covered by Paragraph 2-4.4. In this test, a radiant panel operating as a black body at a temperature of

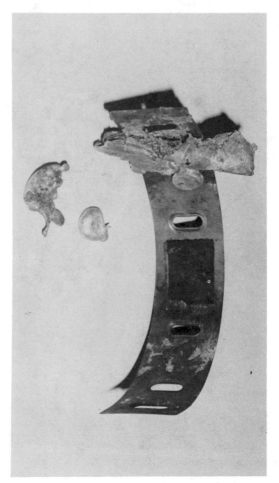

FIG. 9—*Close-up of remains of fireman's self-contained breathing apparatus.*

1000 K (1800 R or 1340°F) shall be permitted to radiate at a portion of the glove approximately 76 mm (3 in. in diameter with an intensity of 1 W/cm² for a period of 1 min. During this time, the temperature of the inner surface of the glove shall not exceed 44°C (111°F). Since the gloves involved in the fire were not reflective, one can assume a maximum emissivity of the surface of the glove of approximately 0.95. Calculation of the radiant interchange between the heating surface and the glove then yields a maximum surface temperature of the glove of 1600 R (1140°F). Since the specification permits this test to run for only 1 min, it is likely that the actual surface temperature of the glove would be somewhat less than this value, which would produce temperatures on the inside less than those produced by the Conductive Heat Resistance Test.

From this analysis, it is evident that National Fire Protection Association Standard 1973, Gloves for Structural Fire Fighters, does not require protective gloves to be exposed to realistic thermal exposure conditions. The tests fail to create requirements which would protect fire fighters when they are exposed to the thermal conditions in a burning structure

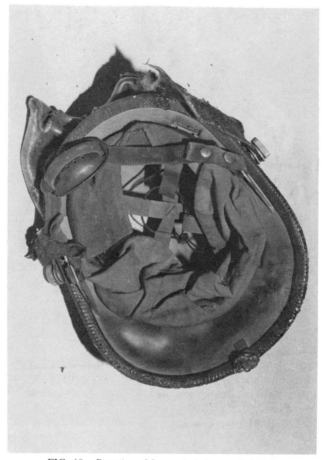

FIG. 10—*Remains of fireman's helmet and shield.*

or building. The 1985 edition of the same Standard does not contain any relevant modifications or improvements in this regard.

Stoll and Chianta [2] document the human tissue tolerance to second-degree burns. Since, as observed in our case study, most of the thermal damage to the fire fighter's hands could have been produced in 7 s, if we further assume that melting of the plastic face shield occurred in 5 s, then the burns to the fire fighter's face would have occurred in less than 2 s. For example, at a temperature of 1092°C (2000°F), corresponding to the temperature of a burning gas cloud, assuming a convective coefficient of 0.5, the absorbed energy in cal/ cm^2/s would have approached 4.0. The corresponding tolerance time would be approximately 0.2 s; that is, in 0.2 s the fireman's face would have been subjected to second-degree burns.

The 1981 NFPA Standard "Self-Contained Breathing Apparatus for Fire Fighters, covers the requirements for SCBA. However, the standard does not provide requirements for testing the apparatus, nor does it cover the requirements for performance at for elevated temperatures. OSHA's Standard 1910.134 on respiratory protection and the ANSI Standards Z-88.2-1969, Practices for Respiratory Protection, similarly do not discuss the requirements of an SCBA for elevated temperatures application.

FIG. 11—*Top view of remains of fireman's helmet and shield.*

Regardless of the deficiencies in the aforementioned standards, it is essential that manufacturers of breathing apparatus test these devices under realistic thermal-exposure conditions and that they make fire fighters aware of the equipment's limitation through suitable warnings and instructions.

Section 1-3.5 titled "Protective Clothing for Structural Firefighting" in the NFPA 1971 Standard relates to

those garments which are worn by fire fighters in the course of performing firefighting operations in buildings. The assembled garment consists of an outer shell, vapor barrier, and lining.

Section 1-4.6 states that

Materials used in garment construction shall not shrink more than 10 percent under heat exposures of 500°F (260°C) in a forced air oven for a period of five minutes.

The test, described in the same standard as the outer shell, does not predict the performance of this protective item in thermal environments of a burning building. In fact, Section 2-2.1.3 states

The outer shell material shall not char, separate, or melt when placed in a forced air laboratory oven at a temperature of 500°F (260°C) for a period of five minutes.

Note that the same temperature exposure as for clothing is permitted for this critical protective shell, with the performance criterion merely changed from shrinkage to charring or melting. The fact is that the maximum temperature specified is unrealistically low.

On 11 June 1986, ANSI and the NFPA revised the 1971 Standard on Protective Clothing for Structural Fire Fighting. The 1986 edition includes more performance requirements. The test methods describing the Thermal Protective Performance Test in Section 5-1, the Thermal Shrinkage Resistance Test in Section 5.2, and the Heat, Char and Ignition Resistance Test

FIG. 12—*Remains of fireman's boots.*

in Section 5.3 demonstrate the NFPA's positive efforts to improve the preexisting standards with more stringent tests.

State of the Art of Technology

Holcombe and Hoschke [3] refer to literature documenting the various heat flux levels of a variety of explosions and chemical fires. A typical building flashover, or mine explosion, reaches a heat flux level of 100 kW/m². Schoppe, Welsford, and Abbott [4], demonstrated that heat-resistant materials such as PBI, Kynol, and Nomex ignite instantaneously at 80 kW/m².

Polybenzimidazole (PBI) fiber, developed by Celanese, as well as DuPont's Nomex SL Aramid fibers, offer a potential for increased protection to fire fighters, as well as the potential for a higher level of flash-fire protection. Nomex and PBI hoods are available as protective apparel for fire fighters. Nomex-lined leather gloves and Nomex/Kevlar blend sleeves are also available, leaving only the face to be protected.

Although there are many self-contained breathing systems on the market available for use in toxic environments, there are no respiratory systems available for high-temperatures exposure in burning buildings or during a flashover.

More than 20 years ago, the aerospace industry developed and used a variety of sacrificial cooling methods involving ablation of materials in the protection of various parts of reentry vehicles. This principle can be applied in conjunction with a combination of reflective materials and new fiber systems. To date, the authors could not find any protective clothing for firemen that contained adequate protective against the heat fluxes typically encountered in burning buildings and flashover exposures.

Recommendations for Improved Material Studies

Although current protective materials cannot withstand the extreme thermal conditions resulting from a flashover, there are methods which can be applied to clothing that have been successfully used in other industries. For example, in the aerospace industry there are specific parts of a missile that must be protected during exposure to hyperthermal environments. Since there is no material available that could withstand the extreme temperatures without destroying the parts to be protected, those parts are protected by materials which ablate during the short-term exposures to the extreme temperatures. The ablation process is nothing more than a sacrificial cooling during the decomposition of the insulation materials. The material exposed to the hyperthermal conditions decomposes and chars while leaving the protected part at extremely low temperatures by comparison. As a result, the protective parts remain intact for the programmed duration of exposure. This principle can be applied to materials being developed for firemen who are entering buildings under conditions where present protective materials would not be able to withstand the foreseeable temperatures of exposure.

Through the successful application of the ablative process in the aerospace industry, the construction of protective clothing for firemen exposed to extreme temperatures could eventually be developed through research. Thermodynamic analysis of ablative materials includes analysis of thermosetting, thermoplastic, and elastomeric insulation materials. Computer programs can be used to make computations for heat of formation of the ablative material and for varying weight fractions of modifiers. Further, enthalpy changes can be determined as a function of temperature for the various materials. The calculations include information of the carbon char residue left on paralysis of the polymeric materials and modifiers. The thermodynamic analysis thus would provide information on a total heat capacity and enthalpy changes of the ablative systems. Also, a comparison can be drawn between the relative amounts of energy absorption available in phase changes and chemical reactions for the ablative systems considered. Once the computer program is created for the materials under study, analysis is computed quickly and at a low cost.

Terms and expressions for material properties can be incorporated in the mathematical derivations for which quantitative values are needed in order to make equations amenable to calculation. The properties are as follows:

1. Emissivity and absorptivity of the char as a function of temperature.
2. Mass fractions of thermal degradation products at a rate at which they are generated from the transition zone as a function of temperature.
3. Heat of chemical decomposition (heat of formation) of the virgin material as a function of temperature.
4. Latent heats of fusion and vaporization of various fillers or modifiers incorporated in the virgin material, or both.
5. Porosity, density, thermal fracture stress, gas permeability, elastic modulus, and thermal conductivity of the char.
6. Thermal conductivity, heat capacity, and density of the virgin material.

Once the computer program is created, one can easily predict and verify the following properties of protective clothing and equipment:

1. Ablation rates of modified and unmodified material with or without reinforcing material.

2. Measurement of the influence of various fire gas chemistries in relationship to the removal rate of the composite material.

3. Determination of the effect of char porosity on the performance of the reinforced composite.

4. Determination of the influence of polymerization on the performance of various polymeric systems used in reinforced or modified composites.

5. Influence of modifying and substitutional polymeric systems on the performance of the basic polymer used in the composite system.

6. Influence of fabricating variables on the performance of the composite material.

7. Effect of various inorganic fillers on the decomposition kinetics of the polymeric system or composite or both.

8. Correlation between the thermal stability of polymeric systems with its structure and thermal conductivity.

9. Correlation of the sintering temperatures of various polymeric systems, with or without modifications.

10. Thermodynamic analysis of the ablative material or composite or both.

The examples described above are only a few of the many studies that could be made by the various manufacturers in order to create higher-performing structural clothing. It is up to the manufacturers of protective clothing to use state-of-the-art technology that has been successful in other areas and to apply it to developing improved protective clothing for firemen. The present standards are minimum standards which are constantly subjected to further revision and improvement as new performance techniques are created and identified and as improved protective clothing is developed.

The test methods recommended in the standards for Thermal Protective Performance Tests are satisfactory for screening purposes. However, the ultimate test for performance would be the use of mannequins wearing the actual protective equipment. These mannequins would be exposed to high-heat fluxes which would simulate the actual exposure conditions of the firemen. The resulting data would allow the manufacturers to optimize the performance of the protective clothing which, in turn, would result in the upgrading of the preexisting standards.

Conclusion

The present requirements specified in existing standards for protective clothing and gear for firemen are unrealistic. They are minimum standards and, understandably, subject to further revision and improvement as improved protective clothing is developed.

The techniques and processes used in other industries such as the aerospace industry could be applied in the development of improved protective clothing. Test mannequins covered with experimental materials and exposed to high-heat fluxes, simulating actual exposure conditions, would yield more realistic information. This information could be used to obtain material performance data which would allow manufacturers to optimize the design of protective clothing and equipment. The improved protective clothing and equipment would result in upgrading the preexisting standards for protective clothing.

There has been little improvement, over the years, in safety for self-contained respiratory systems used in high-temperature environments; the potential for completely protecting a fireman from second- and third-degree burns when he is exposed to heat fluxes and flashovers in burning buildings is limited due to the lack of protection of the fireman's face. The existing standards do not consider the exposure of respiratory systems to the high temperatures that firemen can experience.

Because of the inadequate requirements set forth by the existing standards, as discussed in this paper, including the lack of requirements for necessary warnings and instructions, firemen are given a false sense of security when using present-day protective clothing and equipment. Proper warnings and instructions would allow firemen the choice of exposing themselves to conditions where there would be little or no protection when entering a burning building.

Based upon our study and analysis, it is our conclusion that present-day protective clothing and equipment for firemen are adequate only for fighting fires from without a burning building, but not from within.

References

[1] *1982 Annual Death and Injury Survey*, International Association of Fire Fighters, Washington, DC, 1983.

[2] Stoll, A. M. and Chianta, M. A., "Method and Rating Systems for Evaluation of Thermal Protection," *Aerospace Medicine*, Vol. 40, No. 11, Nov. 1969, pp. 1232–1238.

[3] Holcombe, B. V. and Hoschke, B. N., "Do Test Methods Yield Meaningful Performance Specifications?" *Performance of Protective Clothing, ASTM STP 900*, R. L. Barker and G. C. Coletta, Eds., American Society for Testing and Materials, Philadelphia, 1986, pp. 327–339.

[4] Schoppe, M. M., Welsford, J. M., and Abbott, N. J., "Protection Offered by Lightweight Clothing Materials to the Heat of Fire" in *Performance of Protective Clothing, ASTM STP 900*, R. L. Barker and G. C. Coletta, Eds., American Society for Testing and Materials, Philadelphia, 1986, pp. 340–357.

Janice Huck[1] and Elizabeth A. McCullough[1]

Fire Fighter Turnout Clothing: Physiological and Subjective Evaluation

REFERENCE: Huck, J. and McCullough, E. A., **"Fire Fighter Turnout Clothing: Physiological and Subjective Evaluation,"** *Performance of Protective Clothing: Second Symposium, ASTM STP 989*, S. Z. Mansdorf, R. Sager, and A. P. Nielsen, Eds., American Society for Testing and Materials, Philadelphia, 1988, pp. 439–451.

ABSTRACT: This study collected physical, physiological, and subjective data for different turnout clothing and equipment systems commonly used by fire fighters in the United States. A 3 by 2 by 2 randomized block design was used to determine the effect of (1) garment design; (2) type of moisture barrier; and (3) use or nonuse of a self-contained breathing apparatus (SCBA) on the (1) resistance to dry and evaporative heat transfer provided by the clothing systems measured with a heated manikin, (2) the physiological responses of subjects (that is, rectal temperature, heat rate, weight loss, added energy expenditure, and unevaporated sweat) while performing exercise and wearing the clothing systems, and (3) the subjective responses of subjects after exercise (that is, thermal sensation and wearer acceptability of the clothing systems). Results indicated that the best type of clothing system for structural fire fighting would be either the traditional long turnout coat or the tailed coat worn over waist length pants, constructed with a polytetrafluoroethylene (PTFE) moisture barrier, and worn without a SCBA.

KEY WORDS: fire fighting, protective clothing, physiological measurements, thermal manikin, evaluation, human factors

As an occupation, fire fighting is strenuous and often hazardous. According to the International Association of Fire Fighters [1], there are almost 50 000 injuries to fire fighters in the performance of their jobs each year. Stress (including overexertion) accounts for approximately 20% of all fire fighting injuries [2].

A fire fighter usually approaches a structural fire fighting situation wearing special clothing and equipment for protection and to help him with his fire fighting responsibilities. This protective clothing typically includes a turnout coat and pants, boots, helmet, gloves, and a self-contained breathing apparatus (SCBA). The clothing and equipment system is designed to provide protection from extremes of heat and cold, as well as exposure to fire, steam, impact, toxic fumes, and water.

The heat associated with every fire situation can not only cause tissue damage, but it can also contribute to heat stress in the wearer. The extent to which the clothing affects the heat exchange between the fire fighter and the environment is a crucial parameter that should be considered when evaluating the effectiveness of fire protective clothing. The clothing weight, as well as its stiffness, thickness, and bulkiness, requires additional physical effort on the part of the fire fighter as he performs the already physically stressful tasks necessary to fight a fire. These features can increase the wearer's metabolic heat production

---

[1] Assistant professor and associate professor, Department of Clothing, Textiles, and Interior Design, Kansas State University, Manhattan, KS 66506.

during activity as well as restrict heat flow between the body and the environment. In addition, the nature and thickness of the materials reduce the permeability of the garments and, consequently, inhibit the evaporation of moisture from the body. The permeability of the clothing systems becomes increasingly important in environmental conditions where heat balance can be achieved only by the evaporation of sweat.

Studies with Life-Size Thermal Manikins

Whether or not a fire fighter will suffer from heat stress depends upon the degree to which his clothing alters convective, conductive, radiant, and evaporative heat transfer between the body surface and the environment. The resistance to heat transfer provided by a clothing system can be measured using an electrically heated manikin in a climate-controlled chamber using techniques developed by researchers at the U. S. Army Research Institute of Environmental Medicine [3]. Although insulation values and permeability index values have been reported for military clothing [3,4] and protective work clothing [5], no values could be found for clothing systems worn by fire fighters.

Studies with Human Subjects

Some research studies have evaluated existing or innovative fire protective clothing systems and equipment using human subjects [6,7]. In some studies, fire fighters' physiological responses were measured [8–11]. However, much of this research has not been analyzed statistically, providing only limited information regarding the physiologic costs associated with wearing fire protective clothing systems.

While physiological criteria are important for determination of tolerance to the environment while wearing protective clothing, the wearers' subjective evaluations of thermal comfort and the acceptability of the clothing system must also be considered. For example, a fire fighter who finds his protective clothing very restrictive and uncomfortable may be tempted to remove some of his clothing or wear it incorrectly, increasing his potential for serious injury on the fire ground. Wearers' subjective evaluations of the clothing systems can be measured and compared using different scaling techniques, including semantic differential scales [12]. Prior to this study, little research has quantified subjective evaluations of fire protective clothing systems and analyzed the data with powerful statistical methods.

Purpose

Before new designs of fire protective clothing are developed, it is desirable to systematically evaluate the types of clothing currently available on the market. Such data will provide useful information to designers and manufacturers in the development of more protective, less stressful fire protective clothing systems. Therefore, the purpose of this laboratory study was to evaluate and compare physical, physiological, and subjective data for different turnout clothing and equipment systems that are commonly worn by fire fighters.

Procedure

The research design for this study was a 3 by 2 by 2 randomized block design. Three design variations of turnout clothing were tested (Figs. 1 through 3). Design A was a traditional 89-cm (35-in.) length coat worn over waist-length pants. Design B was a waist-length coat worn over chest-high, bibbed pants. Design C was a "tailed" coat worn over waist-length pants. All designs had identical shell and thermal liner fabrications, with either a Gore-Tex[2] microporous polytetrafluoroethylene (PTFE) film or neoprene moisture barrier layer (Table 1). Finally, each design/fabric combination was tested both with and without subjects carrying—but not breathing through—an SCBA. The dependent variables for this study were: (1) manikin measurements of the resistance to dry and evaporative heat transfer provided by the clothing systems; (2) physiological responses of subjects (that is, rectal temperature, heart rate, weight loss, added energy expenditure, and unevaporated sweat) while performing exercise and wearing the clothing systems; and (3) subjective responses of subjects after exercise (that is, thermal sensation and wearer acceptability of the clothing systems).

Manikin Testing

An electrically heated copper manikin housed in an environmental chamber was used to measure the dry and evaporative heat transfer resistance of the ensembles. A proportional temperature controller was used to keep average skin temperature of the manikin at 33.3 ± 0.5°C and a variable transformer maintained the temperature of the hands and feet at 29.4 ± 0.5°C. For measuring dry thermal insulation values, the environmental conditions in the chamber were controlled as follows: air temperature = 20 ± 0.5°C and air velocity = 0.1 m/s. Relative humidity was not controlled. Temperature measurements from 16 skin thermistors and four air thermistors were recorded at specified intervals on paper tape.

The manikin was dressed in a cotton knit suit and one of the six ensembles (without SCBA) to be tested. After the system reached steady state, a test was conducted [4]. Total

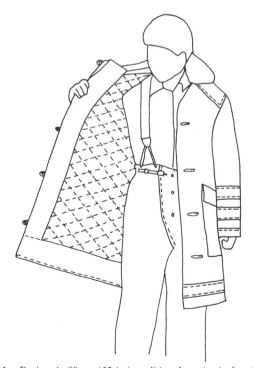

FIG. 1—*Design A: 89 cm (35 in.) traditional coat/waist-length pants.*

[2] Registered trademark of W. L. Gore and Associates, Inc.

FIG. 2—*Design B: 73.7 cm (29 in.) coat/bib pants.*

FIG. 3—*Design C: Tailed coat/waist-length pants.*

TABLE 1—*Fabric characteristics.*

Garment Layer[a]	Outer Shell	Moisture Barrier	Moisture Barrier	Thermal Liner
Fiber content	Nomex III aramid	Gore Tex PTFE laminate bonded to 100% Nomex aramid	Neoprene bonded to 100% cotton	Nomex aramid batting quilted to Nomex facecloth
Fabric count (warp/cm × filling/cm)	19 × 17	32 × 30	23 × 18	29 × 25
Weight, (g/m²)	242.29	139.30	289.46	338.74
Thickness,[b] (mm)	0.51	0.23	0.20	4.52

[a] All fabrics were plain weave constructions.
[b] Measured using a Frazier Compressometer with a 7.6-cm-diameter presser foot and 7.0 g/cm pressure.

hermal insulation (I_T), measured clo units, was calculated by

$$I_T = \frac{K(T_s - T_a)A_s}{H}$$

where

I_T = total insulation value, clo;
K = units constant = 6.45 clo W/m² °C;
T_s = mean skin temperature, °C;
T_a = air temperature, °C;
A_s = body surface area, m²; and
H = power, W.

The total thermal insulation value (I_T) of each ensemble was reported as an average of three independent replications.

The procedure for measuring the permeability index was similar to the procedure just described for measuring thermal insulation. However, the manikin's cotton "skin" was saturated with distilled water to simulate human skin saturated with sweat. The environmental conditions for the wet testing were: air temperature = 26.7 ± 0.5°C, air velocity = 0.1 m/s, and relative humidity = 50%. Again, previous research has documented specific test procedures for measuring the permeability index using a manikin [4]. The permeability index (i_m) for each ensemble was calculated as

$$i_m = \frac{H_d - 1/I(T_s - T_a)}{1/I[2.2\,(P_s - 0_aP_a)]}$$

where

i_m = permeability index, dimensionless;
H_d = power, W;
$1/I$ = watts per degree of temperature difference from dry clo test, W/°C;
T_s = mean skin temperature, °C;
T_a = air temperature, °C;
P_s = saturated vapor pressure at T_s, mmHg;
P_a = saturated vapor pressure at T_a, mmHg, and
0_a = relative humidity, %.

A psychometric table was used to determine P_s and $0_a P_a$ at the mean skin temperature and dew point temperature, respectively. The permeability index (i_m) was calculated as the average of three independent replications.

When clothing ensembles do not have the same insulation values, direct comparison of their permeability index values is not valid. However, i_m/I_T, the coefficient of evaporative heat transfer, can be used to compare different systems [3,4].

Human Subject Testing

Nine professionally trained male fire fighters participated in the human subject evaluation of the clothing systems. The subjects were about the same size as the manikin, about 1.7 to 1.8 m in height and 66 to 75 kg in weight. They were 19 to 33 years of age, and their years in the fire service ranged from 1 to 12.

Exercise/Rest Protocol—The first portion of the testing was done in an environmentally controlled test chamber measuring 3.7 by 7.3 m, with a 3 m ceiling. The chamber was outfitted with a treadmill (including a radiant heat source), a ladder ergometer, a wooden block and axe for chopping, and equipment for monitoring physiological parameters (Fig. 4).

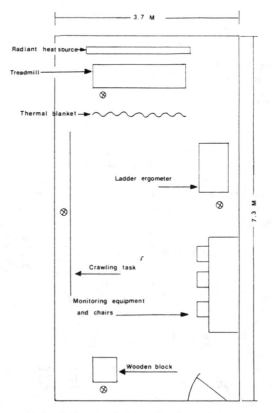

⊗ —   Location of globe and ambient temperature thermometers

FIG. 4—*Environmental chamber and equipment.*

For all tests, the following chamber conditions were used: ambient temperature = 25°C, relative humidity = 50%, and air velocity = 0.1 m/s. Wet bulb globe temperature (WBGT) averaged 21.34 ± 1.0°C for the treadmill station and 18.12 ± 1.0°C all other stations. Treadmill speed was set at 3.2 km/h. The ladder, inclined at a 70° angle, rotated at a rate of 8.5 m/min. The chopping task was done at the subjects' natural rhythmic pace (one chopping motion every 2 to 3 s). All nine fire fighters completed the exercise protocol a total of twelve times, one in each clothing/equipment configuration (that is, 108 observations).

Prior to each testing period, each subject weighed himself nude, inserted a rectal probe 10 cm into the rectum, and dressed in underwear, work pants, work shirt, socks, and boots. The subjects were then seated in the environmental chamber for 30 min prior to testing. At the end of the pretest period and prior to donning the protective clothing ensemble, heart rate and rectal temperature measurements were taken. The subjects then donned the protective clothing and equipment and, on cue from a prerecorded tape, began the exercise protocol (Table 2).

At the end of each exercise/rest protocol, subjects completed a Thermal Sensation Scale (Table 3) and a Wearer Acceptability Scale (Table 4).

After the test period, each subject quickly disrobed and placed all garments (except boots and helmet) in a plastic bag, sealed the bag, and weighed himself nude. The clothing worn by each subject was weighed before and after each test. The weight loss of each subject was calculated, and gain in clothing weight after the exercise/rest protocol was considered to be unevaporated sweat (reported as a percentage of total weight loss). Increase in heart rate was calculated by subtracting each subject's pretest heart rate from that at the end of the exercise/rest protocol; increase in rectal temperature was calculated similarly.

TABLE 2—*Exercise/rest protocol.*

Activity	Time	Variables Measured
Treadmill walking	3.0	
Rest period	2.0	rectal temperature heart rate
Ladder climbing	3.0	
Rest period	2.0	rectal temperature heart rate
Chopping	2.0	
Crawling	1.0	
Rest period	2.0	rectal temperature heart rate
Treadmill walking	3.0	
Rest period	2.0	rectal temperature heart rate
Ladder climbing	3.0	
Rest period	2.0	rectal temperature heart rate
Chopping	2.0	
Crawling	1.0	
Rest period	2.0	rectal temperature heart rate thermal sensation wearer acceptability
Total time	30.0	

TABLE 3—*Thermal sensation scale.*

Please circle the number beside the adjective that best describes how you feel: 9 = very hot 8 = hot 7 = warm 6 = slightly warm 5 = neutral 4 = slightly cool 3 = cool 2 = cold 1 = very cold

Added Energy Expenditure Determination—The second portion of the human subject testing was designed to determine the added energy expenditure associated with wearing the protective clothing system. The technique employed was based on the principle that oxygen intake directly relates to energy expenditure during exercise. To determine the added energy expenditure of wearing each of the ensembles, each subject was tested in each of the twelve protective clothing ensembles and once with the work shirt, work pants, socks, helmet, and boots only. For each test, the subject walked a treadmill at the rate of 4.8 km/h for 8 min. After the first 5 min (to allow the subject to reach a steady-state oxygen consumption level), oxygen consumption was measured once each minute for 3 min. Using open circuit spirometry, exhaled carbon dioxide and oxygen were measured, and oxygen consumption was calculated. Added energy expenditure for each test was calculated as a percent increase in oxygen consumption attributed to wearing the protective clothing and equipment over wearing the station uniform, boots, helmet, and gloves alone.

Results and Discussion

Separate analyses of variance were used to determine the effects of design, moisture barrier, and SCBA on each dependent variable. Although some of the main effects were significant, interactions between the independent variables were not. Fisher's LSD tests were used for mean separation.

Manikin Tests

Total insulation values for the ensembles ranged from 2.80 to 2.85 clo. No statistically significant differences in thermal insulation were found between the three designs or two moisture barrier fabrications tested.

The coefficients of evaporative heat transfer differed significantly as a function of garment design and moisture barrier fabrication (Table 5). The short coat and bibbed pants was found to allow the least evaporative heat transmission; the other two designs were not statistically different from one another. Also, the PTFE moisture barrier was found to allow more evaporative heat transfer than its neoprene counterpart. Even though statistical differences were found in both design and moisture barrier fabrication, the relatively low values of i_m/I_T values measured indicate that all of the ensembles were relatively impermeable.

Human Subject Testing

Table 6 indicates where significant differences in physiological measurements were found as a function of design, moisture barrier fabrication, and use of an SCBA. As shown in this

TABLE 4—*Wearer acceptability scale for fire fighter clothing.*

Place a check between each pair of adjectives at the location that best describes *how you feel*:

Comfortable	9^a / 8 / 7 / 6 : 5 : 4 / 3 / 2 / 1	Uncomfortable
Dry	9 / 8 / 7 / 6 : 5 : 4 / 3 / 2 / 1	Sweaty
Acceptable	9 / 8 / 7 / 6 : 5 : 4 / 3 / 2 / 1	Unacceptable
Fatigued	1 / 2 / 3 / 4 : 5 : 6 / 7 / 8 / 9	Not fatigued

Place a check between each pair of adjectives at the location that best describes *the clothing you are wearing*:

Heavy	1 / 2 / 3 / 4 : 5 : 6 / 7 / 8 / 9	Light
Flexible	9 / 8 / 7 / 6 : 5 : 4 / 3 / 2 / 1	Stiff
Easy to put on	9 / 8 / 7 / 6 : 5 : 4 / 3 / 2 / 1	Hard to put on
Body is adequately covered	9 / 8 / 7 / 6 : 5 : 4 / 3 / 2 / 1	Body is not adequately covered
High visibility color	9 / 8 / 7 / 6 : 5 : 4 / 3 / 2 / 1	Low visibility color
Impractical color	1 / 2 / 3 / 4 : 5 : 6 / 7 / 8 / 9	Practical color
Unattractive design	1 / 2 / 3 / 4 : 5 : 6 / 7 / 8 / 9	Attractive design
Freedom of movement of upper body	9 / 8 / 7 / 6 : 5 : 4 / 3 / 2 / 1	Restricted movement of upper body
Freedom of movement of lower body	9 / 8 / 7 / 6 : 5 : 4 / 3 / 2 / 1	Restricted movement of lower body
Freedom of movement of arms and legs	9 / 8 / 7 / 6 : 5 : 4 / 3 / 2 / 1	Restricted movement of arms and legs
Traditional design	9 / 8 / 7 / 6 : 5 : 4 / 3 / 2 / 1	Nontraditional design
Satisfactory fit	9 / 8 / 7 / 6 : 5 : 4 / 3 / 2 / 1	Unsatisfactory fit
Non-functional design	1 / 2 / 3 / 4 : 5 : 6 / 7 / 8 / 9	Functional design
Non-breathable fabrics	1 / 2 / 3 / 4 : 5 : 6 / 7 / 8 / 9	Breathable fabrics
Comfortable collar design	9 / 8 / 7 / 6 : 5 : 4 / 3 / 2 / 1	Uncomfortable collar design
Dislike	1 / 2 / 3 / 4 : 5 : 6 / 7 / 8 / 9	Like
Would feel good wearing this clothing to fight a fire	9 / 8 / 7 / 6 : 5 : 4 / 3 / 2 / 1	Would not feel good wearing this clothing to fight a fire
Loose	9 / 8 / 7 / 6 : 5 : 4 / 3 / 2 / 1	Tight
Poor protection from radiant heat	1 / 2 / 3 / 4 : 5 : 6 / 7 / 8 / 9	Good protection from radiant heat

^{a} Numbers added for reader reference only. A 9 indicates the best possible rating; a 1 indicates the poorest possible rating.

table, moisture barrier fabrication significantly affected the amount of unevaporated sweat measured in the clothing. This finding is in agreement with previous work [11]. Use of a SCBA significantly affected added energy expenditure, weight loss, increase in rectal temperature, and increase in heart rate. The increased physiological costs of wearing a SCBA are in agreement with the findings of other researchers [13,14]

Results of the analysis of variance procedure for thermal sensation indicated that design, moisture barrier fabrication, and use of an SCBA were significant (Table 7). Subjects reported higher thermal sensation votes for the short coat and bibbed pants than for either of the other two designs. Interestingly, subjects were able to discriminate between the two moisture barrier fabrications although they were not aware as to the specific variations in fabrication; the neoprene moisture barriers were perceived as hotter than their PTFE coun-

TABLE 5—*Fisher's LSD tests for coefficients of evaporative heat transfer (i_m/I_T).*

Variable	Mean Value	LSD Grouping[a]
DESIGN		
Tailed coat, waist-length pants (Design C)	0.07	A
Traditional coat, waist-length pants (Design A)	0.07	A
Short coat, bibbed pants (Design B)	0.05	B
MOISTURE BARRIER		
PTFE film	0.08	A
Neoprene	0.04	B

[a] Means with the same letter are not significantly different from one another.

terparts. As expected, subjects felt hotter when carrying an SCBA than when not carrying one.

The Wearer Acceptability Scale consisted of 23 items, presented in a semantic differential format with nine choice points for each item. A 1 was scored for the least desirable choice point and a 9 was scored for the most desirable choice point. To determine which adjective pairs on the scale were actually measures of the same performance characteristic, a factor analysis procedure was performed on the subjects' raw scores. The specific technique for this statistical procedure has been described by other researchers [12].

From the factor analysis procedure, three factors were identified: (1) Clothing Mobility, (2) Clothing Acceptance, and (3) Clothing Stress. Together, these factors accounted for approximately 34% of the variance in the data set. Using the loading factors generated by the factor analysis procedure for each of the retained adjective pairs, each subject's score was calculated. (A final score of 0 indicated minimum acceptability of the clothing system and a final score of 100 indicated total acceptability of the clothing system.)

TABLE 6—*Fisher's LSD tests for significant physiological measurements.*

Variable	Mean Value	LSD Grouping[a]
MOISTURE BARRIER		
Unevaporated sweat, %		
Neoprene	49.1	A
PTFE film	33.6	B
SCBA		
Added energy expenditure, %		
With SCBA	37	A
Without SCBA	22	B
Nude weight loss of subjects, g		
With SCBA	208.7	A
Without SCBA	176.9	B
Increase in rectal temperature, °C		
With SCBA	0.53	A
Without SCBA	0.39	B
Increase in heart rate, bpm		
With SCBA	72	A
Without SCBA	62	B

[a] Means with the same letter are not significantly different from one another.

TABLE 7—*Fisher's LSD tests for thermal sensation.*

Variable	Mean[a]	LSD Grouping[b]
DESIGN		
Short coat, bibbed pants (Design B)	7.9	A
Tailed coat, waist-length pants (Design C)	7.6	B
Traditional coat, waist-length pants (Design A)	7.5	B
MOISTURE BARRIER		
Neoprene	7.9	A
PTFE film	7.4	B
SCBA		
With SCBA	7.8	A
Without SCBA	7.5	B

[a] The thermal sensation scale ranges from 1 (very cold) to 9 (very hot).
[b] Means with the same letter are not significantly different from one another.

TABLE 8—*Fisher's LSD tests for clothing mobility, clothing acceptance, and clothing stress.*

Factor	Mean[a]	LSD Grouping[b]
CLOTHING MOBILITY, %		
Design		
Tailed coat, waist length pants (Design C)	69.4	A
Traditional coat, waist length pants (Design A)	66.0	A
Short coat, bibbed pants (Design B)	51.2	B
CLOTHING ACCEPTANCE, %		
Design		
Tailed coat, waist length pants (Design C)	64.3	A
Traditional coat, waist length pants (Design A)	59.1	A
Short coat, bibbed pants (Design B)	44.4	B
Moisture Barrier		
PTFE film	61.8	A
Neoprene	50.1	B
CLOTHING STRESS, %		
Moisture Barrier		
PTFE film	57.6	A
Neoprene	45.4	B
SCBA		
Without SCBA	58.2	A
With SCBA	44.8	B

[a] A score of 0% indicates minimum acceptability and a score of 100% indicates maximum acceptability.
[b] Means with the same letter are not significantly different from one another.

The mean scores for each of the three factors were subjected to an analysis of variance procedure. The results of the Fisher's LSD tests for separation of means are presented in Table 8.

The first factor, Clothing Mobility, was dependent upon design. Designs A and C had higher mean values for Mobility than did Design B with the short coat and bibbed pants. Designs A and C also had higher mean Wearer Acceptability scores for Clothing Acceptance than did Design B. Additionally, the ensembles having PTFE moisture barriers were perceived as more acceptable than those with neoprene moisture barriers. Clothing Stress was significant for both moisture barrier fabrication (again, the PTFE garments were perceived as more acceptable than their neoprene counterparts) and use of a SCBA. It would be expected that the ensemble would be more acceptable to the fire fighter if he is not required to wear an SCBA.

Conclusions

From the results of this research, the best possible combination of clothing and equipment was a traditional or tailed coat worn with waist-length pants, fabricated with a PTFE moisture barrier, and worn without an SCBA. Also, this study indicated that the use of a SCBA may be more important in determining the physiological costs of fire fighter protective clothing systems than variations in garment design or moisture barrier fabrication. Finally, this research demonstrated the use of a methodology that can provide important information regarding wearers' subjective evaluations of clothing systems, which may be as important as evaluating the physical characteristics of the clothing systems or physiological responses of subjects wearing the clothing systems.

Acknowledgements

The researchers would like to thank Southern Mills for donating the Nomex quilted thermal liner and the Nomex shell fabric for the ensembles, W. L. Gore & Associates for donating the Gore-Tex laminate, and Morningpride Manufacturing, Inc. for manufacturing the garments.

References

[1] "1983 Annual Death and Injury Survey," International Association of Fire Fighters, Washington, DC, 1983.
[2] Veghte, J. H., "Design Criteria for Fire Fighters Protective Clothing," Janesville Apparel, Dayton, OH, 1981.
[3] Goldman, R. F. in *Transactions,* New York Academy of Science, Series II, Vol. 36, 1974, pp. 531–544.
[4] McCullough, E. A., "An Insulation Data Base for Military Clothing," Technical Report 86-01, Institute of Environmental Research, Kansas State University, Manhattan, KS, July 1986.
[5] McCullough, E. A., Arpin, E. J., Jones, B., Konz, S. A., and Rohles, F. H., *ASHRAE Transactions,* Part I, American Association of Heating, Refrigeration and Air Conditioning Engineers, Vol. 88, 1982, pp. 1077–1094.
[6] Reischl, U. and Stransky, A., *Textile Research Journal,* Vol. 50, 1980, pp. 193–201.
[7] Abeles, F., Bruno, A., Delvecchio, R., and Himel, V., "Project Fires: Firefighters Integrated Response Equipment System," Contract No. NAS-32329, National Aeronautics and Space Administration, Washington, DC, 1978.
[8] Eissing, G., *ASHRAE Transactions,* American Association of Heating, Refrigeration and Air Conditioning Engineers, Vol. 90, 1984, pp. 1099–1115.
[9] Stransky, A. and DeLorme, H. R., *Textile Research Journal,* Vol. 52, 1982, pp. 66–73.

[10] Van deLinde, E. J. G. and Lotens, W. A., "Restraint by Clothing Upon Fire-Fighters Performance," *Proceedings,* 1983 International Conference on Protective Clothing Systems, Stockholm, Sweden, 1981.
[11] Reischl, U. and Stransky, A., *Textile Research Journal,* Vol. 50, 1980, pp. 643–647.
[12] Rohles, F. H., Konz, S. A., McCullough, E. A., and Milliken, G. A., "A Scaling Procedure for Evaluating the Comfort Characteristics of Protective Clothing" in *Proceedings,* 1983 International Conference on Protective Clothing Systems, Stockholm, Sweden, 1981.
[13] Manning, J. E. and Griggs, T. R., *Journal of Occupational Medicine,* Vol. 25, 1983, pp. 215–218.
[14] Duncan, H. W., Gardner, G. W., and Barnard, R. J., *Ergonomics,* Vol. 22, 1979, pp. 521–527.

Richard B. Gaines[1]

The Role of the U.S. Coast Guard Strike Team in Hazardous Chemical Responses

REFERENCE: Gaines, R. B., **"The Role of the U.S. Coast Guard Strike Team in Hazardous Chemical Responses,"** *Performance of Protective Clothing: Second Symposium, ASTM STP 989,* S. Z. Mansdorf, R. Sager, and A. P. Nielsen, Eds., American Society for Testing and Materials, Philadelphia, 1988, pp. 452–460.

ABSTRACT: This paper discusses the U.S. Coast Guard Atlantic Strike Team, covering the general organization, capabilities, and inventory of response equipment and instrumentation. Because the Atlantic Strike Team has become increasingly involved in federal hazardous waste site activities, the paper focuses on chemical response. In the course of such a response, team members must continually evaluate site hazards and thereby play an active role in the determination of adequate personnel protection equipment. The paper discusses procedures used at the Atlantic Strike Team for making these determinations, as well as identifying those resources available to assist in both level of protection and protective suit material selection decision logic.

KEY WORDS: hazardous chemicals, hazardous materials, response, airborne contaminants, health and safety, protective clothing, protective clothing decisions, levels of protection

Public concern about oil spills in the early 1970s demonstrated the need for a federal pollution response program. Congress addressed this concern with the Federal Water Pollution Control Act which, among other things, called for a National Contingency Plan (NCP). The NCP identifies predesignated On-Scene Coordinators (OSC) each of which is responsible for a certain geographic area. The NCP also identifies "special forces" which the OSC may utilize for assistance in a clean-up effort. These include state and local governments, the twelve federal agencies that serve on the National Response Team and Regional Response Teams, Scientific Support Coordinators, and the U.S. Coast Guard National Strike Force (NSF).

The NSF, established in 1973, consists of three elite teams, the Atlantic, Gulf, and Pacific Strike Teams, dedicated solely to pollution response. NSF personnel are specially trained in contemporary techniques needed to contain and remove oil and hazardous materials. They develop cleanup and protection strategies, monitor safety plans and contractor performance, document costs, and prepare cost recovery reports. NSF personnel are involved in an extensive training program including hazardous materials response policy, toxicological problems, flammable and toxic vapor detection and measurement, and hazardous chemical transportation. The NSF maintains an extensive oil and hazardous chemical response and cleanup equipment inventory, and provides training to other federal and industry personnel in the proper cleanup strategies and the use of appropriate equipment.

---

[1] Lieutenant, U.S. Coast Guard, USCG Research and Development Center, Avery Point, Groton, CT 06340.

Atlantic Strike Team

The Atlantic Strike Team (AST), consisting of 25 members located in North Carolina, responds to requests for assistance by an OSC in the northern areas of the United States. This includes the entire northeast coast south to and including North Carolina, westward including West Virginia, Ohio, Michigan, Indiana, Illinois, Wisconsin, and Minnesota. Always on standby, the team can have four men dispatched within 2 h, and a full team with equipment within 6 h. In an emergency, the team can utilize Coast Guard aircraft which are co-located with the AST.

In addition to those items which are basic to any hazardous materials response such as respirators, self-contained breathing apparatus, radios, and decontamination equipment, the AST has an exhaustive inventory of personal protective clothing and air monitoring instruments. Table 1 lists the material composition of suits and accessories available to AST personnel at various levels of protection. Although Teflon[2] coated Nomex[2] has shown a high resistance to permeation by a wide variety of chemicals, it has not been used in the field pending further studies. Table 2 lists the air monitoring instrumentation utilized by the AST. While the use of combustible gas detectors, oxygen-deficient atmosphere detectors, photoionization detectors, and flame ionization detectors have been 'standard' response instruments in the past, the AST has subsequently added three instruments which allow for substantially improved detection of airborne contaminants. The Foxboro Miran 1B is a portable infrared spectrometer capable of detection and quantization of some 120 chemicals programmed into its library. The Microsensor Technology Micromonitor 500 is a prototype four-column portable gas chromatograph modified for field-use by Louisiana State University under contract with the National Oceanic and Atmospheric Association (NOAA). The modification includes interfacing with a microcomputer which allows for the identification of 57 chemicals programmed into its library. An important feature is its capability for separating components of mixtures before identification. The Argonne National Laboratory/ U.S. Coast Guard Chemical Vapor Analyzer is a prototype electrochemical detection device designed specifically for Coast Guard hazardous chemical response. This instrument is currently undergoing a field evaluation at the AST. In its present configuration it can identify and quantify 25 chemicals programmed into its library, as well as detect many organic and inorganic substances. Although the current state of technology is heading towards the ability to identify then quantify a single airborne contaminant, no instrument currently available (with the possible exception of the Micromonitor 500) can do the same for mixtures of contaminants. This is a significant problem because the vast majority of responses by AST personnel involve mixtures.

TABLE 1—*Composition of protective clothing used by the AST.*

Level A[a]	Level B,C	Gloves	Boots
Teflon/Nomex	neoprene	neoprene	neoprene
Viton	Viton	Viton	latex
butyl/Viton	Saranex	nitrile	vinyl
poly(vinyl chloride)	poly(vinyl chloride)	poly(vinyl chloride)	poly(vinyl chloride)
			butyl

[a]Levels of protection are explained in Table 3.

[2] Registered trademarks of E.I. duPont de Nemours & Company.

TABLE 2—*Air monitoring instrumentation at the AST.*

Name	Detector	Linear Concentration Range, ppm	Vapors Detected
Absorption tube	chemical	0 to 500	organics, inorganic gas (compound specific)
Combustible gas	electrochemical	10^4 to 10^6	total combustible organics
HNU	photoionization	0 to 500	most organics, some inorganic
Organic vapor analyzer	flame ionization	1 to 1000	most organics
Miran 1B	IR absorbance	1 to 1000	120 organic/inorganic
Micromonitor 500	thermal conductivity gas chromatograph	(relative only)	identify 57 organic, detect most organic
Chemical parameter spectrometer	electrochemical	1 to 1000	identify 25 organic/inorganic, detect most organic

Response Decisions

Hazardous waste sites pose a multitude of health and safety concerns including chemical exposure, fire, explosion, oxygen deficiency, ionizing radiation, biological hazards, and heat stress. The hazards are a function of the site as well as the work being performed. Several factors distinguish the hazardous waste site environment from other situations involving hazardous substances. One important factor is the uncontrolled condition of the site, which can result in a severe threat to site workers and the general public. Another factor is the large variety and number of substances that may be present at a site. An individual location may contain hundreds of chemicals. Frequently, an accurate off-site characterization (i.e. records search) will yield little or no information concerning the identities of chemicals present. Furthermore, it is impossible to determine what effect the environment has on the chemicals present, or to what degree the chemicals interact. Finally, workers are subject not only to the hazards of direct exposure, but also to dangers posed by the disorderly physical environment present and the stress of working in protective clothing. The combination of all these conditions results in a working environment that is characterized by numerous and varied hazards that may pose an immediate danger to life and health and may not be obvious or identifiable.

An accurate and thorough site characterization is paramount to a safe, effective cleanup because it provides the information needed to identify site hazards and therefore worker protection. Site characterization generally involves three phases. Off-site characterization involves obtaining information about the site from records searches and perimeter searches. Neither usually involves the use of personal protective equipment. On-site surveys verify and supplement information from the off-site characterization. Because team members may be entering a largely unknown environment, suitable protective clothing is a major concern. Once the site has been determined safe for commencement of other activities, continuous site monitoring provides a constant source of information relating to site conditions. This information is critical for the proper reevaluation of protective clothing decisions.

AST members are called upon to perform functions consistent with all three phases of site characterization. These functions involve decisions as to levels of protection, described in Table 3 [1] and selection of appropriate protective clothing. Because off-site characterization generally yields limited information on chemical hazards, the initial site entry requires a conservative approach, indicated in Fig. 1 [2]. Note that the minimum level of protection

TABLE 3—*Minimum equipment for various levels of protection.*

Level A	self-contained breathing apparatus or air line
	fully encapsulating chemical resistant suit
	inner chemical-resistant gloves
	chemical-resistant boots
	chemical-resistant disposable gloves and boots
	two-way radio communications
Level B	self-contained breathing apparatus or air line
	chemical-resistant clothing (splash suit)
	inner and outer chemical resistant gloves
	chemical-resistant boots
	hard hat
	two-way radio communication
Level C	full-facepiece, air-purifying respirator
	chemical-resistant clothing (splash suit)
	inner and outer chemical resistant gloves
	chemical-resistant boots
	hard hat
	two-way radio communications
Level D	coveralls
	safety boots
	safety glasses or chemical splash goggles
	hard hat

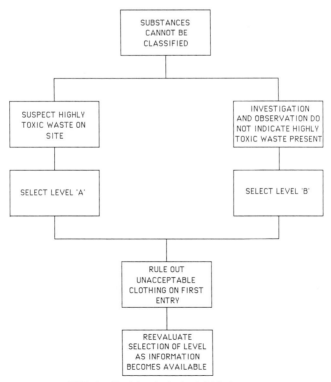

FIG. 1—*Decision logic for initial site entry.*

TABLE 4—*Some guidelines used by the AST for assessing chemical hazards.*

Hazard	Guideline	Explanation	Source for Values[a]
Inhalation of airborne contaminants	threshold limit value, time weighted average (TLV-TWA)	average concentration over an 8-hour day or 40-hour week without adverse effect	ACGIH
	threshold limit value, short-term exposure limit (TLV-STEL)	a 15-minute time-weighted average exposure that should not be exceeded	ACGIH
	immediately dangerous to life or health (IDLH)	the maximum level from which a worker could escape without any escape-impairing symptoms or any irreversible health effects	NIOSH
Dermal absorption by airborne or direct contact	designation 'skin'	substance may be readily absorbed through intact skin; direct contact should be avoided	ACGIH/ OSHA/ NIOSH
	dermal toxicity data	skin penetration, potency, and permissible concentration information	CG
Carcinogens	threshold limit values	some carcinogens have TLVs	ACGIH
	permissible exposure limits (PEL)	individual standards for some specific carcinogens	OSHA
	recommended exposure limit (REL)	recommendations regarding exposure	NIOSH

[a]Sources:

ACGIH, *Threshold Limit Values for Chemical Substances and Physical Agents in the Work Environment Adopted by ACGIH with Intended Changes for 1985–86.* American Conference of Governmental Industrial Hygienists, Cincinnati, Ohio, 1985–1986.

CG, *Policy Guidance for Response to Hazardous Chemical Releases,* U.S. Coast Guard, Washington, D.C., 1984.

NIOSH, *Pocket Guide to Chemical Hazards* (Publication 85-114), National Institute for Occupational Safety and Health, Cincinnati, Ohio, 1985.

OSHA, *Code of Federal Regulations,* Chapter 29, Part 1910.

hazards of the three chemicals and the sources from which they were taken. Acetone shows a high degree of skin penetration. Coupled with a measured concentration at the maximum permissible concentration, a significant risk of adverse skin effects exist. Because work conditions involve the potential for extensive exposure, level A is warranted. Table 6 lists those suit materials that each resource rates as "good" to "excellent" compatibility with the chemicals involved. Of those materials available in the AST inventory, butyl/Viton should provide the best protection. In cases where one suit material would not be adequate for all chemicals involved, it is common practice to wear one suit over top of another. In any case where the probability of contamination is high (for example handling waste materials), it is advisable to wear a disposable covering, such as Tyvek, over the protective suit.

Other Considerations

In addition to material compatibility there are several other factors that must be considered during clothing selection. These affect not only chemical resistance, but also the workers ability to perform the required task. The material must have sufficient durability to withstand the physical stress (tears, punctures, abrasions) of the task, it must not severely interfere with the workers' ability to perform the task, it must maintain its protective integrity and

TABLE 5—*Chemical hazard information for acetone, hexane, and toluene.*

Hazard[a]	Acetone	n-Hexane	Toluene
TLV-TWA, ppm	750	50	100
TLV-STEL, ppm	1000	510	5000
IDLH, ppm	20 000	5000	2000
Target organs	skin, respiratory system	skin, eyes, lungs	central nervous system, skin, lungs, liver, kidney
Skin penetration	high	moderate	low
Potency	moderate	moderate	slight
Category	less serious	less serious	less serious
Permissible concentration	1000 ppm per 8-h	5000 ppm per 8-h	100 ppm per 8-h

[a]Hazard categories are explained in Table 4.

flexibility under hot and cold extremes, and it must be able to be decontaminated (or disposal must be an option). Most importantly, the material must allow a particular task to be accomplished before chemical breakthrough or degradation of the suit becomes significant.

Decontamination of suits on site generally involves the physical removal of contaminants located on the *surface* of the suit. Physical methods involve dislodging, displacement, rinsing, wiping off, and evaporation. Loose contaminants such as dusts and vapors that cling to workers or become trapped in small openings, such as the weave of the fabric, can be removed with water or a liquid rinse. Adhering contaminants are much more difficult to remove. Some compounds such as glues and resins must be scraped or wiped from the clothing. Volatile compounds are normally removed by evaporation. In all cases, the physical removal is followed up with a wash and rinse process using cleaning solutions such as detergents and surfactants. The decontamination of permeated suits is very difficult, if not impossible. A heavily contaminated suit in which permeation is suspected to have occurred will be discarded as hazardous waste. In many cases, workers will wear a protective undergarment until the decision is made to dispose of the suit. Figure 3 [6] depicts the decision logic used for evaluating health and safety aspects of decontamination methods. It should be noted that often, the best protection from permeation hazards is not an effective decontamination, but adequate training on methods of contamination prevention.

TABLE 6—*Material compatibiltiy ratings of "good" to "excellent".*

Chemical	Hazardline[a]	Guidelines[b]
acetone	butyl rubber	butyl chloropolyethylene polyurethane
n-hexane	neoprene nitrile polyurethane Viton poly(vinyl alcohol)	chloropolyethylene nitrile polyurethane Viton poly(vinyl alcohol)
toluene	Viton fluorine/chloroprene	Viton polyurethane

[a]Hazardline, Occupational Health Services, Secaucus, New Jersey.
[b]*Guidelines for the Selection of Chemical Protective Clothing*, A.D. Little, Inc.

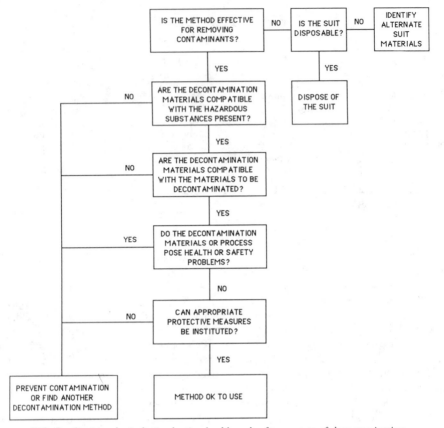

FIG. 3—*Decision logic for evaluating health and safety aspects of decontamination.*

NOTE—The opinions or assertions expressed herein are those of the author and do not necessarily represent the views of the U.S. Coast Guard.

References

[1] *Occupational Safety and Health Guidance Manual for Hazardous Waste Site Activities,* National Institute for Occupational Safety and Health, Cincinnati, Ohio, 1985, pp. 8–14.

[2] *Occupational Safety and Health Guidance Manual for Superfund Activities* (Draft), National Institute for Occupational Safety and Health, Cincinnatti, Ohio, 1984, pp. 8–25.

[3] *Occupational Safety and Health Guidance Manual for Hazardous Waste Site Activities,* National Institute for Occupational Safety and Health, Cincinnati, Ohio, 1985, pp. 6–7.

[4] *Standard On-Line Users Manual,* Occupational Health Services, Inc., Secaucus, New Jersey, p. 1.1.

[5] Schwope, W. D. et al., *Guidelines for the Selection of Chemical Protective Clothing,* 2nd Edition, Arthur D. Little, Cambridge, Massachusetts, 1985, p. 2.

[6] *Occupational Safety and Health Guidance Manual for Hazardous Waste Site Activities,* National Institute for Occupational Safety and Health, Cincinnati, Ohio, 1985, pp. 10–15.

James H. Veghte[1]

Physiologic Field Evaluation of Hazardous Materials Protective Ensembles

REFERENCE: Veghte, J. H., **"Physiologic Field Evaluation of Hazardous Materials Protective Ensembles,"** *Performance of Protective Clothing: Second Symposium, ASTM STP 989,* S. Z. Mansdorf, R. Sager, and A. P. Nielsen, Eds., American Society for Testing and Materials, Philadelphia, 1988, pp. 461–471.

ABSTRACT: Five experienced hazardous material (HAZMAT) fire fighters participated in three cities to evaluate three HAZMAT protective ensembles. The climatic conditions for these field studies were hot/dry (above 40°C), hot/wet (above 32°C) and comfortable (18 to 27°C). Each fire fighter served as his own control and wore a specific HAZMAT protective ensemble once a day for three days. Each test involved an operationally relevant 45-min work session during a total test duration of 55 min. Rectal temperatures (TR), heart rates (HR), blood pressures, respiration rates, clothed weights, and climatic parameters were recorded before and after each test. Test results show TR, HR, and sweat losses increased to 39°C, 208 beats/min, and 2.4 kg, respectively, during the hot/dry, hot/wet exposures. The wet bulb globe temperature levels for the hot/dry conditions exceed the National Institute for Occupational Safety and Health recommended limiting criteria and were marginal for the hot/wet tests. Physiologic parameters measured during comfortable conditions were lower than those measured during the hot/dry or hot/wet conditions. Differences in suit design were clearly reflected in the measured physiologic parameters and the effort required to perform work. Suggested suit modifications are discussed to reduce clothing encumberance and enhance work efficiency.

KEY WORDS: hazardous material, protective clothing, fire fighters' clothing

With increasing frequency during the past 20 years, the fire service has been called upon to respond to and alleviate toxic chemical spills, toxic fumes, and fires involving toxic chemicals. These incidents have become more critical over the past ten years as public awareness has increased to the potential seriousness of the problem. This increased responsibility for fire service personnel has thrust them into a new area that requires different protective clothing than the standard "turnout" clothing. It has also made necessary new procedures for handling these incidents. Although Federal and professional guidelines for handling hazardous material (HAZMAT) incidents are being developed, few studies have addressed protective clothing requirements [1–3]. The obvious problem of chemical permeation through protective clothing has been the first to be addressed. Based upon these chemical data and other criteria, Noll [4] and others developed classes of HAZMAT incidents that require different levels of clothing protection. But the physiologic impact of wearing the impermeable HAZMAT clothing during operationally relevant conditions has seldom been addressed by the fire service. However, pertinent data are available from studies describing the physiologic response of exercising people exposed to temperature extremes dressed in impermeable clothing. The results of some of these studies are described below.

[1] President, Biotherm, Inc., 3045 Rodenbeck Dr., Beavercreek, OH 45432.

Impermeable clothing (chemical resistant) can be defined as clothing which prevents transfer of water and water vapor. Because these materials block water vapor transfer in hot weather, evaporative cooling of body sweat is reduced. Since evaporative cooling is the major physiological protection against overheating, impermeable clothing such as chemical-resistant clothing can present a serious limitation to work in high temperatures. In an early study, Craig [5] exposed exercising men, who were completely covered in an impermeable suit, to a temperature of 27°C. These subjects had a physiological tolerance limit of approximately 30 min. In another study, Darling et al. [6] studied a large number of men marching at 5 km/h (3 mph) while clothed in decontamination suits. The air temperatures ranged from 21° to 29°C. The tolerance time at 21°C was found to be about 100 min. This tolerance time steadily decreased as the temperature increased until at 29°C their tolerance time was only 25 min. Griffin et al. [7] found that subjects dressed in heavy insulation (4 clo) and an exposure suit (impermeable) collapsed in 90 min at temperatures of 32°C. Without the exposure suit, subjects' collapse was delayed until 150 min. In another series of tests using heavy clothing (3.3 clo), Hall [8] found that putting a light, impermeable exposure suit over the clothing doubled the sweating rate and resulted in high skin temperatures. In a report by Robinson et al. [9], a comparison was made among several types of vapor-permeable and impermeable exposure suits. Test temperatures were 27 and 38°C with varying humidities, and the men were exercising either constantly or intermittently. Discomfort was produced wearing all the suits with the criteria being a high skin temperature (35°C or above) and unusual moisture retention in the clothing due to the profuse sweating of the subjects. In one of the few relevant field studies, Smolander et al. [10], determined the heart rates (HR), rectal temperatures (TR), and metabolic levels of fire fighters working for 37 min in a gas protective suit during cold conditions (2.0°C). Their findings showed a HR increase up to 148 beats/min, TR rise of 0.8°C, and sweat loss of 300 g.

In a report by White and Hodous [11], subjects wearing chemical protective clothing exercised on a treadmill at various work levels in a thermally neutral environment (dry bulb 22.6°C, wet bulb 17°C). Their results showed that even at low work intensity (4 METS) tolerance time was limited to 73 min. Tolerance time was defined as 90% of maximal HR, a TR of 39.0°C, or the subject's inability to proceed. At high work levels (7.7 METS), tolerance time working in this clothing was 13 min.

With the tolerance time variously defined in the lower temperature ranges (21 to 32°C) and nonexistent at the higher temperatures (38 to 71°C), it is important to conduct field studies with HAZMAT clothing to quantify the physiologic responses of the fire fighter. This information can then be used to assess the degree of physiologic strain imposed on the fire fighter by his clothing and his work. Clothing design also influences physiologic responses by affecting the level of work involved in performing specific tasks.

Methods

Fifteen experienced fire service HAZMAT trained personnel in designated fire departments across the country participated in this study. These test volunteers were in good physical condition. Field tests were conducted with five fire fighters at three separate geographical locations. The cities that participated in this study were Phoenix, Arizona (hot/dry), Beaumont, Texas (hot/wet), and Memphis, Tennessee (comfortable). The climatic conditions for these three tests were comfortable conditions 18 to 27°C, hot/dry (above 40°C), and hot/wet (above 32°C).

Each fire fighter served as his own control and wore a specific HAZMAT protective ensemble once a day for three days. The type of HAZMAT suit worn on any given day was randomized as much as possible with respect to time of day and which of the three test days it was worn. Each test involved one work session (Table 1). Three different Level A HAZ-MAT clothing ensembles were tested: the Challenge (CHAL) prototype furnished by the U.S. Coast Guard and two commercial ensembles (MSA Chempruf II [MSA] and Trellechem Super Extra [TREL]). Additional tests involved the Sta-Safe (STASAF) and Sijal (SIJAL) suits which were tested by only one subject. The Sijal suit is considered a Level B suit with the self-contained breathing apparatus (SCBA) worn outside the clothing. The STASAF suit is a Level A or totally encapsulated suit. The purpose of evaluating the STASAF suit was to determine the physiologic impact of using a 76-m tethered airline versus a non-tethered airline source to the SCBA.

The experimental procedure was to weigh the person nude and instrument him with a rectal temperature (TR) probe inserted 10 cm into the rectum. Dressing was accomplished in a cool environment to preclude sweating before the test started. After each individual clothing item was weighed dry, the totally clothed subject was weighed. Weights were recorded either on a HeathKit digi scale, Model CD 1186 with an accuracy of 0.04 kg, or on an FWC digital scale, Model DWM IV with an accuracy of 0.009 kg. Each fire fighter was dressed in his undergarments, station uniform, and street shoes. In addition, each wore a HAZMAT protective ensemble, and an MSA 4500 SCBA. The fire fighter used his own personal mask. The total average weight of the entire protective equipment system was approximately 23 kg. The combined weight carried by the fire fighter while carrying the water-filled buckets was 45 kg. TR, HR, blood pressure (BP), and respiration rates (RR) were recorded before and after each test. Blood pressure was recorded either with a Norelco Digital unit, Model HC3030, or manually (blood pressure cuff) by an experienced paramedic fire fighter. Heart rates were obtained with a 1-2-3 Heart Rate monitor. Respiration rates were counted manually. Two inside HAZMAT garment surface temperatures were measured. One site was immediately above the visor, while the second was immediately below the visor. Rectal and suit temperatures were measured with a YSI Model 46 TU telethermometer, using Series 400 thermistor probes. The ambient temperature (TA), ambient humidity (RH), globe temperature (GT), wet bulb globe temperatures (WBGT), and wind velocity (WV) were recorded with a Wibget Heat Stress Monitor Model RSS-212 to describe the physical environment. The fire fighter then proceeded with the prescribed work task.

A realistic HAZMAT work task was selected and used in each test city. The duration of the entire test was 55 min with only 45 min involving active exercise. The event/time frame in Table 1 was followed by each of the 15 subjects in each test city.

TABLE 1—*Work/time sequence.*

Time Frame, min	Events
0 to 5	walking around outside in the sun to simulate prior body heating prior to entering incident site
5 to 10	team briefing
10 to 20	suit up in HAZMAT clothing, on air at time 20 min
20 to 30	walk 152 m to the spill area
30 to 50	walk back 152 m and carry two pails, each filled with 11 kg of water. Repeat this task three times walking between the barrels and the starting point. This activity simulates construction of a moat with absorbent material around an acid spill. Taking the sealing ring off overpack
50 to 55	walk back 152 m to the decontamination site

The physiologic tolerance criterion for this study was a TR of 39°C or a HR of 180 for 3 min. Termination of any test was mandatory upon the request of: the fire fighter, the fire fighter site supervisor, or upon the discretion of the test monitor based on subjective or the aforementioned physiologic criteria. For each test a paramedic team stood by with an ambulance in the event of a medical emergency.

Results

The physiologic responses of fire fighters clearly show the impact of the environmental parameters. Figure 1 shows the effect of the three climatic conditions on the increase in heart rate from resting values while dressed in various HAZMAT protective ensembles. The heart-rate changes are plotted because the beginning control or resting values vary slightly between persons and from day to day. The average resting heart-rate values while wearing each suit and for each climatic condition are given in the table below the figure. Absolute heart-rate values can be obtained from these two data sets. Each bar of tabular data represents the average value for five fire fighters for any given city or while any given clothing was worn unless otherwise specified. These data show the effect of various suit designs on the effort required to perform the specified work.

Table 2 gives the hydration state for each of the Phoenix fire fighters during the course

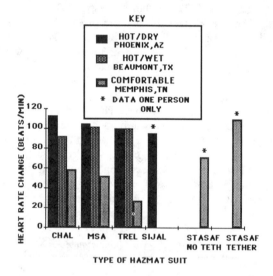

AVERAGE RESTING HEART RATES(BEATS/MIN).

CLOTHING/CITY	CHAL	MSA	TREL	SIJAL	STASAF NO TETH	STASAF TETHER
PHOENIX	64	62	60	51*		
BEAUMONT	63	63	70			
MEMPHIS	69	65	70		61*	57*

FIG. 1—*Comparison of climatic conditions and average heart rates (beats/min).*

TABLE 2—*Individual hydration status during hot/dry conditions (beginning nude weights in kg).*

Fire Fighter, Phoenix	Day 1	Day 2	Day 3	Day 4	Weight Change
1	73.96	73.50	73.18	...	−0.77
2	83.04	81.72	82.54	81.58	−1.45
3	88.12	89.62	89.84	...	+1.82
4	83.76	84.76	84.99	...	+1.23
5	71.19	71.28	71.14	...	−0.05

of the hot/dry study. These weight changes were calculated over the entire study. The data for sweat loss and body temperatures changes are depicted in Figs. 2 and 3. In Fig. 2, the impact of the environmental parameters on sweat production and degree of difficulty in working in the various suits is readily apparent. Similarly, in Fig. 3, the rise in rectal temperatures while working for only short periods of time in various climatic conditions illustrates the severity of the strain placed on the fire fighter.

Table 3 gives the individual physiologic changes for the single fire fighter evaluating the STASAF suit in Memphis. This comparison provides a clear picture of the workload imposed by dragging a tethered air line system when compared with the other nontethered suits tested. Table 4 summarizes the climatic conditions measured during each test. Each data point represents an average value for five tests unless otherwise specified.

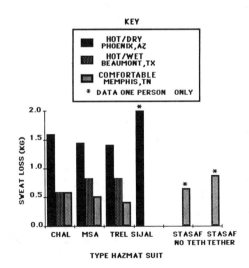

FIG. 2—*Comparison of climatic conditions and average sweat loss (kg).*

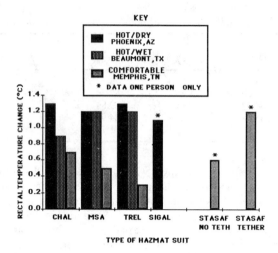

FIG. 3—*Comparison of climatic conditions and average rectal temperatures (°C).*

Discussion

In a totally encapsulated clothing system, the outside surrounding air envelope can play a significant role in the physiologic state of the fire fighter. The impact of thermal transmission from the external environment through the outer clothing shell can be readily seen when the physiological responses are compared to the dry bulb and globe temperatures (reflection of mean radiant temperature) levels. Inside suit temperatures in hot/dry and hot/wet conditions exceeded ambient levels because of the high radiant temperatures (GT). These suit temperatures exceeded 44°C on several occasions, and the fire fighters in hot/dry conditions reported that the inside of the suit was hot to the touch, making it uncomfortable. One person reported "the top of his head was burning" and he had to hold the suit hood off his

TABLE 3—*Physiologic data for fire fighter evaluation of Sta-Safe Suit.*

Suit	Sweat Loss, kg	Rectal, °C	Change Heart Rate, beats/min	Dry Bulb, °C
CHAL	0.78	0.6	98	26.7
TREL	0.52	0.6	62	23.3
MSA	0.47	0.6	50	20.0
STASAF, no tether	0.64	0.6	71	25.0
STASAF, tether	0.88	1.2	110	27.2

TABLE 4—*Summary of climatic conditions.*

Suit	Environmental Parameters					
	Dry Bulb, °C	Wet Bulb, °C	Humidity, %RH	Globe Temperature, °C	WBGT, °C	Wind[a] Yel. (m/min)
HOT/DRY						
1. CHAL	42.2	27.4	32	56.3	34.8	108
2. MSA	40.2	28.2	40	53.1	33.9	92
3. TREL	41.2	27.3	35	52.8	34.4	113
4. SIJAL[b]	39.4	25.6	36	52.8	32.5	161
Average	40.8	27.1	36	53.8	33.9	119
HOT/WET						
1. CHAL	30.1	26.2	75	35.9	28.5	108
2. MSA	32.0	26.3	65	40.0	29.6	74
3. TREL	34.0	26.1	55	41.0	29.8	113
Average	32.0	26.2	65	39.0	29.3	98
COMFORTABLE						
1. CHAL	22.8	16.3	51	30.3	19.8	185
2. MSA	21.1	15.7	58	28.6	18.8	230
3. TREL	21.1	15.7	58	29.4	19.0	150
4. STASAF,[b] no tether	25.2	16.9	43	30.5	20.2	150
5. STASAF,[b] tether	27.2	17.0	37	33.7	21.4	200
Average	23.5	16.3	49	30.5	19.8	183

[a]Conversion factor 1 m/min = 0.03728 mph.
[b]Represents only data from one fire fighter.

head to alleviate the problem. But the humidity in the external surrounding air envelope as well as wind velocity can also influence the heat-transfer rate through the outer shell material. This is so because at any given temperature, the more moisture vapor contained in the air, the greater the heat capacity of that volume of air. Therefore, more heat can be conducted into the clothing system. Also, wind affects the thickness of the boundary layer of air surrounding the clothing, and the greater the air speed, the thinner the boundary layer, which permits more heat transfer into the clothing.

The two major findings of this study are the magnitude of the physiologic impact of external environmental factors and the difference in metabolic effort to perform work while wearing the various protective clothing ensembles.

Physiologic Responses

The three important physiologic parameters for monitoring the thermal state of exercising people (HR, TR, and sweat loss) rise as the ambient temperature increases. Under comfort conditions, wearing the TREL suit resulted in the least body water loss and increase in HR and TR. The increase in these parameters while wearing other protective ensembles above that of TREL reflect to some degree the effort required to work against the clothing or the design deterrent of the other clothing. Simply put, clothing design affects the ease of mobility and therefore effort involved in performing work in the ensembles. In terms of work load, it was described as moderate by the fire fighters, and this is supported by Erb's study [12]. In his study, moderate work is defined as varying from 5.0 to 7.5 kcal/min (4.2 to 6.4 METS) or equivalent to lifting 22.5 kg (50 lb) maximum or with frequent lifting and carrying objects of 11 kg (25 lb). In another study, Webb et al. [13] measured the energy expenditure of a fire fighter carrying 22.5 kg (50 lb) of hose 152 m as 227 kcal/h.

Not too surprising since most of the fire fighters' body is covered by a moisture-impervious shell, the physiologic responses of the one person wearing a Level B protective garment (SIJAL) were similar to that while wearing Level A clothing. The highest individual heart rates were measured during the hot/dry and hot/wet tests with respective values of 190 and 208 beats/min. Similarly, maximal sweat losses for these respective conditions were 2.4 and 1.2 kg. These weight losses are very high considering the short work times, although Kuno [14] has reported higher levels of 4.0 kg sweat loss during very stressful work in the heat. Rectal temperatures rose rapidly and were 1.0°C over resting levels. The maximum individual rectal temperatures recorded at the end of the hot/dry test was 38.5°C and 39.0°C for the hot/wet tests. An overshoot of rectal temperature occurs even when the person is undressed as rapidly as possible and seated at rest in an air-conditioned or cool environment. Rectal temperatures were measured 30 min after the end of each test and in most cases had lowered only a few tenths of a degree Celsius from the end of test levels. Therefore, a person should be closely monitored after HAZMAT incidents in hot/dry or wet environments to prevent possible heat collapse. Gatorade or water or both were given to all subjects upon completion of each test to replenish the water and mineral loss as rapidly as possible. Even with this regimen, three fire fighters showed an overall water deficit at the end of the hot/dry tests (Table 2). Because of the severity of the hot/dry conditions, some subjects complained of nausea or dizziness toward the end of the test cycle.

Although all subjects completed the entire test protocol, at the end of the test period exposed to hot/wet conditions one subject was unable to continue walking to the dressing room. Also, during the hot/dry tests, two subjects were barely able physically to complete the test protocol. Therefore, the test monitors felt that these hot/dry or wet test conditions approached the tolerance limits of the fire fighters. According to National Institute for Occupational Safety and Health (NIOSH) criteria, the hot/dry WBGT of 33.9°C exceeded the permissable exposure for work in hot environments [15]. The WBGT of 29.3°C during the hot/wet tests was marginal at best.

Goldman [16] has suggested the following significant WBGT levels based on physiologic studies for the U.S. Army: 27°C, threshold of concern; 29°C, curtailment of strenuous exercise for unseasoned persons; 31°C, curtailment of all strenuous exercise, and 31 to 32°C, if absolutely necessary, heat-acclimated persons can work for up to 6 h. These suggested limits must be lowered 3 to 5°C or more for persons wearing impermeable clothing.

Design Considerations

These field studies as well as anthropometric measurements revealed design features of these protective ensembles which should be improved. Some features could be potentially hazardous.

CHAL—The major problem with this ensemble is the "ballooning" of the suit. This overinflation restricts movement of the arms, waist flexion, and downward vision. When subjects were working on the barrel or picking up the pails, they had to do a deep knee bend to expel the air from the suit in order to bend over sufficiently. Also, the design of the sleeves tends to push up on the upper arm when inflated so that carrying the pails is uncomfortable and difficult for some subjects. The suit pressure also tends to push the gloves off of the hands. A reduced valve "cracking" pressure is a simple remedy. Visibility through the visor is limited for short persons. At times, the breathing tube snagged on the suit on an obstruction just below the face piece. The inside color of the CHAL suit is light, which is psychologically beneficial since it is perceived by the subjects as conducting "less heat" than the suits with darker interiors.

MSA—The hood design is poor so that the hood pushes down on the inner face mask, forcing the fire fighters to keep pushing the hood up from the outside to maintain adequate vision and to avoid the uncomfortable pressure on the head and neck. The back of the hood presses against the nape of the neck, which is uncomfortable when hot. Visibility is poor because the face piece "rides" too low, making upward vision particularly bad. The hard plastic ring on the top of the boot rubs against the shin, causing blisters and making walking uncomfortable. The Dolman sleeve construction severely limits overhead arm movement. Poor boot fit results in blisters on the feet.

TREL—The hood design is poor: when the wearer looks down while working on the barrel, the suit rests on the nape of the neck, causing uncomfortable pressure. The hands could not be withdrawn from the gloves to reach the SCBA controls because of the inner wrist seal. This problem is considered a major design deficiency by the fire fighters because the likelihood of a SCBA failure is far greater than the need for a secondary chemical barrier at the wrist. Downward visibility is also poor, and the fire fighters were constantly pushing up on the visor to see. The zipper arrangement is poor, and the wearer can't get out of the suit without assistance. The boots fit poorly and rub on the shins.

SIJAL—The straps for the inner face mask became extremely hot and uncomfortable as the test progressed. To remedy this problem, a cap or head cover should be worn over the hood.

General Comments—The gloves on all suits were ruptured repeatedly while the fire fighters were handling the tools and loosening or tightening the protective ring on the overpack barrel. Overgloves must be worn to prevent the suit gloves from rupturing. Rough sizing of the protective ensemble could avoid some of the visibility or hood problem. A simple helmet/suit front pull-down strap would also enhance visibility for a wide height range of fire fighters. Visor fogging was a serious problem at all of the test sites and was particularly severe in the hot/wet tests. In two of these tests, fogging was so severe that the fire fighter could not see, and the test monitor had to lead the fire fighter by the hand in order to complete the work schedule. It was suggested by Memphis fire fighters that a thin film of Prell shampoo be applied on the inside of the HAZMAT suit visor and inner face piece, an innovation which dramatically reduced visor fogging.

Recommendations/Conclusions

1. The work load in hot/dry or hot/wet climatic conditions imposes a serious physiological strain on all fire fighters.
2. Physical conditioning and heat acclimatization is very important for HAZMAT personnel because of the level of physiological strain.
3. The rehydration status for personnel involved in HAZMAT incidents should be closely monitored in hot climates, and personnel should not be allowed to continue fire-service duties until hydration approaches normal level.
4. The impact of these data on rehab is extremely important. For example, people working a HAZMAT incident in hot weather should not be expected to fight a structural fire on the next shift. Body core temperatures can take a long time (hours) to return to resting levels. Ideally, fire fighters should have a full day to recover.
5. Preventive procedures such as periodic hosing off the HAZMAT suits should be considered to reduce the physiologic strain imposed by climatic conditions and the metabolic-generated heat load caused by the workload.

6. Design of the protective ensembles tested could be improved to enhance fire fighters' capability of performing necessary tasks and to reduce the "encumberance" of the clothing.

Acknowledgment

This study was funded by the Federal Emergency Management Agency (FEMA) under Contract EMW-85-C-2130. This contract is monitored by Mr. Tom Smith of the U.S. Fire Administration, Emittsburg, Maryland. My special thanks to Mr. Bob McCarthy of the U.S. Fire Administration; Capt. Jeff Stull of the U.S. Coast Guard; Capt. Steve Storment, Phoenix HAZMAT Section; Mr. Bob Stegall, Texaco Chemical Fort Arthur, Texas; and Capt. John Looney, Memphis F.D., Tennessee, for their assistance in successfully conducting these tests. And foremost, I wish to thank all the fire fighters and the personnel of the Phoenix, Arizona, Beaumont, Texas, and Memphis, Tennessee, Fire Departments who participated in this study and made it all possible. Additional recognition should be given "in toto" to all the members of the HAZMAT Technical Advisory Board for their professional guidance in reviewing and formulating the field testing protocol.

Mention of brand names of various protective garments does not constitute endorsement by FEMA, USFA, or the author. Brand names are mentioned for purposes of identification, to stimulate product improvement, and ultimately to increase the safety for members of the fire service wearing HAZMAT protective clothing.

References

[1] Schwope, H. D., Costas, P. P., Jackson, J. O., and Weitzman, D. L., *Guidelines for the Selection of Chemical Protective Clothing*, Vol. 1, Field Guide, Arthur D. Little for American Conference of Governmental Industrial Hygienists, Cincinnati, OH, 1983.
[2] *1984 Emergency Response Guidebook: Guidebook for Hazardous Materials Incidents*, U.S. Department of Transportation, Washington, DC, 1984.
[3] *Occupational Safety and Health Guidance Manual for Hazardous Waste Site Activities*, National Institute for Occupational Safety and Health, Occupational Safety and Health Administration, U.S. Coast Guard, and U.S. Environmental Protection Agency, Washington, DC, 1985.
[4] Noll, G., "Protective Clothing for Hazardous Material Emergencies," *Fire Engineering*, Vol. 137, No. 7, 1984, pp. 16–23.
[5] Craig, F. N., "Ventilation Requirements of an Impermeable Protective Suit," Medical Division Research Report No. 5, Chemical Corps, Medical Division, Army Chemical Center, MD, 1950.
[6] Darling, R. C., Johnson, R. E., Moreira, M., and Forbes, W. H., "Part I, Physiological Tests of Impermeable Suits. Part II, Improvement of Performance by Wetting the Outside of Impermeable Protective Suits, "Report No. 21, Harvard Fatigue Laboratory, Committee on Medical Research Office of Scientific Research and Development, Boston, MA, 1943.
[7] Griffin, D. R., Folk, G. E., and Belding, H. S., "Physiological Studies of Exposure Suits in Hot and Cold Environments," Report No. 26, Harvard Fatigue Laboratory, Office of Scientific Research and Development, Boston, MA, 1944.
[8] Hall, J. F., "Heat Stress Imposed by Permeable Versus Impermeable Clothing," WCRD TN 52-112, Wright Air Development Center, Wright Patterson Air Force Base, OH, 1952.
[9] Robinson, S., Gerking, S. D., and Newburgh, L. H., "Comfort Limits of Continuous Wear Exposure Suits in the Heat," Report No. 450, National Research Council, Division of Medical Sciences, Washington, DC, 1945.
[10] Smolander, J., Louhevaara, V., and Korhoven, O., "Physiological Strain in Work with Gas Protective Clothing at Low Ambient Temperature," *American Industrial Hygiene Association Journal*, Vol. 46, 1985, pp. 720–723.
[11] White, M. K. and Hodous, T. K., "Reduced Work Tolerance Associated with Wearing Protective Clothing and Respirators," *American Industrial Hygiene Association Journal*, Vol. 48, No. 4, 1987, pp. 304–310.
[12] Erb, D. B., "Applying Work Physiology to Occupational Medicine," *Occupational Health and Safety Bulletin*, Vol. 50, 1961, pp. 20–24.

[13] Webb, P., Annis, J., Crocker, J., and Turner, R., "Design of a One Hour Breathing Apparatus, "Report for Ohio Chemical and Surgical Equipment Co. by Webb Associates, Yellow Springs, OH, 1964.
[14] Kuno, Y., *Human Perspiration,* Thomas, Springfield, IL, 1956.
[15] *Occupational Exposures to Hot Environments,* National Institute for Occupational Safety and Health, Cincinnati, OH, 1986.
[16] Goldman, R., "Environmental Limits, their Prescription and Proscription," *International Journal of Environmental Studies,* Vol. 5, 1973, pp. 193–204.

Protection from Industrial Chemical Stressors

Full Ensemble Performance

Jeffrey O. Stull,[1] *James S. Johnson,*[2] *and Peter M. Swearengen*[2]

Hydrogen Fluoride Exposure Testing of the U.S. Coast Guard's Totally Encapsulated Chemical Response Suit

REFERENCE: Stull, J. O., Johnson, J. S., and Swearengen, P. M., **"Hydrogen Fluoride Exposure Testing of the U.S. Coast Guard's Totally Encapsulated Chemical Response Suit,"** *Performance of Protective Clothing: Second Symposium, ASTM STP 989,* S. Z. Mansdorf, R. Sager, and A. P. Nielsen, Eds., American Society for Testing and Materials, Philadelphia, 1988, pp. 475–483.

ABSTRACT: The U.S. Coast Guard Chemical Response Suit was field tested at the Department of Energy's Nevada Test Site in controlled releases of hydrogen fluoride. Two suits were placed on specially designed mannequins in two separate tests and subjected to hydrogen fluoride vapor concentrations up to 12 000 ppm for a 6-min period. The mannequins contained a pulsed breathing air supply to simulate normal operation of the suit's exhaust valves and four different hydrogen fluoride detection systems. The analytical results of the two tests indicated no penetration of hydrogen fluoride into the suit.

KEY WORDS: chemical protective suit, fluoropolymers, protective suit testing, suit integrity, hydrogen fluoride

The U.S. Coast Guard has developed a new totally encapsulated chemical-protective suit for protection of personnel during chemical spill response. This suit involves a novel fluoropolymer [tetrafluoroethylene (Teflon)]/aramid composite material which has demonstrated a high level of chemical resistance relative to existing commercial protective materials. Most of the suit's exterior components and materials have been evaluated for chemical resistance [1]. Furthermore, the overall physical integrity of the Chemical Response Suit has been assessed using several different methods [2]. However, the ability of the entire suit to maintain its chemical resistance integrity during realistic field exposure conditions has not been tested. Documented evidence from suit failures in a dimethyl amine accident at Benicia, California, demonstrate that chemical protective suit components can fail, exposing the wearer to hazardous chemicals [3].

The U.S. Department of Energy has constructed a large-scale spill test facility for liquefied gaseous fuels and other hazardous materials in the Frenchman Flat Basin on the Nevada Test Site. The Lawrence Livermore National Laboratory (LLNL) assists the Department of Energy with the operation of this facility, which provides data for public safety by studying the controlled spills of hazardous substances. In 1983, large-scale releases of ammonia and nitrogen tetroxide were carried out to measure the atmospheric dispersion of the spilled chemicals [4]. In the summer of 1986, releases of hydrogen fluoride and liquefied petroleum

[1] Senior engineer, Texas Research International, 9063 Bee Caves Road, Austin, TX 78733.
[2] Safety science group leader and principal investigator, respectively, Safety Science Group, Lawrence Livermore National Laboratory, Livermore, CA 94550.

gas of similar magnitude were conducted. Proposed future activities at the spill facility will involve chlorine and other gases.

The U.S. Coast Guard funded the Safety Science Group of LLNL to carry out a small experiment to evaluate the chemical protection of their new Chemical Response Suit in high concentrations of highly corrosive hydrogen fluoride. This evaluation was done as part of the hydrogen fluoride spill series sponsored independently by AMOCO Corp. to develop and test atmospheric dispersion models. This spill test series afforded the Coast Guard and LLNL the opportunity to determine if the new Chemical Response Suit provided protection against high vapor concentrations of hydrogen fluoride. The tests also assessed the feasibility of using high concentrations of hazardous materials to test the performance of chemical protective clothing.

Experimental

Coast Guard Chemical Response Suit

Two Coast Guard Chemical Response Suits were tested in separate hydrogen fluoride spills. The Chemical Response Suit is a totally encapsulating chemical protective suit developed to provide a high level of protection in chemical spill response. This suit is designed to fully enclose both the wearer and his or her breathing apparatus (Fig. 1). Features of this suit include a full body garment with a hood and visor, internal positive pressure operation, a gastight zipper, and integral gloves and boots. The suit uses fluoropolymer-

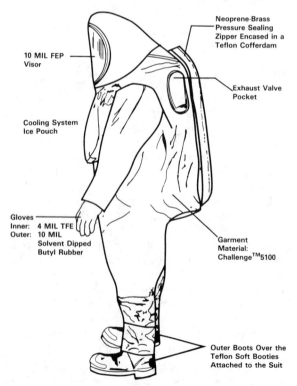

FIG. 1—*Coast Guard Chemical Response Suit.*

based materials for the garment, visor, and gloves: non-fluoropolymer components include the suit zipper and exhaust valves. Only the garment material has been tested against hydrogen fluoride in laboratory testing, and it showed no permeation in a 3-h period.[3] The suit exhaust valves are protected by an inverted pocket to reduce the likelihood of direct contact with chemical splashes. The suit closure is protected by a cofferdam arrangement with two flaps of garment material which are temporarily heat-sealed over the zipper. Positive pressure is achieved within the suit by the exhaust air from a self-contained breathing apparatus (SCBA). This exhaust air is vented through suit exhaust valves adjusted to maintain an average internal suit pressure of 3.8 mm Hg (2.0 in. water).

Suit Mannequin and Instrumentation Package

A mannequin was constructed out of wood to both support the Chemical Response Suit in an upright position and house the instrumentation package (Fig. 2). Figure 3 shows the relative position of equipment on the mannequin. The instrumentation package included both analytical devices to measure hydrogen fluoride intrusion and an air supply system to keep the suit inflated and cool during the experiment. Four separate techniques were used to measure hydrogen fluoride vapor concentrations within the suits. The reason for a fourfold analytical system was to provide redundancy that would assure data collection even if one or more of the individual analytical devices failed. Two techniques were recommended by AMOCO; these included the AMOCO Integrated Field Sampler (IFS) and the GMD Systems AUTOSTEP Model 930 Portable Monitor. Both of these devices were used by the AMOCO spill-site team to analyze hydrogen fluoride concentrations in the spill zone. Two other techniques were added by the Safety Science Group to provide additional analytical information: the Sensidyne SS2000 portable HF monitor and silica gel sorbent tubes. The characteristics of each analytical device are described below.

AMOCO Integrated Field Sampler

The AMOCO IFS is a proprietary air sampling device. The instrument sequentially pulls air through each of ten commercial Air-Sampling Field Monitors (Fisher Scientific: Gelman 4339 styrene filter holder, PN 01-038; Gelman Metricel membrane filters, Grade GN-4, PN 09-730-47). The field monitors contain membrane filters pretreated by a proprietary method specific for retention of hydrogen fluoride. The flow volume through each cassette was precalibrated with an AMOCO data logger designed for use with the IFS. The time of flow through the cassettes is adjustable on a group basis. Once a time interval is selected, every cassette in the series uses the same one. The interval used during this study was 66.6 s. Following use of the IFS, the cassettes were removed and each membrane was analyzed for hydrogen fluoride content by use of ion selective electrodes. The measured detection limit for hydrogen fluoride vapor was 0.03 ppm$_v$. The specific time hydrogen fluoride was first detected is indicated by the number of the cassette which first showed a measurable content.

GMD Systems AUTOSTEP Monitor

This system uses a colorimetric principle in an automatic incremental mode. Color-producing chemicals specific for hydrogen fluoride are impregnated into a paper tape that is stored in a removable cassette. A pump pulls a calibrated air volume sample of the test atmosphere through the tape. The tape is monitored by a light-emitting diode/photodiode

[3] C. E. Garland, private communication, June 1986.

FIG. 2—*Suit mannequin and instrumentation package.*

combination which translates color intensity into a readout. After a programmed interval, the tape is stepped forward and the next sample is taken. At the start of each measuring sequence, a reading is taken of the tape background color intensity, which is stored in memory, and then subtracted from the reading at the end of the sampling interval. The analog output from the AUTOSTEP monitor was sent to a chart recorder within the instrumentation package and also transmitted by field wire to a telemetry station. During each of the suit tests, the instrument was operated in the 0 to 30 ppm_v range. The detection limit calibrated for the specific paper tape used was nominally 3 ppm_v.

Sensidyne SS2000 Portable Toxic Monitor

This device uses an amperometric electrochemical sensor and responds to concentrations of analyte that diffuse across a semipermeable membrane. Calibration of the instrument indicated a repeatable linear response for hydrogen fluoride with a detection limit of 0.4

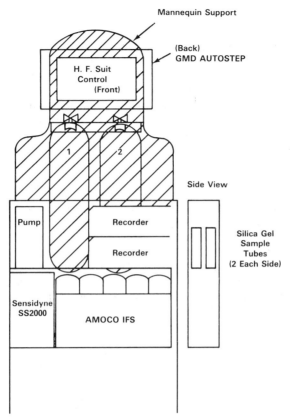

FIG. 3—*Diagram of mannequin equipment layout.*

ppm$_v$ and usable upper range to 10 ppm$_v$. Sensor response was found to be within the 10 s specified by the manufacturer. During this project, an analog output from the Sensidyne was continuously monitored by telemetry in the control room. The signal was also monitored by a strip chart recorder within the suit instrumentation package.

Silica Gel Sorbent Tubes

Four separate SKC, Inc. (Catalog No. 226-10-03) sorbent tubes, two on each side of the mannequin, were used during the tests. A Gillian sampling pump drew air through the tubes at a calibrated flow rate for each tube of 0.2 L/min. Subsequent to the collection period, the tubes were desorbed with eluting solution and analyzed for fluoride by ion chromatography. The measured instrumental detection limit was 1.0 μg. With a controlled flow period of 10 min, the hydrogen fluoride vapor concentration would have to exceed 0.6 ppm$_v$ on a continuous basis to be measured.

Suit Pressurization and Cooling System

Since these experiments were conducted under the high-temperature conditions of the desert, the suit was cooled before and after the experiment to protect the instrumentation package inside the suit. A second requirement was to simulate the operation of an SCBA

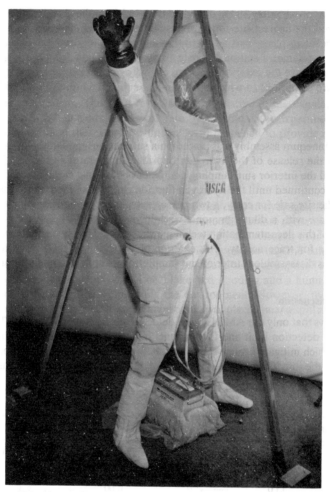

FIG. 5—*Experimental setup prior to hydrogen fluoride exposure.*

TABLE 2—*Summary of hydrogen fluoride measurements inside Chemical Response Suit.*

Detection Method	Detection Limit, ppm	Test 1 Results, ppm	Test 2 Results, ppm
AMOCO IFS	0.03	high: 0.20 low: 0.04 avg: 0.08	high: 0.10 low: 0.03[a] avg: 0.05
Sensidyne SS2000	0.2	ND[b]	ND
GMD Systems AUTOSTEP	3.0	ND	ND
Silica gel sorbent	0.6[c]	ND	ND
Protection factor	9000 with all detection methods		

[a] Low concentration below detection limit of analytical device.
[b] ND = no hydrogen fluoride detected by method.
[c] Actual detection limit is 1 μg mass by ion chromatograph; effective detection limit is 0.6 ppm based on integrated sample over sampling interval.

fluoride concentration at the first two intervals (at 1.1 and 2.2 min, respectively), showing some small quantities of acid, as did those cassettes making hydrogen fluoride measurements during the last 2.2 min of experiment. Comparatively little hydrogen fluoride was detected in the intermediate time intervals of the tests. This indicates a high "blank" (zero) value because there was no hydrogen fluoride vapor outside the suit at initial stage of the experiment. It is known that silica dust will give a false positive for hydrogen fluoride by this method. At an average wind velocity of 3 to 5 m/s, the cloud has insufficient time to move 300 m downwind to the suit location. This observation was confirmed visually for each of the two tests. The second reason against these data showing a suit leak is the observation of IFS precision: measurements appear random throughout its overall operation cycle. For these reasons we feel that the values are so close to the detection limit that they are merely a "blank" reading. If a worse-case position was taken in that the values were true, the measured maximum concentration (0.20 ppm$_v$) of hydrogen fluoride would still be well below the American Conference of Governmental Industrial Hygienists time weighted average (ACGIH TWA) level (3 ppm$_v$) or Short Term Exposure Limit (6 ppm$_v$) [5]. This indicates that the protection offered by the suit is quite high.

The other three analytical techniques showed no measurable hydrogen fluoride at any time during the two field tests. The Sensidyne instrument had the second most sensitive detection limit and, in each test, no measurable signal was generated (in the first test by telemetry, and in the second test by both telemetry and on the chart recorder). The consistency of these data supports our analysis of the IFS data as being variable within the analytical method. Our various monitoring data indicate that the suit maintained complete integrity against a very high external hydrogen fluoride vapor challenge. Furthermore, there was no visual evidence of any suit material or component degradation due to acid exposure.

Conclusions

Our experience with conducting field tests of chemical protective suits under controlled hazardous material spill conditions indicates the feasibility of performing these tests for other protective garments and chemicals. These methods appear useful for determining the performance of protective clothing under actual exposure conditions. The major limitation of these tests would appear to be the lack of dynamic suit use conditions (that is, flex of suit seams and material when worn by a suit subject). While it would be both time-consuming and costly to test a garment against several chemicals, field tests of this type could be conducted on a smaller scale and under more controlled conditions to assess the usefulness of related laboratory garment material testing. In addition, the technique offers a means to test the entire garment to highly toxic and corrosive chemicals without human exposure.

References

[1] Stull, J. O., Jamke, R. A., and Steckel, M. G., "Evaluating a New Material for Use in Totally Encapsulating Chemical Protective Suits," this publication, pp. 847–861.
[2] Johnson, J. S., Swearengen, P. M., Sackett, C., and Stull, J. O., "Laboratory Performance Testing of Totally-Encapsulating Chemical Protective Suits," this publication, pp. 535–540.
[3] Howard, H. A., "Protective Equipment Fails," *Fire Command*, March 1984, pp. 40–43.
[4] Koopman, R. P., "A Facility for Large-Scale Hazardous Gas Testing Including Recent Test Results," Technical Report UCRL-93424, Lawrence Livermore National Laboratory, Livermore, CA, Sept. 1985.
[5] "Threshold Limit Values for Chemical Substances and Physical Agents in the Work Environment with Intended Changes for 1982," American Conference of Governmental Industrial Hygienists, Cincinnati, OH, 1982.

Kenneth S. Ahmie[1]

Propellant Handler's Ensemble: A New-Generation SCAPE

REFERENCE: Ahmie, K. S., **"Propellant Handler's Ensemble: A New-Generation SCAPE,"** *Performance of Protective Clothing: Second Symposium, ASTM STP 989*, S. Z. Mansdorf, R. Sager, and A. P. Nielsen, Eds., American Society for Testing and Materials, Philadelphia, 1988, pp. 484–491.

ABSTRACT: Kennedy Space Center is currently using a self-contained atmospheric protective ensemble (SCAPE) inventory which is more than 20 years old. The present inventory is being replaced with a new-generation SCAPE—the propellant handler's ensemble (PHE). The PHE has design features which allow it to have a leakage rate less than 100 cm³/s (6.1 in.³/s) when tested at an internal positive pressure of 1494.5 N/m² (31.2 lbf/ft²). Manned testing in helium atmospheres shows the PHE to have excellent protection against inward leakage. Extensive testing was performed on the materials that would be exposed to the hazardous atmospheres. Testing with both hypergolic fuel and oxidizer proved the ensemble very capable of meeting both nominal as well as extreme atmospheric hazards.

KEY WORDS: self-contained atmospheric protective ensemble (SCAPE), propellant handler's ensemble (PHE), aerozine-50 (A-50), monomethylhydrazine (MMH), unsymmetrical dimethylhydrazine (UDMH), nitrogen tetroxide (N_2O_4), nitrogen dioxide (NO_2), environmental control system (ECS), emergency air supply (EAS), breathing air, chlorobutyl-coated Nomex fabric, adhesive, thermoplastic rubber, environmental testing, solar radiation, propellant exposure testing

The U.S. Air Force suffered an accident which resulted in injury to their personnel who were in self-contained atmospheric protective ensembles (SCAPE's). This accident was reviewed in depth by the Kennedy Space Center (KSC) and the U.S. Air Force with regard to shuttle operations at KSC and at Vandenberg Air Force Base. These efforts resulted in a decision to jointly share costs and technical expertise toward the development of a new-generation SCAPE.

Background

Martin Marietta Denver was awarded a contract in 1979 to perform an industry-wide technological search for new materials, processes, and designs that could be incorporated into the new-generation SCAPE design. This effort formed the basis for a request for proposal that was advertised in 1981 and awarded in 1981. ILC-Dover Frederica, Delaware, was awarded the contract to provide propellant handler's ensembles (PHEs).

Requirement

The contract was to provide a protective ensemble to be used by a propellant handler for protection against exposure to propellant and chemical vapors and liquids. The ensemble

[1] Section supervisor, Spacecraft and Storables, Engineering Development, National Aeronautics and Space Administration, Kennedy Space Center, FL 32899.

would consist of an outfit and an environmental control system. The outfit would be designed to protect the wearer from exposure to specified propellants: monomethyl hydrazine, unsymmetrical dimethylhydrazine (UDMH), hydrazine, and nitrogen dioxide. The environmental control system (ECS) would provide breathing air, environmental cooling, and an emergency air supply (EAS). The outfit temperature requirements were specified to be 294 K (70°F) in the head area and 299.8 K (80°F) in the torso area, while operating in 266 K (20°F) and 316 K (110°F) temperature ambient environments. The breathing air supply would have an oxygen concentration level between 19.5 and 25% by volume and maintain a carbon-dioxide concentration level between ½ and 5% for standing resting periods and treadmill exercises, respectively. The design weight limitations for the outfit, the environmental control unit (ECU), and EAS were specified as 11.3 kg (25 lb), 18.6 kg (41 lb), and 2.94 kg (6.5 lb), respectively.

PHE Design

The PHE consists of outfits, gloves, boots, and ECSs. The design of the PHE provides the capability to operate in the backpack mode, compressed-air mode, and remote-case mode. The backpack mode requires the operator to wear an ECU on his back, with the PHE outfit totally enclosing both operator and ECU. The compressed-air mode allows the operator to wear the outfit and an EAS with the breathing air supplied from an external supply. The remote-case mode allows the operator to wear the outfit in the low-profile configuration and uses an external ECU enclosed in a case and mounted on a hand cart. In this mode, the ECU supplies breathing air to the outfit through two chemically resistant, butyl rubber hoses, 3.2 cm (1¼ in.) in diameter, one a supply and the other a return. The hoses are connected to the pneumatic quick disconnects of the outfit and the remote case. The remote case is fabricated from the same fabric material used for the PHE.

Outfit Design

The outfit design includes the boots and gloves with their related disconnects, supported chlorobutyl-coated Nomex fabric with a gray wear indicator, unsupported chlorobutyl sheet stock for wear patches and seam covering, a visor with related breast plate, a closure system, pneumatic quick disconnects, an air-distribution system, a low-level liquid warning system, and a communication system.

Boot and Glove Disconnects

The PHE design incorporates a two-component, standard size, boot and glove mechanical disconnect that allows the use of equipment already available from industry. The boots are modified by thermally expanding the upper sleeve to fit over the boot disconnect; the gloves require no modifications. The disconnects are made from an anodized aluminum alloy T6061-T6. The female portion of the disconnect is installed on the suit, and the male portion is installed on the boot and glove. The female portion of the disconnect incorporates a Buna-N O-ring and is slotted to receive alignment pins located on the boot and glove portion of the disconnect. The male portion has three alignment pins located 120 deg apart to retain the boot and glove in position after installation. The boot and glove is mated to the outfit by a push and twist motion. A spring-loaded, slotted collar on the female portion of the disconnect slides over the pins and locks the boot and glove in position. The disconnects are installed using low-profile NAS 1922 type mechanical clamps.

Visor

The visor represents a major advance in SCAPE design. Plexiglas[2]-G with a Dow ARC[3] silicone hardcoating is used. The visor is formed in one plane and is fastened to the outfit with stainless steel screws and cap nuts. The visor is held in place by an anodized aluminum frame, with the fabric material forming the seal between the frame and visor. The inner and outer interface edges between the visor and frame and between the fabric and frame are sealed with RTV 118 sealant. Visors made from polycarbonate materials with and without hardcoatings were originally proposed, but testing showed the materials not to be compatible with hypergols. Hypergolic fuels, aerozine-50, monomethylhydrazine (MMH), hydrazine, and unsymmetrical dimethylhydrazine (UDMH) caused the uncoated polycarbonate to blister, crack, and become translucent. The hypergolic oxidizer, nitrogen tetroxide, caused the material to blister although the material remained optically clear. The coated polycarbonate did not perform any better when exposed to fuels; however, the coated polycarbonate performed quite well when exposed to the oxidizer. Since polycarbonate appeared to be the only material available, consideration was given to using a polycarbonate visor with half the visor having a coating and the other half remaining uncoated, thereby accommodating exposures to both fuel and oxidizer. This proved undesirable and was rejected.

Outfit Fabric

The PHE outfit is fabricated using a chlorobutyl-coated Nomex material. The fabric is 0.043 to 0.053 cm (0.017 to 0.021 in.) thick and has a gray wear indicator. The ratio of gray to white is 2:1. An independent laboratory provided the permeability and compatibility testing to ILC. Permeability results were not as expected: results showed the material to have a breakthrough at 30 min when tested with oxidizers. The maximum permeation rate measured was 0.3 g/min/m^2. This result was contrary to previous test results, which showed the material had at least 2-h breakthrough times. If these results were representative of the true performance of the chlorobutyl-coated Nomex materials, then the National Aeronautics and Space Administration (NASA) and the Air Force were in trouble since both inventories listed SCAPEs made from the same type of material. Re-testing of the material was performed by the KSC laboratories, and results showed that the material met specification requirements. A third testing laboratory repeated the test and verified that the fabric material was acceptable. The question still remained as to why the results should vary from one testing organization to another. This was the basis for The Aerospace Corporation testing, whose test program confirmed the fabric material to be acceptable. Aerospace Corporation also concluded that any test method to be used for future SCAPE applications should control the test program with respect to the test apparatus and to pressure and temperature of the test media. These controls should guarantee results that could be repeated among test agencies.

Outfit Adhesives

The PHE seam is constructed by overlapping the supported chlorobutyl fabric and sewing the fabrics together using a polyester thread. The seam is currently covered with a 2.54-cm (1 in.) unsupported chlorobutyl tape using a two-part STA liquid butyl adhesive (Part No. E-543A/B). During the extended production phase it became necessary to qualify another formulation of the same adhesive. The original formulation was revised to comply with

---

[2] Registered trademark of Rohm and Haas Company.
[3] Registered trademark of Dow Chemical Company.

Occupational Safety and Health Administration (OSHA) environmental regulations. An adhesive 3M-1711 was originally selected and used by ILC-Dover for the prototype PHE's based on the results of their compatibility test program. The testing showed the adhesive to soften, but did not degrade the overall strength of the seam. The Martin Marietta-Denver unmanned fuel and oxidizer exposure tests showed that the exposed surfaces of the adhesive were severely attacked. It was expected that repeated hypergol exposures would continue to degrade the exposed surfaces and eventually lead to seam failure. The 3M-1711 adhesive was rejected on basis of these tests.

Closure

The PHE has demonstrated an overall leakage rate less than 100 standard cm^3/s (6.1 in.^{3}/s) when tested at an internal positive pressure of 1494.5 N/m^2 (31.2 lbf/ft^2) at the time of contract acceptance. The closure system is a major contributor toward the low rate. This system is made up of two components, an inner brass alloy zipper for restraint loading and a double-track tongue-and-groove-type outer closure. The outer closure is molded from Series No. 5290 Uniroyal thermoplastic rubber compound and features a locking vee lip. The closure provides a diagonal opening from the left side of the hood to the right hip with the restraint zipper tab located at the right hip when closed. This location allows the operator to open the PHE if required.

Relief Valves

Each of the three relief valves installed in the outfit consists of a silicone diaphragm in an aluminum valve. The relief valve maintains the outfit pressure at 1.27 to 2.54 cm (½ to 1 in.) of water pressure and closes when the outfit goes into a negative pressure state. Two of the three relief valves were originally located on the lower portion of the backpack envelope. The valve design was found to be susceptible to the extremely cold temperature conditions produced by the ECU. The cold aluminum valve seating surface would condense the moisture expelled by the operator, and over a period of time ice would build up so that the diaphragm would not completely seat. The valves have been relocated to the warmer front chest area to eliminate this problem. A relief-valve cover made of Polysar chlorobutyl rubber is installed over the relief-valve assembly to complete the design. The cover is required to prevent hypergol vapor back migration into the PHE.

Water Impinging Test

A production unit PHE, donned by a SCAPE operator, was subjected to water impinging tests. Water was provided from a fire department pumper at 3.5, 7.0, 8.8, and 10.5 kg/cm^2 (50, 100, 125, and 150 psi) pressures. The water was administered over the entire PHE through a 15.24-m-long, (50 ft), 3.8-cm-diameter (1½ in.) hose using 2.54- and 0.64-cm-diameter (1 and ¼ in.) smooth bore nozzles. In the 2.54-cm (1 in.) tests the distance from nozzle to PHE was 1.5 to 2.4 m (5 to 8 ft); in the 0.64-cm (¼ in.) tests the nozzle was 0.46 to 0.91 m (18 in. to 3 ft) from the PHE. The PHE showed no water intrusion after each test.

ECS Design

The ECS consists of an environmental control unit (ECU) and a emergency air supply (EAS). Each uses a harness assembly for wearing the air supply on the back.

Environmental Control Unit

The PHE uses a portable liquid breathing air supply called an environmental control unit (ECU) that can be worn by the operator (backpack mode) or be located in a case and cart mounted (remote-case mode). The ECU is capable of providing approximately 120 to 180 min of supply air when in the high-flow or the low-flow configuration, respectively. The ECU design has a low-level liquid alarm system that lights up a visual indicator in the hood area when the supply level in the liquid air cylinder reaches the 10% remaining liquid air level. The liquid air storage cylinder is made from 304 stainless steel, is vacuum insulated, and features two swivel tube pickups. The supply swivel is always maintained in the liquid region, while the other swivel is maintained in the vapor region. The orientations are controlled by a weight and a counter-weight, respectively. The swivels were incorporated to allow the operator to work in vertical to horizontal body orientations. The ECU design incorporates a two-circuit, single-valve system which allows the operator to select a warm-air (winter) mode or a cool-air (summer) mode. The winter mode allows the vaporized air to gain more heat before it is mixed with the return air, thereby providing warmer air to the operator. The difference between the two modes is the length of tubing used between the liquid air storage cylinder and the heat exchanger. A manually operated brass valve with a bellow enclosed actuating stem controls the desired configuration. The ECU can supply air at low, medium, or high flows, depending on operator selection. Low flow provides 0.51 L/s (1.1 ft³/min), and high flow supplies a maximum of 0.71 L/s (1.5 ft³/min) of primary air. Once ECU is turned on, it cannot be turned off inadvertently; a spring-loaded lock (pin) in the flow control valve stem engages a valve stop (pin) to prevent the valve from closing. The ECU uses a venturi to recirculate PHE air at 6.136 to 7.55 L/s (13 to 16 ft³/min) for temperature conditioning before returning the air to the outfit proper.

Pneumatic Quick Disconnects

The quick disconnects (QD) are designed to close automatically if an inadvertent disconnect occurs. The quick disconnect design uses a spring-loaded sleeve which seats against the valve in the closed portion. When the hoses are mated to the outfit disconnects, the mating connection forces a sleeve back against the spring which unseats the valve and on further engagement rotates the valve to the open position.

Emergency Air Supply

The emergency air supply (EAS) is worn when the operator is in either the remote-case mode or the compressed-air mode. The EAS provides 4 min of compressed air at 40 L/min. The EAS design uses an Accurex filament wound aluminum cylinder (Part No. 6399-001) which is pressurized to 211 kg/cm² (3000 psi). The compressed air is regulated to provide air on demand to the operator. The EAS is activated at the time the PHE is donned. In the event of an emergency, the operator uses a plastic mouthpiece to receive compressed air. The mouthpiece is located on the right inside surface of the PHE hood. The EAS harness allows the cylinder to be located on the back at the waist. The supply line is routed over the operator's right shoulder.

Physiological Testing

Physiological tests of human subjects wearing the PHE were performed by both the manufacturer and the KSC Bio-Medical Department. The tests were conducted using a

Bruce tread mill protocol. The protocol required two stages of exercise: Stage I for 3 min at 2.73 km/h on a 10% grade, Stage II for 3 min at 4.20 km/h on a 12% grade. A test timeline would consist of 30 min of standing to establish a baseline, followed by Stage I, and Stage II and a 30-min standing recovery. Tests were conducted on 266, 297, 305, 316 K (20, 76, 90, and 110°F) temperature environments using the PHE in all three modes. Respiratory gas concentrations within the helmet area were monitored continuously for oxygen and carbon dioxide. A maximum carbon-dioxide concentration level of 4.2% was reached in the 316 K environment test. The oxygen concentration levels were noted to generally decrease by approximately 4%; that is, if the supply source measured 23.5%, the environment adjacent to the test subject's mouth would measure 19.5%. Temperatures inside the suit were measured at the helmet and upper torso areas. The maximum and minimum helmet and upper torso temperature measured were 309 and 278 K (97.5 and 40°F) for the extreme hot and cold test environments, respectively.

Environmental Testing

One prototype PHE was subjected to an accelerated solar radiation test, with subsequent exposures to liquid and vaporized propellants. The propellant exposures used aerozine-50 (A-50), a 50/50 blend of hydrazine and UDMH fuels, and nitrogen tetroxide (N_2O_4), an oxidizer. The test results provided the basis for determining whether or not to proceed with the production phase.

Accelerated Solar Radiation Test

The PHE was submitted to an independent testing laboratory for an accelerated solar radiation test. The test was performed in accordance with MIL-STD-810C, Method 505, Procedure II. The PHE showed signs of discoloring from white to dull yellow and of some seam separation on the leg seams.

Unmanned Hypergol Propellant Exposure Testing

A prototype PHE was tested by alternately exposing to liquid aerozine-50, liquid nitrogen tetroxide, and to nitrogen tetroxide vapors. The test fixture for all the tests was a 4.5-m³ (160 ft³) chamber that had a temperature conditioning system, a propellant delivery system, a sampling system, and a water wash system. The PHE was supplied with air from the ECU which maintained the outfit at less than 2.54 cm (1 in.) of water. The fuel test was conducted at 299.8 K (80°F). The PHE was sprayed with the fuel for 2 min and then allowed to stand in the chamber for 2 h before being washed with water. The interior air was sampled at 15-min intervals using firebrick coated with sulfuric acid and packed into glass tubes. The samples were analyzed chemically. These samples showed levels of 0.02 ppm hydrozine and 0.06 ppm UDMH were reached inside the outfit. The liquid oxidizer test was performed at 285.9 K (55°F). The PHE was sprayed for 2 min, allowed to stand for 3 min, and then washed down. The suit was allowed to stand for 1 h before it was removed and allowed to dry. In this test the PHE environment reached 0.05 ppm nitrogen dioxide. The gaseous oxidizer test was performed at 305 K (90 ± 5°F) and 80% relative humidity. The chamber nitrogen-dioxide concentration was measured at 26 700 ppm. The suit was allowed to stand in this environment for 2 h before being washed down. After an hour the suit was removed and allowed to dry. The PHE environment reached 0.15 ppm nitrogen dioxide during this test. In both oxidizer tests the interior of the PHE was sampled using a ESI Ecolyzer Propellant Vapor Detector. The PHE passed the permeation requirements in all the tests.

Manned Exposure Tests

Two manned exposure tests were conducted subsequent to the unmanned tests. The first test used helium. The test chamber helium concentration was measured at approximately 0.3%. The second test was conducted using gaseous oxidizer. The chamber nitrogen-dioxide concentration was measured between 0.25% and 0.34%. The duration of each test was 90 min. The interior of the PHE was continuously sampled through a tetrafluoroethylene (Teflon) hose using a nitrogen dioxide Ecolyzer. These tests allowed the PHE to be rated for operational use and provided the final basis for proceeding into the production phase of the contract.

Gaseous Helium Test 90% Atmosphere

This test was just recently conducted. The test was conducted in ambient temperature conditions. A PHE-suited operator was placed inside the test chamber maintained at 0.20 cm (0.08 in.) of water pressure with a greater than 90% helium atmosphere. Via the communications system, the operator was instructed to perform rigorous, exercise-type movements including the rapid deep knee bends (squats). The suit interior was sampled using a 24120 A, CEC Leak Detector. The maximum reading obtained after 65 min was equivalent to 0.3 ppm NO_2.

PHE Fire Test

This test was performed at the White Sands Test Facility (WSTF) in October 1985. The objective of the test was to determine the fire resistance of the PHE butyl rubber components under simulated conditions. A PHE was modified by placing an unsupported chlorobutyl fabric covering over the relief-valve cover located below the visor and on the relief valve located on the right side of the backpack. The covers would perform as flame barriers. Two corner caps on the backpack were also covered. The PHE was placed in a test stand that had a spray nozzle system designed to wet thoroughly the PHE. The stand also included a water deluge system. A thermocouple covered with a fabric material identified as "Refrasil" was attached to the outside of the PHE on the front. One pound (0.45 kg) of Mono-methylhydrazine (MMH) was sprayed onto the suit and manually ignited it. The burning interval was approximately 95 s before the water deluge system was activated. The initial test resulted in the covers over the relief valves catching on fire and partially burning, as did the corner cap covers. The unprotected relief valve and corner covers did not incur any damage. Areas where the fuel puddled incurred some slight surface damage. This was really evident in the area just below the right buttock. The Refrasil fabric retained and channeled the fuel and provided a prolonged burning time. This resulted in some surface scorching where it was in contact with the outfit. The PHE was not penetrated in the initial test. When the test was repeated, the areas that were initially burned degraded more, and there was penetration at the area just below the right buttock. Based on the results of these tests, it was elected not to provide the covers for the relief valves or the corner caps. All other areas on the PHE were not affected by the fire test.

Summary

In 1987, the old inventory was phased out and the new PHE was phased in. Operator comfort and safety are of primary concern since it is the operator who must stay on station for several hours under extreme environmental conditions. With this thought in mind the PHE is always being reviewed and ways for improvement are always being considered. KSC is already looking at the next new generation of SCAPE designs for future use.

Dennis J. Dudzinski[1]

Kennedy Space Center Maintenance Program for Propellant Handlers Ensembles

REFERENCE: Dudzinski, D. J., **"Kennedy Space Center Maintenance Program for Propellant Handlers Ensembles,"** *Performance of Protective Clothing; Second Symposium, ASTM STP 989,* S. Z. Mansdorf, R. Sager, and A. P. Nielsen, Eds., American Society for Testing and Materials, Philadelphia, 1988, pp. 492–500.

ABSTRACT: The Propellant Handlers Ensembles are utilized at the Kennedy Space Center to provide protection to personnel performing tasks in hostile environments related to launch activities, and represent state-of-the-art developments in protective clothing. Problems encountered in attaining a successful maintenance program have been overcome through utilization of available resources and the combined efforts of users, engineers, and maintenance technicians.

This paper provides an overview of the methods used at the Kennedy Space Center to maintain a sizable inventory of Propellant Handlers Ensembles. Discussion includes description of the maintenance facility, suit decontamination, leak testing, cleaning, visual inspection, and repair techniques.

KEY WORDS: Propellant Handlers Ensemble, protective clothing, maintenance program, repair

With the advent of highly sophisticated, reusable protective clothing, it is imperative that organizations establish an end-to-end maintenance program to ensure that the protective clothing continues to provide the original degree of protection throughout its usable life.

The Life Support Facility at the Kennedy Space Center (KSC) presently operated by EG&G Florida, Inc., performs the maintenance and overhaul functions relative to KSC's inventory of Propellant Handlers Ensembles (PHE) (Fig. 1).

The Life Support Department Maintenance Facility contains all the necessary equipment and technical expertise to successfully maintain a sizable inventory of PHEs. The facility has been functioning since 1964, and through the utilization of manufacturing data and input collected from users, engineers, and maintenance technicians, the department possesses a high level of expertise pertaining to the design and maintenance of protective clothing and equipment. The proficiency developed and available through Life Support has resulted in the department being the focal point for maintenance of all protective clothing and other Life Support equipment utilized at KSC.

In addition to the equipment inventories at KSC, the Life Support Department also provides depot repair and overhaul of the Air Force's Rocket Fuel Handler Coveralls (RFHCO), the Air Force functional counterpart of the PHE. Life Support has provided this service for approximately seven years and has processed over 1000 suits. This effort has required the development of various techniques to restore the integrity of each RFHCO

[1] Life support systems engineer, Life Support Engineering, EG&G Florida, Inc., John F. Kennedy Space Center, FL 32815.

FIG. 1—*Propellant handlers ensemble (PHE).*

and encompasses the capabilities for acceptance, inspection, leak testing, repair, fabrication, and replacement of all coverall components, final inspection, and packaging of the completed coveralls for shipment to the respective Air Force facilities.

The PHE is utilized for daily operations relative to Shuttle servicing and is available for use to the wide range of individuals necessary to accomplish these tasks. The frequency of use, ranging up to 5000 recycles per year, dictates that the maintenance program provide an efficient cycle to ensure a consistent, timely, and thorough maintenance sequence in order to guarantee adequate replenishment of equipment in the field. This is accomplished through procedures developed by careful consideration of the expected normal condition of suits received from the field for maintenance or repair or both.

PHE Maintenance

The first step in the maintenance sequence occurs in the field when the user decontaminates the equipment on-site by flushing it in a water-deluge shower (Fig. 2). The potential con-

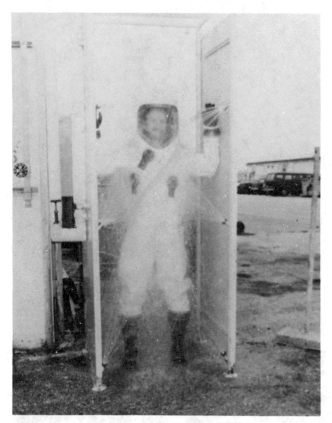

FIG. 2—*Safety shower.*

taminants are adequately removed by thorough exposure to the water shower. Toxic residuals can be detected by smell. This detectable odor is used as a gage to determine whether or not the user can be desuited by technicians at the termination of a particular operation. At this time, the equipment is visually inspected to pinpoint obvious defects to be noted in the operations log for consideration during recycle.

After termination of the operation or at the end of one shift, the used equipment is returned to the Life Support Maintenance Facility for recycle. This action encompasses leak testing, cleaning, visual inspection, repair, and final inspection.

Upon arrival of the equipment at the maintenance facility, the PHE is placed on a trolley to aid technicians in the routing of the equipment through the respective areas during maintenance. Records are then checked to review the operation for which the equipment was utilized and to determine if there were any specific anomalies noted by either the user or the technicians assisting in desuiting. Equipment having anomalies specified in advance is isolated for separate documentation and engineering evaluation.

Each ensemble is then prepared for leak testing by removal of one glove or boot and connection to a pressure test panel via the disconnect/cuff. The pressure test panels are composed of an inlet pressure gage that monitors supply breathing air pressure from a main pressure regulating panel into the ensemble being tested, a flow leakage meter that indicates component leakage, a flow selector valve, and fine flow regulating valve that controls the air supply into the test article. Panel air is routed from the flow leakage meter through tygon tubing to the component pressure test adapter inlet that interfaces with the component being tested through the respective disconnect. The air is routed from the adaptor to the component pressure gage capable of measuring internal pressure in inches of water. The ensemble is pressurized to 4.97×10^2 N/m² (2.00 in. of water pressure) and allowed to stabilize for a period of time (Fig. 3). Upon stabilization of ensemble internal pressure, readings are taken to determine if the leakage rate is acceptable. If an excessive leakage rate is detected, the ensemble then receives a physical inspection utilizing leak detection compound (soap solution) to determine the location and cause of the leak. Ensembles that fail the pressure test are isolated for repair to correct the anomaly. Ensembles passing pressure testing receive quality assurance verification, which is documented on the individual suit maintenance card. Once the ensemble has successfully completed pressure testing, the boots and gloves are removed to be evaluated separately by pressure testing at 2.98×10^3 N/m² (12.00 in. of water pressure) and submersion in water (Fig. 4). Boots and gloves found to be leaking are isolated for replacement or repair.

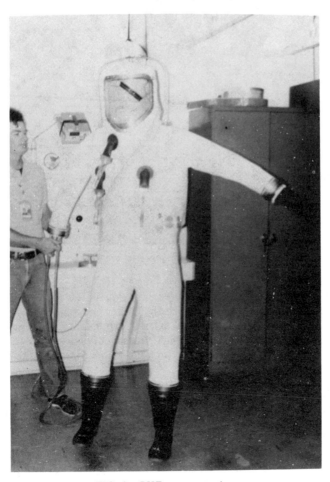

FIG. 3—*PHE pressure testing.*

FIG. 4—*Boot and glove pressure testing.*

Suits passing pressure testing are then routed through a shower system to be cleaned with soap and water (Fig. 5). The shower utilizes hot water and a mild biodegradable soap, directed under pressure to the interior and exterior extremities of each suit. The shower cycle is automatically timed to ensure thorough cleansing of each suit. Once the wash cycle is complete, the suits are then routed to the suit dryer to remove residual moisture (Fig. 6).

Boots and gloves, after passing pressure testing, are cleaned by hand utilizing mild detergent and water. During the boot/glove cleaning process, each boot/glove assembly is visually inspected to ensure that damage has not been incurred that was not indicated during pressure testing. The inspection criteria established include: inspection of the component disconnect to check for any scratches, gouges, broken or bent alignment pins, and loose or damaged clamp bands; inspection of the component (boot or glove) to check for cuts, deep fissures, lifts, and evidence of gum-like deposits that usually are the result of long exposure to high concentrations of hypergol propellants. Depending on the degree of the deposits, a

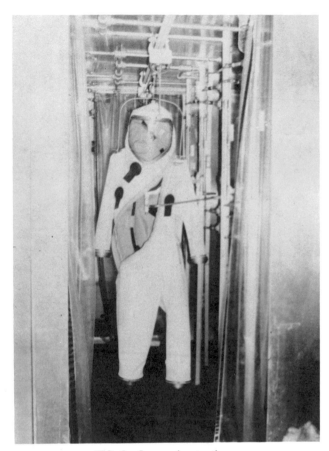

FIG. 5—*Soap and water shower.*

second wash with an alkaline detergent solution or diluted solution of clorox or sodium bicarbonate can be used to remove the sticky feel of the component and restore usability.

After the suits have been cleaned and dried, they are moved to the inspection and repair area of the Maintenance Facility where technicians perform a detailed visual inspection of the suit interior and exterior (Fig. 7). This inspection encompasses checks for any discrepancies to include: loose or broken stitching or both, fabric rips, cuts, punctures, abrasions, delaminations, unusual stains or discoloration, presence of imbedded foreign matter, insecurity of component attachment, lifts measuring 6.35 mm along a lateral edge by 3.175 mm deep into a seam or component, scratches, chips or gouges, sharp edges, burrs, improper lubrication of O-rings, improper cleanliness, any condition of the visor that may impair vision, loose nuts or bolts, improper orientation of air distribution hoses, damaged air distribution hoses, damaged relief valve diaphragms and relief valves seats, proper operation of the suit communication harness, and proper operation of the suit Environmental Control Unit (ECU) low-level warning light. Discrepancies occurring on a frequent basis indicate the need for further evaluation of the suit or work station or both to ascertain the need for a modification to the suit, the work station, or test procedure. Modifications performed on suits and ancillary equipment, indicated by chronic discrepancies or user recommendations or both, serve to promote user confidence and equipment longevity. The evaluation process

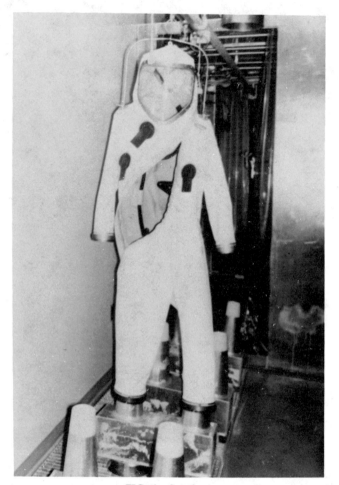

FIG. 6—*Suit dryer.*

also can pinpoint areas of concern relative to the training of individuals required to perform tasks utilizing protective clothing.

Because of the dynamics of the properties of suit fabrics and their interface with other suit components, a suit may possess deficiencies that are not readily detectable by pressure testing or visual inspection. However, these deficiencies are detectable by utilization of a simple, yet very effective technique. Suits having been worn 25 consecutive times are inspected utilizing a high-intensity fluorescent lighting method. This inspection is conducted in a darkened room with the suit torso closure assembly sealed and the visor masked so as not to allow light to escape. A tube-like light is then inserted into one extremity after another with the light being held as closely as possible to the area being inspected. This technique has proven effective for identification of thin areas in the fabric and pin holes. The thin areas, whether from interior or exterior scuffs, are encircled with a black wax marking pencil and holes are identified utilizing a red wax marking pencil. These areas are then repaired according to procedures outlined for the type of deficiency noted.

The next phase of the maintenance process is to perform repairs of anomalies discovered

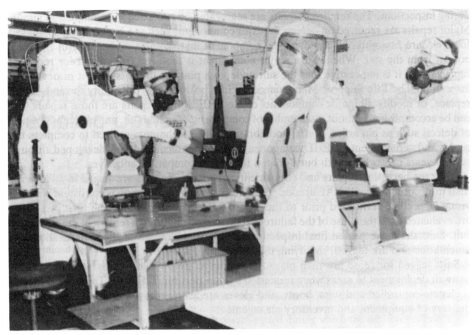

FIG. 7—*PHE repair facility.*

FIG. 8—*PHEs in storage.*

KEY WORDS: ensemble leakage testing, passive sampling, individual protection ensemble (IPE), protective clothing, clothing test method

The protection boundary of greatest concern to the soldier in a contaminated environment is that which is formed by the individual protective ensemble (IPE). This consists of a charcoal-impregnated foam-lined overgarment, gloves, boots, and a protective face mask and hood. Test methods for assessing the protection factor (ratio of ambient-to-internal concentration) afforded the soldier have traditionally focused on static testing of individual components of the ensemble. Factors such as the fitting together of various pieces of the ensemble and leakage caused by motion of the user or simply poor fit have usually gone unexamined. This has been the case because there has not been a test method which could be routinely applied to challenge the total ensemble to evaluate its protection performance. While existing methods permit accurate measurement of the capacity of the clothing materials for uptake of vapors of interest, there has not been a method to measure leakage into the ensemble through the clothing-body interface.

The objective of the work described here has been the development of a new test methodology for assessing the leakage into the individual protective ensemble interior under realistic conditions.

Technical Approach

The major components in the approach include the identification of a challenge vapor and of body motions to use in the performance of testing, identification of principal leakage pathways, and demonstration of the detection method in testing.

Simulant Selection

Isoamyl acetate (IAA) was selected for use as the challenge vapor in this testing from a list of 40 organic compounds which satisfied primary criteria. This compound exhibits sufficient volatility at ambient temperatures to provide a high challenge concentration, has a measured Tenax vapor retention volume of 189 mL/g at 35°C, which is well in excess of this test method's requirements, is detectable using a flame ionization detector (FID) at parts per billion (ppb) concentrations, does not penetrate the protective ensemble material during the test period used (see below), is readily available, easy to handle, and is suitable for use with human subjects under the conditions of the tests performed.

Body Movements Identification

The realistic evaluation of the protection which chemical protective clothing will provide requires that the clothing be stressed in the testing in a fashion which mimics the conditions of actual use. It is clear that an ensemble with closures and joinings will be more subject to leakage when the wearer is moving than when he is not. The likely activities and relevant movements of a soldier wearing an IPE were identified and compiled based on two military manuals [1,2], an existing test procedure, and a human factors handbook [3]. In developing the set of motions for use in the testing it was recognized that there are two possible causes of leakage: (1) pressure differentials induced by volume changes within the garment during joint flexation, or (2) induced stress on component interfaces promoting passive diffusion of the contaminant or an induced pressure differential or both. Each task and activity was assessed for its contribution to either cause of leakage, and appropriate motions were assigned to describe the activity. The body movements were not designed to simulate the appearance of the tasks or activities, but instead were designed to simulate the maximum stresses which would occur on the joints during the task or activity. The body movements were segregated into sets of exercises which concentrate on stressing one specific joint or garment/body interface: (1) head movements, (2) arm movements, (3) turning/bending movements, (4) sitting, (5) kneeling/squatting, (6) leg movements, (7) jumping, (8) crawling, and (9) running.

It is apparent that no two subjects perform a prescribed set of movements in exactly the same way or with the same intensity of movement. For this reason, it was decided to use the same subjects throughout the matrix of tests performed. It was also decided to establish different levels of activity by varying the number of repetitions of the movement scenario rather than by attempting to modify the intensity of its performance. The three levels of activity included in the testing were: (1) rest, which indicates that the subjects were seated throughout the exposure and performed no movements which significantly stressed the suit seals; (2) moderate activity, which required one performance of the movement scenario; and (3) high activity, which entailed three performances of the movements with intervening rest periods.

The movement scenario developed for use in this testing requires approximately 10 min for performance and does not involve very strenuous activity, but does provide stresses for the suit seals of interest. Because of these factors, this scenario may be useful as a standard to be used in other protective clothing testing projects.

Detection Method

The low concentrations of simulant vapor which are likely to occur in the interior air spaces of the IPE, and the fluctuating nature of these concentrations argue for the collection of time integrated samples for subsequent analysis. This provides for the collection of a larger mass of simulant for analysis than would be possible using real-time monitoring inside the IPE, and it damps the fluctuations in concentration which would complicate the interpretation of the protection factor data. The use of gas chromatography with flame ionization detection (GC-FID) of the vapor permits detection of collected masses as low as several nanograms. This level of sensitivity is more than adequate for our purposes here, so the detection method problem reduces to one of choosing the best sampling method which is compatible with the GC-FID.

Sampling Method Selection

Sampling within a worn IPE for intrusion of any simulant is subject to several restrictions that distinguish this operation from external or ambient sampling. Air movement within an ensemble segment is apt to be very slight, so that mixing of the simulant within the ensemble segment under motionless conditions will be primarily by diffusion. Mixing in actual use will be greatly augmented by movements of the wearer.

At any point removed from the leakage site, the concentration of simulant will be reduced by (a) dilution due to mixing with the air within the segment and (b) adsorptive removal of the simulant by the carbon-filled interior of portions of the ensemble. The requirement that the vapor simulant not penetrate the overgarment material has item (b) as a necessary consequence. Any sampler used in the air space within the ensemble, therefore, will be in competition with the internal surface of the overgarment for the intruding species. Thus, collection by the sampler must be essentially irreversible in order to avoid exchange of the simulant with the sorbent contained in the suit. Also, it is clear that a sampler's location will affect its performance, so that sampling at only one location will not yield results that can be used to characterize the concentration of simulant throughout the air space enclosed by the protective ensemble.

Force flow to a sampler may influence the measurements for two reasons. First, disturbance of the normal mixing pattern within the suit will lead to a simultaneous increase in flow across the internal surface of the suit. This may result in increased uptake of simulant by the suit interior. Second, forced flow has the potential to artificially increase in-leakage by reducing the air pressure at key points within the ensemble.

In actual usage of the protective ensemble, the relative humidity of air within the suit is expected to be very high. The sampling system therefore must be capable of operating at relative humidities in the range of 90 to 100%.

Because of the above considerations, it is unlikely that any sampling technique could be expected to quantify the leakage rate of a simulant in terms of mass of simulant entering the ensemble per unit time. Rather, it is suggested that sampling within the ensemble can yield a measure of the dosage that results from leakage.

Passive samplers encompass a range of functionality from qualitative and semiquantitative

FIG. 1—*Photograph of passive sampling device used in study.*

colorimetric indicators to quantitative collectors. In the latter category, both irreversible adsorption using activated carbons and thermally reversible adsorption using porous polymer bead sorbents are employed for the collection process. These quantitative collectors permit more detailed and accurate assessment of the dosage received as a consequence of intrusion. The porous polymer bead sorbents (for example, Tenax) are preferred because of the high relative humidities in the suits. The performance of the activated carbon collectors is adversely affected by high relative humidities [4], but passive devices using Tenax are not [5]. The primary passive sampling device (PSD) selected for use in this project is shown in Fig. 1. This is a Battelle-modified version of the U.S. Environmental Protection Agency PSD [5] which uses Tenax-GC as the collector.

The PSD sampling rates for IAA were experimentally determined over the temperature, concentration, and relative humidity ranges appropriate for this program. Vapor concentrations of IAA used ranged from 30 to 120 ppb; air temperatures were 25 and 38°C; and relative humidities were 10, 50, and 92%.

The apparent sampling rates were determined by comparing amounts of IAA recovered from the PSDs (after a sampling time of 30 min) with amounts of IAA contained in known volumes of gas sampled from the 200-L laboratory chamber, which was precharged with known amounts of IAA using a standard syringe injection technique.

The sampling rates measured at the various concentrations, temperatures, and relative humidities employed are presented in Fig. 2. The data indicate a mean sampling rate of 30.1 cc/min with a relative standard deviation of 10%, which is in excellent agreement with the calculated rate of 31.5 cc/min. Based on these measurements, we conclude that the PSD can be used reliably to measure concentrations of IAA in the ppb range, with an anticipated precision of 10%. If the calculated rate is taken as the true rate, the data indicate an overall accuracy (bias plus twice the coefficient of variation) of 23% for the sampling and analytical procedures. This accuracy is typical of this design PSD performance for vapors exhibiting similar retention volumes on Tenax-GC [6].

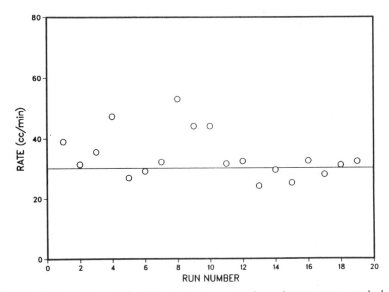

FIG. 2—*PSD sampling rates for IAA at various concentrations, air temperatures, and relative humidities.*

IPE Capacity of IAA

Among the requirements for successful performance of the leakage testing of the IPEs is the requirement that the challenge vapor not penetrate the material of the ensemble. It was clear at the outset of this subtask that IAA is adsorbed by charcoal, but it was necessary to attempt to quantify the capacity of the IPE lining in order to ensure that the experimental conditions would not lead to breakthrough of the vapor through the garments.

The experimental approach used involved preparation of a packed bed of small sections of the carbon-impregnated liner material used in the ensemble. A measured flow of air containing 10 ppm of IAA was continuously passed through the bed, with the outlet gas being monitored with an FID. This arrangement permitted direct measurement of the breakthrough curve and from this one can deduce the capacity of the sorbent for IAA at the test concentration of 10 ppm. The capacity was found to be 52.7 mg/g at 0.5% breakthrough. Using this figure, one can conservatively estimate that the ensemble would become saturated (that is, breakthrough would occur) after an exposure period of 12.6 h. Inasmuch as the actual exposure times were only one hour, this provides a safety factor of at least six in limiting the penetration of the ensemble to less than that corresponding to a protection factor of 200:1. The ensemble material breakthrough data suggest that the breakthrough process in the forced flow laboratory experiment was limited by the kinetics of the sorption process. Thus, under the purely diffusive conditions of ensemble usage, the capacity is expected to be greater than that indicated by the laboratory experiment.

As an additional check on the possibility of loading of the ensembles to the point where breakthrough of the vapor would occur, the data obtained from the mannequin were assessed at the end of the leakage testing. Breakthrough of the material would be indicated by protection factors for a given location which deteriorate with increasing exposure of a given IPE. Each of the six tests in which the mannequin was used resulted in approximately the same dosage for the ensemble. The measured protection factors for several locations were compared against the cumulative exposure. The absence of any discernible trend in these measurements indicates that the capacity of the suit for IAA was not exceeded during the tests performed. As an additional precaution, the suits worn by the human subjects were replaced after the fourth exposure of each suit.

Test Chamber Operations

The test chamber used in this study, shown in Fig. 3, is a well-mixed 17-m³ rectangular cell with interior surfaces of tetrafluoroethylene (Teflon) and aluminum. The interior dimensions of the chamber are approximately 2.5-m height by 2-m width wide by 4-m length. The chamber has viewing ports at either end and an airtight access hatch. The chamber air-handling system permits rapid flushing with filtered ambient air, or a lower flow (~100 L/min) of zero-grade air from a clean air supply system. The chamber also has numerous sampling ports to permit characterization of the test atmosphere and a heated injection manifold for introduction of the test vapor.

Prior to performance of a test the chamber was purged and its atmosphere analyzed to verify that the IAA was not present at a detectable level (>10 ppb). The subjects were then placed in the chamber and equipped with PSDs as described below. Once the chamber was closed, the injection of IAA was begun. It was found in preliminary testing that the rate of loss of IAA from the empty chamber was the same as that for an inert tracer, SF_6. This loss rate, which is a result of leakage and sample withdrawal from the chamber, was 0.033 h⁻¹. With a protective ensemble suspended in the chamber, however, the rate of loss of IAA was approximately 0.39 h⁻¹, while the loss rate of SF_6 was unaffected. Because of this high

FIG. 3—*Photograph of interior of test chamber.*

loss rate, it was necessary during testing to inject IAA continuously, following the initial slug injection. Depending upon the number of subjects in the chamber and their level of activity, different feed rates of IAA were required, and the feed rate was usually adjusted during tests in order to maintain the chamber concentration as near 10 ppm as was practicable.

The IAA concentration in the chamber atmosphere was measured every five minutes. This measurement was performed with a Varian Model 3700 GC, equipped with an FID. The signal from the FID was integrated automatically and printed within 3 min of the sampling time. The temperature and relative humidity of the chamber atmosphere were also monitored during the testing. (These measurements were collected only from considerations of the comfort of the subjects in the chamber and were not part of the varied test parameters.)

Following the 1-h exposure period, the chamber was purged with ambient air to reduce the IAA concentration prior to collection of the samplers. During this period, the chamber concentration was monitored and the chamber remained sealed until the IAA concentration was reduced to less than 100 ppb.

PSD Handling and Analysis

The PSDs were handled only while the test operators wore new vinyl gloves, and they were stored in sealed containers during all storage periods. The placement and collection of the PSDs on the subjects were performed as quickly as possible in order to minimize the PSDs exposure to air external to the protective ensemble. The PSDs were also thoroughly degassed following their analysis in order to prevent cross-contamination of the samples. The analysis of blank PSDs was used to check the success of the handling procedure in avoiding contamination of the samplers.

The analysis of the PSDs was accomplished by thermal desorption of the collected sample into a Varian 3700 GC equipped with a 50-m capillary column, FID, and electronic integrator for recording and preliminary data reduction. One noteworthy modification of the usual thermal desorption procedure which was necessitated by the water vapor collected by the PSDs in their humid sampling environment was the preliminary flushing of the unheated samples to drive off the water vapor. This technique successfully prevented freezing of water in the sample collection loop without loss of the collected IAA.

Typical chromatograms for PSD analyses are shown in Fig. 4, from which the sensitivity of this analytical technique can be inferred. The sensitivity which is achievable is in part determined by how consistent the cleanliness of the blanks can be made. Detection of 10-ng IAA in a sample is possible with this technique, but the variation in the blank sampler analyses results in a higher effective minimum detectable mass. The analysis of 16 unopened blank PSDs in this study resulted in a mean detected loading of 43 ng, with a standard deviation of 44 ng. Thus, the minimum detectable mass for the data reported below was

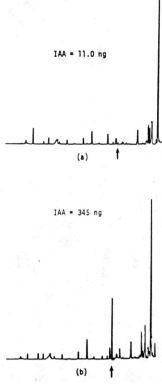

FIG. 4—*Integrator output for a blank and exposed PSDs.*

taken to be 87 ng. Improvement on this figure is possible with increased preparation of the samples prior to each use, but this was not deemed necessary in this study.

Test Subject Considerations

Prior to the performance of each exposure, the subjects were refamiliarized with the measurement scenario to be performed during the test. After entering the test chamber, the subjects were equipped with PSDs immediately prior to their donning the protective ensembles. The objective during this operation was to minimize the length of time the PSDs were exposed to air outside the ensemble. After donning the ensembles, the subjects were assisted by the test operators in obtaining good closure of the ensemble at all points, after which each subject's suit was reinspected to verify its proper use. The subjects were observed during the test period to verify proper performance of the movements and to insure that none of the suit seals were violated during the exposure. At the end of the exposure period, the subjects were directed to rest in the seated or prone position until the chamber was sufficiently purged to permit collection of the PSDs. Throughout the test period, only six of the 263 PSDs deployed on the subjects moved from their intended location and therefore did not obtain valid samples.

Observation of the subjects performing the movement scenarios makes it clear that each performs them somewhat differently, even though each instruction is followed. In some test programs, this may lead to undesired variations in results. Nevertheless, the results obtained make it clear that testing with human subjects yields a different and more realistic assessment of protective ensemble performance than does testing with mannequins.

Test Matrix

The test matrix established for this study addresses two variables only: ensemble size and level of activity. Three ensemble sizes and three levels of activity of the subject were included in the matrix. The tests performed with the mannequin can be considered to be tests in the medium ensemble size, rest activity level. The challenge concentration in the test atmosphere and test duration were maintained at constant values throughout the test matrix, and the same subjects (one of each size) were used throughout the testing to minimize unintentional variations.

Experimental Results

It is not clear at this time how one might use PF values measured at a variety of locations on a subject to arrive at an estimate of the whole-body PF. For this reason, the PFs reported are considered to be applicable only for the location of the sampler, during the test performed.

The parameter actually measured in these tests is the mass of IAA collected by each sampler. This is an integrated measurement of the concentration of vapor near the sampler over the course of the exposure. Because of this, the measured values for each test are compared with a standard value reflecting the mass an unprotected PSD would accumulate during performance of a test. This value is determined by calculating what mass loading results from the average PSD sampling rate and the actual concentration of IAA measured in the test chamber during each test. It should be noted that the chamber concentration does not decrease to zero immediately at 60 min, when the chamber purge begins. The duration of the purging period from the end of the one hour exposure period to the time at which the concentration is lowered to 0.10 ppm is approximately 35 min. The exposure during this period is typically about 10% of the total experienced and is included in the calculation of the standard value used in the PF determinations.

Leakage Test Results

To aid in the interpretation of the results presented here, all measurements are presented graphically, with the PFs shown for the various sampler locations. The data points for each test are identified individually so that test-to-test variability can be distinguished from the variation within a test. Additional data and a more detailed description of test methods used are presented elsewhere [7].

The PF values determined for the small subject in moderate and high activity levels are depicted in Figs. 5 and 6. For these figures, note that the suffixed L or R for locations denote left or right side of the subject, and forearm and mid-back are abbreviated to "F-ARM" and "M-BACK," respectively. In a number of cases, duplicate samplers were used at a particular location. Any PF measured to be in excess of 1000 is reported as 1000 in these figures. This is consistent with the minimum detectable mass for the PSD analysis discussed above.

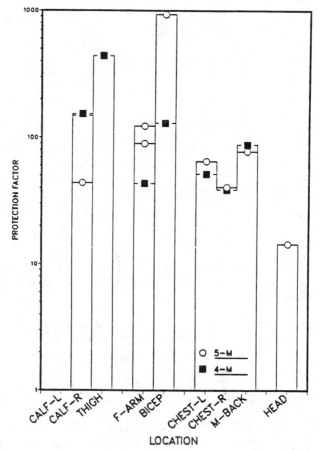

FIG. 5—*Protection factors measured for the small subject in the moderate-activity-level tests.*

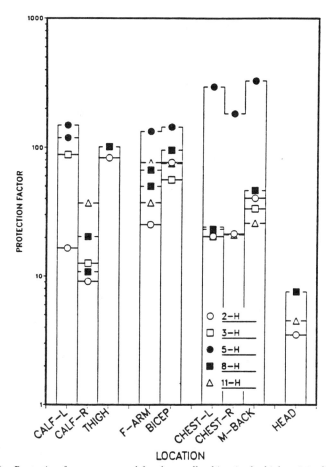

FIG. 6—*Protection factors measured for the small subject in the high-activity-level tests.*

The data for the moderate activity tests how PF values between 40 and 200 for nearly all cases. There is somewhat less scatter in the results for the chest and back sampling locations than elsewhere. This result is obtained in most cases and is believed to reflect the fact that the chest and back sampling locations have a greater amount of sorbent available to the simulant between the point of leakage and the sampler. This would be expected to make these locations less subject to extreme fluctuations in the concentration of IAA which periodically is admitted to the suit interior. It should also be noted in Fig. 5 that the back of the head experiences a fairly low (~14) PF—this indication is consistent throughout the testing.

The protection factors shown in Fig. 6 indicate, in general, lower values for the high activity tests. The greater reproducibility for the chest and back is demonstrated again except for Test 5H results—which are the highest PF values shown in this set of data for all locations measured. The cause of this test's high PF values is not known. The very low (<10) PF values are seen in these tests for the back of the head once again.

The PF values measured for the mannequin in each of the tests it was used are shown in Fig. 7. The mannequin, of course, is at rest for all the tests. In general, the PF values seen

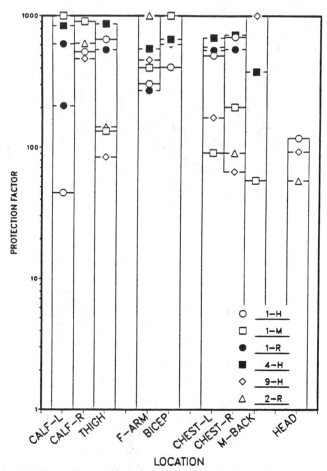

FIG. 7—*Protection factors measured for the mannequin in all tests.*

here are significantly higher than those presented in the previous figures for the active human subject, as one would expect. The values of PF shown for the back of the head are seen to be about an order of magnitude lower than those for the other locations even in the complete absence of wearer movement. Both the inability of the hood material to absorb the IAA and the large perimeter of the hood result in this degraded protection for this location.

It is of interest to compare the PF values measured for the mannequin with those found for the subjects at rest. Figure 8 depicts the measured PF values for each of the subjects at rest and the geometric mean of the data obtained for the mannequin. In most locations, the measured PFs are similar for all subjects and the mannequin. The head position appears to have a higher PF for the mannequin than for the subjects. (The mannequin PF value is based on three different experiments.) This difference is, perhaps, to be expected since the mannequin performs the rest-activity level somewhat better than the seated subjects were able to.

The geometric means of the data obtained for the three subjects in the moderate-activity-level tests are presented in Fig. 9. Several interesting features emerge from this figure. The

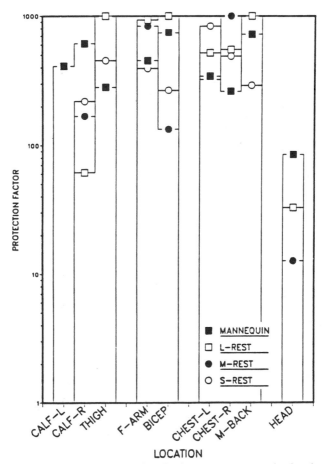

FIG. 8—*Protection factors measured for all subjects at rest-activity level and mannequin.*

medium subject exhibits the poorest PF values at all locations except the mid-back position. We can offer no explanation for this observation, however. One can see in this figure improvement in PF from the calf to the thigh and from forearm to bicep. This is consistent with the absorption of the IAA as it progresses to the suit interior from the leakage location at the wrist or ankle seal. One sees similar results for the right and left side of the chest, and slightly higher PF values for the mid-back, which is also probably due to the greater availability of sorbent between the neck seal and the mid-back sampling position. One also sees very similar and low PF values for the back of the head sampling position for the three subjects.

The results for the high-activity-level tests of the three subjects are shown in Fig. 10. All of the observations made with respect to the previous figure apply here except that the medium subject does not exhibit consistently low PF values in this set. One can note that there is an apparent reduction in the subject to subject variability in PFs at a given location compared to that seen in the moderate-activity-level data. This is probably explained by the fact that these data are based on five tests for each subject, rather than the two used for the moderate activity level. The larger number of tests reduces the apparent variation in

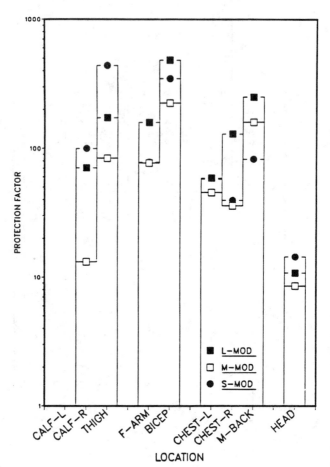

FIG. 9—*Geometric means of protection factors measured for all subjects in moderate-activity-level tests.*

the geometric mean values. Comparison of Fig. 10 with Figs. 8 and 9 indicates a trend towards decreasing PF values with increasing levels of activity of the subjects, but this trend is not as pronounced as was anticipated. If the extent of suit leakage were three times greater in the high activity level, one would expect a corresponding decrease in the measured PF values. The apparent decrease in PF in these limited data is not that large, and we are unable to explain that result at this time.

Blank Test

Given that the PSDs operate by sorption of the IAA vapor and that the ensemble interior surface also sorbs the vapor, there is the possibility that the PSDs might adsorb material collected by the suit interior in previous tests. To ensure that this phenomenon was not occurring during the exposures, we performed a test with one of the subjects performing the high-activity regimen and wearing an ensemble which had been used in four exposures previously. This was intended to provide the greatest likelihood for PSD collection of IAA

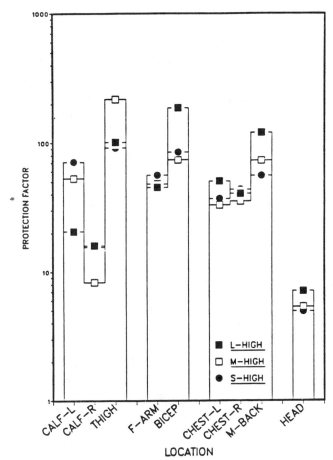

FIG. 10—*Geometric means of protection factors measured for all subjects in high-activity-level tests.*

from the charcoal-impregnated lining of the ensemble. This test was performed in the same manner as the usual leakage tests, except that fewer PSDs were used and no IAA injection was performed. The masses of IAA collected by the samplers used in this test were all below the minimum detectable mass, which indicates that no significant cross-contamination of samplers was occurring during the testing.

Conclusions

To our knowledge, this is the first reported effort involving application of the subject passive sampling devices to this problem. The technique has been demonstrated in this limited testing to be capable of quantitatively determining PF values from 1 to 1000 under the chosen test conditions. The variability in the results obtained at a given sampling location places some limitations on the precision with which one can measure the PF using this technique. We believe that the observed variability in the test results contained in this report are inherent in the phenomenon being measured and are not due to any artifacts of the measurement process. The PSDs are sensitive to the microenvironment in which they are located and do not provide a measurement of the entire volume in which they are placed. They do, however, provide a time-integrated measurement of the mass of vapor transported to the surface of the PSD, which is the parameter of interest in determining the possible dosage received by the wearer of the ensemble.

In addition to the measurement technique developed in this work, a standard set of movements was established which may be useful in any realistic testing of individual protective clothing for the soldier. The use of a standard movement set and activity-level protocol will become a necessity to permit intercomparison of results obtained at different laboratories.

There is a trend towards decreased leakage with decreased subject activity for all monitored locations in these tests. For subjects at rest, the leakage measured is nearly the same as for the mannequin used in these tests in all locations except the back of the head. This was the case since there was always at least some movement of the subjects' heads, even during the rest activity level.

In only one instance was there an apparent difference in the leakage measured for the different subjects, so it does not appear that the different size ensembles perform differently, nor do the subjects themselves have great influence on the amount of leakage detected.

It was found in these tests that the calf and forearm locations received the highest loadings of IAA within the overgarment, but that leakage through the neck of the jacket can lead to similar measurements at the chest and mid-back locations. It was also found that the measured loadings decreased as one moves away from the leakage location. Thus the thigh and bicep locations are characterized by somewhat higher PFs than the calf and forearm. The variability in measured PFs was greatest near the leakage locations and was most consistent—for a given level of activity—at the chest and mid-back locations. This is a result of variation in the quality of the seal obtained at the suit boundary and of the damping affect the sorbent liner of the ensemble has on the vapor which penetrates the leakage pathways.

The basic technique used in this work can be extended to measure higher PF values if required. More effort would be required for pre-test preparation of the PSDs, however. Improved pretreatment of the PSDs may also permit use of shorter duration exposures to achieve measurements of PF $\sim$ 1000.

Acknowledgment

The funding for this work was provided to Battelle by the U.S. Army Chemical Research, Development and Engineering Center, Contract No. DAAA-15-85-D-0010. The body movements scenario was developed by Lisa Jackson of ILC-Dover, Inc. The authors also wish to recognize the expert work of G. William Keigley, of Battelle, in performing the PSD analyses.

References

[1] *Soldiers Manual of Common Tasks,* FM21-2, Skill Level 1, Headquarters, Department of the Army, Washington, DC, 1983.
[2] *Army/Training and Evaluation Program for Mechanized Infantry/Tank Task Force,* ARTEP 71-2, Headquarters, Department of the Army, Washington, DC, Nov. 23, 1981.
[3] Woodson, W. E., *Human Factors Design Handbook,* McGraw-Hill, New York, 1980.
[4] Coutant, R. W., "Evaluation of a Passive Monitor for Volatile Organics," final report from Battelle's Columbus Laboratories to the U.S. EPA, Columbus, OH, Nov. 1982.

[5] Lewis, R. G., Mulik, J. D., Coutant, R. W., Wooten, G. W., and McMillin, C. R., "Thermally Desorbable Passive Sampling Device for Volatile Organic Chemicals in Ambient Air," *Analytical Chemistry,* Vol. 57, 1985, pp. 214–219.

[6] Coutant, R. W., Lewis, R. G., and Mulik, J., "Passive Sampling Devices with Reversible Adsorption," *Analytical Chemistry,* Vol. 57, 1985, pp. 219–223.

[7] Kuhlman, M. R., Coutant, R. W., and Keigley, G. W., "Individual Protection Testing (Task 1— Protective Ensemble Testing)," final report by Battelle Columbus Division to the U.S. Army CRDC, Columbus, OH, Aug. 22, 1986.

Christopher Wiernicki[1]

An Improved Air-Supplied Plastic Suit for Protection Against Tritium

REFERENCE: Wiernicki, C., **"An Improved Air-Supplied Plastic Suit for Protection Against Tritium,"** *Performance of Protective Clothing: Second Symposium, ASTM STP 989*, S. Z. Mansdorf, R. Sager, and A. P. Nielsen, Eds., American Society for Testing and Materials, Philadelphia, 1988, pp. 518–524.

ABSTRACT: A new two-piece, air-supplied, plastic suit has been developed for use in atmospheres containing tritium oxide (HTO). The suit material is 9 mils thick and consists of two 4-mil pieces sandwiched around a polyester scrim for added strength. Each of the 4-mil sections is a laminate of vinylidene chloride (Saran), chlorinated polyethylene (CPE), and ethylene vinyl acetate. The Saran/CPE suit material and 12-mil polyvinyl chloride (PVC), the suit materials previously used in HTO atmospheres, were tested for tritium permeation by the Radian Company. Tritium permeation testing was done in accordance with the ASTM Test Method for Resistance of Protective Clothing Materials to Permeation by Liquids or Gases (F 739-85) using the Radian Microcell. Liquid scintillation counting techniques were used to measure the amount of HTO collected. The two plastic suit materials were challenged with the following: HTO/Duo Seal, 1 mCi/mL. Tritiated water (10 mCi/mL) permeated both materials; however, permeation was at least four times slower, and breakthrough six times longer, for Saran/CPE. HTO in Duo Seal penetrated the 12-mil PVC in 100 min, but did not penetrate the Saran/CPE during the entire 480-min test. No breakthrough was detected with either material when challenged with HTO (1 mCi/mL) in Freon TF or ethanol. The Saran/CPE suit material provides significantly better protection against tritium oxide than 12-mil PVC. In addition, the Saran/CPE plastic suit has an internal noise level of 73 dba, is self-extinguishing, and provides a protection factor of greater than 20 000 when tested against a polydispersed aerosol.

KEY WORDS: air-supplied, plastic suit, tritium, permeation, Saran/CPE, polyvinyl chloride (PVC), improved protection

Tritium, the only radioactive isotope of hydrogen, can exist as elemental hydrogen or combined with oxygen (tritium oxide) as a liquid or vapor. Tritium gas and tritium oxide (HTO) are odorless, tasteless, colorless, and readily dispersed in air. Tritium is a low-energy beta emitter that presents no external hazard because the dead outer layers of skin will completely attenuate the radiation. Tritium that gets inside the body is hazardous because it irradiates the soft tissue. Exposure to HTO is considered to be 10 000 times more hazardous than exposure to elemental tritium. The general assumption is that HTO, whether entry is through the skin, respiratory system, or digestive system, is completely absorbed and mixes freely with body water. It permeates all tissues within a few hours and irradiates the body in a fairly uniform manner. Tritium and HTO exposures are possible in the following operations: light water and heavy water cooled power reactors, fusion reactors, nuclear fuel reprocessing, and tritium production for nuclear weapons.

[1] Senior engineer, E. I. du Pont de Nemours and Company, Savannah River Plant, Aiken, SC 29808; currently with Exxon Biomedical Sciences Inc., Mettlers Road, East Millstone, NJ 08875.

Because HTO readily enters the body via the lungs and skin, total body encapsulation is required to prevent assimilation. In the 1950s, polyvinyl chloride (PVC) was chosen for the fabrication of air-supplied plastic suits for several reasons, including: its advanced stage of development compared with other polymers, extensive industrial experience, and physical properties such as low flammability, good chemical resistance, heat-sealable seams, and good tear resistance [1,2].

Permeation testing conducted in the 1960s and 1970s indicated that vinylidene chloride (Saran)[2] provided up to 150 times more protection against tritium permeation than PVC [1,2]. However, Saran demonstrated poor performance in a number of areas, including drape, flammability, tear resistance, and seam strength. Several unsuccessful attempts were made to produce a Saran plastic suit by laminating Saran to various substrates.

This paper describes a Saran/chlorinated polyethylene (CPE) plastic suit developed at the Savannah River Plant (SRP) for use in tritium and HTO atmospheres.

Saran/CPE Plastic Suit

The plastic of the 9-mil Saran/CPE suit is a lamination of two 4-mil Dow Saran/CPE coextruded XU65505.07 films (Fig. 1). Each of the 4-mil films of a laminate of polyvinylidene chloride (Saran), chlorinated polyethylene (CPE), and ethylene vinyl acetate (EVA). The material is reinforced with a polyester scrim core for added strength and contains flame retardants to make it self-extinguishing when tested per the ASTM Test Method for Rate of Burning and/or Extent and Time of Burning Flexible Plastics in a Vertical Position (D 568-77). The suit is a two-piece design consisting of a jacket and pants (Fig. 2). The jacket has a 35- by 26-cm (14 by 10½ in.) cylindrical 40-mil PVC head piece with a built-in plenum/noise suppressor to deliver air to the wearer's head. Jacket cuffs are constructed from rigid polyethylene to which butyl rubber gloves are attached. The jacket contains an air distribution system/noise suppressor that supplies air to the head, torso, and legs (Fig. 3). The pants have attached boots and are held up with suspenders. The suit comes in two sizes, regular and small; the small suit fits people shorter than 173 cm (5 ft-8 in.). The suit is supplied with 0.0085 to 0.0094 m^3/s (18 to 20 cfm) of air.

Special techniques were developed to heat-seal the suit seams and allow for the heat-sealing of PVC to the Saran/CPE material. These techniques were developed by the suit vendor and are proprietary.

Permeation testing was conducted on the Saran/CPE material and 12-mil PVC that is used in the present tritium plastic suit. In addition, the Saran/CPE suit was vigorously field-tested at SRP and tested by the Personnel Protection Studies Section, Industrial Hygiene Group, of the Los Alamos National Laboratory (LANL) to accept it for use by the U.S. Department of Energy (DOE).

Permeation Testing Methods and Materials

Tritium permeation testing was done by Radian Corporation in accordance with the ASTM Test Method for Resistance of Protective Clothing Materials to Permeation by Liquids or Gases (F 739-85). Test specimens were cut from randomly selected areas. Each specimen was measured at three places to determine the average thickness. A dial gage comparator was used to measure the thickness to the nearest mil (0.001 in.) (0.0254 mm).

Specimens were mounted vertically between two halves of the test cell. Cells were bolted

---

[2] Saran is a registered trademark of Dow Chemical U.S.A.

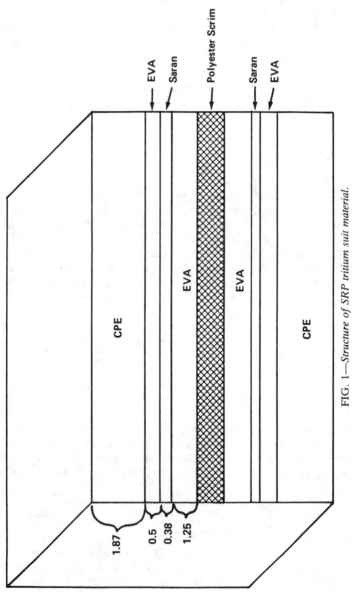

FIG. 1—*Structure of SRP tritium suit material.*

FIG. 2—*SRP tritium suit.*

together to a torque of 2.8 N·m (25 in·lb). Because the cells were manufactured with high-tolerance machine faces, no gaskets were necessary to prevent leakage.

The 9-mil Saran/CPE and 12-mil PVC materials were challenged with the following media:

1. Tritiated water (HTO), 10 mCi/mL.
2. Ethanol/HTO, 1 mCi/mL.
3. Freon[3] TF/HTO, 1 mCi/mL.
4. Duo Seal[4]/HTO, 1 mCi/mL.

The challenge solution was delivered via a 5-mL (0.169-fl oz) disposable syringe. Timing for the test began when the chemical contacted the material. The collection side of the test cell was swept with a stream of zero-grade nitrogen (N^2) at approximately 20 mL/min (0.676

---

[3] Freon TF (trichlorotrifluoroethane) is a registered trademark of Phillips Manufacturing Company.
[4] Duo Seal is a registered trademark of Scott Aviation, a Division of A-T-O Inc.

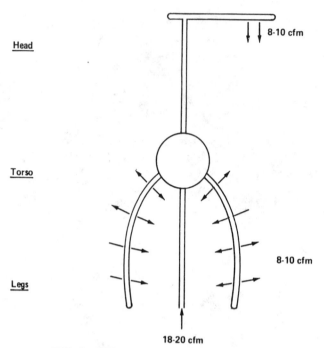

FIG. 3—*SRP tritium suit air distribution system.*

fl oz/min). The nitrogen flowed through a gas dispersion tube into the ethylene glycol collection medium. Samples of this medium were taken every 15 min for 8 h and every 24 h for the neat tritiated water challenges. One-millilitre (0.0338-fl oz) specimens were placed in a low-potassium scintillation vial with 2 mL (0.0676 fl oz) of Aquasol-2, a liquid scintillant. Standards were prepared and a calibration curve generated (alpha = 0.9999) to correct for possible quenching effects in the specimens. The curve was used to convert raw counts per minute to microcuries (μCi) per millilitre. All experiments were carried out in triplicate with a blank cell to check for cross contamination.

Results

Breakthrough times and steady-state permeation rates for all the tests are summarized in Table 1. Tritiated water, 10 mCi/mL, permeated both materials; however, permeation was four times slower, and breakthrough was six times longer, for Saran/CPE. The relative effectiveness of the Saran/CPE suit against tritium permeation is significantly better than the 12-mil suit: more than 75 times greater than 1 h, 13 times better after 2 h, and 7 times better after 3 h. Relative effectiveness is a direct comparison of total tritium that penetrated Saran/CPE and PVC under each set of challenge conditions. The calculation of relative effectiveness is also given in Table 1. The time periods of 1, 2, and 3 h were chosen because they are typical use times at SRP.

Tritium (1 mCi/mL) in Duo Seal oil penetrated the PVC material in 100 min, but did not penetrate the Saran/CPE during the entire 480-min test. Relative effectiveness against tritium was greater than ten times better after 2 h and greater than 28 times better after 3 h for Saran/CPE.

TABLE 1—*Comparison of Saran/CPE versus PVC for tritium permeation.*

Challenge	Material	Permeation Rate, μCi/cm 2 h	Breakthrough Time, min	Relative Effectiveness[a] Against Tritium Permeation		
				1 h	2 h	3 h
10 mCi/mL	Saran/CPE	0.02	95	≥75	13	7
	PVC	0.07	15	1	1	1
1 mCi/mL in	Saran/CPE	0.00	480	1	≥10	≥28
Duo Seal	PVC	0.05	100	1	1	1
1 mCi/mL in	Saran/CPE	0.00	480	1	1	1
Freon TF	PVC	0.00	480	1	1	1
1 mCi/mL in	Saran/CPE	0.00	480	1	1	1
Ethanol	PVC	0.00	480	1	1	1

[a] Relative effectiveness $= \dfrac{\text{SSPR} \times \text{ET for PVC}}{\text{SSPR} \times \text{ET for Saran/CPE}}$

where

SSPR = steady state permeation rate. The lower limit of detection (0.0007 μCi/cm²/h) was used if there was no breakthrough during the time period of the comparison.

ET = elapsed time of tritium permeation = comparison time − breakthrough time.

Example: Saran/CPE versus PVC challenged by 10 mCi/mL for 1 h

$$\text{RE} = \frac{0.07 \times (1\text{ h} - 0.25\text{ h})}{0.0007 \times 1\text{ h}} = 75$$

Both materials were challenged with tritium (1 mCi/mL) in Freon TF and ethanol. No breakthrough was detected for either suit material during the 480-min tests.

DOE Acceptance Testing

The Personnel Protection Studies Section, Industrial Hygiene Group, LANL, tested the SRP Saran/CPE plastic suit in accordance with LA-10156-MS, Acceptance-Testing Procedures for Air-Line Supplied-Air Suits, for evaluation and acceptance by DOE [4]. Suit test procedures examine quality of construction, ability to prevent penetration of aerosols, strength of hose, noise levels, performance in a loss-of-airflow emergency, and skin contamination during removal.

The SRP Saran/CPE plastic suit proved to be of high-quality construction, provided a protection factor of 20 000 against 0.54 ± 0.12-μm mass median aerodynamic diameter polydispersed aerosol, and produced a noise level of 73 dBA inside the suit.

Discussion

The newly developed Saran/CPE plastic suit material offers significantly better protection against HTO penetration and permeation than the 12-mil PVC currently used at SRP and at most DOE and commercial sites where tritium and HTO are exposure hazards. Tritium permeation test results are in general agreement with the published literature. Breakthrough time is an important parameter when evaluating the applicability of protective clothing;

previously published tritium permeation tests did not measure this parameter. Future studies should quantify steady state permeation rate and breakthrough time to more fully evaluate potential tritium protective clothing.

Saran/CPE has successfully been fabricated into a plastic suit because, in addition to its superior tritium resistance, it has all the characteristics required to construct a rugged, dependable, and comfortable suit. The use of the Saran/CPE suit at SRP reactor and tritium production facilities should be a major contribution to the site ALARA (As Low As Reasonably Achievable) program. In addition, several other DOE facilities have expressed an interest in this plastic suit.

Saran/CPE has demonstrated excellent resistance to a wide range of chemical contaminants; therefore, this suit material may have applications in the general chemical industry and hazardous waste site cleanup operations.

Acknowledgment

I would like to thank all the people who contributed to the development of the plastic suit described in this paper, especially Louis Contini of Rich Industries and Michael Ferguson of Dow Chemical Company. The information contained in this paper was developed during the course of work under Contract No. DE-ACO9-76SR00001 with the U.S. Department of Energy.

References

[1] Fuller, T. P. and Easterly, C. E., "Tritium Protective Clothing," ORNL/Tm-6671, Oak Ridge National Laboratory, Oak Ridge, TN, 1979.
[2] Skaggs, B. J., "A Study to Determine the Effectiveness of Personnel Protective Equipment Against Tritium and Tritium/Hydrocarbon Mixtures," LA-UR-86-2015, Los Alamos National Laboratory, Los Alamos, NM, 1986.
[3] Chemical Permeation Report, DEN:86-256-047-01, Radian Corp., Austin, TX, 1986.
[4] Bradley, O. D., "Acceptance-Testing Procedures for Air-Line Supplied-Air Suits," LA-10156-MS, Los Alamos National Laboratory, Los Alamos, NM, 1984.

James S. Johnson[1] and Jeffrey O. Stull[2]

Measuring the Integrity of Totally Encapsulating Chemical Protective Suits

REFERENCE: Johnson, J. S. and Stull, J. O., **"Measuring the Integrity of Totally Encapsulating Chemical Protective Suits,"** *Performance of Protective Clothing: Second Symposium, ASTM STP 989,* S. Z. Mansdorf, R. Sager, and A. P. Nielsen, Eds., American Society for Testing and Materials, Philadelphia, 1988, pp. 525–534.

ABSTRACT: The Lawrence Livermore National Laboratory's (LLNL) Hazards Control Department has recognized for some time that totally encapsulating chemical protective (TECP) suits must be tested to assure that they function properly with a high degree of reliability. Two general types of tests are necessary to properly evaluate a TECP suit: design qualification tests and field use tests. To develop these tests for the evaluation of TECP suit performance, LLNL is participating in a jointly funded project with the U.S. Coast Guard, the U.S. Occupational Safety and Health Administration, and the U.S. Fire Administration. Four tests have been developed to measure TECP suit performance: a quantitative test, a worst-case chemical exposure test, a pressure leakage test, and a chemical leak rate test.

A quantitative TECP suit test method using Freon 12 gas and an aerosol, polyethylene glycol molecular weight 400 (PEG 400), to measure the leakage of both test agents into the suit is being developed. This leak rate is expressed as a "protection factor" that is calculated by dividing the outside test agent concentration by the suit interior test agent concentration. A worst-case chemical exposure test has been demonstrated using the U.S. Coast Guard's Teflon-coated Nomex TECP suit in a hydrogen fluoride gas cloud.

Two types of field use tests have also been developed. The TECP pressure test measures the suit leak rate by inflating the suit to a prescribed pressure and monitoring pressure drop in the suit over time, ASTM Standard Practice for Pressure Testing of Gas Tight Totally Encapsulating Chemical Protective Suits (F 1052-87). The proposed chemical leak rate test consists of generating a known ammonia concentration in a test room, wearing the TECP suit into the room, and doing a series of prescribed light exercises. The interior of the suit is monitored for ammonia after the test subject exists the test room, and the protection factor is calculated in the same manner as described earlier in the Freon 12 test procedure.

By employing the four TECP suit tests in the various stages of suit development and field use, a high degree of safety can be assured.

KEY WORDS: totally encapsulating chemical protective suit, protection factor, design qualification tests, field use tests, quantitative test, worst-case chemical exposure test, pressure leak rate test, chemical leak rate test, protective clothing

The need to provide complete encapsulation of workers to allow them to carry out their jobs safely is becoming very commonplace. Such jobs as hazardous material response, toxic waste dump cleanup, and chemical manufacture and use require complete encapsulation of employees routinely or during accidents. With the increased use of complete encapsulation in the workplace, a high degree of safety is now required of commercially available totally encapsulating chemical protective (TECP) suits. This requirement for chemical protective

[1] Group leader, Lawrence Livermore National Laboratory, Livermore, CA 94550.
[2] Senior engineer, Texas Research International, 9063 Bee Caves Road, Austin, TX 78733.

clothing was characterized by John B. Moran, head, Division of Safety Research, National Institute for Occupational Safety and Health, as "the last line of defense" for the worker.

TECP Suit Testing

A TECP suit is made up of many components (Fig. 1). Most of these components are in themselves individual items of chemical protective clothing for which chemical permeation data is available. Some items however, such as suit closures, vent valves, lens material, suit membranes, and seams are unique to a TECP suit and therefore require individual chemical permeation testing. This type of data does not provide the user with a measure of complete TECP suit integrity. To measure the complete integrity and performance of TECP suits, two general types of tests are necessary: design qualification tests and field use tests. Design qualification tests are normally carried out by manufacturers on a specific TECP suit design or when changes are made to the design that could affect the suit's performance. Results from these tests document the TECP suit's performance at providing complete encapsulation of the wearer in the laboratory and in realistic worst-case chemical exposures. Field use tests are designed to measure the general integrity of the TECP suit in the field, normally after decontamination and before reuse.

Design Qualification Tests

There are three types of design qualification tests that should be carried out to evaluate the performance of a TECP suit. These tests are a quantitative test, a worst-case chemical exposure test, and a pressure leakage test.

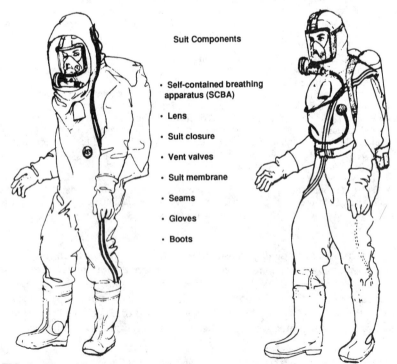

Suit Components

- Self-contained breathing apparatus (SCBA)
- Lens
- Suit closure
- Vent valves
- Suit membrane
- Seams
- Gloves
- Boots

FIG. 1—*The configuration and design of totally encapsulating chemical protective suits.*

Quantitative Test—The quantitative test has been designed to allow one to measure very accurately the level of protection a TECP suit provides the wearer during a series of exercises in a test chamber (Fig. 2). Two types of test agents are used to measure this performance: Freon 12, a gas, and polyethylene glycol molecular weight 400 (PEG 400), in an aerosol form. Either or both of these test agents are generated in the exposure chamber. The test subject enters the chamber wearing a suit and respirator, and the test begins. To test the integrity of the suit, the following exercise protocol is carried out for a 2-min period.

1. Stand in place.
2. Raise the arms above the head, completing at least 15 raising motions per minute.
3. Walk in place, completing at least 15 raising motions of each leg per minute.
4. Touch the toes, making at least 10 complete motions of the arms from above the head to the toes per minute.
5. Perform deep knee bends, making at least 10 complete standing and squatting motions per minute.
6. Stand in place.

Two complete series of tests can normally be conducted using a half-hour-capacity self-contained breathing apparatus (SCBA). By placing the sampling probe in various locations in the suit, localized leak sources can be measured. The overall performance of the suit can be expressed as a protection factor (PF) that is obtained by dividing the chamber concentration by the interior suit concentration. Protection factors of 500 000 can be determined with the Freon 12 system using a gas chromatograph (GC) equipped with an electron capture detector and 20 000 for the aerosol PEG 400 system using a forward light scattering photometer.

To evaluate the performance of the U.S. Coast Guard's new Teflon-coated Nomex TECP suits, a series of quantitative suit tests was carried out. Results from these tests are sum-

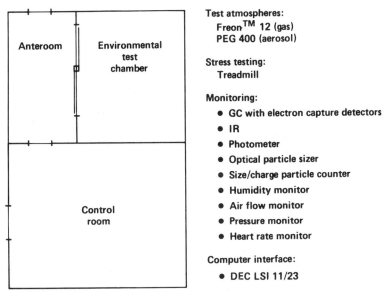

FIG. 2—*Typical test chamber layout and monitoring equipment for carrying out quantitative and chemical leak rate tests.*

marized in Figs. 3 to 6. In analyzing the data, one must remember that the Freon 12 system collects a sample every 2 min, thus providing only a sample analysis every 2 min. The PEG 400 aerosol photometer system, on the other hand, provides a continuous readout of the aerosol concentration in the sample probe area. Two sampling probe locations were chosen to evaluate this suit's performance: a breathing zone location and a vent valve location. No experimental data were left off the bar charts or changed in any way. The data simply reflect the variability that results from man-testing, for example, pinched sampling hose, SCBA out of air, etc.

The bar charts show an obvious difference in the data from the two sampling locations. The breathing zone protection factors average approximately 30 000 for the Freon 12 sampling system and approximately 10 000 for the PEG sampling system. In both sampling systems at the breathing zone area, there is some grouping of concentration data between tests. This indicates a minimal leakage into the TECP suit using the breathing zone air as representative of the entire interior suit air volume.

The vent valve protection factors (PF) averaged approximately 90 000 for the Freon 12 sampling system and approximately 4000 for the PEG sampling system. In both sampling systems at this location, there was more variability in the data. The average Freon 12 PF for the breathing zone location was 30 000, while the PF for the valve location was 90 000. Because of the discrete nature of the sampling system, it is difficult to make any conclusive decisions about this difference in the PFs. It appears, however, that the larger variability in the discrete Freon 12 samples due to rapidly changing Freon 12 concentrations could indicate valve leakage.

The PEG 400 sampling system provides a clearer picture of the difference in the two sampling locations. Since this system is close to a real time monitor, the variation between the two sampling locations is more meaningful. Comparing the average breathing zone PF value of 10 000 to the average vent valve PF value of 4000 indicates a leak in the vent valves. In a preliminary study of the performance of TECP suit vent valves, Swearengen et al.[3] have found that this vent valve and other commercially available valves leak using methane as the challenge agent. Additional quantitative testing of TECP suits is necessary to allow for a better understanding of PF data and its relationship to overall suit performance. A draft ASTM test method for quantitative testing of TECP suits is presently being prepared for ASTM Subcommittee F-23.5 review.

Worst-Case Chemical Exposure Test—The worst-case chemical exposure test has been designed to allow one to determine the survivability and level of protection a TECP suit provides in actual exposures to high levels of hazardous chemicals. An example of a worst-case chemical exposure test has been demonstrated using the U.S. Coast Guard's Teflon-coated Nomex TECP suit in a hydrogen fluoride cloud.[4] The suit was placed on an instrumented mannequin and monitored for leakage as it was engulfed by a large hydrogen fluoride cloud. By using a pressurized valve system on a timer, breathing was simulated so that the suit vent valves operated in a realistic manner. There was, however, no attempt in these experiments to simulate valve performance from such body motions as squatting or touching of the toes. In the two separate experiments using high exposures to hydrogen fluoride, this new TECP suit did not leak.

Due to the complexity of the worst-case chemical exposure test just briefly described, a less complicated approach is being planned for demonstration experiments at a future date. A more controlled chemical suit exposure will be achieved by building a small exposure

[3] Swearengen, P. M., Johnson, J. S., Sackett, C. R., and Stull, J. O., "Evaluation of the Performance of One-Way Valves Used in Chemical Protective Suits," this publication, pp. 535–540.
[4] Stull, J. O., Jamke, R. A., and Steckel, M. G., "Evaluating a New Material for Use in Totally Encapsulating Chemical Protective Suits," this publication, pp. 847–861.

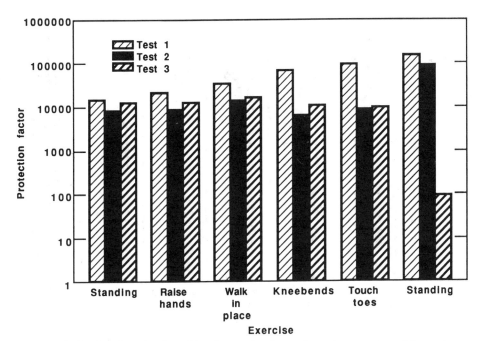

FIG. 3—*Bar chart showing achieved protection factors for various exercises while wearing the U.S. Coast Guard's Teflon/Nomex TECP suit and sampling in the breathing zone for Freon 12.*

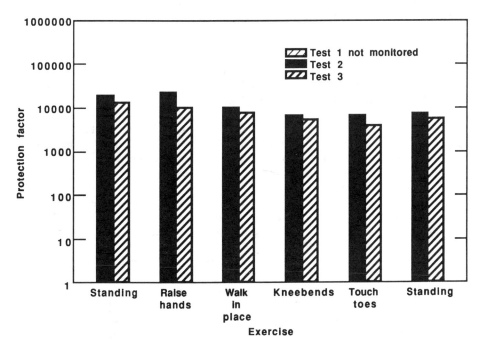

FIG. 4—*Bar chart showing achieved protection factors for various exercises while wearing the U.S. Coast Guard's Teflon/Nomex TECP suit and sampling in the breathing zone for PEG 400.*

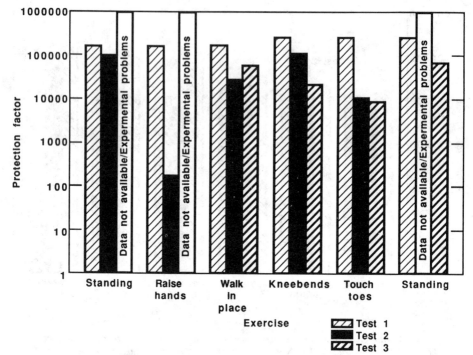

FIG. 5—*Bar chart showing achieved protection factors for various exercises while wearing the U.S. Coast Guard's Teflon/Nomex TECP suit and sampling at the vent valve zone for Freon 12.*

chamber. A test mannequin that bends at the waist will be evaluated to add more realistic body movements to the test protocol. A conceptual schematic of the test is shown in Fig. 7. Results from these tests will be used to develop a draft ASTM F-23.5 test method for worst-case TECP suit chemical exposures.

Pressure Leak Rate Test—The ASTM Standard Practice for Pressure Testing of Gas-Tight Totally Encapsulating Chemical Protective Suits, F 1057–87 is designed to measure the ability of the TECP suit to hold air within the suit at a predetermined pressure. To do this, the vent valves are sealed, along with any other normal component that could cause a leak. The suit is pressurized using an appropriate source of compressed air to a pretest expansion pressure of at least 7.6 cm (3 in.) water gage (wg) for 1 min. The suit pressure is lowered to 5.1 cm (2 in.) wg and the pressure monitored for 3 min (Figs. 8 and 9). If the pressure drops below 80% of the 5.1 cm (2 in.) wg test pressure 4.1 cm (1.6 in.) wg, the suit fails the test and should be removed from service. To find the leak that caused the failure, the suit should be inflated to the pretest expansion pressure and covered with a mild detergent-water solution by brushing or wiping. Formation of soap bubbles indicates the position of the leak, which should then be repaired. After being repaired and successfully passing the pressure test, the suit can be returned to service. Care should be exercised to assure that the inside and outside of the suit are completely dry before it is put into storage.

This practice assumes that if air cannot exit the suit, toxic gases and vapors will not be able to enter the suit. The one shortcoming of this practice is the inactivating of the vent valves and other suit components that could cause outward leaks when the suit is tested.

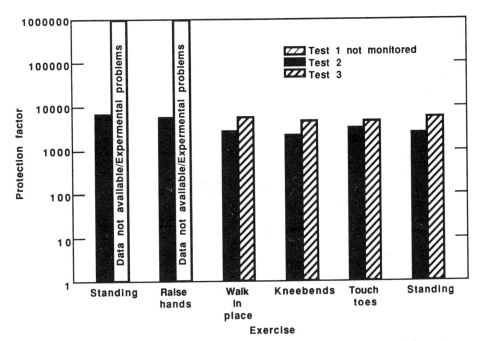

FIG. 6—*Bar chart showing achieved protection factors for various exercises while wearing the U.S. Coast Guard's Teflon/Nomex TECP suit and sampling at the vent valve zone for PEG 400.*

To overcome this problem, a chemical leak test is being developed by ASTM Subcommittee F-23.5.

Field Use Tests

There are two field use tests that should be carried out to evaluate the acceptability of a TECP suit for actual use. These tests are the pressure leak rate test and the chemical leak rate test.

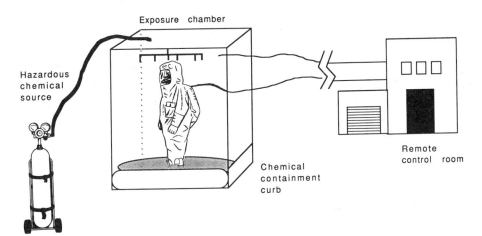

FIG. 7—*A conceptual schematic of a worst-case chemical exposure test.*

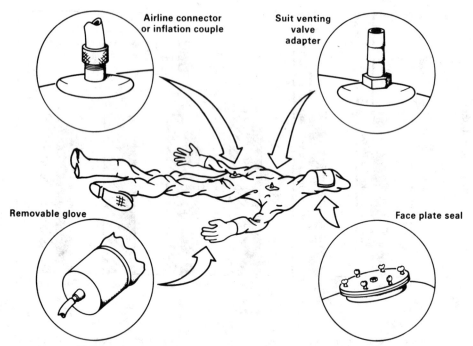

FIG. 8—*Typical examples of TECP suit modifications to permit inflation.*

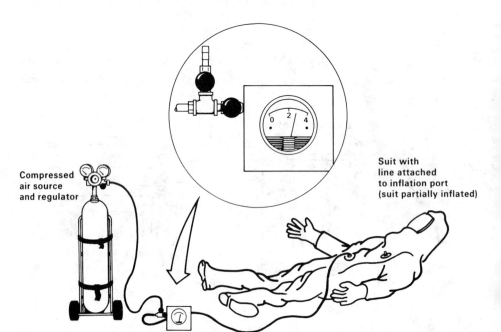

FIG. 9—*Recommended ASTM pressure test apparatus and typical test configuration.*

Pressure Leak Rate Test—The pressure leak rate test that is to be used as the field use test is the same pressure leak rate test described previously in the Design Qualification Tests section.

Chemical Leak Rate Test—The ASTM draft, Practice for Qualitative Leak Testing of Gas-Tight, Totally Encapsulating Chemical Protective Suits with Ammonia Gas, is designed to measure the intrusion coefficient of the TECP suit against ammonia gas. To do this, a test chamber is identified and prepared for the test. In most cases, a small well-ventilated room or simple test chamber will be adequate. The volume of ammonium hydroxide required to produce the test concentration of 1000 ppm ammonia is calculated from a formula that relates room volume to required millilitres of ammonium hydroxide. The airborne concentration of ammonia in the test chamber is determined with length-of-stain colormetric detector tubes. The suit to be tested and the SCBA are donned by the individual. The premeasured volume of ammonium hydroxide, an evaporation pan, detector tubes, and hand pump are taken into the test chamber. The ammonium hydroxide is poured into the evaporation tray and allowed to evaporate for 2 min. A high-range, length-of-stain ammonia detector tube is used to confirm that the airborne concentration of ammonia is 1000 ppm. To test the integrity of the suit, the following exercise protocol has been developed.

1. Raise the arms above the head, completing at least 15 raising motions in a 1-min period.
2. Walk in place for 1 min, completing at least 15 raising motions of each leg in that period.
3. Touch the toes, making at least 10 complete motions of the arms from above the head to the toes in the 1-min period.
4. Perform deep knee bends, making at least 10 complete standing and squatting motions in the 1-min period.
5. Exit the test area.

The low-range length-of-stain ammonia detector tube is inserted through the suit zipper or other appropriate sampling opening to determine the ammonia concentration in the suit. Enough distance from the test area should be provided to prevent a false ammonia reading. The test should be initiated within 1 min after leaving the test chamber to minimize the effect of dilution due to the expired air from the SCBA. After measuring the ammonia concentration within the suit, the test is concluded, and the suit is doffed and the SCBA removed. The test area ventilation fan should be turned on and allowed to run long enough to remove the ammonia gas.

If an ammonia concentration greater than 5 ppm is detected in the suit interior, the suit fails the test. Since 5 ppm is the limit of detection for the ammonia detection tube and the chamber concentration is 1000 ppm, a protection factor of approximately 200 can be measured. To find the leak that caused the failure, determine the gas tightness of the suit by following the ASTM standard practice for pressure testing of gas-tight TECP suits described earlier. The sealing surface of the suit vent valves should also be checked to determine if this was the cause of the leak. After the leaks have been identified and repaired, the suit should be retested for gas tightness and ammonia leakage. This practice allows one to test the complete TECP suit as it is used in the field. Its shortcomings are the use of a corrosive chemical, a delay in sampling, and a protection factor sensitivity of approximately 200.

At the present time, the chemical leak rate test is based on an ammonia test atmosphere. Several people have voiced their strong concern that using a 1000 ppm test atmosphere of ammonia is too hazardous. Other substitutes for ammonia are being investigated by ASTM Subcommittee F-23.5, along with the further evaluation of the ammonia test.

Conclusions

To date, there is one completed ASTM TECP suit test, Standard Practice for Pressure Testing of Gas Tight Totally Encapsulating Chemical Protective Suits (F 1052–87). The other three tests described in this article, a quantitative test, a worst-case chemical exposure test, and a chemical leak rate test, are in various stages of development. When they are finished and available as ASTM standard test methods or practices, a complete and reproducible battery of tests can be completed on commercially available TECP suits. Results from these tests, along with information presently being generated using ASTM test methods for permeation and penetration, will provide the user with a sound technical data base. This will allow the user to effectively evaluate the performance of the TECP suits he or she purchases and uses. The use of these TECP suit tests and related data will assure a high degree of TECP suit safety.

Acknowledgment

This work was performed under the auspices of the U.S. Department of Energy by Lawrence Livermore National Laboratory under Contract No. W-7405-ENG-48. Additional funding was provided by the U.S. Department of Labor, Occupational Safety and Health Administration, The Federal Emergency Management Agency, and the U.S. Coast Guard.

Peter M. Swearengen,[1] *James S. Johnson,*[1] *Carol R. Sackett,*[1] *and Jeffrey O. Stull*[2]

Evaluation of the Performance of One-Way Valves Used in Chemical Protective Suits

REFERENCE: Swearengen, P. M., Johnson, J. S., Sackett, C. R., and Stull, J. O., "**Evaluation of the Performance of One-Way Valves Used in Chemical Protective Suits,**" *Performance of Protective Clothing: Second Symposium, ASTM STP 989,* S. Z. Mansdorf, R. Sager, and A. P. Nielsen, Eds., American Society for Testing and Materials, Philadelphia, 1988, pp. 535–540.

ABSTRACT: We developed a method to test totally encapsulating chemical protective (TECP) suit vent valves. The work reported here is a preliminary investigation of four small valves. Two of the valves were low-pressure vent valves from suits and two others were respirator valves. The latter were a pressure demand type and a standard flapper valve. For testing purposes, a single valve was mounted within a plastic plate. The plate was then installed between two halves of a testing box and consequently served to divide the box into two compartments. Each valve functioned as the only conduit between the separate compartments. We observed that methane gas penetrated through one of the TECP vent valves and the pressure demand respirator valve when each was in a "closed" position (with zero differential pressure across the valve). We also observed that when air pressure on the "inside" of each valve was increased, the leak rate decreased. This effect occurred under both static and dynamic conditions. The dynamic conditions were achieved with the use of a variable rate breathing machine. Further research is needed to allow more general conclusions to be made.

KEY WORDS: protective clothing, vent valves, test methods, leak rate, dynamic tests, static tests, breathing machine, respirator valves, protective covers

We are reporting preliminary results from a study on low-pressure vent valves that we are conducting for the U.S. Coast Guard. Test results from four valves will be discussed here. In service, these valves have protective covers that are often in the form of an inverted pocket made from the suit fabric. In the work we report, covers were not used and the valves were tested in isolation. The valve currently used in the Coast Guard totally encapsulating chemical protective (TECP) suit is made by Stratotech Corporation. A second suit valve that was evaluated is made in Sweden by Trelleborg. To provide a comparison for the evaluation, we included two valves that are used in respirators. These valves were made by MSA Corporation, and included a standard flapper valve and a pressure demand valve.

Background

The U.S. Coast Guard has developed a new totally encapsulating suit for the protection of personnel during chemical spill response. Low-pressure one-way vent valves are used in

---

[1] Principal investigator, group leader, and scientific technologist, respectively, Lawrence Livermore National Laboratory, Livermore, CA 94550.
[2] Senior engineer, Texas Research International, 9063 Bee Caves Road, Austin, TX 78733.

the suit to allow escape of exhaust air from the occupant's self-contained breathing apparatus, and to maintain a small positive pressure (2.5 to 7.6 cm water column pressure) inside the suit. This latter feature is intended to minimize diffusion or penetration of chemical vapors through poor seams, material punctures, or improperly closed zippers. The effectiveness of these valves has not yet been proven. Satisfactory operation of the TECP vent valves (whether positive pressure or flapper type) is critical to the function and protective qualities of encapsulating suits.

While protection factors have been measured for the overall suit in operation, there has been no attempt to exclusively determine suit exhaust valve protection factors. Furthermore, recent overall suit testing has shown differences in suit protection factors when the internal suit probe is located near the breathing zone as compared to locating the probe internally near the exhaust valve. This information indicates that penetration of the challenge agents through the suit exhaust valves may be significant.

Experimental Considerations

We prepared an experimental system that would provide for a high degree of control over the valve environment. A small cast aluminum box (roughly 23 cm long by 13 cm wide by 15 cm high) was fitted with several openings to provide for breathing and test air inputs, analytical sampling ports, and environmental measurements (pressure, temperature). A diagram of the box is shown in Fig. 1. The box was constructed so that a plastic plate could be inserted between the two halves. At the center of the plate, a recessed orifice was machined that allowed the different valves to be inserted with a leak-tight seal. When the box and plate were assembled, the valve was positioned to function as the only conduit between the two resulting compartments. One compartment could then function as the "inside" of a TECP suit, and the other as the "outside."

The complete assembly was tested for leakage with a Stratotech valve installed. A solid cap was threaded onto the inside half of the valve. The outside compartment was filled with methane from a lecture bottle. With the pressure differential between the two chambers at zero, no methane was detected within the second (inside) chamber. We interpreted this data to mean that the test box was leak tight when the insert plate containing a valve was installed. Conversely, with the cap removed, future measurement of methane in the inside chamber would indicate the valve as the source of penetration. A diagram of the Stratotech valve in this testing arrangement is shown in Fig. 2.

A schematic of the complete test assembly is given in Fig. 3. The top left of the diagram

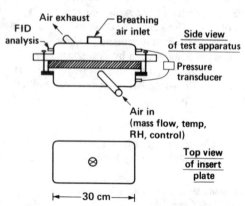

FIG. 1—*A schematic of the cast aluminum box that was used in the valve testing experiment.*

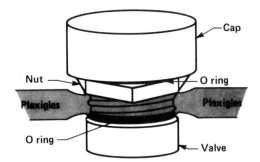

FIG. 2—*A drawing of the Stratotech valve as it appears when installed in the Plexiglas plate.*

shows a source of air that allowed precise control of flow, temperature, and relative humidity (Miller-Nelson Research, HCS-301). At the top center is shown a source (lecture bottle) of test gas (methane) that can be added to the air flow through a mass flow controller. The mixture of air and test gas is passed through a calibrated infrared analyzer (Foxboro Corporation, Miran 1A) to measure test gas concentration. When pure methane was used in this work, the air source and infrared analyzer were disconnected and the methane from the lecture bottle passed through the mass flow controller and then directly to the test chamber.

The previously described test box is shown as a divided box in the lower right of the schematic. Also shown are the probes for differential pressure measurement between the two chambers of the box. In addition, a single pressure transducer could be placed in either part of the box to measure chamber pressure relative to the atmosphere. Finally, the exhaust flow from the lower half of the box was checked for temperature with a thermistor probe (YSI Series 700, Yellow Springs Instrument Company) and a digital thermometer (Cole-Parmer Model 8502-20). We monitored the temperature at the test box exhaust flow and at the controlled air source, and held it constant at $22 \pm 1°C$. The concentration of test gas within the inside chamber of the box was measured with a calibrated total hydrocarbon (THC) analyzer (Beckman, Model 400, FID principle).

We chose methane as a test gas for several reasons. First, under the conditions of this

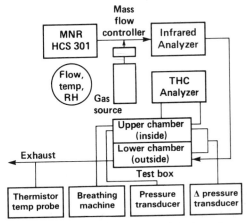

FIG. 3—*Schematic of the experimental test system used in the study on one-way vent valve performance.*

experiment, this gas is inert to the materials used in the different valves. Second, this hydrocarbon can be detected at very low levels with conventional methods. In addition, the THC analyzer can be calibrated to measure methane over a very large linear dynamic range. Finally, the measured diffusion coefficient for methane is on the same order of magnitude as that reported for hydrogen,[3,4] and gaseous diffusion of the compound is therefore quite rapid.

Test Results

Our first test was to observe the valves under static conditions, that is, without use of simulated breathing. A valve was installed in the plastic insert, and the plate was assembled between the box halves. The outside chamber of the test box was filled with pure methane. Leakage rates were determined from the change in the observed concentration of methane in the inside chamber over a specified time period. The calculated volume of the inside chamber is 1616 cm^3 (24.4 by 13.0 by 5.1 cm). If we take the definition of parts per million by volume to be ppmv = [(volume of analyte)/(volume of diluent)] $\times$ 10^6, and then substitute the observed changes of analyte concentration with time, the leak rates can be determined in units of volume.

Two valves were tested in this manner, the Stratotech valve, and the MSA positive pressure valve. After an initial measurement at a pressure differential of zero, compressed air was forced to the inside chamber through a precision valve, and the new concentration recorded over time at the higher pressure. The result of our preliminary testing is shown in Fig. 4. Our technique shows an observable leak of the outside gas into the inside chamber. To provide comparison, we make reference to the current Bureau of Mines Standard for Respiratory Protection Devices.[5] The standard used by the Bureau of Mines is the same as that reported in use by the Chemical Warfare Service during World War II.[6] In this standard,

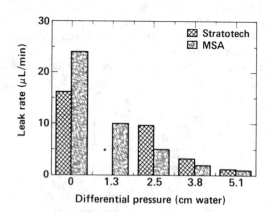

* Stratotech data not available

FIG. 4—*Graphic representation of the leak rates observed during static leak testing of one-way vent valves.*

[3] Wakeham, W. A. and Slater, D. H., *Journal Physics B: Atomic Molecular Physics,* Vol. 6, 1973, pp. 886–896.
[4] *CRC Handbook of Chemistry and Physics,* 61st ed., R. C. Weast, Ed., 1981, p. F-62.
[5] *Federal Register,* Vol. 37, No. 59, Part II, Par. 11.162-2, 1972.
[6] Silverman, L., Lee, R. C., and Lee, G., "Fundamental Factors in the Design of Protective Respiratory Equipment," Office of Scientific Research and Development, Harvard School of Public Health, Boston, MA, Report No. 1864, 1943, p. 6.

the designated respirator exhalation valve leakage is not to exceed 30 mL min^{-1} at a suction of 25 mm of water column height. The implication from this standard is that there is measurable leakage through respirator exhaust valves under normal operating conditions. To provide data comparable to the respirator standard, the suit's one-way vent valves would have to be tested in the same manner.

Our next experiment was to observe the valves during the simulated breathing provided from a breathing machine. We tested four valves at two separate breathing rates, 10 and 20 cycles min^{-1}, respectively. In all cases except one, a constant inside concentration of methane was achieved. Our technique was to observe the background signal of the THC analyzer with the breathing machine on, and then to fill the outside compartment of the box with methane. The internal concentration of methane would rise and then level off at an equilibrium value, which is the data reported in Fig. 5. The exception occurred with the MSA pressure demand valve at the 10 cycle min^{-1} breathing rate. Over the 10-min duration of the test, the internal concentration continued to rise (at a rate of 4.8 μL min^{-1}).

In all the other cases except one, we observed the internal concentration to fluctuate at a rate equal to that of the breathing machine (within a few ppmv). The single exception was that the Trelleborg valve exhibited large oscillations around an average internal concentration. The period of these large oscillations was irregular, but had an average rate of 2.5 cycles min^{-1}. It is these (sawtooth appearing) concentration variations that are shown in the bar graph of Fig. 5. Finally, in addition to the small sinusoidal-type concentration fluctuations seen in the other valves, and the large variation seen in the Trelleborg valve, there was in every case a very small oscillation superimposed on the general trend. This occurred in exact sequence with the cycles of the breathing machine. We could only attribute this fluctuation to the immediate changes that occurred when the valve opened and closed.

We also made observations of the differential pressure during operation of the breathing machine. This was done for each valve and was recorded as a positive pressure within the inner chamber relative to the pressure within the outside chamber. The data are presented graphically in Fig. 6. This data separated the four valves by pairs. The two valves that were controlled by spring tension (to open only after a certain pressure threshold was attained) allowed larger internal chamber pressures. The two flapper-type valves maintained lower pressures. The pressures seen were higher at faster breathing rates, and again the flapper-

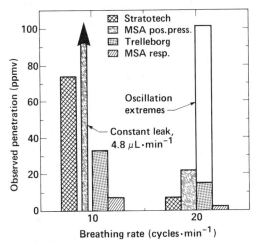

FIG. 5—*Graphic representation of the concentration of methane observed in the "inside" chamber of the test box during simulated breathing.*

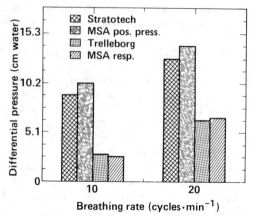

FIG. 6—*Differential pressure observed with different valves during breathing machine operation.*

type valves maintained lower pressure than the spring-tension valves. Suits used with multiple flapper valves typically operate at 0.3 to 0.5 cm of water. Our test with the isolated MSA respirator valve showed a pressure buildup much greater than this.

Conclusions

We have developed a method to test TECP vent valves. This method isolates the valve between two chambers and tests for leakage of the valves by measuring concentration of a test gas in the inside chamber of the test box. The use of a removable plate that contains a valve installed in a leak-tight manner allows for simple and rapid exchange of valves for testing. The valves we studied were tested without the protective covers found in service use. Our preliminary data indicates that there is leakage of the test gas through the valve under normally closed conditions (zero differential pressure). When the pressure on the inside chamber is increased, this leak rate is observed to decrease. One conclusion that follows from these test results is that the vent valves may be a major leak source for the intact suit. Further research is necessary to allow more general conclusions to be drawn.

Acknowledgment

This work was performed under the auspices of the U.S. Department of Energy by Lawrence Livermore National Laboratory under Contract No. W-7405-ENG-48.

Donald F. Doerr[1]

Evaluation of the Physiological Parameters Associated with the Propellant Handler's Ensemble

REFERENCE: Doerr, D. F., **"Evaluation of the Physiological Parameters Associated with the Propellant Handler's Ensemble,"** *Performance of Protective Clothing: Second Symposium, ASTM STP 989,* S. Z. Mansdorf, R. Sager, and A. P. Nielsen, Eds., American Society for Testing and Materials, Philadelphia, 1988, pp. 541–553.

ABSTRACT: Work involved during the preflight preparation of spacecraft involves the handling of materials that are very toxic to humans. These toxins attack the respiratory and skin systems and, therefore, impose the requirement for full suit enclosures. The weight, structure, and operating parameters of such a suit can be expected to have a significant effect upon the metabolic and thermal responses of the user, especially in high workload situations and ambient temperature extremes.

This paper describes the testing of the operational version of the Propellant Handler's Ensemble (PHE). In particular, parameters affecting the physiology of the user were measured during a work-rest regimen performed in three temperature environments: -7, 23, and 43°C (20, 74, and 110°F). Six subjects performed tests in these environments in two versions of the PHE, the autonomous backpack version and the hoseline supplied configuration. Measurements included heart rate, four skin temperatures, rectal temperature, oxygen and carbon dioxide in the helmet area, suit pressure, and interior suit temperature.

It was concluded that the weight and configuration of the suit significantly influenced the physiological stress on the user. The weight, at 29.5 kg (65 lb) for the PHE and backpack, proved to be a primary stressor, as indicated by elevated heart rates. The high workload portion of the protocol also taxed the limit of the environmental control unit because of the increased respiratory requirements. Oxygen levels dropped as much as 4% below resting levels and the carbon dioxide level increased by a similar amount. Finally, thermal stress is clearly evident, especially in the 43°C (110°F) tests.

State-of-the-art design techniques in whole body suits do not provide solutions to these problems. Therefore, it has been necessary to institute operational restrictions and impose medical and physical standards to avoid situations that could adversely affect the well-being of the worker.

KEY WORDS: protective clothing, physiology, totally encapsulating suits, thermal stress, propellant handling

The Kennedy Space Center is the focal point for the preflight checkout and launch of many of this nation's spacecraft. Propulsion systems on these spacecraft rely on a variety of propellants, many of which are extremely toxic to humans. Examples of these toxins are nitrogen tetroxide, hydrazine, and monomethyl hydrazine, all of which have threshold limit values of less than 3 ppm. Despite considerable efforts to institute engineering controls, the potential exists for exposure to workmen during operations such as propellant transfer. Since

---

[1] Chief, Biomedical Engineering, Biomedical Research Laboratory, Kennedy Space Center, FL 32899.

these toxins are damaging to both the skin and respiratory systems, a whole body protective suit must be employed to provide proper protection.

The use of whole body suits can introduce a variety of problems, not the least of which is its effect on the physiology of the user. The weight and encumbrance of the suit contributes an extra workload above the assigned productive work. Thermal loads are imposed and restrictions to vision, mobility, and dexterity are experienced. The cumulative effects of all these factors may prevent the workman from actually working productively, and they can also affect his safety.

As reported earlier (in this publication), the Kennedy Space Center recently embarked on a program to replace the original whole body protective suits called the Self-Contained Atmospheric Protective Ensemble (SCAPE). The SCAPE was tested extensively in our laboratory. The new suit, called the Propellant Handler's Ensemble (PHE), was made available to this laboratory after development of the prototype. Testing of this prototype provided additional data that was fed into the final design of the suit. Many human factor considerations were noted resulting in some rather substantial changes. For example, the emergency air system had to be deleted from the design because it added nearly 9 kg to the overall suit weight of 39.5 kg (87 lb). This burden proved to be too much, especially for the small user.

It is the purpose of this paper to communicate the data resulting from an extensive series of physiological tests on the final production version of the suit hereafter called the qualification (qual) suit.

Methods

The intent in testing the PHE was to examine the factors affecting the physiology of the user during worst-case workloads and extremes of temperature. Therefore, a protocol was developed involving a work-rest regimen that would take place in an environmental chamber.

Initial testing was carried out in normal laboratory conditions to provide a baseline against which data from the cold and hot temperature extremes could be compared.

A bit of background information is necessary to understand the rationale for selection of the test protocol. Experience gained in the field shows that in normal operation, the worker walks several hundred meters to his worksite, sometimes having to climb several flights of stairs. He then performs light plumbing repairs or he adjusts or monitors valve or gage panels. He then walks back after a 2-h work period. Worst-case workload would involve the rescue of a fallen co-worker during a hazardous operation. This would likely be short duration, intensive work.

Laboratory experimentation has shown that even the better conditioned test subjects (VO_2 = 50 mL/kg/min) could not perform treadmill Bruce Stage III while wearing the suit. In the laboratory, the Bruce treadmill protocol has served as a reference for the testing of a wide variety of protective equipment. It also is used as the basic qualification protocol for new subjects and therefore, has, been established as a baseline physiologic load.

A basic description of the suit is necessary to understand other aspects of the protocol origin. The PHE is a completely enclosed whole body suit made of chlorobutyl coated Nomex material. This is one of very few materials that is relatively impervious to the propellants, yet can be joined together into leak-proof seams, and can withstand the rigors of repeated flexing. The suit has two methods of environmental control. One version is called the backpack suit. It contains an Environmental Control Unit (ECU) that is worn on the user's back and is powered by liquid air. The ECU provides gaseous air after expansion through a heat exchanger. The user's body heat contributes to the heat transformation. This primary air is introduced into the ECU's venturi at a rate of approximately 42.5 L/min (1.5

TABLE 1—*Testing procedure.*

Time, min	Activity
−10	Sensor subject, perform final calibrations
0	Start test, collect baseline unsuited data (heart rate, temperatures)
10	Start suiting
20	Suiting complete, enter chamber or laboratory
40	Exercise 3 min on treadmill, 1.7 mph/10% grade
43	End exercise, start recovery period
63	Exercise 6 min: 3 min at 1.7 mph/10% grade and 3 min at 2.5 mph/12% grade
69	End exercise, start recovery period
89	End of test

standard cubic feet/minute (SCFM)). Total flow can reach 425 L/min (15 SCFM), and this flow is divided in an air distribution manifold to allow approximately 60% distribution to the helmet area with the remainder being circulated to the arms and legs. No face mask is worn. This air also provides much needed cooling in the normally experienced hot temperature environments. This version of the suit allows the user to be completely mobile, although the weight penalty, at 29.5 kg (65 lb), is heavy.

The other version of the PHE is the hoseline suit. This suit relies on a hoseline to supply air for respiratory and cooling purposes. A Vortex cooling unit is used. The internal air distribution is identical to the backpack version with the exception that air is not recirculated. Normal flows are about 170 L/min (6 SCFM). This suit has the advantage of relieving the user from carrying the 17.7-kg (39-lb) backpack, but does encumber him with a tether, that is, hoseline.

Considering the foregoing, a protocol was developed and is shown in Table 1.

The first 20 min of suited testing (20 to 40) allowed for the collection of baseline conditions in the suit while the subject stood unsupported. The first exercise period then provided a physical stress for the subject and the suit's ECU. Recovery from this was monitored during the second 20-min standing rest period. If this were satisfactory, that is, the ECU caused the helmet monitored level of oxygen to return to before-work levels, the second and more difficult exercise was imposed. A final 20-min recovery period was then allowed.

This protocol was carried out in three environmental conditions: cold chamber at −7°C (20°F), laboratory at 23°C (74°F), and hot chamber at 43°C (110°F). These extremes were chosen because of the possibility of experiencing the cold during night deservicing in the desert at Edwards Air Force Base and day servicing at the Kennedy Space Center in the summer. Each test was carried out at least four times. The actual test program is shown in Table 2.

In the normal test scenario, the volunteer subject was instrumented for a single channel of electrocardiogram (ECG) using a Hewlett Packard telemetry system. This ECG was received and displayed on a memory scope, fed to a heart-rate counter, and finally recorded on both strip chart and magnetic tape recorders.

TABLE 2—*Actual test program.*

Type	Cold	Laboratory	Hot
Backpack PHE	6	6	6
Hoseline PHE	4	5	7

The subject was also sensored for four skin temperatures and rectal temperature using YSI Series 700 thermistor probes. Skin sites were the forehead, the upper arm, the left chest area, and the right thigh. The suit interior temperature was monitored in the helmet and torso areas. All probes were connected to a Digitec Model 2000 datalogger.

A gas sample line was inserted into the helmet area, just in front of the nose, to monitor the oxygen and carbon dioxide concentrations throughout the test. This line was connected to a Beckman Metabolic Measurement cart that provided continuous analog output to a strip chart recorder and also printed 1-min average value data on a computer.

Suit pressure was monitored using a National Semiconductor integrated pressure chip connected to a locally produced buffer amplifier. Output was recorded on the strip chart.

Finally, the subject was equipped with a Snoopy hat type communications carrier. This afforded communications to the test conductor, safety monitor, and technician via an operational intercommunication system. All voice communication was recorded.

In the laboratory, a Quinton 18-60 treadmill and automatic programmer were used. Inside the Blickman environmental chamber, a Quinton Q-55 treadmill allowed the necessary head clearance and provided the selected workloads.

Results

The subjects used in this test series were all volunteers. Two females participated in some of the tests and four males completed all six configurations. The duration of the test series precluded participation by all subjects throughout the many months of testing. A total of eight subjects ranged in age from 26 to 49 years (mean = 36) and ranged in height from 157 to 193 cm (mean = 178 cm). Weight ranged from 62.1 to 95.2 kg (mean = 80.1 kg). Oxygen uptake, as measured in a standard stress test, with the Bruce protocol ranged from 36.2 to 54.8 mL/kg/min (mean = 42.5 mL/kg/min).

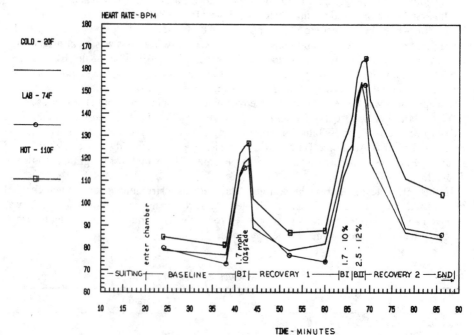

FIG. 1—*Heart rate for the backpack version (average of six subjects).*

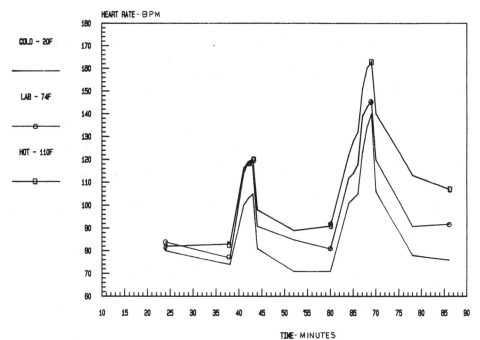

FIG. 2—*Heart rate for the hoseline tests* (*average of two, five, and seven subjects*).

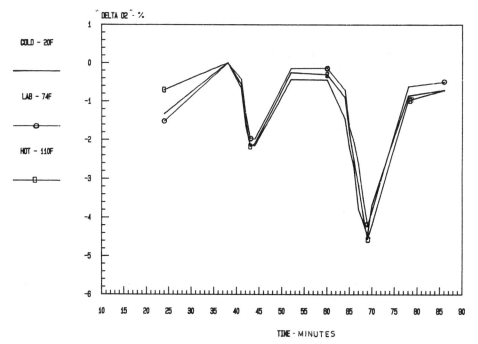

FIG. 3—*Delta oxygen concentration for the backpack suit* (*average of six subjects*).

The data taken were extensive but will be summarized here on a series of figures. Heart rate and average (over 1 min) oxygen and carbon dioxide concentrations were gathered continuously but are plotted at only the most meaningful times. Temperature was recorded every 2 min, but it also was plotted at only the most important times.

The heart rate responded as expected to the imposition of the load and temperature. Figure 1 shows heart rate in beats per minute (bpm) for the backpack version of the suit during all three temperature environments. Notice the rapid increase in heart rate to the first exercise bout at Minute 40. After only 3 min of slow walking, the average rate reached 127 bpm in the hot test. The rate does not recover to the resting value despite 20 min rest. The second exercise period drives the heart rate to 165 bpm. This exercise is very difficult, especially in the heat. One large contributor to the work is the difficulty in actually moving the legs fast enough to keep up with the treadmill at 2.5 mph. One actually pumps air from the leg on each step, a fact that is readily apparent in the suit pressure tracing. Heart rates in the cold and laboratory tests are slightly lower, but still show the difficulty of performing work that would be considered easy without the suit.

The data from the hoseline tests are shown on Fig. 2. Once again, the response to work is apparent although less than that of the backpack suit. The decreased response in this configuration is due to the decreased weight of the suit (11.8 versus 29.5 kg) and the superior gaseous environment of this suit as will be shown soon. Large differences in the heart rate can be noted between the different temperature tests. Since actual physical work is the same in each test, the higher response of the hot test during the second exercise is probably due to heat buildup in the suit.

The next two graphs, Figs. 3 and 4 plot oxygen concentration. This is actually specified as delta oxygen because of the fact that initial oxygen concentration after every backpack

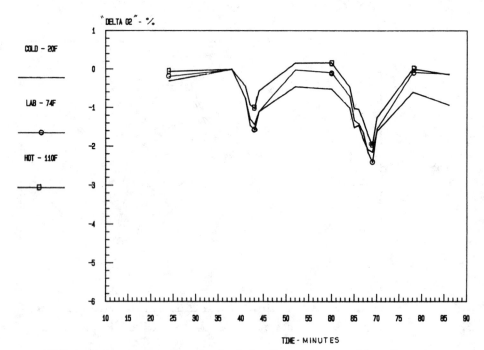

FIG. 4—*Delta oxygen concentration for the hoseline suit (average of four, five and seven subjects).*

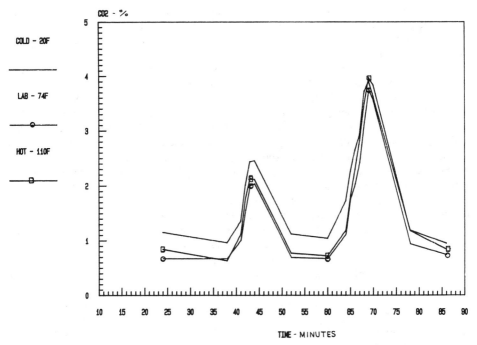

FIG. 5—*Carbon dioxide concentration for the backpack suit* (*average of six subjects*).

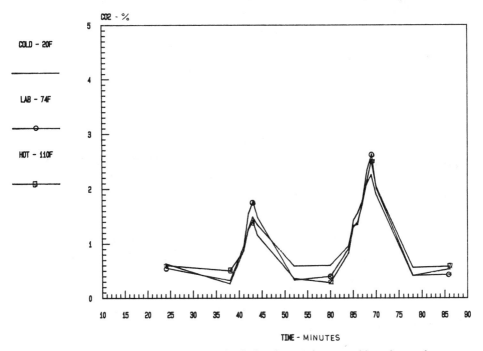

FIG. 6—*Carbon dioxide concentration for the hoseline suit* (*average of four, five, and seven subjects*).

fill is slightly different. Liquid air at the Space Center is manufactured from liquid nitrogen and oxygen. The air becomes oxygen rich after a period of weeks because of nitrogen boiloff. Therefore, a baseline concentration is selected after the first 20 min of baseline acquisition, or Minute 38. All other oxygen concentrations are then expressed as a difference from this value. Figure 3 shows the response of the backpack suit to all three environments. The delta 02 decreases to -2.18% after the first exercise and drops further to -4.56% after the second exercise. This is, of course, a concern when one considers the Occupational Safety and Health Administration (OSHA) limit of 19.5% minimum oxygen. This was not a problem in most of these tests as the absolute value at Minute 38 was 24% or greater. On the other hand, low initial concentrations, due to poor mixing or the use of liquified atmospheric air would most likely result in hypoxic exposure.

The hoseline delta oxygen data are plotted on Fig. 4. Note the general lessened response. Minimum delta oxygen during the first exercise was -1.57% and -2.41% during the second exercise. It was noted that several of the subjects considered the laboratory temperature test to be the most difficult of the three, but no immediate explanation can be found.

The next parameter of interest was the carbon dioxide concentration in the helmet area. Figure 5 shows the backpack suit results. The significant point of this data is that carbon dioxide is always greater than 0.67%, even during rest. Note that the threshold limit value (TLV) for an 8-h working day is 0.5%. However, these suits are infrequently worn more than 2 h, and the maximum duration of the backpack is just about 3 h. The short-term exposure level (STEL), as specified by the American Conference of Governmental Industrial Hygienists (ACGIH) is 1.5%. This value is exceeded during both of the exercise periods. Note the maximum during the second exercise of 3.97%. Data plotted in the Bioastronautics Data Book show that carbon dioxide at twice this concentration produces no detectable effects until the 10 to 15 min exposure is reached. One may suggest that the 1.5% STEL

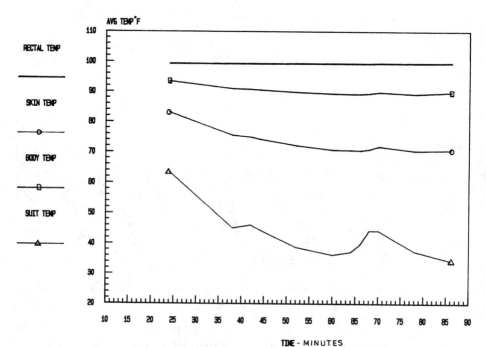

FIG. 7—*Temperature profiles for cold backpack tests at 7°C (20°F) (average of six).*

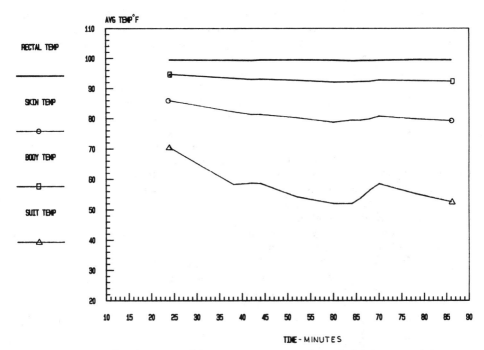

FIG. 8—*Backpack suit test at laboratory temperatures of 23°C (74°F) (average of six).*

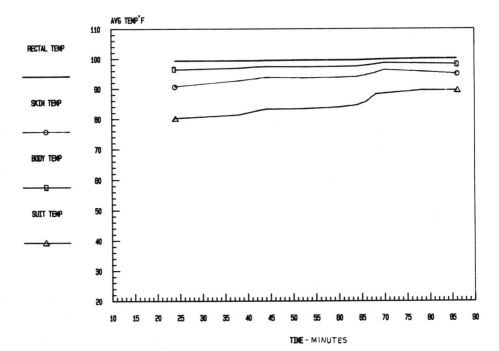

FIG. 9—*Temperature profiles for hot backpack tests at 43°C (110°F) (average of six).*

may not be realistic. Little evidence of adverse reaction to carbon dioxide could be noted in these tests, in spite of an alert posture to a possible response. Inspiration of air containing greater than 2% carbon dioxide is known to trigger significant increases in minute volume. While this was not monitored, particular attention was directed toward detection of other symptoms such as discomfort, fatigue, dizziness, headache, and shortness of breath.

The hoseline suit carbon dioxide data are found in Fig. 6. Once again, the gas concentrations are better. The maximum is 2.62% during the second exercise and was found during the laboratory test. The superior oxygen and carbon dioxide responses of the hoseline suit are due to the introduction of 6 SCFM of new air. Even though the vortex was fed 12 SCFM of new air, full opening of the control valve causes actual flow into the suit to be only 6 SCFM. The vortex was not operating during the cold tests and, as a matter of interest, is seldom used by field forces on hot days as they prefer to have the larger volume of ambient air rather than half that quantity of vortex conditioned air. This may lend some credibility to the thought that the cooling due to evaporation is more effective than cooling due to the lower temperature air being introduced into the suit.

The temperature profiles are now described over a series of six graphs (Figs. 7 to 12). Each of the test blocks has its own graph to avoid confusion of the four temperature profiles of interest. These profiles show suit, body, skin, and rectal temperature. The suit temperature is the average of the readings from the helmet and torso sensors. The sensors were located to avoid contact with the inside surface of the suit and the subject. The skin temperature was the average of four skin sensors located on the forehead, arm, chest, and thigh. The body temperature represented 0.65 times the rectal temperature plus 0.35 times the skin temperature.

Figure 7 shows a plot of these four temperature profiles during the cold backpack tests.

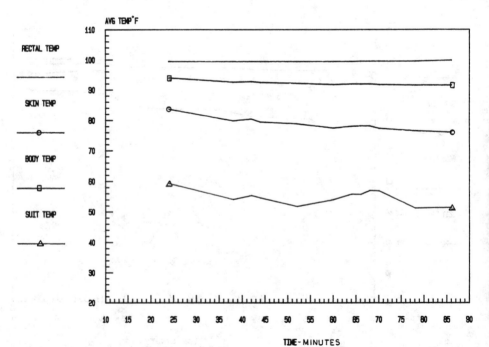

FIG. 10—*Cold hoseline suit temperature test at 7°C (20°F) (average of four).*

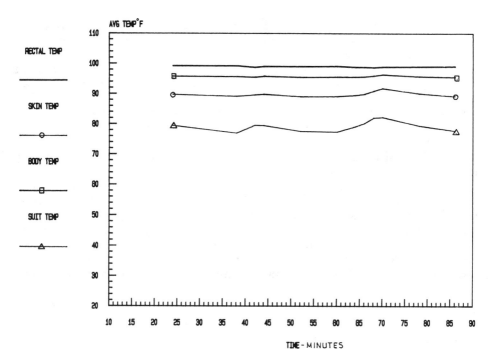

FIG. 11—*Hoseline suit test at laboratory temperatures of 23°C (74°F) (average of five).*

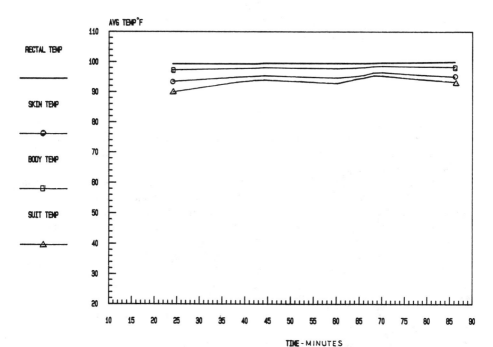

FIG. 12—*Temperature profiles for hot hoseline tests at 43°C (110°F) (average of seven).*

Temperatures are very cold and the subjects were uncomfortable, but none reached the point of uncontrollable shivering by Minute 89. Considering that the suit temperature dipped to 1.4°C (34.5°F) and the normal undergarment for the suit is a single layer of thermal underwear, one can readily understand the problem. If one adds more insulation or clothing to the subject, then less heat is available to the ECU heat exchanger and the air supplied into the venturi becomes colder.

Figure 8 plots the backpack test at laboratory temperatures. One can observe that some cooling capacity is still available as the suit and skin temperatures show a steady decrease, except for the exercise periods.

The backpack suit in the hot environment shown on Fig. 9 can not quite meet the cooling requirements in the hot chamber. A steady rise is apparent throughout.

Comparison of the hoseline suit temperature data in the cold, Fig. 10, with the backpack test shows that the effect of circulating 6 SCFM of ambient air into the suit is not as severe as the cryogenically supplied air of the backpack suit.

Figure 11 shows that the hoseline suit temperatures are basically stable. Figure 12 shows the response to the hot environment. This is subjectively the hottest test. The suit, body, and skin temperatures approach rectal temperature. Little gradient is visible therefore eliminating any real potential for relief on the part of the subject, even during the rest phases of the protocol.

Conclusion

Examination of these data make it clear that the imposition of a whole body suit, such as the PHE, can result in significant physiologic work on the user due to the protective system alone. Heart rates are driven to moderately high levels, supplied respiratory gases are not optimum, and thermal adversities are introduced.

The backpack version of the suit has the important advantage of allowing the user complete freedom to move about without a tether. The ECU provides superior cooling in the normal laboratory and hot environments. It provides nearly complete protection for time periods approaching 3 h. However, it burdens the user with a weight load that may be prohibitive to the small worker.

The hoseline suit offers reasonable conditions in the cold and lower ambient temperatures and provides superior oxygen and carbon dioxide characteristics. The 11.8 kg of suit weight is distributed about the body and mobility is good. The disadvantages are that the worker is tethered to a hoseline and the air supply must be large and of Grade D or better as both respiratory and cooling purposes must be satisfied.

This experience has made it clear to us at the Kennedy Space Center that the purchase and initial introduction of a new suit, even though it was of familiar configuration, is not a simple matter. Should an organization find itself in a position to institute personnel protection in the form of a full body suit, careful consideration must be made of all requirements. Suits that are currently on the market are not universally applicable. It is also impossible for the manufacturer to foresee particular needs and design appropriately.

Perhaps a more fundamental problem is the lack of universal suit testing methods. Suits do not (currently) fall under National Institute for Occupational Safety and Health (NIOSH) certification standards as do self-contained breathing apparatus. The potential for wide application makes creation of a "standard" suit test impractical. However, standard methods for data sampling and collection may be practical. For example, continuous sampling of the breathing zone at a particular location may eliminate testing irregularities. Also, skin and body temperature sampling according to the extensive work done by Ralph F. Goldman could be referenced.

Finally, it is important that the potential user of a whole body protective suit be knowledgeable about the physiologic impact a suit may present to his workmen. When engineering controls cannot eliminate the hazard, then the use of protective equipment is indicated. This use must be appropriate for the hazard and not compromise other aspects of worker safety or health.

Protection from Pesticides

Field Performance

Ann C. Slocum,[1] *Richard J. Nolan,*[2] *Lois C. Shern,*[3]
Sharleen L. Gay,[4] *and A. J. Turgeon*[5]

Development and Testing of Protective Clothing for Lawn-Care Specialists

REFERENCE: Slocum, A. C., Nolan, R. J., Shern, L. C., Gay, S. L., and Turgeon, A. J.,
"Development and Testing of Protective Clothing for Lawn-Care Specialists," *Performance
of Protective Clothing: Second Symposium, ASTM STP 989,* S. Z. Mansdorf, R. Sager, and
A. P. Nielsen, Eds., American Society for Testing and Materials, Philadelphia, 1988, pp. 557–
564.

ABSTRACT: The objective of this study was to develop protective clothing for lawn-care
specialists and to test the effectiveness of the clothing as a barrier to pesticides in comparison
with the regular company uniform. Six volunteers, three wearing the experimental protective
clothing and three the company uniform, sprayed a field dilution of Dursban 4E while carrying
out regular work activities on each of two test days. Volunteers served as their own controls,
wearing a different clothing treatment on the two days. Protective clothing consisted of a
cotton/polyester, long-sleeve knit shirt with woven yoke overlay and work pants lined with a
microporous film laminate in the lower legs and abdominal area. Urinary excretion of the
metabolite 3,5,6-TCP, standardized on the basis of creatinine concentration, measured pesticide absorption. Data indicated that the protective clothing significantly reduced the amount
of pesticide absorbed relative to the regular uniform.

KEY WORDS: protective clothing, lawn care, pesticide, phosphorothioate, Dursban, chlorpyrifos, urinary metabolite, biological monitoring

The lawn-care industry has grown rapidly in recent years [1]. Lawn specialists are employed
to mix, load, and spray pesticides and to maintain and clean equipment. The rapid growth
of the industry, coupled with the potential for considerable dermal exposure of lawn specialists to pesticides [2], makes reducing exposure for this occupational group an important
topic for study. Protective clothing is considered an ideal way to reduce applicators' exposure
to pesticides since the dermal route is typically the route of highest potential exposure,
especially when liquid formulations are used [3–7]. Clothing can provide a physical barrier
which reduces dermal contact with chemicals [8–13]. However, many types of protective
clothing on the market are unacceptable to the wearer because they are unattractive, uncomfortably warm [12,14–17], or do not fit well [14,15]. Thus there is a need to develop
clothing which offers protection and is acceptable to the user.

Laboratory studies [10,18], field studies [18,19], and simulated field studies [9,20,21] have
evaluated the effectiveness of fabrics or garments by quantifying the amount of pesticide

---

[1] Associate professor, Human Environment and Design Department, Michigan State University, East
Lansing, MI 48829.
[2] Health and Environmental Sciences, The Dow Chemical Company, Midland, MI 48674.
[3] Graduate research assistant, Family and Child Sciences Department, Michigan State University,
East Lansing, MI 48824.
[4] Assistant professor, Seattle Pacific University, Seattle, WA 98115.
[5] Formerly vice president, Tru Green Corp.; now professor and head, Department of Agronomy,
The Pennsylvania State University, University Park, PA 16802.

that passes through the clothing barrier. The results have been useful in assessing the relative effectiveness of clothing materials in reducing the amount of pesticide that reaches the underside of the fabric or fabrics and is thus available for contact with the skin. However, the critical question in terms of worker protection is, "Does the clothing reduce the amount of pesticide absorbed by the body?"

Biological monitoring of pesticide metabolites excreted in the urine can be used to evaluate the effectiveness of clothing [22,23], although it has been used more frequently to quantify exposure [5,24–26] and to compare the relative importance of the routes by which pesticides enter the body [6,27,28]. Davies et al. [11] assessed the protection that 100% cotton coveralls afforded citrus-grove workers mixing and applying ethion by quantifying the urinary metabolite, diethyl phosphate. Franklin et al. [23] evaluated the addition of rubberized cotton, when subjects were wearing long-sleeved coveralls over regular clothing, by measuring the urinary output of the active ingredient sprayed.

The objectives of this research were to develop prototype clothing for lawn-care specialists and to evaluate both its acceptability to participants and its effectiveness in reducing bodily absorption of chlorpyrifos when compared with the regular company uniform.

Methodology

Developing the Protective Garment

The design problem was identified using the process described by DeJonge [29], and design criteria were established for protection, motion, aesthetics, and comfort. Protection needs were based on analysis of dye-deposition patterns during timed spray intervals using commercial equipment. The largest exposure occurred on the lower portion of the body, with the lower legs, both front and back, receiving the greatest deposits of spray [30]. The need for free arm extension and arm and shoulder movement was determined from observations of body motions of applicators at work. Aesthetic and comfort criteria were developed through interviews with company officials and employees. Their responses indicated it was important that the protective garment look and fit like regular work clothing, maintain the company color scheme, and be reasonably comfortable when worn during the summer.

Variables tested in the development of the design included ways of providing for arm and shoulder movement, methods of securing lower leg protection, and construction techniques with three different fabrics. Experimental garment designs were constructed in full-scale muslin and judged by a panel of undergraduate students. Two prototypes were constructed and test-worn by a graduate student. Three sketches and two prototype garments were submitted to company officials for consideration.

The prototype chosen for field testing was a two-piece design that inconspicuously incorporated protective features (Fig. 1). The pants were a 50/50 cotton/polyester, green work-weight twill. The pants were lined from the knee down, both front and back, and in the abdominal area, with a 99.5-g microporous film laminate manufactured by Gore Associates. The shirt body and sleeves were constructed in a 50/50 cotton/polyester knit fabric with a yoke overlay of heavy cotton/polyester twill. The shirt had long sleeves, a high roll collar, and stretch mesh underarm gussets. The pants, shirt yoke, collar, and cuffs received three fluorocarbon treatments [31]. Laboratory studies indicate these treatments can reduce pesticide absorption and penetration through fabric [10,32,33]. However, the effectiveness of fluorocarbon treatment under field conditions is uncertain. Davies et al. [11] found no significant differences between plain and treated fabrics worn in field tests.

Wrist-length polyvinyl chloride (PVC) gloves were adapted for use by inserting a breathable mesh panel on the back for ventilation and elastic at the wrist on the palm side to reduce pesticide entry. Results of the deposition study [30] showed that dye was deposited primarily on the palm of the glove.

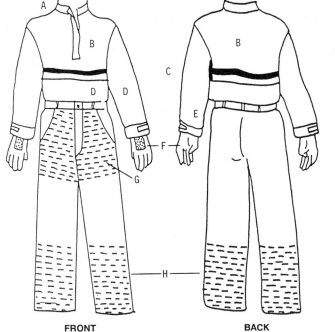

A. Woven stand-up collar
   lined with knit

B. Woven yoke overlay
   attached only at the
   neckline

C. Underarm stretch knit
   mesh gusset

D. Shirt body and sleeves
   of stretch knit

E. Long sleeves with cuffs

F. Wrist length gloves with
   breathable mesh panel

G. Abdominal pocket lining
   of microporous laminate

H. Lower leg lining of
   microporous laminate

FRONT   BACK

FIG. 1—*Two-piece protective uniform.*

Field-Testing the Garment

Participants—Six male Caucasians 21 to 37 years of age participated in the study. Volunteers were obtained through an orientation session presented to applicators employed at two southeastern Michigan branch offices of Tru Green Corp. Their experience as lawn specialists ranged from one to ten years. Participants gave their written informed consent; body measurements for clothing were taken, and specific instructions were given for specimen collection.

Study Design—The study was conducted over a two-week period in June 1985. The pesticide under study, Dursban 4E, was sprayed by all participants on two consecutive Mondays while performing their usual work activities. During the remainder of the study and for one week prior to the study, participants continued to work but did not spray Dursban,[6] Garlon,[6] or other organophosphates. On the first Monday, three participants wore protective clothing and three wore the regular uniform. On the second Monday the clothing treatment was reversed, so the effects of temperature, humidity, and general wind conditions would not be confounded with clothing treatment; thus subjects served as their own control.

Clothing Treatments—The regular uniform consisted of a short-sleeve, polo knit shirt of 50/50 cotton/polyester and 50/50 cotton/polyester work pants, identical to those of the protective garment except without the microporous film laminate lining. In order to remove sizing and increase barrier properties [11,18], new uniforms were prelaundered four times and distributed on each test day. Rubber gauntlet gloves were normally worn by the volunteers and therefore became a part of the regular uniform.

[6] Registered trademark of the Dow Chemical Company.

In addition to the prototype garments (Fig. 1), undershirts were added to the protective ensemble in order to add another layer for protection [11]. The undershirts were new, prelaundered, crew-neck style of 100% cotton. New 100% cotton shorts and socks were prelaundered four times and given to all participants on each test day. New nylon belts, rubber ankle-length boots, and a company baseball-style hat were provided on the first test day and were worn, without decontamination, again on the second test day.

Pesticide and Application Equipment—The insecticide Dursban 4E was chosen for the study because it is frequently used in the lawn-care industry [34] and because data were available on the fate of its active ingredient in the human body [35]. Chlorpyrifos (0,0-diethyl-0-[3,5,6-trichloro-2-pyridyl]) phosphorothioate is the active ingredient in Dursban insecticide. It has been established that chlorpyrifos is rapidly hydrolyzed in the body to 3,5,6-trichloro-2-pyridinol (3,5,6-TCP) [35,36]. A controlled human exposure study revealed that an average of 70% of a 0.5-mg/kg oral dose of chlorpyrifos was excreted in the urine as 3,5,6-TCP, with a half life of 27 h. No unchanged chlorpyrifos was found in the urine [35].

Participants were responsible for mixing and loading their trucks each spray day with the Dursban 4E emulsifiable concentrate, diluted in the ratio of 4.4 kg active ingredient with 378.5 L water. The diluted pesticide was sprayed from a 4560-L (1200 gal) regulation truck with hydrocell pump and Chemlawn hand spray gun which delivered 276 kPa (40 psi) at the nozzle.

Sampling and Analysis—During the field test, participants collected all urine voided during the 24-h interval immediately preceding the test day, the day Dursban 4E was sprayed, and for each of the following three days. This yielded two sets of five 24-h urine collections per subject. The urine was collected in Medi 24-Hour urine containers. As a matter of convenience collections were made in approximately 12-h units, with one unit consisting of time spent at work during the day, and the other unit the time spent after work and until reporting for work the next morning. Urine voided at the beginning of a collection interval was added to the container for the previous collection, except for the initial collection interval when the urine was discarded. Beginning and ending time of the collection period and volume were recorded, and an aliquot was stored under refrigeration for later analysis. Urine was analyzed for creatinine using the Jaffe' reaction [38] and for 3,5,6-TCP using a modification of the method described by Nolan [35]. A small volume of 10-N sodium hydroxide (NaOH) was added to dissolve sediment found in the urine specimens. An aliquot was then removed, treated with sulfuric acid, and placed in a water bath to hydrolyze any conjugates that might be present. The specimen was then extracted with diethyl ether. The extract was derivatized with Bis(trimethylsilyl)trifluoroacetamide (BSTFA) and analyzed by gas chromatography-mass spectrocopy (GC-MS).

Records were also kept of the amount of pesticide sprayed, an estimate of square feet treated, beginning and ending time of the work day, and the time participants showered at the end of the workday. Participants also completed a short questionnaire evaluating the prototype clothing.

Results

Participant Assessment

The participants evaluated four aspects of the protective garment—fit, mobility, appearance, and comfort—on a five-point scale where 1 represented an assessment that the garment was "very good" and 5, "very poor." Participant ratings were higher for the shirt than for the pants for all four criteria, with an overall mean of 2.1 and 2.7, respectively. Thus shirts were evaluated as "good," but the pants were rated about "average." For both shirt and pants, participants rated the individual aspects of fit, mobility, and appearance higher than comfort. For the shirt, the combined mean for fit, mobility, and appearance was 1.7, while the mean comfort rating was 3.0. Similarly, the mean of the combined ratings for fit, mobility, and appearance for the pants was 2.4, compared to a mean of 3.6 for comfort. Participants were not asked to evaluate the regular clothing, so comparisons with the protective clothing cannot be made. However, unsolicited comments indicated participants were also dissatisfied with the comfort of the regular uniform.

Biomonitoring

Based on the volume of the specimen and the amount of creatinine excreted, several of the urine specimens did not represent complete 24-h collections. Therefore, the amount of 3,5,6-TCP in each collection was normalized by dividing the amount of 3,5,6-TCP by the amount of creatinine in that collection. Creatinine is a substance formed endogeneously that is excreted in the urine at a constant rate. It has been previously demonstrated that urinary creatinine is a valid means of normalizing urinary excretion data [35]. The cumulative total (μg) of 3,5,6-TCP excreted per subject was calculated, after normalization, from the urine volume (mL) and the concentration (μg/mL) of 3,5,6-TCP. An adjustment was made for the small amount of 3,5,6-TCP in the pre-spray urine specimens. The cumulative amount of 3,5,6-TCP excreted per subject ranged from 148.6 to 41.6 μg when regular clothing was worn, and from 91.2 to 40.2 μg when protective clothing was worn, with a mean value of 85.88 and 55.05 μg, respectively (Table 1).

It can be observed that in all cases the amount of the metabolite 3,5,6-TCP excreted was less when protective clothing was worn than when the usual company uniform was worn. The differences were statistically significant, $t = 3.443$, $p = 0.009$. The percent reduction in amount excreted when protective clothing was worn ranged from a low of 3% to a high of 54.5%, with an average reduction of 34%. The magnitude of the differences among individuals is striking.

Some of the differences in the amount of pesticide absorbed among subjects can be accounted for on the basis of differences in exposure. An attempt was made to balance work routes so that exposure among applicators would be similar on each of the two test days. However, there was about a twofold variation in the amount of pesticide sprayed across subjects on each test day, and for Volunteers 1 and 2, approximately a 35% difference between the two test days (Table 2).

TABLE 1—*Micrograms of 3,5,6-TCP excreted when regular and protective clothing were worn.*

Volunteer	Regular Clothing, μg	Protective Clothing, μg	Difference, μg	% Reduction
1	101.4	46.7	54.7	53.9
2	148.6	91.2	57.4	38.6
3	103.9	87.2	16.7	16.1
4	61.3	38.4	22.9	37.4
5	58.5	26.6	31.9	54.5
6	41.6	40.2	1.4	3.4
Mean	85.9	55.1	30.8	34.0
Standard deviation	39.5	27.3	21.9	20.5
Paired student: $t = 3.443$, $p = 0.009$				

TABLE 2—*Micrograms 3,5,6-TCP excreted per litre of pesticide sprayed when regular and protective clothing were worn.*

Volunteer	Regular Clothing		Protective Clothing			
	Pesticide Sprayed, L	3,5,6-TCP Excreted, L	Pesticide Sprayed, L	3,5,6-TCP Excreted, L	Difference, μg/L	% Reduction, μg/L
1	3790	0.0268	2843	0.0164	0.0104	38.6
2	2016	0.0737	3127	0.0292	0.0445	60.4
3	3487	0.0298	3638	0.0240	0.0058	19.6
4	2577	0.0238	2843	0.0135	0.0103	43.2
5	1706	0.0343	1546	0.0172	0.0171	49.8
6	2274	0.0183	2464	0.0163	0.0020	10.8
Mean	2642	0.0345	2744	0.0194	0.0150	37.1
Standard deviation	829	0.0200	704	0.0059	0.0153	18.7
Paired student: $t = 2.398$, $p = 0.030$						

When micrograms of 3,5,6-TCP excreted are expressed per litre of pesticide sprayed, the trend continues for more to be excreted when the regular clothing is worn. The differences are statistically significant, $t = 2.398$, $p = 0.030$. There is still approximately 50 percentage points difference between participants with the highest and lowest reduction from protective clothing. When the data are expressed in terms of litres sprayed, Volunteers 3 and 6 continue to have the lowest amount of change in the amount excreted under the two clothing systems, a 20% and 11% reduction respectively. Volunteers 2 and 5 have the highest percent reduction, 60% and 50%, respectively, and also had the highest proportion of micrograms absorbed per litre sprayed. However, Volunteer 1 ranked high in the amount sprayed, but low in the amount of urinary excretion of the metabolite. This phenomenon has been observed by other researchers [37]; the amount absorbed is not simply a function of the amount sprayed.

Discussion

Our study adds to the growing body of research [11,23] that indicates clothing can make a difference in the amount of pesticide absorbed by the body. By analyzing urinary metabolites, Davies et al. [11] found that a 100% cotton coverall significantly reduced absorption of ethion for citrus-grove sprayers. The present study demonstrates that a different fabric and garment design can also succeed in reducing absorption. In this study, regular work clothing was adapted by placing a microporous film laminate in the areas of highest spray deposition and by increasing body coverage through long sleeves and additional layers, for example, an undershirt. These findings are consistent with laboratory studies that indicated additional layers [11] and a microporous fabric [20] provided a good barrier to pesticide penetration.

While the data indicate that modifications in clothing can reduce pesticide exposure, the amount of reduction varies among applicators. Differences among volunteers may be due in part to environmental conditions such as gusts of wind, and to personal work and hygiene practices. Other researchers have observed variability across subjects and hypothesized that it was due to differences in precautions followed by workers [7].

In this study, each participant was observed in an unstructured format for the first two hours of spray time on each test day. Observations of pesticide soiling of clothing during the first two hours are consistent with absorption data. Volunteers 4 and 6, who were neat and careful in their work, had the lowest amounts of metabolite excreted per litre sprayed. However, the participants were not observed during mixing and loading of the insecticide. Libich et al. [6] suggests that a small spill of a concentrated chemical during mixing and loading can be equivalent to a full day of field exposure. The less careful employees could have experienced exposure at this stage that would not have been observed but should have been reported by participants. Daily records were completed in which participants were asked to report any unusual events such as spills or broken hoses. No circumstances were reported that might account for variations.

The problem of designing protective clothing that simultaneously meets multiple criteria is again illustrated. In this project, protection had to be increased while meeting certain aesthetic, motion, and comfort criteria. Since lawn-care specialists act as sales representatives, a professional image is important. Thus, loose-fitting, 100% cotton coveralls were not an acceptable alternative. Although participants' ratings for comfort might have been higher, the company image would not have been maintained. Cost was not considered a primary criterion in this study. It was anticipated that if the clothing successfully reduced absorption, the next phase of the project would involve obtaining cost estimates for large-scale purchases.

Clothing can provide an added margin of safety in the handling of pesticides. Whether the reduction in absorption is large enough to warrant changes in clothing requires a judgment of the long-term health benefits and the immediate cost of implementing a change in clothing. The decision is complicated by the fact that benefits received from protective clothing differ among individuals. However, the experimental uniform did reduce exposure for those who had absorbed the greatest amounts of chlorpyrifos while wearing the regular uniform. This study suggests that in the case of highly toxic substances or large exposures, because an important reduction in absorption of pesticide by the body can be achieved through clothing, protective clothing should be considered.

Acknowledgment

This research was supported by the Dow Chemical Company, Midland, Michigan; Tru Green Corp., Alphretta, Georgia; and Benham Chemical Company, Farmington Hills, Michigan.

References

[1] "Million Dollar List is Up to 39," *Lawn Care Industry*, Vol. 8, No. 5, 1984, pp. 1,32.
[2] Freeborg, R. P., Damiel, W. H., and Konopinski, V. J., *Dermal Exposure Related to Pesticide Use, Discussion of Risk Assessments*. ACS Symposium Series No. 273, R. C. Honeycutt, G. Zweig, and N. N. Ragsdale, Eds., American Chemical Society, 1985, pp. 287–295.
[3] Reinert, J. C. and Severn, D. J., *Dermal Exposure Related to Pesticide Use, Discussion of Risk Assessment*, ACS Symposium Series No. 273, R. C. Honeycutt, G. Zweig, and N. N. Ragsdale, Eds., American Chemical Society, 1985, pp. 357–368.
[4] Dedek, W. in *Proceedings*, 6th International Workshop of the Scientific Committee on Pesticides, International Association of Occupational Health, Elsevier, New York, 1982, pp. 127–207.
[5] Lavy, T. L., Shephard, J. S., and Mattice, J. D., *Journal of Agricultural and Food Chemistry*, Vol. 28, 1980, pp. 626–630.
[6] Libich, S., To, J. C., Frank, R., and Sirons, G. J., *American Industrial Hygiene Association Journal*, Vol. 45, No. 1, 1984, pp. 56–62.
[7] Wolfe, H. R., *Weeds, Trees, and Turf*, April 1973, pp. 12, 36, 37, 52, 53.
[8] Branson, D., Ayers, G. S., and Henry, M. in *Performance of Protective Clothing, ASTM STP 900*, R. L. Barker and G. C. Coletta, Eds., American Society for Testing and Materials, Philadelphia, 1986, pp. 114–120.
[9] Staiff, D. C., Davis, J. E., and Stevens, E. R., *Archives of Environmental Contamination and Toxicology*, Vol. 11, 1982, pp. 391–398.

[10] Laughlin, J. M., Easley, C. B., Gold, R. E., and Hill, R. M. in *Performance of Protective Clothing, ASTM STP 900*, R. L. Barker and G. C. Coletta, Eds., American Society for Testing and Materials, Philadelphia, 1986, pp. 136–150.

[11] Davies, J. E., Freed, V. H., Enos, H. F., Duncan, R. C., Barquet, A., Morgade, C., Peters, L. J., and Danauskas, J. X., *Journal of Occupational Medicine*, Vol. 24, No. 6, 1982, pp. 464–468.

[12] Nigg, H. N., Stamper, J. H., and Queen, R. M., *Archives of Environmental Contamination and Toxicology*, Vol. 15, 1986, pp. 121–134.

[13] Leavitt, J. R. C., Gold, R. E., Holeslaw, T., and Tupy, D., *Archives of Environmental Contamination and Toxicology*, Vol. 11, 1982, pp. 57–62.

[14] DeJonge, J. O. and Munson, D., *Textile Research Institute*, Jan. 1986, pp. 27–34.

[15] Coletta, G. C. and Spence, M. W., *Occupational and Health Safety*, April 1985, pp. 20, 21, 23, 72.

[16] Moraski, R. V. and Nielsen, A. P., *Dermal Exposure Related to Pesticide Use, Discussion of Risk Assessments*, ACS Symposium Series No. 273, R. C. Honeycutt, G. Zweig, and N. N. Ragsdale, Eds., American Chemical Society, 1985, pp. 395–401.

[17] DeJonge, J. O., Vredevoogd, J., and Henry, M. S., *Clothing and Textile Research Journal*, Vol. 2, No. 1, fall/winter 1983/84, pp. 9–14.

[18] Orlando, J., Branson, D., and Henry, M., *Studies in Environmental Science 24, Determinants and Assessment of Pesticide Exposure, Proceedings of a Working Conference, Hershey, PA, 1980*, M. Siewierski, Ed., Elsevier, New York, 1984, pp. 53–66.

[19] Wolfe, H. R., Durham, W. F., and Armstrong, J. F., *Archives of Environmental Health*, Vol. 14, 1967, pp. 622–633.

[20] Orlando, J., Branson, D., Ayres, G., and Leavitt, R., *Journal of Environmental Science and Health*, Vol. 16, No. 5, 1981, pp. 619–628.

[21] Freed, V. H., Davies, J. E., Peters, L. J., and Parveen, F., *Residue Reviews*, Vol. 75, 1980, pp. 159–167.

[22] Davies, J. E. in *Proceedings: Pesticide Residue Hazards to Farm Workers*, U.S. Dept. of Health, Education, and Welfare, May 1976, pp. 19–23.

[23] Franklin, C. A., Fenske, R. A., Greenhalgh, R., Mathieu, L., Denley, H. V., Leffingwell, J. T., and Spear, R. C., *Journal of Toxicology and Environmental Health*, Vol. 7, 1981, pp. 715–731.

[24] Durham, W. F. and Wolfe, H. R., *Bulletin World Health Organization*, Vol. 26, 1962, pp. 75–91.

[25] Davies, J. E., Enos, H. F., Barquet, A., Morgade, C., and Danauskas, J. X., "Pesticide Monitoring Studies: The Epidemiologic and Toxicologic Potential of Urinary Metabolites," *Toxicology and Occupational Medicine*, W. B. Deichmann, Ed., Elsevier-North Holland, New York, Amsterdam, Oxford, 1979, pp. 369–378.

[26] Devine, J. M., Kinoshita, G. B., Peterson, R. P., and Picard, G. L., *Archives of Environmental Contamination and Toxicology*, Vol. 15, 1986, pp. 113–119.

[27] Durham, W. F., Wolfe, H. R., and Elliott, J. W., *Archives of Environmental Health*, Vol. 24, 1972, pp. 381–387.

[28] Manninen, A., Kangas, J., Klen, T., and Savolainen, H., *Archives of Environmental Contamination and Toxicology*, Vol. 15, 1986, pp. 107–111.

[29] DeJonge, J. O., in *Clothing: the Portable Environment*, S. M. Watkins, Iowa State University Press, Ames, IA, 1984, pp. vii–xi.

[30] Slocum, A. C. and Shern, L. C., "Dye Deposition Patterns during Simulated Work Activities by Lawn Care Specialists," unpublished data, Michigan State University, Lansing, MI, 1986.

[31] Gay, S. L., "The Development and Evaluation of a Protective Garment for Lawn Care Specialists," unpublished master's thesis, Michigan State University, Lansing, MI, 1986.

[32] Keaschall, J. L., Laughlin, J. M., and Gold, R. in *Performance of Protective Clothing, ASTM STP 900*, R. L. Barker and G. C. Coletta, Eds., American Society for Testing and Materials, Philadelphia, 1986, pp. 162–176.

[33] Freed, W. H., Davies, J. E., Peters, L. J., and Parveen, F., *Residue Reviews*, Vol. 75, 1980, pp. 159–167.

[34] "State of the Industry: Most LCOs are Independents," *Lawn Care Industry*, June 1984, p. 3.

[35] Nolan, R. J., Rick, D. L., Freshour, N. L., and Saunders, J. H., *Toxicology and Applied Pharmacology*, Vol. 73, 1984, pp. 8–15.

[36] Teitz, N. W., *Fundamentals of Clinical Chemistry*, 2nd ed., Saunders, Philadelphia, 1976.

[37] Grover, R., Cessna, A. J., Muir, N. J., Riedel, D., Franklin, C. A., and Yoshida, K., *Archives of Environmental Contamination and Toxicology*, Vol. 15, 1986, pp. 677–686.

Annette J. Fraser[1] and Vera B. Keeble[2]

Factors Influencing Design of Protective Clothing for Pesticide Application

REFERENCE: Fraser, A. J. and Keeble, V. B., **"Factors Influencing Design of Protective Clothing for Pesticide Application,"** *Performance of Protective Clothing: Second Symposium, ASTM STP 989,* S. Z. Mansdorf, R. Sager, and A. P. Nielsen, Eds., American Society for Testing and Materials, Philadelphia, 1988, pp. 565–572.

ABSTRACT: Use of protective clothing while applying pesticides to crops is considered a deterrent to dermal exposure. A design for a protective garment and hood for fruit orchard workers was developed at Utah State University. The prototype was manufactured in Gore-Tex, Saranex-laminated Tyvek, and an experimental composite structure. Workers wore all garments and personal work clothing in a field setting. After wearing the prototype on the job for an average 2-h period, fruit orchard workers evaluated the garment for functional designing qualities, sizing, and styling features. The protective suit was favorably rated. The hood design was found to be inadequate. Specifications for the garment needed length adjustment in the sleeve area and additional ease through shoulders. The hood design needs complete revision. The prototype evaluation suggested that desirable styling features and appropriate sizing characteristics definitely increase wearability potential of protective garments.

KEY WORDS: suits, protective clothing, pesticide protective clothing, functional clothing design

The use of protective clothing is recommended while applying pesticides to prevent dermal exposure. Pesticide contamination through the skin is the most common cause of poisoning during mixing, loading, application, and equipment maintenance. To limit exposure, a long-sleeved protective suit is recommended to wear over normal work clothes [1].

Protective clothing currently available at retail appears unacceptable to many agricultural workers because of poor thermal comfort qualities. Branson suggested there is a need for clothing that is both comfortable and socially acceptable to individuals occupationally exposed to pesticides [2]. Acceptability of the protective garment's appearance is a factor related to wearing potential and would increase if the protective garment were similar in style to regularly worn work clothing [3]. Optimally, problems associated with pesticide penetration, garment comfort, aesthetic styling, and sizing should be solved in the design stage. Risk assessment and evaluation of design features could be made prior to garment development.

This paper examines the development and evaluation of a design for a protective garment used in a field study involving pesticide applications in Utah fruit orchards. The purpose of this study was to evaluate design qualities and fit of a prototype garment worn while mixing and spraying pesticides. The fruit growers' responses to selected questions after wearing the

[1] Program director, Clothing, Textiles, and Design, University of Wisconsin-Stout, Menomonie, WI 54751.
[2] Extension specialist for Textiles and Clothing, Utah State University, Logan, Utah 84322.

protective garments formed the basis for the initial evaluation of the garment styling and fit. This study was part of a field study in which fabric penetration and thermal comfort were also investigated.

Procedure

Test Garment Development

The prototype for the protective garment was produced in a design laboratory at Utah State University in 1985. Advanced fashion design students utilized theoretically-based functional design strategies [4] and mass-production techniques to produce 14 protective garments.

Investigation of the Problem

A review of pertinent literature that identified the hazards of pesticide application was presented to the design students. Students then examined garments that were available commercially for worker protection and discussed limitations and positive design features.

The students interviewed a panel of workers that included fruit orchard operators, a commercial pest control operator who manages his own business, and a State Extension Horticultural Specialist who operates a family fruit orchard. The students explored fabric qualities and styling features appropriate for protective clothing. The panel concluded that protective clothing needs to be cool enough for use in hot weather; the fabric should be absorbent and breathable, yet form a barrier to agricultural chemicals; bagginess at arm and leg openings in protective garments should be controlled with Velcro closures to avoid catching part of the garment in machinery; high collars and an overlapping front closure would provide protection; and raglan sleeves would afford more movement through the shoulder area. It was stated that during application with airblast sprayers, pesticide contacts the back, shoulders, and head; therefore, a headpiece was considered most desirable. The panel eliminated the idea of a drawstring hood because they thought it inhibited movement. Composite features preferred for protective headgear were described as similar to a French foreign legion hat. The main concern of the panel was comfort. They said that the biggest market for protective clothing for fruit growers was for the man driving the tractor. Fifty dollars was considered a fair market price for a good suit.

Assessment of Critical Design Factors

After several brainstorming sessions, students established design specifications. They were:

1. Garment would cover the body, especially forearms, and fit closely around face and neck.
2. Garment would have adjustable tabs secured by Velcro at wrists.
3. Garment would have a high collar to protect the neck.
4. Garment would have a long zippered closure down the front covered with a flap secured with Velcro.
5. Garment would have a separate hood that could be worn with goggles and respirator. Hood would have a brim to protect the face from drift.
6. Garment would be fabricated in Gore-Tex, a fabric that offers protection and comfort factors.
7. Garment would have seams sealed after construction to prevent leakage.

Each student sketched an original garment, justifying design features. Two designs were selected by the project supervisors to be made up as samples. The selected styles were similar in design except that one had a dropped shoulder with set-in sleeves and the other had raglan sleeves. Basic patterns were derived from commercially available outerwear patterns. Samples were constructed using industrial sewing equipment. The completed samples were tested using simulation of movement involved in pesticide application. Fit was assessed as well as ease in getting in and out of the garment. Styling features from each sample were incorporated into a single protective garment and hood design. The raglan sleeve style was selected because most individuals believed that cut afforded greater movement capability in the shoulder area.

Development of Prototype

The sample patterns were adapted to the prototype style lines. The body of the test garment design combined the side, front, and back in one piece to eliminate the need for joining via a side seam. Thus, front and back pieces were joined at leg inseam, at center front, and center back. A triangular piece was set in the back shoulder area horizontally crossing the center back seam. After underarm seaming, shaped raglan sleeves were set in diagonally to the neck. Finished tabs with Velcro fasteners were preset at center back, underarm, and inseams for width control to adjust for bagginess and leakage potential at the waist, arms and legs. A stand collar for added protection was sewn into the neck. A zipper was set into the front opening, then covered with a lapped panel that extended into the collar area. A Velcro strip was applied to the length of the panel opening for added protection (Fig. 1).

The hood incorporated two sections seamed from center back curving up through the top head area. The pattern was shaped to cover the total head extending over the shoulders for added protection. The hood covered the shoulder seam of the raglan sleeve and continued under the chin area to cover the neck. The center front lapped and was secured with a Velcro fastener. A curved bill or beak stiffened with fusible interfacing was set in the top front opening as additional protection for the face (Fig. 2).

A complete garment was made up in Gore-Tex, the fabric designated to be used for the field study. Gore-Tex, containing a breathable membrane of microporous polymeric film of polytetrafluoroethylene sandwiched between a 100% nylon woven outer layer and a 100% nylon tricot inner layer, has shown to be an effective barrier to penetration of pesticides and is comfortable for workers [5]. As with the earlier samples, the prototype sample was evaluated for fit and comfort factors. An adjustment was made in belt height at insertion point in the center back seam.

Patterns were then drafted in small, medium, and large sizes using commercial men's outerwear patterns as sizing criteria. A marker (cutting layout) was drawn to cut 14 garments with minimum fabric waste.

Garments were cut and sewn using mass production (industrial) techniques. The finished garments were seam sealed according to commercially developed methods.

Research Methods

The purpose of this research was to evaluate the design for a protective garment to be used during mixing and spraying of pesticides in orchards. In addition to the 14 Gore-Tex garments produced for the study, the design was sewn in two other fabrications by commercial manufacturers. One fabric was Saranex-laminated Tyvek, a complex fabric consisting of an outer layer of Saranex 23, a coextruded multilayered film 0.05 mm thick. It has an outside

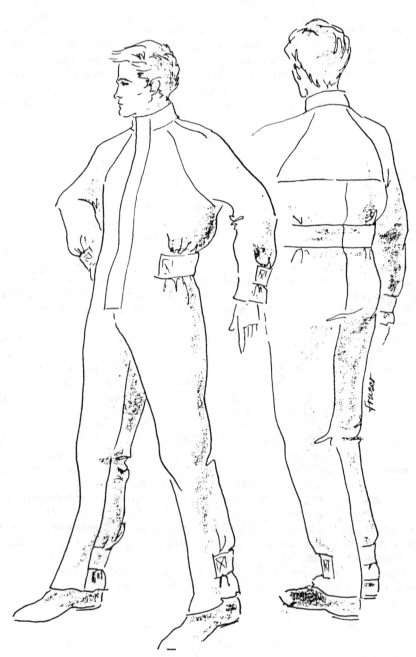

FIG. 1—*Prototype protective garment.*

FIG. 2—*Hood design.*

layer of low-density polyethylene, an inner layer of Saran, a copolymer of vinylidene chloride and vinyl chloride, and the other outside layer of ethyl vinyl acetate/low-density polyethylene copolymer, which is used for bonding to the Tyvek. Tyvek is a spun-bonded sheet structure of high-density polyethylene fibers [6]. The third material was an experimental composite structure. Each thought the same design was used to produce garments from three different fabrications, the design evaluation was independent of the fabrication. This study was part of a field study that investigated fabric penetration by the test pesticide and comfort for the different fabrications.

Workers chose a garment size by selecting and donning garments to determine the best fit. Each worker wore identically sized garments from all three fabrications while routinely mixing and spraying pesticides in fruit orchards. Work periods varied from 1 to 3 h although the mean work time was 2 h. Just before removing each garment, subjects were questioned about how styling features and fit affected the suitability of the garment during wear. Workers were asked if the hood interfered with vision or hearing and if the garment inhibited their work in any way. Subjects were encouraged to comment on any design features that affected the wearability of the garment. Questions also attempted to identify any problems related to length and width measurements of the prototype as well as general fit of the garment and hood.

Results and Discussion

Eighteen fruit growers and workers participated in the field study. Body build and height were variable. Height ranged from 1.57 m with a mean of 1.77 and a standard deviation of 0.11. Weight ranged from 58.96 to 97.51 kg with a mean of 80.27 kg and a standard deviation of 12.12.

To assess the effectiveness of the design in the field situation, subjects were asked questions just before removing the garments, see Table 1. Eighty-eight percent of the workers stated they were able to get in and out of the garment easily. One person said he preferred a two-piece garment and three stated that it was snug getting in and out. One subject said it was

TABLE 1—*Workers' response to questions related to prototype design features.*

	Percentile Response		
Question	N^a	Yes	No
1. Were you able to get in and out of the garment easily?	49	88	12
2. Did any part of the garment get in your way?	45	44	56
3. Could you perform your work easily in this garment?	47	94	6
4. Was there anything about this garment that you disliked?	42	55	45
5. Did the beak on the hood protect your face from drift?	30	63	37
6. Were you able to hear well?	39	87	13
7. Were you able to see well?	46	78	22

[a] N = number of workers asked question. Not all workers participated; therefore, responses are represented by percentile values.

"not bad" compared to what he was used to. Fifty-six percent of the workers did not believe that any part of the garment got in their way during the operation of their work, although some commented that the neck rubbed and was uncomfortable in front. Two individuals found the garment neck too tight, especially when bending over. One person suggested a V-shaped neck.

The subjects were also questioned about the sizing of the garment. Some workers found the sleeves too short and some found them too long. This varied with the size of the garment as well as the size of the worker (Table 2). Approximately 65% of the sample desired longer sleeves, 50% of these were in the largest size garment. Slightly over half the sample was satisfied with leg length. Comments suggested that pant legs rode up while the subject was engaged in driving the tractor. The crotch length related to the body torso of the garment. Even though the workers had a wide variety of body builds, 75% of the workers indicated satisfaction with crotch length.

Sixty-eight percent of the workers indicated there was ample room in the shoulder area of the garment. Eighteen percent felt that the large size needed to be cut with additional ease. The majority of subjects felt that there was plenty of room in the waist, hips, legs, and sleeves. The garment styling for width was adequate except in the shoulder area (Table 3).

The major complaint was the hood. The hood did not fit well or interfered with peripheral vision or both. Although 94% of the subjects stated they could perform their work easily

TABLE 2—*Workers' response to prototype length allowance.*

	Percentile Response[a]								
	Too Short/Small			Just Right			Too Long/Large		
Garment Feature	S^b	M^c	L^d	S	M	L	S	M	L
Is the length of									
the sleeve (N^e = 52)	...	15	50	6	13	4	...	6	6
the leg (N = 52)	...	3	12	6	12	35	...	12	20
the crotch (N = 48)	4	9	4	6	21	48	...	2	6

[a] Percentiles are rounded off to the nearest whole number.
[b] S = small.
[c] M = medium.
[d] L = large.
[e] N = number of workers responding.

TABLE 3—*Workers' response to prototype width allowance.*

| | Percentile Response[a] | | | | | | | | |
| | Too Short/Small | | | Just Right | | | Too Long/Large | | |
Garment Feature	S[b]	M[c]	L[d]	S	M	L	S	M	L
Do you have plenty of room									
through the shoulders (N^e = 50)	2	6	18	6	24	38	...	...	6
around the waist (N = 49)	...	4	...	6	23	57	2	2	6
around the hips (N = 49)	2	...	...	6	23	57	...	6	6
in the legs (N = 47)	...	4	...	6	24	55	...	...	11
in the arms (N = 46)	...	7	2	7	23	54	...	...	7

[a] Percentiles are rounded off to the nearest whole number.
[b] S = small.
[c] M = medium.
[d] L = large.
[e] N = number of workers responding.

while wearing the garment, the exceptions related to the hood. Of the 56% of respondents who stated that there was something about the garment they disliked, most objected to the hood as being unsatisfactory. One person suggested attaching the hood so that it could be pushed back when not needed, but pulled forward when exposure was greatest. When asked if the beak protected the face from drift, 63% of those responding said yes. Those who were able to keep the hood in place believed that it was more protective than working without a head covering. Others stated that their face was protected as long as the hood stayed in place. Some thought the beak should have been longer to better shield the face. Many did not wear the hood because it was a bad fit, crept forward over the eyes, or shifted while they worked. As the workers discontinued wearing the hood as the test progressed, there were fewer respondents to questions related to hood fit. Even though many workers chose not to wear the hood, 95% of those who did use it indicated satisfaction in fit around the face as well as general fit (Table 4).

Eighty-seven percent of those responding could hear well and 78% could see well. Some subjects stated that they preferred not to hear well because their equipment was too loud. Some wear ear plugs and if the hood reduced the noise that would be a plus. Two people

TABLE 4—*Workers' response to hood fit.*

| | Percentile Response[a] | | | | | | | | |
| | Too Short/Small | | | Just Right | | | Too Long/Large | | |
Garment Feature	S[b]	M[c]	L[d]	S	M	L	S	M	L
How well does the hood fit?									
around the face (N^e = 39)	...	...	5	23	18	54	...	...	...
over the head (N = 31)	...	...	2	10	23	65	...	...	...

[a] Response reflects response of workers wearing small, medium and large protective garments who wore hood, which was available only in one size.
[b] S = small.
[c] M = medium.
[d] L = large.
[e] N = number of workers responding.

found the hood did reduce their ability to hear. Those workers who were unable to see well had problems because the hood slipped and blocked their vision.

Workers gave suggestions for improving the garment design. Although they liked the idea of a hood that protected the head and neck, this design was unsatisfactory. They suggested the hood should be redesigned so that it would stay in place during use. Several workers wanted pockets on the outside of the garment as well as slits so that they could reach inside to their jeans pockets.

Conclusions and Recommendations

The design features of the protective suit were acceptable to the workers. However, the hood evaluation suggested that design features need to be readdressed to produce an improved style that will be secure to the head while wearing and offer protection without impeding vision.

Relatively few fit problems were observed. Existing problems related primarily to sizing. Additional width in the back shoulder area would allow for greater ease and more comfort to the worker. The sleeve length was too short for a majority of subjects. One third of workers agreed that the garment pant leg was too long. A number of subjects felt the need of additional length in the crotch area. The limited range of sizes available to a rather wide range of body builds presented some fit problems. Sizing for protective garments that are worn over ordinary clothing should encompass a wider range of sizes than those included in this study. Most problems could be solved with a sizing standard developed to fit a larger portion of the population.

This study also suggests that styling features and appropriate sizing characteristics in protective work wear is important to the wearer and a definite factor to be considered when producing such garments for market consumption.

Acknowledgments

This research was funded in part by the Utah Agricultural Experiment Station, Utah State University. Appreciation is extended to E. I. duPont de Nemours and Company and the Durafab Company for providing the Saranex-laminated Tyvek suits to our specifications. We also acknowledge the assistance of W. L. Gore and Associates in the initial prototype development.

References

[1] "Protective Clothing for Pesticide Users," Cooperative Bulletin by National Agricultural Chemicals Association, U.S. Department of Agriculture Extension Service and U.S. Environmental Protection Agency, Washington, DC, 1986.
[2] Branson, D. H., DeJonge, J. O., and Munson, D., "Thermal Response Associated with Prototype Pesticide Protective Clothing," *Textile Research Journal,* Vol. 56, No. 1, Jan. 1986, pp. 27–34.
[3] Henry, M. S., "User's Perceptions of Attributes of Functional Apparel," Masters thesis, Michigan State University, 1980.
[4] DeJonge, J. O. in *Clothing The Portable Environment,* S. M. Watkins, Ed., Iowa State University Press, Ames, IA, 1984, pp. vii–xi.
[5] Branson, D. H., DeJonge, J. O., and Munson, D., "Thermal Response Associated with Prototype Pesticide Protective Clothing," Textile Research Journal, Vol. 56, No. 1, Jan. 1986, pp. 27–34.
[6] Garland, C. E., Goldstein, L. E., and Campbell, C., "Testing Fully Encapsulated Chemical Suits in a Simulated Work Environment," *Performance of Protective Clothing, ASTM STP 900,* R. L. Barker and G. C. Coletta, Eds., American Society for Testing and Materials, Philadelphia, 1986, pp. 276–285.

Vera B. Keeble,[1] R. Ryan Dupont,[2] William J. Doucette,[2] and Maria Norton[3]

Guthion Penetration of Clothing Materials During Mixing and Spraying in Orchards

REFERENCE: Keeble, V. B., Dupont, R. R., Doucette, W. J., and Norton, M., "**Guthion Penetration of Clothing Materials During Mixing and Spraying in Orchards,**" *Performance of Protective Clothing: Second Symposium, ASTM STP 989*, S. Z. Mansdorf, R. Sager, and A. P. Nielsen, Eds., American Society for Testing and Materials, Philadelphia, 1988, pp. 573–583.

ABSTRACT: Data are presented from a field investigation of the extent of penetration of Guthion wettable powder (azinphos-methyl) through work clothing (regular and Tyvek spray suits) and three protective garments (Saranex-laminated Tyvek, Gore-Tex, and an experimental composite structure) during the typical mixing and spraying activities of fruit orchardists in central and northern Utah. Exterior and interior azinphos-methyl levels were collected using α-cellulose pads and were used by quantify exterior exposure patterns and percent protection afforded by each garment type. No relationship between inside azinphos-methyl pad levels and outside azinphos-methyl loading ($\mu g/cm^2$ of patch), ambient temperature, relative humidity, or wind speed was evident for any of the garments tested based on results of correlation analyses.

KEY WORDS: pesticide application, protective clothing, Guthion, azinphos-methyl, exposure, field monitoring, evaluation

Because dermal exposure is a greater hazard than inhalation or ingestion for most agricultural applications, protective clothing is used to reduce worker contact with toxic substances. Concern has been raised over the interaction and long-term effects of the variety of pesticides, some highly toxic, to which applicators are exposed throughout their working life. Field use of protective clothing has not been extensive, however, due to its lack of thermal comfort and mobility [1,2]. The attention focused on the complex problems of garment design and wearability, thermal comfort, and fabric penetration has not resulted in suitable and effective protective clothing, however, and further research is needed.

The development of protective clothing is often a compromise between protection and comfort. Ideal materials for protective clothing would provide the needed barrier to hazardous chemicals while allowing the wearer to work in comfort. However, few clothing materials are both protective and comfortable, and the current options for cost-effective, comfortable user protection are severely limited.

Laboratory and field studies have been used to determine the effectiveness of various protective clothing fabrications as barriers to the penetration of agricultural chemicals.

[1] Textiles and clothing specialist, Utah State University Cooperative Extension, Logan, UT 84322-2949.

[2] Assistant professors, Civil and Environmental Engineering, Utah State University, Logan, UT 84322-8200.

[3] Computer specialist, Joint Computer Laboratory, College of Family Life, Utah State University, Logan, Utah, 84322.

Laboratory studies identified Tyvek,[4] Gore-Tex,[5] and polyester/cotton and cotton fabrics treated with fluoroaliphatic resins as showing promise as effective barrier fabrics [2–4].

Data documenting the protection afforded by various fabrics in field situations are generally limited and incomplete. Rubberized clothing did not appear to be significantly more protective than a heavy coverall in a field study of Guthion[6] exposure [5]. Tyvek protective suits were shown to reduce total dermal load for workers exposed to dicofol in Florida citrus groves [6]. Further field studies are needed to evaluate the performance of protective clothing under actual working conditions. These conditions may be difficult to reproduce in the laboratory, yet laboratory test methods for evaluating exposure and thermal comfort should be based on typical field conditions because correlation of field and laboratory performance data is often poor [7,8].

The purpose of this study was to evaluate personal clothing (regular work clothing and Tyvek spray suits) and three protective fabrications (Saranex[7]-laminated Tyvek, Gore-Tex, and an experimental composite structure) for penetration of azinphos-methyl in a field situation during mixing and spraying of fruit orchards in northern Utah during July and August 1986. Collection pads placed under and on the surface of the protective clothing were used to estimate percent protection, based on the mass of chemical found on the inside and surface pads. This work was conducted in conjunction with comfort studies to allow the assessment of both garment protection and wearability.

Procedure

Protective suits were produced in three materials (Saranex-laminated Tyvek, Gore-Tex, and an experimental composite structure) based on a design recently developed at Utah State University. These protective garments were used along with work clothing or spray suits normally worn by orchardists as the test garments in this study.

Guthion (azinphos-methyl) was selected as the test pesticide because of its widespread use in the Utah fruit industry. During an orientation meeting for potential subjects, all test procedures were described and demonstrated. Subjects were asked to wear any personal protective equipment usually worn when mixing and spraying azinphos-methyl. Test subjects were asked to wear the same protective equipment with all garments.

Fabric penetration was measured with α-cellulose pads taped on the outside and inside of the garments at seven locations. Pads were placed on lines drawn on the inside of the garments to ensure that pad location would be the same for all garments. Inside pads were placed flush with the garment and covered with glassine paper and tape. Outside pads were attached with tape along all four sides. Placement of pads followed the U.S. Environmental Protection Agency (EPA) guidelines for Applicator Exposure Monitoring [9] with the addition of pads on the front and back of the hood. Figure 1 indicates the pad locations used on all protective garments during the study.

The subjects also wore their own work clothing as the fourth garment evaluated. Pad locations for the subjects' own clothing were the same as on the protective garments, except the two outside and inside chest pads were replaced with a single outside and inside pad at the V of the chest, and a pad was located under the collar on the back of the clothing.

All pads were 6.35 cm by 12.7 cm except those on the outside forearm, which due to sleeve size were limited to 6.35 cm². Pads were attached to garments before they were taken

---

[4] Registered trademark of E. I. duPont de Nemours and Company.
[5] Registered trademark of W. L. Gore Associates, Inc.
[6] Registered trademark of Mobay Chemical Corporation.
[7] Registered trademark of **Dow** Chemical Company.

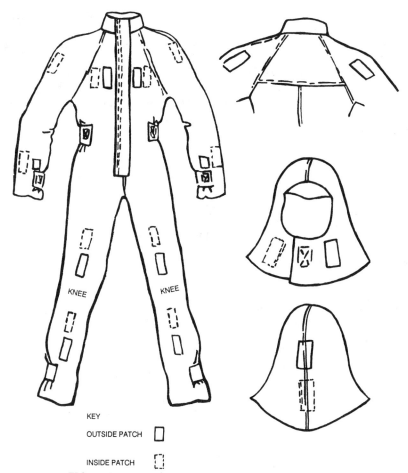

KEY

OUTSIDE PATCH ☐

INSIDE PATCH ⌐⌐

FIG 1—*Protective clothing design and placement of test pads.*

to the exposure site. Protective garments for a single subject were stored in large plastic bags along with all requisite data collection forms prior to use at the field site. The subjects' work clothing was obtained prior to testing to allow proper attachment of exposure pads.

Garments were randomized in a modified Latin Square design, and subjects were issued different garments each day. Following pesticide application, outside exposure pads were removed from the garment and placed in sample containers, the garment was then removed, and the inside pads were collected. After each use, the Gore-Tex garments, which are not disposable, were washed twice in warm water with a heavy-duty liquid detergent and then dried in the sun.

Analysis

After removal from the test garments, exposure pads were placed in clear, widemouth glass jars which were sealed with tetrafluoroethylene (Teflon) lined screw-top lids. For each garment/pad location, pads from opposite sides of the body were placed in the same sample jar and were extracted and analyzed as a composite sample. The samples were transported to the laboratory in ice-filled coolers and stored at 4°C prior to extraction.

Pads were extracted with 100 mL of cyclohexane (Omnisolv, EM Science, Cherry Hill, New Jersey) added directly to the glass jars, followed by 2 h of agitation on an orbital shaker. Inside garment pad extracts were concentrated to 10 mL using a Kadurna-Danish evaporator. Outside pad extracts did not require concentration.

The pad extracts were analyzed for azinphos-methyl using a Shimadzu GC-9A gas chromatograph equipped with an [63]Ni electron capture detector and a 30-m by 0.53-mm inside diameter DB-5 capillary column (J&W Scientific, Folsom, California). The column was operated isothermally at 260°C with helium carrier gas (5 mL/min) and nitrogen makeup gas (35 mL/min). The injection port and detector temperatures were maintained at 300°C.

Field blanks and spikes were prepared in the laboratory and transported along with the empty sample jars to each sample site. These blanks and spikes were later analyzed for azinphos-methyl along with the collected field samples. Blank pads showed no azinphos-methyl contamination. Field and laboratory spikes were not statistically different ($\alpha = 0.05$) based on a Scheffé's multiple comparison test. Azinphos-methyl recovery averaged $82.8 \pm 12.9\%$ (mean $\pm 95\%$ confidence interval) for 41 pads spiked with azinphos-methyl at levels ranging from 1.42 to 72.9 µg azinphos-methyl/pad.

Results and Discussion

General Field Study Data

During July and August 1986, a total of ten subjects participated in this evaluation of azinphos-methyl penetration through work clothing and protective garments. All test subjects were male: five were orchard owners, two were managers, and three were full-time employees. All mixed and applied the 50% wettable powder (50WP) formulation of Guthion (azinphos-methyl) during the field test, while some subjects added other pesticides to their tank mix. Three used Plictran,[7] two added parathion, and one grower used a combination of diazinon, Kelthane,[8] and Vydate[8] in addition to azinphos-methyl. These compounds did not interfere with the chromatographic analysis of azinphos-methyl. All ten subjects sprayed apples; four also sprayed peaches, and two also sprayed pears. One subject applied the azinphos-methyl using a hand spray gun from a tractor, while all others used air-blast sprayers for pesticide application during the field testing period.

As shown in Table 1, each subject wore each of the three protective garments and his own clothing during azinphos-methyl mixing and application. The subjects' own clothing were to be used as the control group; however, comparison between the protective garments and the controls was confounded due to the variability of work clothing worn by the test subjects. Five subjects wore regular work clothing, four wore Tyvek spray suits, and one used a polyvinyl chloride (PVC) spray suit.

The mean number of hours in each suit ranged from 1.8 to 2.6, while the mass of active ingredient applied while wearing the test garments ranged from 1.7 to 4.1 kg (Table 1). Most of the subjects' tanks took approximately one hour to mix and spray. Subjects wore test garments for approximately two mix-and-spray cycles because this pattern reduced interference with their work routine.

Outside and Inside Loading Results

Table 2 summarizes azinphos-methyl mass loading (micrograms of azinphos-methyl per square centimetre) for outside and inside pads for each garment. The percentage of inside pads with measurable amounts of azinphos-methyl varied for the five types of garments.

---

[8] Registered trademark of Rohm and Haas Company.

TABLE 1—*Number of subjects, hours in suit, and kilograms azinphos-methyl applied for each garment.*

Garment	Number of Subjects	Hours in Suit Mean	SD[a]	Azinphos-methyl Applied, kg Mean	SD
Gore-Tex	10	2.3 ±	0.5	2.8 ±	1.8
Saranex-laminated Tyvek	10	2.2 ±	0.5	2.8 ±	1.8
Composite structure	10	2.1 ±	0.4	2.7 ±	1.9
Work clothing	10	2.3 ±	0.4	2.9 ±	1.8
Regular	5	2.6 ±	0.4	1.7 ±	1.3
Tyvek	4	2.2 ±	0.4	4.1 ±	1.8
Spray suit	1	1.8	...	2.8	...

[a] SD = standard deviation.

Regular work clothing had the highest percentage of inside pads with measurable levels of azinphos-methyl (58%). The other garments tested had the following percentages of inside pads with measurable azinphos-methyl: Tyvek work clothing, 30%; Saranex-laminated Tyvek, 21%; composite structure, 21%; and Gore-Tex, 13%. Table 3 gives outside loadings by location for all garments in the study. Table 3 data indicate that outside loading varied with location, with the thighs, forearms, and shins showing consistently higher mass loading than other locations examined.

The mean for measurable inside loadings for regular work clothing was higher than for any other garment tested. When one looks at inside pad loading by location for the work clothing (Table 4), consistent patterns emerge. Upper body locations, specifically those in

TABLE 2—*Outside and inside pad loading ($\mu g/cm^2$) for each garment.*

Garment	n	Mean	Minimum	Maximum	SD[b]	Percent with Measurable Guthion
			OUTSIDE			
Gore-Tex	66	3.85	nd	28.44	6.02	98.5
Saranex-laminated Tyvek	67	3.60	nd	29.63	5.87	98.5
Composite structure	67	9.20	nd	273.85	35.60	98.5
Work clothing						
Regular	30	2.60	0.098	10.18	2.48	100.0
Tyvek	22	7.05	0.004	45.47	12.00	100.0
Spray suit	5	1.44	0.482	3.44	1.31	100.0
Total no. pads	257					
			INSIDE			
Gore-Tex	61	0.000	nd	0.005	0.001	13.0
Saranex-laminated Tyvek	66	0.002	nd	0.025	0.005	21.0
Composite structure	67	0.001	nd	0.015	0.003	21.0
Work clothing						
Regular	29	0.038	nd	0.296	0.078	58.0
Tyvek	22	0.001	nd	0.012	0.003	30.0
Spray suit	4	0.002	nd	0.006	0.003	50.0
Total no. pads	249					

[a] nd = nondetectable.
[b] SD = standard deviation

TABLE 3—*Outside pad loading ($\mu g/cm^2$) by location for all test garments.*

Outside	n	Mean	Minimum[a]	Maximum	SD[b]
Hood neck	25	1.39	nd	4.31	1.22
Hood seam	25	3.03	0.090	27.73	5.51
Shoulders	40	2.00	0.003	16.71	3.43
V of chest	7	1.34	0.059	2.88	1.05
Chest	35	2.50	0.022	23.25	4.29
Forearms	40	5.45	0.004	33.36	7.13
Thighs	40	16.98	0.258	273.86	46.05
Shins	39	4.06	nd	38.93	7.48
Neck back	6	1.95	0.792	3.51	0.92
Total no. pads	257				

[a] nd = nondetectable.
[b] SD = standard deviation.

a horizontal orientation, show higher inside pad levels than other pad locations. The combination of generally thinner and lighter-weight fabrics used for shirts as compared to trousers and the increased potential for maintaining pesticide/fabric contact in a horizontal orientation are thought to be the causes of this result. Inside loadings for the other test garments were much lower, and no obvious patterns were evident relating to pad location.

Analysis of Variance Results

To identify differences in subjects, garments, and locations, an analysis of variance (ANOVA) was performed on the outside and inside pads for the three protective garments (Gore-Tex, Saranex-laminated Tyvek, and experimental composite structure). Only results from these three protective garments were included in the statistical analysis because the growers wore more than one type of clothing as their usual outfit for mixing and applying the test pesticide. These ANOVA results are given in Table 5. There were no differences in the outside loading of azinphos-methyl for subjects, garments, or the interaction between garment and location. Only location was statistically significant at an observed significant level of 0.007. Table 6 gives the number of pads for each garment, estimated means for outside loading, and the standard deviation of the mean for each garment included in the analysis. Although the experimental composite had a higher mean outside loading, it was not significantly different from the other protective garments. Table 7 provides the number of pads for each location, the estimated means, and the standard deviation of the mean for the seven locations on the three protective garments. Azinphos-methyl loadings on the thigh pads were significantly greater than at all other pad locations using an α-level of 0.05 as

TABLE 4—*Mean inside pad loading ($\mu g/cm^2$) for regular work clothing by location.[a]*

Subject	Shoulders	V of Chest	Forearms	Thighs	Shins	Neck Back
1	np	0.006	0.273	0.021	nd	0.062
2	0.015	nd	0.012	0.009	nd	np
3	0.011	0.005	0.099	nd	nd	0.009
4	0.017	0.003	0.223	nd	nd	nd
5	0.174	nd	0.296	nd	nd	nd
Total	0.217	0.014	0.903	0.030	nd	0.071

[a] np = no pad; nd = nondetectable.

TABLE 5—*Adjusted ANOVA for azinphos-methyl loading ($\mu g/cm^2$) for each of seven outside pad locations on the three protective garments.*

Source	Degrees of Freedom	Mean Squares	F	Significance Level
Subjects	9	638.7	1.60	0.119
Garments	2	519.9	1.30	0.275
Location	6	1239.6	3.10	0.007
Garment × Location	12	601.6	1.51	0.126
Error	170	399.5	...	...

TABLE 6—*Estimated means for outside pad loading ($\mu g/cm^2$) for the three protective garments.*

Garment	n	Mean	Standard Deviation of Mean
Gore-Tex	66	4.16	2.48
Saranex-laminated Tyvek	67	3.75	2.46
Composite structure	67	8.79	2.46

TABLE 7—*Estimated means of outside loading ($\mu g/cm^2$) for the seven pad locations on the three protective garments.*

Location	n	Mean	Standard Deviation of Mean
Hood neck	26	1.60	3.97
Hood seam	25	3.90	4.06
Shoulders	30	1.51	3.65
Chest	30	2.03	3.65
Forearms	30	5.44	3.65
Thighs[a]	30	19.82	3.65
Shins	29	4.64	3.72

[a] Thigh loadings significantly greater than all other locations.

TABLE 8—*Adjusted ANOVA for azinphos-methyl loading ($\mu g/cm^2$) for each of seven inside pad locations on the three protective garments.*

Source	Degrees of Freedom	Mean Squares	F	Significance Level
Subjects	9	0.000016	1.28	0.251
Garments	2	0.000046	3.72	0.026
Location	6	0.000004	0.31	0.933
Garment × Location	12	0.000007	0.58	0.859
Error	173	0.000012	...	...

TABLE 9—*Estimated means for inside pad loading ($\mu g/cm^2$) for the three protective garments.*

Garment	n	Mean	Standard Deviation of Mean
Gore-Tex	69	0.00038	0.00042
Saranex-laminated Tyvek[a]	67	0.00196	0.00043
Composite structure	67	0.00074	0.00043

[a] Saranex-laminated Tyvek inside loadings significantly greater than all other garments.

measured by the Least Significant Difference Test. The two-way interaction between garment and location was not significant (Table 5), indicating that high azinphos-methyl thigh pad loading was consistent for all garments.

An ANOVA was performed on inside pad levels for subjects, garments, locations, and the interaction of garment by location. Table 8 summarizes the ANOVA results for the inside pads and identifies garments to be different at an observed significance level of 0.026. The number of pads by garment, the estimated mean of azinphos-methyl loading, and standard deviation of the mean are given in Table 9. The means for Gore-Tex and the composite structure were not significantly different; however, Saranex-laminated Tyvek was significantly different from the other protective garments when tested with the Least Significant Difference test at the 0.05 level.

Correlation Analyses

Correlation analysis was performed to identify the relationship between inside and outside azinphos-methyl loading for work clothing and the three protective garments. All work clothing worn by all ten subjects were grouped for this analysis. Results of the analysis are given in Table 10. As azinphos-methyl levels increased on the outside pads, there was no corresponding increase in inside pesticide levels; that is, there were no high, positive correlation coefficients for any of the fabrics tested. When inside and outside azinphos-methyl loadings are plotted as shown in Fig. 2, no linear relationship is obvious. Environmental conditions, including ambient temperature, relative humidity, and wind speed, also failed to show a high correlation with inside azinphos-methyl levels (Table 10).

TABLE 10—*Results of correlation analysis of inside pad loading ($\mu g/cm^2$) versus outside pad loading ($\mu g/cm^2$) and environmental conditions for work clothing and the three protective garments.*

Garment	n	Correlation Coefficient Inside Loading Versus Independent Variables			
		Outside Loading	Temperature	Relative Humidity	Wind Speed[a]
Gore-Tex	53	0.157	−0.154	0.209	0.019
Saranex-laminated Tyvek	61	−0.073	−0.244	0.107	−0.188
Composite structure	67	0.507	0.070	−0.218	0.233
Work clothing	47	−0.086	−0.103	−0.024	−0.021

[a] Difference between maximum and minimum wind speed during exposure period.

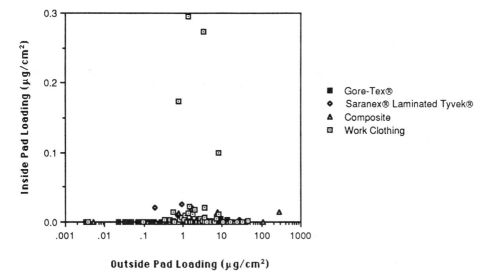

FIG. 2—*Plot of inside versus outside pad loading (µg azinphos-methyl/cm²) for the three protective garments and work clothing.*

Percent Protection Data

Table 11 provides the percent protection data [(Outside Loading-Inside Loading)/Outside Loading × 100] for all garments. The assumption made in estimating protection afforded by the test garments is that the same levels of pesticide exist on the outside pads and the surface of the garment directly above the inside pad. The three protective garments all exhibited mean protection above 99%. Regular work clothing provided 97.58% protection with a standard deviation of 5.78%, while for individuals wearing Tyvek work clothing, protection was also above 99%.

The small differences in percent protection provided by all test garments was not expected based on previous literature for other pesticides studied. In particular, results from a laboratory evaluation of methyl parathion migration through Tyvek and Tyvek-composite specimens [10] indicate that while penetration through regular Tyvek was nearly instantaneous, Saranex-laminated Tyvek provided a complete barrier over a 4-h test period. Staiff et al. [11], however, found that there was no significant difference in protection against azinphos

TABLE 11—*Percent protection by garment.*

Garment	n	Mean	Minimum	Maximum	SD[a]
Gore-Tex	57	99.99	99.86	100.0	0.02
Saranex-laminated Tyvek	66	99.74	89.35	100.0	1.36
Composite structure	65	99.94	98.44	100.0	0.21
Work clothing					
Regular	29	97.52	77.60	100.0	5.78
Tyvek	22	99.93	99.08	100.0	0.21
Spray suit	4	99.83	99.48	100.0	0.24

[a] SD = standard deviation.

methyl penetration afforded by rubberized cotton, regular Tyvek, and two types of poly-ethylene-coated Tyvek materials.

Results obtained in this study seem to support the findings by Schwope [10] and Staiff et al. [11] that pesticide penetration is affected by the active ingredient, pesticide formulation, application regiment, and composition of protective fabric. Further laboratory and field research is necessary to determine the role these variables play in the design and evaluation of protective clothing used in various exposure situations.

Conclusions

The following conclusions were reached regarding the protection provided by Gore-Tex, Saranex-laminated Tyvek, and an experimental composite structure fabric against azinphos-methyl penetration:

1. No significant differences were observed in outside azinphos-methyl loading as a function of subject and garment type (Table 5).
2. Outside thigh pads were shown to have a significantly higher loading than all other patch locations based on ANOVA results (Table 5 and 7).
3. No significant differences were observed in inside azinphos-methyl levels as a function of pad location and subject (Table 8).
4. Significant differences in inside pad levels were observed based on garment type when tested with the Least Significant Difference test. Results showed inside levels for Saranex-laminated Tyvek to be significantly greater than for Gore-Tex and the experimental composite structure (Tables 8 and 9).
5. The percent protection for each of these garments was greater than 99.7%.

Based on results from both the work clothing and protective garment studies, the following conclusions can be reached:

1. Inside azinphos-methyl pad levels were not highly correlated with outside pad levels or environmental conditions (Table 10). This lack of correlation between inside and outside pad levels was unexpected in light of the assumed concentration driven garment penetration behavior used to describe the movement of chemicals through protective clothing.
2. Greater penetration of azinphos-methyl was observed through regular work clothing in the upper body region, in terms of mean inside pad levels and percent of inside pads with measurable azinphos-methyl concentrations (Tables 2 and 4).

Acknowledgments

This research was funded in part by the Utah Agricultural Experiment Station and the Utah Water Research Laboratory, Utah State University. Appreciation is extended to E. I. duPont de Nemours and Company, and the Durafab Company for providing the Saranex-laminated Tyvek suits to our specifications.

References

[1] Keeble, V. B., Norton, M. J. T., and Drake, C. R., "Clothing and Personal Equipment Used by Fruit Growers and Workers When Handling Pesticides," *Clothing and Textiles Research Journal*, Vol. 5, No. 2, 1987, pp. 1–7.

[2] Orlando, J., Branson, D. H., Ayres, G. S., and Leavitt, R., "The Penetration of Formulated Guthion Spray Through Selected Fabrics," *Journal of Environmental Science and Health,* Vol. 16, No. 5, 1981, pp. 617–628.

[3] Branson, D. H., Ayres, G. S., and Henry, M. S., "Effectiveness of Selected Work Fabrics as Barriers to Pesticide Penetration," *Performance of Protective Clothing, ASTM STP 900,* R. L. Barker and G. C. Coletta, Eds., American Society for Testing and Materials, Philadelphia, 1986, pp. 114–120.

[4] Freed, V. H., Davies, J. E., Peters, L. J., and Parveen, F., "Minimizing Occupational Exposure to Pesticides. Repellency and Penetration of Treated Textiles to Pesticide Spray," *Residual Reviews,* Vol. 75, 1980, pp. 159–167.

[5] Franklin, C. A., Fenske, R. A., Greenhalgh, R., Mathieu, L., Denley, H. V., Leffingwell, J. T., and Spear, R. C., "Correlation of Urinary Pesticide Metabolite Excretion with Estimated Dermal Contact in the Course of Occupational Exposure to Guthion," *Journal of Toxicology and Environmental Health,* Vol. 7, 1981, pp. 715–731.

[6] Nigg, H. N., Stamper, J. H., and Queen, R. M., "Dicofol Exposure to Florida Citrus Applicators: Effects of Protective Clothing," *Archives of Environmental Contamination and Toxicology,* Vol,. 15, 1986, pp. 121–134.

[7] Lloyd, G. A., "Efficiency of Protective Clothing for Pesticide Spraying," *Performance of Protective Clothing, ASTM STP 900,* R. L. Barker and G. C. Coletta, Eds., American Society for Testing and Materials, Philadelphia, 1986, pp. 121–135.

[8] Nielsen, A. P. and Moraski, R. V., "Protective Clothing and the Agricultural Worker," *Performance of Protective Clothing, ASTM STP 900,* R. L. Barker and G. C. Coletta, Eds., American Society for Testing and Materials, Philadelphia, 1986, pp. 95–102.

[9] "Pesticide Assessment Guidelines," EPA Draft, Subdivision U, Applicator Exposure Monitoring, Office of Pesticide Programs, Hazard Evaluation Division, U.S. Environmental Protection Agency, Washington, DC, Jan. 1986.

[10] Schwope, A. D., "The Effectiveness of Tyek and Tyek Composites as Barriers to Methyl Parathion" in *Distributor Handbook Tyek Spunbonded Olefin in Limited-Use Apparel,* DuPont Co., Wilmington, DE, April 1981.

[11] Staiff, D. C., Davis, J. E., and Stevens, E. R., "Evaluation of Various Clothing Materials for Protection and Worker Acceptability During Application of Pesticides," *Archives of Environmental Contamination and Toxicology,* Vol. 11, 1982, pp. 391–398.

Wilhelm Batel[1] and Torsten Hinz[1]

Exposure Measurements Concerning Protective Clothing in Agriculture

REFERENCE: Batel, W. and Hinz, T., **"Exposure Measurements Concerning Protective Clothing in Agriculture,"** *Performance of Protective Clothing: Second Symposium, ASTM STP 989,* S. Z. Mansdorf, R. Sager, and A. P. Nielsen, Eds., American Society for Testing and Materials, Philadelphia, 1988, pp. 584–596.

ABSTRACT: Risk assessments associated with the handling and spraying of pesticides are based on both exposure and toxicological data. The quality of this assessment depends, to a large degree, on the circumstances at the time of worker exposure. Therefore, in addition to field exposure measurements, it is important to develop both predictive methods for estimating worker exposures and for establishing the causal relationships between exposure and adverse health effects. Using data derived from both approaches, it is possible to estimate the degree of risk associated with various pesticide application scenarios and make recommendations for appropriate protective clothing. Actual field measurements can be used to test the suitability and effectiveness of protective clothing requirements. In this report, we describe the use of two protective clothing indicators: the penetration rate and water vapor permeability with respect to thermal comfort.

KEY WORDS: pesticide application, exposure measurements, predictive models, risk assessment, protective clothing, penetration measurements, thermal comfort

The potential risks of personal exposure from spraying and handling pesticides are based on the possible routes of exposure and the toxicological characteristics of the pesticide and can be expressed as follows:

$$\text{risk factor} = \frac{\text{inhalative exposure}}{\text{relevant tox data}} + \frac{\text{dermal exposure}}{\text{relevant tox data}}$$

If the risk factor is greater than 1 ($R_i > 1$), it is suggested that protective clothing is used to eliminate or reduce this risk.

If the toxicological data for pesticides are known, two exposure-related questions remain:

1. What levels and duration of exposure during spraying of pesticides are to be expected?
2. What are the appropriate means of providing personal protection based on the exposure and toxicological data?

The answers to these questions will serve as the subject of this report.

[1] Director and scientist, respectively, Institute for Bio-System Engineering, Federal Agricultural Research Center, Braunschweig-Völkenrode, West Germany.

Determination of Exposure Levels During Spraying and Handling of Pesticides

There are two possible ways to determine personal exposure levels while spraying and handling pesticides:

1. Measurement of exposures in field conditions.
2. Prediction of potential exposure to pesticides.

In this report, we will concentrate on predictive methods for estimating exposure, since methods of field exposure measurements have already been described by the U.S. Environmental Protection Agency (EPA) [1].

To date, there are only a few methods described in the literature for predicting personal exposures. A predictive method has several advantages over actual field measurements including: (1) the prediction of worker exposure associated with various application scenarios, and (2) the estimation of the reduction in exposures based on the use of protective clothing.

Field studies have shown that it is possible to distinguish between several exposure scenarios using different types of application methods or mixing and loading operations or both (Fig. 1).

Personal exposure is caused by transporting the pesticides from the source to the applicator and can be described by the following chain of events

$$\text{emission} - \text{transmission} - \text{exposure}$$

This chain of events serves as the basis for our predictive method (Fig. 2).

The emissions by sprayers can be described by the flow rate of active ingredients, F_E, or their concentration, c_E, at the source of the emission. These values are obtained from the application rate, A_a, concentration of active ingredients, c_a, sprayers velocity, v, a working band width, b, and air flow, Q.

Pesticide transmission refers to the dispersal from the site of application, and can be considered a function of propagation, k_1, k_3, or a dilution factor, k_2, of the pesticide. The

symbols	description
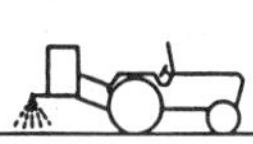	field or row crop spraying
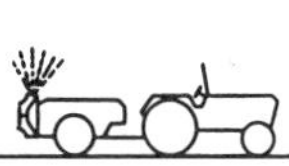	orchard spraying — airblast spraying
	spraying in greenhouses
	mixing and handling

application by aircraft is not considered

FIG. 1—*Application and handling of agricultural chemicals.*

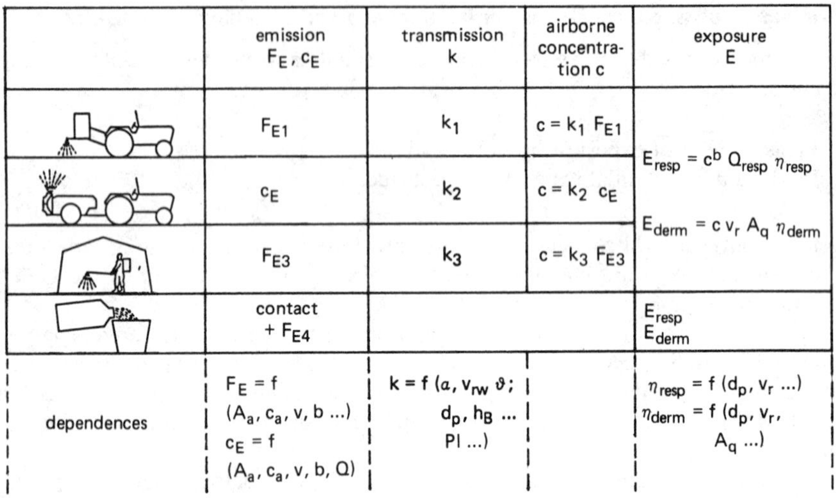

	emission F_E, c_E	transmission k	airborne concentration c	exposure E
(tractor)	F_{E1}	k_1	$c = k_1 F_{E1}$	$E_{resp} = c^b Q_{resp} \eta_{resp}$
(sprayer)	c_E	k_2	$c = k_2 c_E$	
(greenhouse sprayer)	F_{E3}	k_3	$c = k_3 F_{E3}$	$E_{derm} = c\, v_r\, A_q\, \eta_{derm}$
(mixing/loading)	contact $+ F_{E4}$			E_{resp} E_{derm}
dependences	$F_E = f$ $(A_a, c_a, v, b ...)$ $c_E = f$ (A_a, c_a, v, b, Q)	$k = f (a, v_{rw}\, \vartheta;$ $d_p, h_B ...$ $Pl ...)$		$\eta_{resp} = f (d_p, v_r ...)$ $\eta_{derm} = f (d_p, v_r,$ $A_q ...)$

FIG. 2—*Exposure prediction chart.*

transmission coefficients are a function of the angle of wind incidence, α, relative velocity of the wind, v_{rw}, the temperature of ambient air ϑ, droplet size, d_p, the physical dimensions of the application device, for example, height of nozzle, h_B, and crop population. Detailed information regarding the development and measurement of these parameters are given in Refs 2–4. By combining the transmission and emission factors, the airborne concentrations, c, are obtained for a given application scenario. From this information, when combined with the area flow patterns around the workplace and the pesticide deposition pattern, applicator exposures can be estimated. This approach is analogous to predictive methods based on fluid dynamics that include parameters such as droplet size, d_p, deposition velocity, v_r, deposited areas, A_q, etc., that are then combined to give ratios of inhalation and dermal deposition. The results of this predictive method combined with results from our field measurements and from the literature are shown in Fig. 3.

These results show that dermal exposure is consistently higher than inhalation exposure especially while spraying tall crops. Inhalation exposures are considerably lower while spraying ground plants. However, inhalation exposure can be significant while spraying tall plants and in greenhouses. When mixing and loading pesticides, dermal exposure is predominant.

In estimating the portion of the applied pesticide that reaches a worker's target organ, a penetration coefficient of 1 is assumed for inhalation exposures and a value of 0.1 for dermal exposure, providing no other specific experimental data are available.

Figure 4a shows the maximum permissible penetration rate associated with four application scenarios. These penetration rates are given for hand protection, body protection, and respiratory protection for the three levels of increasing toxicity (note: protection rate $s = 100 - $ penetration rate P). From the derived penetration rates, appropriate recommendations for protective clothing can be derived and are shown in Fig. 4b. The selection of the protective clothing ensembles is based on the penetration (P) factor and the thermal comfort (acceptance) factor given by the water vapor resistance factor, R_{et}, which has been chosen according to local climatic conditions (see section on Thermal Comfort for Protective Clothing).

For hand protection, protective gloves with either short-term or long-term resistant char-

application handling	application rate of active ingredient kg/ha	a.i. concentration at working place (mg/m³) breathable c^b range / standard value	total c range / standard value	exposure (mg/h) respiratory E_{resp} range / standard value	dermal E_{derm} range / standard value
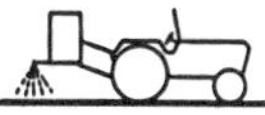	0.3	0.002–0.04 / 0.008		0.004–0.08 / 0.015	0.2–4 / 1
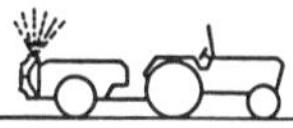	0.3	0.01–0.2 / 0.04	0.05–1 / 0.2	0.02–0.4 / 0.08	5–100 / 20
	0.2	0.05–0.15 / 0.03		0.015–0.45 / 0.09	0.2–6 / 1.2
	—	0.001–0.1 / 0.01		0.003–0.3 / 0.03	1–40 / 15

FIG. 3—*Exposure and concentration of active ingredient at working place—range and standard values. Values obtained from field measurements and prediction. Dermally affected area (head, neck, and hands) A = 0.22 m². Air temperature 20 to 25°C. Under comparable conditions, exposure is roughly directly proportional to the active-substance application rate.*

acteristics are indicated. Body protection can be obtained by using protective suits with varying degrees of permeability. For certain applications, for example, pouring and mixing, an apron with good resistant characteristics is recommended. Respiratory protection is achieved using either a filter face piece or a half-face respirator, depending on the degree of protection indicated. From the data in Fig. 4b, the protective clothing requirements and recommended clothing ensembles can be roughly classified. For individual pesticides, other recommendations may be required by the manufacturer or by the Federal Government. In addition, more specific recommendations may be made if additional toxicological data are available and taken into account.

With regard to the protective clothing recommendations, it is evident that they do not necessarily correspond to the degree of pesticide toxicity. Several reasons for this discrepancy can be presented. While an estimation of risk could include total body exposure, only dermal exposures are considered because it is well known that the hands represent the greatest amount of exposure received by the body. Only in the case of air blast spraying is head exposure approximately the same as hand exposure.

If worker acceptance of protective clothing is desired, it is important to specify protective clothing recommendations in simple terms. The recommendations given in Fig. 4b provide an excellent guide to pesticide applicators and can give direction to developers of new pesticides and manufacturers of packing, spray, and worker protection devices. Finally this predictive approach gives pesticide applicators and various safety advisors a simple method to estimate potential risks associated with handling pesticides and to follow safe application procedures. Based on the recommendations just given, we feel that there would be no need for personal protective clothing if pressurized cabs equipped with filtered air were used during pesticide spray applications.

requirements of protection measures

application handling	hand protection			body protection			respiratory protection		
	LD50 derm. [mg/kg]			LD50 derm. [mg/kg]			LC50 [mg/l] 4 h		
	400–4000	50–400	< 50	400–4000	50–400	< 50	0,5–5	0,1–0,5	< 0,1
 *(tractor with rear sprayer)*	30	20	10	50	25	10	–	–	30
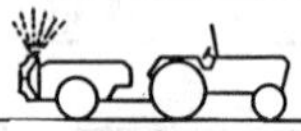 *(tractor with front sprayer)*	10	5	5	10	5	5	–	20	5
 *(greenhouse spraying)*	20	10	5	20	10	5	30	20	5
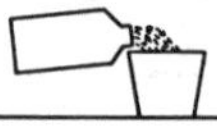 *(pouring from container)*	10	5	5	20	10	5	–	20	10

and therefrom derived protective means

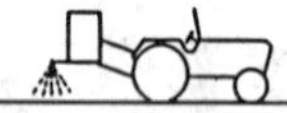 *(tractor with rear sprayer)*	+	+	+	A	A	B	–	–	filtering face piece
 *(tractor with front sprayer)*	+	++	++	B	B	B	–	filtering face piece	 half mask
 *(greenhouse spraying)*	+	++	++	A	B	B	filtering face piece	filtering face piece	 half mask
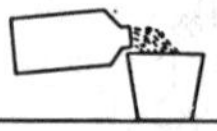 *(pouring from container)*	+	++	++	(A, B) + S			–	filtering face piece	 half mask

symbols
+   protective gloves with short term resistance
++   protective gloves with long term resistance
A   protective suits e.g. cotton coverall
B   protective suits (permeability < 5 %)
    e.g. Goretex, Propylen-Kimberly-Clark, Tyvek
S   apron of good resistance

filtering face piece

half mask with filter e.g. A2 + P2

FIG. 4—*Protection required during the application and mixing of pesticides of various toxicity, represented* (top) *by the maximum permissible penetration rate P in % and* (bottom) *by symbols for personal protection gear satisfying the penetration requirements.*

Acceptability and Rating of Personal Protective Clothing

The acceptability of personal protective clothing ensembles can be described by two specific properties of the clothing material:

1. The protective clothing material provides a barrier against pesticides and is characterized by a specific penetration rate.
2. An adequate level of thermal comfort is provided by the particular protective ensemble.

Body Protection

Protective clothing used for pesticide application generally consists of tightly woven cotton fabric material or PVC-coated fabric materials. Worker acceptance of this clothing would be low if there is relatively low thermal comfort under particular climatic conditions. One reason for the lack of thermal comfort is the low water-vapor permeability of the material. During the last several years, new fabrics have appeared on the market that exhibit a higher water-vapor permeability combined with low penetration characteristics against airborne particles.

In principle, these characteristics are in opposition to each other. In order to identify materials with these contrasting properties, special test methods have been designed to measure these two characteristics.

Measurement of Penetration—The measurements of penetration through protective clothing materials are carried out by the test apparatus shown in Fig. 5.

A manikin (1:1) was used as the test subject and was placed in a typical greenhouse structure. An air sample was passed through a separator that was cooled by liquid oxygen in order to condense the gaseous material. Airborne particles collected in the sample were separated by a filter system. A typical experiment consisted of about 10 min. The concentration of the active ingredient (a.i.) in the ambient air, c_G, was determined by the same method.

The conditions of a typical experiment are as follows. A greenhouse (7 m wide, 11 m long, and 2.5 m high with a corridor down the middle) is used as the experimental site. The temperature of ambient air is 25°C with relative humidity 85 to 92%. A knapsack sprayer, equipped with a cone-nozzle at a distance of 0.8 m above the ground, sprayed liquid water containing 0.025% LINDAN. The application rate is 20 to 60 mg/m^2. The flow rate against the model, which consisted of a sheffield head, is about 0.2 m/s. The penetration of the active ingredient is measured in a circular area of 0.45 m^2 between a height of 0.8 and 1.2

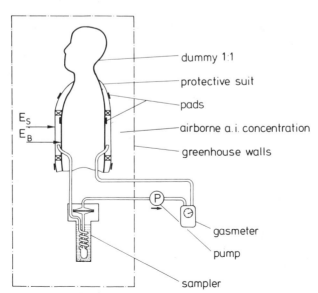

FIG. 5—*Test arrangement to measure active-ingredient flow against* E_S *and through* E_B *protective suit.*

m (that is, between the chest and trunk region). The volume between the manikin and the protective suit had been pre-determined and the protective suit had no physical contact between the undergarments and the body of the manikin.

The penetration results are given as

$$P_S = \frac{E_B}{E_S}$$

The amount of the active ingredients, E_S, on the surface of the protective suit is determined by the use of exposure pads that are attached to the surface of the manikin.

If the gaseous fraction of the test pesticide is considerable, the amount of the active ingredient that passes through a suit cannot be adequately determined using exposure pads. Therefore, in this case, a certain amount of air from inside the suit is removed and the active ingredient separated and determined. From an estimate of the active ingredient deposited on the manikin measured by pads, a calculation of the volume between the protective suit and the model, and taking the sample results into account, the amount of active ingredient on the body, E_B, can be calculated.

The penetration rate, as measured by the test apparatus, assumes that a reservoir of active ingredient exists inside the protective suit and that the degree of penetration of the skin by the active ingredient is significantly higher than that observed experimentally. In other words, the active ingredient penetrating the skin of the person can never be higher than the penetration values measured by the test equipment.

The characteristics of selected test fabrics are given in Table 1. Figure 6 shows the results of penetration tests conducted on three types of materials, in terms of the measured flow of active ingredient and the airborne concentration of active ingredient in the greenhouse.

The penetration characteristics (P_S) of the protective suits made of K, N, and U materials are shown in Fig. 7 as a function of airborne concentrations of active ingredient. The penetration of Fabrics U and N is nearly constant over the range of airborne active ingredient concentration. For Fabric K (Kimguard), a significant increase in penetration of the active ingredient is observed.

The barrier effect of air-permeable textile fabrics with agriculture pesticides is due mainly to the diffusion and capillary flow through material. There is considerable literature describing these phenomena. These factors are determined by the intensity and duration of the exposure and the contact between the protective material and the wearer.

For the particular clothing material, the diffusion process is dominant, which explains the course of corresponding curves of Fig. 6.

TABLE 1—*Characteristics of protective suit fabrics investigated.*

Symbol	Name Company	Material Structure	Weight, g/m²	Water Vapor Resistance, m² mbar/W	Water Vapor Permeability, g/m²h mbar
U	Avilastik, Nordfaser Neumünster	polyester filament-yarn with polyurethane coating	200	$2466 \cdot 10^{-3}$	0.6
K	Kimguard, Kimberly-Clark	3 layer polypropylene material	60	$25.6 \cdot 10^{-3}$	58.1
N	Tyvek 1422 A, Dupont	spunbonded olefine	40	$142 \cdot 10^{-3}$	10.4

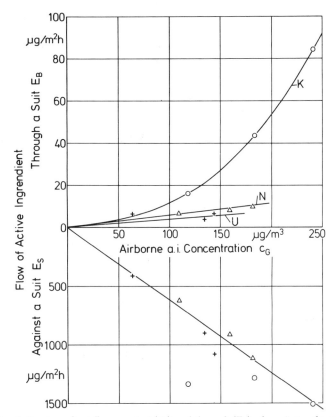

FIG. 6—*Active-ingredient flows against* (E_S) *and through* (E_B); *the suit as a function of the airborne active-ingredient concentration* (c_G) *measured at a height,* h = 1 m.

Regarding worker exposure, the penetration results represent pesticide applications in greenhouses and while spraying low crops. To assess the exposure situation associated with applying agriculture pesticides to higher crops, tests have been made using additional spray devices directed towards the surface of the protective garment. The results (not shown here) suggested that the penetration rate decreases, remains constant, or increases at a lesser rate than results shown in Fig. 7. This can be explained in part because the pesticide contains no active ingredient in the gaseous form and the barrier effect of the material is more effective for larger size droplets. The test results suggest that adequate barrier performance is achieved using protective Suits K and N. According to other measurements, Gore-tex® has characteristics between Fabrics N and K, and therefore has acceptable penetration characteristics.

Thermal Comfort for Protective Clothing—Thermal comfort is probably the most important factor affecting wearer acceptance of a protective clothing. It is well known that at high ambient temperature and humidity, the evaporation of sweat becomes a significant factor, particularly if the wearer of protective clothing is doing physical work. Therefore, water-vapor permeability of protective clothing ensembles must be high in order to achieve some degree of thermal comfort.

Based on research work carried out by the Bekleidungsphysiologisches Institut [5], a functional and subjective relationship exists between the subjective sensation of thermal

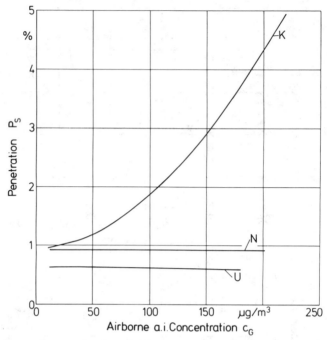

FIG. 7—*Penetration as a function of airborne active-ingredient concentration for Suits K, N, and U, calculated from the values plotted in Fig. 6.*

comfort and the water-vapor resistance (R_{et} according to the German Standard DIN 54101) of protective clothing fabrics. This relationship between the subjective comfort rating as a function of water-vapor resistance for three types of fabrics is shown in Fig. 8.

The results are shown for moderate ambient conditions (20°C 60% relative humidity) and moderate physical work, for an individual walking on even, level ground, and at a speed of 4.5 km/h. This effort results in a total body heat production of approximately 300 W.

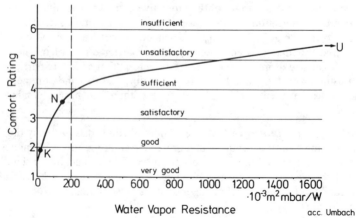

FIG. 8—*Subjective comfort rating of wearing comfort of protective suits as a function of water-vapor resistance.*

Thermal comfort under these particular work conditions is achieved when the water-vapor resistance (R_{et}) of the fabrics is lower than 200×10^{-3} m²mbar/W (as indicated by the dotted line in Fig. 8).

The protective Suit U has a water vapor resistance, $R_{et} = 2466 \times 10^{-3}$ m²mbar/W, that is far above this limit. Therefore, the wearer will experience thermal discomfort. Protective suits with Fabric N show a water-vapor resistance of 142×10^{-3} m²mbar/W, indicating that a satisfactory thermal comfort can be expected.

Fabric K provides the highest level of thermal comfort ($R_{et} = 25.6 \times 10^{-3}$ m²mbar/W). Tests by Umbach [5] show that other fabrics, for example, two- or three-layer fabrics with one polytetrafluoroethylene layer, exhibited similar low water-vapor resistance values (between 21 and 135×10^{-3} m²mbar/W). As a result, this material has a thermal comfort rating that is between Fabrics N and K.

Textile Penetration Test Equipment—The penetration measurements described earlier can be quite cumbersome in evaluating the performance characteristics of a large number of test fabrics. Therefore, efforts have been made to develop a simple and standardized test method. Such a test protocol has been proposed and laboratory tested (Fig. 9).

A predetermined amount of pesticides (dilute or concentrate) is dispersed by a two-phase nozzle. The pesticide mixture is injected through the nozzle by syringe that is activated by a microprocessor-controlled step motor. To evaluate the amount of the injected material applied to the target area, calibration of the apparatus is required. A sorption agent that receives the penetrating pesticide material is placed beneath the textile sample. Based on an analysis of the pesticide on the sorption agent, the penetrated amount of the material can be determined. Efforts to achieve optimal performance of the test equipment and to

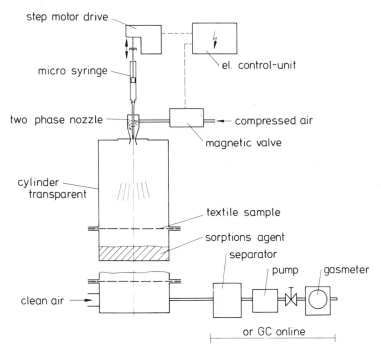

FIG. 9—*Design for test apparatus for determination of penetration of substances through protective suits.*

obtain standardized test procedures are underway. A yet unsolved problem is the degree of penetration caused by mechanical contact between the clothing material and the wearer's body.

Evaluation of the Penetration Rate of Respirators With and Without Head Protection— Under some conditions, respiratory protection is needed while applying pesticides. Therefore, it is important to measure the penetration of respirators when the worker is equipped with or without protective hoods. The penetration rate, P_R, of the pesticide material is calculated from the concentration of a.i. in inhaled air, c_{resp}, and ambient air, c_G

$$P_R = \frac{c_{resp}}{c_G}$$

The test equipment designed to measure these parameters is shown in Fig. 10.

The test conditions for measuring respiratory penetration rates were the same for the greenhouse described earlier. The test equipment is basically the same as the respiratory protection testing devices described in the literature [6]. In this instance, the entire inhaled air passes through the sampling device in order to record the volume of pesticide passing through the respiratory device. However, since this method caused higher pressure losses, an artificial lunge with sufficient pumping power has been developed. To achieve a predetermined breathing profile, the inhale and exhale valves are microcomputer controlled. Eighteen air exchanges per minute with an air volume of about 1.5 m³/h are used. Gaseous chemicals are measured using a Flame Ionization Detector (FID). The result obtained with this apparatus is comparable with the official test results used for respiratory equipment. It is also possible to directly record the amount of leakage around the face seal. In this case, the respirator without filter is supplied with pure air.

The concentration of airborne active ingredient at the breathing zone and the concentration

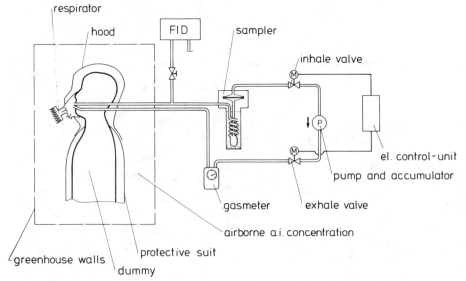

FIG. 10—*Apparatus for measurement of inhaled quantities of active ingredient for various respirators and protective suits; the active ingredient inside the hood is assayed by an arrangement as illustrated in Fig. 5.*

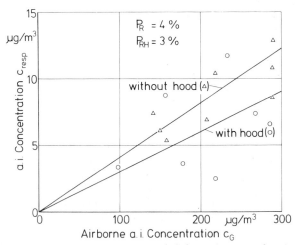

FIG. 11—*Active-ingredient concentration in inhaled air, c_{resp}, as a function of airborne active-ingredient concentration in greenhouse; half mask S (combination filter A2 + P3) with and without hood.*

of the active ingredient of the inhaled air using the respirator with and without the hood is shown in Fig. 11.

The comparatively high deviation of the observed values in Fig. 11 is due to the variation of leakages around the face seal. It is apparent that without a hood, penetration ($P_R = 4\%$) for pesticides is slightly higher than for pure gas (for example, propane, $P_R = 1.2\%$). Again the reason for this penetration is leakage around the face seal. In all tests using pure gas, the mask had been adjusted to give the lowest values. This adjustment was not performed for tests using pesticides. However, as a consequence of these test results (not presented in

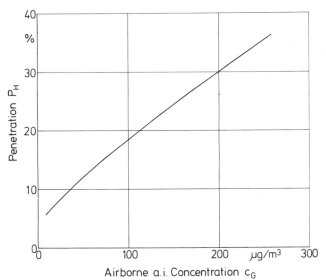

FIG. 12—*Penetration of a Tyvek® hood.*

this paper), it can be concluded that the penetration rates from the official respiratory protective device tests for gases are equally valid for agricultural pesticides. Tests with the hood in place show that the penetration rate can be slightly diminished, again depending on the leakage around the face seal.

For some applications of agricultural pesticides, a full-face respirator is required. With regard to respiratory risk, if real exposure data are taken into account, the requirement for a full-face respirator is only necessary under special conditions.

Another reason for using full-face respirators is that a degree of eye protection can also be obtained. In this case, the issue of whether a similar degree of protection can be achieved by the hood was investigated by measuring the penetration rate of the hood material. The test equipment, as shown in Fig. 5, was used for the testing. The exposure pads were placed on the hood and the results of exposure are shown in Fig. 12.

The protective hood provided no special seal against the protective suit but was merely lying against it. The results showed that at an airborne concentration of 100 $\mu g/m^3$, as found in typical greenhouses, a penetration rate of less than 20% is observed. Using improved hood designs, an even lower penetration rate can be observed. As a result, dermal exposure can be reduced by protecting the entire head and neck region with a protective hood, as compared to wearing a full-face mask.

Acknowledgment

The authors appreciate the support given by J. D. Sakura, Ph.D., in preparing the final paper and acknowledge the support given by D.-I. H. Speckmann in designing and building the electronic test and control devices.

References

[1] "Pesticide Assessment Guidelines," Subdivision U: Applicator Exposure Monitoring, Environmental Protection Agency, Jan. 1986 (draft).
[2] Batel, W., "Exposure of the User During the Application of Plant Protection Products by Spraying—Summary of Up To Now Results," *Grundlagen Landtechnik,* Vol. 34, No. 2, 1984, pp. 33–53.
[3] Batel, W., "Inhalative and Dermal Exposure of the User During Plant Protection Work With Blower Sprayer," *Grundlagen Landtechnik,* Vol. 35, No. 3, 1985, pp. 65–70.
[4] Batel, W., "Exposure of the User During Application of Plant Protection Products in Greenhouses," *Grundlagen Landtechnik,* Vol. 35, No. 6, 1985, pp. 177–182.
[5] Umbach, K. H., "Physiological and Technical Aspects of Water Repellent but Water-Vapor-Permeable Fabrics, *Schlußbericht zum Forschungsvorhaben,* AIR-Nr. 5296, 1985, Bekleidungsphysiologisches Institut, 7124 Bönnigheim, Schloß Hohenstein.
[6] Riediger, G., Tobis, H. U., and Kunst, H., "Investigations of Leakage of Filtering Respirators," *Staub-Reinhaltung der Luft,* Vol. 45, No. 11, 1985, pp. 525–531.

Rinn M. Cloud,[1] David J. Boethel,[1] and Steven M. Buco[1]

Protective Clothing for Crop Consultants: Field Studies in Louisiana

REFERENCE: Cloud, R. M., Boethel, D. J., and Buco, S. M., **"Protective Clothing for Crop Consultants: Field Studies in Louisiana,"** *Performance of Protective Clothing: Second Symposium, ASTM STP 989*, S. Z. Mansdorf, R. Sager, and A. P. Nielsen, Eds., American Society for Testing and Materials, Philadelphia, 1988, pp. 597–604.

ABSTRACT: Factors affecting the level of foliar residue transfer of permethrin to the clothing of crop consultants working in aerially sprayed soybean fields were studied. Significant variation occurred by subject in the degree of exposure of specimens taken from specific locations on garments. Results indicated that ultra-low-volume oil spray techniques increase the level of residue transfer to clothing. Delayed entry is highly effective in reducing the degree of clothing contamination. Fiber content of fabrics worn into fields had some effect, but presence or absence of applied surface finishes had little or no effect.

KEY WORDS: pesticides, insecticides, toxicology, contamination, textile fibers, cotton fibers, polyester fibers, woven fabrics, nonwoven fabrics, protective clothing

Previous research has established that the principal means of pesticide exposure is dermal [1] and that clothing serves with varying degrees of success as a barrier to dermal exposure [2,3]. Studies of pesticide contamination of clothing have considered three situations in which exposure may occur. During mixing of pesticide and loading into application equipment, undiluted or field-strength pesticide may be spilled directly on clothing. Such spills have been simulated in the laboratory by submerging fabric in pesticide [4] or pipetting pesticide onto the surface of the fabric [5]. Contamination of clothing during application of pesticide involves exposure of fabric to a forceful spray. Laboratory equipment has been developed by some researchers to simulate the contamination and penetration of fabric by spray [6]. In the third situation, dislodgeable pesticide residues transfer from plant foliage to the clothing of harvesters and other field workers as they work in treated fields or orchards [7,8]. A review of the literature indicates no studies that attempt to simulate this type of contamination in a laboratory other than by pipetting similar levels of pesticide onto fabric.

The importance of field studies in pesticide exposure research is well supported in the literature [8,9]. Much of the work that has been done in field studies of dermal exposure/clothing contamination has been concerned with applicators and the penetration of clothing by pesticide spray. Work that has been conducted related to foliar residue transfer to workers has primarily involved tree crop harvesters [10,11]. In Louisiana, research has focused on the exposure of individuals whose occupations require that they enter fields of soybeans or cotton shortly after the fields have been aerially sprayed with pesticides. The pioneer work in this area was conducted in 1967 by Finley and associates at Louisiana State University (LSU) through funding from the Louisiana Agricultural Experiment Station (LAES) [12].

---

[1] Associate professor, professor, and assistant professor, respectively, Louisiana State University Agricultural Center, Baton Rouge, LA 70803-4300.

Their findings indicated that the clothing of commercial cotton scouts entering insecticide-treated cotton fields absorbed significant levels of methyl parathion and DDT and that even after laundering these residues were biologically active. Since that time, the Louisiana Agricultural Experiment Station has continued to support research to determine factors affecting levels of foliar transfer of pesticide residue to the clothing of individuals who enter pesticide-treated fields and the potential use of clothing as a barrier to dermal exposure.

Crop Consultants as Test Subjects

There are numerous challenges in conducting field research. Soliciting workers as test subjects is one of the first problems a field researcher faces. Often workers are unconcerned about their exposure to pesticides and are thus unwilling to participate in studies. In Louisiana, crop consultants have served as an alternative test group to other farm workers. A questionnaire survey conducted of licensed crop consultants in Louisiana in 1982 indicated that crop consultants face some of the same problems that other farm workers face with regard to exposure [13]. They spend a high percentage of the work week in pesticide-treated fields, and they desire clothing that maximizes comfort in the semi-tropical Louisiana climate. In addition, crop consultants have unique exposure problems related to the types of activities conducted in the field.

Field Conditions

Field studies of crop consultants' exposure to pesticides have been conducted in conjunction with research projects of the LSU/LAES Entomology Department, which in recent years have focused on the effectiveness of synthetic pyrethroids, permethrin and fenvalerate, on insect control in soybean fields. Greater than 800 000 hectares throughout Louisiana are planted in soybeans annually, and synthetic pyrethroids are being widely used by soybean farmers. In the 1982 survey of crop consultants, permethrin was identified as the most commonly encountered pesticide, with 71% of the respondents indicating recurring exposure to fields treated with this synthetic pyrethroid [13]. The insecticide has a moderate level of mammalian toxicity.

Aerial spraying is the common method of insecticide application used on soybeans in Louisiana. Recent interest has been generated concerning the use of ultra-low-volume oil spray techniques where insecticide is mixed in an oil medium rather than the conventional water medium. Such techniques are being studied to determine if they reduce insecticide drift, improve canopy penetration, and increase persistence [14].

An additional field condition under study by entomologists has been the experimental use of narrow row spacing to increase crop yield. Cooperative research efforts have provided opportunities for textile researchers to determine possible effects of new or different application technology or field conditions on increasing or decreasing hazard related to residue transfer to clothing.

Fabric Characteristics

Some studies concerning pesticide exposure have made reference to the use of clothing without a clear description of the numerous variables that exist in any given piece of cloth. In most cases, the objective has been to use whatever clothing the individual worker normally wears rather than to compare different types of clothing. It is the intent of our work to compare fabrics of varying fiber contents, weave structures, weights, and surface finishes to determine the effects of these variables on limiting dermal exposure. Fabrics may vary significantly in these areas and still be very similar to the fabrics worn by the farm worker.

To the worker, the degree to which a fabric protects him may be less important than the comfort of that fabric, especially in the type of climate that is encountered in Louisiana. In the 1982 survey, consultants indicated that the most common attire worn into fields was a woven, light-weight, short-sleeved sportshirt and either khaki or denim work pants [13].

Procedure

Fabric Exposure Tests

In order to expose several different fabrics at once, a patchwork of fabrics was pinned to the outside of the right front thigh area of participants' pants. Previous research indicated this area received the greatest amount of exposure for crop consultants working in low-crop fields [12]. Each patchwork contained three replications of five shirt-weight test fabrics, which were arranged randomly. The five test fabrics were unfinished 100% cotton, unfinished 100% polyester, unfinished 50% cotton/50% polyester (50/50 blend), durable-press finished 50/50 blend, and acrylic acid soil-release finished 50/50 blend. For this study, the field variables that were investigated were row spacing, use of nonconventional ultra-low-volume oil spray techniques, and day of entry.

Wear Study

In the wear study, subjects wore one of two types of coveralls: a 35% cotton/65% polyester twill weave work coverall sold by a major department store chain, and a nonwoven disposable garment produced from a spunbonded olefin. To determine the amount of foliar residue transfer to the garments, test specimens were taken from the thigh, arm, and chest areas of the suits after exposure. Field variables in the wear study were use of nonconventional ultra-low-volume oil spray techniques and method of insect collection.

Test Subjects

Individuals employed by the LAES to conduct insect counts for other research purposes served as the subjects for fabric exposure tests and wear studies. The activities of the subjects were governed by the other research needs and, for the wear study, included two methods of collecting insect counts. The sweep net method involved 50 sweeps with a 38-cm diameter sweep net so that the opening of the net passed through the foliage with each step [15]. The ground cloth method involved placing a 91-cm length sheet between two adjacent soybean rows and forcefully displacing the insects by shaking the plants over the sheet [16]. These activities required that the subjects be in a particular plot for approximately 15 min. Clean patchworks or suits were put on each time a new plot was entered. For the fabric exposure study, subjects entered plots within 2 h of spraying and again 48 h later. Wear tests were conducted two days after spraying.

Field Treatment

Equal-sized plots in experimental soybean fields were sprayed using three field treatments: (1) permethrin in water carrier (water treatment), (2) permethrin in soybean oil carrier (oil treatment), (3) permethrin in soybean oil/water mixture carrier (oil/water treatment). Treatment parameters were controlled to produce 0.084 kg active ingredient/ha.

TABLE 1—*Worker variability in mean permethrin residues absorbed during fabric exposure tests.*

Worker	n	Residue, ng/cm^2		
		Cis	Trans[a]	Total[a]
1	30	81.4	87.2	168.6
2	30	108.8	123.7	232.6
3	30	122.1	174.9	297.0

[a] $p \leq 0.05$.

Residue Analysis

Immediately after exposure, suits or patchworks were placed in aluminum foil and put on ice for transportation to the laboratory where they were held in a freezer until residue analysis was performed. Cis and trans permethrin were extracted from fabric specimens using hexane and were analyzed on Perkin Elmer 3920 and Hewlett Packard 5880A gas chromatographs equipped with Nickel 63 electron capture detectors. Concentrated samples were calculated using the peak height measurements of standards and samples. Data were analyzed using analysis of variance for main effects and Duncan's Multiple Range Test for mean separation.

Results

Subject Variables

In conducting field studies with human subjects, numerous factors that cannot be controlled in the research design may affect results. Size and shapes of individuals may be similar but are never the same. Body movements and behavioral habits vary considerably among individuals. While these factors may confound the results, they will, in fact, be present in real-life situations and thus should not necessarily be avoided. Most field studies are limited in the numbers of subjects available for any given study and this again may affect results. In our studies, individual variance by subject was significant only for the fabric exposure tests conducted on the day sprayed (Table 1). The results indicated that fabric swatches worn by a particular subject were consistently lower in residue absorbed than those worn by the other two subjects. Ex post facto comparison of the subjects revealed a difference in size and body proportions, which would have resulted in the plant foliage striking the leg of that individual in a different location than other workers. This finding suggests that placement of fabric swatches or collection pads or location of specimens taken from clothing for individuals working in these situations should be based on a specified distance from the ground related to plant height at the time of exposure.

TABLE 2—*Mean permethrin residues by location of specimen on suits worn in wear study.*

Location	n	Residue, ng/cm^2		
		Cis	Trans	Total
Arm	48	...[a]	...[a]	...[a]
Chest	48	...[a]	...[a]	...[a]
Leg	48	22.3	23.8	46.0

[a] Less than 0.9 ng/cm^2.

TABLE 3—*Mean permethrin residues during fabric exposure tests by field treatment.*

Treatment	n	Residue, ng/cm²		
		Cis	Trans	Total
Oil	30	192.3[a]	214.8[a]	407.1[a]
Oil/Water	30	89.8[b]	125.5[b]	215.3[b]
Water	30	30.3[c]	45.5[c]	75.7[c]

[a,b,c] Means with the same letter in the same column are not significantly different ($p \leq 0.05$).

In wear studies, an attempt was made to determine effects of differences in worker activity on residue levels on various areas of the body. Workers used the two methods described previously to collect insect counts. However, a single collection was made before suits were removed, and no measurable residue was found in arm or chest specimens (Table 2). Measurable residue was located only in the specimens taken from the thigh area of the suits and did not differ significantly by worker activity, indicating that the greatest amount of foliar residue transfer to clothing occurred as workers entered and exited the fields. These results may have implications for the design of protective wear for individuals working in these situations. If a lower garment of high protective value is worn, lighter weight upper garments designed to be replaced after a few uses may provide sufficient protection during insect collection activities. Such an alternative is likely to be well received by field workers. Individuals participating in our studies complained of discomfort when wearing either the woven cotton/polyester suit or the spun/bonded olefin suit.

Field Variables

It is not unexpected that alterations in pesticide treatments designed to improve level and persistence of pesticide on plants may also lead to increased levels of pesticide residue that transfer to clothing. The use of ultra-low-volume oil spray techniques did result in higher levels of foliar residue transfer to fabrics in both the fabric exposure studies and the wear studies. Table 3 shows the effects of field treatment on residues absorbed during fabric exposure studies. Fabrics worn into fields sprayed using oil or a combination of oil and water as the mixing medium had significantly higher residues than those worn into fields sprayed using the conventional water medium.

Effects of row spacing were the opposite of those expected (Table 4). Nonconventional narrow row spacing resulted in significantly lower residue on fabrics in the fabric exposure studies than did the conventional wider spacing of rows. Crop consultants indicated that plant canopies were heavily lodged in narrow spaced fields, resulting in less penetration of insecticide to lower foliage. This suggests that while garments may have come in contact with more foliage, less of the contacted foliage had dislodgeable insecticide residue.

TABLE 4—*Mean permethrin residues absorbed during fabric exposure tests by row spacing.*

Row Spacing	n	Residue, ng/cm²		
		Cis	Trans	Total
16 in. (narrow)	45	89.7[a]	108.4[b]	198.1[b]
32 in. (conventional)	45	118.6[a]	148.8[a]	267.3[a]

[a,b,c] Means with the same letter in the same column are not significantly different ($p \leq 0.05$).

TABLE 5—*Mean permethrin residues absorbed during fabric exposure tests by day of entry.*

Day of Entry	Residue, ng/cm^2
Day sprayed	232.7
48 h later	73.8

Day of entry is clearly recognized as an important means of reducing hazard associated with entering treated fields. Table 5 indicates the effects of day of entry on foliar residue transfer during fabric exposure studies. These results also indicate that studies not conducted immediately after spraying should incorporate longer exposure times for fabrics in order to have measurable residue levels. While several subject, field, and fabric variables were significant factors for data collected on the day sprayed, no significant variables were identified for data collected 48 h after spraying.

Fabric Variables

Conflicting results have been reported regarding the effects of fiber content on absorption of pesticide residues. Some of the studies have used fabrics that differed in aspects other than fiber content. In the fabric exposure study, fabrics used were produced by the U.S. Department of Agriculture (USDA) Southern Regional Research Center in New Orleans to be comparable in all aspects except fiber content. Fiber content was not found to be a significant variable, although mean residue levels tended to be higher for the 100% cotton and 100% polyester than for the 50/50 blend (Table 6). Neither was there a significant difference in residue absorption between the two types of suits used in the wear study, which differed not only in fiber content but also in fabric structure and weight.

Fabric finishes were investigated as both a potential problem related to the wearing of normal attire and as a potential means of making normal attire more protective. Commonly used fabric finishes such as durable-press finishes may present a problem if they enhance residue absorption. On the other hand, soil-release finishes could improve the resistence of fabric to foliar residue transfer or to subsequent pesticide penetration. Durable-press and acrylic acid soil-release treated blend specimens were not significantly different from unfinished blend specimens in level of residue absorbed (Table 7). Penetration was not measured in these studies. The results only indicate that neither finish increased nor decreased the amount of foliar residue transfer.

TABLE 6—*Mean permethrin residues absorbed by unfinished fabrics and suits.*

Fabric	n	Residue, ng/cm^2		
		Cis	Trans	Total
FABRIC EXPOSURE TEST				
100% cotton	27	136.5	171.9	308.4
50/50 blend	27	81.5	97.2	178.6
100% polyester	27	115.1	151.0	266.1
WEAR STUDY				
Nonwoven suit	24	26.0	27.6	53.6
Woven suit	24	18.6	19.9	38.5

TABLE 7—*Mean permethrin residues absorbed by blend fabric on day sprayed by type of finish.*

		Residue, ng/cm^2		
Finish	n	Cis	Trans	Total
Unfinished	18	81.5	97.2	178.6
Durable press	18	87.9	115.3	203.2
Soil release	18	99.6	107.6	207.2

Conclusions

The overall results of our studies suggest that foliar residue transfer is affected more by subject and field variables than by fabric variables. There were statistical interactions that indicated that fabrics of some fiber contents were less affected by type of field treatment than others, but these results are inconclusive. Penetration of the transferred residues through fabrics was not measured. It is expected that differences in penetration would be strongly related to fabric variables.

Much remains to be done in the area of foliar insecticide residue transfer to the clothing of field consultants in order to make comprehensive recommendations regarding the design and use of appropriate protective clothing. Numerous other subject, field, and fabric variables remain to be studied. Models for predicting foliar residue transfer when entering and exiting fields are needed since this activity appears to represent the greatest hazard related to insecticide residues in the clothing of crop consultants.

Acknowledgments

Portions of this research were conducted as part of Southern Regional Research Project S-163. Other financial support for these studies was provided by the Southern Region Pesticide Impact Assessment Program and the American Soybean Association. Soybean oil, soybean oil surfactant, and nonwoven coveralls were provided by LouAna Foods, Stoller Chemical Co., Inc., and Dupont, respectively. Appreciation is also expressed to Mary Lynn Zimpfer, Jamie Yanes, and Claire Harmon for technical assistance.

This paper was approved for publication by the Director of the Louisiana Agricultural Experiment Station as manuscript No. 87-25-10044.

References

[1] Wolfe, H. R., Armstrong, J. F., and Durham, W. F., *Archives of Environmental Health*, Vol. 13, 1966, pp. 340–344.
[2] Davies, J. E., Freed, V. H., Enos, H. F., Duncan, R. C., Barquet, A., Morgade, C., Peters, L. J., and Danauskas, J. X., *Journal of Occupational Medicine*, Vol., 24, No. 6, 1982, pp. 464–468.
[3] Lillie, T. H., Livingston, J. M., and Hamilton, M. A., *Bulletin of Environmental Contamination and Toxicology*, Vol. 27, 1981, pp. 716–723.
[4] Laughlin, J. M., Easley, C. B., Gold, R. E., and Tupy, D. R., *Bulletin of Environmental Contamination and Toxicology*, Vol. 27, 1981, pp. 518–523.
[5] Kim, C. J., Stone, J. F., and Sizer, C. E., *Bulletin of Environmental Contamination and Toxicology*, Vol. 29, 1982, pp. 95–100.
[6] Easter, E. P. and DeJonge, J. O., *Archives of Environmental Contamination and Toxicology*, Vol. 14, 1985, pp. 281–287.
[7] *Pesticide Residues and Exposure*, J. R. Plimmer, Ed., American Chemical Society Series, Series No. 182, 1982.

[8] *Dermal Exposure Related to Pesticide Use,* R. C. Honeycutt, D. Zweig, and N. Ragsdale, Eds., American Chemical Society, 1985.

[9] *Field Worker Exposure during Pesticide Application,* W. F. Tordoir and E. A. H. Heemstra-Le-quin, Eds., Elsevier Scientific Publishing Company, New York, 1980, pp. 17–19.

[10] Stamper, J. H., Nigg, H. N., and Queen, R. M., *Bulletin of Environmental Contamination and Toxicology,* Vol. 36, 1986, pp. 693–700.

[11] Nigg, H. N., Stamper, J. H., and Queen, R. M., *Journal,* American Industrial Hygiene Association, Vol. 45, 1984, pp. 182–186.

[12] Finley, E. L. and Rogillio, J. R. B., *Bulletin of Environmental Contamination and Toxicology,* Vol. 4, No. 6, 1969, pp. 343–351.

[13] Cloud, R. M., Hranitzky, M. S., Day, M. O., and Keith, N. K., *Louisiana Agriculture,* Vol. 26, No. 4, Summer 1983, pp. 20–21.

[14] Southwick, L. M., Clower, J. P., Clower, D. F., Graves, J. B., and Willis, G. H., *Journal of Economic Entomology,* Vol. 76, 1983, pp. 1442–1447.

[15] Kogan, M. and Pitre, H. N., Jr., *Sampling Methods in Soybean Entomology,* Springer-Verlag, New York, 1980.

[16] Boyer, W. P. and Dumas, W. A., *Cooperative Economic Insect Report,* Vol. 13, 1963, pp. 91–92.

Curt Lunchick,[1] *Alan P. Nielsen,*[1] *and Joseph C. Reinert*[1]

Engineering Controls and Protective Clothing in the Reduction of Pesticide Exposure to Tractor Drivers

REFERENCE: Lunchick, C., Nielsen, A. P., and Reinert, J. C., **"Engineering Controls and Protective Clothing in the Reduction of Pesticide Exposure to Tractor Drivers,"** *Performance of Protective Clothing: Second Symposium, ASTM STP 989,* S. Z. Mansdorf, R. Sager, and A. P. Nielsen, Eds., American Society for Testing and Materials, Philadelphia, 1988, pp. 605–610.

ABSTRACT: Reduction of exposure received during occupational use of pesticides is one of the principal strategies available to the Environmental Protection Agency in reducing risk. The Exposure Assessment Branch of the Office of Pesticide Programs has evaluated applicator exposure studies in which pesticides were applied by ground boom or airblast spray methods. The dermal exposure to the applicator in an enclosed tractor cab during ground boom application was estimated to be one-sixth the exposure to the applicator conducting similar operations with an open tractor cab. During airblast applications, exposure to applicators in enclosed vehicles was more than thirtyfold less than to airblast applicators in open vehicles. At present, the major emphasis in reducing applicator exposure to pesticides has been to require the use of protective clothing. Surveys of applicators indicate that they frequently do not use the required protective equipment. The use of engineering controls such as enclosed vehicles is another risk-reduction strategy available to the Agency. This paper reviews the effects on exposure by tractor-cab design and the advantages and disadvantages of protective clothing and engineering controls in reducing total pesticide applicator exposure.

KEY WORDS: airblast application, ground boom application, enclosed cabs, protective clothing, parathion

The U.S. Environmental Protection Agency (EPA) has been empowered by the Federal Insecticide, Fungicide, and Rodenticide Act (FIFRA) to register chemicals for pesticidal use that do not cause unreasonable adverse effects on the environment when used in accordance with widespread and commonly recognized practices. The Exposure Assessment Branch (EAB) of the Office of Pesticide Programs has the task of estimating the exposure received by individuals handling pesticides. The risk from a given pesticide is a function of the pesticide's toxicity and an individual's exposure to the pesticide.

Recently, EAB has begun assembling generic databases for estimating the exposure received during various application methods. The use of generic databases to estimate exposure is an accepted practice [1,2]. The first generic data base assembled by EAB was for airblast applicators [3]. During the past year, EAB has reviewed six ground boom applicator exposure studies available in the published literature [4–9]. When adjusted to an application rate of 0.45 kg (1.0 lb) active ingredient per acre (0.4 hectare [ha]) and normal work attire of long

[1] Chemist, biologist, and chemist, respectively, U.S. Environmental Protection Agency, Office of Pesticide Programs, Hazard Evaluation Division (TS-769C), Exposure Assessment Branch, 401 M Street, SW, Washington, DC 20460.

pants and long sleeve shirt, the average exposure was estimated to be 6.3 mg/h. This is a geometric mean of the 92 replicates in which exposure ranged from 0.33 mg/h to 146 mg/h. The replicates at the lower end of the range involved application from enclosed tractor cabs, while open tractor cabs were used by individuals receiving the higher exposures. Other factors, including personal work habits and boom placement, were also believed to influence applicator exposure.

Effect of Enclosed Cabs on Exposure

Because of the wide range of exposures received by ground boom applicators and the apparent effect of cab structure on such exposure, EAB evaluated additional exposure studies in which the effect of open and enclosed cabs could be compared. Airblast applicators were included in these evaluations.

Wojeck et al. [8] evaluated the exposure received by applicators applying paraquat by ground boom to tomatoes. Applicators in the open tractor cabs received six times the exposure received by applicators in the enclosed cabs. Four persons applied paraquat in open tractor cabs at an application rate of 0.57 kg acid equivalent/ha and received an average exposure of 169 mg/h. Two operators applying paraquat at 0.57 kg acid equivalent/ha in enclosed air-conditioned cab tractors received an average exposure of 27 mg/h. In both instances, the applicators sat approximately 0.3 m above the horizontal spray boom.

In an exposure study submitted in support of the registration of a herbicide, the average exposure to individuals applying the herbicide from open tractor cabs was five times greater than that received by applicators in enclosed cabs. A total of eight replicates were measured in which two individuals applied the herbicide by ground boom sprayer drawn by an enclosed tractor cab. Four replicates of a third individual measured exposure when the ground boom sprayer was drawn by an open-cab tractor. The booms were 0.4 to 0.5 m above the ground, and the application rates were nearly identical.

Carmen et al. [10] conducted an extensive study in which the effect of cab design on applicator exposure was examined. The exposure to applicators spraying citrus by airblast from open tractors, enclosed cabs with open windows, enclosed cabs with closed windows, tractors with open-cage canopies, and trucks with open or closed windows were studied. The results are presented in Table 1. The estimates presented in the table were calculated by EAB from the unit exposures expressed by Carmen in mg/cm^2/h.

Table 2 provides the assumed dermal protection factors for airblast applicators in truck or tractor cabs when compared to airblast applicators on tractors with no cabs. Tractor cabs with all windows closed provided an assumed dermal protection factor of 72 when compared with tractors with no cab. Canopies did not reduce the exposure. Applying pesticides from either a truck cab or a tractor cab with the driver window open completely negated any protection. Exposures to these individuals were similar to the airblast applicators on open tractors.

Examining data on the deposition of pesticides presented by Carmen indicates that the left sides of the applicators received almost all the exposure.

Dermal exposure to the applicators in a cab with open passenger windows was essentially the same exposure as to applicators in totally enclosed cabs. However, this result is deceiving. Based on the deposition of pesticide upon applicators near the open window on the driver's side, it would be reasonable to assume that a passenger in a cab with an open passenger window would receive little protection. In addition, the inside of the cab is being contaminated as the pesticide is drawn through the open window.

Both trucks and tractor cabs afforded protection to the airblast applicator. The authors stated that the exposure received by the applicators in enclosed vehicles represents exposure

TABLE 1—*Dermal exposure to airblast applicators.*[a]

Vehicle	Number of Replicates	Exposure Range,[b] mg/h	Mean Exposure, mg/h
Tractor:			
no cab, no canopy	8	2.2 to 49	23
no cab, with canopy	2	17 to 28	23
with cab			
both windows open	4	2.7 to 17	8.8
driver window open	1	21	21
passenger window open	3	0.29 to 0.93	0.50
passenger window open, with grid	4	0.053 to 0.96	0.58
both windows closed	4	0.009 to 0.74	0.32
Truck:			
both windows open	1	16	16
driver window open	1	6.6	6.6
passenger window open	1	0.92	0.92
both windows closed	1	0.092	0.092

[a] All exposure adjusted to an application rate of 2.25 kg (5.0 lb) active ingredient/acre to eliminate variation from different application rates (that is, the exposure for application at 6.75 kg [15 lb] active ingredient/acre was multiplied by $\frac{5}{15}$ or 0.33).

[b] The exposure in $mg/cm^2/h$ presented by Carmen for each body part was multiplied by body part surface area to express exposure in mg/h. The following surface areas were used. The shoulders represented the head, 650 cm^2; upper areas, 1320 cm^2; forearms, 1210 cm^2; thighs, 2250 cm^2, front, 3550 cm^2; and back, 3550 cm^2.

from contamination of the cab with pesticide residues. This contention is reasonable. The EPA believes that the assumed dermal protection factors are maximum values because the applicators were not permitted to leave their vehicles during application. Since it is sometimes necessary for the applicator to leave the vehicle to repair the equipment during the spray operation, the reduction in dermal exposure afforded by the enclosed cab could be negated without proper precautions.

Protective Suit Concerns

From the limited exposure data available to the Agency, it is clear that enclosed application vehicles offer a means of reducing applicator exposure to pesticides. Traditionally, protective clothing has been used as the Agency's method of reducing applicator exposure. However, several distinct disadvantages exist with protective clothing requirements.

Lack of compliance is a major disadvantage. Waldron [11] surveyed pesticide applicators in Ohio and determined that compliance with stringent protective clothing requirements as stated on parathion and paraquat labels is minimal. Both products require protective gloves, waterproof spray suits, and respiratory protection during spraying. Despite these require- ments, 83% of applicators wore protective gloves, 39% wore spray suits, and 57% used a dust mask or respirator with parathion. When using paraquat, 45% wore gloves, 24% wore spray suits, and 26% used a dust mask or respirator.

Keeble [12] surveyed fruit growers and workers in Virginia. Of those applicators applying parathion, 19% wore water-repellent pants, 26% water-repellent coats, 9% water repellent coats with hoods, 51% protective gloves, 45% respirators, and 0.6% disposable coveralls. The low compliance was not a result of ignorance of the pesticide's acute toxicity. A total of 93% of the users considered parathion to be very hazardous. The effect of discomfort and heat stress appear to be factors in the users' decision not to wear the protective clothing.

TABLE 2—*Assumed dermal protection factor for various types of cabs during airblast application.*

Vehicle	Applicator Dermal Exposure, mg/h[a]	Assumed Dermal Protection Factor[b]
Tractor:		
no cab, no canopy	23	1.0
no cab, with canopy	23	1.0
with cab		
both windows open	8.8	2.6
driver window open	21	1.1
passenger window open	0.50	46
passenger window open,		
with grid	0.58	40
both windows closed	0.32	72
Truck:		
both windows open	16	1.4
driver window open	6.6	3.5
passenger window open	0.92	25
both windows closed	0.092	250

[a] The exposures are adjusted to an application rate of 2.25 kg (5.0 lb) active ingredient per acre and are not corrected for clothing.
[b] The assumed dermal protection factor is defined as the dermal applicator exposure for each vehicle type divided into the dermal applicator exposure received during application from an open tractor with no cab and no canopy.

Heat stress with protective suits is a real problem. Arthur D. Little, Inc. [13] evaluated the heat stress issue for the Agency and developed a heat stress model. The model relates to a hypothetical 1.7-m, 70-kg, 25-year-old male applicator doing light work (tractor driving) in an 29°C and 50% relative humidity environment. Based on this model, a nude individual would obtain an equilibrium rectal temperature of 37.7°C. An individual in long-sleeve shirt and long pants would achieve an equilibrium rectal temperature of 38.0°C. A similar temperature is predicted for an individual in a Tyvek coverall with no clothing underneath. The use of the Tyvek coverall over a shirt and pants was predicted to produce an equilibrium rectal temperature of 38.3°C.

The EPA Standard Operating Safety Guides [14] state that internal body temperature in excess of 38.3°C is a sign of heat stress in which excessive fatigue, physical exhaustion, and dizziness can occur. The EPA is concerned that such symptoms may reduce the alertness of the applicator and are a definite hazard during pesticide applications, especially with Toxicity Category I pesticides. The California Department of Food and Agriculture has recognized the threat of heat stress from chemical-resistant protective suits such as Tyvek suits. Their 1986 amendments to Title 3 of the Administrative Code prohibits the use of this type of protective clothing at ambient temperatures in excess of 29°C (85°F) [27°C (80°F) in sunlight] unless a cooling source is provided.

Enclosed Cabs as a Regulatory Option

In conducting a special review of a pesticide's registration, the Agency has three options after determining the risks and benefits of the pesticide's use. One option is to retain the pesticide's uses without amending the existing registration. A second option is to cancel some or all of the registrations of that pesticide. This second option occurs when the risks outweigh the benefits and the risks cannot be mitigated. A third option is to amend the existing registration to reduce the risks to a level at which the benefits of use are greater.

As the Agency re-registers Toxicity I formulations, the acute toxicity hazard to persons handling these pesticides will be examined. Parathion, a Toxicity I pesticides, is an example of a pesticide for which an extensive combination of protective clothing (waterproof pants, coat, hat, rubber boots, rubber gloves, goggles, and respirator) has been required since 1971 (by PR Notice 71-2). Despite these stringent label requirements, "ground applicators" was the job classification with the highest number of systemic illnesses due to parathion in California between 1976 and 1981 [15]. A review of the California data suggested that some poisonings occurred even when all of the required protective clothing was worn. Parathion poisoning to ground applicators may occur from sudden wind shifts blowing the spray across the face, improper use of protective clothing, or not using protective clothing [16].

The parathion poisoning incidents suggest that protective clothing may not provide adequate protection. Therefore, requiring enclosed application systems might be needed to adequately protect ground applicators.

Several factors must be considered before a regulation requiring such systems can be recommended. The capital outlay for users not currently owning such a system could be prohibitive. A second concern is that the size of some enclosed-cab tractors may preclude their use in certain areas such as orchards or vineyards. Pickup trucks may be able to fit into more confined spaces, but may not be capable of hauling large airblast sprayers. Finally, there is the problem of decontaminating the interior of an enclosed cab should it become contaminated. Contamination could occur, for example, if the applicator must leave the cab during application to repair the spray equipment. Even if the applicator wears protective clothing while making repairs, the possibility of contaminating the cab exists when the applicator reenters the cab. Even if the use of enclosed cabs reduces applicator exposure to acceptable levels and closed loading systems likewise reduce mixer/loader exposure, the issue of hazardous field residues to field workers will still remain. These concerns will have to be addressed during any Agency consideration of enclosed cabs as a regulatory option.

Conclusions

A review of applicator dermal exposure data indicates that the use of totally enclosed vehicles will reduce the applicator's dermal exposure to pesticides. The limited data available estimated that enclosed cabs reduced ground boom applicator dermal exposure fivefold to sixfold. Airblast applicator dermal exposure was reduced seventy-fold by totally enclosed cabs.

Worker compliance with extensive protective clothing requirements is poor. Discomfort and heat stress are two reasons for the poor compliance. Therefore, protective clothing may not provide adequate protection against extremely toxic pesticides such as parathion or dinoseb, even when such clothing is used.

Requiring ground application from totally enclosed vehicles will be considered by the Agency as an approach to mitigating the potential risks of some Toxicity Category I pesticides. Data on the economic and practical impacts of such a requirement will be needed to assess the appropriateness of such an action.

References

[1] Honeycutt, R. C. in *Dermal Exposure Related to Pesticide Use,* R. C. Honeycutt, G. Zweig, and N. H. Ragsdale, Eds., American Chemical Society, Washington, DC, 1985, pp. 369–375.

[2] Hackathorn, D. R. and Eberhart, D. C. in *Dermal Exposure Related to Pesticide Use,* R. C. Honeycutt, G. Zweig, and N. H. Ragsdale, Eds., American Chemical Society, Washington, DC, pp. 341–355.

[3] Reinert, J. C. and Severn, D. J. in *Dermal Exposure Related to Pesticide Use,* R. C. Honeycutt, G. Zweig, and N. H. Ragsdale, Eds., American Chemical Society, Washington, DC, 1985, pp. 357–368.

[4] Abbott, I. et al., *American Industrial Hygiene Association Journal,* Vol. 48, No. 2, 1987, pp. 167–175.

[5] Dubelman, S. et al., *Journal of Agriculture and Food Chemistry,* Vol. 30, No. 3, 1982, pp. 528–532.

[6] Maitlen, J. C. et al. in *Pesticide Residues and Exposure,* J. R. Plimmer, Ed., American Chemical Society, Washington, DC, 1982, pp. 83–103.

[7] Staiff, D. C. et al., *Bulletin of Environmental Contamination and Toxicology,* Vol. 14, No. 3, 1975, pp. 334–340.

[8] Wojeck, G. A. et al., *Archieves of Environmental Contamination and Toxicology,* Vol. 12, 1983, pp. 65–70.

[9] Wolfe, H. R., Durham, W. F., and Armstrong, J. F., *Archieves of Environmental Health,* Vol. 14, 1967, pp. 622–633.

[10] Carmen, G. E. et al., *Archieve of Environmental Contamination Toxicology,* Vol. 11, 1982, pp. 651–659.

[11] Waldron, A. C. in *Dermal Exposure Related to Pesticide Use,* R. C. Honeycutt, G. Zweig, and N. H. Ragsdale, Eds., American Chemical Society, Washington, DC, 1985, pp. 413–425.

[12] Keeble, V. B., Norton, M. J. T., and Drake, C. R., *Clothing and Textiles Research Journal,* in press.

[13] "Manual for Selecting Protective Clothing for Agricultural Pesticide Operations," Environmental Protection Agency Project No. 68-03-3293, Work Assignment 0-09, Arthur D. Little, Inc., Cambridge, MA, 1986.

[14] *EPA Standard Operating Safety Guides,* Office of Emergency and Remedial Response, Hazardous Response Support Division, U.S. Environmental Protection Agency, Edison, NJ, Nov. 1984.

[15] Maddy, K. T., Winter, C. K., and Ochi, E. T., "Occupational Illnesses and Injuries Due to Exposure to Parathion as Reported by Physicians in California in 1981," Report HS-993. California Department of Food and Agriculture, 1982.

[16] Peoples, S. A., Maddy, K. T., and Topper, J., *Veterinary and Human Toxicology,* Vol. 20, No. 5, 1978, pp. 327–329.

William Popendorf[1]

Mechanisms of Clothing Exposure and Dermal Dosing during Spray Application

REFERENCE: Popendorf, W., **"Mechanisms of Clothing Exposure and Dermal Dosing during Spray Application,"** *Performance of Protective Clothing: Second Symposium, ASTM STP 989,* S. Z. Mansdorf, R. Sager, and A. P. Nielsen, Eds., American Society for Testing and Materials, Philadelphia, 1988, pp. 611–624.

ABSTRACT: This report focuses on the mechanisms underlying the various sources of exposure during two replicate exposure studies of two ground crews applying Difolatan 80 Sprills (80% captafol) in central Florida orange groves.

Aerosolized captafol concentrations averaged 56 $\mu g/m^3$ for mixer-loaders and 34 $\mu g/m^3$ for spray applicators but were not statistically different.

Dermal "exposures" (outside the coveralls) were consistent within participants but not within operations for reasons that are compatible with observed differences in work conditions. The density of exposure ranged roughly from 1 to 10 $\mu g/h/cm^2$ to the hands, legs, and arms (but up through 20 when direct contact with captafol solutions was evident) to 0.1 to 1 $\mu g/h/cm^2$ for other locations. Whole-body exposures had a mean of 40 mg/h and ranged from 15 to 116 mg/h, with the hands accounting for around 40% of these totals.

Dermal "doses" (inside the coveralls) were reduced and generally more uniform than either dermal exposures or aerosols. Penetration through the cotton-polyester coveralls differed from a mean of 2% for the spray applicators to 7% for the mixer-loaders, for reasons believed to be related to wet versus dry deposition, respectively. This protection reduced skin dosing densities under the coveralls to 0.01 to 0.1 $\mu g/h/cm^2$ except to the largely horizontal upper legs and lower arms that ranged from 0.1 to 0.5 $\mu g/h/cm^2$. Whole-body doses had a mean of 19 mg/h and ranged from 10 to 30 mg/h, with hands accounting for around 90% of these totals. Laboratory methods to predict the clothing penetration rates in similar field settings are needed.

KEY WORDS: dermal deposition, skin deposition, dermal exposure, pesticide application, clothing protection, clothing penetration, field studies, protective clothing

The goal of this project was to assess operator exposure and the effectiveness of their clothing protection during the application of Difolatan 80 sprills in Florida citrus groves. The study design was to monitor the exposure of two crews for each of two application periods of at least 4 h. Each crew consisted of one mixer-loader and one spray applicator. Both participating crews were experienced, clearly interested in the project, and their degree of cooperation was excellent. The locations, equipment, and operating conditions of each study are recorded in Table 1. Both dermal exposure and skin deposition as well as airborne concentrations of captafol (the active ingredient of Difolatan) were monitored. Close observation of the setting permitted later correlations to be made between physical mechanisms and measurement results within this report.

The agricultural pesticide application process begins with the arrival of the crew, a tractor with air-blast oscillating nozzle pesticide application machine, a tank truck to transfer the

---

[1] Associate professor, University of Iowa, Institute of Agricultural Medicine, Iowa City, IA 52242.

TABLE 1—*Study design and performance parameters.*

	Studies 1 and 2		Studies 3 and 4	
Tractor	1950 Oliver			
Sprayer	FMC 9100 speed sprayer		FMC 957 speed sprayer	
Spray tank	1000 gal/tank		500 gal/tank	
Speed, km/h (mph)	2.4 (1.5)		2.4 (1.5)	
kg/ha (lb active/acre)	5.5 (5)		5.5 (5)	
L/ha (gal H_2O/acre)	4670 (500)		4670 (500)	
Mixture concentration, %	0.12%		0.12%	
Total ha (acres)	14 (34)		14 (35)	
Tree spacing, m (ft)	4.3 (14)		6.1 (20)	
Row spacing, m (ft)	8.5 (28)		6.1 (20)	
Trees/ha (trees/acre)	274 (111)		269 (109)	
	Studies		Studies	
	1	2	3	4
Dates, 1983	27 June	27 June	29 June	29 June 30 June
Starting times	10:20 am	3:10 pm	8:00 am	2:00 pm 9:30 am
Ending times	2:45 pm	7:50 pm	1:10 pm	5:50 pm 12:20
Exposure period				
M-L h:min	3:42	4:47	5:23	6:30
S-A h:min	4:15	4:45	5:08	6:35
Average lb active/h	19	19	16	15

spray mixture to the application machine in the field, and some form of trailer to carry the formulated pesticide to the mixing site. The mixing site is selected to be near a water source (usually surface water in Florida). A gasoline-powered water pump is used to draw water into the tank truck. As the tank is being filled, the mixer-loader measures out the specified amount of pesticide (in this case 2.835 kg (6.25 lb) of 80% active ingredient sprills per 500 gal per acre into a bucket, climbs upon the truck, and adds it to the water in the tank through a top-loading hatch.

This mixture is then usually agitated, driven into the grove to the point at which the previous spray load was completed, connected to the application machine, and transferred to its spray tank via a 10.16 cm (4-in.) rubber hose. The spray applicator manually operates the transfer hose, drives the tractor towing the application machine, regulates the hydraulic spray controls, and is responsible for equipment calibration and maintenance.

Field Methods

Dermal monitoring was conducted using gauze pad dosimeters as originally described by Durham and Wolfe [1] and more recently by Popendorf and Leffingwell [2]. The construction of these particular dosimeters consisted of a 12-ply 7.62 by 7.62 cm (3 by 3 in.) surgical gauze sponge backed by an impervious sheet of aluminum foil, all contained within a glassine-coated paper envelope with a 6-cm diameter hole cut in its face exposing 28 cm² of the gauze pad. These dosimeters were then attached to both the outside of the cotton coveralls worn by each participant and against their skin at the locations listed in Table 2.

The placement of dosimeters outside of the clothing measures the dermal exposure or potential deposition of material onto the body (as if no protective clothing were worn). Those inside the coveralls against the skin measure actual dermal deposition or dose. The

TABLE 2—*Dermal sample identifying code in tables, and areas used to extrapolate from dosimeter pesticide density to deposition rate on body location. Values in parentheses indicate areas for each dosimeter on two sides of body.*

Location (Comment)	Identifying Code	Location Area [2], cm^2
Head (each side)	HD	(650)
Chest (mid-chest)	CH	2190
Back (mid-back)	BA	2190
Upper leg (each thigh)	UL	(1730)
Lower leg (each calf)	LL	(1300)
Upper arm (each bicep)	UA	(930)
Lower arm (each forearm)	LA	(645)
Hand (glove on each hand)	GL	(1075)

ratio of inner-to-outer pesticide collected at the same location is a measure of clothing protection, provided care is taken not to occlude the inner dosimeter by physically overlapping the two dosimeters.

Only one set of dosimeters were necessary for unclothed areas, that is, the head and hands. The set of dosimeters to assess head exposure was attached to both sides of a cap worn by each participant. A set of thinly knit white cotton gloves without any backing materials was used to assess hand exposure. The physical integrity of these dosimeters sometimes needs to be closely monitored because of exposure to both high moisture and physical contact during these operations.

The duration of exposure for each set of 14 dosimeters was recorded as listed on Table 1. After use, these dosimeters were removed, folded in half to enclose the exposed sponge within the envelope, placed into individual zip-lock polyethylene bags (except the gloves that were placed into individual 236.5-mL (8-oz glass jars), and shipped at ambient temperature to Chevron's Residue Laboratory for extraction and analysis.

The resulting chemical mass data was first divided by the exposure time and collection area of each dosimeter to calculate the rate-density of collection at each body location. These values were then multiplied by the skin surface area of the body part represented by each dosimeter (also listed in Table 2) to extrapolate to both the local and whole-body dose rates. For the purpose of comparing nominal exposures and doses among operations and participants, a single, standard body size is assumed [2].

Total aerosol samples were collected into open-faced 37-mm A/E glass fiber filters (Gelman Corp., Ann Arbor, Michigan). One filter cassette was placed on each shoulder of each participant and aspirated at near 1.85 L/m. To avoid overloading filters by what was expected to be higher aerosol levels on the spray applicator, one of his filters was changed approximately every hour while a tank truck load was being transferred to the spray machine. All other filters were worn for the duration of the study period. At the end of each study, these filters were removed from their plastic (Tyril) cassettes, folded, and placed into 20 mL foil-sealed glass vials for shipment. Analyses of several batches of back-up filter support pads indicated an average of 99% of the recovered captafol was contained within the filter with no breakthrough.

Among the other data not reported here in detail were spike recovery samples from each collection media, and hourly wind (averages 0 to 1.5 m/s with gusts to 5), temperature (27 to 32°C), and relative humidity (60 to 90%). The weather was typical in terms of temperature but locally dry for this time of year. Thunderstorms threatened to terminate Studies 2 and 4 between 3 and 4 p.m. each afternoon but in both cases passed without a major disruption (Study 4 was delayed approximately 40 min due to lightning).

Results

The physiologic effect of the combined heat and humidity upon the crew cannot be fully assessed based on these data, but the "effective temperature" above 27°C in the afternoons was near the prescriptive occupational exposure limit of 28 to 30°C [3]. Exposures to additional solar radiation may well have exceeded ACGIH TLV guidelines depending upon metabolic work rates. Clearly, people in this region and industry routinely work in these conditions; but these data indicate the limited potential of relying on additional protective clothing during the summer, should such an option be considered.

The results of both the field and laboratory spike recovery tests were highly variable and often low when spiked below 100 μg. After due consideration, it was concluded not to correct any dermal data for recovery losses because of the following: (1) the high variability in gauze recovery efficiencies, (2) its unclear relationship to the deposition density, (3) the probable differences in the recoveries or media retention of dry versus aqueous captafol, (4) the small potential mean correction (1.56) relative to the variability between dermal exposure and dosing levels ($\sim \times 2$ to $\times 3$), and (5) the potential for similar losses on the skin.

Similar low recoveries from cloth of 30% for parathion and 70% for kelthane were reported by Serat et al. [4] and were attributed to an initial vaporization of even such low volatility chemicals. Winterlin et al. [5] found recoveries of only 20% for captan, when applied in an aqueous solution onto cotton or even glass slides, versus nearly complete recovery when applied in an organic solution; they attribute this loss to "codistillation." There seems to be some important mechanism at work during the absorption and evaporation of thin films on environmental surfaces that is not well understood but can have a strong bearing on evaluations of dermal exposure and the performance of protective clothing.

Throughout the remainder of this report, sets of data are assumed to be log-normally rather than normally distributed. All statistical tests and conclusions are based on the logarithmic transformation of the data sets. Thus, geometric means and geometric deviations are reported. The latter is a multiplying and dividing factor expressed herein as a percentage, for example, ±50% means that the limits of one geometric deviation above and below the mean are calculated by multiplying and dividing the mean by 1.50, respectively. Given the small sample sizes of many of the following groups of data, the power of any inferences based only on statistical tests may be limited; however, the statistical significance is presented in most cases to reinforce the importance of the physical differences between settings as described in the Observations and Discussion section of this report.

Aerosolized captafol concentrations are summarized by study and operation in Table 3. Overall and uncorrected, the mixer-loaders were exposed to a geometric mean airborne concentration of 56 μg/m³ (95% confidence limits of 20 to 160 μg/m³); the spray-applicators to 34 μg/m³ (with corresponding limits of 5.3 to 215 μg/m³). Using various tests, no statistical significance was attributed to either this difference or to those between individuals or between replicated studies within each operation.

Dermal exposure rate-density (μg/h/cm² outside clothing), dose rate-density (inside clothing), and clothing penetration results are grouped in Table 4 by operation, location, and by study. Density data are useful to compare exposures or doses among body locations independent of the skin area of each location and to assess potential dermatologic responses. These same data are adjusted for the anatomical site-specific area and grouped as before in Table 5 as exposure rates and dose rates (mg/h), respectively, then summarized and compared statistically in Table 6. Rate data can be summed and interpreted relative to potential systemic (internal) toxicologic responses.

A statistical analysis of the replicate measurements (that is, left and right at each location other than the chest and back) was conducted using an SAS general linear models procedure.

TABLE 3—*Aerosol summary*, $\mu g/m^3$.

Study	Mixer-Loader	Spray Applicator
1	$91 \pm 36\%$ $n = 2$	69
2	$56 \pm 54\%$ $n = 2$	33
3	$31 \pm 96\%$ $n = 2$	64
4	$63 \pm 86\%$ $n = 2$	9
Geometric mean, $\mu g/m^3$	56	34
Pooled geometric deviation	$\pm 70\%$	...
Geometric deviation between	$\pm 79\%$	$\pm 157\%$
Analysis of variance on multiple samples within a study: probability of no difference between studies within operations	$F_{3,4} = 1.48$ $p < 0.4$	... ...
Student's t test between operations: probability of no difference between mixer-loader and spray applicator	$t_6 = 1.77$ $p < 0.1$	

It was concluded that differences between sides within any given participant and study were statistically insignificant in comparison to other factors such as location, participant, and operation. Therefore, both sides were arithmetically averaged (as if they had been pooled for laboratory extraction) and only a single value for each location was used in all subsequent analyses.

It was estimated from the right-hand column in Table 4 that the pooled geometric deviation within paired dosimeters (right and left sides) was between 110 and 135%, except the spray applicators inside dosimeters that typically differed by about 225%. It was also observed using this study design, that for exposures or doses at any given location, differences between studies were not statistically larger than differences within studies unless the paired deviation was in the range of 50 to 100%, or less. From this analysis, it can be expected that differences at any location would be detected more frequently if multiple dosimeters were co-located at each location, that is, the use of two or four dosimeters would reduce the expected geometric deviation by the square-root of two to four, respectively. As it was, significant differences among whole-body dermal depositions reported later within this report were probably the result of different predominant exposure mechanisms in each setting that affected multiple locations on a given person.

The penetration of pesticide through the cotton-polyester blend coveralls was assessed by comparing the dose inside to the exposure outside, as shown by the rate-density data in the bottom of Table 4. (Penetration based on dose and exposure rate data (Table 5), of course, would follow the same pattern having only been adjusted for skin area by location.) Penetration data are generally quite consistent within each operation at each location, exceptions being relatively low penetration onto the lower leg during both operations and one mixer-loader's upper leg being somewhat high; any consideration of these exceptions as "outliers" would reduce slightly the magnitude of the difference between mixer-loaders and applicators but would simultaneously increase its statistical significance. As it was, the percent penetrations are quite different ($p < 0.01$) between operations being 7.2% for mixer-loaders versus 2.4% for spray applicators (geometric errors of the mean are ± 27 and $\pm 39\%$, respectively). There is no definitive explanation for this difference, but it may relate to a larger amount of dry-dust exposure by the mixer-loader versus wet-mist exposure by the spray applicator, as will be discussed later.

The variability among all whole-body exposure rates (as represented by overall geometric deviations in Table 6) was $\sim\pm100\%$, but only $\pm40\%$ within replicates by participants. Separate Student's t tests indicate that the variability between the two mixer-loader's whole-body exposure rates was no greater than the variability within their two replicates, but the corresponding variability between the two spray applicators was significantly greater than within their two replicated studies ($p < 0.02$). Similar tests on the overall geometric means for whole-body dermal exposure rates of 25 mg/h for mixer-loaders and 66 mg/h for spray applicators indicate they were also significantly different from each other ($p < 0.02$).

The hand exposures accounted for 50% of the totals during mixer-loader operations and 35% during spray application; although in absolute terms the spray applicators' mean hand exposure was essentially twice as high as the mixer-loaders'. These patterns resulted in the exposures to the body other than the hands ("not hands") to be statistically different both

TABLE 4—*The mean rate density and penetration of deposited captafol grouped by operation, location, and study, with geometric mean and pooled geometric deviation within all sampled pairs.*

	Study 1	Study 2	Study 3	Study 4	Geometric Mean ± Geometric Deviation within
	MIXER-LOADER EXPOSURE (OUTSIDE) RATE DENSITY, $\mu g/h/cm^2$				
HD	0.033	0.043	0.625	0.372	0.099 ± 240%
CH	0.082	0.022	0.202	0.236	0.096
BA	0.045	0.022	0.129	0.085	0.057
UL	0.380	0.597	2.476	7.190	1.314 ± 76%[b]
LL	0.365	0.228	2.184	2.720	0.799 ± 57%
UA	0.111	0.181	1.015	0.890	0.329 ± 98%
LA	1.665	0.132	2.459	2.429	0.859 ± 178%[b]
GL	3.81	6.62	3.52	11.5	5.59 ± 22%[b]
	MIXER-LOADER DOSE (INSIDE) RATE DENSITY, $\mu g/h/cm^2$				
HD	0.033	0.043	0.625	0.372	0.099 ± 240%
CH	0.0058	0.000	0.0046	0.079	0.013
BA	0.0058	0.0030	0.0066	0.0049	0.005
UL	0.200	0.636	0.171	0.805	0.257 ± 275%
LL	0.0068	0.0059	0.023[a]	0.361	0.023 ± 60%[b]
UA	0.011	0.012	0.034	0.061	0.021 ± 125%
LA	0.051	0.043	0.157	0.134	0.077 ± 65%
GL	3.81	6.62	3.52	11.5	5.59 ± 22%[b]
	MIXER-LOADER CLOTHING PENETRATION (DOSE × 100/EXPOSURE), %				
CH	7.2	0.0	2.2	33.5	<8.2 ± 290%
BA	13.	13.3	5.3	5.9	8.6 ± 64%[c]
UL	52.	106.	6.9	11.2	26. ± 260%[d]
LL	1.8	2.6	1.0	13.2	2.8 ± 200%
UA	10.2	6.7	3.8	6.2	6.4 ± 50%
LA	3.0	32.7	6.4	5.5	7.7 ± 180%
	SPRAY APPLICATOR EXPOSURE (OUTSIDE) RATE DENSITY, $\mu g/h/cm^2$				
HD	0.433	0.286	0.132	0.221	0.215 ± 112%
CH	0.490	1.105	0.676	1.276	0.827
BA	0.773	0.693	0.175	0.085	0.299
UL	9.705	7.523	1.693	1.169	3.386 ± 37%[e]
LL	22.386	10.409	3.178	0.971	3.843 ± 223%
UA	5.546	1.737	0.507	1.178	1.373 ± 105%
LA	7.084	5.451	1.357	2.757	3.303 ± 57%
GL	16.09	13.08	10.40	8.615	11.61 ± 22%

TABLE 4—*Continued*

	Study 1	Study 2	Study 3	Study 4	Geometric Mean ± Geometric Deviation within
	SPRAY APPLICATOR DOSE (INSIDE) RATE DENSITY, $\mu g/h/cm^2$				
HD	0.433	0.286	0.132	0.221	0.215 ± 112%
CH	0.416	0.0135	0.0007	0.020	0.017
BA	0.011	0.0135	0.0042	0.018	0.010
UL	0.789	0.213	0.027	0.078	0.123 ± 99%[e]
LL	0.060	0.137	0.0038	0.008	0.015 ± 325%
UA	0.165	0.093	0.017	0.031	0.046 ± 138%
LA	0.549	0.103	0.767	0.077	0.159 ± 340%
GL	16.09	13.08	10.40	8.615	11.61 ± 22%
	SPRAY APPLICATOR CLOTHING PENETRATION (DOSE × 100/EXPOSURE), %				
CH	85.0	1.2	0.10	1.6	4.3 ± 1800%[f]
BA	1.4	2.0	0.49	22.0	2.3 ± 390%
UL	8.1	2.8	1.6	6.6	3.9 ± 115%
LL	0.27	1.3	0.12	0.83	0.4 ± 195%
UA	3.0	5.3	3.4	2.6	3.5 ± 36%
LA	7.7	1.9	53.0	2.8	6.8 ± 340%

[a] Value based on only one dosimeter.
[b] Differences between studies was significantly greater than differences within duplicate dosimeters at this location.
[c] The two mixer-loaders were statistically different with 13.1 ± 2% and 5.6 ± 8%, respectively.
[d] The two mixer-loaders were statistically different with 74 ± 65% and 8.8 ± 41%, respectively.
[e] Differences between studies was significantly greater than differences within duplicate dosimeters at this location.
[f] Without Study 1, statistics would have been 0.6 ± 94%.

within each operation between the two participants ($p < 0.02$) and between operations ($p < 0.02$).

The variability among dosing patterns (inside the coveralls plus head and hands) was ~55% both within operations and within participants. The only statistically significant difference within an operation is between the whole-body doses for the two spray applicators ($p < 0.02$). As with preceding exposures, the overall mean whole-body doses of 14 mg/h

TABLE 5—*Mean dermal dose-rate and exposure rate of deposited captafol grouped by operation, location, and study, with geometric mean and geometric deviation of the listed values.*

	Study 1	Study 2	Study 3	Study 4	Geometric Mean ± Geometric Deviation between
	MIXER-LOADER DERMAL EXPOSURE (OUTSIDE) RATE, mg/h				
HD	0.043	0.057	0.813	0.484	0.176 ± 340%
CH	0.180	0.047	0.443	0.516	0.210 ± 200%
BA	0.099	0.049	0.282	0.186	0.126 ± 115%
UL	1.316	2.067	8.568	24.88	4.907 ± 285%
LL	0.950	0.594	5.678	7.074	1.091 ± 250%
UA	0.206	0.335	1.889	1.655	0.681 ± 206%
LA	2.148	0.171	3.172	3.133	0.691 ± 310%
GL	8.190	14.226	7.565	24.738	12.15 ± 75%[d]
Total	13.132	17.546	28.41	62.67	25.3 ± 98%
Not hands	4.94	3.32	20.85	37.93	10.7 ± 220%
% of total	(38%)	(19%)	(73%)	(61%)	(42% ± 82%)

TABLE 5—*Continued*.

	Study 1	Study 2	Study 3	Study 4	Geometric Mean ± Geometric Deviation between
MIXER-LOADER DERMAL DOSE OR DEPOSITION (INSIDE) RATE, mg/h					
HD	0.043	0.057	0.813	0.484	0.176 ± 340%
CH	0.013	0.000	0.010	0.173	0.028 ± 385%[a]
BA	0.013	0.0065	0.015	0.011	0.011 ± 44%
UL	0.685	2.201	0.593	2.785	1.256 ± 120%
LL	0.017	0.0155	0.058[b]	0.937	0.061 ± 580%[c]
UA	0.021	0.0226	0.072	0.103	0.044 ± 135%
LA	0.065	0.056	0.202	0.173	0.106 ± 95%
GL	8.190	14.226	7.565	24.738	12.15 ± 75%[d]
Total	9.047	16.585	9.328	29.404	14.2 ± 75%
Not hands	0.814	2.36	1.76	4.67	1.99 ± 106%
% of total	(9.0%)	(14%)	(19%)	(16%)	(14% ± 38%)
SPRAY APPLICATOR DERMAL EXPOSURE (OUTSIDE) RATE, mg/h					
HD	0.563	0.372	0.172	0.287	0.319 ± 64%
CH	1.073	2.421	1.480	2.79	1.810 ± 56%
BA	1.693	1.518	0.384	0.185	0.654 ± 195%
UL	33.582	26.028	5.859	4.05	12.01 ± 188%
LL	58.205	27.065	8.264	2.525	13.47 ± 295%
UA	10.316	3.230	0.944	2.19	2.882 ± 170%
LA	9.139	7.032	1.851	3.55	4.472 ± 110%
GL	24.588	28.126	22.36	18.52	25.19 ± 31%
Total	139.2	95.792	41.314	34.097	65.8 ± 95%
Not Hands	114.6	67.67	18.95	15.58	38.9 ± 165%
% of total	(82%)	(71%)	(46%)	(46%)	(59% ± 35%)
SPRAY APPLICATOR DERMAL DOSE OR DEPOSITION (INSIDE) RATE, mg/h					
HD	0.536	0.372	0.172	0.287	0.319 ± 64%
CH	0.911	0.030	0.0015	0.044	0.037 ± 1275%
BA	0.024	0.030	0.0019	0.040	0.027 ± 90%
UL	2.730	0.737	0.093	0.268	0.473 ± 323%[e]
LL	0.156	0.356	0.010	0.021	0.058 ± 435%
UA	0.306	0.172	0.032	0.058	0.099 ± 180%
LA	0.708	0.133	0.990	0.099	0.310 ± 220%
GL	24.59	28.13	22.36	18.52	25.19 ± 31%
Total	29.96	29.95	23.66	19.337	25.31 ± 24%
Not Hands	5.40	1.83	1.30	0.817	1.795 ± 125%
% of total	(18%)	(6.1%)	(5.5%)	(4.2%)	(8% ± 90%)

[a] Statistics based on three nonzero values.
[b] Value based on only one dosimeter.
[c] Without Study 4, statistics would have been 0.025 ± 109%.
[d] Without Study 4, statistics would have been 9.59 ± 41%.
[e] Differences between studies was significantly greater than differences within studies, but no single outlier can be identified.

for mixer-loaders and 25 mg/h for spray applicators are significantly different from each other at $p < 0.02$.

The hand doses are taken as equal to the hand exposures because no protective gloves are normally worn during these operations. Because of previously discussed differences in the protective effect of the coveralls, the hand doses represented 85% of the total to the mixer-loaders versus 92% of the whole body doses to the spray applicators (a difference significant at $p < 0.02$). If hand exposures were eliminated (as by the use of impervious

TABLE 6—*Dermal exposure-rate (mg/h) and dose-rate (mg/h) summaries with log-normal parametric statistical tests of significant differences.*

	Whole-Body Dermal		Not-Hands Dermal	
	Mixer-Loader	Spray Applicator	Mixer-Loader	Spray Applicator
DERMAL EXPOSURE RATES				
Study 1	13.00	139.00	4.90	115.00
Study 2	17.50	96.00	3.30	68.00
Geometric mean	15.00	116.00	4.00	88.00
Geometric deviation	±23%	±30%	±32%	±45%
Study 3	28.00	41.00	21.00	19.00
Study 4	63.00	34.00	38.00	16.00
Geometric mean	42.00	37.00	28.00	17.00
Geometric deviation	±77%	±14%	±52%	±13%
Overall				
Geometric mean	25.00	66.00	11.00	39.00
Geometric deviation	±100%	±96%	±220%	±160%
t statistic between participants ($df = 2$)	−3.35 $p < 0.10$	7.70 $p < 0.02$	−7.74 $p < 0.02$	8.31 $p < 0.02$
t statistic between operations ($df = 6$)	−3.45 $p < 0.02$		−2.99 $p < 0.03$	
DERMAL DOSE RATES				
Study 1	9.00	30.00	0.81	5.40
Study 2	17.00	30.00	2.40	1.80
Geometric mean	12.00	30.00	1.40	3.10
Geometric deviation	±57%	±0%	±115%	±115%
Study 3	9.30	22.00	1.80	1.30
Study 4	29.00	19.00	4.70	0.82
Geometric mean	16.00	20.00	2.90	1.00
Geometric deviation	±125%	±11%	±97%	±40%
Overall				
Geometric mean	14.00	25.00	2.00	1.80
Geometric deviation	±75%	±26%	±105%	±125%
t statistic between participants ($df = 2$)	−0.62 $p < 0.60$	7.40 $p < 0.02$	−1.43 $p < 0.40$	2.62 $p < 0.10$
t statistic between operations ($df = 6$)	−3.18 $p < 0.02$		0.37 $p < 0.80$	

gloves), the overall mean "not hands" doses of 2.0 and 1.8 mg/h would have been virtually identical in both operations.

Observations and Discussion

The exposures observed during these studies were generally equivalent to other results in comparable settings reported in the literature, for example, Wojeck et al. [6] with ethion, Leavitt et al. [7] with carbaryl, Carman et al. [8] with parathion, and Nigg and Stamper [9] with chlorobenzilate. Considerable variation in mixer-loader exposures was reported both within and among studies by other investigators, for example, exposures within individual body locations on mixer-loaders seem to vary over two orders of magnitude. The initial pesticide formulation (for example, EC, WP, or sprills) would be expected to have some effect on their exposure, but its effect cannot be discerned given the other possible influences of pesticide packaging, package handling, pesticide transfer, in-field exposures to the spray, to wet foliage and to contaminated equipment, individual technique, and investigative methodologies.

The variability of reported exposure rates within selected body locations on spray applicators seems to be at least as great as that for mixer-loaders but can be reduced somewhat by adjusting for the tank concentration and gallons per acre application rate; the former ranges over a factor of 5 (0.045 to 0.225%) among these other tests, and the latter over a factor of 4 (250 to 1000 gal/acre).

By and large, the present study of only four individuals lies within all of the preceding bounds (in terms of both concentration and gallonage (Table 1) and exposure rates). The following field observations offer alternatives to random fluctuations to explain some of the variability discussed in the previous section.

First in the operational sequence is the mixer-loader opening and handling the pesticide package. During these studies, the mixer-loader had to cut or tear away a major section of the carton to access the spout adequately for pouring. Although the sprill formulation significantly simplifies pouring (compared to WP or EC formulations), it is not totally dust free. Dust (sprill fragments) was observed during the transfer of pesticide from the carton into the labeled measuring cup and from the cup into a bucket, especially in Studies 1 and 2 when the carton was supported on a nearly 1.219 m (4-ft) high flat-bed trailer.

An even larger impact on mixer-loader's exposure is believed to have resulted from the combined effects of tank size and tree spacing. Because the mixer-loader must drive into the citrus grove each time a load is transferred to the sprayer applicator, the smaller tank-truck used in Studies 3 and 4 caused there to be twice as many trips through the grove as in Studies 1 and 2. In addition, the mixer-loader in Studies 3 and 4 received more exposure to recently sprayed (wet and dripping) foliage on each trip due to the closer row spacing and overhanging foliage in this grove versus Studies 1 and 2. Thus, it is not surprising that the mean "not hands" M-L exposure in replicate Studies 3 and 4 was seven times higher than in Studies 1 and 2. Hand exposures were more uniform and of course large, causing the total exposures in the second pair of studies to be only three times higher than in the first.

Mixer-loader dermal doses (compared to exposure) were even more uniform than their exposures due to an apparent combined effect of clothing deposition and penetration dominated by the upper legs. It can be seen in Table 7 that the heaviest "not hand" exposures to the mixer-loader occur onto the upper legs probably because they were horizontal when exposed to spray and tree droplets, and for the first mixer-loader (Studies 1 and 2) because he was standing downwind (such as it was) of the aerosol generated while pouring the dry sprills from the carton into the cup and bucket.

As shown in Table 4, the percent penetration onto the upper legs was over eight times higher for the first mixer-loader than for the second (74%/8.8%). This difference in penetration does not appear to correlate with exposure rate-densities. It is conjectured that much of this difference may be due to the intrinsically higher rate of clothing penetration for dry versus wet aerosols. In support of this conjecture are the following observations:

1. The mean penetration (7.2% ± 27% geometric error of the mean) of the mixture of wet and dry exposures for mixer-loaders was earlier shown to be significantly higher

TABLE 7—*Comparison of mean exposure rates (mg/h) and mean dose rates (mg/h), assuming a RMV of 16.7 L/min or 1 m³/h corresponding to a moderate metabolic work rate of 4 kcal/min.*

	Airborne	Whole Dermal	% Air/ total	Not Hands	% Air/ total
	MEAN EXPOSURE RATES, MIXER LOADER				
Study 1	0.091	13.0	0.7%	4.9	1.80%
Study 2	0.056	17.5	0.3%	3.3	1.70%
Study 3	0.031	28.0	0.11%	21.0	0.14%
Study 4	0.063	63.0	0.10%	38.0	0.17%
Overall					
Geometric mean	0.056	25.0		11.0	
Geometric deviation	±70%	±100%		±220%	
	MEAN EXPOSURE RATES, SPRAY APPLICATOR				
Study 1	0.010	139.0	0.007%	115.0	0.009%
Study 2	0.010	96.0	0.011%	68.0	0.015%
Study 3	0.013	41.0	0.031%	19.0	0.067%
Study 4	0.006	34.0	0.017%	16.0	0.036%
Overall					
Geometric mean	0.012	66.0		39.0	
Geometric deviation	±123%	±96%		±160%	
	MEAN DOSE RATES, MIXER-LOADER				
Study 1	0.091	9.0	1.00%	0.81	10.0%
Study 2	0.056	17.0	0.30%	2.40	2.2%
Study 3	0.031	9.3	0.33%	1.80	1.7%
Study 4	0.063	29.0	0.22%	4.70	1.3%
Overall					
Geometric mean	0.056	14.0		2.00	
Geometric deviation	±70%	±75%		±105%	
	MEAN DOSE RATES, SPRAY APPLICATOR				
Study 1	0.010	30.0	0.033%	5.40	0.18%
Study 2	0.010	30.0	0.034%	1.80	0.56%
Study 3	0.013	22.0	0.058%	1.30	0.97%
Study 4	0.006	19.0	0.031%	0.82	0.70%
Overall					
Geometric mean	0.012	25.0		1.80	
Geometric deviation	±123%	±26%		±125%	

than of the totally wet exposures for spray applicators (2.4% ± 39% geometric error of the mean).

2. Penetration of dry pesticide residues through more-or-less horizontal body surfaces has been reported ranging from 30 to 50% [10].

3. Because of the physical setting and wind, the first mixer-loader is believed to have been exposed to more captofol dust in the area of his upper legs and had penetration rates there ~75%. The wearing of coveralls would make deposition here via other than penetration unlikely.

Whatever the complete reason, the differences seen between the two mixer-loaders in terms of exposures largely disappears in terms of both the total and "not hands" doses.

A rather similar trend toward a less significant difference between doses versus exposures can be discerned between the two spray applicators. In this case, the first spray applicator's "not hands" exposures were over five times higher than the second; including hand exposures similarly reduced this difference to three times. The high exposure rate during the first set of studies is believed to have resulted primarily from the need for this spray applicator to frequently remove and clean nozzles while transferring tank loads. This maintenance was attributed by him to recent kelthane applications with this equipment, causing the rapid deterioration of the sprayer's rubber hoses and gaskets that would slough off and clog the spray nozzles, forcing the driver to stand on a small platform near the rear of the speed-sprayer to remove, clean, and replace nozzles. During this time, he would lean with his legs against the wet spray machine, receive residual droplets carried out by the idling fan, and thoroughly wet his hands. This experience was apparently not unusual, at least for this crew.

Penetration of lower leg exposures was consistently low throughout all studies (consistent with the wet dosing mechanism just noted), thus negating its influence upon dose. Thus, the trend toward a smaller difference between the doses among the two spray applicators than between their exposures, is similar to the trend for mixer-loaders but appeared to have a different causation.

The differences in the exposure mechanisms both between and within operations may also explain why there is no overall correlation between, for instance, aerosol and dermal "not hands" exposure, as might be expected if mist aerosol deposition were the singular predominant mechanism in all cases. In order to simplify this comparison, a nominal respiration rate of 1 m³/h is assumed, corresponding to a moderate metabolic work-rate of 4 kcal/min and creating units of equivalent airborne exposure rate, mg/h, as included in Table 7. The fraction of the total exposure attributed to the airborne route, as indicated by the ratio "%Air/Total," is consistent within a multiplying factor of two within the four pairs of replicated studies by each participant but is not equal among the four participants.

It is noted that the mixer-loader during the first two studies was not observed to be exposed to spray mist, but had a much higher airborne fraction compared to other participants, one very close to the 0.8 to 1.7% found among harvesters exposed to dry pesticide residues [11]. The second mixer-loader's ratio indicates that a larger fraction of his exposure resulted from liquid-contact routes, a conclusion also compatible with the low penetration observations just described. The low airborne fraction for most of the spray applications is compatible with the time spent cleaning nozzles (in the case of the first participant), with bodily contact with contaminated surfaces, and with the large droplet size of the mist itself.

The distribution of doses between airborne and dermal routes with and without hands (equivalent to without and with gloves, respectively) shows patterns similar to the foregoing with proportionate increases in the airborne fraction attributable to the measured protective qualities of the coveralls and hypothetical qualities of gloves.

Conclusions

The overall implications of these findings can be viewed in terms of (1) the density of dermal exposure for dermatological considerations, (2) the relative magnitude of dermal versus respiratory dose rates for toxicological considerations, (3) the potential for dermal dosimeters and physical observations to identify and explain differences in sources and routes of exposure, and (4) the importance and potential for clothing or other forms of personal protection to affect the preceding considerations.

When considering either exposure or dose rate densities, the hands are a major exception compared to all other parts of the body. The spray applicators' hands were frequently wet, with reported densities of 10 to 15 μg/h/cm²; while the mixer-loaders' hands ranged from 4 to 11 μg/h/cm². The highest density of doses at other locations only approached 1 μg/h/cm² to the upper legs, head, and the applicators' forearms while most other sites ranged generally $\leq$0.1 μg/h/cm².

Aerosols and dermal exposures were positively correlated between replicate studies within each participant but demonstrated no consistency between participants, probably because of differences in the predominant exposure mechanisms among the four working environments. In general, it would be overly optimistic to expect consistent correlations among other application settings without taking these exposure mechanisms into consideration. Based on the calculations noted in Table 7, at least 99% of the total 15 to 25 mg/h doses was delivered to the skin versus the lungs and upper respiratory tract. If effective glove usage could be enforced, the dermal doses would drop by nearly an order-of-magnitude and the importance of the airborne route would take on a larger but not necessarily predominant intrinsic importance. In order for the airborne route to approximate the importance of the dermal route for mixer-loaders, dermal absorption must be no more than about 1% without gloves or 5% if gloves are to be worn. For the airborne route to approach the importance of the dermal route for spray applicators, both glove protection and less than 1% dermal absorption would have to be assumed. The actual importance of these two routes of dosing for captafol or any other chemical used in a similar setting requires further predictive knowledge concerning the absorbed fraction of dermally deposited formulated and or mixed chemical.

The coveralls used during these studies substantially reduced the exposures of captafol onto the outside of the clothing from reaching the skin. The mean coverall penetration was ~2% for spray applicators and 7% for mixer-loaders. It appeared that coveralls are more effective against mist than dry materials, and when covering vertical rather than horizontal body surfaces. Due to the heat and humidity, it is unlikely that more occlusive forms of dermal protection or respirators would be feasible in this setting. It is uncertain whether any proposed laboratory protocols would adequately predict these field penetration conditions.

References

[1] Durham, W. F. and Wolfe, H. R., "Measurement of the Exposure of Workers to Pesticides," *Bulletin*, World Health Organization, Vol. 26, 1962, pp. 75–91.

[2] Popendorf, W. J. and Leffingwell, J. T., "Regulating OP Pesticide Residues for Farmworker Protection," *Residue Reviews*, Vol. 82, 1982, pp. 125–201.

[3] *Heating and Cooling for Man in Industry*, 2nd ed., American Industrial Hygiene Association, Akron, OH, 1975.

[4] Serat, W. F., VanLoon, A. J., and Serat, W. H., "Loss of Pesticides from Patches Used in the Field as Pesticide Collectors," *Archives of Environmental Contamination and Toxicology*, Vol. 11, 1982, pp. 227–234.

[5] Winterlin, W., Kilgore, W., Mourer, C., and Schoen, S., "Worker Re-entry Studies for Captan Applied to Strawberries in California," *Journal of Agricultural Food Chemistry*, Vol. 32, No. 3, 1984, pp. 664–672.

[6] Wojeck, G. A., Nigg, H. N., Stamper, J. H., and Bradway, D. E., "Worker Exposure to Ethion in Florida Citrus," *Archives of Environmental Contamination and Toxicology*, Vol. 10, 1981, pp. 725–735.

[7] Leavitt, J. R. C., Gold, R. E., Holcslaw, T., and Tupy, D., "Exposure of Professional Pesticide Applicators to Carbaryl," *Archives of Environmental Contamination and Toxicology*, Vol. 11, 1982, pp. 57–62.

[8] Carman, G. E., Iwata, Y., Pappas, J. L., O'Neal, J. R., and Gunther, F. A., "Pesticide Applicator Exposure to Insecticides During Treatment of Citrus Trees with Oscillating Boom and Airblast Units," *Archives of Environmental Contamination and Toxicology*, Vol. 11, 1982, pp. 651–659.

[9] Nigg, H. N. and Stamper, J. H., "Exposure of Spray Applicators and Mixer-Loaders to Chlorobenzilate Miticide in Florida Citrus Groves," *Archives of Environmental Contamination and Toxicology*, Vol. 12, 1983, pp. 477–482.

[*10*] Popendorf, W. J., Spear, R. C., Leffingwell, J. T., et al., "Harvester Exposure to Zolone (Phosalone) Residues in Peach Orchards," *Journal of Occupational Medicine,* Vol. 21, 1979, pp. 189–194.
[*11*] Popendorf, W., "Exploring Citrus Harvesters' Exposure to Pesticide Contaminated Foliar Dust," *Journal,* American Industrial Hygiene Association, Vol. 41, 1980, pp. 652–659.
[*12*] Davies, J. E., Freed, V. H., Enos, H. F., et al., Reduction of Pesticide Exposure with Protective Clothing for Applicators and Mixers, *Journal of Occupational Medicine,* Vol. 24, 1982, pp. 464–468.

R. Grover,[1] A. J. Cessna,[1] N. I. Muir,[2] D. Riedel,[2]
and C. A. Franklin[2]

Pattern of Dermal Deposition Resulting from Mixing/Loading and Ground Application of 2,4-D Dimethylamine Salt

REFERENCE: Grover, R., Cessna, A. J., Muir, N. I., Riedel, D., and Franklin, C. A., **"Pattern of Dermal Deposition Resulting from Mixing/Loading and Ground Application of 2,4-D Dimethylamine Salt,"** *Performance of Protective Clothing: Second Symposium, ASTM STP 989*, S. Z. Mansdorf, R. Sager, and A. P. Nielsen, Eds., American Society for Testing and Materials, Philadelphia, 1988, pp. 625–629.

ABSTRACT: Regional dermal deposition of the dimethylamine salt of 2,4-(dichlorophenoxy) acetic acid (2,4-D DMA) on farmers when handling, mixing, and spraying the herbicide was determined following 30 separate exposures. Before each spraying operation, the farmers were issued a standardized set of laundered cotton clothing that covered most of the body with two layers of clothing. Dermal samplers were placed under the clothing at nine locations and also on the neck and head regions. Two additional dermal samplers, attached over the clothing in the chest and elbow regions, provided direct comparison of outside/inside deposits.

Three distinct levels of dermal deposits could be clearly ascertained. The lowest median 2,4-D (acid equivalent; a.e.) deposit densities occurred under two layers of cotton clothing. These densities were quite uniform and indicated a greatly reduced but nevertheless general permeation of the herbicide through two layers of protective clothing. Somewhat higher median deposit densities were found on exposed body regions less likely to be contaminated during the mixing process, such as the head, neck, and outside elbow regions. The highest median deposit densities occurred on regions of the body most likely to be contaminated during the mixing process, that is, the wrist and chest regions. In the two body regions (chest and left elbow) where direct comparison could be made between deposits outside and beneath the clothing, a significant protective effect was observed, with the median deposit density suggesting a three- to eight-fold protective effect. It may, however, be pointed out that the overall body deposits, minus the hand regions, were only 10 to 20% of the total body deposits and thus, when protective garments equivalent to those used in this study are worn, hand protection must remain a major concern when spraying herbicides with ground-rigs.

KEY WORDS: herbicides, spraying, exposure, 2,4-D, 2,4-dichlorophenoxyacetic acid, protective clothing

The dermal route has been identified as the main route of exposure to those handling and spraying pesticides in the field [1,2]. In a recent but more comprehensive study in which farmers, using tractor-pulled ground-rigs to apply 2,4-(dichlorophenoxy) acetic acid (2,4-D DMA), were monitored for inhalation and dermal exposure during 30 separate spraying operations, the dermal route was again identified as the main route of exposure, with the

[1] Environmental Chemistry of Herbicides Section, Research Station, Agricultural Canada, Regina, Sask. S4P 3A2, Canada.
[2] Environmental Health Directorate, Health & Welfare Canada, Ottawa, Ont. K1A OL2, Canada.

hands contributing the greatest portion (80 to 90%) of the total dermal exposure [3]. Hand-washes were used to monitor exposure to the hands, whereas dermal samplers, placed at several regions of the body both under and over a standardized set of clothing, were used to determine exposure to the remainder of the body, as described earlier [3]. Total body deposition, calculated from the amount of 2,4-D (acid equivalent; a.e.) found on the dermaĺ samplers and the appropriate body surface areas plus that recovered in the hand-washes and the total expressed as microgram of body weight (BW), ranged from 0.1 to 552 µg/kgBW [3]. This paper describes in detail the regional dermal deposition of 2,4-D (a.e.) on these farmers and assesses the protective value of the standardized clothing under typical farm spraying conditions.

Materials and Methods

Field Design

Farmers were monitored for exposure to 2,4-D DMA during 30 individual spraying operations during the 1981 and 1982 spraying seasons. In all 30 exposures, the farmers performed the same work practices, which involved the hand transfer, mixing, and spraying of the herbicide using tractor-pulled ground-rigs. From 6.7 to 88.3 kg of 2,4-D (a.e.) were applied to 16 to 194 ha of cereal crops at rates of 350 to 630 g/ha. These operations were carried out under typical spraying conditions over time periods varying from 55 to 870 min.

Estimation of Dermal Deposition

Each farmer was issued a laundered short-sleeved cotton T-shirt, long work pants, and cotton coveralls that were worn throughout each spraying operation. Hand deposition was determined by rinsing the farmer's hands with 750 mL of a 1% sodium bicarbonate solution after each spraying operation. Exposure to the remainder of the body was determined using dermal samplers that consisted of ethylene glycol impregnated glass fiber filter papers (47 mm) sandwiched between two layers of surgical gauze and attached over and under the clothing. Under the clothing, one patch was placed on the left side of the chest and one just above each elbow, wrist, knee, and ankle [3]. One dermal sampler was placed at the neck and another on the left side of the head on a baseball-type cap. Two dermal samplers, mounted in Petri dishes, were attached on the left side of the chest and above the left elbow outside the clothing. In this way, deposition on the outside of the clothing could be compared with that under one (at the elbow) or two (at the chest) layers of clothing. From these results, deposition (ng/cm²) of 2,4-D (a.e.) on the various regions of the body was determined and compared, and the protective effect of two layers of cotton calculated for chest and upper arm regions of the body. The anatomic proportions used were those of Spear et al. [4], whereas the total body surface area was determined using the height-weight formula of DuBois and DuBois [5].

Analysis of Dermal Samplers and Hand-Wash Solutions

The 2,4-D (a.e.), deposited on the dermal samplers and recovered in the hand-wash solutions, was determined using procedures for which percent recoveries and detection limits have been previously described [6].

Results and Discussion

All sampling patches had detectable levels of 2,4-D (a.e.) indicating that the herbicide had penetrated the clothing regardless of whether one or two layers of clothing was worn.

The median dermal sampler concentrations of 2,4-D (a.e.) by body area and number of layers of clothing are summarized in Table 1.

The median values of 2,4-D (a.e.) deposition on different regions of the body, minus the hands, revealed that the observed differences in 2,4-D (a.e.) deposition resulted from two factors:

1. The presence or absence of two layers of clothing over the dermal sampler.
2. The location of the dermal sampler with respect to body areas that are likely to become contaminated during the mixing processes.

In general, three categories of deposition could be defined. The lowest median 2,4-D (a.e.) deposit densities occurred under two layers of clothing and were quite uniform, ranging from 32 ng/cm^2 on the right elbow to 63 ng/cm^2 on the right thigh. That multiple layers of fabric assist in limiting pesticides from contaminating the skin has been shown in laboratory studies [7]. Somewhat higher median deposit densities were found on body regions that were exposed, but less likely to be contaminated during the mixing processes; that is, head (134 ng/cm^2), outside left elbow (122 ng/cm^2), and neck (71 ng/cm^2). The highest deposit densities occurred on regions of the body that were likely to become contaminated during the mixing processes; that is, right wrist (852 ng/cm^2), left wrist (539 ng/cm^2), and outside chest (396 ng/cm^2).

Two body regions (left elbow and chest) permitted a direct comparison between 2,4-D (a.e.) deposit densities on the dermal samplers that were located on the outside as well as

TABLE 1—*Estimated median amounts and the range of 2,4-D DMA (a.e.) deposited on various regions of the body using dermal samplers situated below one or two layers of cotton clothing or exposed outside the clothing.*

Body Region	Area,[a] (%)	Patch Location	Deposit Densities and Range, ng/cm^2		
			Median	Min	Max
UNDER TWO LAYERS OF CLOTHING					
Arm, upper (r)[b]	4.85	above elbow	32	2	2 100
Arm, upper (l)	4.85	above elbow	45	6	167 000[c]
Thorax[d]	22.80	chest	45	11	880
Thigh/hip (r)	13.55	above knee	63	8	50 600
Thigh/hip (l)	13.55	above knee	43	6	1 490
Calf/foot (r)	9.95	ankle	48	7	4 520
Calf/foot (l)	9.95	ankle	47	8	3 940
UNDER ONE LAYER OF CLOTHING					
Forearm (r)	3.35	wrist	852	14	37 600
Forearm (l)	3.35	wrist	539	26	5 670
OUTSIDE THE CLOTHING					
Neck	1.20	neck	71	4	1 340
Head	5.60	head	134	20	1 480 000[c]
Arm, upper (l)	4.85	elbow	122	36	17 300
Thorax	22.80	chest	396	25	28 400
Hands[e]	7.00	hand-wash	7750	7	33 900

[a] Area expressed as percent of total body surface (mean body surface area was 19 791 cm^2 and mean body weight was 82.3 kg [3]).
[b] r = right and l = left.
[c] Includes one outlier each.
[d] Includes chest, back, and shoulders.
[e] Sampled as hand-wash after each exposure (calculated from data in Ref 3).

on the inside of clothing. These values were compared using Wilcoxon's nonparametric two-sample sign rank tests [8] that showed a significant protective effect of the clothing at the 0.05% level. On the basis of the median deposit densities for the outside and inside samplers at these two locations (Table 1), this protective effect was in the range of three- to eight-fold.

It is encouraging that a protective effect of this magnitude was provided by laundered garments typically worn in the field while spraying. However, since dermal samplers in all regions of the body under two layers of fabric received low levels of 2,4-D (a.e.) deposition, it would indicate that both outer and inner garments were also contaminated with the herbicide. Thus, if protective clothing was reused without laundering, the protective effect may be markedly reduced, emphasizing the need to launder clothing regularly after each spraying operation.

Inhalation exposure was less than 2% of the total body exposure, the remainder being dermal exposure, of which 80 to 90% was on the hands [3]. However, the hands were not protected from exposure, whereas the remainder of the body was protected by one or two layers of cotton fabric, which provided a significant protective effect. In extreme cases, where an applicator may not wear any protective clothing, then the dermal exposure to the remainder of the body may be in the same order of magnitude as exposure to the hands. Therefore, the use of protective garments on the remainder of the body could be equally as effective in minimizing dermal exposure as the use of protective gloves. In terms of the absorbed dose, dermal exposure to the remainder of the body may be more important because the chemical is distributed over a much larger surface area.

However, problems remain that must be solved regarding the effectiveness of materials used in the manufacture of protective clothing and gloves to minimize exposure. For example, methods to adequately test the protective effect of materials exposed to pesticide splashes on an intermittant basis must be developed. Materials for use with emulsifiable concentrate, water soluble, and dry formulations should be available to the pesticide applicator. As well, any limitations of protective gloves or garments (for example, can they be washed and reused?) should be transmitted to field users.

Clearly, the results of material permeation tests have aided in the selection of glove materials with characteristics that allow very toxic chemicals to be used in the industrial setting. The challenge of the next few years will be to adapt and expand garment and glove performance studies to aid in the selection of reliable hand and body protection for those handling pesticides.

Acknowledgments

The authors wish to express their appreciation to Jane Duncan for recruiting subjects and explaining to them the necessary conditions for obtaining the appropriate samples. Technical assistance of Lorne Kerr and Verne Becke of the Regina Research Station for analysis of dermal samples is acknowledged. Dr. D. T. Spurr, statistician, Saskatchewan region, Agriculture Canada, provided the statistical support.

References

[1] Nash, R. G., Kearney, P. C., Maitlin, J. C., Sell, C. R., and Fertig, S. N., "Agricultural Applicators Exposure to 2,4-Dichlorophenoxy Acid" in *Pesticide Residues and Exposure*, J. R. Plimmer, Ed., American Chemical Society Symposium Series No. 182, Washington, DC, 1982, pp. 119–132.
[2] Nigg, H. N. and Stamper, J. H., "Exposure of Spray Applicators and Mixer-Loaders to Chloro-benzilate Miticide in Florida Citrus Groves," *Archives of Environmental Contamination and Toxicology*, Vol. 12, 1983, pp. 477–482.

[3] Grover, R., Cessna, A. J., Miur, N. I., Riedel, D., and Franklin, C. A., "Factors Affecting the Exposure of Ground-Rig Applicators to 2,4-D Dimethylamine Salt," *Archives of Environmental Contamination and Toxicology,* Vol. 15, 1986, pp. 677–686.

[4] Spear, R. D., Popendorf, W. F., Leffingwell, J. T., Milby, T. H., Davies, J. E., and Spencer, W. F., "Fieldworkers' Response to Weathered Residues of Parathion," *Journal of Occupational Medicine,* Vol. 19, 1977, pp. 406–410.

[5] DuBois, D. and DuBois, D., "A Formula to Estimate the Approximate Surface Area if Height and Weight be Known," *Archives of Internal Medicine,* Vol. 17, 1916, pp. 863–871.

[6] Grover, R., Cessna, A. J., and Kerr, L. A., "Procedure for the Determination of 2,4-D and Dicamba in Inhalation, Dermal, Hand-Wash, and Urine Samples from Spray Applicators," *Journal of Environmental Science and Health,* Vol. B20, 1985, pp. 113–128.

[7] Laughlin, J. M., Easley, C. B., Gold, R. E., and Hill, R. M., "Fabric Parameters and Pesticide Characteristics that Impact on Dermal Exposure of Applicators" in *Performance of Protective Clothing,* R. L. Barker, and G. C. Coletta, Eds., *ASTM STP 900,* American Society for Testing and Materials, Philadelphia, 1986.

[8] Sokal, R. R. and Rohlf, F. J., *Biometry,* W. H. Freeman and Co., San Francisco, 1969, p. 400.

Richard A. Fenske[1]

Use of Fluorescent Tracers and Video Imaging to Evaluate Chemical Protective Clothing During Pesticide Applications

REFERENCE: Fenske, R. A., "Use of Fluorescent Tracers and Video Imaging to Evaluate Chemical Protective Clothing During Pesticide Applications," *Performance of Protective Clothing: Second Symposium, ASTM STP 989*, S. Z. Mansdorf, R. Sager, and A. P. Nielsen, Eds., American Society for Testing and Materials, Philadelphia, 1988, pp. 630–639.

ABSTRACT: The addition of fluorescent tracers to agricultural spraying systems allows the direct evaluation of protective clothing under actual field conditions. Deposition of the tracer beneath clothing is quantified by means of a video imaging system. Workers are illuminated under long-wave ultraviolet light following exposure, and the fluorescence on the skin is measured. This exposure assessment technique was tested with 25 workers conducting mixing and high-volume airblast application procedures with the organophosphate pesticide malathion in citrus groves. Three types of protective clothing were compared: cotton/polyester workshirts, cotton/polyester coveralls, and nonwoven coveralls (untreated Tyvek). All workers had measurable hand exposure, despite the use of neoprene gloves. Hand exposure to mixers was more than twice that of applicators. Substantial deposition of tracer occurred beneath clothing. The applicator group wearing workshirts was more highly exposed than either of the coverall groups (Analysis of Variance, $p < 0.05$). Exposure between the two coverall groups did not differ significantly for either mixers or applicators. The heaviest depositions for the coverall groups occurred near the openings of the garments (collar and sleeves), suggesting that studies of penetration through fabric do not provide a complete evaluation of potential exposure under field conditions. This study demonstrates that fluorescent tracers and video imaging can be used to provide a quantitative index of protective clothing field performance.

KEY WORDS: exposure assessment, evaluation, protective clothing, pesticide application, malathion, fluorescent tracer, video imaging

Dermal exposure to pesticides during mixing and application procedures can constitute a serious occupational health hazard. Chemical protective clothing has been employed with increasing frequency as a means of reducing worker exposure in the agricultural workplace. This paper offers a new approach to evaluating the effectiveness of protective clothing under realistic field conditions [1,2]. A fluorescent compound is employed as a tracer of pesticide deposition on the skin. Deposition levels are quantified with a video imaging system. This method allows a direct evaluation of the field performance of various protective clothing alternatives.

The field of industrial hygiene has as its stated goal the recognition, evaluation, and control of health hazards in the workplace [3]. Industrial hygiene recognizes a hierarchy of control strategies in which engineering and administrative alternatives are considered first.

[1] Assistant professor, Department of Environmental Science, New Jersey Agricultural Experiment Station, Rutgers University, New Brunswick, NJ 08903; field work was conducted at the University of California Lindcove Field Station.

Protective clothing is considered to be a control of "last resort," acceptable only for temporary activities, or if emergency conditions arise. This approach to exposure reduction is outlined explicitly in the standards adopted under the Occupational Safety and Health Act of 1970 [4,5].

A control strategy based on protective clothing must include a systematic evaluation program of actual performance, and requires continuous training of personnel in the proper use and maintenance of the clothing. Use of chemical protective clothing often reduces comfort, agility, and dexterity for the worker, and may contribute to heat stress under environmental conditions commonly encountered in agriculture.

Nonetheless, protective clothing is an important control option for the reduction of dermal exposure among agricultural workers, since traditional engineering and administrative controls are either not appropriate for outdoor environments or are not relevant to the dermal route of exposure. Engineering approaches such as closed cab systems and closed mixing systems can reduce occupational exposures substantially [6], but such systems are not feasible for many pesticide application conditions. As a result, federal and state regulatory agencies are focusing on protective clothing as a primary strategy for exposure and risk reduction [7].

Current knowledge of the efficiency of protective clothing performance in agriculture is very limited [8]. Recent studies have concentrated on laboratory techniques for determining the movement of pesticides through materials [9–11]. Field studies have been hampered by lack of an adequate methodology for characterizing the performance of protective clothing under realistic conditions. The most common field method is an adaptation of the patch technique, involving attachment of collection devices both inside and outside clothing [12,13]. This approach has been employed to evaluate direct penetration of clothing, but the accuracy of the technique is severely limited by the assumption that exposure will be uniform over entire body regions. This method is further limited by an inability to monitor deposition at the openings of garments, and by an inability to detect contamination due to temporary removal or inappropriate use of clothing.

The use of fluorescent tracers in conjunction with video imaging offers an alternative methodology for examining protective clothing performance. This technique provides a means of visualizing patterns of exposure on skin surfaces beneath clothing and can assist in determining the source of such exposures. It also allows direct comparison of different types of protective clothing employed under similar exposure conditions. It is hoped that the use of fluorescent tracers and video imaging will clarify the appropriateness of particular types of clothing for a wide range of work activities.

Procedure

The research design for this project differed from traditional monitoring studies by strictly controlling a number of application parameters. The aim was to create a series of exposure episodes in which each worker's "exposure potential" was similar. Reduction of variability due to factors extraneous to the issue of protective clothing allows a direct comparison of exposure values among the workers. The standardized procedures implemented in this study are summarized in Table 1.

Separate individuals conducted the mixing and application activities throughout the study. All workers employed the same 1892 L (500 gal) air blast sprayer for mixing and application. Each mixer added 5.6 kg (15 lb) of a 25% wettable powder (WP) formulation of malathion [O,O-dimethyl-S-(1,2-dicarbethoxyethyl)phosphorodithioate] per tank. He was also asked to follow the same procedures in adding 300 g of a fluorescent tracer [4-methyl-7-diethyl-aminocoumarin] to the tank. Each applicator sprayed four tanks. The clothing worn by the

TABLE 1—*Pesticide exposure assessment study design.*

1. Separation of mixer and applicator activities
2. Standardized procedures
a. Protective clothing
b. Application equipment
c. Number of work cycles
d. Amount of pesticide/tracer handled
3. Uncontrolled variables
a. Wind/environmental factors
b. Time
c. Work practices

workers is outlined in Table 2. All workers wore baseball caps, half-mask respirators with pesticide cartridges, neoprene gloves, workboots, 100% cotton T-shirts, and 65/35 cotton/polyester workpants. The protective garments employed were new and double-washed. Group 1 wore nonwoven coveralls made of untreated Tyvek.[2] Group 2 wore woven coveralls made of a 65/35 cotton/polyester fabric (weight = 27.2 mg/cm^2). Group 3 wore workshirts made of a 50/50 cotton/polyester fabric (weight = 11.0 mg/cm^2).

Several variables could not be controlled during the study. Environmental factors such as temperature and humidity were monitored and did not change greatly over the study period. Wind, however, was highly variable and unpredictable, substantially affecting the exposure potential of the individual subjects. The range of time worked for mixers was 50 to 61 min, while applicators displayed a broader range, 72 to 122 min. Work practices also appeared to influence exposure potential; for example, differences in mixing techniques, allowing the tank to fill to overflow, continuing applications during turns in the groves, and application procedures in response to changes in wind.

The study was conducted in citrus groves in the Central Valley of California during May and June. The study group included eight mixers and 17 applicators, with one mixer working with two applicators each day. The mixer was supplied with a total of 22.4 kg (60 lb) of malathion 25% WP and 1200 g of fluorescent tracer. The resulting tank ratio was 5.7:1, pesticide to tracer. A high-volume spray schedule typical of that followed by growers in

TABLE 2—*Protective clothing worn by workers.*

Worker Group	Body Region	Personal Protection
All workers	head	baseball cap
	face	respirator
	hands	rubber gloves [neoprene]
	feet	boots
	torso	T-shirt [cotton]
	legs	workpants [cotton/polyester]
Group 1	torso + legs	nonwoven coveralls [untreated Tyvek]
Group 2	torso + legs	woven coveralls [cotton/polyester, 65/35]
Group 3[a]	torso	cotton workshirt [cotton/polyester, 50/50]

[a] Applicators only.

[2] Registered trademark of I. I. duPont de Nemours & Company.

the area was conducted. The air blast sprayer was calibrated at 189 L/min (50 gal/min) for an application rate of 18 690 L/hectare (2000 gal/acre). Tractor speed was 1.6 km/h (1 mile/h).

Fluorescent Tracer/Video Imaging Methodology

The fluorescent compound, 4-methyl-7-diethylaminocoumarin, is a fluorescent whitening agent (FWA) commonly used in commercial products. The use of FWAs as tracers of chemical deposition on the skin has been discussed previously [2]. This compound is not acutely toxic, does not produce dermal irritation, and is nonmutagenic in bacterial assays. It has been employed previously in the evaluation of worker exposure during pesticide applications in agriculture [14]. In the current study, the deposition of the tracer compound and malathion in the vicinity of the worker was highly correlated ($r = 0.94$) [2].

The instrument employed to quantify dermal fluorescence was a computer-based imaging system interfaced with a television camera. A single scan of the camera is captured by the computer and written onto disk as a digitized video image. The design and testing of this system have been described in detail elsewhere [1]. Subjects were examined in a darkened room under long-wave ultraviolet light (320 to 400 nm) both prior to work and immediately following the exposure period. Workers wore only black athletic shorts during the examination. Ultraviolet-A illumination was approximately 200 μW/cm^2 at the subject plane (20% of the American Conference of Governmental Industrial Hygienists Threshold Limit Value) [15]. The subjects were able to view themselves on television throughout the examination. Video images were recorded of the following body parts; head, hands, left and right forearms, left and right upper arms, torso, upper legs, and lower legs. Four views of each part were taken so that all sides were included.

Standards were read before and after each examination. Analysis of the video images consisted of several steps: (1) comparison of before and after images to determine if exposure had occurred for a particular part/view; (2) adjustment of exposed images for the nonplanar aspects of the specific body region by means of geometric models; and (3) use of a standard curve to relate the fluorescent intensity detected by the camera to the amount of fluorescent tracer resident on the skin surface.

Results and Discussion

Several general exposure patterns were evident. First, little exposure occurred to the legs of either the mixers or applicators, regardless of the type of clothing worn. The results presented later, therefore, pertain only to exposure occurring above the waist. Second, all workers were exposed on the face and neck, these being the only unprotected body regions. Third, all workers exhibited measureable exposure to the hands despite the use of rubber gloves. Finally, 24 of the 25 workers received exposure to the arms or torso, regions covered by protective clothing.

Exposure values are reported as micrograms of fluorescent tracer deposited on the skin. Studies are currently underway to determine the appropriate calibration factors for relating the two compounds both in cases of direct deposition and for penetration through protective clothing. The exposure values reported for mixers represent exposure received during four tank loadings, while values for applicators represent exposure from four applications.

Hand Exposure

Summary statistics of hand exposures for mixers and applicators are presented in Table 3. For each work group, hand exposure exhibited high variability, with an overall range of

TABLE 3—*Hand exposure beneath rubber gloves during mixing and application.*

Worker Type	N	Mean, μg	Range, μg	Standard Deviation
Mixers	8	138.5[a]	52 to 263	66.6
Applicators	17	58.0[a]	4.9 to 221	60.8

[a] Mean values are significantly different (ANOVA: $F = 9.0$; $p < 0.01$).

5 to 263 μg of tracer deposition. The mixers had much higher exposure levels, averaging 138 μg, whereas the mean for the applicators was only 58 μg. The mean values are significantly different when compared by Analysis of Variance (ANOVA) ($p < 0.01$). These results are in accord with many earlier studies that have found higher exposures during mixing activities. Clearly, exposure potential is highest when working with concentrated rather than diluted active ingredient.

Field observations of work practices indicated that both the mixers and applicators removed their gloves during breaks in work. Mixers nearly always removed the gloves immediately following the loading activity, and applicators kept their gloves off while mixing occurred. Removal of gloves during work in agriculture is a normal and predictable behavior. Considering the physical discomfort of rubber gloves, it is not surprising that the gloves were removed several times over the 2-h work period. Thus, hand exposure can be expected during mixing and application regardless of the type of glove worn. The difference in magnitude between the mixer and applicator exposures may be due either to glove failure during mixing or to contact with the concentrated or dilute residues on the gloves themselves.

Regional Variability of Exposure

The body can be separated into three regions for comparative purposes: hands, face, and neck, and protected regions (above the waist). Comparisons of exposure to each of these regions may assist in developing appropriate control strategies. The relative importance of each region is often expressed as a percentage of total exposure. These values are presented in Table 4. The overall results show that the contribution of hand exposure to total exposure can vary substantially due both to worker activity and to other types of protection used by the workers. The high exposure to mixers' hands, for example, represents 41% of their total exposure, while applicator hand exposure is a much smaller 13%. These values demonstrate that hand exposure does not necessarily dominate total exposure. Caution should be taken in applying generalizations to different types of work activities and exposure situations.

TABLE 4—*Percent fluorescent tracer exposure by body region.*

Worker/Clothing Type	N	Hands	Head	Protected Regions[a]
Mixers	8	41	22	37
Nonwoven[b]	2	33	26	41
Coveralls[b]	6	43	21	36
Applicators	17	13	28	59
Nonwoven	4	16	47	36
Coveralls	7	17	31	52
Workshirt[b]	6	10	20	70

[a] Includes arms and torso; does not include regions below the waist.
[b] Nonwoven = untreated Tyvek coveralls; coveralls = cotton/polyester coveralls (65/35); and workshirt = cotton/polyester workshirt (50/50).

The alteration of arm and torso protection can affect the percentage values of the other body regions, as is illustrated by differences among the applicators. Exposure to protected regions ranged from 36% for workers wearing the nonwoven coveralls to 70% for those wearing only the workshirt. Thus, the higher levels of exposure to protected regions among the workshirt group tends to diminish the relative importance of hand and face exposure. However, the actual exposure values for the hands and face differ very little among the three applicator groups. Thus, data expressed solely as percentages can be misleading when making comparisons across studies if attention is not paid to the impact of differences in overall protection.

Performance of Protective Clothing

The effect of specific clothing types on dermal exposure is an important aspect of this study, and the change in percentages for protected regions among applicator groups in the last column of Table 4 is suggestive of marked differences. Table 5 provides a comparison of exposure levels to protected regions by worker activity and by the type of clothing worn.

Mixers wearing nonwoven coveralls and cotton coveralls show little difference in exposure levels. The mean values of 124.9 and 125.8 μg are virtually identical. There is also high variability in the individual exposure values. Among the applicator groups, however, a clear pattern emerges indicating a reduction in exposure due to clothing type. The low mean exposure level of 94 μg for the nonwoven coverall group more than doubles to 205 μg for the cotton coverall group, and increases again to 432 μg for the workshirt goup. Within each group, however, there is very high variability, as indicated by the ranges and standard deviations. The differences in mean values are not statistically significant.

This study was designed to produce equal exposure potential for each clothing type group, but clearly this aim was not achieved. Wind conditions created a wide range of exposure values, requiring adjustment of the data to normalize exposure potential among workers. The amount of exposure that each worker received to the unprotected head region provides a direct indication of the variability in wind effects during mixing and application. The video imaging values for total head exposure were therefore employed for normalizing exposure potential. The protected region exposure value for each worker was divided by the individual's head exposure value. This adjustment reduces variability within the clothing type

TABLE 5—*Fluorescent tracer exposure beneath protective garments.*

Clothing Type	N	Mean, μg	Range, μg	Standard Deviation	Percent Reduction[a]
Mixers					
Nonwoven[b]	2	124.9	39 to 210	121	1
Coveralls[b]	6	125.8	41 to 216	63	…
Applicators					
Nonwoven	4	94[c]	0 to 248	109	78
Coveralls	7	205[d]	21 to 555	191	52
Workshirt[b]	6	432[c,d]	68 to 1371	478	…

[a] Reduction in exposure for mixers compared to wearing coveralls; for applicators compared to wearing workshirt.

[b] Nonwoven = untreated Tyvek coveralls; coveralls = cotton/polyester coveralls (65/35); workshirt = cotton/polyester workshirt (50/50).

[c,d] Mean values with matching letters are significantly different when data are normalized for exposure potential (ANOVA, Student-Newman-Keuls: $p < 0.05$).

groups, allowing the effects of clothing type to be isolated from the confounding variables contributing to unequal exposure potential.

These adjusted data were submitted to statistical analysis. The two mixer groups still did not exhibit a significant difference in exposure. However, among the applicators, those wearing workshirts had significantly higher exposure than either the cotton coverall or the nonwoven coverall group (ANOVA, Student-Newman-Keuls: $p < 0.05$). The apparent difference between the two coverall groups is not statistically significant. Thus, the cotton workshirt material clearly provides a lower level of protection than do either of the coverall materials.

Evidence of the actual penetration of workshirt material is depicted in Fig. 1. Here the fabric has become wet and has adhered to the skin for a short period of time, producing a pattern of distinct spots of tracer deposition. Forearm exposure patterns of this kind were common in the workshirt group, whereas they did not occur with either of the coverall materials. These results suggest that workshirts are not an appropriate protective clothing for airblast applications.

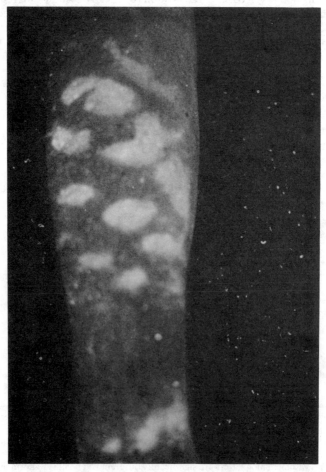

FIG. 1—*Fluorescent tracer exposure to applicator forearm through workshirt material.*

If the workshirt is considered minimal protection, the relative protection afforded by the other types of clothing can be calculated. These values, expressed as percent reduction in exposure, are presented in the final column of Table 5. The cotton coverall provides more than 50% reduction in exposure to protected regions, and the nonwoven coverall provides nearly 80% exposure reduction.

Several explanations can be offered in regard to the lack of differences between the two coverall materials. First, the sample size for the groups was small, and particularly so for the mixers. The difference in exposure between the two applicator coverall groups is intriguing, and may prove to be real in a larger study. A second explanation suggests that once a reasonably high level of protection is achieved (that is, coveralls of any kind), then the major source of exposure to protected regions is not by penetration through fabric but by the entrance and deposition of material through openings in the clothing. Figure 2 illustrates the deposition pattern seen on the neck and back of an applicator. The tracer is deposited far below the coverall collar. Deposition occurs on the shoulder blades, decreasing with increasing distance from the clothing opening. This pattern is not in accord with penetration through the clothing, but rather with material being drawn inside the clothing before deposition. This "bellows" effect has been described in studies of air movement within protective garments, and may well play an important role in the field performance of clothing.

Conclusion

In conclusion, this study demonstrates that substantial spray penetration of protective clothing can occur during high-volume airblast applications. The forearm receives the most

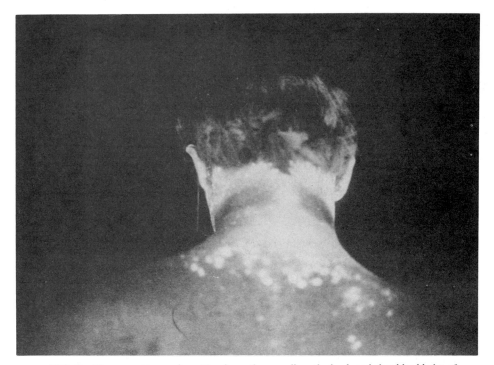

FIG. 2—*Fluorescent tracer deposition beneath coverall on the back and shoulder blades of applicator.*

exposure among protected regions, and cotton coveralls and nonwoven coveralls provide significantly greater protection than workshirts during application. It appears that a major portion of exposure to the torso and forearms can occur through openings in clothing rather than through fabric penetration. This finding should be examined in a larger study comparing woven and nonwoven coveralls. The results of this study clearly demonstrate that the use of fluorescent tracers and video imaging provides a quantitative procedure for developing a relative index of protection under realistic field conditions. Finally, the qualitative use of fluorescent tracers has great potential for the rapid evaluation of the effectiveness of protective equipment and clothing, and can serve as an important tool in the education and training of workers.

Acknowledgments

The excellent cooperation of the administration and staff of the University of California Lindcove Field Station is acknowledged and greatly appreciated. The field work was supported by the National Institute for Occupational Safety and Health (Grant No. 1 RO1 OH 01234-01 A1) and the U.S. Environmental Protection Agency (Cooperative Agreement CR-810691-01-0). The data analysis portion of this research was funded in part by the U.S. Department of Agriculture (Grant No. 83-CRSR-2-2121), and was also supported by state funds through the New Jersey Agricultural Experiment Station (Publication No. K-07905-1-87).

References

[1] Fenske, R. A., Leffingwell, J. T., and Spear, R. A., *Journal,* American Industrial Hygiene Association, Vol. 47, No. 12, Dec. 1986, pp. 764–770.
[2] Fenske, R. A., Wong, S. M., Leffingwell, J. T., and Spear, R. A., *Journal,* American Industrial Hygiene Association, Vol. 47, No. 12, Dec. 1986, pp. 771–775.
[3] *Patty's Industrial Hygiene and Toxicology, 3rd Ed., Vol. I: General Principles,* G. D. Clayton and F. E. Clayton, Eds., Wiley, New York, 1978.
[4] Schulte, H. F. in *The Industrial Environment—Its Evaluation and Control,* National Institute for Occupational Safety and Health, Washington, DC, U.S. Government Printing Office, 1973.
[5] Moran, J. B. in *Proceedings,* 2nd International Symposium on the Performance of Protective Clothing, American Society for Testing and Materials, Philadelphia, 1987.
[6] Lunchick, C., Reinert, J., and Nielsen, A., this publication, pp. 605–610.
[7] Nielsen, A. P. and Moraski, R. V. in *Performance of Protective Clothing, ASTM STP 900,* R. L. Barker and G. C. Coletta, Eds., American Society for Testing and Materials, Philadelphia, 1986, pp. 95–102.
[8] Freed, V. H., Davies, J. E., Peters, L. J., and Parveen, F., *Residue Review,* Vol. 75, 1980, pp. 159–168.
[9] Branson, D. H., Ayers, G. S., and Henry, M. S. in *Performance of Protective Clothing, ASTM STP 900,* R. L. Barker and G. C. Coletta, Eds., American Society for Testing and Materials, Philadelphia, 1986, pp. 114–120.
[10] Hobbs, N. E., Oakland, B. G., and Hurwitz, M. D. in *Performance of Protective Clothing, ASTM STP 900,* R. L. Barker and G. C. Coletta, Eds., American Society for Testing and Materials, Philadelphia, 1986, pp. 151–161.
[11] Leonas, K. K. and DeJonge, J. O. in *Performance of Protective Clothing, ASTM STP 900,* R. L. Barker and G. C. Coletta, Eds., American Society for Testing and Materials, Philadelphia, 1986, pp. 177–186.
[12] Davis, J. E., *Residue Reviews,* Vol. 75, pp. 33–50.
[13] Davies, J. E., Enos, H. F., Barquet, A., Morgade, C., Peters, L. J., and Danauskas, J. X. in

Pesticide Residues and Exposure, ACS Symposium Series, Vol. 182, American Chemical Society, Washington, DC, pp. 169–182.

[*14*] Fenske, R. A., Leffingwell, J. T., and Spear, R. C. in *Dermal Exposure Related to Pesticide Use,* ACS Symposium Series, Vol. 273, American Chemical Society, Washington, DC, pp. 377–393.

[*15*] *TLV—Threshold Limit Values for Chemical Substances in the Work Environment,* American Conference of Governmental Industrial Hygienists, Cincinnati, OH, 1986–87.

Charles B. Hassenboehler, Jr.,[1] *Herbert N. Nigg,*[2] *and Jacquelyn Orlando DeJonge*[1]

Comparison of a Thermal Test Battery Analysis and Field Assessments of Thermal Comfort of Protective Apparel for Pesticide Application

REFERENCE: Hassenboehler, C. B., Jr., Nigg, H. N., and DeJonge, J. O., "**Comparison of a Thermal Test Battery Analysis and Field Assessments of Thermal Comfort of Protective Apparel for Pesticide Application,**" *Performance of Protective Clothing: Second Symposium, ASTM STP 989,* S. Z. Mansdorf, R. Sager, and A. P. Nielsen, Eds., American Society for Testing and Materials, Philadelphia, 1988, pp. 640–648.

ABSTRACT: Protective apparel fabrics designed to reduce dermal exposure during pesticide spraying procedures were evaluated relative to thermal comfort under sweating/high heat stress conditions. Preliminary results from field tests conducted in Florida were compared to thermal comfort predictions inferred from a battery of thermal tests on the fabrics.

The thermal battery consisted of thermal transmittance with simultaneous moisture transport, a profile of air permeability over a range of differential pressure, and a radiant temperature parameter introduced as the clothing radiant temperature. The performance of a fabric in each specific test was graded on a 10-point scale system. Higher points were awarded according to the fabrics' ability to relieve or lower heat stress associated with thermal discomfort. Individual and cumulative thermal comfort scores were compared to field test results. The effect of heat, moisture, and air transport on thermal comfort as inferred from the thermal test battery data provides a reasonable indicator of actual field study thermal comfort testing.

KEY WORDS: pesticide, protective clothing, dermal exposure, thermal comfort, sweating heat stress

The development of protective apparel for pesticide application has long been hampered by the often contradictory requirements to provide adequate protection against pesticide penetration and thermal comfort in conditions of high temperature and humidity. As pesticide penetration work continues to identify acceptable fabrics, attention must be paid to the thermal characteristics of these fabrics in order to make final recommendations of the fabrics that provide relative comfort in heat stress. If this is not accomplished, the agricultural workers will continue to choose not to wear the available protective apparel.

Thermal comfort testing has been conducted in several ways: (1) in a controlled laboratory setting by testing the fabric alone [1], (2) in a controlled environment chamber using human

[1] Graduate research assistant and department head, respectively, Department of Textiles, Merchandising and Design, College of Human Ecology, The University of Tennessee, Knoxville, TN 37996-1900.

[2] Professor, Agricultural Research and Education Center, University of Florida, Lake Alfred, FL 33850.

subjects [2], and (3) in actual human subject field testing conditions. The latter two are extremely expensive and should be used only after a fabric has been properly screened for expected thermal performance.

This study attempts to compare a battery of laboratory thermal tests with field assessments of thermal comfort for four prospective fabrics. The laboratory testing was conducted at The University of Tennessee, Knoxville, and the field assessments were conducted by The University of Florida, Lake Alfred field station. This paper compares initial data obtained in laboratory and field setting situations, in anticipation of refining the laboratory techniques to pre-assess fabrics for field economic study assessment.

Experimental Procedure

Fabric

Four fabrics were selected for this study that have been found to be good barriers against pesticide penetration: 50/50 cotton polyester twill; 50/50 cotton polyester twill with a fluorocarbon finish; Gore Tex[3] (a fabric with 50/50 cotton polyester woven face and polytetrachloroethylene film back); and SMS (a composite nonwoven polypropylene) [3]. In addition, 100% cotton chambray was selected as a thermal comfort reference fabric for the laboratory evaluation. Standard physical test methods were used to characterize the fabrics (Table 1).

Laboratory Thermal Test Battery

A battery of comfort related physical tests were used to assess the ability of experimental protective clothing fabrics to relieve the heat stress developed during summer-time pesticide application work. Three physical measurements were performed on the fabrics: thermal transmittance (U-values), wind penetration potential (WPP), and the clothing radiant temperature (Clort).

U–Values—Fabric thermal transmittance (U-value) is a measure of the total heat flow through the fabric per unit temperature difference across the air layers on each side of the fabric. U-value units are W/m²K or Btu/h/ft²/°F. U-values were measured during simultaneous moisture transport to reflect the ability of the fabrics to transport heat under simulated sweating/heat stress conditions [4].

A guarded heat and mass transmittance test instrument measures thermal transmittance

TABLE 1—*Fabric physical characteristics.*

	Yarn Count, yarns/in.		Weight, g/m²	Thickness, mm	Air Permeability, m³/s · m²
	Warp	Fill			
Chambray	58	52	117	0.30	0.075
Twill	50	58	235	0.43	0.075
Twill F1	50	57	234	0.36	0.094
Gore Tex	60	57	165	0.25	0.00
SMS	NA	NA	65	0.18	0.048

[3] Registered trademark of W. L. Gore and Associates, Inc.

of flat fabrics under a mixed flow state of free convection [1]. The mixed flow transmittance tester (MFTT) uses solid-state heat flow transducers situated on large flat plate heat sinks that are critically spaced, guarded, and temperature controlled. Mixed flow is a distinctive state of heat transfer for which the ratio of total transmittance to the conductive heat flow across an air space is constant. This controlled, natural convection feature makes it possible to standardize thermal transmittance results under defined free-convective conditions.

Water reservoirs made with silicone caulk levees were placed on the heat source plate to serve as a moisture generator. The water surface area was one half the plate area. The cold plate served as a condensing chill plate for the transported moisture as its temperature was always less than the dew point of the adjacent air mass. The moisture-generating system was designed to produce a constant vapor supply for transport through the air and textile being tested. The mixed flow convective state produced in the apparatus was expected to prevent vapor stratification in the chambers.

At perspiring heat stress conditions, U-values of apparel fabrics can range from 8.5 to 14.2 W/m²K. On initial exposure to "sweating" heat stress conditions in the MFTT, fabrics generally condense some moisture while transmitting heat and moisture through the fabric. The maximum U-value attained is affected by the ability of the fabric to maintain high moisture flux and on its physical characteristics related to effective thickness. Hairiness, texture, and wicking affect surface temperatures, and the location and type of moisture condensation produced.

In a U-value determination, water was added to reservoirs on the heat source plate, excess condensation was removed from the chilled top plate, and the fabric specimen was inserted between the MFTT halves. Temperature and heat flux signals were converted to U-values by a data acquisition system that also plotted the results. Values of the initial U-value and equilibrium values were recorded. The initial value indicates the initial comfort response (high U implying greater relief of heat stress). Equilibrium U-values were expected to reflect the ongoing ability of the fabric to transfer heat away from a sweating subject.

Wind Penetration Potential—The wind penetration potential (WPP) was designed to characterize the heat stress relief benefit of wind. The WPP of the fabric is defined as the area under a graph of the air flow rate through a fabric over a range of differential pressure across the fabric. The WPP for a fabric assesses the potential for wind (or tractor movement) to reduce "on the job" heat stresses with respect to the chambray. The WPP parameter is supported by other work on wind effect reviewed by Fourt and Hollies [5] and the recent areal penetration work of Harter et al. [6] and others [7–9] using tracer gas techniques to assess microclimate air exchange due to wind. A Gurley Densiometer/Permeometer was modified and automated to plot air flow as the differential pressure across the fabric was varied.

Clothing Radiant Temperature (Clort)—The test battery further includes an "instrumental operative temperature" parameter (after Gagge [10]) that assesses the radiant temperature field to which the skin of the clothed subject in a controlled environment is exposed. The Clort is affected by the clothing and the microclimate to which the wearer's skin (heated plate) is exposed. The parameter is named the clothing radiant temperature or Clort, as it assesses the effect of the clothing on the radiant field to which the "skin" is exposed for a controlled external environment. Conventional woven fabrics transmit radiant heat as a function of their cover. In a given environment, the Clort parameter simulates the mean radiant temperature (heat) signal felt by the skin as seen through the clothing, as in Fig. 1. This thermal stimulus activates one's physiology to react to the thermal environment. As an element of the thermal test battery, a higher Clort value is interpreted as a hotter

ENVIRONMENT   CLORT

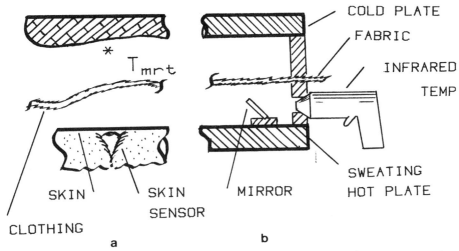

FIG. 1—*Relation between* (a) *physical environment and* (b) *experimental measurement of the clothing radiant temperature* (*Clort*).

environment to the skin, potentially contributing to or signalling the onset of thermal discomfort.

The infrared thermometer and front surface mirror assembly (Fig. 1) is built into one side of the MFFT instrument. Clort measurements are taken concurrent with thermal transmittance testing. The infrared radiant temperature of the fabric is measured from the warm air side of the fabric.

Florida Field Assessment

Skin temperatures were measured on the upper arm, the upper chest, and the outside lower leg of each subject and the mean skin temperature was calculated according to Burton's equation [11]:

$$Tmsk = 0.50 \ Tskc + 0.36 \ Tskl + 0.14 \ Tska$$

where msk = mean skin temperature, skc = chest temperature, skl = leg temperature, and ska = arm temperature.

Temperatures were measured with thermo-couples taped to the skin with 3M surgical tape and an Omega digital thermometer Model 871. Connectors were Model SMP-K-M miniature Type T. Thermocouples were constructed from copper constantan wire ~0.01 mm (Omega' Engineering, Stanford, CT) by silver soldering the junction and protecting bare areas near the tip with heat-shrink tubing. All thermocouples were checked daily by immersion in an ice bath and also in a heated bath at 37°C. The connectors were taped outside the clothing for easy reading. Outside temperature and relative humidity were measured with a Fisher Digital Hygrometer/Thermometer.

All subjects wore coveralls constructed of the test fabrics used in the laboratory over their

50/50 polyester cotton long-sleeved shirts and long pants each monitoring day. Shoes varied from sneakers to cowboy boots. All subjects wore hats of various descriptions. No gloves were worn. Subjects applied or loaded pesticides. This work is light and consists of riding a tractor with canopy or an open vehicle with a 3785 L (1000 gal) water tank. About every 25 min, workers reloaded the spray machine by connecting one hose from the water tank to the spray machine. Each subject was asked to rate a coverall from 1 to 10 (1 is low) in the categories: comfortable, temperature, pleasant, ventilation, acceptable, and satisfied. These scores were added for a subjective rating. A later rating scale was also used in which the subject rated the comfort of the suit relative to seven numbered verbal descriptors of its judged comfort. Suit materials were individually tailored for fit.

Experimental Results

Thermal Test Battery

A thermal comfort reference fabric (100% cotton chambray) was added to the laboratory testing for data comparison and for developing rating scales. The results of all three tests, along with field tests, are summarized in Table 2. They will be discussed individually and then as a total thermal comfort test battery.

U-Values

Thermal transmittance U-value responses ranged from 11.1 to 12.7 W/m^{2}K. The standard cotton chambray received a high U-value of 12.7 that relieves the most heat stress. Of the four test fabrics, Gore Tex performed best in U-value, followed by untreated twill, treated twill, and SMS.

Table 3 reconciles the general effects of fabric thickness, texture, and condensation behavior on the U-values expected for the fabrics. Fabric thickness and surface texture (for example, hairiness) are combined as an effective thickness defining the boundary layer on each side of the fabric. Greater thickness creates decreased thermal transmittance (U-value). Also, the location of condensation that formed on the face or back of the fabric and its formation mode (film or drop wise) on the fabric affect the magnitude of the heat transported via conduction and latent heat with vapor transport.

The chambray was selected as a standard for high thermal comfort for the Florida climate. Chambray exhibited dripwise condensation on the face (exterior) surface hairs of the fabric. This probably allows an unimpeded moisture vapor flux for maximum latent heat transport. The low reduction in U-value expected from its relatively high effective thickness is possibly negated by its high optical openness. Openings at yarn crossings would provide a direct path

TABLE 2—*Instrumental test results.*

| | | | | Skin Temperature, °C | |
| | | | | | |
Sample	U-Value, W/m^{2}K	WPP, m^3/kPa/m^{2}min	Clort, °C (°F)	High Stress, °C (°F)	Low Stress, °C (°F)
Chambray	12.7	24.0	24.1 (75.5)	...	...
Twill	11.7	7.1	25.0 (77.0)	34.4 (94.0)	...
Twill F1	11.2	8.4	25.2 (77.4)	35.0 (95.0)	32.8 (91.0)
SMS	11.1	5.6	25.8 (78.4)	36.1 (97.0)	...
Gore Tex	11.9	0.0	25.0 (77.0)	37.2 (99.0)	34.7 (94.5)

TABLE 3—*Fabric property and effect on* U-*value.*

Fabric	Thickness	Texture	Location and Form of Condensation	Measured Relative U-Value
Chambray	thin/high	hairy/low	face, drops/high	high
Twill	thick/low	rough/low	back, wicks/medium	medium
Twill F1	thick/low	rough/low	back, wicks/medium	low
SMS	thin/high	smooth/high	back, film/low	low
Gore Tex	thin/high	smooth/high	back, drops/high	high

for radiant heat transfer from the heat source to heat sink. Each of the four experimental samples displayed condensation on the back (moisture source) side of the fabric. Initially, the fabrics displayed dropwise condensation on the back that altered in time.

For the twill material, the condensing moisture wicked into the back surfaces resulting in a slightly dampened surface. Treated twill developed a wetter, heavier water film layer that possibly reduced the moisture flux through the treated fabric and reduced its U-value. No wetness was observed on the face of the fabrics.

The dropwise condensation formed on the continuous Gore Tex polymer film remained unchanged throughout the hours of testing in the MFTT. The unwetted surface provided a dry material path for water vapor transport that aided latent heat transfer. In addition, dropwise condensation on a flat plate typically increases the heat transfer to the surface compared to filmwise condensation. The drops possibly provide greater surface area to absorb heat via convection. The relative thinness of the Gore Tex material, along with the sustained vapor flux, resulted in a relatively high U-value.

Condensation on the SMS appeared to develop into a film of water on the fibrous back S component in time. The SMS and Gore Tex were judged to have equivalent effective thicknesses; therefore, the continuous water film on the SMS is blamed for the lower U-value of SMS. Although this condensing water film is supplying latent heat to the fabric back, the heat must then transfer through the M and S layers. Also, water probably doesn't wick into the M layer. These considerations are interpreted as lowering the U-value of SMS.

WPP Values

The WPP values reflect the "airiness" or potential benefit of light breezes to improve the thermal comfort of the fabric. The area under the WPP curve was used as the WPP-value. With standard cotton chambray having the highest rating of 24, the four test fabrics ranged from 0 (Gore Tex) to 8.4 (treated twill).

The twill fabrics had low-flow properties characteristic of their tightly woven staple yarn construction. The SMS had lower air permeability values over the pressure range due to the microporous texture of the melt blown M component. Gore Tex film has very low air permeability due to submicron pore structure. In tests on wet fabric samples, all WPP values were well below the dry SMS value (except chambray) and thus would contribute little wind effect relief of high heat stress if the fabrics were wet.

Clort

The clothing radiant temperature, or Clort, was defined as a reference mean radiant temperature that the skin sees (comfort sensation with the clothing). The lower the Clort value, the lower the heat stress on the skin. With the chambray standard having the lowest value of 24.15°C, the four fabrics tested ranged from 25.0 to 25.75°C. Untreated twill and

Gore Tex had equal values of 25.00°C; treated twill was slightly higher at 25.20°C; and SMS provided the highest heat stress to the skin at 25.75°C.

The Clort values were probably influenced strongly by the condensation behavior of the fabric. Condensation creates localized absorptive heating [12], possibly accounting for the higher Clort values for fabrics that exhibited condensation on the fabric back. The SMS and treated twill had the higher Clorts and they produced continuous water layers on the heated-sweating side of the MFTT.

Field Test

Outdoor field tests were mostly done during a season of high outdoor temperature; thus, incomplete appraisal of all four samples resulted.

Only one field observation was taken for the Gore Tex suit. At 34.4°C (94°F) outside temperature, the skin temperature for Gore Tex was 37.2°C (98.6°F), extremely high skin temperature for the external temperature. In comparison, treated twill skin temperatures ranged from 34 to 37°F in outdoor conditions of 33 to 37.8°F. Likewise, untreated twill skin temperatures never rose above 35.6°F, with the range of outdoor temperature was 32 to 38°C. The SMS skin temperature tended to rise with increased outdoor temperature. The highest was 37.2°F for an outdoor temperature of 42.2°C (108°F).

During the high-temperature field tests, the untreated twill yielded the lowest skin temperatures for a range of outdoor temperatures. It was the only fabric judged tolerable for the hot testing conditions. Gore Tex yielded intolerable skin temperature in the initial test at the lower end of the outdoor temperature range, 34.4°C.

The subjective comfort ratings for these four fabrics reflect the same results as the mean skin temperature. Ratings for each test period were averaged. Gore Tex had the lowest rating of 30 with 60 being the most comfortable. The other three fabrics were very close in their ratings and were indistinguishable by statistical intepretation. Untreated twill had a rating of 38, SMS a rating of 39, and treated twill 40 points. In ongoing field tests, the field study participants are testing all fabrics over the same test period to control temperature and other variables affecting thermal comfort. In lower stress tests at temperatures in the mid 70's (mid 20°C), Gore Tex compared equally with the Twill F1 fabric.

Relative Thermal Comfort

The relative thermal comfort of the test fabrics was assessed by their overall performance in the physical test battery as they relate to comfort under high heat stress/sweating conditions. Table 4 is a record of points awarded on a 10-point system, based on their relative

TABLE 4—*Instrument discomfort relief and field test point scoring.*

Sample	U-Value	WPP	Clort	Total	Field Ratings Low	High Stress
Chambray	10.0[a]	10.0[a]	10.0[a]	30.0	(7)	(60)
Twill	7.6	3.0	8.0	18.6	...	38
Twill F1	6.1	3.5	7.6	17.2	7	40
SMS	5.8	2.3	6.5	14.6	...	39
Gore Tex	8.0	0.0	8.0	16.0	7	30

[a] Scoring: 0 to 10, 10 = best relief.

ability to relieve thermal discomfort for each cell or test category. Field ratings are included for data taken at both high and moderate heat stress field conditions. For the U-values, a linear point award scale was derived based upon extreme U-values, 9 (0 points for a water vapor impermeable fabric) through 1.3 W/m^{2}K (10 points).

The WPP values in Table 2 show the clustering of treated twill and SMS, with Gore Tex and chambray being on the extremes of the readings. Chambray was the most comfortable (comfort standard), with Gore Tex allowing the least amount of relief from heat stress under wind influenced conditions. The WPP points awarded were based on the percent area under the WPP curve relative to chambray equaling 10 points.

The Clort values ranged on a scale from 29 (0 points) to 24°C (10 points) for cotton chambray. The fabrics tested were awarded points from 6.5 to 8.0. The combination of these three tests with the point system gave untreated twill the greatest amount of points, with 18.6, followed by treated twill with 17.2, Gore Tex with 16.0, and SMS with 14.6. The chambray laboratory reference fabric for thermal comfort accumulated 30 points, the maximum possible score.

The thermal comfort collective rating from the laboratory test battery and the preliminary field test data generally agree. Only the ratings from WPP tests related directly to the high stress field test results. The good success of this ranking parameter is probably due to the necessity of wind effect to obtain thermal comfort in the field tests under very hot conditions. The Gore Tex was rejected by subjects after initial field tests. If the initial minimal data on Gore Tex (zero WPP) is ignored then the U-value and Clort rankings reflect those of the high stress field data.

If wind had been absent in the hot field test situation, then field worker thermal comfort ratings might have been equal for all fabrics as predicted by the U-value and Clort parameter (while WPP values would fail to independently rank the field results). Individual cells in the three-part test battery appear to indicate additive levels of heat stress relief.

In the lower heat stress field tests with temperatures in the mid 70's, Gore Tex compared equally with the treated twill (F1) fabric. The seven-point award scale was used. The high U-value of Gore Tex provided sufficient heat removal to maintain thermal comfort in the lower heat stress environment.

Conclusions

Individual cells of the test battery were selected to represent mechanisms for both heat removal and comfort sensations effected by the fabric. Laboratory test point award schemes appear to reconcile the limited field findings. A closer analysis of field and test battery results require the collection of more controlled data from the field.

The observed condensation difference between the chambray and the four test fabrics indicated the potential importance of establishing the effect of condensation during heat and mass transport through fabrics at sweating/heat stress conditions.

Acknowledgment

Although the research described in this article has been funded wholly or in part by the U.S. Environmental Protection Agency under Assistance Agreement (CR 812846-01-0) to Dr. Jacquelyn Orlando DeJonge of The University of Tennessee, it has not been subjected to the Agency's peer and administrative review and therefore may not necessarily reflect the views of the Agency. No official endorsement should be inferred.

References

[1] Fourt, L. and Harris, M. in *Physiology of Heat Regulation and the Science of Clothing,* L. H. Newburgh, Ed., Hafner, New York, 1968.

[2] Branson, D. H., DeJonge, J. O., and Munson, D., *Textile Research Journal,* Vol. 56, No. 1, 1986, pp. 27–34.

[3] Leonas, K. K., "Apparel Fabrics as Barriers to Pesticide Penetration," Doctoral thesis, The University of Tennessee, Knoxville, 1985.

[4] Hassenboehler, C. B., Jr., and Vigo, T. L., *Textiles Research Journal,* Vol. 52, 1982, pp. 510–517.

[5] Fourt, L. and Hollies, N. R. S., "The Comfort and Function of Clothing," Technical Report 69-74-CE, The U.S. Army Natick Laboratories, Natick, MA, 1969.

[6] Harter, K. L., Spivak, S. M., Yeh, K., and Vigo, T. L., *Textile Research Journal,* Vol. 51, 1981, pp. 345–355.

[7] Stuart, I. M. and Denby, E. F., *Textile Research Journal,* Vol. 53, 1983, pp. 655–660.

[8] Shivers, J. L. in *Clothing Comfort,* N. R. S. Hollies and R. F. Goldman, Eds., Ann Arbor Science Publishers, Ann Arbor, MI, 1977.

[9] Spencer-Smith, J. L., *Clothing Research Journal,* Vol. 5, 1977, pp. 82–100.

[10] Gagge, A. P., *American Journal of Physiology,* Vol. 120, 1937, p. 277.

[11] Burton, A. C. and Edholm, O., *Man in a Cold Environment,* Williams and Wilkins, London, 1954.

[12] Kreith, F., *Principles of Heat Transfer,* 2nd ed., International Textbook Company, Scranton, PA, 1965.

Protection from Pesticides

*Laboratory Test Methods for Materials Resistance
and Decontamination*

Donna H. Branson[1] and Seema Rajadhyaksha[1]

Distribution of Malathion on Gore-Tex Fabric Before and After Sunlight Exposure and Laundering as Determined by Electron Microscopy

REFERENCE: Branson, D. H. and Rajadhyaksha, S., **"Distribution of Malathion on Gore-Tex Fabric Before and After Sunlight Exposure and Laundering as Determined by Electron Microscopy,"** *Performance of Protective Clothing: Second Symposium, ASTM STP 989,* S. Z. Mansdorf, R. Sager, and A. P. Nielsen, Eds., American Society for Testing and Materials, Philadelphia, 1988, pp. 651–659.

ABSTRACT: The distribution of malathion residue on Gore-Tex fabric was examined using electron microscopy. The percent malathion retained in laundered and unlaundered fabric specimens and in the individual Gore-Tex layers was determined with gas chromatography. Concentrations of malathion were found on the surface of nylon fibers and in the interstices between fibers in both the face and the backing fabric layers as well as in the membrane of both the unlaundered and laundered samples. These findings were verified and the percent residue retained by fabric layer was quantified. Thus, the Gore-Tex fabric tested was not an effective barrier to the undiluted malathion solution. Also, the distribution of the pesticide changed after laundering. The Gore membrane retained the highest proportion of pesticide of the three layers, suggesting the ineffectiveness of an aqueous laundry system for removal under these conditions.

KEY WORDS: malathion, pesticide, Gore-Tex, decontamination, electron microscopy, protective clothing

Previous research has shown Gore-Tex[2] fabric to be an effective barrier to field-strength concentrations of a limited number of pesticides including parathion, paraquat, Guthion, and dinoseb [1,2]. In a controlled laboratory thermal analysis study, subjects wearing ensembles of Gore-Tex fabric experienced a similar thermal response (assessed by skin temperature and subjective comfort measures) as subjects wearing 100% cotton chambray ensembles [3]. Higher mean skin temperatures and greater perceived thermal discomfort were found for subjects wearing polyethylene-coated Tyvek ensembles in the same study. The combination of good barrier properties to a limited set of pesticides and acceptable thermal comfort properties prompted studies focusing on the decontamination of Gore-Tex [4,5].

Park [5] investigated the pesticide residue retained in Gore-Tex fabric as a function of a sunlight exposure treatment, detergent treatments, and repeated contamination and subsequent launderings. The pesticide of interest was a field-strength emulsifiable concentrate

---

[1] Professor and graduate research assistant, respectively, Department of Clothing, Textiles and Merchandising, Oklahoma State University, Stillwater, OK 74078-0337.
[2] Registered trademark of W. L. Gore and Associates.

commercial-grade mixture of parathion and methyl parathion (6-3 Parathion-Methyl). Several results of this study prompted the present research. Specifically, Park found between 36 and 43% of the parathion, and 16 to 23% of the methyl parathion was retained by the Gore-Tex fabric following the various treatments. Secondly, the simulated sunlight exposure treatment contributed to improved removal rates. Lastly, no pesticide build-up was found after five contaminations and subsequent launderings. The combination of the relatively high percent residue retained and no build-up seemed to require further information on the location of the pesticide within the three-layer fabric structure.

The purpose of the present study was to investigate the microscopic distribution of the organophosphorus pesticide malathion on Gore-Tex fabric before and after laundering.

Procedure

A controlled laboratory study was conducted with three independent variables: pesticide, simulated sunlight exposure, and laundering. A 2 by 2 by 2 experimental design with three replications was used. Half of the samples were analyzed using gas chromatography (GC) and half using the scanning electron microscope (SEM).

Test Fabrics

The second generation Gore-Tex test fabric was a three-layer fabric consisting of a microporous membrane of polytetrafluoroethylene laminated between an outer fabric of 100% nylon ripstop and an inner layer of nylon tricot. Fabric weight was determined to be 85.45 g/m^2 in accordance with ASTM Test Methods for Weight (Mass) per Unit Area of Woven Fabric (D 3776-84).

All of the fabric specimens (14.6 by 8.9 cm) were initially given a prewash treatment in the Launder-Ometer. Three fabric specimens, 500 mL of distilled water at 49°C, and 50 steel balls were placed in stainless steel cannisters and operated in the Launder-Ometer for 0.3 ks. This was followed with a single rinse consisting of soaking three fabric specimens and 500 mL of distilled water at 49°C in beakers for 0.3 ks. The fabric specimens were then allowed to air dry overnight in the conditioning room.

Pesticide

A commercial-grade emulsifiable concentrate formulation of malathion formulated for Fords Chemical and Service Inc. was used to contaminate the test fabric specimens. Active ingredients were malathion 57% (00-dimethyle dithiophosphate of diethyl mercaptosuccinate) and xylene 30%. The use of the undiluted pesticide simulated an accidental spill scenario.

Contamination Procedure

Before contamination, fabric specimens with machine zigzag-stitched edges were conditioned for 86.4 ks, in accordance with the ASTM Practice for Conditioning Textiles for Testing (D 1776-79). The pesticide solution was placed on a magnetic stirrer prior to and during the contamination process to facilitate the solution being held in suspension. A 100 µL Hamilton syringe was used to pipette 100 µL of the full-strength pesticide onto the fabric specimen surface. All specimens were stored in foil-lined glass desiccators containing magnesium acetate. The contaminated and control samples remained in the sealed desiccators in the conditioning room for 54 ks.

Simulated Sunlight Exposure

After the 54-ks storage period, half of the dried contaminated test specimens and the controls were given a simulated sunlight exposure treatment in an Atlas [35]Ci Fade-Ometer. Test conditions included temperature of 55°C ± 2°C, 55% ± 2 relative humidity, and an irradiance band of 1.5 W/m². Fabric specimens were secured to cardboard backing with double-stick tape and placed in Fade-Ometer stainless steel holders such that the entire fabric surface was exposed to the zenon light source.

Laundry Procedure

Each specimen for the pesticide retention analysis was individually laundered for 0.6 ks in the Atlas Launder-Ometer with 3 mL detergent, 200 mL of distilled water at 38°C, and ten steel balls (included to simulate the abrasive action of a single-home laundering). The detergent (Fresh Start) contained nonionic surfactant, sodium tripolyphosphates protease, and amylase. The formula averaged 14.7% phosphorous in the form of phosphates. The detergent was selected since previous research [5] had found a trend for it to be more effective than other detergents tested. After two 0.3 ks rinses, each with 200 mL of distilled water at 25 ± 2°C, the specimens were allowed to air dry for 1.44 ks in the conditioning room.

Determination of the Pesticide Residue

Both unlaundered and laundered fabric specimens were individually extracted in 10 mL of nanograde acetone on an Eberbach shaker for 0.9 ks. The samples were then stored in the refrigerator for 8.64 ks, and subsequently shaken again on a Yankee Pipette Shaker. The fabric specimen was removed and acetone was added to bring the volume to mL.

The quantity of malathion on each fabric specimen for each test condition was determined using a Tracor 560 gas chromatograph with a flame ionization detector. The chromatograph column was glass packed with liquid phase 5% OV-1 on 80/100 mesh size Supelcopert. The carrier gas was helium set at a flow 20.7 kPa, air 8.3 kPa, and hydrogen 206.9 kPa. The selected conditions were oven temperature 200°C, injection post 250°C, and flame detector 230°C.

Electron Microscopy

Pesticide distribution on the laundered and unlaundered fabric specimens was determined using the technique developed by Obendorf [6,7] for X-ray microanalysis. Yarns that were extracted from the center of the top-fabric layer of 100% nylon ripstop, and from the center of the 100% nylon tricot backing fabric, were placed in 2 mL of 2% (weight/volume) osmium tetroxide (OsO_4) in water for 10.8 ks. The specimens were then rinsed in water for three 200-s periods and air dried overnight. Five (less than 3 μm in width) fabric strips from the center of the test specimens were also prepared in the same manner. Longitudinal yarns were mounted on carbon stubs and carbon-coated in a Veeco VE400 high vacuum evaporator.

Cross-sections of the OsO_4-treated fabric strips were prepared for microscopical analysis in the following way. The specimens were dehydrated in ethanol and infiltrated using a mixture of one part ethanal to one part Spurr resin, capped, and allowed to remain overnight. The following day, the specimens were uncapped under a vacuum for 2.52 ks, embedded in 100% Spurr low viscosity resin, and the resin was cured for 2.88 ks in an oven at 70°C.

TABLE 1—*Mean amount (mg) of malathion residue retained by test fabrics.*[a]

Sunlight Treatment	Laundering Treatment		Mean
	Laundered	Unlaundered	
Sunlight exposure	13.5	34.1	23.8
No sunlight exposure	13.6	35.8	24.7
Mean	13.6	34.9	24.3

[a] Three replications.

Cross-sectional specimens of the fabric were cut (10 μm) using glass knives on an ultramicrotome. Embedded cross-sections were placed on a carbon stub and coated with carbon.

Energy dispersive X-ray data were obtained using a Japan Electron Optics Laboratory (JEOL) JSM 35-U scanning electron microscope. Counts of X-ray emission from osmium were recorded at 0.304 fJ (1.90 keV) for five locations on the nylon fibers (both layers) and for three locations within each layer of the fabric strip cross-sections. The locations for the face and backing fabrics included fibers near the Gore membrane, fibers in the center of the fabric layer, and fibers located near the fabric surface. Within the Gore membrane, the three sites included one central location and two locations near each surface boundary.

Results and Discussion

Since pesticides breakdown in the natural environment, it was hypothesized that exposure to simulated sunlight, heat, and humidity might represent an effective means of reducing the pesticide residue retained in fabrics. As Table 1 indicates, the simulated exposure treatment did not influence pesticide residue found in either the laundered or in the unlaundered samples. This finding is not in agreement with Park [5], possibly because a lower temperature was maintained in the present research.

Laundering once with a phosphate nonionic detergent resulted in a mean residue of approximately 24 mg of malathion retained by the Gore-Tex fabric (Table 1). This was significantly less (at the 0.00001 level) than the residue found in the unlaundered samples. Table 2 presents these data as relative percentages of the initial contamination. The relative percentage takes into account that the percent active ingredient varied by replication. Thus, almost 27% was retained by the laundered fabrics regardless of the exposure to simulated sunlight treatment.

The fairly high percent of malathion retained by the fabric was somewhat surprising since malathion is highly water soluble. Park [5] found between 36 and 43% parathion and 16 to

TABLE 2—*Pesticide residue remaining, calculated as a relative percentage of initial contamination.*

Treatment	Rep 1, %	Rep 2, %	Rep 3, %	Mean, %
PSL[a]	29.6	24.7	26.2	26.8
PNL[b]	27.2	23.8	29.2	26.8
PSN[c]	75.6	56.9	71.6	68.0
PNN[d]	80.3	57.3	75.4	71.0

[a] PSL = simulated sunlight exposure and laundered.
[b] PNL = no simulated sunlight exposure and laundered.
[c] PSL = simulated sunlight exposure and not laundered.
[d] PNN = no simulated sunlight exposure and not laundered.

23% methyl parathion retained in Gore-Tex when laundered at 49°C. Easter [4] found a 14% Guthion residue and 27% captan residue in Gore-Tex fabrics when laundered at 38°C. Both of these studies had used field-strength pesticide solutions, whereas the present study used a full-strength pesticide solution.

The location of pesticide residue within the complex three-layer Gore-Tex structure, which has not previously been determined, was investigated using electron microscopy. High concentrations of malathion were found on the surface of the nylon fibers in both the face and the tricot backing fabrics of the unlaundered fabrics (Figs. 1 and 2). Pesticide residue appears to be distributed throughout the Gore membrane, as shown in Fig. 3, a cross-section of an unlaundered Gore-Tex fabric strip. Concentrations of malathion were also located in the spaces between the nylon fibers in the face fabric and the Gore membrane (Fig. 3). Thus, the findings from this portion of the study suggest that the undiluted malathion penetrated the Gore membrane to the tricot backing.

Examination of the laundered fabrics indicated that smaller concentrations of malathion were present on the surface of the nylon fibers for both the face and the tricot backing fabrics (Figs. 4 and 5). Cross-sectional views of the three-layered fabric specimens (Fig. 6) also suggest malathion concentrations were trapped in the Gore membrane, in the interstices between the Gore membrane and the two fabric layers.

Thus, the results of the GC analysis indicated approximately 70% residue retained in the unlaundered fabrics and 27% residue in the laundered fabrics. The SEM analysis detected that malathion residue was present in all three layers of the unlaundered Gore-Tex specimens. The SEM analysis further indicated a similar malathion distribution throughout the three layers in the laundered specimens but with a lower concentration present.

In order to rule out the possibility that the rinsing step in the SEM procedures was responsible for finding relatively high counts on the backing fabric, a follow-up study using GC was conducted in the same manner as indicated in the methods section of this paper. There were two notable exceptions: (1) the simulated sunlight exposure treatment was not used; and (2) before the extraction process for GC analysis, the three layers of the Gore-Tex samples were separated and each layer was prepared for GC analysis individually. Three replications were performed.

The results of this follow-up study are given in Table 3. Slightly over 47% of the malathion contamination (100 μL) was found in the face fabric, slightly over 14% in the Gore membrane, and almost 20% had penetrated the Gore membrane and was detected in the nylon tricot backing fabric. Thus, the malathion concentrations observed in the SEM phase of the study were substantiated with the GC analysis. The full-strength malathion used in this study, which simulated a spill, did penetrate the protective Gore membrane.

Examination of the results for the laundered specimens (Table 3) indicated that the aqueous laundry system was most effective in reducing the pesticide residue found in the face fabric from approximately 47% to 8.2%, almost a six-fold decrease. The tricot backing fabric retained almost 6% of the initial contamination, which represents more than a three-fold reduction.

The Gore membrane was the most difficult layer to decontaminate with 9% remaining in this membrane. It is apparent that an aqueous medium was not effective in pesticide removal from this membrane. A solvent-based decontamination system would likely be required to reduce this residue.

Conclusions

This study was designed originally to investigate the location and concentration of malathion residue in laundered and unlaundered Gore-Tex fabric specimens using electron

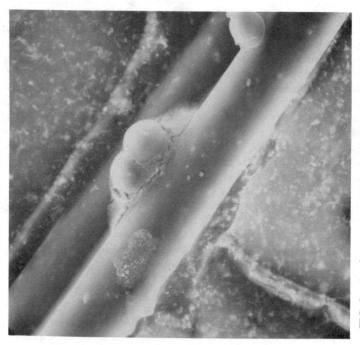

FIG. 2—An electron micrograph of malathion residue located on the surface of nylon fibers taken from the tricot-backing fabric layer of unlaundered Gore-Tex fabric (×400).

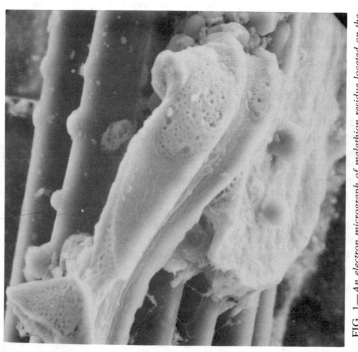

FIG. 1—An electron micrograph of malathion residue located on the surface of nylon fibers taken from the face-fabric layer of unlaundered Gore-Tex fabric (×400).

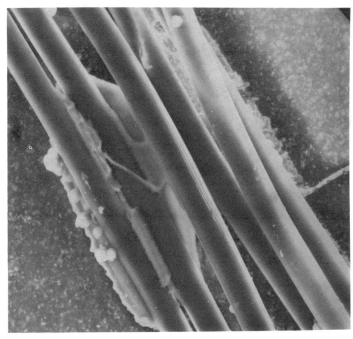

FIG. 4—An electron micrograph of malathion residue located on the surface of nylon fibers taken from the face-fabric layer of laundered Gore-Tex fabric (×400).

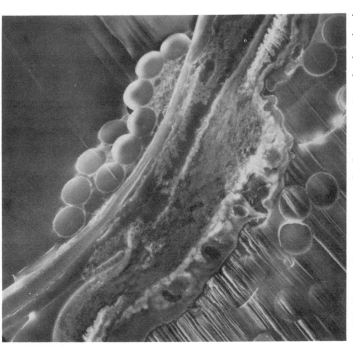

FIG. 3—An electron micrograph of a cross-section of unlaundered Gore-Tex fabric showing the distribution of malathion throughout the three layers (×400).

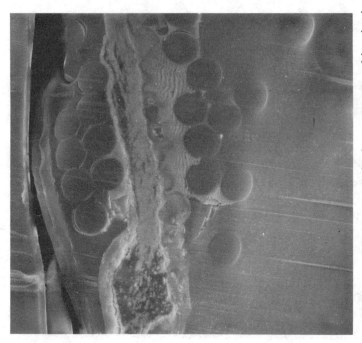

FIG. 6—An electron micrograph of a cross-section of laundered Gore-Tex fabric showing the distribution of malathion throughout the three layers (×400).

FIG. 5—An electron micrograph of malathion residue located on the surface of nylon fibers taken from the tricot-backing fabric layer of laundered Gore-Tex fabric (×400).

TABLE 3—*Pesticide residue remaining in each fabric layer, calculated as a relative percentage of initial contamination.*[a]

Fabric Layer	Laundering Treatment		Mean, %
	Laundered, %	Unlaundered, %	
Ripstop face fabric layer	8.2	47.4	27.8
Gore membrane	9.0	14.3	11.7
Tricot backing layer	5.9	19.7	12.8
Mean	8.4	27.1	17.5

[a] Three replications.

microscopy. Results indicated that pesticide residue was present in all three layers of the Gore-Tex fabrics both before and after laundering. This finding, of course, suggested that Gore-Tex was not an effective barrier to the full-strength malathion emulsifiable concentrate used in this study to simulate a spill situation. A second study was conducted in which the three-layered fabric was separated and the individual layers were analyzed using gas chromatography. The results of this study confirmed the presence of malathion residue in all three layers of both the laundered and unlaundered specimens. The distribution pattern differed, however, for the laundered and unlaundered samples.

Acknowledgments

The authors thank the Oklahoma State University Agricultural Experiment Station and the Center for Energy Research for their financial support of this research. This study was part of the U.S. Department of Agriculture Cooperative Regional Research NC-170. We also wish to thank Dr. S. Burks and Greg Smith for technical assistance and use of their laboratory for GC analysis.

References

[1] Orlando, J., Branson, D. H., Ayers, G. S., and Leavitt, R., *Journal of Environmental Science and Health*, Vol. 5, 1981, pp. 617–628.
[2] Branson, D. H., Ayers, G. S., and Henry, M. S. in *Performance of Protective Clothing, ASTM STP 900*, R. L. Barker and G. C. Coletta, Eds., American Society for Testing and Materials, Philadelphia, 1986, pp. 114–120.
[3] Branson, D. H., DeJonge, J. O., and Munson, D., *Textile Research Journal*, Vol. 56, No. 1, Jan. 1986, pp. 27–34.
[4] Easter, E., *Textile Chemist and Colorist*, Vol. 47, No. 3, March 1983, pp. 29–33.
[5] Park, J., "Pesticide Decontamination from Fabric by Laundering Following Repeat Contamination and Simulated Weathering," unpublished doctoral thesis, Oklahoma State University, Stillwater, 1986.
[6] Obendorf, S. K. and Klemash, N. A., *Textile Research Journal*, Vol. 52, No. 7, 1982, pp. 434–442.
[7] Solbrig, C. M. and Obendorf, S. K., *Textile Research Journal*, Vol. 55, 1985, pp. 540–546.

Karen K. Leonas,[1] *Jacquelyn Orlando DeJonge,*[2] *and*
Kermit E. Duckett[2]

Development and Validation of a Laboratory Spray System Designed to Contaminate Fabrics with Pesticide Solutions

REFERENCE: Leonas, K. K., DeJonge, J. O., and Duckett, K. E., **"Development and Validation of a Laboratory Spray System Designed to Contaminate Fabrics with Pesticide Solutions,"** *Performance of Protective Clothing: Second Symposium, ASTM STP 989,* S. Z. Mansdorf, R. Sager, and A. P. Nielsen, Eds., American Society for Testing and Materials, Philadelphia, 1988, pp. 660–670.

ABSTRACT: A chamber designed to contaminate fabric samples by delivering controlled amounts of pesticide solution has been constructed. The pesticide solution is applied in spray form simulating droplets as found in actual field pesticide spraying of crops. This chamber provides a systematic method for evaluating barrier properties of fabrics under consideration for use in protective clothing for agricultural workers. The design and operation procedures are discussed in terms of its applicability as a standard test method for pesticide penetration of textiles and protective fabrics.

KEY WORDS: protective clothing, fabric penetration, pesticide, spray chamber

Currently, three methods of fabric contamination are commonly used to evaluate the ability of a fabric to act as a barrier against pesticide penetration: field studies, drop methods, and laboratory spray systems. Field studies involve using actual field conditions and monitoring the amount of pesticide that penetrates the garment or fabric. The penetrating pesticide is collected on gauze pads that are later analyzed. Field studies are invaluable in ascertaining the acceptability of a garment but they can be costly, time consuming, and it is difficult to control variables that may influence penetration. These variables include droplet size, velocity, and environmental conditions such as wind and sun. The total amount of pesticide one comes in contact with is undeterminable, making reproducibility unlikely [1].

The drop method involves using a pipette to apply a known amount of pesticide to the fabric [2,3]. This method provides for application of a controlled amount of pesticide in a reproducible manner; however, it more closely simulates mixing, cleaning up, and spill situations than application situations.

The third method of exposing fabrics to pesticide solutions is a laboratory spray system. Orlando et al. [4] used a Beltsville sprayer initially designed to spray vegetation. Its size made it cumbersome for fabric testing and control of environmental conditions was difficult because it had to be housed in a large room. Easter [5] used a spray chamber in her studies

---

[1] Assistant professor, Division of Textiles, Apparel, and Interior Design, University of Illinois, Urbana, IL 61801.
[2] Department head and professor, respectively, Department of Textiles, Merchandising and Design, College of Human Ecology, The University of Tennessee, Knoxville, TN 37996-1900.

designed to spray one fabric sample at a time. The sample was manually inserted into the chamber, exposed to a known amount of pesticide (based on time), and then removed. These spray systems were refined and expanded, evolving into the spray system discussed here.

This spray system design allows for monitoring the influence of droplet size, as in field studies, while controlling the amount of pesticide delivered to the fabric as in the drop method. The droplets, formed from a spray nozzle, more closely simulate the actual field condition of a pesticide spray impinging on the test fabric. The chamber can be housed in a laboratory hood, providing localized control of environmental conditions. The amount of pesticide that each fabric is exposed to is mechanically controlled.

Pesticide Spray Chamber Design

The equipment is designed for observation of the spraying operation, operator protection, pesticide concentration and liquid flow rate control, and areal distribution uniformity of liquid sprayed across the fabric sample. The integrated system consists of specific components involving the: (*a*) spray chamber, (*b*) fabric sample holders, (*c*) spray nozzle and mechanical control, and (*d*) pesticide reservoir and liquid feed line.

The chamber housing is constructed in the form of a rectangular box from 1.2-cm-thick clear plexiglass sheeting (Fig. 1). The outside dimensions measure 115 cm long, 60 cm high, and 30 cm deep. Contacting edges are attached with a solvent contact adhesive, with the exception of the top plate. This plate is attached to the four sides with machine screws, thereby providing access to the interior in the event of necessary repair or top replacement. For the purpose of general cleaning and routine access to the inside of the chamber, two 20-cm-diameter portholes have been machined into the front panel. These are fitted with

FIG. 1—*Spray chamber.*

circular plexiglass window panes, each held in place by frictional contact of an O-ring attached to the outside edge of the window pane and the edge surface of the porthole in the chamber. Handles on the panes provide easy removal and replacement.

The top plate also has a slot machined lengthwise through the middle to within about 10 cm of the ends. The slot width accommodates a stainless steel feed line to which is attached a spray nozzle directed downward into the chamber interior. The purpose of the lengthwise slot is to provide clearance for a sweeping motion of the spray nozzle across fabric samples mounted beneath. Sample-nozzle separation distance adjustments are set by lowering or raising the stainless steel tubing to which the nozzle is attached. This tube, itself, passes through a matched hole drilled through a plexiglass block carriage constructed to ride on two stainless steel rods mounted above the top plate of the chamber and parallel to the open slot through which the spray head projects. Left and right carriage motion is handled by a positive-drive chain driven by a variable speed, reversible Dayton gearmotor. Appropriate electrical switching and voltage controls provide variable speeds of the nozzle and, hence, rate of pesticide coverage when used in conjunction with known rates of liquid volume feed. A sweeping spray can be obtained from either direction. The bottom of the chamber is configured with a V-shape geometry, giving drainage to the center of the chamber where liquids are removed by gravity feed through a drain pipe into collector cans for appropriate and safe disposal. Plexiglass panels are mounted at each end of the chamber on the inside in a configuration that protects the fabric specimens from direct pesticide spray when the nozzles reach the end of their sweeping motions. This allows the spraying to continue unabated, eliminating variability in spraying rates that are present when starting and stopping the spray feed operation. The complete spray chamber unit is placed within a laboratory fume hood for obvious safety reasons associated with spraying toxic pesticides.

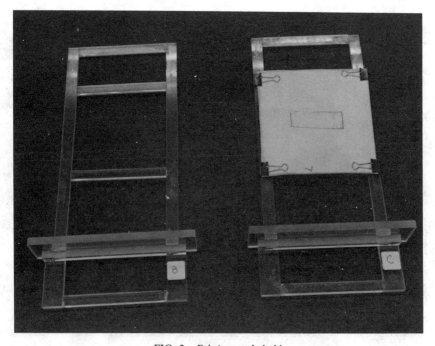

FIG. 2—*Fabric sample holder.*

The fabric holders are similarly fabricated of 1.5-cm-thick plexiglass in a tenter frame style with spring-loaded clamps to hold the fabric specimens in place and aid in ease of securing and removing fabric samples (Fig. 2). A tenter frame is commonly used in textile processing to hold the fabric firmly at the edges, leaving the center area taut and unconstricted. The overall holder dimension is 37 cm long by 18 cm wide, with the sample area 12.8 cm by 12.8 cm. At the opposite end to the sample area is the handle, with a stop plate for setting the distance to which the fabric holder can be inserted through a slot milled in the front panel of the spray chamber. There are four slots in the front panel to accomodate four fabric holders. A plexiglass strip on the back, inside panel acts as a rest stop for each holder when set in place for a spray treatment. A splash guard is attached to the inside front panel above the slots to eliminate pesticide leakage during spraying.

The pesticide solution is delivered from a 2-L poly(ethylene terephthalate) bottle through a reinforced polyethylene standard spray hose. This 6-mm internal diameter (ID) hose is coupled with another similar sized flexible pneumatic hose connected to the nozzle tube that is fixed in the traveling carriage. The flexibility of the hose minimizes drag as the nozzle carriage traverses its sweep directions. Connections between hoses and other points in the feed line are made with Kwik Change[3] Connectors (Milton Couplet). The pesticide reservoir bottle rests on the plate surface of a magnetic stirrer, operated during spraying to provide a uniform, constant concentration of pesticide solution to the spray feed line. Helium gas, a chemically inert gas, is used to provide the driving pressure and is fed from a regulator-

TABLE 1—*Rate of nozzle movement in two directions at each of eleven dial settings.*

Dial Setting	Direction of Nozzle Movement	Mean Rate,[a] cm/s	SD,[a] cm/s
0–3	L to R[b]	NM[c]	...
0–3	R to L[d]	NM	...
4	L to R	1.468	1.360
4	R to L	2.083	0.059
5	L to R	3.099	0.026
5	R to L	3.472	0.019
6	L to R	4.163	0.007
6	R to L	4.425	0.013
7	L to R	4.981	0.006
7	R to L	5.080	0.002
8	L to R	5.522	0.005
8	R to L	5.639	0.007
9	L to R	5.773	0.002
9	R to L	5.048	0.002
10	L to R	5.908	0.006
10	R to L	6.350	0.003

[a] Mean and standard deviation (SD) based on five replications.
[b] L to R = left to right.
[c] NM = no movement.
[d] R to L = right to left.
NOTE—Chamber parameters:
  Driving pressure = 27.5 N/cm².
  Distance between nozzle and sample = 30.48 cm.

[3] Registered trademark of Milton Industries, Inc.

TABLE 2—*Spray chamber validation of consistency of liquid delivery.*

Dial Setting	Amount of Liquid Delivered, gm/cm²					
	Slot					
	A	B	C	D	Mean[a]	SD[b]
6	0.020	0.020	0.021	0.021	0.021	0.00057
8	0.015	0.016	0.016	0.016	0.016	0.00044
10	0.014	0.015	0.015	0.015	0.015	0.00040

[a] Mean based on three replications.
[b] Standard deviations between locations (slots) at three dial settings.
NOTE—Chamber parameters:
Driving pressure = 27.5 N/cm².
Distance between nozzle and sample = 30.48 cm.

controlled tank through a rubber hose to the space in the reservoir above the pesticide solution. The volume flow rate of solution delivered to the spray chamber is controlled by a combined choice of gas driving pressure and spray nozzle.

The nozzle is passed across the length of the chamber, above the fabric samples, by a motor mounted on the outside of the chamber. A control box, outside the chamber, is used to start and stop the motor, change nozzle direction, and control the nozzle speed. The nozzle speed is adjusted by a variable dial allowing for settings from 0 (lowest) to 10 (highest).

Stop latches at each end of the chamber automatically shut the motor off when the nozzle reaches the end of the chamber. This allows for continued feed of the pesticide solution while removing or inserting fabric samples. Continuous spray of the pesticide solution ensures no fluctuation in pressure due to starting and stopping of the liquid flow.

The nozzle can be changed easily to allow for varying spray patterns and amount of pesticide delivered. The distance between the nozzle tip and fabric sample is also adjustable ranging from 25.4 to 38.1 cm. The amount of total spray delivered to the sample depends on choice of nozzle, the hydraulic driving pressure in the feed line at the nozzle, separation distance between the nozzle and the fabric, and the time interval over which the test fabric is subject to spraying.

Calibration and Validation

The spray chamber has been evaluated using several different methods to ensure consistent and accurate spray application to each test specimen location. Spray coverage uniformity, consistency in the amount of liquid delivered, the effect of pressure on droplet size, and the effect of separation distance between the nozzle and the test specimen location were evaluated.

To determine the uniformity of the spray (a well-dispersed pattern with equal amounts of liquid/particle solution striking all areas of the test specimen), a fluorescein dye solution (0.12%) was used in the chamber. This method is similar to one used by DeJonge et al. [6] and provides an orange-green solution by which to visually evaluate the spray pattern. A 130 g/m², 100% cotton plain weave fabric was used for this portion of the validation. The fabric samples were exposed to one pass of the nozzle at Dial Setting 8 with separation distance of 30.5 cm, were allowed to dry, sprayed with a potato starch solution to activate the fluorescein dye, and evaluated under black lights to determine deposition distribution of the dye particles.

A wide variety of nozzle types are commonly found on agricultural sprayers. The nozzle chosen for this study was of the hydraulic type, where a pesticide solution is pressure forced through an orifice and the liquid is sheared into droplets. Generally, there are five classes of hydraulic operated nozzles found in agricultural sprayers: (*a*) side-entry hollow cone, (*b*) disk-type solid cone, (*c*) core-insert hollow cone, (*d*) fan spray, and (*e*) flooding. Solid cone, hollow cone, and fan spray nozzles were initially evaluated for use with this spray system. The solid and hollow cone nozzles produce circular spray patterns. In this spray system, the nozzle travels in a sweeping motion causing extensive spray overlap when the cone nozzles are used making them unacceptable. In addition, the circular pattern resulted in reproducibility difficulties at the end chamber specimen positions. The fan spray nozzle was chosen for the chamber because of the uniformity in pesticide distribution that can be obtained across fabric specimens, when used in conjunction with the sweeping motion of the nozzle from one end of the chamber to the other. The fan spray nozzle produces a flat slot-like pattern that works well with this system easily overlapping the intended fabric test specimen width of 15.2 cm but not overlapping between specimens. Sweeping the fan at a constant rate past the specimen provides uniform coverage. TeeJet[4] nozzle 720023 was selected as standard because the pattern produced by this particular nozzle gave reproducible results in spray uniformity across each sample. Currently, this TeeJet nozzle (Model 730023) is

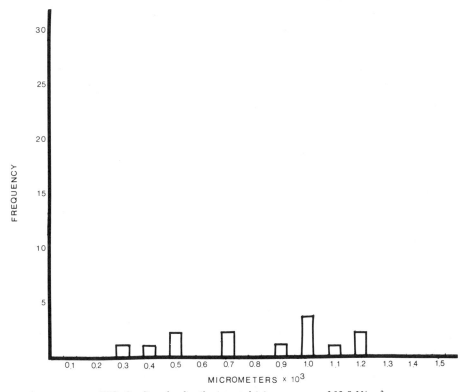

FIG. 3—*Droplet distribution at driving pressure of 13.8 N/cm².*

[4] Registered trademark of Spraying Systems Company.

standard equipment. This nozzle, manufactured by Spraying Systems Company, provides a 52° angle of discharge at 0.023 gal/min when operated at 27.5 N/cm². The volume mean diameter (vmd) of water spray droplets is 150 μm, under 27.5 N/cm² of driving pressure. The rate of nozzle movement was determined by timing how long it took for the nozzle to travel a predetermined distance. Rate determination was based on five replications.

The speed of the nozzle in traveling the length of the chamber is adjustable (Table 1). Except at the lowest dial settings where movement occurred (4 and 5), these rates are consistent (indicated by the low standard deviation) while the nozzle travels in one direction. However, there are inconsistencies when traveling in opposite directions under the same Dial Setting 9. In all cases, except at the dial setting, the left-to-right movement had a lower rate than the right-to-left movement. This inconsistency is due to drag effects as the hose is pulled by the nozzle unit as it moves from right to left. Consistency of results required choosing one or the other of the two directions during actual testing. For the remaining calibration and validation testing, the left-to-right nozzle movement was used and only Dial Settings 6 and above were evaluated.

Consistency of the amount of liquid delivered to each sample was determined by weight. Validation of the consistency of the total coverage (volume/unit area) was determined in each of the four slots and was made following exposure to one pass of the spray nozzle. Water was the liquid used during this testing procedure. Precision was determined by replication. The collector plates were weighed before and after each spray exposure to determine the amount of liquid delivered at each location. Using this method (volume/unit area), a calibration was made between nozzle sweep speed and the amount of liquid delivered at

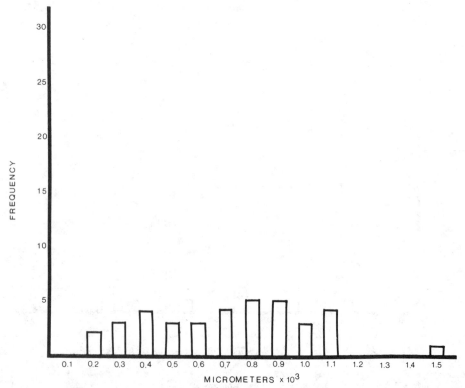

FIG. 4—*Droplet distribution at driving pressure of 20.6 N/cm².*

three dial settings with separation distance between the nozzle and collector plate set at 30.5 cm. No dial setting below 6 was chosen for testing because of the high volume of liquid delivered due to the slow rate of nozzle movement and the increased standard deviation described earlier. Only dial settings of 6, 8, and 10 were evaluated to show significant differences in rates. Results showed reproducibility and consistency from position to position in passes made in the same direction (left to right) of the moving nozzle at the three dial settings (Table 2).

Pressure is controlled using a regulator and the possible variation in pressure is continuous. The effect of pressure on droplet size and spray pattern was evaluated. Droplet size is affected by the amount of hydraulic driving pressure used. Droplet size distribution was measured at pressures of 13.8 N/cm² (20 psi), 20.6 N/cm² (30 psi), 27.5 N/cm² (40 psi), 34.4 N/cm² (50 psi), and 41.3 N/cm² (60 psi). This range provided the maximum and minimum measurable range available. The droplet size was determined by covering a microscope slide with a thin layer of castor oil and exposing it to the spray (water) from the TeeJet 730023. The separation distance between nozzle and slide was constant at 30.5 cm. The nozzle remained stationary and the slide was inserted for 1.5 s then removed. The slide was immediately observed using a Projectina 4002 (Heerburgs) and the droplets within a predetermined area were measured and counted to give a droplet-frequency distribution. Five replications at each pressure were completed to compensate for the change in droplet size due to heat of the microscope light and movement from the chamber to the microscope. Results of the distribution are in Figs. 3 to 7 and Table 3. According to Pratt et al. [7], the size of the droplets are classified as aerosols and fogs. As the driving pressure increases,

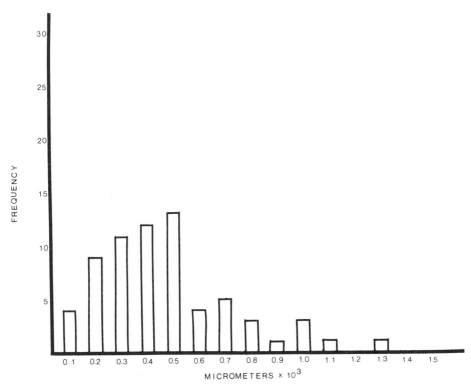

FIG. 5—*Droplet distribution at driving pressure of 27.5 N/cm².*

the size of the droplet decreases and the droplet size range also decreases. However, the number of droplets collected on the slide peaked at a pressure of 27.5 N/cm². As the pressure was increased after this point, the spray began to hang in the air and drift, producing a fog rather than aerosol, and resulting in a reduced number of droplets seen on the slide and reproducibility decreased as indicated by the standard deviation at 41.3 N/m².

Pressure also affects the spray pattern the fabrics are exposed to. Pressures of 13.8 N/cm², 27.5 N/cm², and 41.3 N/cm² were used to evaluate the fan spray pattern. At 13.8 N/cm², a 15° angle fan was produced but did not provide complete coverage of the fabric sample area. The outer edges of the spray were more heavily concentrated than liquid delivery in the middle of the spray as indicated using the fluorescein dye technique. The driving pressure of 27.5 N/cm² produced a fan of 52° angle and good dispersion of spray pattern. This pattern covered the fabric sample area. When using 41.3 N/cm², a fan of 87° was produced and drifting of spray within the chamber was present, resulting in the inability to reproduce the amount of pesticide that would accumulate on each sample.

The spray system was designed so that the distance between the nozzle tip and test speciman location could be adjusted. Three distances were chosen to evaluate spray nozzle-fabric separation and liquid delivery. Data was completed using the previously described technique of using collector plates to quantify coverage (volume/unit area). Results in Table 4 show how liquid coverage is affected by the separation of test sample location and nozzle tip. Liquid coverage can be changed by approximately 60% simply by adjusting height of the spray head in the system from the lowest to highest positions.

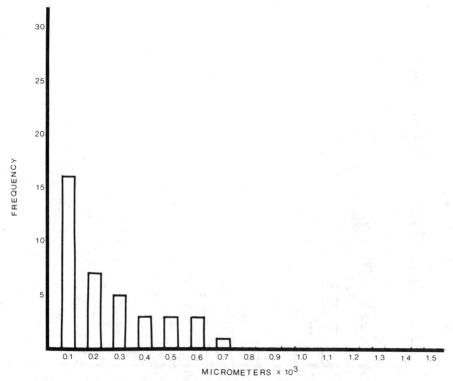

FIG. 6—*Droplet distribution at driving pressure of 34.4 N/cm².*

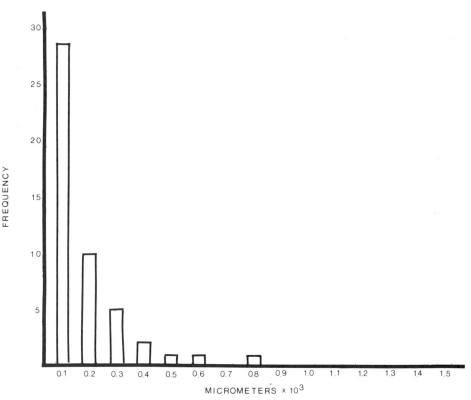

FIG. 7—*Droplet distribution at driving pressure of 31.3 N/cm².*

Conclusion

In this study, a laboratory spray system was developed that delivers a controlled amount of pesticide solution of fabric samples. Tests of the equipment indicated that it could be adjusted so that the spray pattern was uniform and the amount of liquid delivered to each sample was consistent. This chamber can be used under a fume hood to minimize laboratory research exposure.

TABLE 3—*Effect of pressure on droplet size and distribution.*

		Most Frequent			
Pressure, N/cm²	Droplet Size Range, μm × 10³	Total Number of Droplets	Size, μm × 10³	Mean Size[a] μm × 10³	SD[a] μm × 10³
13.8	0.1 to 1.5	59	1.0	0.861	0.882
20.6	0.1 to 1.5	152	0.8	0.753	1.045
27.5	0.1 to 1.5	201	0.5	0.475	0.276
34.4	0.1 to 0.8	142	0.1	0.232	0.176
41.3	0.1 to 0.6	144	0.1	0.232	0.233

[a] Mean size and standard deviation (SD) based on five replications.
NOTE—Chamber parameters:
Distance between nozzle and sample = 30.48 cm.

TABLE 4—*Effect of separation distance between fabric sample and nozzle tip on amount of liquid delivered.*

Distance, cm	Liquid Amount, Mean[a] g/cm^2	SD,[a] g/cm^2
25.40	0.012	0.00040
30.48	0.016	0.00045
38.10	0.019	0.00007

[a] Mean size and standard deviation (SD) based on three replications.
NOTE—Chamber parameters:
Dial setting = 8.
Driving pressure = 27.5 N/cm^2.

This chamber allows for reproducible laboratory testing of fabric samples exposed to pesticide droplets similar to actual field situations. Cost and hazards are reduced by moving testing from the field to the laboratory, and through the closed system design and incorporation into a chemical exhaust hood. This system provides a method for laboratory screening of fabrics for pesticide penetration. Data collected will aid in the understanding of fabric and pesticide variables in the penetration process. Verification can be made through field studies on fabrics that perform well in the less costly laboratory testing. These combined efforts will increase the dependability and speed of results leading to fabric recommendations for protection apparel for pesticide application.

Acknowledgment

This research was funded in part by the Agricultural Experiment Station, The University of Tennessee, Knoxville, a part of the Southern Region Project S-163. In addition, the U.S. Environmental Protection Agency provided funding for this project.

References

[1] Serat, W. F., Van Loon, A. J., and Serat, W. H., "Loss of Pesticides from Patches Used in the Field as Pesticide Collectors," *Archives of Environmental Contamination and Toxicology*, Vol. 11, 1982, pp. 227–234.
[2] Raheel, M. and Gitz, E. C., "Effect of Fabric Geometry on Resistance to Pesticide Penetration and Degradation," *Archives of Environmental Contamination and Toxicology*, Vol. 14, 1985, pp. 273–279.
[3] Lillie, T. H., Livingston, J. H., and Hamilton, M. A., "Recommendations for Selecting and Decontaminating Pesticide Application Clothing," *Bulletin of Environmental Contamination and Toxicology*, Vol. 27, 1981, pp. 716–723.
[4] Orlando, J., Branson, D., Ayres, G., and Leavitt, R., "The Penetration of Formulated Guthion Spray Through Selected Fabrics,"*Journal of Environmental Science and Health*, Vol. 5, 1981, pp. 617–628.
[5] Easter, E. P., "Decontamination of Pesticide Contaminated Fabrics by Laundering," unpublished thesis, The University of Tennessee, 1982.
[6] DeJonge, J. O., Ayres, G., and Branson, D., "Patterns of Pesticide Deposition of Clothing During Air Blast Spraying," *Home Economics Research Journal*, Vol. 14, No. 2, 1985, pp. 262–268.
[7] Pratt, H. D. and Littig, K. S., "The Relationship Between Insecticide Application and Particle Size," *Insecticide Application Equipment for the Control of Insects of Public Health Importance*, U.S. Department of Health, Education, and Welfare, Publication (CDC) 76-8273, 1974.

Cynthia J. Goodman,[1] *Joan M. Laughlin,*[2] *and Roger E. Gold*[3]

Strategies for Laundering Protective Apparel Fabric Sequentially Contaminated with Methyl Parathion

REFERENCE: Goodman, C. J., Laughlin, J. M., and Gold, R. E., "**Strategies for Laundering Protective Apparel Fabric Sequentially Contaminated with Methyl Parathion,**" *Performance of Protective Clothing: Second Symposium, ASTM STP 989,* S. Z. Mansdorf, R. Sager, and A. P. Nielsen, Eds., American Society for Testing and Materials, Philadelphia, 1988, pp. 671–679.

ABSTRACT: Fabrics were contaminated daily with methyl parathion for up to five days, laundered, and residues before and after laundering determined. Two fabrics (100% cotton and 50/50% cotton/polyester) of two finishes (unfinished and fluorocarbon soil-repellent) were studied with one-half the specimens laundered daily following contamination and the other half laundered only on the last day following daily contamination.

Gas chromatographic analysis showed that methyl parathion soiling was additive over the five-day period at each additional contamination. The soil-repellent finish was effective through two launderings in limiting sorption of pesticide. In specimens laundered daily, post-laundering residues were similar across each of the five days. Differences in laundered fabric attributable to the fiber content of the specimen and to the functional finish were observed.

Soiling increased across the five days with no laundering such that the washing process was not as effective in removal of the methyl parathion residues. When specimens were contaminated repeatedly and laundered only once, the analysis of wash water showed sizable contaminate; however, when the specimens were laundered daily, proportionately less contamination was evident.

KEY WORDS: pesticide, pesticide residue, protective clothing, laundering, functional finish

Effective laundering procedures that reduce pesticide residues in contaminated clothing already have been studied [1–5]. Several studies have been conducted comparing the effectiveness of various laundry procedures in reducing pesticide residues in contaminated clothing. From this work, laundering recommendations have been published for minimizing direct and indirect exposure to workers and family members through appropriate laundering procedures. With few exceptions [6–8], laboratory laundering has been used to duplicate the home laundering procedures. The situation modeled in the laundering studies is that, when contaminated, the protective garment should be quickly removed and quickly laundered. However, a garment may be worn more than one day (without laundering) during repeated mixing, handling, and application of the pesticide. Researchers have examined a single application of pesticide to fabric and have made recommendations [1–8]; however, no studies to date have examined the role of pesticide recontamination of protective-apparel

[1] Florida Cooperative Extension Service, St. Augustine, FL 32085.
[2] Professor and chairman, Textiles, Clothing, and Design, 234 Home Economics Building, University of Nebraska-Lincoln, Lincoln, NE 68583-0802.
[3] Head, Department of Entomology, University of Nebraska-Lincoln, Lincoln, NE 68583-0818.

fabrics. As the concentration of active ingredient in the pesticide mixture increases, the completeness of removal of pesticide residues in laundering decreases. In general, repeated launderings help to remove pesticides [2,9,10]. Laboratory procedures for waste water from washing and rinsing involved neutralization with sodium hydroxide. In the home-laundering situation, waste water is evacuated from equipment and delivered to a common sewage system. Although the detergent, additives, or pre-wash treatment used in laundry may degrade pesticides, or water volume may dilute contamination to negligible levels, no data published to date have addressed this issue.

The need for reduction of dermal exposure to pesticides among mixer/handlers and applicators has prompted widespread recommendations for protective clothing [11]. Occupational Safety and Health Administration Standard 1910.267a suggests that protective clothing include a "washable fabric" [12]. However, recent investigations at the University of Nebraska [1–5,13–15] have confirmed that pesticide residues remain in these washable fabrics after laundering. In addition, indirect exposure to members of the applicators' families may occur in the home laundry situation [1,4].

The usual processes of soiling and the resultant mechanisms of soil removal from textile substrates are complicated by the chemical nature of pesticides. Generally, soiling depends upon the chemical nature of the textile; the characteristics (geometry) of the fiber, yarn, and fabric; the chemical treatments of the textile; and the conditions of liquid soil, whether of an oil type or a water type [16]. Since soil removal is a displacement process wherein the soil, such as pesticide/water/oil mixture, is displaced by water or a detergent solution, the diffusion of contamination may be in large part from the fabric to the wash water [17]. Even though a large portion of contamination is released into the water, a small portion of the released pesticide is deposited onto the fabric. Leveling—generalized redeposition of residues through the fabric—has been reported [8].

Fluorocarbon soil-repellent (SR) finishes decrease pesticide absorption; however, these finishes may hinder pesticide removal in laundering [15]. Fluorocarbon polymers have very low surface tensions and therefore show very good oily soil repellency in air, but form high-energy surfaces in water during fabric laundering, enabling hydrophobic soil dispersion in the aqueous wash medium [18].

Thus, a study of repeated contamination of fabric with pesticide, such as might occur when a garment is worn repeatedly with and without laundering, is justified. The determination of methyl parathion in effluent wash and rinse water is needed to elucidate the soil-redeposition situation.

Experimental Procedure

Fabrics were contaminated with methyl parathion (MeP) insecticide on successive (up to five) days with and without daily laundering. A second part of this study assessed the pesticide levels in the wash and rinse waters. The research design included two phases. Phase I was a $2 \times 2 \times 5$ design and included two fabrics (all-cotton and cotton/polyester) of two finishes (unfinished and fluorocarbon finished) which were sequentially contaminated with MeP on one through five days (Fig. 1). Specimens were laundered on the last day of contamination. Phase II was a $2 \times 2 \times 5 \times 2$ design, examining the same fiber contents (two), fabrics finishes (two), and number of days (five) as in Phase I. Specimens for Phase II were split into two groups; one half of the paired specimens were laundered daily after each contamination, and the other half were held overnight and recontaminated and laundered the subsequent day. Controls for each time period and for each phase were prepared using specimens that were contaminated and then extracted (Fig. 1). Waters from wash and rinse procedures were extracted and analyzed for pesticide content.

	1	2	3	4	5
PHASE I	C - Ex				
	C - L-Ex				
	C - H	C-Ex			
	C - H	C - L-Ex			
	C - H	C - H	C-Ex		
	C - H	C - H	C - L-Ex		
	C - H	C - H	C - H	C - Ex	
	C - H	C - H	C - H	C - L-Ex	
	C - H	C - H	C - H	C - H	C - Ex
	C - H	C - H	C - H	C - H	C - L-Ex

	1	2	3	4	5
PHASE II	C - Ex				
	C - L-Ex				
	C - L-H	Ex			
	C - L-H	C-Ex			
	C - L-H	C - L-Ex			
	C - L-H	C - L-H	Ex		
	C - L-H	C - L-H	C-Ex		
	C - L-H	C - L-H	C - L-Ex		
	C - L-H	C - L-H	C - L-H	Ex	
	C - L-H	C - L-H	C - L-H	C-Ex	
	C - L-H	C - L-H	C - L-H	C - L-Ex	
	C - L-H	C - L-H	C - L-H	C - L-H	Ex
	C - L-H	C - L-H	C - L-H	C - L-H	C - Ex
	C - L-H	C - L-H	C - L-H	C - L-H	C - L-Ex

C = Contaminate   H = Hold

Ex = Extract   L = Launder

FIG. 1—*Research design for Phase I (recontaminated daily and laundered on day of extraction) and Phase II (recontaminated and laundered daily).*

Analysis included the amounts of MeP ($\mu g/cm^2$) in the specimens at initial contamination, amounts of residue in the fabric specimens after successive recontaminations with and without laundering, and the amount of contamination in the wash and rinse waters (ppm). All work was replicated a minimum of three times.

Fabrics

Fabrics under study were an unfinished 100% cotton poplin and a 50% polyester/50% bleached and mercerized cotton poplin. Both fabrics were 106 cm in width and were undyed. Fabrics such as these were commonly used for work garments. The all-cotton fabric, TestFabric Style No. 407, had a thread count of 44×20 yarns/cm and weighted 231 g/m^2. The blend fabric of 50% cotton/50% polyester, TestFabric Style No. 7428, had a thread count of 48×20 yarns/cm and a weight of 210 g/m^2. Fabrics were studied unfinished and finished with Scotchgard,[4,5] a consumer-applied aerosol fluorocarbon renewable soil-repellent finish. The soil-repellent was applied at 1.5% weight/weight (w/w).

---

[4] Use of brand name does not imply product endorsement.
[5] Registered trademark of 3 M Company.

Specimen Preparation

Specimens, 8 by 8 cm, were randomly selected from yardages of fabric that had been stripped of warp sizing and softeners through a series of five launderings, using American Association of Textile Chemists and Colorists (AATCC) Test Method for Dimensional Changes in Automatic Home Laundering of Woven and Knit Fabrics (135-1978, revised 1982). The outer 10% of the yardage was removed prior to preparation of test specimens as described in the ASTM Test Methods for Breaking Load and Elongation of Textile Fabrics [D 1682-64 (1975 R-82)] to ensure consistency of warp yarns under evaluation.

Spiking the Specimens

Methyl parathion (MeP) (O,O-dimethyl O-*p* nitrophenyl phosphorothioate) dilutions were prepared at 1.25% active ingredient (AI) field strength concentration from emulsifiable concentrate formulation. Solutions were held in suspension during the contamination process by placing them on a magnetic stirrer. Two-tenths of a millilitre were pipetted onto the center of the specimen using a MicroLab P programmable micropipette. Contaminated specimens were allowed to air dry 4 to 6 h. Following contamination and drying, specimens were laundered and evaluated, or were evaluated unlaundered. A 0.2-mL aliquot of the 1.25% AI MeP was placed in glass for each replication for each formulation, allowed to air dry, and prepared in acetone as a baseline for determining recovery rate.

Laundering the Specimens

Prior to laundry, contaminated specimens were treated with 0.225 mL of a pre-wash product (Spray-n-Wash[6]). Specimens were laundered with a non-ionic heavy-duty liquid detergent (Dynamo[7]) (0.13%) using procedures modified from AATCC Test for Colorfastness to Washing, Domestic; and Laundering, Commercial: Accelerated (61-1875). The accelerated method was adjusted to simulate a single laundry cycle. All specimens were laundered for 12 min followed by 5- and 3-min rinses, respectively. Agitation was provided by 25 stainless steel balls. All the cycles used 49°C (120°F) distilled water.

Residue Analysis

MeP was extracted from fabric specimens utilizing acetone as a solvent as per procedures previously reported [14], with results calculated in $\mu g/cm^2$. MeP was double extracted from the wash and rinse water with 100 mL of dichloromethane [19] and reported in ppm.

The extracts were concentrated with rotary and nitrogen stream evaporation, and made up in hexane, which was analyzed on a Varian Vista 6000 gas chromatograph with automatic injection, electron capture detector, and 4270 dedicated microprocessor. Separation was achieved on a 2-m by 2-mm glass column packed with 100% OV-101 on 80/100 mesh Chromasorb W-HP with a nitrogen flow of 30 mL/min. Injection, detector, and oven temperature were 220°C (428°F), 270°C (518°F), and 220°C (428°F), respectively. Total amount of MeP residue in each specimen was expressed in $\mu g/cm^2$.

Statistical Analysis of Data

The statistical analyses of data were based on computed amount of pesticide residue remaining in the fabric after treatment and as a proportion that the after-treatment residue

---

[6] Registered trademark of Texize Company.
[7] Registered trademark of Colgate Palmolive Company.

was as compared with the initial contamination. Arc sine transformations were applied to the percent residue remaining, and statistical analysis system (SAS) general linear models analysis was used to test for main effects and all interactions of the main effects. Least significant (LS) means tests were performed to separate means where significant differences were observed. The decision level was $p \leq 0.05$.

Findings and Discussion

Phase I: Successive Contaminations With Laundering on Terminal Day Only

Phase I modeled the wearing of a contaminated garment for one day to a maximum of five days prior to laundering. Laundering occurred immediately following the last contamination. Fabric contamination on successive days ranged from 1.67 $\mu g/cm^2$ on Day 1 for the cotton/polyester soil-repellent finished specimens to 93.05 $\mu g/cm^2$ on Day 4 for the cotton/polyester unfinished specimens (Table 1). Mean amounts of MeP on the unfinished and soil-repellent finished·specimen were 65.08 $\mu g/cm^2$ and 12.68 $\mu g/cm^2$, respectively. Soil-repellent finished specimens used in this study retained only 20% of initial contamination of an unfinished specimen. The contamination levels were similar to those determined by Keaschall et al. [14] and Laughlin et al. [15]. Soil-repellent treated specimens absorbed less MeP due to the low surface energy of the textile imparted by the fluorocarbon finish which, in effect, repelled liquid soil [20].

Based on the analysis of variance (ANOVA) performed on amount ($\mu g/cm^2$) of initial contamination, significant differences due to the main effect of day ($F = 69.46$, $df = 4$, $p \leq 0.05$), finish ($F = 756.76$, $df = 1$, $p \leq 0.05$), and an interaction effect of day with finish were observed ($F = 25.81$, $df = 7$, $p \leq 0.05$). Scrutiny of the interaction data showed that amount of MeP on unfinished specimens ranged from 20.99 $\mu g/cm^2$ on Day 1 to 88.98 $\mu g/cm^2$ MeP on Day 4 (Table 2) while the amount of contamination was heaviest on Day 5 for the soil-repellent finished specimens.

Laundering significantly decreased the amount of MeP residue (Table 1). Amount of pesticide residue remaining on the specimens after laundering generally increased through the successive days (Table 1). Thus, it appeared more difficult to remove pesticides following

TABLE 1—*Amount ($\mu g/cm^2$) of MeP on specimens for Phase I.*

	Unfinished			Soil-Repellent Finish		
Day	Initial	After Laundering	% Residue	Initial	After Laundering	% Residue
		COTTON				
1	19.33	0.24	1.39	3.05	0.23	4.99
2	51.34	0.15	0.74	7.63	0.21	3.39
3	73.39	1.61	8.61	16.82	0.52	12.37
4	84.90	2.07	10.16	20.54	0.33	8.01
5	90.80	4.05	19.85	20.67	0.25	7.71
		COTTON/POLYESTER				
1	22.64	0.32	1.49	1.67	0.20	10.92
2	55.13	0.16	1.49	9.25	0.19	9.94
3	75.37	0.67	3.10	15.88	0.46	23.91
4	93.05	1.23	5.62	13.82	0.48	24.76
5	84.78	1.37	6.19	17.44	0.42	20.36

multiple contaminations. These amounts represented as little as 0.7% of initial contamination for the unfinished cotton (Day 2) to as much as 24.8% of initial contamination for the soil-repellent finished blend (Day 4). The mean amount of MeP residues on the specimens after laundering was greater after the third day for the cotton than the blend and was greater for the unfinished specimens than for the SR finished specimens. In comparison to the findings of an earlier work [5], increased quantities of pesticide (AI) are more difficult to remove in laundering than less concentrated pesticide.

Generally, larger percentages of residue remaining after laundering were found for the soil-repellent finished specimens; however, the amount ($\mu g/cm^2$) was lower for SR than for unfinished specimens. The soil-repellent finished specimens had absorbed less MeP at initial contamination, and the amount of pesticide remaining on the soil-repellent finishes was lower than for the unfinished specimens. This finding is important because the critical concern in laundering studies is for lowering the post-laundering residues to the minimal amount possible.

For the cotton, the amount of contamination on the soil-repellent finished specimens was significantly less than the amount of contamination on the unfinished cotton. A significant difference was not observed for the cotton-polyester blend specimens. Obendorf and Solbrig [21] postulated that this phenomenon was due to the site of soiling on fiber. Through an electron scanning microscopic analysis, these researchers documented MeP soil on the surface of polyester and cotton fibers (in the convolutions and crenulations) and in the lumen of the cotton fibers of an unfinished fabric. Laughlin et al. [15] found that the after-laundering residues were similar across the unfinished, soil-repellent, and durable press finished cotton and cotton/polyester specimens even though the initial contamination differed markedly. The soil-repellent finish inhibited soil removal in laundering.

ANOVA of residue remaining after laundering resulted in main effects due to day ($F = 8.77$, $df = 4$, $p \leq 0.05$), finish ($F = 25.69$, $df = 1$, $p \leq 0.05$), and fiber content ($F = 6.02$, $df = 1$, $p \leq 0.05$); however, there were interaction effects of day with finish ($F = 6.88$, $df = 4$, $p \leq 0.05$) and of finish with fiber content ($F = 7.33$, $df = 1$, $p \leq 0.05$). The important contribution of the soil-repellent fabric finish to after-laundering residues was reflected in these interaction effects; that is, the contribution of soil-repellent finish was dominant initially but diminished with time. At initial contamination, interaction effects were found for finish and day.

There was a significant difference in after-laundering residue between the cotton unfinished and cotton/polyester blend unfinished fabrics, with greater amounts of MeP remaining in the cotton unfinished fabrics. This may be due to the MeP entrapment in the cotton lumen, convolutions, and crenulations, making it more difficult to remove than is true of removal from the polyester fiber surfaces and capillary spaces [21]. This finding is unlike that of Easley et al. [2,3] but supportive of Finley et al. [9,22].

TABLE 2—*Interaction effect of fabric finish and day of initial contamination for Phase I.*

| Day | Amount, $\mu g/cm^2$ | |
	Unfinished	Soil-Repellent
1	20.99	2.36
2	53.24	8.44
3	74.38	16.35
4	88.98	17.18
5	87.79	19.06

Phase II: Successive Contaminations with Laundering

Phase II modeled the wearing of a contaminated garment for one day to a maximum of five days with daily laundering. Laundering occurred every day after the specimens had been exposed to MeP.

Specimens contaminated and laundered on successive days ranged in contamination from 2.86 μg/cm² on Day 1 for the cotton/polyester soil repellent specimen to 29.29 μg/cm² on Day 3 for the cotton/polyester unfinished specimen (Table 3). Unfinished specimens absorbed more MeP initially than the soil repellent finished specimens. This was consistent with Phase I of the experiment (Table 2). The functional finish lowered the surface energy of the textile which repelled the MeP; however, the beneficial properties of the functional finish were lost after two launderings, perhaps because of the synergistic effect of the pesticide and the laundering. After two launderings, enough of the soil-repellent finish had been removed that the specimen performed similar to the unfinished specimens.

ANOVA comparing the amount of MeP at the initial contamination across specimens showed there were significant differences due to the main effects of day ($F = 9.26$, $df = 4$, $p \leq 0.05$) and of fabric finish ($F = 98.93$, $df = 1$, $p \leq 0.05$). There was an interaction effect of day with fabric finish ($F = 4.85$, $df = 4$, $p \leq 0.05$).

For specimens contaminated, then laundered daily, post-laundering residues ranged from 0.09 μg/cm² on the soil-repellent cotton/polyester (Day 1) to 0.88 μg/cm² on the unfinished cotton (Day 3) (Table 3). Mean amount of MeP residue for unfinished specimens was 0.31 μg/cm² and for soil-repellent finished specimens was 0.15 μg/cm².

Statistical analysis showed a significant difference in amount of residue remaining after daily laundering due to the main effect of fabric finish ($F = 9.13$, $df = 1$, $p \leq 0.05$). In explaining the residue remaining due to fabric finish, it is important to remember that the initial contamination significantly differed; that is, there was less contamination on the soil repellent finished fabrics.

Contrasts were performed between the amount of residue in specimens from Phase I (Table 1) and amount of residues in specimens from Phase II (Table 3). Residue amounts ranged from 4.05 μg/cm² for UN cotton on Day 5 (Phase I) to 0.15 μg/cm² on Day 1 (Phase II). Analysis of variance revealed significant differences due to daily laundering ($F = 35.56$, $df = 1$, $p \leq 0.01$). Based on this finding it is recommended that MeP-contaminated clothing be laundered daily.

TABLE 3—*Amount (μg/cm²) of MeP on specimens for Phase II.*

	Unfinished			Soil-Repellent Finishes		
Day	Initial	After Laundering	% Residue	Initial	After Laundering	% Residue
			COTTON			
1	19.88	0.15	0.72	4.22	0.60	1.74
2	26.82	0.27	1.92	12.69	0.22	4.14
3	22.51	0.88	4.56	20.12	0.18	2.75
4	22.42	0.27	1.33	15.27	0.19	4.59
5	20.69	0.22	0.88	16.81	0.12	2.87
			COTTON/POLYESTER			
1	22.64	0.21	0.95	2.86	0.09	4.01
2	18.81	0.36	1.32	8.78	0.15	12.29
3	29.29	0.33	1.45	14.09	0.17	6.91
4	26.88	0.41	1.89	16.03	0.22	11.42
5	19.36	0.33	1.20	15.55	0.16	8.87

TABLE 4—*MeP residues (ppm) in water from laundering unfinished cotton.*

	Phase 1			Phase II		
Day	Wash, ppm	Rinse 1, ppm	Rinse 2, ppm	Wash, ppm	Rinse 1, ppm	Rinse 2, ppm
1	60.37	24.69	6.36	86.58	25.19	5.84
2	89.05	17.20	20.06	54.07	21.71	10.06
3	129.08	55.98	13.00	51.91	7.99	1.54
4	237.12	68.10	13.29	55.48	8.12	7.21
5	309.33	74.62	14.15	62.00	14.84	5.29

Analysis of MeP Residues in Wash and Rinse Water

Since the residue levels from Phase I were the greatest for unfinished cotton, that fabric/finish combination was selected for water analysis. There was a daily increase in the amount of MeP in wash and rinse waters from Day 1 to Day 5 (Table 4) for Phase I. For the specimens from successive contaminations without daily laundering, an interaction effect of day and water source (wash, rinse one, rinse two) was found ($F = 2.70$, $df = 8$, $p \leq 0.05$). For the specimens from successive contamination with daily laundering, only a main effect due to water sources was found ($F = 34.00$, $df = 2$, $p \leq 0.05$). The greatest amount of MeP was found in the wash water, followed by the first rinse which was followed by the second rinse. These findings confirm the need for daily laundering of MeP-contaminated clothing.

Conclusions

The contribution of soil-repellent finish to initial contamination and to residues after treatment was of major importance. The soil-repellent finish lowered the amount of initial contamination. However, the soil-repellent finish was effective in lowering the amount of MeP absorbed into the fabric only through two launderings. This factor may have been due to a synergistic effect of soil plus laundering in diminishing the effectiveness of the soil-repellent finish. Based on these findings, a recommendation is made that renewable soil-repellent finishes be reapplied after every second laundering.

MeP contamination increased across the five days without daily laundering such that the washing process was not as effective in removing residues. Given the MeP levels in the fabric and in the waste wash and rinse waters, the recommendation to launder protective clothing daily is supported.

Acknowledgments

This research was supported in part by the Nebraska Agricultural Research Division Project 94-012 and it contributes to North Central Research Project NC-170, "Limiting Dermal Exposure to Pesticide Through Effective Cleaning Procedures and Selection of Clothing." This paper is published as Paper No. 7981, Nebraska Agricultural Experiment Station.

References

[1] Laughlin, J. M., Easley, C. B., Gold, R. E., and Tupy, D., "Methyl Parathion Transfer from Contaminated Fabrics to Subsequent Laundry and to Laundry Equipment," *Bulletin of Environmental Contamination and Toxicology*, Vol. 27, No. 4, Oct. 1981, pp. 518–523.

[2] Easley, C. B., Laughlin, J. M., Gold, R. E., and Schmidt, K., "Detergents and Water Temperature as Factors in Methyl Parathion Removal from Denim Fabrics," *Bulletin of Environmental Contamination and Toxicology,* Vol. 28, No. 2, Feb. 1982a, pp. 239–244.

[3] Easley, C. B., Laughlin, J. M., Gold, R. E., and Hill, R. M., "Laundry Factors Influencing Methyl Parathion Removal from Contaminated Denim Fabric," *Bulletin of Environmental Contamination and Toxicology,* Vol. 29, No. 4, Oct. 1982b, pp. 461–468.

[4] Easley, C. B., Laughlin, J. M., Gold, R. E., and Tupy, D., "Laundering Procedures for Removal of 2,4-Dichlorophenoxyacetic Acid Ester and Amine Herbicides from Contaminated Fabrics," *Archives of Environmental Contamination and Toxicology,* Vol. 12, No. 1, Jan. 1983, pp. 71–76.

[5] Laughlin, J. M., Easley, C. B., and Gold, R. E., "Methyl Parathion Residues in Contaminated Fabrics After Laundering," *American Chemical Society Symposium Series,* No. 273, 1985, pp. 177–188.

[6] Finley, E. L. and Rogillio, J. R. B., "DDT and Methyl Parathion Residues Found in Cotton and Cotton-Polyester Fabrics Worn in Cotton Fields," *Bulletin of Environmental Contamination and Toxicology,* Vol. 4, No. 6, June 1969, pp. 343–351.

[7] Lillie, T. H., Livingston, J. M., and Hamilton, M. A., "Recommendations for Selecting and Decontaminating Pesticide Applicator Clothing," *Bulletin of Environmental Contamination and Toxicology,* Vol. 27, No. 6, June 1981, pp. 716–723.

[8] Laughlin, J. M. and Gold, R. E., "Removal and Redeposition of Methyl Parathion During Laundering of Functionally Finished Textiles," *Technical Papers, The First International Symposium on the Impact of Pesticides, Industrial and Consumer Chemicals on the Near Environment,* B. Reagan, D. Johnson, and S. Dusaj, Eds., sponsored by USDA-CSRS, 1988, pp. 12–14.

[9] Finley, E. L., Graves, J. B., and Hewitt, F. W., "Reduction of Methyl Parathion Residues on Clothing by Delayed Field Re-Entry and Laundering," *Bulletin of Environmental Contamination and Toxicology,* Vol. 22, No. 4, July 1979, pp. 598–602.

[10] Easter, E., "Removal of Pesticide Residues from Fabrics by Laundering," *Textile Chemist and Colorist,* Vol. 47, No. 3, March 1983, pp. 29–33.

[11] Baker, D. E. and Bradshaw, W., "Homeowner Chemical Safety," *Science and Technology Guide,* University of Missouri-Columbia Extension Division, Bulletin No. 1918, May 1979.

[12] Occupational Safety Health Standard 1910.267a, FR 38, 10715, 1 May 1973.

[13] Easley, C. B., Laughlin, J. M., Gold, R. E., and Tupy, D., "Methyl Parathion Removal from Work Weight Fabrics by Selected Laundry Procedures," *Bulletin of Environmental Contamination and Toxicology,* Vol. 27, No. 2, July 1981, pp. 101–108.

[14] Keaschall, J. L., Laughlin, J. M., and Gold, R. E., "Effect of Laundering Procedures and Functional Finishes on Removal of Insecticides Selected from Three Chemical Classes," *Performance of Protective Clothing, ASTM STP 900,* R. L. Barker and G. C. Coletta, Eds., American Society for Testing and Materials, Philadelphia, 1986, pp. 162–176.

[15] Laughlin, J. M., Easley, C. B., Gold, R. E., and Hill, R. M., "Fabric Parameters and Pesticide Characteristics that Impact on Dermal Exposure of Applicators," *Performance of Protective Clothing, ASTM STP 900,* R. L. Barker and G. C. Coletta, Eds., American Society for Testing and Materials, Philadelphia, 1986, pp. 136–150.

[16] Morris, M. A. and Prato, H. H., "The Effect of Wash Temperatures on Removal of Particulate and Oily Soil from Fabrics of Varying Fiber Content," *Textile Research Journal,* Vol. 52, April 1982, pp. 280–286.

[17] Kissa, E., "Kinetics of Oily Soil Release," *Textile Research Journal,* Vol. 41, No. 9, Sept. 1971, pp. 760–767.

[18] Das, T. K. and Kulshreshtha, A. K., "Soil Release Finishing of Textiles: A Review," *Journal of Scientific and Industrial Research,* Vol. 38, Nov. 1979, pp. 611–619.

[19] Goodman, C. J. S., "Removal of Methyl Parathion Through Laundering Recontaminated Fabrics," unpublished master's thesis, University of Nebraska-Lincoln Libraries, 1985.

[20] Berch, J., Peper, H., and Drake, G. L., Jr., "Wet Soiling of Cotton. Part II: Effect of Finishes on the Removal of Soil from Cotton," *Textile Research Journal,* Vol. 34, Jan. 1964, pp. 29–34.

[21] Obendorf, S. K. and Solbrig, C. M., "Distribution of the Organophosphorous Pesticide Malathion and Methyl Parathion on Cotton/Polyester Fabrics After Laundering as Determined by Electron Microscopy," *Performance of Protective Clothing, ASTM STP 900,* R. L. Barker and G. C. Coletta, Eds., American Society for Testing and Materials, Philadelphia, 1986, pp. 187–204.

[22] Finley, E. L., Bellon, J. M., Graves, J. B., and Koonce, K. L., "Pesticide Contamination of Clothing in Cotton Fields," *Louisiana Agriculture,* Vol. 20, No. 3, Spring 1977, pp. 8–9.

Charles J. Kim[1] and Jong-Ok Kim[1]

Dispersion Mechanism of a Pesticide Chemical in Woven Fabric Structures

REFERENCE: Kim, C. J. and Kim, J.-O., **"Dispersion Mechanism of a Pesticide Chemical in Woven Fabric Structures,"** *Performance of Protective Clothing: Second Symposium, ASTM STP 989*, S. Z. Mansdorf, R. Sager, and A. P. Nielsen, Eds., American Society for Testing and Materials, Philadelphia, 1988, pp. 680–691.

ABSTRACT: The dispersion mechanism of a DDT solution in a variety of woven fabric structures is described. The dispersion is illustrated by scanning electron microscope (SEM) micrographs and related to fiber morphology, fabric geometry, weight, and finish. The kinetics of dispersion is related to the size and shape of the dispersion area and to the concentration of radioactive-labeled DDT. Findings show that final location of DDT within textile substrates is related to fiber content and morphology, fabric geometry, and finish. Fiber irregularities serve as deposit sinks for the chemical soil. The finish influences soil dispersion by making the fiber surface smoother. Fabric weight and geometry determine the size and shape of the dispersion area. The maximum concentration of DDT deposit is found to be at the outermost boundary of the dispersion area, not at the center.

KEY WORDS: capillary radius, chemical soil, cotton, cotton/polyester blend, DDT, deposit sink, dispersion areas, dispersion mechanism, fabric and yarn geometry, fiber morphology, finish, pesticide, polyester, protective clothing, radioactive-labeled, scanning electron micrographs

Concern for safety with respect to exposure to pesticide has increased in recent years. Human exposure and absorption of pesticide can be minimized through the appropriate use of clothing as a body covering. When the contaminated clothing comes in contact with the skin, the chemical transfers from the fabric to the skin [1,2]. If the contaminant chemical is not removed from the clothing, the skin may absorb the chemical from the clothing fabric [3]. Appropriate care methods for washable, contaminated clothing, therefore, would reduce the health risk from pesticide exposure [4].

To select effective methods of removing chemical soils from contaminated fabrics, it is necessary to understand the soiling process and the distribution of the soil in a fabric structure. The literature has shown that soiling and soil-release processes are controlled by many of the same parameters [5]. The purpose of this research is to describe the soil-dispersion mechanism of an industrial/pesticide chemical in fabrics having different geometries and characteristics, in terms of the final location of the soil, the size and shape of the dispersion area, and the soil concentration in specific locations of the dispersion area.

The main soiling mechanism for liquid soil is reported as capillary penetration [6]. According to Raheel and Gitz [7], the rate of penetration and transport of a pesticide-soil solution from a garment to the underlayers depends on fabric geometry. For all-cotton fabrics, the ease of wettability is higher in fabrics with larger interfiber and interyarn cap-

---

[1] Associate professor and graduate student, respectively, Department of Textiles and Clothing, Iowa State University, Ames, IA 50011.

illaries, whereas the level and rate of wicking are higher in fabrics with smaller interfiber and interyarn capillary radii [7]. Brown et al. [8] showed in electron micrographs that oily soil deposits heavily at the fiber crossings in spun polyester yarns, whereas filament polyester yarns have fewer fiber junctions and soil is distributed uniformly over the fiber surfaces. According to Smith and Sherman [5], irregularities in the fiber surface serve as deposit sinks for the soil. Protruding fiber ends and irregularities in fabric surface due to weave also contribute to the retention of soil.

Finley and Rogillio [9] reported more absorption of methyl parathion by all-cotton fabrics than by cotton/polyester blend fabrics. Lillie et al. [10] found that all-cotton fabrics are less penetrable than all-polyester fabrics and that, in the two fabric types, pesticide penetrability is inversely proportional to absorbability. Fort et al. [11] showed a more rapid oily-soil diffusion at elevated temperatures for synthetic fibers than for cellulosics. Easter [12] showed similarly that oleophilic synthetic fabrics retain more oil-based emulsion pesticide than pesticide particles in an aqueous suspension.

The hydrophobic nature of fabric surfaces is related to the ease of soiling and soil release. When a fabric is hydrophobic, it is generally oleophilic and can absorb and retain large quantities of oily soil [13,14]. Bowers and Chantrey [15] stated that the original differences in fiber hydrophobicity are completely negated when the fiber is made into yarns and fabrics; chemical finishes, however, change the hydrophobic nature of the fabric. Venkatesh et al. [16] reported that mercerization eliminates most of the surface irregularities of cotton, decreases the surface area, and thereby provides an improvement in resistance to soiling.

Obendorf et al. [17] found that fiber hydrophilicity and fabric weight are the dominant parameters that affect triolein retention. Kim et al. [18,19] also cited fabric weight as a factor in pesticide retention. To show oily soil distribution and specific location of trapped residual soils in fiber, yarn, and fabric structure, scanning electron microscope (SEM) micrographs are used [5,14,17,20–22].

Procedure

Test Fabrics

Fabrics used in this research were selected on the basis of fiber content, yarn type, weave construction, weight, and finish. They were obtained from Testfabrics Inc. The fabric names and their characteristics are presented in Table 1. The construction characteristics of the fabrics were determined in accordance with the ASTM Test Method for Yarn Number Based on Short-Length Specimens (D 1059-76), ASTM Test Method for Fabric Count of Woven Fabric (D 3775-79), and ASTM Test Method for Weight (Mass) per Unit Area of Woven Fabric (D 3776-79). The fabrics were cut into 15 by 15-cm specimens along the warp and filling directions for pesticide contamination.

Chemical Soil

As contaminant chemical soil, technical-grade DDT (1,1,1-trichloro-2,2-bis(p-chloro-phenyl)ethane) supplied in a crystalline form, was used with certified-grade hexane (99.9%) as solvent. An 8% DDT solution was prepared by dissolving the crystalline-form DDT in hexane at 40°C and used for the SEM-micrograph analysis of soiling and soil-dispersion mechanisms.

For the radioactive-tracer analysis, a ^{14}C-labeled DDT solution (3.14 GBq/mmol) in hexane was diluted to 1/400 (0.06 μCi/mg) to reduce the radioactivity, so that clearer autoradiogram images could be obtained. The concentration of DDT in the diluted solution was 10 mg/cm^2.

TABLE 1—*Construction characteristics of test fabrics.*

Fabric Symbol	Fabric Name and Weave	Fabric Finish	Weight, g/m²	Count,[a] W × F	Yarn No. cc W	Yarn No. cc F	Yarn Type
			100% COTTON				
F1	Print cloth, plain	bleached	108	83 × 74	33	41	spun, single
F2	Army duck, plain	bleached	337	55 × 40	8	7	spun, ply
F3	Cotton twill, twill	bleached, mercerized	247	108 × 54	16	15	spun, single
			100% POLYESTER				
F4	Plain	none	128	66 × 52	17	28	spun, single
F5	Taffeta, plain, rib	none	139	87 × 65	35	20	plain filament
F6	Twill	none	161	59 × 43	16	16	textured filament
			50/50% COTTON/POLYESTER				
F7	Print cloth, plain	bleached	91	81 × 53	32	33	spun, single
F8	Poplin, plain	bleached, mercerized	189	107 × 46	19	20	spun, single

[a] Per 2.54 cm; W = warp yarn and F = filling yarn.

Soiling Procedure

For the SEM-micrograph analysis, 15 µL of the 8% DDT solution in hexane was applied onto the center of the taut surface of a fabric swatch mounted on an embroidery hoop. The tip of a fixed-volume capillary micropipet was used and held 1 cm above the fabric surface during the application. The contaminated swatch was air dried. One swatch from each fabric type was contaminated.

For the radioactive-tracer analysis, 100 µL of the diluted ^{14}C-labeled DDT solution was applied, using a 0.1-mL graduated pipet in the same manner as in the SEM-micrograph analysis. The swatch was then air dried for 20 min. Three swatches from each fabric type were contaminated.

Procedure for Soil-Dispersion Analysis

SEM Micrograph—A specimen, 1.2 by 1.8 cm, was cut from the center of the contaminated swatch, mounted on a carbon stub, 2.5 cm in diameter, and coated with gold to make it more conductive of the electrical charge. Electron images were obtained using a Joel JSM-U3 SEM at 7-15 keV accelerating voltage with a specimen current of 10^{-10} A and at a 32-mm working distance. Three micrograph locations were selected for each specimen at a position in the plane of the fabric: one above, below, and right or left of the center drop point.

Radioactive Tracer—Two different tests were involved in the radioactive-tracer analysis of soil dispersion: autoradiogram and liquid-scintillation counting. Autoradiograms provide information on the size (area) and shape of the spread of the soiling DDT solution over the fabric specimen. Liquid-scintillation counting shows the concentration of the ^{14}C-labeled DDT in a specific location of the fabric surface.

For autoradiograms, the soiled swatch was placed directly in contact with an X-ray film; the film was exposed to the labeled DDT contained in the swatch for 48 h. From the developed film, the size and shape of the dispersion area of the labeled DDT were analyzed. The data on size as well as height-to-width ratio of the dispersion area were average values of three replications. The shape of the dispersion area, however, was taken from one specimen typical of the three replications.

For measuring the concentration of the labeled DDT in the contaminated swatches, two of the three ^{14}C-labeled swatches prepared for autoradiograms were selected randomly and used. Square specimens, 0.64 by 0.64 cm (a ¼-in. square), were taken from the swatch (Fig. 1) and measured for contaminant concentration. Each square was cut and placed in a scintillation vial with a 5-mL multi-purpose scintillation solution. The vials were placed in a LKB Rackbeta Model 1217 liquid-scintillation counter and disintegrations per minute (dpm) were recorded for each vial. The data on dpm were the average of two replications.

Results and Discussion

When fabrics with different fiber content and morphology, weave, count, yarn number, weight, and finish were contaminated with a DDT solution, they allowed spread of the liquid-DDT solution from the center drop point toward the outer edges of the fabric, leaving

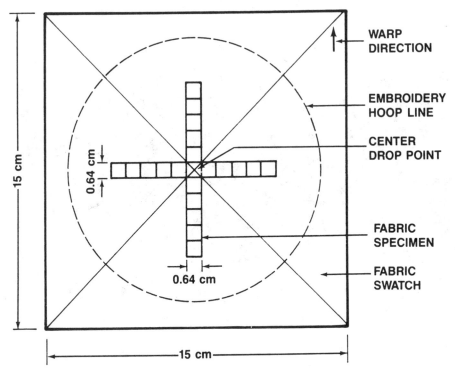

FIG. 1—*Specimen marking for measuring the concentration of the radioactive DDT.*

oval stains. To understand the mechanism of soil dispersion, analyses were made on the specific locations of the non-radioactive DDT in the fabric structure, the size and shape of the dispersion area of the radioactive DDT, and the concentration of the radioactive DDT in the dispersion area.

Location of DDT in the Fabric Structure

The SEM micrographs showed that the main locations of pesticide deposits are fiber surfaces, small capillaries formed between closely spaced fibers within the yarn, and fiber irregularities such as the crenulation due to convolution of the cotton fiber. Fiber and fabric surface characteristics also influence dispersion of the DDT solution.

100% Cotton Fabrics—The SEM micrographs show that, regardless of weave type, the crystal soil deposits mainly on fiber convolutions, protruding fiber ends that reach into voids between yarn interlacings, yarn cross-over points, and interfiber spaces within the yarn structure (Fig. 2a through 2d). The cotton print cloth, F1, constructed with fine yarns in balanced count (Table 1), shows large pores between yarns. The chemical soil, how-

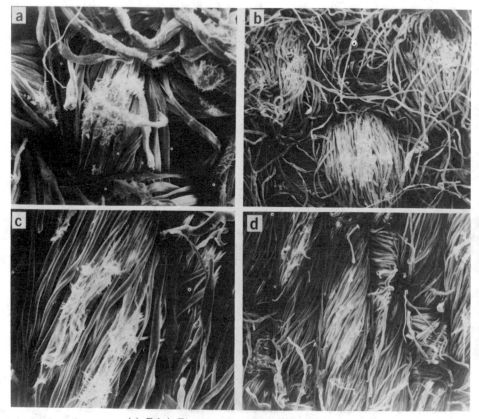

(*a*) Fabric F1, contaminated (original magnification ∼ ×180).
(*b*) Fabric F2, contaminated (original magnification ∼ ×60).
(*c*) Fabric F3, contaminated (original magnification ∼ ×180).
(*d*) Fabric F3, contaminated (original magnification ∼ ×60).

FIG. 2—*SEM micrographs of DDT dispersion in all-cotton fabrics.*

ever, does not fill the pores, but deposits mainly on the fiber and yarn cross-over points (Fig. 2a).

With coarse two-ply yarns that provide deep weave crimps as in F2 (Fig. 2b), the crest portion of the warp cross-over points shows a larger portion of DDT than the filling. In the twill fabric, F3, the DDT crystals spread along the warp floats (Fig. 2c and 2d). These results support findings of Smith and Sherman [5] that fabric surface irregularities due to weave crimps and fiber irregularities appear to serve as soil-deposit sinks, especially near the edges of the contaminated area where the rate of dispersion of the liquid soil is lowest.

The bleached and mercerized cotton fabric, F3, shows more even spread of DDT (Fig. 2c and 2d) than the bleached cotton fabrics, F1 (Fig. 2a) and F2 (Fig. 2b). This is consistent with findings of Venkatesh et al. [16] that fabric finish influences soil dispersion in fabrics. Our results, however, may be partly due to the difference in weave structure, the plain weave versus more warp-faced twill weave, and in yarn geometry such as twist. Fabric F3 (Fig. 2d) shows more compact yarns than those of F1 (Fig. 2a) and F2 (Fig. 2b).

100% Polyester Fabrics—Although F4 is constructed with spun yarns, it has fewer protruding fiber ends, irregularities, or crevices than the all-cotton fabrics due to the smoother, more uniform polyester staple fibers (Fig. 3b). The pesticide solution, consequently, flows

(a) Fabric F5, control (original magnification $\sim \times 180$).
(b) Fabric F4, contaminated (original magnification $\sim \times 60$).
(c) Fabric F5, contaminated (original magnification $\sim \times 180$).
(d) Fabric F6, contaminated (original magnification $\sim \times 60$).

FIG. 3—*SEM micrographs of DDT dispersion in all-polyester fabrics.*

along the fiber surface rather than being absorbed. As the solvent, hexane, evaporates, it leaves behind a more uniform and less concentrated deposit of DDT crystals.

The smooth, uniform, filament-fiber surface of the polyester taffeta control fabric, F5, (Fig. 3a) has no visible fiber protrusions nor surface irregularities so that the DDT solution spreads uniformly over all fiber surfaces as shown in Fig. 3c. This is in contrast to the all-cotton fabrics where the chemical soil deposits heavily in crenules and crevices. Obendorf and Klemash [14] observed similar differences in soil-distribution pattern, which suggest different soil-removal mechanisms for the two fabric types.

On the textured-filament yarn fabric, F6, the DDT solution deposits primarily in the textured crimp, which serves as sinks (Fig. 3d). The pattern of DDT distribution in the textured-filament polyester fabric, F6, resembles more the pattern of the spun-polyester fabric, F4 (Fig. 3b), than that of the plain-filament polyester fabric, F5 (Fig. 3c). In F6, heavy deposits are observed along the crimped filaments as well as on weave crimps, especially along the fillings (Fig. 3d). Unlike spun yarns, there are no protruding fiber ends showing DDT deposits. This suggests that yarn geometry affects soiling and soil-removal mechanisms. Brown et al. [8] reached a similar conclusion between polyester filament-yarn fabrics and cotton/polyester blended-yarn fabrics.

In comparing the two twill-weave fabrics, F3 and F6, fabric geometry seems to play a role in the dispersion mechanism. Fabric F3 in a 3/1 twill construction is more warp-faced than F6 in 2/1 twill, due to the longer warp float and the higher warp count (Table 1). The warp yarns in F3 are, therefore, packed more closely together and cover the fabric surface more completely than in F6. No appreciable DDT deposit is shown on the filling yarns of F3 that surface to the technical face of the fabric (Fig. 2d), whereas the filling yarns of F6 show heavy deposits of DDT (Fig. 3d). This effect of fabric geometry seems true regardless of fiber content and yarn geometry, since F3 is made of soil-attracting cotton fiber in a spun-yarn structure and F6 of polyester fiber in a textured-filament structure.

50/50% Cotton/Polyester Fabrics—The surfaces of the two blended-yarn fabrics, F7 and F8, show a thin coating of DDT along the smooth polyester fibers and heavier deposits of DDT crystals on cotton-fiber protrusions, crossings, and convolutions (Fig. 4a and 4b). The mercerization finish provides cotton fibers in F8 with more cylindrical and smoother surfaces and makes the fibers more like polyester fibers in their external morphology. Protrusions, crossings, and convolutions, however, are less affected by the finish. Major DDT deposits are noted on these irregularities, but the concentration of deposit is less in the mercerized fabric, F8 (Fig. 4b), than in the nonmercerized fabric, F7 (Fig. 4a). Changes in fiber surface morphology due to mercerization and their consequent effect on DDT distribution is noted earlier in the analyses of the all-cotton fabrics F3 versus F1 and F2.

Although Kim et al. [18,19] reported significantly higher amounts of residues of alachlor and fonofos in heavier-weight fabrics than in lighter-weight fabrics after laundering, fabric weight does not seem to affect the pattern of DDT distribution on the surfaces of the eight fabrics studied. The focus limitations of the SEM prevent meaningful observations of soil distribution below the fabric surface. For thick fabrics, fabric interior cross-sections may provide a better observation of soil-deposit pattern within the fabric depth.

Size and Shape of the Dispersion Area

Size of the Dispersion Area—The two plain-weave, all-cotton fabrics, F1 and F2, show a significant difference in the size of their dispersion areas (Fig. 5a versus Fig. 5b). Fabric F1 is a lighter, higher-count, finer-yarn fabric than F2 (Table 1). This indicates that F1 has smaller interfiber and interyarn capillary radii, which tend to increase the transporting

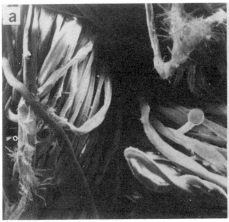

(*a*) Fabric F7, contaminated (original magnification ~ × 180).
(*b*) Fabric F8, contaminated (original magnification ~ × 180).

FIG. 4—*SEM micrographs of DDT dispersion in cotton/polyester fabrics.*

capacity of the liquid DDT [7]. The thinner fabric, F1, seems to spread the DDT solution in a wider area, whereas the thicker fabric, F2, retains the soil in its depth. The heavier the all-cotton fabric is, the smaller the size of the dispersion area is (Table 1 and Fig. 5). This explains partly why, in the studies by Kim et al. [18,19], the heavier fabrics retained more residues of alachlor and fonofos than the lighter fabrics after laundering. Similar analysis of fabric geometry explains the difference in size of dispersion areas between F2 and F3, as well as between F1 and F3. The mercerization finish of F3 further helps the fabric to retain the liquid soil in a smaller area, due to the increased absorbency of the cotton fiber and the polar bonding between the hydroxyl groups in the cellulose polymer chain and the DDT solution.

Among the three all-polyester fabrics, the heaviest fabric, F6, also shows the smallest dispersion area, as for the all-cotton fabrics (Table 1 and Fig. 5*f*). The low-count construction of F6 and thus larger interyarn capillary radii, in addition to textured-filament crimps, contributes to the retention of the liquid soil in a small area. Fabric F5 exhibits the largest area of dispersion as can be expected from its fabric and yarn geometries: high fabric count and fine, smooth, filament yarns (Table 1 and Fig. 5*e*). The liquid soil appears to spread unhampered along the plain-filament surfaces, except in the fabric crimps caused by yarn interlacings. Brown et al. [8] observed a similar phenomenon with filament polyester and nylon fabrics in comparison to spun-yarn polyester and polyester/cotton fabrics. The spun-yarn polyester fabric, F4, has similar construction characteristics as F1, although F4 is made of coarser yarns in a lower-count construction and is heavier (Table 1); accordingly, F4 shows a smaller area of DDT distribution (Fig. 5*d* versus 5*a*).

Between the two cotton/polyester blended-yarn fabrics, F7 (Fig. 5*g*) exhibits a larger area of dispersion than F8 (Fig. 5*h*) due to the finer yarn size and lighter fabric weight (Table 1). The mercerization finish of F8 does not seem to help the DDT solution spread, but rather retains it due to the same reasons as in F3. Fabric F8 shows a similar dispersion area (Fig. 5*h*) as F3 (Fig. 5*c*), the other mercerized fabric, due to similar construction characteristics, except that F3 is heavier (Table 1). Likewise, F7 has almost the same construction characteristics as F1 and thus has a similar size of the dispersion area as F1 (Fig. 5*g* versus 5*a*).

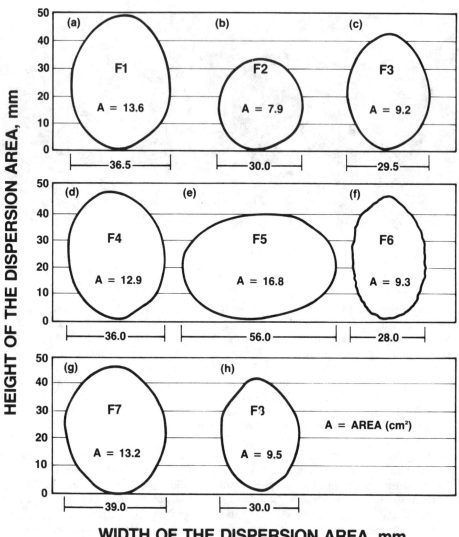

FIG. 5—*Size and shape of the dispersion area of the radioactive DDT.*

(a) Autoradiogram images of F1.
(b) Autoradiogram images of F2.
(c) Autoradiogram images of F3.
(d) Autoradiogram images of F4.
(e) Autoradiogram images of F5.
(f) Autoradiogram images of F6.
(g) Autoradiogram images of F7.
(h) Autoradiogram images of F8.

Shape of the Dispersion Area—Figure 5 illustrates that all fabrics show oval dispersion areas with a larger warpwise dimension than fillingwise except F5. Table 2 shows that F5 is the only fabric that has a height-to-width ratio below one (0.68) and that F5 has the highest ratio of warp yarn size to filling yarn size: the warps are 1.75 times finer than the fillings.

The fabric balance ratio of the number of warps to that of fillings, however, is 1.34 (Table 2). This combination of fabric and yarn geometries of F5 indicates smaller interyarn capillary radii between the flatter, ribbon-shaped, smooth-filament fillings. In addition, F5 is a ribbed taffeta and thus has more warpwise crimp than fillingwise crimp. The DDT solution, consequently, spreads more along the filling direction.

In all other fabrics, the warp and filling yarn sizes are similar or warps are slightly bigger than fillings (Table 2). This, in addition to the larger number of warps than fillings, indicates smaller warpwise interyarn capillary radii, and thus a greater spread in the warp direction. In addition, warp yarns usually have less crimp than fillings. Generally, the larger the balance ratio is, the more asymmetrical dispersion the fabrics show along the warp and filling directions, as expected by theory that the moisture-transport mechanism is largely through the capillaries formed between parallel fibers and yarns in warp and filling directions [6]. Large capillaries increase liquid-holding capacity of fabrics rather than transport the liquid to other surfaces [7].

Concentration of DDT in the Dispersion Area

Figure 6 illustrates generally symmetrical distribution of concentration of the ^{14}C-labeled DDT along the warp and filling directions in the 0.64 by 0.64-cm squares (Fig. 1) for F3, F4, F5, and F7. In both the warp and filling directions, there is a small drop in concentration at the first 0.64-cm distance from the center drop point, except in the warp direction for F5 (Fig. 6c). After this point, the concentration rises, culminating at the edges of the dispersion area.

Figure 6 also illustrates that DDT concentration is higher in the filling direction than in the warp except for F5, which shows higher warpwise concentration. This is exactly the opposite to the shape of dispersion (Fig. 5): more warpwise spread than fillingwise except, again, in F5. This indicates that concentration of DDT is inversely related to the shape of spread.

Fabric F7 shows nearly the same concentration along the warp and filling directions (Fig. 6d), suggesting the same mechanism of spread along the two directions. Observation of the flow of the DDT solution at the time of contamination suggests that speed of the dispersion

TABLE 2—*Height-to-width ratio of the dispersion area and related construction characteristics of test fabrics.*

Fabric Symbol	Yarn No. Ratio,[a] (W-Nc)/(F-Nc)	Balance Ratio, W/F	Height-to-Width Ratio of the Dispersion Area
	100% COTTON		
F1	0.81	1.12	1.29
F2	1.14	1.38	1.10
F3	1.07	2.00	1.36
	100% POLYESTER		
F4	0.61	1.27	1.31
F5	1.75	1.34	0.68
F6	1.00	1.37	1.57
	50/50% COTTON/POLYESTER		
F7	0.97	1.53	1.18
F8	0.95	2.33	1.40

[a] W-Nc = warp yarn number in cotton count. F-Nc = filling yarn number in cotton count.

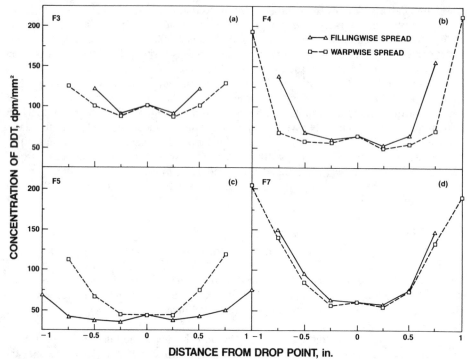

FIG. 6—*Concentration of the radioactive DDT in the dispersion area.*

influences DDT deposit concentration. At the initial contamination, the speed of movement is higher and thus the solution is more transported than retained within the depth of the fabric. When the spread reaches about the 0.64-cm distance from the center drop point, the speed begins to fall and the soil consequently tends to deposit or be retained. At the outermost boundary of the stain area, frictional resistance and evaporation of solvent halt dispersion, leaving the maximum concentration.

Conclusions

The specific location of the deposited DDT crystals in a fabric structure, the size and shape of the dispersion area of the radioactive DDT, and the concentration of the radioactive DDT in specific locations of the fabric provide information on the relationships between the three soil-distribution mechanisms and the construction and other characteristics of the test fabrics. Soil-removal mechanisms may be related to similar factors, although there is the need for further systematic research on the relationship between the two.

An understanding of the triangular relationships among fabric characteristics, soil-distribution mechanisms, and soil-removal mechanisms seems essential as a basis for designing fabrics that provide maximum protection to individuals against exposure to toxic chemicals.

Acknowledgments

The authors express their appreciation to the Monsanto Company for providing funds for this research. This paper was published as Journal Paper No. 405 of the Family and Consumer

Sciences Research Institute, College of Family and Consumer Sciences, Iowa State University, Ames, IA.

References

[1] Hollies, N. R. S., Krejci, V., and Smith, I. T., *Textile Research Journal,* Vol. 52, No. 6, June 1982, pp. 370–376.
[2] Laughlin, J. M., Easley, C. B., Gold, R. E., and Hill, R. M., "Fabric Parameters and Pesticide Characteristics That Impact on Dermal Exposure of Applicators," *Performance of Protective Clothing, ASTM STP 900,* R. L. Barker and G. C. Coletta, Eds., American Society for Testing and Materials, Philadelphia, 1986, pp. 136–150.
[3] Wolfe, H. R., Durham, W. F., and Armstrong, J. F., *Archives of Environmental Health,* Vol. 14, April 1967, pp. 622–633.
[4] Stone, J. F., Koehler, K. J., Kim, C. J., and Kadolph, S. J., *Journal of Environmental Health,* Vol. 48, No. 5, March/April 1986, pp. 259–264.
[5] Smith, S. and Sherman, P. O., *Textile Research Journal,* Vol. 39, No. 5, May 1969, pp. 441–448.
[6] Miller, B., Coe, A. B., and Ramachandran, P. N., *Textile Research Journal,* Vol. 37, No. 11, Nov. 1967, pp. 919–924.
[7] Raheel, M. and Gitz, E. C., *Archives of Environmental Contamination and Toxicology,* Vol. 14, No. 3, May 1985, pp. 273–279.
[8] Brown, C. B., Thompson, S. H., and Stewart, G., *Textile Research Journal,* Vol. 38, No. 7, July 1968, pp. 735–743.
[9] Finley, E. L. and Rogillio, J. R. B., *Bulletin of Environmental Contamination and Toxicology,* Vol. 4, No. 6, 1969, pp. 343–351.
[10] Lillie, T. H., Livingston, J. M., and Hamilton, M. A., *Bulletin of Environmental Contamination and Toxicology,* Vol. 27, No. 5, Nov. 1981, pp. 716–723.
[11] Fort, T. Jr., Billica, H. R., and Grindstaff, T. H., *Textile Research Journal,* Vol. 36, No. 2, Feb. 1966, pp. 99–112.
[12] Easter, E., *Textile Chemist and Colorist,* Vol. 15, No. 3, March 1983, pp. 29–33.
[13] Berch, J., Peper, H., and Drake, G. L., Jr., *Textile Research Journal,* Vol. 35, No. 3, March 1965, pp. 252–260.
[14] Obendorf, S. K. and Klemash, N. A., *Textile Research Journal,* Vol. 52, No. 7, July 1982, pp. 434–442.
[15] Bowers, C. A. and Chantrey, G., *Textile Research Journal,* Vol. 39, No. 1, Jan. 1969, pp. 1–11.
[16] Venkatesh, G. M., Dweltz, N. E., Madan, G. L., and Alurkar, R. H., *Textile Research Journal,* Vol. 44, No. 5, May 1974, pp. 352–362.
[17] Obendorf, S. K., Namaste, Y. M. N., and Durnam, D. J., *Textile Research Journal,* Vol. 53, No. 6, June 1983, pp. 375–383.
[18] Kim, C. J., Stone, J. F., and Sizer, C. E., *Bulletin of Environmental Contamination and Toxicology,* Vol. 29, No. 1, July 1982, pp. 95–100.
[19] Kim, C. J., Stone, J. F., Coats, J. R., and Kadolph, S. J., *Bulletin of Environmental Contamination and Toxicology,* Vol. 36, No. 2, Feb. 1986, pp. 234–241.
[20] Fort, T., Billica, H. R., and Sloan, C. K., *Textile Research Journal,* Vol. 36, No. 1, Jan. 1966, pp. 7–12.
[21] Breen, N. E., Durnam, D. J., and Obendorf, S. K., *Textile Research Journal,* Vol. 54, No. 3, March 1984, pp. 198–204.
[22] Solbrig, C. M. and Obendorf, S. K., *Textile Research Journal,* Vol. 55, No. 9, Sept. 1985, pp. 540–546.

Maria Thompson Anastasakis,[1] *Karen K. Leonas,*[2] *Carol Dimit,*[1]
John Brothers,[1] *and Jacquelyn Orlando DeJonge*[1]

Effect of Temperature and Humidity on Laboratory Pesticide Penetration Studies

REFERENCE: Anastasakis, M. T., Leonas, K. K., Dimit, C., Brothers, J., and DeJonge, J. O., **"Effect of Temperature and Humidity on Laboratory Pesticide Penetration Studies,"** *Performance of Protective Clothing: Second Symposium, ASTM STP 989*, S. Z. Mansdorf, R. Sager, and A. P. Nielsen, Eds., American Society for Testing and Materials, Philadelphia, 1988, pp. 692–696.

ABSTRACT: The purpose of this study was to evaluate the effect of three conditions of temperature and humidity on pesticide penetration through the fabric. Two fabrics were used, a 100% cotton and a 50/50 cotton/polyester. The three conditions used were 29°C (85°F) with 75% relative humidity (RH), 22°C (72°F) with 65% RH, and 18°C (65°F) with 55% RH. The application of Dicofol pesticide was completed using an enclosed spray chamber that simulated actual field conditions encountered during air blast spraying. Fabric type had a greater influence on pesticide penetration than atmospheric conditions. Consistent standard laboratory conditions of 21.1°C (70°F) with 65% RH are recommended for use in future laboratory penetration tests.

KEY WORDS: pesticide, penetration, humidity, temperature, 4,4-alpha-trichloromethylbenzyhydrol, protective clothing

The effect of temperature and humidity on the performance of fabrics has been documented in textile research and has been shown to affect tensile strength, elastic recovery, electrical resistance, and rigidity [1]. A study evaluating soiling at different levels of relative humidity (RH) revealed that differences in soiling occurs by a wicking mechanism. The sorption of cotton plotted against relative humidity creates a Sigmoid Curve [2]. The mechanism of this interaction is hydrogen bonding. At low humidity, water molecules are present in a compact localized layer and will provide high resistance to wicking. Increases in humidity will induce a dipolar field in this layer of water. Thus, another layer is attracted and continues into a mobile three-dimensional layer, and the resistance to wicking will be lowered. Soiling increased with increases in relative humidity up to 65% and then decreased with further increases in humidity [2].

Laboratory methods for evaluating pesticide penetration[3] have been used as a pre-screening for field study testing as documented by Leonas [3] and by Staiff et al. [4]. The use of both drop and spray system methods are usually conducted in an indoor laboratory setting.

---

[1] Graduate research assistant, research assistant, senior research associate, and department head, respectively, Department of Textiles, Merchandising and Design, College of Human Ecology, The University of Tennessee, Knoxville, TN 37996-1900.

[2] Assistant professor, Department of Textiles, Apparel, and Interior Design, University of Illinois, Urbana, IL 61801.

[3] The term, penetration, is used throughout this paper to describe the concepts of both permeation and penetration.

The conditions commonly accepted for standard textile testing are 21°C (70°F) and 65% RH. A few studies have documented the use of these standard conditions, although in the majority of cases there is no documentation of the conditions during the experiment. The primary objective of this study was to evaluate the effect of temperature and humidity on the penetration of pesticide through fabrics. A secondary objective was to determine the conditions for laboratory pesticide penetration testing for consistent and valid results.

Actual pesticide application is more likely to occur in the hot and, in some cases, humid summer months. Consequently, this study chose three conditions that simulated summer temperatures and humidities found in different geographical areas. The conditions chosen were 29°C (85°F) with 75% RH, 22°C (72°F) with 65% RH, and 18°C (65°F) with 55% RH. The fabrics selected simulated the type of clothing worn by pesticide applicators and field workers during the summer months. The fabrics were low to medium weight, as would be expected for summer shirts and pants.

Experimental Work

The two fabrics chosen were a plain weave 50/50% cotton/polyester and a plain weave 100% cotton. The polyester/cotton weighed 105.67 g/m^2 (3.12 oz/yd^2) and was 0.2794 mm thick (0.011 in.). The cotton weighed 250.32 g/m^2 (7.38 oz/yd^2) and was 0.4572 mm thick (0.018 in.). The thickness, weight, yarn count, air permeability, and spray rating are shown in Table 1.

The fabrics were prewashed and dried according to an adaptation of the AATCC Test Method 135-78. They were then randomly cut in accordance with ASTM Test Methods for Breaking Load and Elongation of Textile Fabrics (D 1682-64) into 15 by 15 cm squares (6 by 6 in.). Each test specimen was composed of the test fabric, a collector layer, and two foil layers. The four layers were assembled with the test layer on top, followed by the collector layer and two foil layers. The test layer was the fabric being evaluated; the collector layer was a 50/50% cotton/polyester blend tee-shirt knit fabric; and the foil layer served to collect any pesticide penetrating through the collector layer. Five replications of each fabric were sprayed with the pesticide under the three conditions. The collector layer was analyzed for all specimens to determine the amount of pesticide penetrating through the test fabric. To validate consistency in the total spray quantity, 20% of the test fabric and foil were randomly analyzed for the presence of pesticide. The total quantity of pesticide from the test, collector, and foil layers should be equivalent for all specimens.

Exposure

The pesticide used in this study was Dicofol (4,4-alpha-trichloromethybenzhydrol) a chlorinated hydrocarbon with an active ingredient of 42%. An emulsifiable concentrate formulation was diluted in water to make a 0.12% solution, the same concentration used in field spraying. An enclosed chamber designed to simulate field conditions of air blast spraying was used [5]. After spraying, the specimens were removed from the chamber and allowed to dry for 1 h. A 7.5 by 2.5 cm rectangle (3 by 1 in.) was cut from the middle of each layer and placed in individual test tubes, in preparation for the extraction.

The three conditions tested on three separate dates were 29°C (85°F) with 75% RH, 22°C (72°F) with 65% RH, and 18°C (65°F) with 55% RH. They were maintained at a constant ±2° and ±2%, through the use of a humidifier, dehumidifier, heater, and air conditioner. The temperature and humidity were monitored every half hour using a hygrometer with wet and dry bulb thermometers. They were also recorded by a Honeywell chart recorder.

TABLE 1—Selected fabric characteristics.

Fabric	Thickness[a]		Weight[b]		Air Permeability[c]		Spray[d] Rating	Yarn Count[e]	
	in.	(mm)	g/m²	(oz/yd²)	m³/s/m²	(ft³/min/ft²)		Warp	Fill
50/50 polyester/cotton	0.011	(0.2794)	105.67	3.12	1.113028	219.10	0	88	55
100% cotton	0.018	(0.4572)	250.32	7.38	0.033528	6.60	0	112	55

NOTE—The following test methods were used:

[a] ASTM Method for Measuring Thickness of Textile Materials (D 1777-64).
[b] ASTM Test Methods for Weight (Mass) per Unit Area of Woven Fabric (D 3776-79).
[c] ASTM Test Method for Air Permeability of Textile Fabrics (D 737-75).
[d] AATCC 22-1980.
[e] ASTM Test Method for Fabric Count of Woven Fabric (D 3775-79).

Analysis

Each 7.5 by 2.5 cm rectangle (3 by 1 in.) was analyzed with gas chromatography (GC). These GC techniques calculate the amount of pesticide residue found on each respective layer. The pesticide was extracted from the fabric layers using the extraction method outlined by Easter et al. [6]. Two 30-min extractions were done with 50 mL of Hexane for a total of 100 mL of Hexane for each specific 7.5 by 2.5 cm (3 by 1 in.) layer. A 5 mL aliquot of the extraction fluid was poured in a small vial and stored in a freezer until it was analyzed. The GC used was a Varian 3700. The glass column was 1.8 m by 2 mm with 3% SP-2100 on 100/120 Supelcoport packing. The operating conditions used were 52 mL/min of nitrogen gas with an electron capture detector (ECD). The column temperature was at 195°C (383°F), the injector at 210°C (410°F), and the detector at 300°C (572°F).

Results and Discussion

The independent variables analyzed in the investigation were temperature/humidity conditions and fabric. Pesticide penetration was the dependent variable.

Statistical Analysis

A two-way analysis of variance was used to evaluate the data. No significant differences were found among the three temperature/humidity combinations. The analysis also showed that fabric type had a significant effect on pesticide penetration (Table 2). Penetration was consistently greater for the lighter weight cotton/polyester blend than for the heavier weight cotton fabric.

Fabrics

The 100% cotton fabric, the heavier fabric of the two, consistently showed less penetration than the cotton/polyester blend. Besides weight, the difference in penetration could also be due to the denseness of yarns per inch; the lower warp yarn count may have contributed to the blended fabric's inferior performance.

In addition to weight and yarn count, fiber content and air permeability of the fabrics may have influenced penetration. The results indicated that 100% cotton provided better resistance to penetration than the 50/50 cotton/polyester blend. These results are in agreement with those of previous studies [3,7]. Also, the fabric with the higher air permeability, the 50/50 cotton/polyester, allowed greater penetration. The results relative to fabric air permeability and pesticide penetration concur with those reported by Leonas [3].

TABLE 2—*ANOVA of the effects of temperature and humidity on pesticide penetration through cotton and cotton/polyester fabrics.*

Source	Degrees of Freedom	Sum of Squares	Mean Square	F Value
Temperature/humidity	2	0.6600	0.6600	2.80
Fabric	1	156.3497	156.3497	663.33[a]
Temperature/humidity × fabric	8	1.2904	0.1613	0.68

[a] Significant at 0.0001.

TABLE 3—*Mean scores of pesticide penetration through cotton and cotton/polyester fabrics at three temperature/humidity conditions.*[a]

Temperature/Humidity	100% Cotton	50/50 Cotton/Polyester
29°C (85°F) at 75% RH	0.00080	0.05186
22°C (72°F) at 65% RH	0.00052	0.07636
18°C (65°F) at 55% RH	0.00082	0.08284

[a] Measured in ng/μL.

Temperature and Humidity Conditions

Although there were no statistical differences in the effect of temperature/humidity conditions on penetration, there was a slight variation at different conditions. This was more apparent for the cotton/polyester fabric. No apparent trend was evident in mean scores of penetration through the 100% cotton fabric for the three conditions (Table 3).

Conclusion

The results of the study showed that fabric type had a greater influence on pesticide penetration than atmospheric conditions. Both fabrics in the study performed consistently regardless of temperature/humidity combinations. Differences in fabric weight, yarn count, fiber content, and air permeability could account for the significant difference in fabric performance.

Though minor differences may occur in pesticide penetration, comparisons of fabric performance can be effectively made if conditions are held constant. Therefore, standard laboratory conditions of 70°F and 65% RH are suggested for pesticide penetration laboratory work. The use of standard laboratory conditions will ensure consistency in procedure and comparability of results with other areas of textile research.

Acknowledgment

Although the research described in this article has been funded wholly or in part by the U.S. Environmental Protection Agency under Assistance Agreement (CR 812846-01-0) to Dr. J. O. DeJonge of The University of Tennessee, it has not been subjected to the agency's peer and administrative review and therefore may not necessarily reflect the views of the agency. No official endorsement should be inferred.

References

[1] Lyle, D. S., *Performance of Textiles,* Wiley, New York, 1977, p. 23.
[2] Kumar, R., Dave, A. M., and Srivastava, H. C., *Textile Research Journal,* Vol. 54, No. 9, Sept. 1984, pp. 585–589.
[3] Leonas, K. K., "Apparel Fabrics as Barriers to Pesticide Penetration," doctoral thesis, The University of Tennessee, Knoxville, 1985.
[4] Staiff, D. C., Davis, J. E., and Stevens, E. R., *Archives of Environmental Contamination and Toxicology,* Vol. 11, 1982, pp. 227–234.
[5] Leonas, K. K. and DeJonge, J. O. in *Performance of Protective Clothing, ASTM STP 900,* R. L. Barker and G. C. Coletta, Eds., American Society for Testing and Materials, Philadelphia, 1986, 177–186.
[6] Easter, E. P., Leonas, K. K., and DeJonge, J. O., *Bulletin of Environmental Contamination and Toxicology,* Vol. 31, Dec. 1983, pp. 738–744.
[7] Freed, V. H., Davies, J. E., Peters, L. J., and Parveen, F., *Residue Reviews,* Vol. 75, 1980, pp. 159–167.

Karen Pedersen Ringenberg,[1] *Joan M. Laughlin,*[2]
and Roger E. Gold[3]

Chlorpyrifos Residue Removal from Protective Apparel Through Solvent-Based Refurbishment Procedures

REFERENCE: Ringenberg, K. P., Laughlin, J. M., and Gold, R. E., **"Chlorpyrifos Residue Removal from Protective Apparel Through Solvent-Based Refurbishment Procedures,"** *Performance of Protective Clothing: Second Symposium, ASTM STP 989,* S. Z. Mansdorf, R. Sager, and A. P. Nielsen, Eds., American Society for Testing and Materials, Philadelphia, 1988, pp. 697–704.

ABSTRACT: Dry-cleaning solvent (tetrachloroethylene) and the dry-cleaning procedures were studied in the refurbishment of chlorpyrifos-contaminated clothing. Also evaluated were solvent-based pretreatments followed by laundering. Fabrics were a 100% cotton and a 50% cotton/50% polyester each either with a renewable, soil-repellent finish, or unfinished. Field strength chlorpyrifos (0.5% active ingredient), prepared from emulsifiable concentrate formulation, was pipetted onto specimens which were then dry-cleaned or laundered following solvent pretreatment. Transfer of pesticide from contaminated to uncontaminated fabric specimens was also studied.

Dry-cleaning treatments were most effective, with less than 1% chlorpyrifos residue remaining on fabric specimens. Mechanical agitation was an aid in chemical removal. Pesticide transfer to the uncontaminated fabrics was minimal, but did occur. Pretreatments, followed by laundering, were less effective than dry-cleaning procedures.

KEY WORDS: pesticide, pesticide soiling, pesticide residue, laundering, dry cleaning, solvent pretreatment, protective clothing, soil-repellent finish

Pesticides are chemicals intentionally applied in the environment by man to improve quality of life. Direct exposure to these chemicals may occur during handling, mixing, and application of pesticides [1]. Clothing worn during direct-exposure situations may limit dermal contact. Although several researchers have studied refurbishment of protective apparel through laundering [2–8], no work in the literature has addressed the effectiveness of an organic solvent in refurbishment of protective apparel. Fluorocarbon water-repellent finishes reduce the levels of initial soil deposition [8,9]

Recommendations for laundering and refurbishment vary with the chemical class of pesticide under evaluation [10]. Easley et al. [11] found the emulsifiable concentrate formulation of methyl parathion to be more difficult to remove than either a wettable powder or an encapsulated formulation. The emulsifiable concentrate formulation was the most difficult to remove since oil-based soil became entrapped between the fibers and within the cotton lumen [12].

[1] Elwood, NE 68937.

[2] Professor and chairman, Textiles, Clothing, and Design, 234 Home Economics Building, University of Nebraska-Lincoln, Lincoln, NE 68583-0802.

[3] Head, Department of Entomology, University of Nebraska-Lincoln, Lincoln, NE 68583-0818.

Detergent type and water temperature have also been investigated as parts of the laundering procedure. A nonionic heavy-duty liquid detergent and "hot" water temperature (60°C) were recommended for maximum pesticide residue removal [3]. Easley et al. [4] and Laughlin et al. [13] examined the transfer of pesticides from contaminated to clean fabrics during the laundering process. Although the amounts of pesticides transferred were small, recommendations were made for dedicated laundering and cleanup of laundering equipment.

Pesticide soiling has been paralleled to oily soiling of fibers. Easter [14] labeled some pesticide soils "oil soils" and indicated that the difficulty of removing these oily soils involves not only the geometric entrapment of soils on the surface of the fiber but also fiber-soil interactions. Easter [14] theorized that components of pesticide diffuse below the fiber surface and become molecularly entangled in the body of the filaments or possibly the polymer. However, Solbrig and Obendorf [15] demonstrated that the location of malathion and methyl parathion were on the surface of polyester fibers, in the lumen, secondary wall, and on the surface and crenulations of cotton fiber. They found that one laundering greatly reduced the amount of malathion and methyl parathion on the surfaces of the cotton fibers, but laundering did not reduce the concentration of pesticide in the cotton lumen [12].

Most research to date on refurbishment of protective apparel has examined laundering procedures to determine the most effective practices. Keaschall et al. [10] examined eleven pesticides within three classes: organochlorines, carbamates, and organophosphates. The after-laundering residues of chlorpyrifos were the greatest (35.8%) of the eleven pesticides studied. With specific pesticides such as chlorpyrifos, alternative refurbishment practices must be evaluated, including dry cleaning.

The synthetic solvents used in dry cleaning are chlorinated hydrocarbons. Unlike water, dry-cleaning solvents such as perchloroethylene ($Cl_2C = CCl_2$) do not cause fibers to swell, thus aiding removal of oil from within the fiber structure [16]. Some organic solvents can "wet" and penetrate faster and to a greater extent than water, and the low contact angle, θ, of these halogenated solvents allows the liquid to penetrate the intermicellar spaces within the fiber more thoroughly than water [17]. The use of synthetic solvents enhances oil-soil removal and may increase pesticide-soil removal.

The solvents used in a standard dry-cleaning process are reused. Dry cleaners control impurities in these solvents by one of four methods or some combination of these: continuous solvent filtration to remove insoluble impurities; dilution by adding new solvent; absorption by filtering through activated charcoal; or distillation by converting a liquid to a gas and then condensing to a liquid, leaving impurities behind in the process [18].

The International Fabricare Institute advised its members to avoid dry cleaning garments containing pesticides and chlorinated contaminants because the result can be highly toxic residues and contaminated filter wastes [19]. Dry cleaning at home is not recommended (except localized spot removal) because it requires special precaution due to toxicity and flammability of the solvents.

The purposes of this study were to evaluate the refurbishment of chlorpyrifos-contaminated fabrics, including refurbishment by (1) dry cleaning or solvent soak or both and (2) laundering following pretreatment with solvent-based products. Transfer of pesticide from contaminated specimens to "clean" specimens during the standard dry-cleaning procedure was also evaluated.

Methods and Procedures

Experimental Design

The refurbishment procedures were designed to simulate the commercial dry-cleaning process (Phase I) and laundering following solvent pretreatment (Phase II). Two fabrics, two finishes, and one pesticide formulation were examined.

Fabrics

The fabrics for this study were 100% cotton (C) and 50% cotton/50% polyester (CP). Both undyed fabrics were a 474.7 gm/m^2 (14 oz/yd^2) (trouser weight), 3/1 left-hand twill. The fabrics were stripped of all manufacturing-applied softeners and warp sizing by laundering them as stated in the American Association of Textile Chemists and Colorists (AATCC) Test Method for Dimensional Change in Automatic Home Laundering of Woven and Knit Fabrics (135-1978, R-85). The fabrics were laundered five times with 90 g AATCC Detergent 124 and followed by one laundering with a non-precipitating phosphate water softener.

Fabric Preparation

The fabrics included two finishes, unfinished (UN) and soil-repellent finish[4] (SR). Matched sets of specimens were prepared. One half were treated with a renewable fluorocarbon finish.[5] To control the amount of finish applied to each yardage, we measured the percentage dry weight gain of the fabric as well as the net amount used from each aerosol can. An average of 0.78% add-on of fluorocarbon soil-repellent finish was applied to each specimen.

Specimens 8 by 8 cm^2 (64 cm^2) were cut from each yardage of each finish and were randomly assigned to treatments in the research design. All specimens were conditioned at 21 ± 1°C and 65 ± 2% relative humidity (RH) prior to pesticide contamination.

Spiking the Fabric Specimens

A field-strength solution of 0.5% active ingredient (AI) chlorpyrifos (0,0-diethyl 0,3 5,6-trichloro-2-pyridyl-phosphorothioate) was prepared from an emulsifiable concentrate formulation.[6]

A Microlab P programmable pipette was used to deliver 0.2 mL chlorpyrifos onto the fabric specimen surface. After spiking, all specimens were allowed to air dry. Control specimens of each fabric type and finish were included in the experimental design.

Phase I: Dry Cleaning

The research design for dry-cleaning with tetrachloroethylene ($Cl_2C = CCl_2$) [16] solvent included three treatments: Tx$_1$ solvent tumble, solvent rinse, and air dry; Tx$_2$ solvent presoak, solvent tumble, and solvent rinse, and air dry; and Tx$_3$ solvent soak, and air dry. The dry cleaning was done in an AATCC Atlas Launder-Ometer (Model LEF). Twenty stainless-steel balls were added to each canister to provide agitation during the solvent tumble and solvent rinse cycles.

For the first treatment (Tx$_1$), specimens were solvent tumbled in 150 mL tetrachloroethylene for 10 min at 40°C. The solvent was decanted and 150 mL fresh tetrachloroethylene (40°C) added for the 3-min rinse. Following rinsing, specimens were air-dried under a fume hood. An all-cotton, uncontaminated specimen was placed in each canister during all replications to assess possible pesticide transfer.

In the second treatment (Tx$_2$), the specimens were placed in a 250-mL beaker and pre-soaked in 150 mL tetrachloroethylene for 10 min at 20°C, after which time the solvent was decanted and 150 mL fresh tetrachloroethylene was placed in each canister, tumbled, solvent-rinsed for 10 min at 40°C, and dried following the same procedures as for Tx$_1$.

---

[4] Use of brand name does not imply product endorsement.
[5] Scotchgard Fabric Protector, 3-M Company.
[6] Dursban 4E, Dow Chemical Corporation, EPA Registration No. 464-360.

Treatment three (Tx_3) consisted of a solvent soak for 10 min (no agitation) in 150 mL tetrachloroethylene at 20°C. The solvent was decanted and specimens were dried under the fume hood.

Phase II: Laundering Following Solvent Pretreatment

The research design for the laundering process included two treatments: tetrachloroethylene and a commercially available pre-spotter.[7] Treatment four (Tx_4) included prespotting the specimens with 1 mL tetrachloroethylene placed on the spiked area. After 10 min, the specimens were laundered in an AATCC Atlas Launder-Ometer (Model LEF) as per AATCC Test Method Colorfastness to Washing, Domestic, and Laundering, Commercial: Accelerated (61-1985). Twenty-five steel balls provided agitation during the laundry cycles. A nonionic, heavy-duty liquid detergent[8] at 0.13% in distilled water was used for the laundry procedure. Fabric specimens were washed for 12 min at 49°C in 150 mL detergent solution. Following the wash cycles, the wash liquor was decanted and 150 mL distilled rinse water were added for a 5-min rinse, followed by decanting and second rinse of 150 mL distilled water for 3 min. The fabric specimens were removed from the canisters and allowed to air dry.

Treatment 5 (Tx_5) included prespotting with a pre-spotter. Following this pretreatment, the specimens were laundered using the same procedures as for Tx_4.

Extraction and Gas Chromatographic Analysis

To determine the amount of pesticide remaining on the fabric following treatment, we added 200 mL hexane to a glass bottle containing a specimen. The bottles were placed on Precision Shaking Water Bath (Model 50) at 120 cpm for 30 min. Extracts were either concentrated or diluted to facilitate gas chromatographic analysis. Concentration of samples was done with a Brinkmann Buchi Rotavapor rotary vacuum evaporator and a Meyer-N-Evap Analytical Evaporator, water bath nitrogen stream.

A Varian Vista 6000 Gas Chromatograph with automatic injection and a thermionic specific detector were used. A 2-m glass column was used with 4.0 mm inside diameter. Packing was 6% OV-210 plus 4% OV-101 on 80/100 mesh Chromosorb.[9] Nitrogen was used as a carrier gas at a flow rate of 30 mL/min. Hydrogen gas and compressed oxygen were passed over the detector base at 4.5 mL/min and 175 mL/min. The injector, column, and detector temperature were 220, 200, and 300°C, respectively. The Varian 4720 Integrator was programmed to compute retention times and area counts.

Each sample was doubly injected on the gas chromatograph and a mean computed for the sample. The apparatus was recalibrated after every six samples or 12 injections with an external reference standard of known concentration (2 ppm). The reference standard was obtained from the Environmental Protection Agency, Health Effects Research Laboratory, with a purity of 99.93%.

Statistical Analysis

The percentage of residue remaining after treatment was calculated, and since the percentage of residue remaining ranged from less than 1 to 15%, thus were not normally

---

[7] Carbona, Carbona Products Company; contains inhibited 1,1,1-trichloroethane, perchlorethlane, and petroleum hydrocarbons.
[8] Dynamo is a registered trademark of Colgate-Palmolive Company.
[9] Registered trademark of Manville Company.

distributed, the percentages were arc sine converted prior to analysis (general linear model) with means separated by a Least Significant Means Test. The probability level as a criterion for significance was set at 0.05. All work was replicated a minimum of three times.

Findings and Conclusions

Initial Contamination

It was necessary to determine the initial level of contamination of the fabric specimens for comparison to after-treatment levels. Three control fabric specimens for each replication of each fabric and finish were used as an indication of initial contamination. Chlorpyrifos extracted from the control fabrics were: UN cotton, 15.87 $\mu g/cm^2$; SR cotton, 15.17 $\mu g/cm^2$; UN cotton/polyester, 16.56 $\mu g/cm^2$; and SR cotton/polyester, 16.38 $\mu g/cm^2$. There were no significant differences attributable to fabric or finish ($F = 0.877$, $df = 3.8$).

It was anticipated that the SR finish would decrease the level of initial contamination as reported by several other researchers [8,10,18]; however, this was not found for this study. The average percent add-on of SR was similar to Goodman [20] and Keaschall et al. [10], but the fabric was much heavier with a more complex surface structure. Therefore, a heavier application of the fluorocarbon soil-repellent topical finish appears to be necessary for heavier weight fabrics and fabrics with a more complex surface geometry.

Phase I: Dry Cleaning

The amount of chlorpyrifos residue remaining after dry-cleaning treatment ranged from 0.08 to 2.51 $\mu g/cm^2$ (Table 1). There were no significant differences due to fiber content of the specimen or finish. Therefore, only differences due to treatment were examined ($F = 79.40$, $df = 1$, $p \leq 0.05$) (Table 2). The agitation and solvent volume (Tx_1, Tx_2) were the factors most responsible for the significant differences in residue removal. For Tx_1 and Tx_2, there was less than 1% chlorpyrifos residue remaining on the fabric. Soaking in the dry-cleaning solvent (Tx_3) without agitation was not as effective as the dry-cleaning procedure that included solvent plus agitation (Tx_1 and Tx_2). These results were a considerable improvement over those of Keaschall et al. [10], who found chlorpyrifos residue as high as 35.8% after laundering.

During Tx_1, chlorpyrifos residue transferred from the contaminated specimens to uncontaminated fabric specimens when the specimens were refurbished simultaneously. Residue transferred in this study were minute, ranging from 0.002 to 0.006 $\mu g/cm^2$ (Table 3). These numbers are very small and would present a minimal risk; however, they raise some ethical questions. Commercial applicators or commercial dry cleaners might invest in a dedicated dry-cleaning apparatus. In this situation, solvent would only be used to clean contaminated fabrics, and the solvent could be discarded or distilled after each cleaning cycle, reducing the risk of transfer.

Phase II: Laundering Following Solvent Pretreatment

The amounts of chlorpyrifos residue remaining after Tx_4 and Tx_5 were examined (Table 1). There were no significant differences in residue due to fiber content and finish. Across the fiber contents and fabric finishes, the average $\mu g/cm^2$ was 2.77 for Tx_4 and 4.01 for Tx_5 (Table 2). Based on these data, the dry-cleaning solvent (tetrachloroethylene) was significantly better as a pretreatment before laundering than was the commercially available

TABLE 1—Initial contamination levels and the amount and percent residue remaining of chlorpyrifos after refurbishment procedures.

Fabric and Finish	Initial Contamination, µg/cm²	Tx_1[a]		Tx_2[b]		Tx_3[c]		Tx_4[d]		Tx_5[e]	
		µg/cm²	%[f]	µg/cm²	%[f]	µg/cm²	%[f]	µg/cm²	%[f]	µg/cm²	%[f]
100% cotton (UN)	15.87	0.14	0.85	0.12	0.73	2.04	12.87	1.82	11.14	4.87	30.72
100% cotton (SR)	15.17	0.13	0.85	0.29	1.88	2.51	16.54	2.53	16.68	3.66	24.12
50% cotton/50% polyester (UN)	16.56	0.11	0.64	0.08	0.48	1.68	10.16	2.35	14.21	3.55	21.42
50% cotton/50% polyester (SR)	16.38	0.08	0.50	0.08	0.48	1.51	9.34	4.39	26.77	3.96	24.19

[a] Tx_1 = Solvent tumble, rinse, dry.
[b] Tx_2 = Solvent presoak, tumble, rinse, dry.
[c] Tx_3 = Solvent soak, dry.
[d] Tx_4 = Tetrachloroethylene pretreatment, launder.
[e] Tx_5 = Solvent based pretreatment, launder.
[f] Percent residue remaining after treatment.

TABLE 2—*Mean values and LS differences among the dry-cleaning procedures based on amounts and percent residue remaining.*

Treatment	$\mu g/cm^2$	$\%^a$
Tx$_1$ (dry-cleaning procedure)	0.12a[b]	0.71e
Tx$_2$ (presoak plus dry cleaning)	0.14a	0.89e
Tx$_3$ (solvent soak)	1.94b	12.20f
Tx$_4$ (solvent pretreatment, launder)	2.77c	17.20g
Tx$_5$ (commercially available		
pretreatment, launder)	4.01d	25.11h

[a] Percent residue remaining after treatment.
[b] Groups with same letters are not significantly different at $p \leq 0.05$.

product (Carbona) ($F = 6.75$, $df = 1$, $p \leq 0.05$). The percentage of residue remaining for Tx$_4$ was 17.20%, while the percentage for Tx$_5$ was 25.11% of the initial contamination. These results are also an improvement over those of Keaschall et al. [*10*], who observed that a pre-wash surfactant-based product plus laundering produced less chlorpyrifos residues than laundering alone, but residue were still sizeable (27.13%). This study, using a solvent based pre-spotter, produced lower residue, an improvement over the Keaschall work, which used a surfactant based pre-wash product.

When comparisons were made between the dry-cleaning processes (Tx$_1$, Tx$_2$, and Tx$_3$) and laundering (Tx$_4$ and Tx$_5$), it was concluded that the dry-cleaning was more effective in reducing chlorpyrifos residue than either of the laundering procedures; however, the use of commercial dry-cleaning should be discouraged due to possibility of pesticide residue accumulating in the reused solvents and transferring to clean fabrics. It is unknown at this point whether pesticide residue may be separated from other filter wastes through the clarification or distillation of dry-cleaning solvents.

Acknowledgments

The authors would like to acknowledge the contribution of Dow Chemical, North Central Regional Project 170 and Nebraska 94-012: Limiting Dermal Exposure to Pesticides through Effective Cleaning Procedures and Selection of Clothing. This paper is published as Paper No. 8226, Nebraska Agricultural Research Divison.

TABLE 3—*Amount transferred to uncontaminated specimens (transfer specimens) through simultaneous refurbishment with contaminated specimens.*

	Contaminated Specimen		Transfer Specimen Amount Transferred, $\mu g/cm^2$
Fabric	Initial Contamination Level, $\mu g/cm^2$	After Refurbishment Residues, $\mu g/cm^2$	
Cotton (UN)	15.87	0.14	0.005
Cotton (SR)	15.17	0.13	0.006
50% cotton/50% polyester (UN)	16.56	0.11	0.002
50% cotton/50% polyester (SR)	16.38	0.08	0.004

References

[1] Matsumura, F. and Madhukar, B. V., "Exposure to Insecticides," *Pharmacology and Therapeutics,* Vol. 9, 1980, pp. 27–49.

[2] Easley, C. B., Laughlin, J. M., Gold, R. E., and Hill, R. M., "Laundry Factors Influencing Methyl Parathion Removal from Contaminated Denim Fabric," *Bulletin of Environmental Contamination and Toxicology,* Vol. 29, No. 4, Oct. 1982, pp. 416–468.

[3] Easley, C. B., Laughlin, J. M., Gold, R. E., and Schmidt, K., "Detergents and Water Temperature as Factors in Methyl Parathion Removal from Denim Fabrics," *Bulletin of Environmental Contaminatioin and Toxicology,* Vol. 28, No. 2, Feb. 1982, pp. 239–244.

[4] Easley, C. B., Laughlin, J. M., Gold, R. E., and Tupy, D., "Laundering Procedures for Removal of 2,4-Dichlorophenyl Oxyacetic Acid Ester and Amine Herbicide From Contaminated Fabrics," *Archives of Environmental Contamination and Toxicology,* Vol. 12, No. 1, Jan. 1983, pp. 71–76.

[5] Finley, E. L., Graves, J. B., Hewitt, F. C., Morris, H. F., Harmon, C. W., Iddings, F. A., Schilling, P. E., and Koonce, K. L., "Reduction of Methyl Parathion Residues on Clothing by Delayed Field Re-Entry and Laundering," *Bulletin of Environmental Contamination and Toxicology,* Vol. 22, 1979, pp. 590–602.

[6] Kim, C. J., Stone, J. F., Coats, J. R., and Kadolph, S. J., "Removal of Alachlor Residues from Contaminated Clothing Fabrics," *Bulletin of Environmental Contamination and Toxicology,* Vol. 36, 1986, pp. 234–241.

[7] Laughlin, J., Easley, C., and Gold, R. E., "Methyl Parathion Residues in Contaminated Fabrics After Laundering," *Dermal Exposure Related to Pesticide Use,* ACS Symposium Series No. 273, 1985, pp. 177–187.

[8] Laughlin, J. M., Easley, C. B., Gold, R. E., and Hill, R. E., "Fabric Parameters and Pesticide Characteristics that Impact on Dermal Exposure of Applicators," *Performance of Protective Clothing, ASTM STP 900,* R. L. Barker and G. C. Coletta, Eds., American Society for Testing and Materials, Philadelphia, 1986, pp. 136–150.

[9] Davies, J. E., Enos, H. F., Barquet, C., Morgade, L., Peters, J., and Danaukas, J. X., "Protective Clothing Studies in the Field—An Alternative to Reentry," *American Chemical Symposium,* 1982, pp. 169–182.

[10] Keaschall, J. L., Laughlin, J. M., and Gold, R. E., "Effect of Laundering Procedures and Functional Finishes on Removal of Insecticides Selected from Three Chemical Classes," *Performance of Protective Clothing, ASTM STP 900,* R. L. Barker and G. C. Coletta, Eds., American Society for Testing and Materials, Philadelphia, 1986, pp. 162–176.

[11] Easley, C. B., Laughlin, J. M., Gold, R. E., and Tupy, D., "Methyl Parathion Removal From Denim Fabrics by Selected Laundry Procedures," *Bulletin of Environmental Contamination and Toxicology,* Vol. 27, 1981, pp. 101–108.

[12] Obendorf, S. K. and Solbrig, C. M., "Distribution of Organophosphorus Pesticides Malathion and Methyl Parathion on Cotton/Polyester Fabrics After Laundering as Determined by Electron Microscopy," *Performance of Protective Clothing, ASTM STP 900,* R. L. Barker and G. C. Coletta, Eds., American Society for Testing and Materials, Philadelphia, 1986, pp. 187–204.

[13] Laughlin, J. M., Easley, C. B., Gold, R. E., and Tupy, D., "Methyl Parathion Transfer from Contaminated Fabrics to Subsequent Laundry and to Laundry Equipment," *Bulletin of Environmental Contamination and Toxicology,* Vol. 27, No. 4, Oct. 1981, pp. 518–523.

[14] Easter, E. P., "Removal of Pesticide Residues from Fabrics by Laundering," *Textile Chemist and Colorist,* Vol. 15, No. 3, 1983, pp. 29–33.

[15] Solbrig, C. M. and Obendorf, S. K., "Distribution of Residual Pesticide Within Textile Structures as Determined by Electron Microscopy," *Textile Research Journal,* Vol. 55, No. 9, 1985, pp. 540–546.

[16] Davidsohn, A. and Milwidsky, B. M., *Synthetic Detergents,* Wiley, New York, 1978.

[17] Drexler, P. G. and Tesor, G. C., "Materials and Processes for Textile Warp Sizing," *Handbook of Fiber Science and Technology: Vol. 1 Chemical Processing of Fibers and Fabrics, Part B Fundamentals and Preparation,* M. Lewin and S. B. Selo, Eds., Marcel Dekker, New York, 1984, pp. 1–90.

[18] *How Dry Cleaners Keep Solvents Clean,* PC Report Bulletin PCR-40, International Fabricare Institute, Silver Spring, MD, 1982.

[19] *Processing Contaminated Loads,* Technical Bulletin T-559, International Fabricare Institute, Silver Spring, MD, 1984.

[20] Goodman, C. J., "Removal of Methyl Parathion through Laundering Recontaminated Fabrics," unpublished Master's thesis, University of Nebraska-Lincoln, 1985.

Joan M. Laughlin,[1] *Jana Lamplot,*[1] *and Roger E. Gold*[2]

Chlorpyrifos Residues in Protective Apparel Fabrics Following Commercial or Consumer Refurbishment

REFERENCE: Laughlin, J. M., Lamplot, J., and Gold, R. E., "**Chlorpyrifos Residues in Protective Apparel Fabrics Following Commercial or Consumer Refurbishment,**" *Performance of Protective Clothing: Second Symposium, ASTM STP 989*, S. Z. Mansdorf, R. Sager, and A. P. Nielsen, Eds., American Society for Testing and Materials, Philadelphia, 1988, pp. 705–714.

ABSTRACT: This study examined the effectiveness of home laundering with a fabric softener and commercial laundering with starch on the removal of pesticide residues from fabric specimens. A 0.2-mL aliquot of chlorpyrifos (0.5% AI, emulsifiable concentrate) was pipetted onto fabric specimens cut from two fabrics (100% cotton and 50% cotton/50% polyester) with two finishes (unfinished or fluorocarbon soil-repellent finished). Laundering treatments included commercial laundering with and without starch, and home laundering with and without fabric softener. The most significant factor in removal by laundering and in contamination level before laundering was the soil-repellent (SR) finish. The soil-repellent finish inhibited absorption of chemical on the controls; however, any treatment that involved laundering diminished the effectiveness of the SR finish. Pesticide removal was similar between the unfinished specimens and the SR finished specimens.

Additional aqueous solutions in commercial laundering did not result in greater residue removal. A single application of starch or fabric softener did not affect pesticide absorption. Starch and fabric softener were not shown to be effective laundry auxilaries in lowering pesticide residues and did not affect after-laundering residue levels when specimens had been laundered with these auxiliaries prior to contamination.

KEY WORDS: chlorpyrifos, pesticide residue, home laundering, commercial laundering, fabric softener, starching, protective clothing

Soiling of agricultural workers' clothing with pesticides has been an issue since Wolfe et al. [1] reported that dermal exposure is of greater concern than oral or respiratory exposure. Researchers have focused attention on protective clothing, functional design, and efficiency of refurbishment. Studies of refurbishing procedures have centered on home laundering; however, no work published to date has discussed the effects of fabric softeners, estimated to be used by 85% of all consumers. In addition, many professional pesticide applicators wear career apparel that is commercially laundered; however, no research to date has examined removal of pesticide residues from fabrics by commercial laundering. Commercial laundering differs from automatic home laundering in type of detergents, temperatures, rinse cycles, and auxiliaries used in the laundering procedure. Such differences may affect the pesticide absorption and removal from uniforms.

An important textile auxiliary often applied as a textile finish, is a fabric softener. With machine washing, laundered clothing would become "hard" to the feel, due to buildup of

[1] Professor and chairman and graduate research assistant, respectively, Textiles, Clothing, and Design, Home Economics Building, University of Nebraska-Lincoln, Lincoln, NE 68583-0802.
[2] Head, Department of Entomology, University of Nebraska-Lincoln, Lincoln, NE 68583-0818.

salts on the fibers that were not being rinsed away in the final machine rinse when relatively hard water was used [2]. Fabric softeners, adsorbed during the final rinse, add softness probably due to an interfiber lubricating effect [2]. Fabric softeners are cation-active surface-active compounds, usually amine salts, quarternary ammonium [3], or pyridinium derivatives [4]. These cationics adsorb strongly onto most solid surfaces, which in contact with water at pH ≤ 5 assume a negative charge [5]. This fact is used as the basis for attachment of surface films on fibers [4] or to impart special characteristics to the substrate [6].

The fabric softener tends to diminish production of static electricity [2]. Static electricity may contribute to particulate soil attraction. Quarternary ammonium salts are also used as emulsifying agents where adsorption of the emulsifying agent onto the substrate is desirable (for example, insecticidal emulsions) [6]. Given the lipophillic fabric softener, an emulsifiable concentrate pesticide may be miscible in this auxiliary [6].

Fabric softeners are dispersed in water at approximately 0.1% concentration [weight/ weight (w/w) of clothes]. However, cationic fabric softeners and anionic detergents are mutually incompatible because the anion and cation precipitate each other [2]. Subsequent washing of clothes treated with fabric softeners does not tend to neutralize the softener already adsorbed on the fiber; rather, the softener tends to accumulate with consecutive washes [2]. "Softener buildup" is the term used for this phenomenon in which softness increases and absorbency decreases. Repeated use or high concentrations may render a fabric moisture repellent. It is postulated that a softener may function to decrease pesticide absorption during soiling.

Commercial launderers use starch as a temporary textile finish. At very low add-ons, starch will give a fabric crispness and body, impart a stiff and smooth appearance for collars and cuffs, and facilitate soil removal. Chemically, starch is a naturally occurring poly-α-glucopyranase [7] in a water-soluble, film-forming polymer to meet the requirements of sizing compositions. Since starches in unmodified form are not effective as sizes for synthetic fibers, lubricants and softeners have been added, and derivatives have been developed through specific substituents. Among the advantages of starches and derivatives are easy removal from fabrics as starches become soluble in an aqueous alkali [8] and absence of serious problems with biodegradability. Refurbishment in an alkaline medium can be accelerated by anionic or nonionic wetting agents that promote wetting, swelling, or diffusion; however, ionic surfactants inactivate enzymes that might be selected for a starch-removing laundering process [7].

Soil is more easily removed from starched fabrics because the soil attaches to the starch rather than to the fabric and is removed with the starch during subsequent laundering. Stout and Schiermeier [9] found that a 4 to 5% add-on (w/w) of starch assisted in soil removal in subsequent laundering. Unstarched specimens required as much as three washings to obtain a level of cleanliness equivalent to one post-starch washing. Utermohlen et al. [10] examined starch used at 0.5 to 5% add-ons. They found that, while amounts of starch in the 5% range assisted in subsequent soil removal, lesser concentrations either did not aid or hindered soil removal. Starch products used by commercial launderers recommend application at rates of less than 1% add-on, where the starch is used to improve handle and appearance, rather than to retard soiling and to assist in soil removal.

Keaschall et al. [11] examined differences among pesticides in soiling potential and soil removal in laundering attributable to chemical class. They found that chlorpyrifos, an organophosphate, was the most difficult to remove of the eleven pesticides examined. Urban commercial pest control operators use chlorpyrifos to manage a number of insect problems around the home, lawn, and garden. Chlorpyrifos, as well as other organophosphates and carbamates, have become the products of choice for pest control and have replaced the organochlorines once used in urban pest management.

Purpose

This study was undertaken to determine if laundering that included the temporary functional finishes of starch or fabric softener affected pesticide absorption and pesticide soil removal. The application of these temporary functional finishes occurs during the laundering procedure. The research design included laundering with these auxiliaries and the contribution of previous laundering with these auxiliaries to pesticide residue retention.

Methods and Materials

The study was designed to assess the role of laundry auxiliaries in inhibiting pesticide absorption and in completeness of removal in laundering under two conditions: (1) when the laundry product had been a part of a laundering cycle prior to contamination, and (2) when the laundry product had been used in laundering following pesticide exposure. The research design included: (a) commercial laundering, without starch (Phase I); (b) commercial laundering, with starch (in a fourth immersion) (Phase II); (c) home laundering without fabric softener (Phase III); and (d) home laundering with fabric softener (in the second rinse) (Phase IV). Control specimens of each fiber content and fabric finish were also included in the research design. All work was replicated at least three times.

Fabric Specimens

Two fabrics of similar construction were studied. A poplin of 100% cotton and a poplin of 50% polyester/50% cotton were obtained from TestFabrics, Inc. (Table 1). The fabrics were initially stripped of warp sizing and manufacturer applied fabric finishes by washing fives times as per the American Association of Textile Chemists and Colorists (AATCC) Test for Dimensional Changes in Automatic Home Laundering of Woven and Knit Fabrics (135-1978, R-85). The outer 10% of the fabric was removed for preparation of test specimens as described in the ASTM Test Methods for Breaking Load and Elongation of Textile Fabrics (D 1682-64, R-75) to assure consistency of the warp yarns under evaluation. A fluorocarbon soil-repellent (SR) finish (SR)[3,4] was applied to paired specimens of all-cotton and the blend

TABLE 1—Description of fabric.

Fabric	Designation		Test Fabric Number	Fabric Count Yarns/10 cm	Weight, g/m²
100% cotton, bleached and mercerized	C	UN	407	440 by 200	231
100% cotton, bleached and mercerized with soil-repellent finish	C	SR	407	440 by 200	231
50% Fortrel Polyester/ 50% cotton poplin, bleached and mercerized	C/P	UN	7428	480 by 200	210
50% Fortrel polyester/ 50% cotton poplin, bleached and mercerized with soil-repellent finish	C/P	SR	7428	480 by 200	210

[3] Scotchgard Fabric Protector is a registered trademark of 3-M Company.
[4] Use of product name does not imply endorsement.

fabrics, at an average of 1.0% add-on (w/w). Unfinished (UN) and finished portions of the fabric yardages were cut into specimens of 8 by 8 cm. Using a plot diagram, we randomly assigned specimens to treatments.

Spiking of the Specimens

A 0.5% active ingredient concentration of chlorpyrifos (0,0-diethyl 0,3, 5,6-trichloro-2-pyridyl-phosphorothioate) was prepared from Dursban 4-E.[5] A 0.2-mL aliquot of chlorpyrifos dilution was applied to each specimen using a MicroLab P* programmable micropipette. Specimens were placed on a raised surface to minimize contact points during contamination. The micropipette unit was held in a padded ring stand, allowing a constant distance of 5 cm between the pipette tip and the specimen surface. The pesticide was allowed to absorb into the unfinished specimens (<1 s). Pesticide spiked onto SR specimens was allowed to remain on the fabric surface for 10 s. Any solution not absorbed by the specimen was rolled into a waste container. All specimens were then air-dried before further treatment.

Refurbishment

The AATCC Test for Colorfastness to Washing, Domestic; and Laundering, Commercial: Accelerated (61-1985) was used to simulate the home laundering or commercial laundering procedures. Solutions were prepared at a volume of 150 mL wash liquor per fabric specimen. Detergent and laundry additives were prepared at concentrations recommended by the product manufacturers. For home laundering, a heavy duty nonionic liquid detergent (Dynamo[6]) was used at 0.13%, and a fabric softener (Downy[7]) at 0.125% was used in the second rinse cycle. For commercial laundering, a 6.5% industrial phosphate detergent (Factor[7]) was used at 0.05%, and vegetable starch (H.K. Instant) at 0.05% was used in an additional step following laundering. All work was done using distilled water at 60°C in an Atlas Launder-Ometer (Model LEF). Agitation was provided by 25 steel balls during the 12-min wash cycle and the two rinses of 5 and 3 min each. For the home laundering treatments, fabric softener was added to the second rinse; for commercial laundering treatments, the three rinses were followed by immersion in a starch solution. Dedicated starch solutions were used for each contaminated specimen. Following laundering treatments, specimens were allowed to air dry, and prepared for analysis.

Extraction and Gas Chromatography Analyses

Laundered fabric specimens were placed in individual glass bottles with 200 mL glass-distilled hexane, and extracted on a Precision shaker at 120 cpm for 30 min. Recovery rates were 79.8 to 83.3%. Following extractions, the hexane/chlorpyrifos mixture was concentrated to approximately 10 mL on a Bucchi Rotavapor, then further evaporated on a Meyer N-Evap evaporator with a nitrogen stream. Specimens were analyzed on a Varian Vista 6000 Gas Chromatograph equipped with a thermionic specific detector. The column was 2 by 4 mm, glass packed with 6% OV-210 + 4% OV-101 on 80/110 mesh Chromsorb W,[8] with a nitrogen flow of 30 mL per min. Temperatures were: injection, 200°C; column, 220°C; and detector, 300°C. Retention times and area counts were computed with the Varian

---

[5] Registered trademark of Dow Chemical Company.
[6] Registered trademark of Colgate Palmolive Company.
[7] Registered trademark of Proctor and Gamble Company.
[8] Registered trademark of Manville Company.

dedicated computer integrator. Pesticide standard solutions, at 2 ppm, were made up from 99.97% pure chlorpyrifos, obtained from the Health Effects Laboratory, Environmental Protection Agency. Residues were expressed in $\mu g/cm^2$, and residues remaining after treatments were calculated as percentages of initial contamination.

Statistical Analyses

Statistical differences were computed using General Linear Models with level of significance at $p \leq 0.05$. Since the percentages of residues remaining after treatment were not normally distributed, arc sine transformations were performed on these data prior to analyses. Comparisons examined fiber content, fabric finish, laundry treatment, and interactions associated with these variables. Least significant difference (LSD) mean comparison tests were performed to determine where differences occurred, and least significant means were computed for the examination of interactions.

Findings and Discussions

The factorial experiment for assessing the residue of chlorpyrifos in unfinished (UN) and soil-repellent (SR) finished all-cotton and cotton/polyester-blend fabrics was completed for home laundering with fabric softener or commercial laundering with starch. Prior to examination of residue remaining in the specimens after treatment, it was necessary to establish baseline levels of contamination before treatments. Initial contamination ranged from 3.73 $\mu g/cm^2$ for SR finished cotton/polyester fabrics to 12.25 $\mu g/cm^2$ for unfinished cotton fabrics (Table 2). One-way analysis of variance revealed significant differences attributable to finish on unlaundered specimens ($F = 32.76, df = 3,11, p \leq 0.05$), but no differences attributable to fiber content of the unfinished specimens ($F = 0.63, df = 1,14$) which supported previous research results [12]. The fluorocarbon finish (SR) limited initial contamination through altered surface tension of the fabric. The observed differences in initial pesticide levels must be considered when interpreting after treatment residues.

However, as noted in Table 2, a single laundering (with or without starch) prior to contamination diminished the effectiveness of the SR finish significantly. Thus, there were no significant differences attributable to fiber content or finish in comtamination levels following the initial laundering procedure. This resulted from the inhibition or removal of the functional SR finish during the laundering process prior to contamination. This was not due to a masking of the functional finish by the starch (Table 2). Confirmation of this was obtained through Analysis of Variance (ANOVA), which revealed no significant differences between the specimens laundered before contamination resulting from the presence or absence of starch. Goodman [18] found that the soil repellency of the fluorocarbon finished fabric was ineffective after two exposures to methyl parathion followed by home laundering.

TABLE 2—*Initial chlorpyrifos levels ($\mu g/cm^2$) in test specimens for commercial laundering.*

| | Cotton | | Blend | |
| | UN, $\mu g/cm^2$ | SR, $\mu g/cm^2$ | UN, $\mu g/cm^2$ | SR, $\mu g/cm^2$ |
Pre-Spiking Condition				
No treatment (control)	12.12a[a]	5.88b	11.29a	3.73b
Laundered with starch	11.84a	11.96a	10.82a	12.53a
Laundered without starch	11.63a	11.12a	11.57a	12.50a

[a] Means with the same letter are not statistically different from each other.

In this study, only one laundering resulted in an amount of chemical similar to the amount of chemical in unfinished specimens.

Commercial Laundering Following Contamination (Phase I)

Commercial laundering reduced chloropyrifos residues to 23 to 49% of initial contamination (Phase I, Table 3). After-laundering residue was significantly different from pre-laundering levels (Table 2) of chlorpyrifos regardless of functional finish (initial versus laundered, $F(2) = 75.99$, $p \leq 0.05$). This finding that refurbishment made a difference in levels of chemical in the specimen is supportive of earlier work [12–17]; however, the percentages of after-laundering residue are much greater than those found for other pesticides, but are similar to the findings of Keaschall et al. [11] who also examined chlorpyrifos.

Comparison among the laundering treatments (Table 3) was performed to determine if starch, as a part of the commercial laundering process, made a contribution to residue removal. A main effect attributable to finish and an interaction effect of fiber content with finish (Table 4) elucidated the significant contribution of the SR finish. The SR finish had more of an impact on the cotton/polyester blend (Table 3) than on the all-cotton; however, one laundering cycle diminished the effectiveness of the SR finish. Starch did not make a difference in after-laundering residues, even though the laundering-starching process involved two more exposures to an aqueous medium than did the laundering without starching. Thus, based on these data, there was no evidence to support the use of starch in laundering to enhance pesticide removal.

Commercial Laundering as a Pretreatment (Phase II)

Starch is a renewable finish and thus remains in the textile product during its use through the next refurbishment cycle. The presence or absence of starch may affect subsequent contamination or subsequent cleanup efforts. The research design addressed this possible contribution of the pretreatment with starch.

The effect of pretreatment with commercial laundering prior to chlorpyrifos spiking was examined. Paired specimens were commercially laundered with starching and without starching prior to spiking. As a result of the subsequent laundering, significant reduction in chlorpyrifos residue occurred (Table 3). Paired comparison of the specimens commercially laundered with or without starch were made to detemine whether the presence of starch on the specimen at the time of spiking contributed to the completeness of chlorpyrifos removed

TABLE 3—*Contribution of commercial laundering (with and without starch) to after-laundering chlorpyrifos residues.*

	Cotton				Blend			
	UN,		SR,		UN,		SR,	
Treatment	$\mu g/cm^2$	%	$\mu g/cm^2$	%	$\mu g/cm^2$	%	$\mu g/cm^2$	%
Phase I								
Cont,L[a]	2.94	25.10	2.86	31.04	5.17	46.79	0.59	29.49
Cont,L,S	2.82	23.48	2.75	49.09	5.33	47.81	0.67	28.49
Phase II								
L,Cont,L	2.25	19.12	5.67	55.04	4.60	39.63	4.24	32.86
L,S,Cont,L	4.25	34.75	5.35	44.31	4.54	40.94	5.45	42.97

[a] Cont = contaminate; L = launder; S = starch.

TABLE 4—*Comparison of chlorpyrifos residue between specimens laundered with starch and laundered without starch.*

Source	df	F value	PR ≤ F
Fiber content	1	0.03	n.s.
Finish	1	19.37	*
Treatments[a]	1	0.00	n.s.
Fiber content × finish	1	18.22	*
Fiber content × treatments	1	0.05	n.s.
Finish × treatments	1	0.00	n.s.
Fiber content × finish × treatment	1	0.00	n.s.

[a] Treatments = Tx_1 = Contaminate, launder; Tx_2 = contaminate, launder with starching.

in subsequent laundering. Based on ANOVA, the presence or absence of starch did not contribute to more complete removal in laundering. Although this finding appears to conflict with the observations of Stout and Schiermeier [9] that the presence of starch on a fiber assists soil removal in laundering, it must be remembered that Utermohlen [10] observed that, at very low add-ons (less than 1%), starch hindered subsequent soil removal. Using commercial launderers' procedures, the add-on was less than 0.05%. At this percent add-on (w/w), starch as a pretreatment neither inhibited nor assisted in removal of chlorpyrifos. Additional work is needed to explore the contribution of starch at higher add-ons.

Home Laundering Following Contamination (Phase III)

Home laundering reduced chlorpyrifos residue by 9 to 42% of initial contamination (Table 5). These after-laundering residue (Phase III, Table 6) were significantly different ($F = 215.61$, $df = 2,3$, $p ≤ 0.05$) from prelaundering levels (Table 5). The SR finish inhibited absorption of chemical on the controls; however, any pretreatment that involved laundering (with or without fabric softener) had rendered of the SR finish less effective. These findings (Table 5) were consistent with the findings for commercial laundering pretreatments (Table 2).

One cycle of home laundering (with or without fabric softener) significantly ($F = 231.61$, $df = 2,3$, $p = 0.05$) reduced chlorpyrifos levels in fabric specimens (Table 6). In order to ascertain whether there was a contribution of fabric softener to pesticide removal in the laundering process, a contrast between laundered without fabric softener and laundered with fabric softener was completed. An interaction of fiber content and finish was revealed by ANOVA (Table 7). The residue levels on the SR cotton laundered with fabric softener were significantly greater than the residue levels on the SR blend specimen ($p ≤ 0.05$).

The best combination was the cotton/polyester blend specimens with the SR finish as

TABLE 5—*Initial chlorpyrifos levels ($\mu g/cm^2$) in test specimens (for home laundering).*

	Cotton		Blend	
Pre-Spiking Condition	UN, $\mu g/cm^2$	SR, $\mu g/cm^2$	UN, $\mu g/cm^2$	SR, $\mu g/cm^2$
No treatment (controls)	12.25a[a]	5.88b	11.29a	3.73b
Laundered with fabric softener	11.80a	10.90a	10.35a	12.42a
Laundered without fabric softener	12.77a	11.14a	10.08a	13.09a

[a] Means with the same letter are not significantly different.

TABLE 6—*Contribution of home laundering (with and without fabric softener) to after-laundering chlorpyrifos residues.*

	Cotton				Blend			
	UN,		SR,		UN,		SR,	
Treatment	$\mu g/cm^2$	%	$\mu g/cm^2$	%	$\mu g/cm^2$	%	$\mu g/cm^2$	%
Phase III								
Cont,L[a]	1.42	11.98	1.85	32.46	2.55	22.82	0.72	11.47
Cont,L,FS	1.13	9.48	2.40	42.28	1.83	16.40	0.69	19.12
Phase IV								
L,Cont,L	1.60	13.90	4.35	39.22	2.26	21.58	1.72	13.72
L,FS,Cont,L	2.41	19.08	4.50	41.85	2.51	22.97	3.49	21.88

[a] Cont = contaminate; L = launder; and FS = fabric softener.

measured by after-laundering residues (Phase III, Table 6). Based on these data, use of a fabric softener is optional in home laundering since the chlorpyrifos residues are neither more nor less completely removed in laundering with a fabric softener.

Home Laundering as a Pretreatment (Phase IV)

Fabric softener as a pretreatment might affect the absorbency of the fabric specimen. Given that a single laundering cycle limited the functionality of the SR finish (Table 5), laundering (with and without fabric softener) as a pretreatment was examined. There was a slight but nonsignificant increase in residue retention when the specimen had been previously laundered with a fabric softener (Phase IV, Table 6).

Noteworthy was the observation that when the SR specimens had been laundered prior to contamination, greater residue was found than when the SR specimen had not been laundered (Table 6). These findings were consistent with the result of the commercial laundering phases of this study (Table 3). The SR finish limited the level of chlorpyrifos on the specimens; however, any pretreatment that involved laundering inhibited the effectiveness of the SR finish.

Based on these data, additional investigations of repeated exposure to fabric softener (and the possibilities of fabric-softener buildup) are needed. The lipophillic nature of fabric softener and miscibility of emulsifiable concentrates in these auxiliaries [6] may contribute to increased retention of chemical. With a single application of fabric softener, a slight

TABLE 7—*ANOVA on amount of chlorpyrifos on home-laundered specimens with and without fabric softener.*

Source	df	F value	$PR \leq F$
Fiber content	1	0.71	n.s.
Finish	1	1.13	n.s.
Treatments[a]	1	0.16	n.s.
Fiber content × finish	1	15.48	*
Fiber content × treatments	1	0.70	n.s.
Finish × treatments	1	1.69	n.s.
Fiber content × finish × treatments	1	0.02	n.s.

* Significant at $p \leq 0.05$
[a] Treatments = Tx_1, Contaminate, launder; Tx_2, Contaminate, launder with fabric softener.

increase in residue retention was found. Additional work needs to be done to elucidate the observations of this study.

Conclusions

The most significant factor in pesticide contamination and removal by laundering was the soil-repellent finish. The SR finish inhibited absorption of chemicals on the specimens (controls); however, any treatment that involved laundering diminished the effectiveness of the SR finish. Chlorpyrifos was not more completely removed from the SR specimens than from the UN specimens. Starch and fabric softener were not effective laundry auxiliaries in lowering pesticide residues and did not affect after-laundering residue levels when specimens had been laundered with starch or with fabric softener prior to contamination.

Starching did not make a difference in after-laundering residue, even though the laundering-starching process involved two more immersions in an aqueous medium than did laundering without starching. Additional work is needed to explore the contribution of starch at higher add-ons.

A trend was observed for increased after-laundering residues when fabric softener had been used in the laundering prior to contamination than when it had not been a part of the pre-laundering. Given the situation of fabric softener "buildup," additional work is needed on its potential for miscibility with pesticides.

In general, the use of starch or fabric softener in laundering is not a major factor in initial contamination or retention of chlorpyrifos. The findings of this study indicate that either can be used at the launderers' discretion without concern for adverse effects.

Acknowledgments

This research was supported in part by the Nebraska Agricultural Research Division Project 97-012 and it contributes to North Central Research Project NC-170, "Limiting Dermal Exposure to Pesticide Through Effective Cleaning Procedures and Selection of Clothing." This paper is published as Paper Number 8225, Nebraska Agricultural Experiment Station.

References

[1] Wolfe, H. R., Durham, W. F., Armstrong, J. F., "Exposure of Workers to Pesticides," *Archives of Environmental Health*, 1967, Vol. 14, pp. 622–633.
[2] Davidsohn, A. and Milwidsky, B. M., *Synthetic Detergents*, 6th ed., Wiley, New York, 1978.
[3] Egan, R. R., "Cationic Surface Active Agents as Fabric Softeners," *Journal of American Oil Chemists' Society*, Vol. 55, Jan. 1978, pp. 118–121.
[4] Trotman, E. R., *Dyeing and Chemical Technology of Textile Fibers*, 6th ed., Wiley, New York, 1984.
[5] Berg, J. C., "The Role of Surfactants," *Textile Science and Technology, Vol. 7: Absorbency*, Chatterjee, P. K., Ed., Elsevier, New York, 1985, pp.149–196.
[6] Rosen, M. J., *Surfactants and Interfacial Phenomena*, Wiley, New York, 1978.
[7] Drexler, P. G. and Tesor, G. C., "Materials and Process of Textile Warp Sizing," *Handbook of Fiber Science and Technology: Vol. 1. Chemical Processing of Fibers and Fabrics. Part B: Fundamentals and Preparation*, M. Lewin and S. B. Sello, Eds., Marcel Dekker, New York, 1984, pp. 1–84.
[8] Stannett, V. T., Fanta, G. F., and Doane, W. M., "Polymer Grafted Cellulose and Starch," *Textile Science and Technology, Vol. 7: Absorbency*, Chatterjee, P. K., Ed., Elsevier, New York, 1985, pp. 257–279.
[9] Stout, L. E. and Schiermeier, K. F., "Effect of Previous Starching Upon Ease of Washing Cotton Fabrics," *Industrial and Engineering Chemistry*, pp. 1403–1405.

[10] Utermohlen, W. P., Jr., Ryan, M. E., and Young, D. O., "Improvement of Cotton Clothing in Resistance to Soiling and in Ease of Washing," *Textile Research Journal*, July 1951, pp. 510–521.

[11] Keaschall, J. L., Laughlin, J. M., and Gold, R. E., "Effect of Laundering Procedures and Functional Finishes on Removal of Insecticides Selected from Three Chemical Classes," *Performance of Protective Clothing, ASTM STP 900*, R. L. Barker and G. C. Coletta, Eds., American Society for Testing and Materials, Philadelphia, 1986, pp. 162–176.

[12] Easley, C. B., Laughlin, J. M., Gold, R. E., and Tupy, D., "Methyl Parathion Removal from Work Weight Fabrics by Selected Laundry Procedures," *Bulletin of Environmental Contamination and Toxicology*, Vol. 27, pp. 101–108.

[13] Easley, C. B., Laughlin, J. M., Gold, R. E., and Schmidt, K., "Detergents and Water Temperature as Factors in Methyl Parathion Removal from Denim Fabrics," *Bulletin of Environmental Contamination and Toxicology*, Vol. 28, No. 2, Oct. 1982, pp. 239–244.

[14] Easley, C. B., Laughlin, J. M., Gold, R. E., and Hill, R. M., "Laundry Factors Influencing Methyl Parathion Removal from Contaminated Denim Fabric," *Bulletin of Environmental Contamination and Toxicology*, Vol. 29, No. 4, Oct. 1982, pp. 461–468.

[15] Easley, C. B., Laughlin, J. M., Gold, R. E., and Tupy, D., "Laundering Procedures for Removal of 2,4-Dichlorophenoxyacetic Acid Ester and Amine from Contaminated Fabrics," *Archives of Environmental Contamination and Toxicology*, Vol. 12, pp. 71–76.

[16] Laughlin, J. M., Easley, C. B., and Gold, R. E., "Methyl Parathion Residues in Contaminated Fabrics After Laundering," *Dermal Exposure to Pesticide Use*, ACS Symposium Series No. 273, 1985, pp. 177–187.

[17] Laughlin, J. M., Easley, C. B., Gold, R. E., and Hill, R. M., "Fabric Parameters and Pesticide Characteristics that Impact on Dermal Exposure of Applicators," *Performance of Protective Clothing ASTM STP 900*, R. L. Barker and G. C. Coletta, Eds., American Society for Testing and Materials, Philadelphia, 1986, pp. 136–150.

[18] Goodman, C. J., "Removal of Methyl Parathion Through Laundering Recontaminated Fabrics," unpublished Master's thesis, University of Nebraska-Lincoln, 1985.

James R. Fleeker,[1] Cherilyn Nelson,[2] Mohamad F. Wazir,[2] and Marilyn M. Olsen[2]

Effect of Formulation on Removal of Carbaryl and Chlorothalonil from Apparel Fabrics by Dry Cleaning, Aqueous Extraction, and Vaporization

REFERENCE: Fleeker, J. R., Nelson, C., Wazir, M. F., and Olsen, M. M., "**Effect of Formulation on Removal of Carbaryl and Chlorothalonil from Apparel Fabrics by Dry Cleaning, Aqueous Extraction, and Vaporization,**" *Performance of Protective Clothing: Second Symposium, ASTM STP 989,* S. Z. Mansdorf, R. Sager, and A. P. Nielsen, Eds., American Society for Testing and Materials, Philadelphia, 1988, pp. 715–726.

ABSTRACT: Several methods were examined for efficacy in removal of carbaryl and chlorothalonil contamination from indigo-dyed cotton twill, undyed cotton twill, polyester/cotton broadcloth, and cotton weft knit terry. Fabrics were contaminated with commercial formulations of carbon-14 labeled carbaryl and chlorothalonil, as well as pure carbaryl and chlorothalonil. Vaporization of the pesticides over a 21-day period resulted in a loss of up to 35% of the carbon-14. A hydrocarbon-based dry cleaning solvent was ineffective in removing the pesticides, while a perchloroethylene-based solvent removed ≥76% of the pesticides. More of the pure form of the pesticides was extracted than the formulated form. Repeated extraction with warm water removed 80 to 98% of the pesticides except for pure chlorothalonil of which 40 to 60% was removed. The inert ingredients in the commercial formulation of chlorothalonil appear to increase the efficiency of aqueous extraction.

KEY WORDS: pesticide, pesticide residue, laundering, dry cleaning, carbaryl, chlorothalonil, vaporization, protective clothing

Commercial pesticides are a diverse group of chemicals not only because of different chemical classes, for example, organophosphates and carbamates, but also because of different formulations. Literature on the reduction of exposure to these chemicals has included the examination of protective clothing as well as techniques for pesticide removal. To date, pesticide removal has focused mainly on aqueous extraction (laundering).

Pesticides examined in extraction studies have included some of the more toxic pesticides such as methyl parathion, fonofos, and azinphos methyl, which are used on agronomic crops [1]. Two other pesticides that have wide use in turf, ornamental, agronomic, and vegetable crop pest control are chlorothalonil and carbaryl [2,3]. These pesticides have a low LD_{50} [1], but their use patterns suggest a higher potential for contamination of the public as compared to pesticides used strictly on agronomic crops [4–7]. Removal of these pesticides

---

[1] Professor of Biochemistry, Biochemistry Department, North Dakota State University, Fargo, ND 58105.

[2] Assistant professor and graduate students, respectively, Apparel, Textiles and Interior Design Department, North Dakota State University, Fargo, ND 58105.

has not been adequately studied. Laundering to remove carbaryl residues has been examined [8,9], but these experiments did not include recovery determinations to validate the analytical method.

The efficacy of laundering for pesticide removal from apparel fabrics has recently been reviewed [10,11]. Because no single method has been described that will effectively remove all pesticide contamination from fabrics, the objectives of this study were to determine the efficacy of vaporization, aqueous extraction, and dry cleaning-solvent extraction on the removal of chlorothalonil and carbaryl residues from four apparel fabrics.

Procedures

Fabrics

The fabrics used in this study are characterized in Table 1. The blue cotton denim (CD) was obtained from Cone Mills, Greensboro, North Carolina. Mercerized cotton twill, No. 423 (CT), and Dacron polyester/cotton broadcloth, No. 7409 (PC), were obtained from Testfabrics, Inc., Middlesex, New Jersey. The cotton knit terry (CK) was obtained from Acme-McCrary Corp., Asheboro, North Carolina. Sizing was removed by American Association of Textile Chemists and Colorists (AATCC) Test Method 135-1978. Fabric specimens of 5 by 5 cm or 4 by 16 cm were prepared from the fabric after removal of the outer 10% of the fabric in the warp direction as per ASTM Test Method for Breaking Load and Elongation of Textile Fabrics (D 1682-64).

Pesticides

Four commercial formulations were used: (1) chlorothalonil, flowable formulation (F), EPA Registration No. 50534-8; (2) carbaryl, emulsifiable concentrate (EC), EPA Registration No. 239-2356-AA; (3) carbaryl, wettable powder (WP), EPA Registration No. 1016-43; and (4) carbaryl (F), EPA Registration No. 264-333. Analytical standards of the pesticides were obtained from the U.S. Environmental Protection Agency (EPA) Pesticides and Industrial Chemicals Repository, Research Triangle Park, NC.

Chlorothalonil[^{14}C], specific activity 10 mCi/mmole, and carbaryl[^{14}C], 21 mCi/mmole, were checked for radiochemical purity by thin-layer chromatography using two solvent systems [12,13]. Both compounds were >98% radiochemically pure by this criterion.

The contamination solutions of the pesticides contained 0.6% (weight/volume) of active ingredient and 0.8 μCi/mL of carbon-14. The commercial formulations were prepared with

TABLE 1—Fabric characterization.

Fabric Name	Fabric[a] Structure	Weight, g/m²	Dye	Yarns/cm	
				Warp	Filling
100% cotton denim (CD)	3/1 twill weave	475	indigo	28	18
100% cotton twill (CT)	3/1 twill weave	258	undyed	43	23
65% polyester 35 % cotton broadcloth (PC)	plain weave	95	undyed	53	28
100% cotton weft pile knit (CK)	terry knit	407	undyed	5.3[b]	28[c]

[a] All yarns were single yarns.
[b] Wales/cm.
[c] Courses/cm.

distilled water. Solutions of pure chlorothalonil and carbaryl (P) were prepared in toluene-methanol (1:1, volume/volume).

Vaporization

Specimens (5 by 5 cm) were treated with 50 µL of the contamination solution and hung in a fume hood for various time periods. The temperature range was 23 to 26°C and the air-flow rate was 44 m³/min/m². The specimens were cut in half and each piece assayed for carbon-14 by scintillation counting as described later. Control specimens were measured for radioactivity 20 min after application of the pesticide.

Dry Cleaning Solvents

A perchloroethylene-based solvent was prepared with surfactant and water according to AATCC Test Method 158-1985. A hydrocarbon-based solvent was obtained from the extraction chamber of a local dry cleaning establishment. Surfactant and water were present in this solvent according to the manufacturer's directions.

Specimens (4 by 16 cm) were held horizontal with tweezers, treated with 50 µL of the contamination solution, and then hung vertically to dry. The pesticide drop was confined to the 4 by 4 cm area at the lower end of the specimen. The swatches were conditioned at 22 ± 2°C and 65 ± 3% relative humidity for 24 h prior to extraction. Controls were assayed for carbon-14 just prior to the dry cleaning treatment.

Each specimen was swirled at 250 rpm for 20 min in 150 mL of the dry cleaning solvent. A rotary shaker with a 1-cm displacement was used with 250-mL flasks (Kimax No. 26500). The starting temperature of the solvent was 32°C and the final temperature 29 ± 1°C. Excess solvent was gently squeezed out of the fabric and the specimen dried at room temperature. While drying, the specimens were hung so the original contaminated area was downward. The specimens were cut across the width in 2 by 4 cm sections and each section assayed by scintillation counting as described later.

Aqueous Extraction

Specimens were contaminated and conditioned as described for the dry cleaning experiments. Controls were assayed for carbon-14 just prior to aqueous extraction. The specimens were swirled in water at 250 rpm under various conditions of time, volume, and temperature. In some experiments, the specimen was removed from the flask, the excess solvent squeezed out, and the specimen transferred to another flask with fresh solvent for continued extraction. Where indicated, AATCC standard detergent 124 (12.4% phosphorus) was used at a concentration of 1.8 g/L. Distilled water was used throughout except where indicated. A solution of 10% isopropyl alcohol (volume/volume) in water was used in one experiment and municipal water was used as a hard water source in another. The hardness range of the water source was 85 to 137 mg/L as measured by the standard sodium salt of ethylenediaminetetracetic acid (EDTA) titrimetric method [14]. The conductivity of the water at the time of the experiment was 500 µmhos/cm. The specimens were dried and sectioned as described for the dry cleaning experiments.

Radioactivity Determination

Carbon-14 was measured with a liquid scintillation spectrometer. Sections of white fabric were placed in 1 mL of water and 10 mL of scintillation solvent No. 1 [15]. This solvent contained 10 g of 2,5-diphenyloxazole (PPO) and 1 g of 1,4-bis(4-methyl-5-phenyloxazole-

2-yl)benzene (dimethylPOPOP) per litre of toluene/Triton x-100 (2:1, volume/volume). Counting efficiency was determined with toluene[^{14}C] as an internal calibration standard (New England Nuclear, NES-006) [12].

Colored specimens could not be assayed by the method just described. The total specimen was assayed by oxidizing 200-mg portions (Harvey biological oxidizer, Harvey Instrument Co., Hillsdale, NJ) and collecting the resulting radioactive carbon dioxide in 20 mL of scintillation solvent No. 2. This solvent contained 10 g of PPO, 1 g of dimethylPOPOP per liter of 2-aminoethanol/toluene/2-methoxyethanol, 3:10:7 (volume/volume/volume). Methyl acrylate[^{14}C] tablets (New England Nuclear, NES-009) were used to calibrate the oxidizer for oxidation and counting efficiency.

Statistical Analysis

The data from three replicates were analyzed by using the Analysis of Variance (ANOVA) and Duncan's New Multiple Range Statistical Test. Significance between the means of pairs was determined by least significant means tests at $P \leq 0.05$.

Results and Discussion

The carbon-14 label allowed direct measurement of the pesticide on fabric. The method allows for rapid quantitation, but does identify the chemical form of the residue remaining. Some chemical change of the pesticide may have occurred on the specimen during treatment, although this was not expected due to the chemical stability of the compounds [6,16]. Chlorothalonil is stable to aqueous alkali and acid and to ultraviolet light [16]. Carbaryl is stable in water at pH less than 9 [6].

Pure chlorothalonil and carbaryl were used for comparison to the commercial formulations. Most pesticide applicators do not use the pure forms of pesticides, although chlorothalonil is used in a nearly pure form (90% active ingredient) to fumigate greenhouses [1,16].

TABLE 2—*Loss of ^{14}C-labeled chlorothalonil and carbaryl from fabrics exposed to moving air at room temperature.[a]*

| | Radioactive Remaining, % | | | | | |
| | Chlorothalonil | | | Carbaryl | | |
Exposure Time	F	P	EC	F	WP	P
	POLYESTER/COTTON (PC)					
2 days	90.3	99.0	96.0	95.0	93.3	96.0
7 days	82.3	96.0	94.7	82.3	85.0	93.3
21 days	80.0a	91.3b	80.3a	65.3c	74.3d	84.0e
	COTTON TWILL (CT)					
2 days	99.3	98.7	96.3	91.7	98.0	95.3
7 days	97.3	97.3	94.7	91.0	90.3	93.3
21 days	78.3a	94.3b	95.3b	87.0c	93.7b	90.0d

[a] Formulations: F = flowable, P = pure compound, EC = emulsifiable concentrate, and WP = wettable powder. 300 μg of each pesticide was applied to 4 by 5 cm swatches. Means for 21 days and within a fabric type that have the same letter are not statistically different at $P \leq 0.05$. The means for 21 days were all statistically different at $P \leq 0.05$ between fabrics treated with the same formulation.

Vaporization of Residues

Both chlorothalonil and carbaryl were slowly vaporized from the fabrics (Table 2). There was not a large difference in the vaporization rates of the two pesticides, although chlorothalonil has a much higher vapor pressure than carbaryl [1]. The PC fabric lost more of the pesticides after 21 days than the CT. Humidity may affect the dissipation rate but was not controlled in this study. Although vaporization was unsuccessful as a refurbishing technique in this case, Kim et al. [11] found that heat and drying reduced the levels of alachlor on fabrics after laundering.

Chlorothalonil has been found to vaporize from the walls of a room covered with paint containing the compound [7]. The rate of vaporization was slow but significant. Laughlin and Gold [17] examined loss of methyl parathion from apparel fabrics exposed to air. The rate of dissipation of the insecticide was similar to that observed here, although long-term exposure (six months), was effective in removing essentially all of the insecticide in moving air. Chemical alteration of pesticide on the fabric, as well as vaporization, may account for some of the loss of pesticide observed by Laughlin and Gold, and by Kim et al., and may also have occurred in our studies.

Drycleaning Solvents

Pesticides that are not salts are usually more soluble in organic solvents than in water. Therefore, dry cleaning would appear to be an effective method to remove these pesticides from apparel fabrics. Table 3 shows the effectiveness of two common dry cleaning solvents in the extraction of chlorothalonil and carbaryl from fabrics. The pure pesticides were removed more efficiently than the commercially formulated pesticide. Inactive materials in the formulations appear to influence the extraction efficiency. The perchloroethylene-based

TABLE 3—*Efficacy of drycleaning solvents on the removal of carbaryl and chlorothalonil formulations from apparel fabrics.*

		Radioactivity Remaining, %					
		Chlorothalonil		Carbaryl			
Fabric	Swatch[a]	F	P	EC	F	WP	P
		PERCHLOROETHYLENE-BASED SOLVENT					
CT	first	23.7a	8.7b	15.8c	8.8b	11.1d	5.0e
	second	1.2	1.2	0.8	0.5	0.6	0.6
PC	first	2.5a	2.1a	4.2b	2.1a	3.2c	1.5d
	second	0.9	1.0	0.7	0.7	0.9	0.7
		HYDROCARBON-BASED SOLVENT					
CT	first	80.6a	38.7b	97.8c	76.8a	73.9d	53.4e
	second	0.4	1.2	0.1	0.4	0.6	0.8
PC	first	81.3a	39.4b	97.9c	76.0d	72.1e	35.6f
	second	0.2	0.7	0.1	0.4	0.6	0.8

[a] The first swatch contained the original contamination. The second swatch was not treated with pesticide and was added to the dry cleaning solvent after removal of the first swatch.

See Table 2 for formulation designations. The differences between means within a solvent system and fabric with the same letter were not statistically different at $P \leq 0.05$. The differences in the means between CT and PC for the perchloroethylene solvent and first swatches were all significant at $P \leq 0.05$.

solvent was more effective than the hydrocarbon-based solvent and this may reflect the greater polarity of perchloroethylene. The PC fabric retained significantly less pesticide than CT when perchloroethylene was the solvent.

The presence of moisture and surfactant in the perchloroethylene-based solvent influenced the efficacy of the extraction. Fleeker et al. [18] found that perchloroethylene without moisture and detergent, and under similar experimental conditions as used here, was much less effective in removing the same pesticides than the dry cleaning solvent system.

The extracted pesticides were redeposited onto fabric that was originally uncontaminated. The solvent continued to contaminate fresh specimens even after removal of the original contaminated swatch (Table 3). The amount of pesticide redeposited was low, but was uniformly distributed over the fabric. For these reasons, the authors recommend commercial dry cleaning not be used to remove pesticide contamination from clothing. It is not known if filtration of the solvent will remove dissolved or suspended pesticide.

Ringenberg et al. [19] have found that perchloroethylene-based dry cleaning solvent was effective in removing chlorpyrifos from apparel fabrics. Kim et al. [11] lowered residues of alachlor on contaminated apparel fabrics by using a pretreatment consisting of perchloroethylene suspended in water. Although perchloroethylene can decrease pesticide residues on apparel fabrics, fumes of this solvent are hazardous, and household use of the solvent for this purpose may not be appropriate.

Aqueous Extraction

Tables 4 and 5 show the effects of temperature, time of extraction, volume, and solvent composition on the aqueous extraction of carbaryl and chlorothalonil from fabrics. Increasing

TABLE 4—Effect of aqueous extraction conditions on removal of chlorothalonil and carbaryl from white-cotton twill and cotton/polyester.[a]

			Radioactivity Remaining, %					
			Chlorothalonil		Carbaryl			
	Extraction Conditions	Fabric	F	P	EC	F	WP	P
I.	2 min, 24°C, 150 mL	CT	21.1	53.7	5.1	36.8	6.2	26.0
		PC	16.4	63.2	5.2	17.9	3.0	20.1
II.	2 min, 49°C, 150 mL	CT	11.5	50.5	4.0	29.1	4.9	4.7
		PC	8.9	55.8	3.6	5.6	2.6	7.3
III.	5 min, 49°C, 150 mL	CT	8.0	45.8	4.3	12.1	4.5	4.1
		PC	7.5	50.4	3.3	3.7	3.4	4.8
IV.	5 min, 49°C, 300 mL	CT	7.7	44.7	2.6	12.9	2.6	3.2
		PC	6.6	46.6	3.3	3.4	2.6	2.9
V.	Extracted twice for 5 min, 49°C, 150 mL	CT	3.5	38.6	1.3	3.3	0.9	1.6
		PC	4.2	42.7	2.8	2.5	0.8	1.2
VI.	5 min, 49°C, 150 mL of isopropyl alcohol-water (1:9)	CT	8.1	43.3	3.0	6.8	3.4	2.4
		PC	5.9	47.3	2.2	3.1	2.2	3.4
VII.	5 min, 49°C, 150 mL of municipal water	CT	11.3	47.8	3.7	12.0	3.9	3.7
		PC	7.8	51.8	3.2	3.8	2.1	4.6

[a] Each specimen received 300 µg of pesticide. For formulation designations see Table 2. Statistical treatment of data is reported in Table 5. All specimens were extracted once with distilled water unless otherwise stated.

TABLE 5—*Statistical treatment of the data in Table 4.*

Analysis of Variance Test

Source of Variations	Degrees of Freedom	Sum of Squares	F Values	PR > F
Fabric (Fab)	1	151	110	0.01
Formulation (For)	5	64 127	9321	0.01
Treatment (Tre)	6	4 598	557	0.01
Fab × For	5	1 294	188	0.01
Fab × Tre	6	145	17.5	0.01
For × Tre	30	2 115	51.2	0.01
Fab × Tre × For	30	610	14.7	0.01

Duncan's Multiple Range Test:
mean of percent recovery[a]

Item	N	Mean	Grouping
	FORMULATION		
Chorothalonil (P)	42	48.7	A
Carbaryl (F)	42	10.9	B
Chlorothalonil (F)	42	9.2	C
Carbaryl (P)	42	6.4	D
Carbaryl (EC)	42	3.4	E
Carbaryl (WP)	42	3.1	E
	TREATMENT		
Treatment I	36	22.9	A
Treatment II	36	15.7	B
Treatment VII	36	13.0	C
Treatment III	36	12.7	C
Treatment IV	36	11.6	D
Treatment VI	36	10.9	E
Treatment V	36	8.6	F
	FABRIC		
CT	126	14.4	A
PC	126	12.9	B

[a] Means followed by common letter are not significantly different at $P \leq 0.05$ by Duncan's New Multiple Range Test.

the water temperature and length of extraction time were generally effective in increasing the extraction efficiency. Two extractions were more effective than one. Similar results have been reported by other investigators for different pesticides [10,11,20–22]. Variations in extraction efficiencies were noted with the different formulations. This trend was also expected from other studies. For example, the F formulation of carbaryl was found, in this study, to have slightly greater resistance to multiple extractions compared to the WP, P, and EC preparations, while Easley et al. [20] found an EC formulation of methyl parathion to have greater resistance to extraction than WP or encapsulated formulations. The composition of the inactive substances in the formulations must account for the variation in the extraction efficiency within a single fabric type and pesticide.

Surfactants are common components of pesticide formulations. A commercial surfactant used with pesticides (Ortho X77) was added to a portion of the pure chlorothalonil solution to the extent of 0.1% (weight/volume). Both solutions were applied to CD and extracted as in method III of Table 4. With pure chlorothalonil, $47.6 \pm 4.3\%$ was retained, while

35.7 ± 2.6% of the fungicide was retained on the fabric when contaminated with the solution containing the surfactant. A control was used in which specimens treated with the pure chlorothalonil solution were extracted with water containing an amount of X77 surfactant equivalent to that placed on the chlorothalonil-X77 treated specimens. Extraction of the control specimens left 45.1 ± 3.1% of the chlorothalonil applied. The presence of the surfactant in the contamination solution increased the extraction efficiency, but not to the extent observed when the F formulation was used (Table 4).

Pure chlorothalonil was resistant to aqueous extraction, while pure carbaryl was readily removed. Carbaryl has a solubility in water over 100 times that of the fungicide [1], and thus carbaryl may be expected to be removed from the fabric with greater efficacy. While 40 to 50% of the chlorothalonil residues were removed with one extraction, the remaining residue resisted further extraction. The commercial formulation of chlorothalonil was removed much more efficiently than the pure compound. This is opposite of the effect observed with the dry cleaning solvents (Table 3) and again must reflect solubility of the chlorothalonil and the influence of the inactive substances in the formulation. The chlorothalonil formu-

TABLE 6—*Effect of three aqueous extractions, fabric type, and detergent on the removal of chlorothalonil and carbaryl from apparel fabrics.*[a]

		Radioactivity Remaining, %					
		Chlorothalonil			Carbaryl		
Detergent	Fabric area	F	P	EC	F	WP	P
		COTTON TWILL (CT)					
No detergent	Total swatch	4.0	40.4	1.5	3.0	1.0	1.1
	Remainder	0.2	0.5	0.1	0.1	0.1	0.1
Detergent added	Total swatch	3.2	35.6	0.9	2.5	0.5	0.6
	Remainder	0.1	0.4	<0.1	0.2	<0.1	<0.1
		COTTON KNIT (CK)					
No detergent	Total swatch	3.6	31.3	4.3	4.7	4.1	2.7
	Remainder	0.9	2.8	1.3	1.4	1.6	1.4
Detergent added	Total swatch	3.6	31.5	4.1	3.6	4.4	2.6
	Remainder	0.9	0.8	0.7	0.7	0.4	0.6
		BLUE COTTON DENIM (CD)					
No detergent	Total swatch	10.7	57.8	2.3	7.7	2.1	2.9
	Remainder	0.8	0.6	0.1	0.5	0.2	0.3
Detergent added	Total swatch	4.6	57.7	2.0	7.6	1.1	2.2
	Remainder	0.6	0.5	0.1	0.2	0.1	0.2
		POLYESTER/COTTON BROADCLOTH (PC)					
No detergent	Total swatch	4.1	40.7	1.1	2.9	0.7	1.2
	Remainder	0.2	0.4	0.1	0.1	0.1	0.2
Detergent added	Total swatch	3.8	36.6	0.9	2.7	0.7	1.1
	Remainder	0.1	0.2	<0.1	0.1	0.1	0.1

[a] The extraction was carried out three times for 5 min each, with distilled water at 49°C and a volume of 150 mL. When detergent was used, the second extraction used AATCC detergent at 1.8 g/L. Each 4 by 16 specimen received 300 μg of pesticide on a 4 by 4 cm section at one end of the swatch. Remainder refers to the remaining 4 by 12 cm area. Statistical treatment of the data is reported in Table 7.

lation has in it substances such as surfactants that facilitate the aqueous extraction of the pesticide. This may explain the effectiveness of laundering in the removal of such water-insoluble pesticides as DDT, toxaphene [21], chlordane, aldrin, and lindane [10] from fabrics when commercial formulations of these insecticides were studied.

The use of 10% (volume/volume) isopropyl (rubbing) alcohol or municipal water caused small changes in extraction efficiency when used in place of distilled water, although in some instances there were statistically significant differences (Tables 4 and 5). Kim et al. [11] also found that dilute aqueous ethyl alcohol, when used as a pretreatment, had about the same effect in removing alachlor as distilled water.

TABLE 7—*Statistical treatment of the data in Table 6.*

Analysis of Variance Test

Variable	Degrees of Freedom	Sum of Squares	F Values	PR > F
Fabric (Fab)	3	359	517	0.01
Formulation (For)	5	15 213	13 138	0.01
Detergent (Dt)	1	25.68	111	0.01
Area (Ar)	1	5 723	24 710	0.01
Dt × For	5	13.1	11.3	0.01
Dt × Fab	3	1.12	1.61	0.189
For × Fab	15	917	264	0.01
Dt × Ar	1	7.35	31.7	0.01
For × Ar	5	14 590	12 600	0.01
Fab × Ar	3	378	545	0.01
Dt × For × Fab	15	30.7	8.84	0.01
Dt × For × Ar	5	9.81	8.47	0.01
Dt × Fab × Ar	3	11.9	17.1	0.01
For × Fab × Ar	15	960	276	0.01
Dt × Fab × For × Ar	15	33.4	9.63	0.01

Duncan's Multiple Range Test:
mean of percent recovery[a]

Item	N	Mean	Grouping
	FORMULATION		
Chlorothalonil (P)	48	21.1	A
Chlorothalonil (F)	48	2.6	B
Carbaryl (F)	48	2.4	B
Carbaryl (EC)	48	1.2	C
Carbaryl (WP)	48	1.1	C
Carbaryl (P)	48	1.1	C
	DETERGENT		
Deleted	144	5.2	A
Added	144	4.6	B
	FABRIC		
CD	72	6.8	A
CT	72	4.8	B
CK	72	4.1	C
PC	72	4.0	C

[a] Means followed by a common letter are not significantly different at $P \leq 0.05$ by Duncan's New Multiple Range Test.

A significant difference in the removal of paraquat from apparel fabrics was noted by Olsen et al. [22] when distilled and municipal water were compared. In this case, however, the differences were attributed to ion exchange between the paraquat cation and anionic groups on the cotton fibers. It was proposed that the low salt concentration of the municipal water was effective in reducing the ionic interaction between paraquat and fabric. Few pesticides are cations in the normal pH range of ground and municipal water, so that this ion-exchange effect should be confined to a limited group of pesticides such as paraquat, diquat, and difenzoquat.

The influence of fabric types and the effect of added detergent were examined using a sequence of three aqueous extractions (Tables 6 and 7). The pesticides were redeposited during extraction onto uncontaminated areas of the specimens. This is consistent with other studies and was expected. The amount redeposited was considerably less than that found at the original area of contamination, and it was spread over a much greater area. The use of detergent resulted in a small but significant increase in efficacy of extraction, although this was not observed in all cases. The standard AATCC detergent used here is not considered representative of present-day commercial detergents and may not have the same cleaning ability as commercial detergents. Some investigators have observed significant differences in the efficacy of commercial detergents in removing pesticide residue [10,11] while other investigators have seen essentially no differences [3]. This may reflect the properties of the different pesticides studied and their formulations as well as fabric and detergent type.

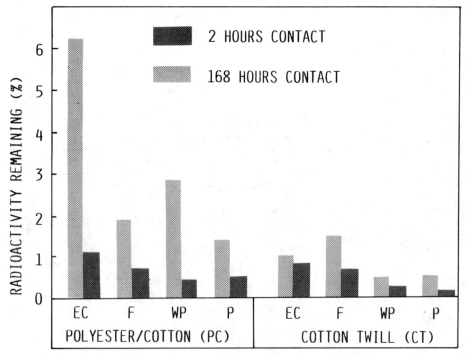

FIG. 1—*Effect of contact time on efficacy of carbaryl extraction from fabrics. Specimens maintained 22°C and 65% relative humidity. Extraction protocol is described in Table 5 and included detergent. The differences in the mean values between contact time, for each formulation, were all statistically different at P ≤ 0.05 (F = 523.6). Formulation designations are given in Table 2.*

Of the four solutions of carbaryl used, the F formulation of carbaryl was slightly more resistant to aqueous extraction than the others (Tables 6 and 7). The CD fabric retained more pure chlorothalonil and more pesticide from the F formulations than the other fabrics, while CK fabric retained slightly more residue from the WP and EC formulations of carbaryl. If the data in Table 6 are expressed on the basis of pesticide weight per weight of fabric, the PC fabric material generally retained more of the pesticide than the all-cotton fabric, however, the PC fabrics also received more pesticide on a weight/weight basis than the other fabrics. An earlier study on carbaryl reported complete removal of the insecticide after laundering [8,9], however, the investigators used an extraction solvent that was found to be ineffective in removing carbaryl from fabric for chromatographic purposes [18].

Effect of Exposure Time—In general, prolonged exposure of fabric to soil reduces the removability of the soil [23]. This was also observed here using carbaryl and exposure times of 2 and 168 h (Fig. 1). Overall, the retention effect was greater with the PC fabric than with the all-cotton fabric. The EC formulation of the carbaryl showed the largest retention effect, but only with the PC fabric.

Conclusions

Inactive materials in pesticide formulations appear to favorably affect aqueous extraction of pesticides that are insoluble in water. The inactive ingredients also reduce the effectiveness of dry cleaning solvents in extracting pesticide residues. Dry cleaning solvents can transfer pesticide from contaminated garments to those not contaminated, even after removal of the original contaminated article. Therefore, dry cleaning of pesticide contaminated clothing is not recommended.

Previous studies have indicated that multiple extractions of a few garments with large volumes of warm water is most effective in reducing pesticide contamination on apparel fabrics. Our studies confirm the efficacy of this protocol for removal of chlorothalonil and carbaryl residues. Laundering of contaminated fabrics should be carried out as soon as possible after contact with pesticide.

Acknowledgments

The authors extend their appreciation to Ann Bratten and Jody Fleeker who assisted in the studies.

References

[1] Worthing, C. R. and Walker, S. B., *The Pesticide Manual*, 7th ed., Lavenham Press, England, 1983.
[2] Leavitt, J. R. C., Gold, R. E., Holeslaw, T., and Tupy, D., "Exposure of Professional Pesticide Applicators to Carbaryl," *Archives of Environmental Contamination and Toxicology*, Vol. 11, 1982, pp. 57–62.
[3] O'Brian, R. D. and Dannelly, C. E., "Penetration of Insecticide Through Rat Skin," *Journal of Agriculture and Food Chemistry*, Vol. 13, 1965, pp. 245–247.
[4] Gold, R. E., Leavitt, J. R. C., Holeslaw, T., and Tupy, D., "Exposure of Urban Applicators to Carbaryl," *Bulletin of Environmental Contamination and Toxicology*, Vol. 11, 1982, pp. 63–67.
[5] Kawar, N. D., Gunther, G. A., Serat, W. F., and Iwata, Y., "Penetration of Soil Dust Through Woven and Nonwoven Fabrics," *Journal of Environmental Science and Health*, Vol. 13, 1978, pp. 401–415.

[6] Mount, M. E. and Oehme, F. W., "Carbaryl: A Literature Review," *Residue Reviews*, Vol. 80, 1981, pp. 1–64.

[7] Zimmerli, B. and Marek, B., "Transfer of the Biocidal Material Tetrachloroisophthalonitrile (Chlorothalonil) from Paints into the Gaseous Phase," *Chemosphere*, Vol. 6, 1977, pp. 215–221.

[8] Lillie, T. H., Livingston, G. M., and Hamilton, M. A., "Recommendations for Selecting and Decontaminating Pesticide Applicator Clothing," *Bulletin of Environmental Contamination and Toxicology*, Vol. 27, 1981, pp. 716–723.

[9] Livingston, J. M., "Evaluation of Cotton and Polyester Coveralls for Protection for Pesticide," TR 78-75, USAF Occupational and Environmental Health Laboratory Technical Report, U. S. Air Force, Wright-Patterson Air Force Base, OH, Aug. 1978

[10] Keaschall, J. L., Laughlin, J. M., and Gold, R. E., "Effect of Laundering Procedures and Functional Finishes on Removal of Insecticides Selected from Three Chemical Classes," *Performance of Protective Clothing, ASTM STP 900*, R. L. Barker and G. C. Coletta, Eds., American Society for Testing and Materials, Philadelphia, 1986, pp. 162–176.

[11] Kim, C. J., Stone, J. F., Coats, J. R., and Kadolph, S. J., "Removal of Alachlor Residues from Contaminated Clothing Fabrics," *Bulletin of Environmental Contamination and Toxicology*, Vol. 36, 1986, pp. 234–241.

[12] *Radiotracer Techniques and Applications*, Vol. I, E. A. Evans and M. Muramatsu, Eds., Marcel Dekker, New York, 1977.

[13] Sherma, J., "Thin-layer Chromatography: Recent Advances," *Analytical Methods for Pesticide and Plant Growth Regulators*, G. Zweig, Ed., Vol. 7, 1973, pp. 3–87.

[14] *Standard Methods for the Examination of Water and Wastewater*, 15th Ed., American Public Health Association, Washington, DC, 1980.

[15] Patterson, M. S. and Greene, R. C., "Measurement of Low Energy β-Emitters in Aqueous Solution by Liquid Scintillation Counting of Emulsions," *Analytical Chemistry*, Vol. 37, 1965, pp. 854–857.

[16] Ballee, D. L., Duane, W. C., Stollard, D. E., and Wolfe, A. L., "Chlorothalonil," *Analytical Methods for Pesticides and Plant Growth Regulators*, G. Zweig, Ed., Vol. 8, 1976, pp. 263–274.

[17] Laughlin, J. M. and Gold, R. E., "Vaporization of Methyl Parathion Soil from Unfinished and Soil Repellent Finished Cotton and Cotton/Polyester Twill Fabrics," *AATCC Book of Papers*, American Association of Textile Chemists and Colorists, 1986, pp. 294–298.

[18] Fleeker, J. R., Nelson, C. N., Braaten, A. W., and Fleeker, J. B., "Quantitation of Pesticides on Apparel Fabrics," this symposium, pp. 745–749.

[19] Ringenberg, K., Laughlin, J. M., and Gold, R. E., "Removing Chlorpyrifos Residues from Clothing Fabrics Through Drycleaning or Laundering," *Research Abstracts*, American Home Economics Association, 1986, pp. 84–87.

[20] Easley, C. B., Laughlin, J. M., Gold, R. E., and Tupy, D., "Methyl Parathion Removal from Denim Fabrics by Selected Laundry Procedures," *Bulletin of Environmental Contamination and Toxicology*, Vol. 27, 1981, pp. 101–108.

[21] Finley, E. L., Metcalfe, G. I., and McDermott, F. G., "Efficacy of Home Laundering in Removal of DDT, Methyl Parathion and Toxaphene Residues from Contaminated Fabrics," *Bulletin of Environmental Contamination and Toxicology*, Vol. 12, 1974, pp. 268–274.

[22] Olsen, M. M., Janecek, C., and Fleeker, J. R., "Removal of Paraquat from Contaminated Fabrics," *Bulletin of Environmental Contamination and Toxicology*, Vol. 37, 1986, pp. 558–564.

[23] Wentz, M., Lloyd, A. C., and Watt, A., "Experimental Removal of Stain," *Textile Chemists and Colorist*, Vol. 7, 1975, pp. 30–34.

Daniel J. Ehntholt,[1] *Robert F. Almeida,*[1] *Kevin J. Beltis,*[1]
Deborah L. Cerundolo,[1] *Arthur D. Schwope,*[1] *Richard H. Whelan,*[1]
Michael D. Royer,[2] *and Alan P. Nielsen*[3]

Test Method Development and Evaluation of Protective Clothing Items Used in Agricultural Pesticide Operations

REFERENCE: Ehntholt, D. J., Almeida, R. F., Beltis, K. J., Cerundolo, D. L., Schwope, A. D., Whelan, R. H., Royer, M. D., and Nielsen, A. P., **"Test Method Development and Evaluation of Protective Clothing Items Used in Agricultural Pesticide Operations,"** *Performance of Protective Clothing: Second Symposium, ASTM STP 989,* S. Z. Mansdorf, R. Sager, and A. P. Nielsen, Eds., American Society for Testing and Materials, Philadelphia, 1988, pp. 727–737.

ABSTRACT: Protective clothing is the principal defense of millions of farm workers and pesticide mixers/applicators against skin contamination by pesticides. Unfortunately, few data are available in the public domain to permit the identification of appropriate protective items for specific categories and types of pesticides and pesticide formulations. The U.S. Environmental Protection Agency (EPA) has therefore sponsored an effort to produce, assemble, and critically evaluate data on the effectiveness of protective clothing for agricultural uses. This paper describes (1) efforts to develop a rapid screening method for determining the compatibility of various glove materials with selected pesticides and (2) comparative analyses of collection media for use with the ASTM Test Method for Resistance of Protective Clothing Materials to Permeation by Liquids and Gases (F 739–85) permeation cell in tests of low-volatility, low-water-solubility active ingredients in pesticide formulations.

Two methods for the rapid screening of protective glove-pesticide formulation compatibilities were tested. Results obtained with a draft ASTM Test Method for Evaluating Protective Clothing Materials for Resistance to Degradation by Liquid Chemicals suggest that inappropriate formulation-glove combinations may be identified by their excessive weight increase after 1-h exposure.

Permeation testing to determine breakthrough times and measure the permeation of formulation components using the ASTM F 739–85 cell and selected gloves and pesticide formulations is a major focus of this program. However, the collection medium to be used in this test when evaluating low-volatility, low-water-solubility pesticides poses a particular program. Various media (water, aqueous surfactant solutions, aqueous alcohol solutions, and solid phase collection materials) were tested using actual pesticide formulations spiked with ^{14}C-labeled active ingredients. A solid silicone rubber material was identified as the most efficient material for collection of parathion and ethion. It has now been used with several protective glove materials and two concentrated pesticide formulations.

[1] Chemical Measurement Technology unit manager, chemist, chemist, chemist, Applied Polymer Science unit manager, and pesticide applicator, respectively, Arthur D. Little, Inc., Cambridge, MA 02140.
[2] Project officer, Release Control Branch, Hazardous Waste Engineering Research Laboratory, U.S. Environmental Protection Agency, Edison, NJ 08837-3679.
[3] Biologist, Exposure Assessment Branch, Hazard Evaluation Division (TS-769C), U.S. Environmental Protection Agency, Washington, DC 20460.

KEY WORDS: pesticides, pesticide formulations, organophosphorus compounds, protective clothing, gloves, test methods, collection media, sampling and analysis, permeation tests

The potential ability of protective clothing items to significantly reduce the exposure of workers to pesticides is widely recognized. However, insufficient data are available in the public domain to permit the identification of appropriate materials for protection against specific categories and types of pesticides and pesticide formulations. For example, it is not known how the concentration of the active ingredient or the concentration and composition of other "inert" components affects breakthrough times and permeation rates of formulated pesticide products through protective clothing items.

The U.S. Environmental Protection Agency's (EPA's) Office of Pesticide Programs (OPP) is responsible for protecting human health and the environment from the adverse effects of pesticides, pursuant to its authority under the Federal Insecticide, Fungicide, and Rodenticide Act (FIFRA). This responsibility is carried out through regulations and training. It is the objective of this work to produce, assemble, and critically evaluate data on the effectiveness of protective clothing for agricultural uses. The information will then be used to support the development and evaluation of information and regulations by OPP. Two specific areas of interest are being pursued:

1. Evaluation of the effectiveness of polymer gloves for protecting against pesticide exposures.
2. Identification and quantification of correlations between characteristics of protective clothing materials, pesticide formulations, and protective clothing effectiveness.

Primary laboratory emphasis in the program is on the measurement of breakthrough times of concentrated and application-strength pesticide formulations through a variety of rubber gloves. Initial work focused on the identification of "rapid screening" test procedures (that is, compatibility or degradation tests) that might be used to identify and eliminate inappropriate pesticide-glove combinations from further testing (or field use). Other efforts on this program have addressed the identification of an efficient and reproducible collection medium for use in the permeation testing of pesticide formulations containing low-volatility, low-water-solubility active ingredients.

Eventually, this program is expected to identify and verify correlations in the permeation and compatibility data generated in this work and in the literature that will significantly improve the process of selecting or recommending appropriate protective clothing for agricultural uses.

Rapid Screening Tests

To minimize the number of time-consuming permeation tests that must be conducted, a test was sought that would rapidly eliminate inappropriate pesticide-glove combinations at the outset. Such a test could also be considered for use or specification by the EPA for investigatory or regulatory purposes.

Two approaches to be the development of a rapid screening test protocol were examined:

1. The draft ASTM Test Method for Evaluating Protective Clothing Materials for Resistance to Degradation by Liquid Chemicals" (NIOSH 200-84-2702 DEG, Revision 3, 6/5/85) uses a modified "sandwich" configuration consisting of a one-chamber cell (Fig. 1). The cell is modeled after those described in ASTM Test Method for Rubber

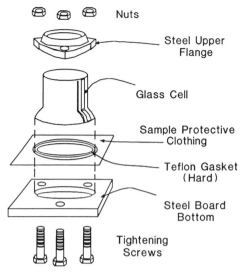

FIG. 1—*Modified ASTM test cell.*

Property—Effect of Liquids (D 471-79), and is designed to bring the normal outside surface of a material specimen in contact with a test chemical. The exposed area is 20 cm^2, and sufficient pesticide formulation was applied to cover that area to a 1.0 to 1.5 cm minimum depth.

To determine the effects of exposure, the weight and thickness of the specimen were measured before and after the test.

2. Degradation tests were also run in a cell that was designed by Arthur D. Little, Inc., for use in testing the effects of chemicals for the U.S. Coast Guard. Like the ASTM cell, it exposes one side of a material to a chemical, but it accommodates a sufficiently large specimen (44 cm^2) to permit standard ASTM tension test "dog bones" to be obtained. These were used to measure the elongation of the specimens by subjecting them to a specific tensile stress.

Most of the tests used the proposed ASTM method, since it may become a standard, but some were duplicated on the Coast Guard cell to permit elongation testing. Weight, thickness, and elongations changes were measured with the intention of correlating the results with those of the permeation testing. The initial measurements were compared with others made 1 and 3 h after exposure.

Twenty-one different concentrated pesticide formulations were selected to represent active ingredients having a range of solubilities in water, a variety of structural categories (for example, amides, organophosphates, and organochlorines), and different carrier solvent types (for example, alcohols, xylenes, ketones, and oils).

The glove materials included those normally recommended on pesticide labels and material safety data sheets; other protective glove materials, both more and less resistant, were also tested. Table 1 lists the manufacturers, designations, and thicknesses of the ten types of gloves examined. In several cases, different "brands" of the same generic materials were tested to ensure that the results were generally applicable. All tests were done in duplicate, and measurements were made after both 1 and 3 h of exposure.

TABLE 1—*Characteristics of protective glove samples tested.*

Glove Description	Manufacturer	Model	Nominal	Actual[a]
Nitrile #1	Edmont	Sol-Vex	15	20
Nitrile #2	Pioneer	Stanslov A-15	16	20
Natural rubber #1	Granet	541	18	26
Natural rubber #2	Pioneer	ivory white	18	23
Polyethylene	Fisher	polyethylene	[b]	2
PVC	Pioneer	Stan-Flex	20	19
Neoprene #1	Ansell	neoprene	18	21
Neoprene #2	Edmont	neoprene	18	19
Butyl	North	butyl	16.5	19
Silver shield[c]	North	silver shield	[b]	4

[a] Average of 6 to 10 samples.
[b] Not specified by manufacturer.
[c] Proprietary laminated structure.

Table 2 lists the weight changes (in percent) as measured by the draft ASTM degradation test when the various glove materials were exposed for 1 h to the 21 pesticide formulations. The increases range up to 130%. There are a few negative values, which indicate extraction of glove components by the formulation. A similar variety of weight change percentages was observed on the smaller group of specimens tested in the Coast Guard cell.

In all the test results, the weight increases displayed clear and consistent trends, that is, if the weight increased after 1 h of exposure, a greater percentage increase was measured after 3 h. This was not true of the thickness and elongation measurements; the absolute direction and magnitude of the change that occurred between 1 and 3 h of exposure was not always consistent or predictable.

Table 3 compares the weight-change percentages in Table 2 with the total comulative amounts of carrier solvent (xylenes measured as a meta-xylene) and active ingredient that permeated the specimen for three cases of Parathion 8-E and Methyl Parathion 4E concentrated pesticide formulations and glove materials. The Parathion 8E formulation was composed of 72% parathion and 10% xylenes whereas the Methyl Parathion 4E contained 43% methyl parathion and 48% xylenes carrier solvent. The amounts of carrier solvent and active ingredient were measured over the 8-h permeation test period. In each case, when results of challenge of a single glove type by each formulation is compared, an increase in the percentage weight change after a 1-h exposure of glove material to the formulation corresponded to an increase in the cumulative amount of active ingredient and carrier solvent permeating the glove over the 8-h test period. Although only a few permeation tests have been conducted to date, the degradation test thus shows promise for use as a screening device to select resistant glove materials.

Collection Media for Low-Volatility, Low-Water-Solubility Pesticides

To accurately measure the permeation of active ingredients through protective gloves, it is necessary to select an appropriate medium for collecting the permeant. As currently used, ASTM F 739-85 typically involves collecting the permeant in either water or a continuous gas stream. In many pesticide formulations, however, the active ingredient of interest is neither water-soluble nor very volatile. Therefore, it is necessary to identify a collection medium that will efficiently dissolve and collect the active ingredient (as well as other permeating materials). In addition, it should not itself be absorbed into the challenged glove

TABLE 2—Results of ASTM degradation tests (percent by weight change following 1-h exposure).

Formulation	Active Ingredient/ Carrier Solvent	Natural #1 (Granet)	Natural #2 (Pioneer)	Nitrile #1 (Edmont)	Nitrile #2 (Pioneer)	Neoprene #1 (Ansell)	Neoprene #2 (Edmont)	PE	PVC	Butyl	Silver Shield
Bidrin 8WM	dicrotophos	5.3	...	4.6	2.8	3.7	...	...	1.1	...	-2.1
Azodrin 5WM	monocrotophos	8.2	...	16.4	...	3.2	...	...	0.3	-0.1	5.4
Phosdrin 4EC	mevinphos xylenes	57.1	45.9	...	48.1	...	56.2	...	...	10.6	...
Monitor 4 Spray	methamidophos	0.3	...	...	...	0.2	...	...	...	...	...
Monitor 4	methamidophos	1.0	...	...	...	0.6	...	...	...	...	...
Metasystox-R Spray Concentrate	oxydemeton-methyl petroleum distillates	39.2	33.0	...	42.7	...	38.0	...	...	3.0	...
Parathion 8-E	ethyl parathion (76%) xylenes (10%)	15.6	11.6	15.5	9.5	26.5	16.4	12.1	10.4	0.9	0.9
Phoskil Spray	ethyl parathion dry powder	0.4	...	-0.6	...	-0.4	...	...	...	...	...
Methyl Parathion 4E	methyl parathion (43%) xylenes (48%)	64.6	...	19.1	...	74.5	...	...	...	...	...
Aqua 8 Ethion	Ethion	2.5	...	0.9	...	8.1	...	...	2.3	-1.0	...
Ethion Superior 70 Oil	Ethion oil	11.6	...	...	...	0.0	...	...	...	7.4	...
Diazinon 4E	diazinon xylenes	34.7	33.9	...	3.4	29.7	...	14.7	7.2	2.2	-3.3
Diazinon 14G	diazinon dry	...	0.2	...	...	0.6	...	...	...	...	...
Diazinon AG500	diazinon xylenes	...	49.0	...	...	39.7	...	...	...	...	...
Diazinon 2MEC	diazinon xylenes	...	10.1	...	...	10.9	...	...	3.2	...	-4.0
Dibrom 8E	naled petroleum distillates	65.5	45.1	31.1	16.5	62.8	34.4	15.4	26.8	8.3	0.5
Cygon 2-E	dimethoate	61.3	...	21.3	...	64.5	...	...	...	...	...
Zolone EC	phosalone	...	46.2	...	...	...	38.9	...	71.2	8.0	6.0
DEF 6	DEF	...	40.2	...	...	...	23.9	...	...	...	...
Thiodan 3EC	endosulfan xylenes	...	112.8	...	6.3	...	...	...	...	...	...
Thiodan 3EC	endosulfan xylenes	...	130.7	...	-5.2	...	...	...	...	...	...

TABLE 3—*Comparison of percent by weight change from ASTM test (1 h) to permeation data.*[a]

	Parathion 8-E (76% parathion, 10% xylenes) Glove			Methyl Parathion 4E (43% methyl parathion, 48% xylenes) Glove		
	Natural	Nitrile	Neoprene	Natural	Nitrile	Neoprene
Percent by weight change	15.6	15.5	26.5	64.6	19.1	74.5
Cumulative amount of meta-xylene from carrier solvent permeated at 8 h (μg/cm^2)	558	ND[b]	172	4690	2170	6130
Cumulative amount of active ingredient permeated at 8 h (μg/cm^2)	5.0	ND	1.0	230	33	262

[a] Average of three tests.
[b] ND = none detected.

material, thereby altering the barrier characteristics of the glove, and it should be compatible with the analytical chemistry method(s) used to quantify the collected compounds.

In this program, various collection media were evaluated in a series of tests. The candidates included aqueous isopropanol solutions, aqueous solutions containing surfactants, and solid collection media such as gauze pads, filter paper, and solid polymeric materials. The aqueous isopropanol solutions were chosen because the alcohol should solubilize many of the active ingredients as well as carrier solvents. Initial tests were conducted with 5% isopropanol in water to determine its relative utility compared with an aqueous medium. Isopropanol was chosen as a result of information in the literature that suggested that it is absorbed less quickly than other organic solvents into the material found in many protective clothing items.[4]

Similarly, the surfactants Sur-ten (dioctyl sodium sulfosuccinate) and ammonium lauryl sulfate were chosen for study in the solubilization of typical active ingredients. A 1% solution of Sur-ten has been used by the California Department of Food and Agriculture to wash the inside of protective gloves in a study designed to identify residual pesticide contamination.[5] A 6% solution of aqueous ammonium lauryl sulfate is a commonly used surfactant solution.

Absorbent solid materials were also of interest, although removal of such a collection medium would terminate the permeation test (that is, there would be only one data point for each test, unless the standard permeation test cell were slightly modified). A gauze pad and filter paper were initially chosen as candiate solid collection media.

To evaluate the selected media, ^{14}C-labeled parathion and ^{14}C-ethion (from Pathfinder Laboratories) were used in spiking commercial concentrated pesticide formulations of those low-water-solubility active ingredients. Solubility studies conducted with the concentrated ^{14}C-labeled parathion formulation showed that neither aqueous isopropanol nor the Sur-ten

[4] Sakura, J. D., Schwope, A., Cotas, P., and Goldman, R. F. "Interim Guidance Manual for Selecting Protective Clothing for Agricultural Pesticide Operations," U.S. Environmental Protection Agency, Contract 68-03-3293, Sept. 1986.

[5] Schwope, A., Costas, P., Jackson, J., and Weitzman, D. "Guidelines for the Selection of Chemical Protective Clothing," American Conference of Governmental Industrial Hygienists, Cincinnati, 1985.

TABLE 4—*Solubility test results using ^{14}C-labeled parathion 8-E.*

Candidate Medium	Disintegrations per Minute in Supernatant Solution (corrected for background)[a]
Water	18 500[b]
Isopropyl alcohol in water	
5%	14 000[c]
10%	10 500[b]
20%	11 500[b]
30%	34 000[b]
50%	103 000[b]
1% Sur-Ten in water	18 000[c]
6% ammonium lauryl sulfate in water	22 000[c]

[a] Theoretically available disintegrations per minute were 104 000.
[b] Average of four determinations.
[c] Average of two determinations.

solution was significantly better than a water solution in dissolving the parathion. Table 4 shows the results obtained when 5 µL of radiolabeled parathion formulation were added to 5 mL of the collection medium of interest. This table also shows results obtained with a range of isopropanol concentrations. Only when a significant amount of alcohol was used (isopropanol constituted 30 to 50% of the total amount of the solution) did the solubility of the active ingredient increase substantially. However, the use of such a high concentration of an alcohol for up to 8 h in a permeation test cell might distort permeation results, since isopropanol is known to be absorbed by many protective glove materials. It was therefore rejected as a generally applicable collection medium.

Subsequently, four solid collection materials were tested using a modified ASTM F 739-85 permeation test procedure:

1. filter paper (Whatman #1),
2. synthetic gauze (polyester/rayon),
3. cotton gauze, and
4. silicone rubber sheeting, 0.102 and 0.05 cm (0.04 and 0.02 in.) thicknesses (Silastic[6] silicone rubber).

The gauzes were selected for evaluation based on the recommendations of workers in this field who have used this type of material as a collection medium in permeation tests with pesticides. Filter paper was chosen as an alternate collection medium since it is frequently used in surface wipe or "swipe" tests. The silicone rubber sheeting was selected on the basis of previous work in which this material was found useful for sampling surfaces contaminated with chemicals.[7] Use of these media required modification of the standard ASTM F 739-85 permeation cell to allow the insertion and removal of the collection media. Briefly, the collection chamber of the cell was modified to allow placement of the solid in intimate

[6] Registered trademark of Dow-Corning, Inc.
[7] Schwope, A. D., Masucco, A., and Wilson, D. "Development of Chemical Warfare Agent Simulants and Methods for their use in Evaluating the Effectiveness of Decontaminating Components," Report N62269-85-C-0243, Naval Air Development Center, Warminster, PA, Aug. 1986.

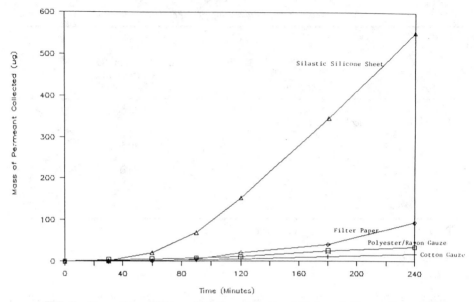

FIG. 2—*Permeation of ^{14}C-parathion from Parathion 8-E through Pioneer (Ivory White) natural rubber material to several solid collection media.*

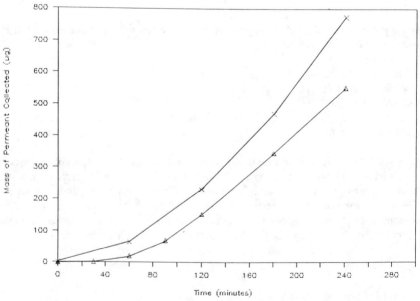

FIG. 3—*Permeation of ^{14}C-parathion from Parathion 8-E through Pioneer (Ivory White) natural rubber to 0.102 and 0.05 cm (0.02 and 0.04 in.) of silicone rubber sheeting as the collection media.*

contact with the downstream surface of the glove material. Periodically, the solid is removed and then extracted in order to measure the mass of chemical(s) that permeated during the contact period. The solid is replaced with fresh collection media and the cycle repeated for the duration of the test. The collection media can be replaced as frequently as desired; when more often, the more definitive is the resulting permeation curve and the detection of breakthrough.

Figure 2 summarizes the results of a series of permeation tests using a natural rubber glove material (Pioneer) and radiolabeled Parathion 8E concentrated formulation. For this series, the permeation tests were conducted using 0.102 cm (0.04 in.) silicone rubber sheeting as the collection medium. The silicone rubber sheeting shows a significantly greater efficiency for collecting the radiolabeled parathion than the other three solid collection media.

Permeation tests using a second thickness of the silicone rubber sheeting, 0.05 cm (0.02 in.), were also conducted with the natural rubber glove material using the radiolabeled Parathion 8-E. The 0.05 cm (0.02 in.) sheet is pliable, readily conforms to the glove surface, and thus appears to be more adaptable for use in the modified ASTM permeation test cell although its capacity for permeant would be less than that of the 0.102 cm (0.04 in.) sheeting. (In some cases, this capacity may be required for collection of rapidly permeating carrier solvents.) Results of permeation tests with the two silicone rubber thicknesses are shown in Fig. 3.

TABLE 5—*Permeation data summary for concentrated pesticide formulations using silicone rubber sheeting as the collection medium[a] (8-h permeation test).*

Glove Material	BTR,[b] min	Total Mass Permeated, $\mu g/cm^2$	BTR, min	Total Mass Permeated, $\mu g/cm^{2c}$
	PARATHION 8-E Active: Parathion (72%)		Carrier: Xylene (10%)	
Natural rubber	30 to 60	4.7	15 to 30	523
	120 to 180	7.1	15 to 30	593
	120 to 180	3.0	15 to 30	559
Nitrile	ND[d]	ND	ND	ND
Neoprene	240 to 360	1.6	120 to 180	186
	360 to 480	0.3	120 to 180	163
	240 to 360	1.0	120 to 180	166
	METHYL PARATHION 4E Active: Methyl Parathion (43%)		Carrier: Xylene (48%)	
Natural rubber	0 to 15	202	10 to 15	4972
	15 to 30	214	0 to 15	5–35
	15 to 30	278	0 to 15	4065
Nitrile	180 to 240	32.7	120 to 180	2068
	180 to 240	24.4	120 to 180	2180
	180 to 240	42.3	120 to 180	2260
Neoprene	30 to 60	266	15 to 30	6385
	15 to 30	181	15 to 30	6125
	60 to 120	340	30 to 60	5884

[a] Samples were taken at 0, 15, 30, 60, 120, 180, 240, 360, and 480 min.
[b] BTR = breakthrough time range (first sample showing presence of permeant was collected over this time period).
[c] Values reported are for meta-xylene component of carrier solvent.
[d] ND = none detected up to 8 h.

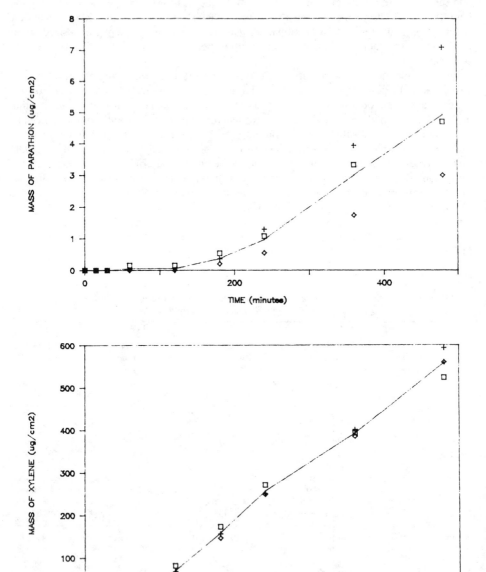

FIG. 4—*Permeation of parathion and xylenes from Parathion 8-E through natural rubber to silicone rubber sheeting. Xylene data are for meta-xylene isomer.*

In consideration of all of the preceding findings, the silicone rubber sheeting was selected as the collection medium for subsequent permeation tests with active ingredients having low water solubility. These tests are in progress and are providing further insight into the utility and limitations of the collection medium. Some initial results are described in the following discussion.

Initial Permeation Test Results for Concentrated Parathion and Methyl Parathion Formulations

Permeation tests were conducted with two concentrated pesticide formulations, Parathion 8-E and Methyl Parathion 4E. Each contains low-water-solubility, low-volatility active ingredients (parathion and methyl parathion), and xylene range aromatics as carrier solvents. As noted earlier, the Parathion 8-E formulation contains approximately 76% parathion as the active ingredient and 10% xylenes as carrier solvent. The Methyl Parathion 4E formulation contains approximately 43% methyl parathion and 48% xylenes.

All permeation tests were performed using the modified ASTM F 739-85 with the 0.05 cm (0.02 in.) silicone rubber sheeting as the collection medium. The permeation of both the active ingredient and the xylene carrier solvent was measured in each test.

The tests were conducted by regularly removing and replacing the silicone rubber sheeting on the collection side of the protective glove material, extracting permeated compounds from the silicone rubber, and analyzing the resulting solution. Table 5 contains the measured breakthrough time range and cumulative mass permeated over 8 h for each of the triplicate tests conducted with three glove materials. These data were shown earlier to correlate with percent by weight change results from the degradation tests.

Figure 4 depicts the permeation of parathion and xylenes from Parathion 8-E through a natural rubber glove material, as measured using the silicone rubber sheeting as a collection medium.

We are currently examining the applicability of the collection method to a wide range of pesticide formulations. As the rapid screening test results discussed earlier and actual permeation data for various pesticide formulations and glove materials become available, we will continue to examine the data for useful correlations. If it is possible to rapidly eliminate incompatible pesticide formulation-protective glove combinations, this would be a useful screening tool. We also anticipate that analysis of the permeation data will yield information about the permeation of active and inert ingredients, which may permit better specification and formulation of such ingredients as well as more appropriate specification of protective clothing requirements.

Acknowledgment

Although the research described in this article has been funded wholly or in part by the United States Environmental Protective Agency Contract No. 68-03-3293 to Arthur D. Little, Inc., it has not been subject to the Agency's review and therefore does not necessarily reflect the views of the Agency, and no official endorsement should be inferred. Mention of trade names or commercial products does not constitute endorsement or recommendation for use.

Nancy E. Hobbs[1] and Billie G. Oakland[2]

Use of Methylene Blue Dye to Predict Fabric Penetration by Malathion

REFERENCE: Hobbs, N. E. and Oakland, B. G., **"Use of Methylene Blue Dye to Predict Fabric Penetration by Malathion,"** *Performance of Protective Clothing: Second Symposium, ASTM STP 989*, S. Z. Mansdorf, R. Sager, and A. P. Nielson, Eds., American Society for Testing and Materials, Philadelphia, 1988, pp. 738–744.

ABSTRACT: This study developed and verified a test method for the screening and selection of fabric finishes proposed for use in protective garments for agricultural workers exposed to pesticide sprays. Six nonwoven and four woven fabrics with and without barrier finishes were tested for aerosol spray penetration using 0.1% methylene blue dye as an indicator of pesticide penetration with three spray emulsions: (1) water; (2) water/surfactant (48:1); and (3) unrefined cottonseed oil/surfactant (4:1). The procedure was repeated using the following pesticide emulsions: (1) water/malathion (100:4) and (2) unrefined cottonseed oil/malathion (100:8).

Analysis of variance was used to compare the results of the water/malathion and the results of the cottonseed oil/malathion spray tests between categories of dye penetration. In each case, there was significant difference between group variation. Duncan's Multiple Range Test ($p < 0.05$) indicated that fabrics that did not allow penetration by the methylene blue dye had significantly lower levels of malathion present in the extracts. The spray tests indicated that aerosol spray penetration was reduced by the presence of fluorocarbon barrier finishes on the fabrics.

KEY WORDS: aerosol spray, pesticides, protective clothing, fluorocarbon finish, disposable garments, nonwoven fabrics, malathion

The purpose of this study was to develop and verify a test method for the screening and selection of fabrics and fabric finishes proposed for use in protective garments for pesticide handlers and sprays. The fabrics in question included fabrics for use in reusable garments, semi-disposable or limited-use garments, and disposable garments. The development of this test method was prompted by the need to identify fabrics, fabric finishes, and seaming techniques that could be used in the manufacture of protective garments for mixers and applicators of aerosol pesticides. The criteria for the test method were that (1) the test be easy to administer with a minimum of equipment, (2) the results should be easily interpreted, and (3) the test be inexpensive to run.

Another aspect of the study was to examine the effectiveness of fluorocarbon finishes as barriers to aerosol penetration. Pesticide sprays are applied in either a water-based emulsion of the pesticide or an oil-based emulsion of the pesticide dissolved in an organic solvent. Clothing sprayed with a wet pesticide spray (aerosol) is not only a source of direct dermal contamination but also a source of vapor that when diffusing toward the skin also results in dermal contamination from the pesticide.

[1] Assistant professor, School of Home Economics, East Carolina University, Greenville, NC.
[2] Professor of Clothing and Textiles, School of Human Environmental Science, The University of North Carolina at Greensboro, Greensboro, NC 27412.

The application of a barrier finish coats the fabric surface, filling in crevices and surface rugosities on fibers. Finishes also fill in the interstices between fibers in nonwoven fabrics and between fibers in yarn bundles in woven fabrics. This polymetric coating limits entrapment of soils and liquids on the fabric surface [1].

Both fabric structure and surface free energy affect the extent of fabric penetration by impinging aerosol droplets. Phobicity between a liquid and a solid film is dependent on contact angles of 90° or more [2]. The critical surface energy for wetting of a solid is the amount of surface energy required by a liquid to ensure perfect wetting. A liquid of equal or lower surface energy of a given solid would exhibit a contact angle of zero. A liquid of higher surface energy would yield a finite contact angle on that solid [3].

Barrier finishes, such as fluorocarbon-based finishes impart oleophobic and hydrophobic properties by lowering fiber/fabric free surface energy, thus limiting wetting by any solvent [4]. Zisman (1964) stated that finishes such as poly 1, 10 dihydroperfluoroctyl methacrylate have very low surface energy values ($10.6/\gamma$) and impart both oil and water repellency [3]. Silicon-based finishes do not lower the fabric free surface energy to a value low enough to render the fabric oil repellent [5].

Materials

Two general classes of fabrics, woven and nonwoven, were selected for this study. The nonwoven fabrics were tested both in the finished and the unfinished state. Unfinished nonwoven fabrics were laboratory finished with a Scotchgard[3] fluorocarbon finish. The nonwoven fabrics selected for the study were:

(a) spun-lace 100% polyester (90 g/m^2), unfinished and laboratory finished with Scotchgard;

(b) spun-lace rayon/polyester (70 g/m^2), commercially finished with a fluorocarbon-based finish by Johnson & Johnson for surgical gown use;

(c) spun-bonded 100% olefin (40 g/m^2), unfinished and laboratory finished with Scotchgard;

(d) spun-bonded polypropylene with melt-blown fibers (50 g/m^2), unfinished and finished with Scotchgard;

(e) spun-bonded polypropylene with melt-blown fibers (50 g/m^2), commercially finished by Kimberley Clark for industrial use (chemical base of finish is not known); and

(f) polyethylene coated spun-bonded olefin (40 g/m^2).

The woven fabrics selected for the study were:

(a) twill weave, denim, 100% cotton (283 g/m^2), unfinished;

(b) shirting-weight muslin, 100% cotton, commercially finished with Scotchgard (157 g/m^2);

(c) shirting-weight muslin, 100% cotton with a Quarpel® (fluorocarbon) finish (113 g/m^2); and

(d) shirting-weight muslin, 35% cotton/65% polyester, with a Quarpel (fluorocarbon) finish (117 g/m^2).

The nonwoven and woven fabrics were subjected to the following aerosol spray ratios by

---

[3] Registered trademark of 3M Company.

volume: (1) water, (2), water/surfactant (48:1), and (3) unrefined cottonseed oil with/surfactant (4:1). All carriers contained 0.1% methylene blue dye as an indicator. The surfactant was obtained from a pesticide manufacturer and is used in pesticide formulations. In addition, the test fabrics were subjected to aerosol sprays containing a commercial malathion emulsifiable concentrate containing 50% active ingredients, a surfactant, and other inert ingredients. The following volume ratios were used: (1) water/malathion (100:4) and (2) unrefined cottonseed oil/malathion (100:8). Filter paper (18.5 cm in diameter) was used as the backing substrate.

Procedure

The methodology reported earlier by Hobbs et al. proposed the use of methylene blue dye spray to predict pesticide penetration [6,7]. The test method developed was based on the modification of ASTM Method of Salt Spray (Fog) Testing (B 117-73 (1979)), developed primarily for testing metallic and metallic-coated specimens.

Methylene Blue Aerosol Spray Penetration

The spray test procedure was designed to assess the resistance of fabrics to penetration of aerosol sprays. The testing procedure consisted of mounting a fabric sample (18.5 by 18.5 cm) in a frame over an absorbent backing (18.5 cm diameter filter paper) and placing the framed sample upright in a containment box 30.4 cm (12 in.) in front of an airless spraying device. A 12.5 cm diameter area was exposed to the aerosol spray. Three replications were completed for each fabric tested with each aerosol spray formulation. The spray time for the water-based solution was 50 s. The spray time for the oil-based solution was 30 s. The increased concentration of surfactant in the oil-based formulation resulted in a superior wetting system that reduced the spray time for sample differentiation to 30 s. The average amount of spray applied to the surface of the test fabric per square centimeter was 0.25 g/cm^2 for the water and the oil-based solutions. Scotchgard aerosol spray manufactured by the 3M Corporation was applied to the surface of unfinished nonwoven fabric. The Scotchgard spray was applied to thoroughly wet the fabric surface following label directions and allowed to dry for 24 h prior to testing.

Methylene blue dye was selected for use as the indicator of fabric penetration for the following reasons: (1) it is visible on the backing paper; (2) it is highly soluble in water and easily emulsifiable in the cottonseed oil and remained in emulsion over a 12-h period; (3) the molecular weight was within the range of the molecular weights of the organo-phosphate pesticides (methylene blue = 319.86, malathion = 330.36, and parathion = 291.27); and (4) methylene blue was easy to store and offered few clean-up problems.

The results of the spray tests were recorded as either pass or fail with no relative ranking of the degree of dye penetration. The presence of any methylene blue dye on the backing fabric within the test area was categorized as "fail." This procedure was selected since the objective of the research was the identification of fabrics that offered "complete" protection; therefore, graduation of aerosol dye spray test results were considered of little value.

Malathion Aerosol Spray Penetration

The aerosol pesticide spray test followed the same spray procedure used for testing the water-based dye spray solutions. However, to prevent the contamination of the filter paper, aluminum foil was placed under the filter paper. The test fabric was placed on top of the filter paper and the aluminum foil was folded over the edges and creased to seal the filter paper between the foil and the test fabric. The malathion spray did not contain the methylene blue dye indicator. Following the spray test, the filter paper was immediately removed, sealed in a self-sealable freezer bag, and placed in a container in the freezer until extraction procedures could be followed.

Each filter paper containing malathion was extracted for 15 min at room temperature using 25 mL of methanol as the solvent. This was followed by a 5-min rinse of 25 mL methanol for a total of 50 mL solution/extract. Following extraction, the solution/extract was frozen in individual glass containers until the gas chromatographic analyses could be performed. Prior to analysis the samples were allowed to warm to room temperature.

Analyses were performed using a Tracor® 560 gas chromatograph equipped with a flame photometric detector operating in the phosphorus specific mode (526-mm filter). Separations were achieved at an isothermal oven temperature of 210°C using 1.8 m by 2 mm glass column packed with 10% DC-200 on 80/100 GCQ (Applied Science Laboratories, Inc.). A detector operating temperature of 200°C and an injection port temperature of 225°C were used.

The standard stock solution used was a 1 to 20 dilution of the stock spray solution and contained 700 μg/mL as malathion. All analyses were based upon standards generated from this stock solution. When necessary, a small portion of a test sample solution/extract was diluted with methanol to achieve an appropriate concentration for gas chromatographic (GC) analysis.

TABLE 1—*Pass/fail results of the methylene dye spray penetration test.*

Fabrics Tested by Category	Water	Water/ Surfactant	Cottonseed Oil/Surfactant
CATEGORY I			
Scotchgard finished spun-lace polyester	pass	pass	pass
Commercially finished spun-lace rayon/polyester	pass	pass	pass
Scotchgard finished spun-bonded olefin	pass	pass	pass
Scotchgard finished spun-bonded/ MB polypropylene (50 g/m²)	pass	pass	pass
Polyethylene coated spun-bonded olefin	pass	pass	pass
CATEGORY II			
Commercially finished (Scotchgard) cotton	pass	pass	fail
Quarpel finished cotton	pass	pass	fail
Quarpel finished 35/65 cotton/ polyester	pass	pass	fail
CATEGORY III			
Spun-bonded olefin	pass	fail	fail
Spun-bonded/MB (50 g/m²)	pass	fail	fail
Commercially finished spun-bonded/MB (50 g/m²)	pass	fail	fail
CATEGORY IV			
Denim-unwashed	fail	fail	fail
Denim-3 washes	fail	fail	fail
Spun-lace polyester	fail	fail	fail

Statistical Analysis

The results of the aerosol dye spray test were evaluated visually and recorded as pass or fail. Failure was determined by noting the presence of methylene blue dye on the backing substrate. The fabrics were grouped according to performance (pass/fail) on the three aerosol dye penetration tests. The four groupings that resulted with the three spray tests were pass-pass-pass, pass-pass-fail, pass-fail-fail, and fail-fail-fail (Table 1).

Following extraction and analysis, the mean penetration of pesticide per cm^2 was computed for each fabric in test runs with malathion (Table 2). A one-way ANOVA was performed to examine between group variability by fabric category (Table 1) for both the malathion water test results and the Malathion/cottonseed oil test results. Duncan's Multiple Range tests, ($p < 0.05$) results were performed as post hoc tests. In addition, a two-way ANOVA was performed to compare the results of the two malathion test solutions by dye-spray test fabric category to examine similarities by malathion penetration between the water and oil-based solutions.

TABLE 2—*A comparison of aerosol dye spray penetration of water/malathion and cottonseed oil/ malathion penetration by test fabric test performance.*

Dye Test Performance Category by Fabrics Tested	Water/Malathion[a] Penetration, $\mu g/cm^2$	Cottonseed Oil/Malathion[a] Penetration, $\mu g/cm^2$
CATEGORY I, PASS-PASS-PASS		
Scotchgard finished spun-lace polyester	5.9	0.05
Commercially finished spun-lace rayon/polyester[b]	3.7	14.5
Scotchgard finished spun-bonded olefin[c]	5.1	0.04
Scotchgard finished spun-bonded/MB polypropylene (50 g/m²)[d]	16.5	0.07
Polyethylene coated spun-bonded olefin	4.7	0.03
CATEGORY II, PASS-PASS-FAIL		
Commercially finished (Scotchgard) cotton	5.6	0.04
Quarpel finished cotton	16.8	0.55
Quarpel finished 35/65 cotton/ polyester	16.4	45.3
CATEGORY III, PASS-FAIL-FAIL		
Spun-bonded olefin	17.8	114.9
Spun-bonded/MB (50 g/m²)	145.1	98.5
Commercially finished spun-bonded/MB (50 g/m²)	218.3	229.8
CATEGORY IV, FAIL-FAIL-FAIL		
Denim-unwashed	259.9	75.2
Denim-3 washes	156.1	14.9
Spun-lace polyester	177.9	207.6

[a] Based on an exposed circular area of 126.7 cm^2.

[b] Commercially finished with a fluorocarbon-based finish by Johnson & Johnson for surgical gown use.

[c] The fabric was manufactured by du Pont and was finished with fluorocarbon and flame retardant finish.

[d] The finished fabric was manufactured by Kimberley Clark; the composition of the finish is proprietary.

TABLE 3—*Duncan's Multiple Range Test ($p < 0.01$) of water/malathion penetration ($\mu g/cm^2$) by fabric category.[a]*

	Fabric Categories			
	I	II	III	IV
Malathion penetration ($\overline{X}$), μ/cm^2	7.18	12.90	101.46	197.96

[a] Based on an exposed area of 126.1 cm².

Results and Discussion

The results of the dye spray test as summarized in Table 1 indicate that the fabrics in Category I (pass-pass-pass) consisted of nonwoven fabrics either commercially or laboratory finished with fluorocarbon finish. The woven fabrics with fluorocarbon finishes are located in Category II (pass-pass-fail) due to the fact that the finish did not prevent the oil-based spray from penetrating the fabric interstices. Fabrics without fluorocarbon barrier finishes failed to pass either dye spray test that contained the surfactant.

Analysis of variance was used to test for differences in pesticide levels on the backing substrate between the dye spray tests by fabric categories. The results of the ANOVA of the water/malathion spray test indicate a significant difference ($p < 0.01$) between dye spray penetration fabric categories. This difference, as illustrated in Table 3 using Duncan's Multiple Range Test, indicates that at the 0.01 level of significance, Categories I and II fabrics differ from the fabrics in Categories III and IV. Similarities in malathion penetration exist between the fabrics in Categories I and II; and similarities in malathion penetration exist between the fabrics in Categories III and IV. It should be noted that fabrics in Categories I and II are finished with fluorocarbon finishes or coated with polypropylene whereas the fabrics in Categories III and IV are unfinished fabrics.

A one-way ANOVA of the results of the cottonseed oil/malathion spray test indicates a significant difference between fabric categories at the 0.01 level of significance. The Duncan's Multiple Range Test indicate that at $p < 0.05$ differences exist between the fabrics in Categories I and II and fabrics in Categories III and IV (Table 4).

There was no difference between fabric categories when the Duncan's Multiple Range Test was performed at the 0.01 level of significance. This is due to the large standard deviation found in the cottonseed oil/malathion penetration in the fabrics in Category IV ($s = 98.56$). This is primarily due to the low level of malathion found in backing substrates of the washed

TABLE 4—*Duncan's Multiple Range Test ($p < 0.05$) of cottonseed oil/malathion penetration ($\mu g/cm^2$) by fabric category.[a]*

	Fabric Categories			
	I	II	III	IV
Malathion penetration ($\overline{x}$), $\mu g/cm^2$	2.94	15.186	99.23	147.74

[a] Based on an exposed area of 126.1 cm².

TABLE 5—*Results of two-way ANOVA of water/malathion and cottonseed oil/malathion penetration by dye spray test category.*

Effects	Degrees of Freedom	F Value	$p < 0.01$
Fabric Category (H$_2$O × oil)	7	6.91	...
Spray media (H$_2$O × oil)	1	0.76	n.s.
Between fabrics	3	14.36	...
Interaction	3	1.50	n.s.

denim samples. The washed denim samples absorbed much of the cottonseed oil spray but little was released to the backing substrate in the short time allowed by the testing protocol.

A two-way ANOVA was performed to compare the results of the pesticide spray media (water/malathion and cottonseed oil/malathion) penetration as grouped by fabric spray test category (Table 5).

As expected, the overall differences between fabric group variation was significant ($p < 0.01$). Variation between fabric categories was also significant ($p < 0.01$). There was no significant difference between spray media nor was the interaction significant. The results of the two-way ANOVA indicate that methylene blue dye penetration can be used to predict malathion penetration using either a water or cottonseed oil spray medium.

Conclusions

Conclusions are drawn as follows: (1) the methylene blue dye may be used as an indicator for pesticide penetration in prescreening fabrics for protective clothing; (2) nonwovens finished with fluorocarbon finishes acted as a barrier to aerosol spray (in the laboratory testing); and (3) disposable nonwoven fabrics with fluorocarbon barrier finishes show promise for inexpensive protective clothing for agricultural workers exposed to pesticide sprays.

Field exposure studies, based on the laboratory results, are planned.

References

[1] Warburton, L. E. and Parkhill, F. T., *Textile Chemist and Colorist*, Vol. 5, 1973, pp. 41–45.
[2] Pittman, A. G., Roitman, T. M., and Sharp, D., *Textile Chemist and Colorist*, Vol. 3, 1971, pp. 175–180.
[3] Zisman, W. A., *Advances in Chemistry Series*, R. F. Gould, Ed., American Chemical Society, No. 43, 1964, pp. 1–51.
[4] Ellzey, S. E., Connick, W. J., Drake, G. L., and Reeves, W. A., *Textile Research Journal*, Vol. 39, 1969, pp. 209–815.
[5] Berch, T., Peper, H., and Drake, G., *Textile Research Journal*, Vol. 35, 1965, pp. 252–260.
[6] Hobbs, N. E., Oakland, B. G., and Hurwitz, M. D., "Effects of Barrier Finishes on Aerosol Spray Penetration and Comfort of Woven and Disposable Nonwoven Fabrics for Protective Clothing," *Performance of Protective Clothing, ASTM STP 900*, R. L. Barker and G. C. Coletta, Eds., American Society for Testing and Materials, Philadelphia, 1986, pp. 151–161.
[7] Hobbs, N. E., "Pesticide Penetration and Comfort Properties of Protective Clothing Fabrics," unpublished dissertation, University of North Carolina, Greensboro, 1985.

James R. Fleeker,[1] Cherilyn N. Nelson,[2] Ann W. Braaten,[2] and Jody B. Fleeker[1]

Quantitation of Pesticides on Apparel Fabrics

REFERENCE: Fleeker, J. R., Nelson, C. N., Braaten, A. W., and Fleeker, J. B., **"Quantitation of Pesticides on Apparel Fabrics,"** *Performance of Protective Clothing: Second Symposium, ASTM STP 989*, S. Z. Mansdorf, R. Sager, and A. P. Nielsen, Eds., American Society for Testing and Materials, Philadelphia, 1988, pp. 745–749.

ABSTRACT: The efficiency of common organic solvents in extracting carbon-14-labeled carbaryl, dicamba, chlorothalonil, and glyphosate from cotton twill was examined. The degree of extractions was influenced by pesticide formulation, pesticide polarity, and solvent polarity. In most but not all instances, higher solvent polarity resulted in greater pesticide extraction. The data have implications in extraction methods selected for quantitation of pesticide residue on fabrics following various refurbishing techniques.

KEY WORDS: chlorothalonil, carbaryl, glyphosate, dicamba, pesticide residue, laundering clothing

Pesticide-contaminated garments are a potential source of chronic exposure to pesticide applicators [1]. Exposure may be lowered by refurbishing techniques which would remove or reduce pesticide residue. The efficiency of a refurbishing technique can be determined by several methods.

Radioisotope levels have been used to measure the level of soil on textiles after laundering [2]. The advantage of this approach is the direct measurement of soil on the fabric without extraction. The use of carbon-14-labeled pesticides would be expected to have a similar advantage. However, soiling with radioisotopically labeled pesticides is inconvenient for many laboratories because of regulatory constraints and lack of suitable equipment. Thus, chromatographic or spectrophotometric methods are more often used to measure pesticide residue.

Chromatographic analysis entails extraction and, in some cases, derivatization before quantitation. For such analysis to be accurate, the efficiency of the extraction must be known, that is, what percentage of the pesticide on the fabric is being removed for quantitation [3,4]. Not all published studies have indicated that this analysis was completed prior to assessing the degree of pesticide removal by various refurbishing techniques.

This study examined the efficacy of several solvents in removing pesticides from apparel fabric and factors which affect the extraction efficiency

[1] Professor and laboratory assistant, respectively, Biochemistry Department, North Dakota State University, Fargo, ND 58105.
[2] Assistant professor and graduate research assistant, respectively, Department of Apparel, Textiles and Interior Design, North Dakota State University, Fargo, ND 58105.

Materials and Methods

Fabric

Undyed cotton twill, No. 423, was obtained from TestFabrics, Inc. (Middlesex, NJ). The fabric weight was 258 g/m², and the fabric count was 43 yarns/cm in the warp and 23 yarns/cm in the filling.

Pesticides

The following pesticide formulations were used: carbaryl, emulsifiable concentrate (EC), Environmental Protection Agency (EPA) Reg. No. 239-2356-AA; carbaryl, wettable powder (WP), EPA Reg. No. 1016-43; carbaryl, flowable (F), EPA Reg. No. 264-333; chlorothalonil (F), EPA Reg. No. 50534-8; glyphosate, water-soluble concentrate (WS) of the isopropylamine salt, EPA Reg. No. 524-308-AA; and dicamba, WS formulation of the dimethylamine salt, EPA Reg. No. 876-25-AA. Analytical standards of the pesticides were obtained from the U.S. EPA Pesticides and Industrial Chemicals Repository, Research Triangle Park, NC.

Carbon-14-labeled compound were obtained either from commercial suppliers or from the pesticide manufacturers. These were carbaryl[^{14}C], 21 mCi/mmole; chlorothalonil[^{14}C], 10 mCi/mmole; glyphosate[^{14}C], 2.3 mCi/mmole; and dicamba[^{14}C], 9.9 mCi/mmole. Radiochemical purity was checked by thin-layer chromatography (TLC) using two solvent systems for each compound [5]. Purification was necessary for dicamba[^{14}C], and this was done by silica gel TLC [5]. The labeled compounds were all >98% radiochemically pure when used in the experiments.

All solutions used for contamination contained 0.6% weight/volume (w/v) of the active ingredient prepared by diluting the pesticide formulation with distilled water. Radioactive pesticide was then added to this solution to a level of 0.8 μCi/mL. The weight of the radioactive pesticide was less than 1% of the total weight. The analytical samples of carbaryl and chlorothalonil were dissolved in methanol/toluene, 1:1 volume/volume (v/v). The contamination solution of analytical-grade dicamba was prepared in water as the dimethylamine salt, and the contamination solution of analytical grade glyphosate was prepared in water as the isopropylamine salt.

Sample Preparation and Contamination

The fabrics were stripped of warp sizing and manufacturer-applied fabric softeners by washing as per American Association of Textile Chemists and Colorists (AATCC) 135-1978 [6]. Specimens were cut from the inner 80% of the fabric width to ensure construction consistency. Each specimen was held horizontally, and 50 μL of the contamination solution was applied with a pipet. The specimens were conditioned 24 h at 22 ±2°C and 65 ±3% relative humidity prior to extraction.

Extraction

The extraction method was adapted from Easter et al. [7]. The specimens were placed in 250-mL flasks (Kimax No. 26500), to which 100 mL of solvent (reagent grade) was added. The flasks were swirled on a rotary shaker (New Brunswick, Model G-2) at 250 rpm for 60 min at room temperature. The excess solvent was gently squeezed from the specimen, which was then placed in a second flask containing the same volume of fresh solvent. The agitation procedure was repeated. The specimen was then squeezed to remove excess solvent, held vertically, and air-dried.

Radioisotope Measurement

Each specimen was cut in half and each section placed in a counting vial with 1 mL water and 10 mL scintillation solvent. This solvent contained 10 g of 2,5-diphenyloxazole and 1 g of 1,4-bis(4-methyl-5-phenyloxazole-2-yl)benzene per litre of toluene/Triton x-100 (2:1, v/v) [8]. Counting efficiency was determined with standard toluene[^{14}C] used as an internal calibration standard [9]. Time required for collection of 10 000 counts was used to compute counts per minute.

Statistical Analysis

The data from three replicates for each contamination solution in each extracton solvent were analyzed using the Analysis of Variance (ANOVA) and Duncan's New Multiple Range Statistical Test. Least Significant Means tests were applied to determine significant differences between pairs of means within each pesticide treatment.

Results and Discussion

Four common agronomic pesticides of diverse solubility were used in the study. The water solubilities of the pure pesticides increase in the following order: chlorothalonil, carbaryl, dicamba, and glyphosate [10,11]. Dicamba and glyphosate are present as salts of amines in the commercial pesticide formulations and are both very water soluble [11]. The pure pesticides were used to indicate the effect of the inactive ingredients present in the commercial pesticide formulations on the extraction efficiency. The solvents used increased in polarity in the following order: hexane, perchloroethylene, benzene, acetone, methanol, and water [12]. All except perchloroethylene have been used for pesticide extraction from laundered fabric prior to chromatographic analysis. Neat perchloroethylene is used for stain removal, as well as for commercial dry-cleaning when mixed with moisture and detergent. The extraction method employed here [7] has been widely used with variations in solvent and extraction time.

Table 1 lists the residue remaining after extraction of the four pesticides from twill fabric. Pure chlorothalonil and carbaryl were extracted from the fabric more efficiently than the corresponding formulations. The inactive ingredients in the commercial formulations may account for the difference, although the different solvents used for preparing the contamination solutions cannot be excluded from having an effect. The solvents may influence the deposition of the residues on the fabric surface as well as into yarn interiors and lumen of cotton fibers.

There were considerable differences in the extraction efficiency between the commercial formulations of carbaryl. Since a fixed amount of carbaryl was applied in water, differences in the inactive ingredients appear to have considerable influence on the extraction efficiency.

Salts of glyphosate and dicamba are considerably more polar than carbaryl or chlorothalonil. This could account for the ineffectiveness of the less polar solvents to remove the residues of glyphosate and dicamba (Table 1). The largest differences in extraction efficiency were between pure carbaryl and chlorothalonil and their formulations. These substances also have the lowest water solubility of the four pesticides, and chlorothalonil is much less water soluble than carbaryl.

The data in Table 1 suggest that in the analysis of pesticide residue on fabrics, extraction efficiency be determined with the selected pesticide formulation rather than the pesticide analytical standard. The results indicate that extraction efficiency cannot be confidently extrapolated from the pure or analytical standard of the pesticide to pesticide formulation.

TABLE 1—^{14}C-labeled pesticides remaining on cotton twill after extraction with organic solvents.

Pesticide	Formulation	Radioactivity Remaining, %[a]				
		Hexane	Perchloroethylene	Benzene	Acetone	Methanol[b]
Chlorothalonil	F	37.6a	84.8c	38.9a	25.8b	9.1d
Chlorothalonil	P	31.9e	41.3g	30.0e	10.8f	3.1h
Carbaryl	EC	83.1a	94.5d	90.7b	36.2c	ND[c]
Carbaryl	F	40.6e	85.2g	42.7e	11.0f	ND
Carbaryl	WP	62.3i	74.5m	69.9j	22.0k	ND
Carbaryl	P	21.8n	46.7q	17.5o	7.2p	ND
Dicamba	WS	100.6a	98.5a	101.6a	83.5b	0.3c
Dicamba	P	97.6a	98.2a	99.4a	79.2d	0.3c
Glyphosate	WS	96.6a	95.7a	93.0a	95.5a	47.6b
Glyphosate	P	97.7a	95.6c	98.2a	95.6a	45.4b

[a] Mean values for a single pesticide followed with the same letter are not statistically different at $P = 0.05$.
[b] Solvents have been placed in order of increasing polarity [12].
[c] ND = not determined.

If laundering or dry-cleaning techniques are being conducted prior to solvent extraction and quantitation, an additional analytical problem exists. Table 2 gives the amount of pesticide remaining on fabric after water extraction alone (control) and water extraction followed by organic solvent extraction. This experiment simulates a situation in which organic extraction is done in order to determine the efficiency of water extraction (laundering). The percent of the pesticide removed by the organic solvent is considerably different in some instances than that given in Table 1. Water extraction may selectively remove inactive ingredients present in the pesticide formulation, resulting in an extraction efficiency different from that observed with extraction by organic solvent alone.

The analytical problem is greatly reduced if the solvent removes >95% of the pesticide. Pesticides are so diverse in solubility properties that no single solvent can be recommended that will effectively extract all pesticides from fabric. The investigator must determine experimentally the most effective solvent. Laundering has been shown to be effective in removing many pesticides from fabric [13–15]; therefore the analyst may wish to consider using small amounts of water in the extracting solvent in order to increase extraction efficacy.

TABLE 2—Efficacy of solvent extraction in removing pesticide after preliminary extraction with water.

	Radioactive Pesticide Remaining, %[a]					
	Chlorothalonil			Carbaryl		
	F	P	ED	F	WP	P
Water alone (control)	100(19.0)a	100(45.9)b	100(2.6)a	100(23.4)b	100(3.0)a	100(4.0)c
Acetone	11.5c	1.7c	50d	12.8a	40d	35d
Hexane	46.6d	13.7e	88a	56.8e	50d	78f
Benzene	36.1f	10.5g	77g	63.4h	67ad	95c

[a] Formulation designations are given in the text. The values in parentheses represent the percent of pesticide remaining after water extraction. Mean values for a single pesticide followed by the same letter are not statistically different at $P = 0.05$.

Water can be removed before chromatography by forming a suitable azeotrope system in advance of the concentration step [*16*].

Conclusions

Accurate chromatographic analysis of pesticide residue on fabrics requires that the efficiency of extraction be determined. Accuracy of the data is questionable in experiments which have not included this analytical quality control. Extraction efficiency based on only the analytical grade pesticide is inadequate for quantitation purposes if the pesticide formulation was used for contamination. Use of an extraction solvent which results in essentially complete removal (>95%) should reduce quantitation error in measuring pesticide residue on fabrics.

References

[*1*] Wolfe, H. R., Durham, W. F., and Armstrong, J. F., "Exposure of Workers to Pesticide," *Archives of Environmental Health,* Vol. 14, 1967, pp. 622–633.
[*2*] Gordon, B. E., "Radiotracers in Fabric-Washing Studies," *Journal of the American Oil Chemical Society,* Vol. 45, 1968, pp. 367–373.
[*3*] Leng, M. L., "Governmental Requirements for Pesticide Residue Analysis and Monitoring Studies," *Analysis of Pesticide Residues,* Vol. 58, 1981, pp. 395–448.
[*4*] Garber, M. J., "Statistical Evaluation of Results," *Analytical Methods for Pesticides and Plant Growth Regulators,* G. Zweig and J. Sherma, Eds., Vol. 15, Academic Press, New York, 1986, pp. 19–65.
[*5*] Sherma, J., "Thin-layer Chromatography: Recent Advances," *Analytical Methods for Pesticides and Plant Growth Regulators,* G. Zweig, Ed., Vol. 7, Academic Press, New York, 1973, pp. 3–87.
[*6*] *AATCC Technical Manual,* American Association of Textile Chemists and Colorists, 1978.
[*7*] Easter, E. P., Leonas, K. K., and DeJonge, J. O., "A Reproducible Method for the Extraction of Pesticide Residues from Fabrics," *Bulletin of Environmental Contamination and Toxicology,* Vol. 31, 1983, pp. 738–744.
[*8*] Patterson, M. S. and Greene, R. C., "Measurement of Low Energy β-Emitters in Aqueous Solution by Liquid Scintillation Counting of Emulsions," *Analytical Chemistry,* Vol. 37, 1965, pp. 854–857.
[*9*] *Radiotracer Techniques and Applications,* Vol. 1, E. A. Evans and M. Muramatsu, Eds., Marcel Dekker, New York, 1977.
[*10*] Ballee, D. L., Duane, W. C., Stollard, D. E., and Wolfe, A. L., "Chlorothalonil," *Analytical Methods for Pesticides and Plant Growth Regulators,* G. Zweig, Ed., Vol. 8, Academic Press, New York, 1976, pp. 263–274.
[*11*] Worthing, D. R. and Walker, S. B., *The Pesticide Manual,* 7th ed., Lavenham Press, U.K., 1983.
[*12*] Snyder, L., "Classification of the Solvent Properties of Common Liquids," *Journal of Chromatography,* Vol. 92, 1974, pp. 223–230.
[*13*] Easley, C. B., Laughlin, J. M., Gold, R. E., and Tupy, D., "Removal of 2,4-Dichlorophenoxyacetic Acid Ester and Amine Herbicide from Contaminated Fabrics," *Archives of Environmental Contamination and Toxicology,* Vol. 12, 1983, pp. 71–76.
[*14*] Keashchall, J. L., Laughlin, J. M., and Gold, R. E., "Effect of Laundering Procedures and Functional Finishes on Removal of Insecticides Selected from Three Chemical Classes," *Performance of Protective Clothing, ASTM STP 900,* R. L. Barker and G. C. Coletta, Eds., American Society for Testing and Materials, Philadelphia, 1986, pp. 162–176.
[*15*] Kim, C. J., Stone, J. F., Coates, J. R., and Kadolph, S. J., "Removal of Alachlor Residues from Contaminated Clothing Fabrics," *Bulletin of Environmental Contamination and Toxicology,* Vol. 36, 1986, pp. 234–241.
[*16*] "Azeotropes," *CRC Handbook of Chemistry and Physics,* R. C. Weast, Ed., Vol. 50, 1969, pp. D1–D44.

Protection from Pesticides

User Attitudes and Work Practices

Chris L. Maas[1]

Impact of Labor Protection in the Registration Process of Pesticides in the Netherlands*

REFERENCE: Maas, C. L., **"Impact of Labor Protection in the Registration Process of Pesticides in the Netherlands,"** *Performance of Protective Clothing: Second Symposium, ASTM STP 989*, S. Z. Mansdorf, R. Sager, and A. P. Nielson, Eds., American Society for Testing and Materials, Philadelphia, 1988, pp. 753–764.

ABSTRACT: Pesticide registration in the Netherlands is the responsibility of four governmental departments dealing with (1) agriculture, (2) public health, (3) environment, and (4) occupational health and safety. It is the responsibility of the Directorate-General of Labour (DGA) within the Ministry of Social Affairs and Employment to direct the evaluation of occupational toxicity and exposure data. Further, DGA maintains via the Labour Inspectorate standards of application technology; procedures for storage, mixing and loading; sound hazard communication; and the quality of protective measures. The ultimate goal is to achieve a quality of health and safety in agriculture comparable to other work places where toxic materials are used. Therefore, consistency with premanufacturing notification policy, for example, is obligatory.

As far as labor protection is concerned, the registration process of pesticides starts with toxicology. The relevant adverse effects determined in experimental animals are further evaluated if dose levels at which these effects occur indicate serious risk during or after occupational exposure via the dermal, inhalation, or secondary oral route. Quantitative exposure assessment is then carried out. Currently, a systematic approach for gaining exposure data is being developed in the Netherlands. This process is a multistep system consisting of (1) theoretical estimation of exposure, (2) using mathematical models and generic data, (3) personal dermal monitoring, (4) environmental (air) monitoring and personal air monitoring, and (5) biomonitoring.

Higher levels in the system are triggered if results at lower levels seem to have serious implications. In certain situations, a combination of elements from different levels within one study may have advantages. The exposure data then are challenged by an extrapolated human threshold dosage-level in order to rationally decide on registration. This approach of weighing possible risks is carried out for all relevant differential applications. Special attention is paid to indoor applications, like thermally isolated warehouses.

When the decision to register a pesticide has been made, different instruments may be used to control the actual exposure by defining the application technology and using warnings on the label, ergonomic packaging, and protective clothing. The Labour Inspectorate may add special protective measures to these demands for separate locations. During the registration, DGA may ask for further testing either in toxicology or exposure issues. The field data reported by the Labour Inspectorate provide feed-back on registration decisions. Efforts to answer questions concerning pesticide application and regulation are preferentially to be improved by a joint effort of the pesticide industry, the farmers, their organizations, and the registration authorities, all cooperating for better labor protection.

[1] Toxicologist, Ministry of Social Affairs and Employment, Directorate-General of Labour (DGA), P.O. Box 69, 2270 MA Voorburg, the Netherlands.

* This paper expresses the personal view of the author and does not necessarily reflect the formal policy of DGA.

KEY WORDS: protective clothing, pesticide registration, exposure analysis, toxicology, occupational health

Pesticides are biologically active agents, selected for their toxicity. Although manufacturers strive for a great safety-interval between the toxicity for the target species and human toxicity, most pesticides are potentially hazardous to man. Adverse effects to the human being, however, will only occur if exposure takes place. Human exposure might result from residues in food or drinking-water, from pollution in the environment and from application at the working-place. The highest levels of exposure are encountered at the working-place. This paper describes the labor protection policy as part of pesticide registration in the Netherlands.

The Use of Pesticides in the Netherlands

The Dutch pesticides Act [1] regulates both: agricultural and industrial pesticides. As is normally the case, in the Dutch Pesticides Act agricultural pesticides are defined in such a way that all agents (substances, mixtures of substances or microorganisms) causing a herbicide, insecticide, fungicide, or rodenticide action are covered. Furthermore, the legislation covers agents against ectoparasites on cattle, growth regulators for plants and agents to control pests in buildings used to house animals, in means of transport for animals or in milking equipment. "Industrial pesticide" has been defined slightly broader in the Dutch Pesticides Act than internationally accepted. The definition includes: wood preservatives, antifouling paints, the classical disinfectants, biocides used in cooling-water systems or in air-conditioning systems, as well as preservatives for metal-working fluids.

The distribution of pesticides, the concurring occupational exposure and the inspection regime acted upon by the Dutch Labour Inspectorate are shown in Fig. 1. This figure shows the complete life-cycle of a pesticide from production to application. In some cases this complete cycle is present in the Netherlands. In most cases however, the production of the active ingredients takes place in other countries, whereas formulation is frequently carried out in the Netherlands. The Pesticides Act does not regulate the production of active ingredients nor the formulation of pesticides. Regulation under the purview of the Pesticides Act starts with the registration of a certain pesticide-formulation, which authorizes the selling and using of that particular formulation, and goes downwards in the scheme of Fig. 1 via trade to application and re-entry situations.

On the level of trade, prescriptions are in force regarding the stocking and packaging of the pesticide, whereas on the level of application and re-entry situations protective clothing and equipment or certain application technology may be required. In the Dutch pesticides industry, where active ingredients and formulations are produced either for export purposes or for use in the Netherlands, labor protection regulations are based on the general Act on Labour Conditions [2]. In this way, it is ensured that, for pesticides in production or formulation, equal labor protective standards are maintained, whether or not these pesticides have been registered in the Netherlands.

An essential difference between the production or formulation of pesticides on the one hand and the use of pesticides on the other, is the impossibility (especially in agricultural applications) of total containment of the pesticide during and after application. Thus, whereas in production or formulation situations one may strive to a closed system in which the harmful compounds are contained, application of pesticides includes a more or less open system in which workers have to be protected against harmful compounds. This difference together with the fact that many pesticides show a biological effect of a more or less toxic

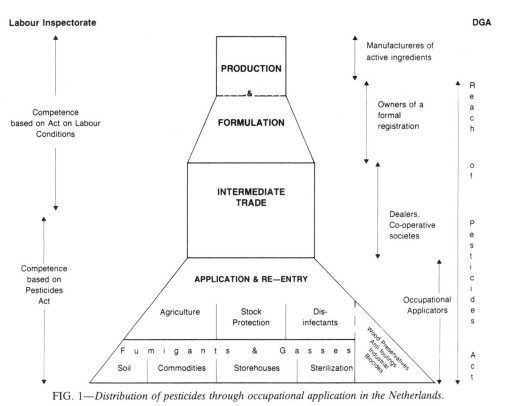

FIG. 1—*Distribution of pesticides through occupational application in the Netherlands.*

nature, form the fundamental reason for a specific approach to labor protection in the field of pesticide application.

Pesticide Registration in the Netherlands

The Pesticides Act: History and Predominant Features

Pesticide regulation in the Netherlands goes back to 1947, when the Pesticides and Fertilizers Act came into force. The purpose of this act was to prevent inferior agricultural pesticides and fertilizers from being put on the market. Given the limited objectives of this legislation the possible risks resulting from the differential toxicity of otherwise reliable pesticides could not be weighed. To overcome this omission, therefore, the Pesticides and Fertilizers Act was replaced by the Pesticides Act in 1964. This act was adapted in the mid-seventies to implement the purpose of environmental protection. At the present time the following criteria have been laid down in the Pesticides Act:

1. Efficacy.
2. No damage to public health or the quality of food.
3. No damage to the health and safety of the applicator, provided he is complying with the prescribed protective measures.
4. No unacceptable damage to the environment.

A pesticide should meet all of these criteria, otherwise it will not be registered. An essential feature of the Pesticide Act is, that the use of any pesticide, which has not been registered, is prohibited. Four Ministers of State take responsibility for the decision whether a certain pesticide meets the criteria mentioned: The Minister of Agriculture & Fisheries, The Minister of Public Health, The Minister of Environmental Hygiene and The Minister of Social Affairs and Employment. The latter Minister considers the aspect of the health and safety of the applicators.

The Pesticide Registration Process

The Ministers concerned are advised on registration decisions by the Pesticide Registration Committee. This committee is composed of high officials from the ministries concerned and is responsible for a balanced advice. In this committee the Director of Health of the Directorate-General of Labour is in charge of labor protection policy. The Pesticide Registration Committee is assisted by specialized working parties, which review the dossiers and prepare a decision. The registration process is shown in Fig. 2.

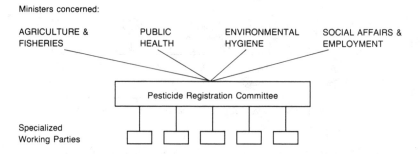

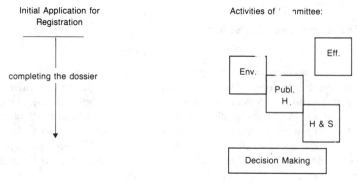

Eff.	:	Research & evaluation concerning efficacy
Env.	:	Environmental hygiene
Publ.H.	:	Hazards resulting from residues in food
H & S	:	Occupational Health and Safety

FIG. 2—*Pesticide registration process in the Netherlands.*

Review activities are carried out in a cascade system. They start with the evaluation of and research into pest combatting efficacy and continue, if the advice on efficacy is positive, with the review of ecotoxicology, the hazard resulting from residues in food and the evaluation of the occupational hazard respectively. During the reviews, outside experts including experts from companies applying for registration, may be consulted. Depending on the time needed to complete the dossiers, the review procedure for new active ingredients will take 5 months to 2½ years. Registration is valid for a given period of time with maximum of ten years registration. Registration may be extended but will be reevaluated. If necessary, for older pesticides additional research has to be carried out before the end of a registration period to comply with the standards for new pesticides. A significant effort is made to balance the objectives of the different ministries, to obtain an integrated policy.

Labor Protection Policy as Part of Pesticide Regulation in the Netherlands

General Strategy on Occupational Hygiene

In the late seventies, a structural policy for health protection at workplaces, where toxic substances are used, was developed by the Directorate-General of Labour. This new policy fits in the new framework of labor protection as has been laid down in the new Act on Labour Conditions, which is currently entering into force. One of the basics of the new policy is that employers have to meet standards of healthy and safe working places in conjunction with their employees.

At the present time, labor protection policy in pesticide regulation forms an integral part of the structural policy on toxic substances at the workplace. This structural policy consists of a general strategy which applies all situations (including the application of pesticides), supplemented by specific regulation for defined categories of substances as is outlined in Table 1.

Basically the general strategy provides for an effective reduction of harmful exposure to toxic agents. Emission may be prevented, for example, by replacing a toxic substance by a less toxic one, by applying cleaner technology, by using alternative methods or by adapting methods, and by adapting the workplace or the maintenance and cleaning operations. Exhausting should be preferentially implemented at the primary source. If this cannot be realized, general ventilation may be applied. By adapting the work organization, the population at risk as well as the duration of exposure per worker should be minimized. If these measures fail to reduce exposure to an acceptable level, protective clothing or respiratory equipment must be used. Superimposed on this general strategy, specific regulation for well-defined groups of substance as has been developed.

TABLE 1—*Structural policy on toxic substances at the working place in the Netherlands.*

A. General Strategy
  1. Prevention of harmful emission
  2. Exhaust of polluted air
  3. Work organization designed to minimize exposure
  4. The use of protective clothing and/or respiratory equipment
B. Supplementary Specific Regulation
  1. lead, asbestos and new chemicals (implementation of EC directives)
  2. carcinogenic substances
  3. teratogenic substances (in preparation)
  4. laying down of limit values
  5. pesticides

Specific Regulations on Toxic Substances

The European Community (EC) puts forward directives for a harmonized regulation in the member states on different substances which cause a substantial health hazard at the workplace. At present, such directives entered into force for lead and asbestos. Furthermore, the EC adopted a directive on the notification of new chemical substances, which is fully operative. The latter type of regulation is known in the United States as "premanufacturing notification" (PMN). However, significant differences do exist between the European and the American systems. Roughly, in the EC the notification concentrates on the toxicological and physico-chemical properties rather than on exposure, whereas in the United States the PMN-system demands a detailed description of the processes in which the new chemical is produced or used rather than of the toxicological profile of the substance. Therefore, the European system (unlike the American one) is more or less comparable with the world-wide accepted system of pesticide registration. Special measures for carcinogenic substances have been carefully prepared during recent years and are currently under review by the Dutch Council on Labour Conditions, which advises the Minister on labor conditions policies [3]. Within the framework of the proposed policy, carcinogens are defined as: "substances which may cause the development of a maligne tumour." The IARC (WHO) list of proven human and proven experimental animal carcinogens has been adoped, and it is suggested to enforce the policy on carcinogens to this group of substances as a body. The proposed regulation primarily includes labeling conform EC rules and the obligatory registration at company level of data relating to the carcinogenic substance and to exposure profiles. Further regulation should be governed by a specific classification scheme, which is shown in Table 2.

The proposed classification is based upon the mechanism of carcinogenic action and potency, respectively. In case of a nongenotoxic mechanism of action, a limit value of exposure may be defined below which no adverse health effects have to be encountered. For genotoxic substances such a limit value without carcinogenic risks cannot be given. Therefore, for these substances strick measures are foreseen, depending on the potency of carcinogenicity. The measures may rank from replacing the substance to notification of use of the substance to the Dutch Labour Inspectorate, compulsory membership of an occupational health service, compulsory monitoring of the substance, and strict codes for hygiene as well as the obligation to bring other measures into force if certain exposure-limits of action have been reached.

A specific policy of similar structure is in preparation for teratogenic and maternotoxic substances at the working place. Another specific policy, superimposed on the general strategy, is the current laying down of threshold limit values, defined in the Netherlands

TABLE 2—*Proposed specific regulation on carcinogenic substances in the Netherlands.*

Classification	Measures
A. Nongenotoxic	limit value minimizing exposure
B. Genotoxic with:	
1. Very powerful potency	replacing the substance (possible exemptions)
2. Moderately powerful potency	notification occupational health service
3. Slightly powerful potency	monitoring hygiene code and protocols limits of action

as Maximal Accepted Concentrations (MAC-values) at the workplace. The statuting of MAC-values has been facilitated by setting up a tripartite committee, composed of representatives of employers, labor organisations and the Dutch government.

Pesticide Registration as a Specific Regulation on Toxic Substances at the Workplace

The objective of specific regulation superimposed on the general strategy must be: reducing risks at critical occupational situations to the risk level of "normal" workplaces. Working situations may be denoted as critical either if substances with extreme harmful potencies are present or if extensive exposure to toxic substances may occur. Both criteria apply during or after the occupational application of pesticides. For this reason pesticide registration, in fact, is the oldest specific occupational regulation on a group of substances. With the other specific regulations fading in, there is a need to implement their useful elements into pesticide regulation and further to balance costs and benefits of the different specific regulations on toxic substances to each other. Some of these consequences will be addressed next.

Decision-making on Registration or Extension of the Registration

The most striking feature of pesticide regulation is that the use of any pesticide is prohibited, unless the pesticide has been registered. The decision making on registration from the view-point of labor protection involves a risk-assessment regarding application and re-entry situations.

For *novel pesticides,* this assessment of possible, occupational risks has to be based primarily on physico-chemical properties and on the results of toxicological studies conducted in experimental animals. Typical basic toxicological data requirements regarding the formulation and the active ingredient(s) of a pesticide have been presented in Table 3.

The basic toxicological data set comprises relevant acute toxicity and irritation studies carried out with both the formulation and its active ingredient(s). Further, skin sensitization, mutagenicity, and subacute oral toxicity of the active ingredient(s) should be investigated. If significant occupational exposure has to be anticipated, subacute dermal or inhalatory studies as well as a Segment II teratology and a fertility study may be required also.

TABLE 3—*Basic toxdata requirements for the risk assessment concerning application and reentry.*

Active ingredient
  acute oral
  acute dermal
  dermal irritation
  skin sensitization
  subchronic oral (90 d) or subacute oral (28 d)
  mutagenicity
  eye irritation
  exp: subacute dermal (28 d)
     subacute inhalation (28 d)
     teratogenicity
     fertility
Formulation
  acute oral
  acute dermal
  dermal irritation
  eye irritation
  exp: acute inhalation

In addition to the basic data requirements, further testing may be indicated. For instance, if positive mutagenicity results have been confirmed in different test models, the results of chronic carcinogenicity studies should be evaluated before an advice on registration is given. Other additional studies which are frequently encountered are: investigation on the mechanism of toxic action, dermal penetration, parallel studies in other animal species to obtain insight in interspecies variation and toxicokinetics.

Results of such additional studies may as well be of importance for the risk-analysis concerning possible residues of active ingredients in food. In many cases therefore, the pesticide industry will conduct this kind of research on its own behalf, in order to advance registration procedures.

A positive advice on registration will be given if sufficient evidence has been obtained indicating that no harmful effects at the workplace need to be encountered. However, if potential harmful effects have been determined somewhere along this line of testing, a specific risk assessment has to be carried out.

The same applies for "old" pesticides (pesticides which have been registered previously, while data requirements were incomplete compared to the present state of the art in toxicology), if new knowledge indicates potential harmful effects which had not been taken into account during earlier registration procedures. Currently, this situation has to be dealt with for some "old" pesticides, as a result of the updating programs on the data files enforced by registration authorities of different countries including the Netherlands. In the Netherlands, for example, the last four years the mutagenicity item has been cleared for almost all the older pesticides. New knowledge may come up also in the open literature or result from research projects subsidized by the Dutch government.

Having identified the most relevant adverse effects, risk assessment continues with extrapolation from that animal toxdata to the applicator and the worker at re-entry situations. In most cases extrapolation from oral studies has to be performed. Risk assessment then includes both extrapolation from oral animal data to estimates for dermal and inhalatory exposure and extrapolation from these estimates to the human being. As described in depth elsewhere [1], extrapolation needs expert judgments, since dose relationships (if present), mechanisms of action (if understood), and toxicokinetics should be considered. Available data on worker health from industrial health services concerned with production-sites for the active ingredients under review may be of significant help in obtaining pragmatic human exposure-limits of action.

At this stage risk assessment advances with exposure analysis. Before describing this part of the evaluation process, first, a short excursion into the interrelationships between pesticide regulation and the other specific regulations on toxic substances will be made. At first, there exists a strong interrelationship between pesticide regulation and the system for notification of new chemicals in the EC, because the scientific system of gaining insight into the toxicological profile of new chemicals or pesticides, basically, is similar. Also the basic toxdata requirements in both regulations are very similar, the only essential difference being the formulation-toxicology for the pesticides. At first sight, there seems to be another significant difference, in that the toxicological profile of new pesticides has to be clarified *before* marketing, whereas that of new chemicals will be developed gradually if the market for the particular chemical is growing. However, the specific regulation for new chemicals includes an *optional* requirement for advanced completing of the data file, if harmful effects have been established. Further, restriction on the use of the new chemical (including complete removal from the market, if necessary) may be set forth. Those restrictions will be guided by pertinent toxdata on harmful effects or by medical observations of workers applying the chemical. Therefore, the notification system for new chemicals, in fact, is very similar to the procedures for "old" pesticides. Thus, it may be concluded that enough conditions have been fulfilled to obtain a flexibile and balanced policy for both new chemicals and pesticides, as far as meeting equal standards for occupational health and safety in agriculture and industry is concerned.

Another interrelationship does exist between the specific regulation proposed for carcinogenic substances in industry and pesticide regulation. In principle, the new regulation for carcinogenic substances will introduce a similar regulation for known nongenotoxic industrial chemicals, as has been in place already in pesticide regulation. For proven genotoxic carcinogens, however, there remains some of a divergence, in so far, that genotoxic carcinogenic pesticides have not been registered in the past and unlikely will be registered in The Netherlands. This strict policy results a good deal from the great prudence regarding possible residues in food or drinking water. In case of a nonpersistent pesticide which will be used either in nonfood crops or in industry, there will be made an evaluation of carcinogenic potency versus occupational exposure. Thus, as far as worker protection is concerned, pesticide registration at least meets the standards proposed in the specific regulation of carcinogens at industrial working places. In fact, the pesticide-registration procedure is more strict as a result of the active data requirements, which will possibly balance the greater difficulties in many branches of agriculture with implementation of enclosed application systems, compared to industry.

After having established a human limit of action based on toxicology, exposure analysis is needed. A general scheme for the analysis of occupational exposure has been shown in Table 4. As a routine, exposure analysis should imply the breakdown of working processes into tasks performed. Whereas a general breakdown into application tasks is easily given, this certainly is not the case for re-entry situations. Thus, for re-entry situations the different sources of pesticide exposure, as well as of the tasks performed per worker, should be analyzed carefully. A nonlimitative list of possible sources has been presented in Table 4.

Per task, the exposure via the differential routes should be either estimated or determined. A tricky pitfall may be the overconcentration on the quantitative determination of the deposits on the skin, without proper understanding of the bioavailability of such deposits

TABLE 4—*Analysis of occupational exposure during and after the application of pesticides.*

Tasks and Sources	Routes of Exposures	Bio-Availability Resulting From	Response as a Function Of

Application
  Preparation
  Mixing and Loading
  Use
  Cleaning
  Clearing
  Maintenance
Re-entry
  Soil/floor/walls
  Air/water
  Crop
  Reused packagings
  Machines and
    technology
  Solvents
  Products and
    materials
	Dermal
	
	Inhalatory
	
	Secondary Oral
		dermal penetration
		
		deposit in and uptake via lungs
		
		uptake via alimentary canal
			human toxico-kinetics
			
			and
			
			human toxico-dynamics

TABLE 5—*Proposed system for exposure-assessment in Dutch pesticide regulation.*

1. Theoretical estimation.
2. Model calculation using some empirically determined
   parameters.
3. Personal dermal monitoring in
   addition to 1 or 2 (if dermal exposure is predominant).
4.1 Environmental (air) monitoring in addition to 1, 2
    or 3.   } (If significant inhalatory
4.2 Personal air-monitoring in addition to 1, 2, 3 or 4.1.   exposure can be anticipated.)
5. Biomonitoring.
   ┌──→ increasing
   │   liability
   ↓ increasing
cost and time

under real world practices. Secondary oral exposure may result from the "hand-face shunt." The significance of this route of exposure varies with working practices, which are influenced by physical parameters like humidity and temperature and by the tasks performed. Inhalatory exposure can be estimated for operations in closed workplaces. For outdoors situations a worst case for metereological conditions may be assessed. Currently, in the Netherlands a working party from industry and DGA is reviewing the possibility of a flexible approach of exposure assessment, implicating a multi-step system as shown in Table 5.

An urgent need does exist for priority setting, to answer the questions:

1. In which case do theoretical estimations or model-calculations of exposure apply?, and
2. Where and during which stage of the application or re-entry tasks should exposure be analyzed?

Furthermore, for agricultural pesticides worst case cultures and crops should be identified. It seems worthwhile to base priority setting both on the present status in the field and on current trends in working practices and application technology. Current trends for pesticide application in agriculture have been presented in Table 6. Obviously, new exposure assess-

TABLE 6—*Current trends in application technology for the use of agricultural pesticides in the Netherlands.*

Outdoors	Greenhouses and Other Indoor Applications
PAST	
Open tractors	knapsack pulse
Unfolding and adjustment of booms by handling	
Poor quality of protective clothing (seldomly used)	idem, respiratory equipment seldomly used
PRESENT	
Closed tractorcabs with ventilation	rapid increase in automatic application via
Remote control of spray/blast apparatus	raining system or via pulse fog in
Closed loading system	combination with Co_2 fertilizer
Cleaning systems for packaging	
Better quality of protective clothing, protective	idem, respiratory equipment more
clothing more easily available	frequently in use
FUTURE	
Automation and remote control of the application, emphasis on protective measures for re-entry situations.	

ment priorities will appear for indoor re-entry situations like greenhouses and growing facilities for mushrooms rather than for outdoor application of pesticides.

One of the spotted worst case indoor cultures is the taking of chrysanthemum cuttings, because of frequent use of different pesticides, extensive (dermal) contact of workers with the crop, and the nonavailability of comforting protective gloves for the fine-handling tasks to be performed. A possible framework for the final risk-assessment concerning such a "re-entry" situation, based upon toxicology and anticipated dermal exposure, has been shown in Fig. 3. In case of a complete theoretical exposure analysis many estimations have to be made for deposition characteristics, stability, uptake by the crop, percentage of dislodgeable residue, working practices et cetera.

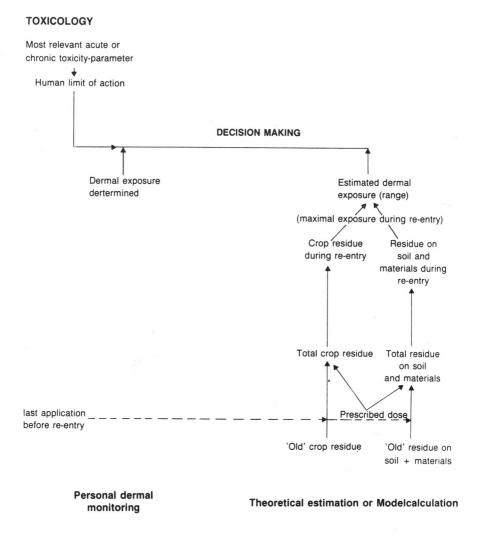

FIG. 3—*Risk assessment on the occupational, dermal exposure during re-entry tasks performed in agricultural indoor situations with regular pesticide application.*

If personal monitoring results have been obtained, it should be reviewed critically whether or not conditions and working practices in that exposure study were typical. Accepted values or ranges for exposure should be then challenged by the toxicological limit of action and an advice on registration may be given.

For novel pesticides the advice on registration has to be based entirely on a risk-assessment procedure as described previously. In case of the reevaluation of the registration of old pesticides, the experience obtained in different application situations and in production and formulation facilities may be weighed. If withdrawal of the registration of an old pesticide has to be considered, pertinent attention has to be paid to the question which other pesticide(s) will replace the one to be withdrawn. As a part of the final decision making it has to be determined whether the benefits for occupational health and safety resulting from the withdrawal of the one pesticide will outweigh the possible risks of a greater use of the replacing pesticide(s).

Conclusions and Recommendations

The application technology for the use of agricultural pesticides is rapidly improving and many indoor applications, for instance, in horticulture, may be carried out by remote control in the near future. As a result, worker exposure during the application of pesticides may decrease. On the other hand, current pest combatting programs imply more frequent and intensive use of many different pesticides in order to meet the extremely high quality standards which the competitive agricultural world market is demanding. Furthermore, the use of pesticides does imply an open system and, consequently, causes distribution of the pesticide over the working-place. Thus, workers performing "re-entry tasks" may be exposed to increased quantities of pesticides. Therefore, pertinent attention in the risk assessment has to be paid to the re-entry issue. In the working-places with significant exposure during re-entry, a sound hazard communication between employer and employees should be developed. Further, basic research into the effects of multiple exposure, which are at present poorly understood, should be encouraged. Also the involvement of Agricultural Hygiene and Health Services is recommended. Such services may play a key role in the systematic investigation of exposure patterns and in the tracing of early health effects of applicators and "re-entry workers."

In order to improve risk assessment, internationally accepted protocols for the different types of exposure studies should be developed, and the international adoption of standards for the quality assurance of these studies is recommended. It should be considered whether the inspectorates for Good Laboratory Practice might inspect the quality assurance of exposure as well. Finally, it seems worthwhile if registration bodies commence requiring advices on the choice of protective clothing, gloves, and respiratory protection for the pesticides of which the use implicates relevant health hazard. Such advices should be based upon breakthrough time and comfort tests, carried out to conform with internationally adopted test methods.

References

[1] Stemers, W., *Bestrijdingsmiddelenwet 1962*, W. E. J. Tjeenk Willink, Zwolle, 1982, pp. 1–200.
[2] Aarts, A. C. M. and van Riet Paap, K. W., *Arbeidsomstandighedenwet*, W. E. J. Tjeenk Willink, Zwolle, 1983, pp. 1–522.
[3] Balemans, A. W. M., Regulatory Aspects of Carcinogenic Substances in The Netherlands, Med. Lav. 77, 1986-4, pp. 399–401.
[4] Kroes, R., "Food Additives and Risk Assessment, Current Status and Future" in *FEST-Toxicology in Europe*, Elsevier, Amsterdam, 1986, pp. 32–34.

Curt Lunchick,[1] *Alan P. Nielsen,*[1] *Joseph C. Reinert,*[1] *and Dea M. Mazzetta*[1]

Pesticide Applicator Exposure Monitoring: EPA Guidelines

REFERENCE: Lunchick, C., Nielsen, A. P., Reinert, J. C., and Mazzetta, D. M., **"Pesticide Applicator Exposure Monitoring: EPA Guidelines,"** *Performance of Protective Clothing: Second Symposium, ASTM STP 989,* S. Z. Mansdorf, R. Sager, and A. P. Nielsen, Eds., American Society for Testing and Materials, Philadelphia, 1988, pp. 765–771.

ABSTRACT: The Environmental Protection Agency's Office of Pesticide Programs has published final guidelines for monitoring occupational pesticide exposure. This paper describes various aspects of these guidelines. The advantages and disadvantages of quantifying exposure both by passive dosimetry and biological monitoring are discussed. Requirements for monitoring dermal exposure when protective clothing is worn are described. When field exposure monitoring studies must be carried out and the number of individuals that must be monitored in each study are also specified.

KEY WORDS: pesticides, applicator exposure, passive dosimetry, biological monitoring, EPA guidelines

Under the aegis of the Federal Insecticide, Fungicide, and Rodenticide Act (FIFRA), the Office of Pesticide Programs (OPP) of the Environmental Protection Agency (EPA) registers chemicals for use as pesticides provided that, among other criteria, "when used in accordance with widespread and commonly recognized practice, [they] will not generally cause unreasonable adverse effects on the environment." Protection of agricultural workers involved in the application of pesticides is considered under the umbrella of this risk criterion, and the estimation of human exposure is an integral part of OPP's risk assessment procedures.

On 2 January 1986, EPA issued draft guidelines designed to aid pesticide registrants and others in designing and carrying out field studies which measure potential dermal and respiratory occupational exposure to pesticides, using personal monitoring techniques. These draft guidelines underwent a review by the FIFRA Scientific Advisory Panel (SAP) as well as a public comment period.

There are two basic approaches to estimating occupational exposure to pesticides. Passive dosimetry estimates the amount of the chemical contacting the surface of the skin or the amount of the chemical available for inhalation through the use of appropriate trapping devices. Biological monitoring estimates internal dose from either a measurement of body burden in selected tissues or fluids or from the amount of pesticide or its metabolites eliminated from the body. There are theoretical and practical advantages and disadvantages to each approach.

---

[1] Chemist, biologist, chemist, and environmental protection specialist, respectively, U.S. Environmental Protection Agency, Office of Pesticide Programs, Hazard Evaluation Division (TS-769C), Exposure Assessment Branch, 401 M. Street SW, Washington, DC 20460. Please address all inquiries to Mr. Lunchick.

Passive Dosimetry

Two major advantages of direct entrapment procedures are, first, the ability to differentiate exposure received during discrete work activities within a workday, such as mixing/loading versus applying the pesticide, and, second, the ability to compare the relative contributions of the dermal and respiratory exposure routes for each separate work activity [1]. As seen in Table 1, for many field applications, potential respiratory exposure does not contribute significantly to total exposure. Table 2 indicates that for most applications, a high proportion of total dermal exposure occurs to unprotected hands. These advantages of passive dosimetry are extremely important for evaluating exposure reduction safety practices such as personal protective equipment for the various work activities associated with pesticide application. For example, if a substantial fraction of total exposure for a workday occurs to the hands, forearms, and face during the short period of time a worker pours a concentrated pesticide formulation into a mix tank, risk may be mitigated significantly by requiring chemical-resistant gloves, a face shield, a closed mixing system, or some combination of these, for this short-duration/high-exposure period.

Traditionally, applicator exposure monitoring studies carried out for the Agency have employed passive dosimetry techniques. Because of this history of use, experimental design and execution of exposure studies by passive dosimetry are routine for many investigators. This in itself is an advantage, but leads to a much more important advantage of passive dosimetry, which is the ability to create large generic data bases from studies carried out with different pesticides. This generic approach is discussed in detail in another paper in this volume.[2]

A final advantage in using passive dosimetry is that the study participants are typically under the supervision of the investigator during the entire period when exposure data are being collected, thus helping to ensure that good quality-assurance practices are conducted and that sample integrity is not compromised.

A major disadvantage of passive dosimetry is that these techniques measure only the

TABLE 1—*Contribution of respiratory exposure to total potential exposure for pesticide applicators.*

Activity	Physical Form of Pesticide Used	Respiratory Exposure as % of Total Exposure	Reference
Drivers of tractors equipped with canopies during air-blast spraying of citrus	spray	0.029	[12]
Bulk spray suppliers for above application	unspecified	0.004	[12]
Drivers of ordinary tractors during boom spraying of tomatoes	spray	0.04	[13]
Applicators using handgun to spray aquatic weeds from airboat	spray	ND[a]	[13]
Drivers of airboats for above application	spray	ND[a]	[13]
Applicators using aerosol generator for mosquito control	aerosol	9.1	[14]
Applicators using hand knapsack mister for spraying tomatoes	mist	3.1	[15]

[a] None detectable.

[2] Eberhart, D. C., Day, E. W., Knarr, R. D., Zimmerman, D. M., Nielsen, A. P., Reinert, J. C. and Muir, N. I., "Progress in the Development of a Generic Data Base for Mixer/Loader-Applicator Exposures," this publication, pp. 772–775.

TABLE 2—*Contribution of hand exposure to total potential dermal exposure for pesticide applicators.*

Activity	Hand Exposure as % of Total Dermal Exposure	Reference
Drivers of tractors equipped with canopies during air-blast spraying of citrus	41	[12]
Bulk spray suppliers for above application	52	[12]
Drivers of ordinary tractors during boom spraying of tomatoes	25	[13]
Applicators using hand guns to spray aquatic weeds from airboats	47	[13]
Drivers for airboats for above application	89	[13]
Aerial applicators	55	[16]
Flaggers for above application	39	[16]
Applicators spraying lawns, trees, and gardens with power sprayers	37	[17]
Applicators spraying lawns with hose-end sprayers	98	[18]

amount of chemical potentially available for absorption; independent estimates of dermal and lung absorption are required to estimate dosage for hazard assessment purposes. These absorption estimates are generally difficult to derive and interpret using data available from either *in vivo* or *in vitro* penetration studies. In addition, when passive dosimetry is used, assumptions concerning the value of the clothing worn on the interception of pesticide residues must also be made.

Another disadvantage of or potential source of error in the use of patches for residue collection is the extrapolation from the residues on the relatively small surface area of the trapping devices to entire body surface areas. Also, when the patch technique is used, crucial areas of exposure may be missed, depending on the location of the trapping devices [2].

Biological Monitoring

The distinct advantage of the biological monitoring approach is that, under proper conditions, actual internal dose may be estimated from the results of the monitoring study. Biological monitoring implies that a biological measurement is intrinsic to the experimental design of an exposure assessment study. The use of the information derived from biological monitoring studies varies from the early detection of a health effect to establishing a correlation between concentration of a chemical in fluids to absorbed dose.

For pesticides, urinary metabolites have been used to detect exposure during field operations. Swan [3] measured paraquat in the urine of applicators; Gollop and Glass [4] and Wagner and Weswig [5] measured arsenic in timber applicators; Lieben et al. [6] and Durham and Wolfe [7] measured paranitrophenol in urine after parathion exposure. A chlorobenzilate metabolite was detected in citrus workers by Levy et al. [8]; phenoxy acid herbicide metabolites in farmers by Kolmodin-Hedman et al. [9], and organophosphate metabolites in the urine of people exposed to mosquito treatments by Kutz and Strassman [10]. Davies et al. [11] used urinary metabolites of organophosphates and carbamates to confirm poisoning cases.

While biological monitoring offers a distinct advantage, there are inherent difficulties in study design, execution, and data interpretation. Prior to establishing specific testing procedures, the experimenter must have a broad knowledge of the chemical being tested. The pharmacokinetics of the chemical of interest must be known and fully understood so that the appropriate tissue, fluid, or excretion pathway, as well as the appropriate time periods for monitoring, can be chosen with regard to this information. Because this level of detailed information is currently unavailable for many pesticides, extrapolations back to the actual dose, essential in the risk assessment process, are difficult.

When measuring exposure via human substrate one must consider the different ways through which a substance may be eliminated (sweat, urine, saliva, etc.) or that it may be stored in adipose and other tissues. Because the measurements relevant to the estimation of workers' exposure to pesticides must be carried out under field conditions, the Agency believes that urine analysis is a realistic and practical approach to implement biological monitoring. This emphasis on a particular method is based on a consideration of the type of information desired and what is feasible to obtain under field conditions. It should be understood, however, that this observation is not intended to exclude other methods which are demonstrably effective.

Biological monitoring clearly lacks the definition of source of exposure provided by the passive dosimetry method. The results of an appropriate biological monitoring study provide an integrated exposure picture. Biological monitoring should be considered a chemical specific approach. The potential for extrapolating results of an exposure study to other pesticides appears limited by probable differences in absorption, metabolism, and excretion profiles. Thus the opportunity for developing generic data bases, such as those from passive dosimetry studies, does not appear promising. This difficulty, however, may be overcome by the application of structure activity principles which would allow the grouping of pesticides according to chemical descriptors, thus providing a reasonable foundation for a useful, predictive data base.

A potential difficulty that may be encountered when using a biological method is obtaining strict cooperation and adherence to protocols from study participants when collecting specimens (urine, saliva, etc.). A plan for intensive study oversight is necessary to ensure participant compliance in collection, storage, and handling of urine specimens before, during, and after exposure to pesticides. It is not always practical, however, since investigators are usually on the site during the work activities only, while exposure data are collected beyond the supervised period.

When a registrant believes that the problems associated with biological monitoring as described here can be overcome for a particular pesticide and chooses to monitor worker exposure using biological monitoring, the Agency will carefully evaluate the results and, if judged valid, use the results in the risk assessment process. In these circumstances, the advantage of biological monitoring discussed earlier would naturally dictate the use of such data.

An examination of the limitations of the trapping devices presently used in passive dosimetry has suggested exposure scenarios where it would be necessary to utilize biological monitoring to estimate exposure. Examples of specific situations not amenable to current passive dosimetry methodologies include measuring exposure to pesticides in swimming pools or in dishwashing detergents, or measuring dermal exposure to volatile organic chemicals.

Advantages and disadvantages of passive dosimetry and biological monitoring are summarized in Table 3. The points delineated may be considered as initial guidance in determining the feasibility and objectives of conducting an exposure study by either method.

Monitoring Dermal Exposure by Passive Dosimetry when Protective Clothing is Worn

If the pesticide being studied is or will be registered for use only when protective garments are worn, dermal monitoring is slightly different. In addition to the standard external patches, the Agency may require that investigators monitor penetration. These pads should be located in regions that are expected to receive maximum exposure, such as beneath garment seams. Since the maximum exposure or area of penetration cannot always be accurately anticipated,

TABLE 3—*Advantages and disadvantages[a] of estimating occupational exposure with passive dosimetry and biological monitoring.*

Method	Advantages	Disadvantages
Passive dosimetry	* routes and areas of exposure clearly defined routine experimental design and execution participant under supervision of investigator genetic data bases may be created	* dermal and respiratory absorption must be estimated extrapolation from patch to body surface area must be made not all exposure scenarios are amenable
Biological monitoring	* actual dose may be estimated unnecessary to adjust for value of garment/protective clothing	* pharmacokinetics must be known routes of exposure cannot be distinguished difficult to ensure participant cooperation potential problems when using invasive techniques required for specimen collection

[a] Those advantages and disadvantages highlighted with an asterisk are considered most important.

additional pads should be placed under unseamed areas of protective garments in all the same locations as specified for dermal exposure monitoring of workers not wearing protective clothing. Pads under protective garments should be near, but not covered by, any pads on the outside of the clothing. Even if the label specifies that protective gloves must be worn, hand exposure under the gloves must be assessed. Maddy et al. [16] as well as other investigators have clearly demonstrated that workers, even when wearing rubber gloves, can receive significant hand exposure.

When Exposure Monitoring Field Studies Are Required

Dermal

Field studies of potential dermal exposure during the various work activities involved in the application of pesticide products under actual use conditions will be required of registrants when each of the following criteria is met:

1. The toxicological evaluation of a pesticide product indicates that the use of the product may pose an acute or chronic hazard to human health.
2. Dermal exposure is likely to occur during use.
3. The Agency lacks data to estimate with an acceptable degree of confidence the magnitude of exposure for a particular work activity.

Inhalation

Field studies of potential inhalation exposure for the various work activities involved in the application of pesticides under actual use conditions will be required when each of the following criteria is met:

1. The toxicological evaluation of a pesticide product indicates that the use of the product may pose an acute or chronic hazard to human health.

2. Respiratory exposure is likely to occur during use.

3. The Agency lacks data to estimate with an acceptable degree of confidence the magnitude of exposure for a particular work activity.

4. Depending on the physicochemical properties of the pesticide product, the conditions of use, and the toxicological effects, it can be expected that exposure by the route of inhalation will be of significant concern as compared with exposure by the dermal route. Field inhalation exposure studies will not be required in all cases where dermal exposure studies are required.

Number of Replicates—Under most conditions, 15 replicates each for total dermal and inhalation exposure will be required for each study. In this context, a replicate is defined as a measure of potential dermal (or inhalation) exposure using a personal monitoring device for an individual over the course of one work cycle or portion thereof. Studies may be required at indoor and outdoor sites.

Biological

Biological monitoring studies may be employed to estimate exposure as an alternative to passive dosimetry when each of the preceding four criteria is satisfied and will be required when all five of the following criteria are met:

1. The toxicological evaluation of a pesticide product indicates that the use of the product may pose an acute or chronic hazard to human health.

2. Exposure is likely to occur during use.

3. The Agency lacks data to estimate with an acceptable degree of confidence the magnitude of exposure for a particular work activity.

4. The pharmacokinetics of a pesticide (or other component in a pesticide product) are sufficiently understood so that a back-calculation to actual dose is possible.

5. Passive dosimetry techniques are determined not to be applicable for a particular exposure scenario.

Number of Replicates—Normally, 15 replicates are required for each biological monitoring study, with a replicate defined as a measure of a total exposure to an individual over the course of a workday or portion thereof.

Summary

Subdivision U (Applicator Exposure Monitoring) of the Pesticide Assessment Guidelines provides direction for conducting acceptable worker exposure studies, including criteria for the reporting of data, quality assurance, and for determining when the Agency may require the studies to be done.

Subdivision U has been finalized, is available to the public, and can be purchased through the National Technical Information Service (NTIS). Orders should be addressed to

National Technical Information Services
Attn: Order Desk
5285 Port Royal Road
Springfield, VA 22161
(Tel. 703-487-4850)

Subdivision U has been assigned the accession number PB87-133286 (EPA Report No. 540/9-87-127). Use this number when inquiring about the current purchase price.

It should be noted in closing that before any worker exposure monitoring studies are initiated by registrants which will be used to support registration actions, it is imperative that the registrant receive Agency approval of the study protocol. It should also be noted that the Agency will accept exposure monitoring studies using alternative methodologies, if given appropriate scientific justification. The Agency strongly encourages new methods development and stresses that the methodologies in Subdivision U are subject to change.

For further information, contact:

> Mr. Curt Lunchick
> USEPA
> Office of Pesticide Programs
> Hazard Evaluation Division (TS-769C)
> 401 M. St. SW
> Washington, DC 20460

References

[1] Durham, W. F. in *Dermal Exposure Related to Pesticide Use,* R. C. Honeycutt, G. Zweig, and N. N. Ragsdale, Eds., American Chemical Society Symposium Series 273, Washington, DC, 1985, pp. xiii–xiv.

[2] Franklin, C. A., Fenske, R. A., Greenhalgh, R., Mathieu, L., Denley, H. V., Leffingwell, J. T., and Spear, R. C., *Journal of Toxicology and Environmental Health,* Vol. 7, No. 5, 1981, pp. 715–731.

[3] Swan, A. A. B., *British Journal of Industrial Medicine,* Vol. 26, 1969, pp. 322–329.

[4] Gollop, B. R. and Glass, W. I., *New Zealand Medical Journal,* Vol. 89, 1979, pp. 10–11.

[5] Wagner, S. L. and Weswig, P., *Archives of Environmental Health,* Vol. 28, 1974, pp. 77–79.

[6] Lieben, J. R., Waldman, K., and Krause, L., *Industrial Hygiene and Occupational Medicine,* Vol. 7, 1953, pp. 93–98.

[7] Durham, W. F. and Wolfe, H. R., *Bulletin of the World Health Organization,* Vol. 26, 1962, pp. 75–91.

[8] Levy, K. A., Brady, S. S., and Pfaffenberger, P., *Bulletin of Environmental Contamination and Toxicology,* Vol. 27, No. 2, 1981, pp. 235–238.

[9] Kolmodin-Hedman, B., Hoglund, S., and Okerblom, M., *Archives of Toxicology and Contamination,* Vol. 54, 1983, pp. 267–273.

[10] Kutz, F. W. and Strassman, S. C., *Mosquito News,* Vol. 37, No. 12, pp. 211–218.

[11] Davies, J. E., Enos, H. F., Barquet, A., Morgade, C., and Danauskas, J. X. in *Toxicology and Occupational Medicine,* W. B. Deichman, Ed., Elsevier, New York, 1979, pp. 369–380.

[12] Wojeck, G. A., Nigg, H. N., Braman, R. S., Stamper, J. H., and Rouseff, R. L., *Archives of Environmental Contamination and Toxicology,* Vol. 11, 1982, pp. 661–667.

[13] Wojeck, G. A., Price, J. F., Nigg, H. N., and Stamper, J. H., *Archives of Environmental Contamination and Toxicology,* Vol. 12, 1983, pp. 65–70.

[14] Culver, D., Caplan, P., and Batchelor, G. S., *American Medical Association Archives of Industrial Health,* Vol. 13, 1956, pp. 37–50.

[15] Simpson, G. R. and Beck, A., *Archives of Environmental Health,* Vol. 11, 1965, pp. 784–786.

[16] Maddy, K. T., Wang, R. G., and Winter, C. K., "Dermal Exposure of Mixers, Loaders, and Applicators of Pesticides in California," Worker Health and Safety Unit Report No. HS-1069, California Department of Food and Agriculture, Sacramento, 1983.

[17] Leavitt, J. R. C., Gold, R. E., Holcslaw, T., and Tupy, D., *Archives of Environmental Contamination and Toxicology,* Vol. 11, 1982, pp. 57–62.

[18] Davis, J. E., Stevens, E. R., Staiff, D. C., and Butler, L. C., *Environmental Monitoring and Assessment,* Vol. 3, No. 1, 1983, pp. 23–38.

D. C. Eberhart,[1] E. W. Day,[1] R. D. Knarr,[1] D. M. Zimmerman,[2]
A. P. Nielsen,[3] J. C. Reinert,[3] and N. I. Muir[4]

Progress in the Development of a Generic Data Base for Mixer/Loader-Applicator Exposures

REFERENCE: Eberhart, D. C., Day, E. W., Knarr, R. D., Zimmerman, D. M., Nielsen, A. P., Reinert, J. C., and Muir, N. I., **"Progress in the Development of a Generic Data Base of Mixer/Loader-Applicator Exposures,"** *Performance of Protective Clothing: Second Symposium, ASTM STP 989,* S. Z. Mansdorf, R. Sager, and A. P. Nielsen, Eds., American Society for Testing and Materials, Philadelphia, 1988, 772–775.

ABSTRACT: In recent years, occupational exposure to pesticides by users of these materials has received increasing emphasis from manufacturers, users, and regulatory agencies. To quantitate exposures, either a field study is performed, or data from a study of a similar product are used as a surrogate on a "worst-case" basis. In 1984, a new concept for estimation of occupational exposure to pesticides was presented at the 187th National Meeting of the American Chemical Society in St. Louis. This proposal called for the creation of a database including all the available mixer/loader-applicator data, submitted in generic form without identifying the product or producer. The data would be compiled according to formulation and application technique, and made available to the public. Since 1984, a joint task force of the National Agricultural Chemicals Association, the U.S. Environmental Protection Agency, and Health and Welfare Canada has worked to create the database, which is now accepting data for compilation. This paper reports progress to date on the database.

KEY WORDS: protective clothing, pesticides, exposure, mixers, loaders, applicators, database

In recent years, exposure assessment has received increasing emphasis in risk assessments and in the registration process for pesticides. The exposures referenced here are occupational exposures to mixer-loaders and applicators. Exposures to the general public have long been addressed through the tolerance-setting process.

The ideal approach to exposure monitoring would be to conduct an exposure study of every use of every product. Then directly applicable data would be available as a basis for decisions on application methods and work practices in every situation. The cost of such an approach, however, would be prohibitive, since a single product might be registered for use on a dozen or more crops, and an exposure study can cost $50 to $150 thousand. Feasibility, then, dictates that a "worst-case" approach be taken. Under this approach, a single exposure study of perhaps four to eight exposure episodes is conducted on the crop and application practices that are likely to yield the highest exposure values. The results of this study are

---

[1] Members, Explosive Assessment Subcommittee, National Agricultural Chemicals Association, Washington, DC 20005.

[2] Environmental protection specialist, U.S. Environmental Protection Agency, Chicago, IL 60604.

[3] Agronomist and biochemist, respectively, U.S. Environmental Protection Agency, Washington, DC 20460.

[4] Chemist, Health and Welfare Canada, Ottawa, Canada.

then used to specify work practices and equipment for all uses of the product. In an effort to improve the exposure-asessment process, scientists from Mobay Chemical Corporation reviewed published mixer/loader-applicator exposure data in 1983, to determine whether it could be used to predict exposure. This report describes the results of that review, and the progress to date on the mixer/loader-applicator exposure database that was proposed by the reviewers.

Development of the Database Concept

The 1983 review of published mixer/loader-applicator exposure data yielded three principal observations:

1. Exposure data are not uniformly reported. Units of milligram/hour (mg/h), milligram/kilogram (mg/kg) active ingredient, and milligram/kilogram (mg/kg) body weight have all been used.
2. In cases where data are comparable, correlations between studies are good. This observation supports the assumption that exposures depend on the physical parameters of the application process and provides the basis for a generic database.
3. The published literature does not contain enough adequately-detailed data to predict exposures.

These findings were presented at the symposium on Dermal Exposure Related to Pesticide Use at the American Chemical Society National Meeting in April 1984.[5]

Included in this presentation was a proposal to create a generic database on mixer/loader-applicator exposure. The concept of a generic database is based on the assumption that exposure to pesticide applicators is governed by the physical parameters of the application, that is, by the type of formulation and the application method, and is independent of the chemical nature of the active ingredient. Under this assumption, a 4 EC formulation of a herbicide applied with a tractor-mounted boom-spraying rig, for example, would give a similar exposure to the applicator to a 4 EC insecticide applied by the same technique. All exposure data from ground-spray application of 4 EC products could be combined into a single data set of perhaps 100 observations, rather than the ~10 usually available from a single study. Similarly, data from other combinations of formulation type and application technique could be combined.

Data from individual companies, universities, and regulatory agencies would be submitted on a generic basis, that is, without identifying the product or the submitter, and compiled under a uniform reporting format to ensure comparability of data. The database would be available to data submitters and the general public.

The benefits of a generic database would include:

1. Scientists would be able to conduct mixer/loader-applicator risk assessments with a much greater degree of certainty than is currently possible, since their conclusions will be based on a much larger number of observations than the four to eight commonly used today.
2. The influences of different physical parameters on exposure could be determined and compared.

---

[5] Hackathorn, D. R. and Eberhart, D. C., "Data-base Proposal for Use in Predicting Mixer-Loader-Applicator Exposure," in *Dermal Exposure Related to Pesticide Use*, R. C. Honeycutt, G. Zweig, and N. N. Ragsdale, Eds., American Chemical Society Symposium Series No. 273, 1985, pp. 341–355.

3. The conflict and confusion resulting from the use of several different databases would be eliminated by a common database accepted by the scientific community.
4. The cost of mixer/loader-applicator exposure evaluations could be reduced through better use of scientific resources.

Since the fall of 1984, a task force comprised of representatives from National Agricultural Chemicals Association (NACA) and the U.S. Environmental Protection Agency (EPA), with technical support from Health and Welfare Canada, the California Department of Food and Agriculture, and public-interest groups, has addressed the technical and administrative issues pertinent to the database. Accomplishments of the task force to date are:

1. Criteria for acceptance of data into the database and for grading the data have been established.
2. Forms for submitting data in a uniform format have been developed.
3. A contractor to computerize and administer the database has been chosen.
4. NACA member companies have agreed to submit data from studies conducted in support of registrations to the database.
5. Funding has been secured. The EPA has allocated funds in Fiscal Year (FY) '86 and FY '87, and Health and Welfare Canada has financed the initial development of the database.
6. It has been agreed that the database will be available through the National Technical Information Service (NTIS).

The extent of the cooperation on this project is testimony to the perceived usefulness of the database to all the participants.

Use of Database for Protective Clothing Recommendation

The connection between the use of the database and protective clothing occurs in the risk management area. The database could be used to make decisions on protective-clothing recommendations. When the exposures associated with a formulation type and application method are known, the recommended protective clothing can be specified accordingly. Consider the following hypothetical example:

Given conditions:

Product:	50-WP
Application Method:	ground-spray, tractor-mounted
Application Crew:	single mixer/loader-applicator
Oral NOEL rat:	100 mg/kg/day

In many instances, neither dermal toxicity nor percutaneous absorption information are available. In such cases, the oral NOEL must be used and 100% absorption assumed.

Database Information:
M/L-A exposure for application of 50-WP is 12 mg/h.

Predicted Exposure:

$$\frac{(12 \text{ mg/h})(8 \text{ h/day})}{70 \text{ kg}} = 1.4 \text{ mg/kg/day}$$

If SF = 100, acceptable exposure = 1 mg/kg/day.
Conclusion: borderline.

Database Information:
For 50-WP formulations, M/L-A exposure is reduced by 80% when protective gloves are worn.

Recalculate:
(1.4 mg/kg/day)(0.20) = 0.28 mg/kg/day

Options:
1. Accept database estimate, put glove statement on label.
2. Conduct study to confirm.
3. Conduct dermal penetration study.
4. Conduct biomonitoring study.

Conclusion

We don't expect the database to solve all our exposure-estimation problems. Certainly, more studies will need to be done, and improved methodology, particularly in dermal-exposure monitoring, is critical. But we are confident, as our hypothetical example showed, that access to existing exposure data, when organized and compiled in a readily-accessible form, will allow us to make decisions in some cases, and to direct our research and monitoring efforts in the most productive directions in others.

Herbert N. Nigg[1] and James H. Stamper[1]

Field Evaluation of Protective Clothing: Experimental Designs

REFERENCE: Nigg, H. N. and Stamper, J. H., "**Field Evaluation of Protective Clothing: Experimental Designs,**" *Performance of Protective Clothing: Second Symposium, ASTM STP 989,* S. Z. Mansdorf, R. Sager, and A. P. Nielsen, Eds., American Society for Testing and Materials, Philadelphia, 1988, pp. 776–787.

ABSTRACT: Field evaluation of the protection afforded pesticide workers by apparel lis an emerging research area. The design of experiments is critical for obtaining unequivocal data upon which to base recommendations.

The simplest design is a penetration study using collection devices inside and outside the clothing. The residues on inside devices are compared with the residues on outside devices to calculate "protection." The residue proportion between inside and outside devices may be as large as 1:99. Although proportions this large are helpful for establishing statistical differences, this design still presents certain problems for the analyst.

A second design utilizes urine only. A 24-h urine specimen is collected. The worker wears the garment for one week and removes it the next. Changes in the concentration of residues in urine allow calculation of protection. This design may be flawed by extraneous exposure, uncertainties about urinary excretion kinetics, or nonurinary excretion routes.

A third design combines the first two, but unfortunately a new set of experimental considerations then emerges.

This manuscript discusses the statistical, chemical, metabolic, and human problems inherent in the three designs, with special reference to the experimental goal of each design.

KEY WORDS: protective clothing, experimental design, pesticide protection, field experiments, worker protection, exposure estimation

The goals of a field exposure study dictate methods, designs, and subsequently the difficulties encountered in the field portion of the study. Determination of pesticide penetration through fabrics might require that collection devices be placed under and on the surface of the worker's suit. To determine the *actual* protection provided by the suit might require that urine specimens be collected each day. Twenty-four-hour urine specimens are preferable for applicator/mixer-loader studies; daily timed urine specimens may suffice for a harvester experiment. Depending on urine excretion kinetics for the compound used, a protective garment might be worn on alternate days or on alternate weeks. Field methods are also influenced by the chemical extraction and analytical techniques employed. The length and complexity of any individual design will depend on the variation among replications reported by previous researchers. Exposure experiments are very expensive and time-consuming; no researcher wants to discover belatedly that only a few more specimens collected and a little more effort expended would have salvaged a statistically unanalyzable experiment.

Many of the methods mentioned have been previously reviewed [1,2]. It is our purpose

[1] Professor and research associate, respectively, Citrus Research and Education Center, University of Florida, IFAS, 700 Experiment Station Road, Lake Alfred, FL 33850.

here to consider experimental designs for monitoring worker exposure so that prospective researchers understand some of the sources of error or ambiguity in field experiments.

Penetration Assessment

Penetration of chemicals through protective clothing may be assessed with dermal exposure pads. Dermal exposure pads have been constructed of alpha-cellulose, cloth, polyurethane foam, and combinations of these materials. They are generally designed to collect spray materials and have been used for emulsifiable concentrates and wettable powders mixed with water. Almost never are these collection devices assessed for pesticide loss. That is, if the pad is left on the worker for 6 h, how much material evaporated, degraded, or otherwise disappeared in those 6 h? Davis [1] reviewed the practice of fortifying pads with the same pesticide-water mixture as in the field experiment. Alternatively, unfortified pads could be applied to the worker, with some left for 1 h and then removed, some left for 2 h and removed, and so forth. This latter test assumes that the exposure to the pads is equal for all exposure periods and ignores temperature effects. Fortified pads placed in an out-of-doors holding device appears to be the better approach. By removing and analyzing these pads at intervals, one can determine the approximate length of time a pad should be worn by a worker in the field. This is one important requirement for obtaining "unquestionable" worker data.

Once the time of exposure and laboratory recovery studies are completed, storage stability should be determined. Although a separate experiment is possible, the simple expedient of storing one or two fortified pads with each worker's pad set will determine storage stability as extraction and analyses proceed. The resulting recoveries also serve as a check on the accuracy of laboratory technical help. The required number of these fortified pads depends on the size of the experiment. The criterion is to allow for enough measurements to validate the quality of both storage and extraction. We use a minimum of three fortified pads per exposure day.

The analyst must be prepared to work throughout a wide range of sensitivities. The amount of chemical on an outside collection pad depends on its location on the worker and on job type, work practice, the pad area which is extracted, exposure time, application rate, and wind. Under a worst circumstance, 1 mg may be an upper limit for pad quantity. Twenty nanograms might be a lower limit. If the pad is inside and a 95% reduction occurs, the limits are correspondingly lower. Operating at high sensitivity may lead to "peaks" from contaminated glassware and needles or through various extraction procedures. It is important to understand analytical limitations so that we understand what observation of chemical on an inside pad means to the worker.

Placement of the pads on the body of the subject has many ramifications. If a total body exposure estimate is to be made, the calculation method should be considered. Davis [1] and Popendorf and Leffingwell [3] are good sources for these methods. Pads will then be placed so as to optimize the total body estimate. The location of pads on the subjects should not be decided upon hastily. For those areas of the United States where the climate is very warm or humid or both during a spray season, data on where the greatest exposure occurs to the worker's body would be very helpful. These data might allow development of protective suit designs which do not attempt total coverage and which therefore might prove more comfortable. For instance, the exposure of an applicator or mixer-loader on the backs of the arms and legs is not known. Whether the lower arms receive more exposure than the upper arms is seldom monitored. In many of our measurements, the lower arms received significantly more exposure than the upper arms, but the generality of this result is unconfirmed. If these data were available, comfortable protective suits utilizing relatively open mesh areas might be certifiably protective at this time. Certainly, taking these additional data adds extra work and expense to a study, but the long-term benefits might be substantial.

Exposure Versus Work Practice

Applicators and mixer-loaders receive different levels and types of exposure than do harvesters. Mixer-loaders have an opportunity for exposure to concentrated materials as well as drift exposure. An applicator is primarily exposed to drift and the diluted, tank-mixed material. Both groups may also be exposed by working on or around contaminated machinery and in or around contaminated loading areas. A harvester is exposed to a presumably homogeneous application of pesticide on fruit, leaf, and soil surfaces. Theoretically, harvesters are exposed only to the residues remaining in the field.

The experiment appears simple. For the design, however, there are all sorts of possible constructions. Will clothing chosen by the worker suffice, or should standard clothing be issued? Is the field to be sprayed over one hour, one day, or three days? If the application takes longer than one day, where in the field should the harvesters start working? Will they overlap sprayed sections as the work progresses? How many daily residue specimens should be taken as a consequence? Should pads with a surgical gauze front be used or would polyurethane foam be satisfactory? How should these pads be assessed for residue loss? How long should the worker wear the pads? Is the pesticide converted in the field into a toxicologically important metabolite? Can it be extracted and analyzed? How should the urine be collected: 24-h urines or a timed grab specimen? And finally, how many sampling periods (days) are necessary to determine the statistical significance of the results?

For harvesters, the following suggestions are offered: The pads should be placed inside the clothing for lower and upper arm, chest, back, shoulders, and shin exposure. For the upper body, the pads can be conveniently pinned inside an issued shirt. They can also be pinned inside the pants, but it should be noted whether or not the worker wears the same pants each day. Intuition is a fine tool here, especially as enhanced by experience; but for the first experiment, we recommend the above placement of pads. For later experiments, a reduction in the number of pads may be possible. It is, however, a mistake to simply observe a harvest operation and decide *a priori* that only leg patches are necessary.

Knowing the pesticide application date can be crucial. An experiment of this type should begin at the legal reentry time and extend through at least two pesticide "half-lives." This insures variable doses of the pesticide to both the outside and inside of clothing. This sampling time may last one week or longer, and the area to be sprayed may be necessarily large. We have, for instance, used three spray machines simultaneously in order to assure a one-day application. All harvesters are then exposed to the same daily residue over the sampling period.

When a "blind" harvester experiment is conducted and the pesticide application is made over a few days, the number of residue specimens (for a half-life calculation) per substrate should be doubled and taken from where the harvesters are working that particular day. Even if the experiment is only a one- or two-day experiment, reentry should commence as soon as possible after application. This assures some results at least, from an analytical standpoint. If workers reenter a field after ten days and the analytical chemist detects no residues because of low levels, little has been accomplished except the expense of time and money.

The most commonly used exposure pad for worker *reentry* experiments is faced with surgical gauze, backed with alpha-cellulose and glassine weighing paper. This pad has proven uncomfortable for the worker, difficult to attach, and time-consuming to prepare. We know of one instance in which polyurethane foam pads were used.[2] Polyurethane foam pads are very convenient and may be efficient. However, there is no good method for assessing the residue collecting efficiency of any of these collection devices for a harvester exposure experiment. In spite of years of research in this area, the transfer process of field surface residues to the body of the harvester is not known with certainty. Probably foliar and field dust are primarily involved. How then is the efficiency of a collection device for a harvesting operation measured? The experimenter is presently confined to the application of pesticide-laden dust or a pesticide solution to the exposure pad, followed by a disappearance study. Although the disappearance study may indicate a 50% loss from a pad in, say, two hours, the pads may have to be worn longer. The reason is simple: Exposure for a harvester is generally low, and enough residue must be collected for analyses. We attach the pads just before workers enter the field in the morning and remove them 4 to 5 h later at the noon break. The chemical residue level on a pad can be corrected according to the disappearance experiment, but this correction is not entirely reliable since the pesticide may disappear at a different rate when attached to dust.

The production of a toxic metabolite on foliage or in soil and the possible consequence to harvesters have been reviewed [4,5]. We mention this consideration because of its importance to harvesters and because urine analyses may have to account for the excretion products of these metabolites. Urine collection from harvesters is not difficult, but consistently reliable urine sampling is another matter. We have attempted to collect 24-h urine specimens; this did not work well. Harvesters in Florida are more mobile, may or may not appear for work, and are less likely to understand instructions than the applicator, mixer-loader group. However, a timed grab specimen from the start of work until the noon break has provided excellent correlations between residue levels on foliage and urinary metabolites in harvesters [6].

Harvesters and applicators represent a large difference in exposure. Similar considerations must be factored into an experiment for different types of application methods as well.

Work Practices and Work Rates

For the applicator/mixer-loader group, the type of equipment used, the number of tanks applied per unit time, the concentration of the tank mix, and the loading method all affect the exposure process. This has been known for years and is described in many published reports [1,6].

For harvesters, only a few field experiments are described in the literature. The harvesting method and crop have been studied, and some reports exist which can be compared. What seems apparent from these reports is that the exposure process is the same for the harvesting of both citrus and apples. At least, the proportion of harvester exposure to pesticide on the leaf surface is the same. For other types of crops this proportion may be different.

Regardless of crop type, the work rate appears to be related to exposure. This means that the number of boxes picked, crates loaded, tassels removed, etc., is confounded with residue levels in affecting exposure. The worker's production affects the amount of contact with the plant, a subject which has been studied using movies and time analysis [7], and estimated with surveys [8]. Therefore, work rate data should be gathered for each subject; they may explain variation in urinary or dermal exposure unaccounted for by field residues.

[2] Brady, E., University of Georgia, Athens, GA,, personal communication.

Analytical Considerations

For applicators and mixer-loaders, methods can be developed as needed with defined substrates. For the extraction of leaf, fruit, and soil surface residues, peculiar to harvester exposure studies, a standard methodology has been adopted by many researchers [9,10]. Fruit and leaf surface residues are recovered with organic solvents from a mild soap solution in which they have been shaken. Soil surface residues are recovered by vacuuming surface soil through a 100-mesh screen. However, for foliar residues, some experimenters shake leaves in organic solvents [11,12]. These organic solvent residue data are generally higher and usually lead to slower calculated rates of disappearance, making it *appear* that the worker is exposed to higher residues of longer duration.

Should these differences in extraction methods really be a concern? Regardless of the volume of research on the harvester exposure *process*, that is, the transport mechanism from foliage or soil to the worker, it is not precisely understood now, nor is a better understanding likely to be forthcoming in the near future. Nor are there enough researchers or funds available to investigate every exposure situation. But models of worker exposure as a function of foliar residues recovered by the dislodgeable method (nonsolvent) have been and are being produced. A model developed for one chemical is then used for another. Solvent residue data for a chemical could be used in these models once the relationship between the solvent and dislodgeable methods is understood and quantified.

Acetylcholinesterase Monitoring

One of the first applicator exposure studies was undertaken by Quinby et al. [13]. Cholinesterase activity was measured in aerial applicators together with residues on worker clothing and in respirator pads. In spite of physical complaints by the pilots exposed to organophosphate, either normal or only slightly depressed cholinesterase values were reported. Cholinesterase values were compared with the "normal" range for the U.S. population rather than the pilots' own individual "normal" values. In 1952, Kay et al. [14] measured cholinesterase levels in orchard parathion applicators. These were compared with cholinesterase levels taken from the same workers during non-spray periods. Plasma cholinesterase for workers reporting physical symptoms was 16% lower during the spray period. The corresponding reduction for symptomless workers was about 13%. Red blood cell cholinesterase was depressed 27% for the symptom group versus 17% for the nonsymptom group, but these means were not statistically different. Roan et al. [15] measured plasma and erythrocyte cholinesterase and serum levels of ethyl and methyl parathion in aerial applicators. Serum levels of the parathions could not be correlated with cholinesterase levels. However, serum levels did correlate with the urine concentration of p-nitrophenol. Drevenkar et al. [16] measured plasma and erythrocyte cholinesterase levels and urine concentrations of organophosphate and carbamate pesticides in plant workers who formulate these materials. No correlation existed between urinary metabolites and cholinesterase depression. Bradway et al. [17] studied cholinesterase, blood residues, and urinary metabolites in rats. Even under controlled conditions and known doses, correlations of cholinesterase activity with blood residues and urine metabolite levels were poor. Eight organophosphates were included in the Bradway et al. [17] experiment. The overriding general conclusion from the above is that cholinesterase inhibition as an exposure indicator contains too many variables, known and unknown, to be of use for monitoring the protective effect of personal equipment.

Urine Monitoring

Urine monitoring to determine protective clothing efficacy necessarily depends on assumptions and disparate pieces of available data. The excretion kinetics of the pesticide employed must be known. If the total dose is excreted by small animals in 24 to 48 h, the same may also be true of humans, and a simple experimental design may suffice. If the dose is excreted over the period of a week, a simple design correlating dose with an *immediate* effect on urine pesticide levels is not possible. The difficulty with long sampling periods because of slow excretion kinetics derives from the variation normally observed in the urinary exposure estimation for field experiments. It is common to see a 100% coefficient of variation among urinary excretion specimens. As a consequence, a major change in urinary pesticide levels comparing clothing versus no clothing would have to occur before a significant difference could be detected statistically. Part of this unfortunate variation results from reckless work practices. Workers spill concentrated materials. They service contaminated machinery with bare hands. In one experiment, we observed a worker remove his gloves, roll up his protective suit, and retrieve a crescent wrench from the tank. These increases in exposure variation decrease the statistical significance of observed mean differences in urinary metabolites between workers wearing protective gear and those not wearing protective gear.

The prior knowledge of the excretion routes is also a key piece of data. Many pesticides in common use are not excreted in urine in proportion to dose. Organochlorines are excreted more in feces than in urine. An example of current interest is dicofol. With multiple doses this organochlorine rapidly reaches a plateau in urine, while excretion levels steadily increase in the feces [18]. While dicofol is an exception, the excretion kinetics available for most compounds have resulted from studies using only one or a very few oral doses. Fieldworkers, however, receive daily dermal and respiratory doses. There is no comparable animal model for these multiple routes of exposure. There are kinetic models which could represent pesticides recycled into the bile [19]. These models do not account for the differences in the route of the dose for workers, their frequency of dose, and subsequent excretion differences in workers compared with those for small animals.

Urinary metabolites of pesticides have been used for a variety of experimental goals. Swan [20] measured paraquat in the urine of spraymen, Gollop and Glass [21] and Wagner and Weswig [22] measured arsenic in timber applicators, Liehen et al. [23] measured para-nitrophenol in urine after parathion exposure, as did Durham et al. [24]. Chlorobenzilate metabolite (dichlorobenzilic acid) was detected in citrus workers [25], phenoxy acid herbicide metabolites in farmers [26], and organophosphate metabolites in the urine of ordinary citizens exposed to mosquito treatments [27]. Davies et al. [28] used urinary metabolites of organophosphates and carbamates to confirm poisoning cases. These studies document exposure, but no estimate of exposure can be made from urinary metabolite levels alone. Other studies have used air sampling and monitored hand exposure in combination with urine levels [29], and air sampling plus cholinesterase inhibition plus urine levels [30].

The exposure pad method, combined with measurement of urinary metabolites, has been used to compare worker exposure for different pesticide application methods [31,32], as well as to monitor formulating plant worker exposure [33] and homeowner exposure [34].

Several studies have used the exposure pad method to estimate total dermal exposure, attempting to correlate urine levels with the exposure estimate [31,34,35–39]. Lavy et al. [38,39] found no such correlation with 2,4-D or 2,4,5-T. Wojeck et al. [31] found no paraquat in urine and consequently no relationship between dose and urine levels. However, the *group* daily mean concentration of urinary metabolites of ethion and the group *mean* total dermal exposure to ethion on that day correlated positively, with significance at the 97% confidence level [37]. For arsenic, the cumulative total exposure and daily urinary arsenic concentration correlated positively, with significance at the 99% confidence level [36]. Franklin et al. [34] found a positive correlation between 48-h excretion of azinphosmethyl metabolites and the amount of active ingredient sprayed. A significant correlation could not be made, however, between 48-h excretion and an exposure estimate. In the Franklin et al. [34] experiment, a fluorescent tracer had been added to the spray mixture. Qualitatively, unpatched areas (face, hands, neck) also received significant exposure, perhaps leading to a weak correlation between the patch estimate and urinary metabolites. Winterlin et al. [40] monitored the dermal exposure of applicators, mixer-loaders, and strawberry harvesters to captan using exposure pads. Although the applicator, mixer-loader group showed higher dermal exposure, no metabolite was detected in their urine, whereas harvester urine had detectable levels.

We return, consequently, to the problem of the excretion kinetics of pesticides, the complexity of which may render useless any search for a simple linear correlation between dose and urinary metabolites. Drevenkar et al. [16] studied the excretion of phosalone metabolites in one volunteer. Excretion reached a peak in 4 to 5 h, but was not complete in 24 h. Funckes et al. [41] exposed the hand and forearm of human volunteers to 2% parathion dust. During exposure, the volunteers breathed pure air and placed their forearm and hand into a plastic bag which contained the parathion. This exposure lasted 2 h and was conducted at various temperatures. There was an increased excretion of paranitrophenol with increasing exposure temperature. More important, paranitrophenol could still be detected in the urine 40 h after exposure. In another human experiment, Kolmodin-Hedman et al. [42] applied methylchlorophenoxy acetic acid (MCPA) to the thigh. Plasma MCPA reached a maximum in 12 h, and MCPA appeared in the urine for five days with a maximum after about 48 h. When MCPA was given orally, urinary MCPA peaked in 1 h with about 40% of the dose excreted in 24 h. In a rat experiment, seven different organophosphates at two different doses were fed to two rats [17]. Rats were removed from exposure after the third day and blood and urine collected for the next ten days. The average percents of the total dose excreted in urine over the ten days were: dimethoate, 12%; dichlorvos, 10%; ronnel, 11%; dichlofenthion, 57%; carbophenothion, 66%; parathion, 40%; and leptophos, 50%. Very little of this excretion occurred beyond the third day post-exposure. Intact residues of ronnel, dichlofenthion, carbophenothion, and leptophos were found in fat on Day 3 and Day 8 post-exposure. In another rat experiment, animals were dosed once dermally and intramuscularly with azinphosmethyl [43]. About 78% of the dermal dose had been excreted in urine in 24 h. Its rate of excretion peaked in 8 to 16 h, continued at about the same rate for another 16 h, and declined to a steady level 16 h thereafter. There was a linear relationship between dermal dose and urinary excretion. The intramuscular dose was excreted much more rapidly than the dermal dose. No apparent relationship existed between the intramuscular dose and urinary excretion. In a worker exposure experiment with azinphosmethyl, Franklin et al. [44] concluded that without adequate metabolism information for the parent compound, urine is useful only as a screening tool for indicating exposure or as a predictor of overexposure.

These experiments illustrate the excretion differences between dermal, intramuscular, and oral dose excretion, the excretion differences between compounds, and also problems about which urinary metabolite to monitor [43]. In a very perceptive study, Akerblom et al. [45] clearly showed that humans excrete dermal and oral doses of MCPA at quite different rates. A very comprehensive experimental design would consequently be necessary to model dermal exposure, absorption, and urinary metabolite levels versus protection afforded by protective gear. Statistical problems, centered on replicate variation and the resulting necessity for large numbers of replications, make the cost of either a human or small-animal experiment too expensive.

There may be solutions to this difficulty. A small-animal excretion study with multiple doses could be done. However, which dosing method should be used? Should the dose be dermal, as the fieldworker has mostly dermal exposure, or should a mixture of oral, dermal, and respiratory dosing be used? What if the compound has associated data suggesting skin penetration is low? Since regulations are designed to protect humans, real-world estimates, relevance to work exposure, and accuracy in the preliminary small-animal studies are extremely important.

Another possibility is to perform an actual human excretion study. At the end of the spray season, with the workers removed from further exposure, a series of 24-h urine specimens might yield the necessary kinetic data. We suggest splitting the daily sampling periods into two 12-h or four 6-h segments. For compounds which are very rapidly excreted, this division into timed segments will help to obtain the necessary data points. Removal of workers from exposure is paramount, even if the experimenter must closely monitor their work activities and wash the workers' clothing over this period. We have discovered, to our surprise, that workers have interpreted "removal from exposure" as "stop spraying," and then proceeded to clean and maintain heavily contaminated equipment over our sampling period. Even if all of these factors are controlled, the chemical type, its metabolism, its variability among subjects, or some combination of these may prohibit the drawing of statistically valid inferences.

Some of the other factors which invalidate a clothing efficacy experiment are loosely fitting respirators, workers with highly variable hand exposure, varying wind speed and direction, and spraying when the wind velocity exceeds 16 km/h (10 mph).

Let us assume that all problems associated with monitoring urine for pesticides and metabolites have been solved. We have eliminated exposure from the hands, face, feet, and respiration. The worker wears issued shirt and pants which are exchanged for another decontaminated shirt-and-pants set each day. We now design an experiment where workers are completely protected during one week and the protective garment is removed the next. The worker continues to protect the hands, face, feet, and respiration when the protective garment is removed in the second week. Further, assume that the level of exposure (dose) is the same in both weeks. Can we compare the level of urinary metabolites in the second week with the first week and determine the protective clothing efficacy? Rosenberg et al. [46] used sweat patches to collect dichlorobenzilic acid (a metabolite of chlorobenzilate). Urine analyses were also performed. The estimated dose ratio excreted in sweat and urine was approximately 1:1. Sell et al. [47] conducted a human exposure experiment with 2,4-D and monitored urine and sweat. Of the total 2,4-D excreted over one week after a single hand dermal dose, 15% was excreted in urine and 85% in sweat (average for three workers). Although the sweat monitoring data for pesticides are preliminary, we have overlooked sweat excretion in our design. Unfortunately, rats do not sweat, so rat excretion data are probably not applicable. If a pesticide or metabolite or both are excreted in sweat, the sweat/urine ratio will change with the human physiological response to heat. Since protective suits add a heat factor, the estimate of protection will be higher because of increased sweat production while wearing the suit. When the excretion of pesticide metabolites and parent compounds in human sweat is understood, we may be able to determine the magnitude of this overestimation. For comments on the future of this technology, see Peck et al. [48].

If dermal dose urinary excretion kinetics are ignored in monitoring urine, the results may be up to 100% wrong. If sweat excretion is ignored, 100% to 300% mistakes appear possible. If our dose estimation from urine alone is 100% wrong, is the worker 100% sick?

Replications, Statistical Considerations

Here we discuss in greater detail the statistical considerations in experimental design for the field assessment of protective clothing efficacy in reducing dermal pesticide exposure. Data analysis for current experiments usually necessitates comparing two mean values, each deriving from about the same number of replications. One mean represents exposure to a pad placed outside the clothing while the other represents exposure to a pad placed inside. The number of replications (n) per mean is typically the number of separate exposures monitored.

The question must be addressed as to how many replicates per mean (subjects × days) will be necessary to provide a reliable estimate of the difference between means and its statistical significance. Two parameters must be estimated first: What difference in means is expected to arise from the data, and what variation among replicates is expected? A large estimate for the first expectation or a small estimate for the second expectation would reduce the number of replicates necessary per mean. Our experience has shown, for example, that protective clothing reduces mean pad exposure by about 90% [49]. Similarly, the use of gloves reduces handwash residues by about 85%, according to a recent study of dicofol by Nigg et al. [50]. As for variation among replicates, the coefficient of variation is typically about 100% for exposure pad and handwash residues [6,49].

Standard statistical procedures then show [51] that, on the basis of the above estimates, two exposure pad means may be judged significantly different at the 95% confidence level by taking at least ten replications per mean. The (approximate) calculation is $n > 8$ (100% ÷ 90%)^2, or $n > 10$ replications. For gloves, the corresponding calculation would be $n > 8$ (100% ÷ 85%)^2, or $n > 11$ replications. These numbers represent an absolute minimum based on the above two expected values. An increase in n of 50% to provide some margin for error, in the above cases to $n > 15$ and $n > 17$, is certainly warranted in view of the guesswork involved about sample means and variances.

If more than two means are to be compared concurrently, as with comparing residues at various body locations, the situation becomes somewhat more complicated. While it is now harder to generalize, the optimum number of replications per mean can be roughly adduced by the same calculation, with the final result that the means are grouped into significantly different categories, at some confidence level.

Whether to utilize, for example, three subjects for six sampling periods each or six subjects for three sampling periods each, to obtain, say, 18 replications, is usually dictated by factors other than statistical ones. A good rule to follow is not to overload the design too heavily in favor of either variable. If many subjects are used for a very few sampling days, and the data indicate large differences from subject to subject (admittedly an unlikely result), the available number of replicates now decreases to the number of sampling days alone. This experiment makes a very unreliable statement about each of many subjects; no valid overall conclusion may emerge.

Remember that these statistical calculations show significant differences between means, but do not estimate the difference. Suppose one mean is 90% less than the other, as above, but that one wishes to validate the claim that 50% of that 9% is statistically significant. The approximate calculation for the requisite number of replications per mean is now $n > 8$ [100% ÷ (90% − 50%)]^2 = 50, or $n > 75$ with the safety factor.

A Combined Design

The third design combines urine data and exposure collection device data. One consideration is the amount of chemical removed by a collection device which subsequently is *not* excreted in urine. If twenty 103-cm^2 (4 in. by 4 in.) pads are used for monitoring a worker, about 2000 cm^2 (320 in.^{2}) of body surface area would be covered. The average human has a surface area of about 20 000 cm^2 (3200 in.^{2}). Since 10% of the body surface area is covered, will this reduce the urinary metabolite by 10%? Hardly. The pads are worn no longer than four hours, and usually only one or two hours, over the eight-hour workday. The maximum percentage reduction is thus 5%. We challenge any researcher to show a statistically valid 5% difference in urinary excretion in any field experiment.

However, the use of a hand rinse or cotton gloves to monitor hand exposure may invalidate urine results. Suppose 24-h urines were collected from 20 workers during different periods; one with a protective device and one without a protective device. The urine excretion means for the two periods would have to differ by 35% to be statistically different at the 95% level, assuming a coefficient of variation of 75%. For studies using fewer than 20 subjects (replicates), differences in urinary excretion means become evermore insignificantly statistically.

Suppose we measured 1000 units of a pesticide urinary metabolite in an experiment where hand exposure contributed 50% of the urinary metabolite. If we removed (rinse or gloves) hand residue after 2 h, we would remove one quarter of the hand contribution for an 8-h day. The new urine value would be 875 units, $1000 - (0.25 \times 500)$, or a urinary metabolite reduction of 12.5%. If we monitored hands during an ungloved period, measured 1000 urinary metabolite units, and otherwise totally excluded other sources of exposure, then protected the hands with gloves, what then? Gloves have afforded 85% protection [50]. Now we see only 150 urinary units. If we removed one quarter with a hand rinse, we would see about 113 urinary units. We could easily describe this difference statistically, but we would overestimate the protection afforded by gloves by above 4%.

Unfortunately, such arguments are superfluous because we cannot use urinary metabolites to monitor either dose or protective device efficacy. As previously stated, sweat excretion and dermal dose urine excretion kinetics must be understood in order to use urine. But, when we understand *human* excretion routes and their interrelationships, passive monitoring devices and biological fluid monitoring may be combined.

General Conclusions

The complexity of monitoring urine, blood, and other biological fluids precludes their use for assessing protective clothing or for determining dose. Further development of "biological" monitoring techniques is a challenging scientific frontier for estimating internal doses and should be vigorously pursued. Frequently, the failure of a protective clothing experiment results from the lack of basic information on relevant human biochemistry and physiology. A thorough literature search for information on mammalian excretion kinetics could be a very worthwhile prior investment.

Otherwise unexplainable sources of variation within replicate specimens might become evident if the work practices of the subjects are considered. This may require a period of careful observation in the field extending over a week or longer.

The level of and reasons for variation encountered in previous published studies can be helpful for experimental design. Once this anticipated variation is estimated, numbers of replications per specimen can be rationally established.

Monitoring the efficacy of protective clothing under field conditions is the ultimate test. Because we are the subjects, every effort to perform well-designed experiments must be made, regardless of cost.

Acknowledgment

This paper is Florida Agricultural Experiment Station Journal Series No. 7946.

References

[1] Davis, J. E., *Residue Reviews,* Vol. 75, 1980, pp. 33–50.
[2] Nigg, H. N. and Stamper, J. H. in *Dermal Exposure Related to Pesticide Use,* R. C. Honeycutt, G. Zweig, and N. Ragsdale, Eds., American Chemical Society Symposium Series 273, Chapter 7, Washington, DC, 1985, pp. 95–98.
[3] Popendorf, W. J. and Leffingwell, J. T., *Residue Reviews,* Vol. 82, 1982, pp. 125–201.
[4] Nigg, H. N. and Stamper, J. H. in *Pesticide Residues and Exposure,* J. Plimmer, Ed., American Chemical Society Symposium Series 182, Chapter 6, Washington, DC, 1982, pp. 59–73.
[5] Gunther, F. A., Iwata, Y., Carman, G. E., and Smith, C. A., *Residue Reviews,* Vol. 67, 1977, pp. 1–139.
[6] Nigg, H. N., Stamper, J. H., and Queen, R. M., *American Industrial Hygiene Association Journal,* Vol. 45, 1984, pp. 182–186.
[7] Wicker, G. W. and Guthrie, F. E., *Bulletin of Environmental Contamination and Toxicology,* Vol. 24, 1980, pp. 161–167.
[8] Wicker, G. W., Stinner, R. E., Reagan, P. E., and Guthrie, F. E., *Bulletin of the Entomological Society of America,* Vol. 26, 1980, pp. 156–161.
[9] Iwata, Y., Knaak, J. B., Spear, R. C., and Foster, R. J., *Bulletin of Environmental Contamination and Toxicology,* Vol. 18, 1977, pp. 649–655.
[10] Spencer, W. F., Kilgore, W. W., Iwata, Y., and Knaak, J. B., *Bulletin of Environmental Contamination and Toxicology,* Vol. 18, 1977, pp. 656–662.
[11] Ware, G. W., Estesen, B., and Cahill, W. P., *Bulletin of Environmental Contamination and Toxicology,* Vol. 14, 1975, pp.606–609.
[12] Ware, G. W., Estesen, B., and Buck, N. A., *Bulletin of Environmental Contamination and Toxicology,* Vol. 25, 1980, pp. 608–615.
[13] Quinby, G. F., Walker, K. C., and Durham, W. F., *Journal of Economic Entomology,* Vol. 51, 1958, pp. 831–838.
[14] Kay, K., Monkman, L., Windish, J. P., Doherty, T., Pare, J., and Raciot, C., *Industrial Hygiene and Occupational Medicine,* Vol. 6, 1952, pp. 252–262.
[15] Roan, C. C., Morgan, D. P., Cook, N., and Paschal, E. H., *Bulletin of Environmental Contamination and Toxicology,* Vol. 4, 1969, pp. 362–369.
[16] Drevenkar, V., Stengl, B., Tralcevic, B., and Vasilic, Z., *International Journal of Environmental Analytical Chemistry,* Vol. 14, 1983, pp. 215–230.
[17] Bradway, D. E., Shafik, T. M., and Loros, E. M., *Journal of Agricultural and Food Chemistry,* Vol. 25, 1977, pp. 1353–1358.
[18] Brown, J. R., Hughes, H., and Viriyanondha, S., *Toxicology and Applied Pharmacology,* Vol. 15, 1969, pp. 30–37.
[19] Colburn, W. A., *Journal of Pharmaceutical Sciences,* Vol. 73, 1984, pp. 313–317.
[20] Swan, A. A. B., *British Journal of Industrial Medicine,* Vol. 26, 1969, pp. 322–329.
[21] Gollop, B. R. and Glass, W. I., *New Zealand Medical Journal,* Vol. 89, 1979, pp. 10–11.
[22] Wagner, S. L. and Weswig, P., *Archives of Environmental Health,* Vol. 28, 1974, pp. 77–79.
[23] Lieban, J., Waldman, R. K., and Krause, L., *Industrial Hygiene and Occupational Medicine,* Vol. 7, 1953, pp. 93–98.
[24] Durham, W. F., Wolfe, H. R., and Elliot, J. W., *Archives of Environmental Health,* Vol. 24, 1972, pp. 381–387.
[25] Levy, K. A., Brady, S. S., and Pfaffenberger, C., *Bulletin of Environmental Contamination and Toxicology,* Vol. 27, 1981, pp. 235–238.
[26] Kolmodin-Hedman, B., Hoglund, S., and Akerblom, M., *Archives of Toxicology,* Vol. 54, 1983, pp. 257–265.
[27] Kutz, F. W. and Strassman, S. C., *Mosquito News,* Vol. 37, 1977, pp. 211–218.
[28] Davies, J. E., Enos, H. F., Barquet, A., Morgade, C., and Danauskas, J. X. in *Toxicology and Occupational Medicine,* W. B. Deichmann, Ed., Elsevier, New York, Vol. 4, 1979, pp. 369–380.
[29] Cohen, B., Richler, E., Weisenberg, E., Schoenberg, J., and Luria, M., *Pesticides Monitoring Journal,* Vol. 13, 1979, pp. 81–86.
[30] Hayes, A. L., Wise, R. A., and Weir, F. W., *American Industrial Hygiene Association Journal,* Vol. 41, 1980, pp. 568–575.
[31] Wojeck, G. A., Price, J. F., Nigg, H. N., and Stamper, J. H., *Archives of Environmental Contamination and Toxicology,* Vol. 12, 1983, pp. 65–70.
[32] Carman, G. E., Iwata, Y., Pappas, J. L., O'Neal, J. R., and Gunther, F. A., *Archives of Environmental Contamination and Toxicology,* Vol. 11, 1982, pp. 651–659.

[33] Comer, S. W., Staiff, D. C., Armstrong, J. F., and Wolfe, H. R., *Bulletin of Environmental Contamination and Toxicology,* Vol. 13, 1975, pp. 385–391.

[34] Franklin, C. A., Fenske, R. A., Greenhalgh, R., Mathieu, L., Denley, H. V., Leffingwell, J. T., and Spear, R. C., *Journal of Toxicology and Environmental Health,* Vol. 7, 1981, pp. 715–731.

[35] Staiff, D. C., Comer, S. W., Armstrong, J. F., and Wolfe, H. R., *Bulletin of Environmental Contamination and Toxicology,* Vol. 14, 1975, pp. 334–340.

[36] Wojeck, G. A., Nigg, H. N., Braman, R. S., Stamper, J. H., and Rouseff, R. L., *Archives of Environmental Contamination and Toxicology,* Vol. 11, 1982, pp. 661–667.

[37] Wojeck, G. A., Nigg, H. N., Stamper, J. H,. and Bradway, D. E., *Archives of Environmental Contamination and Toxicology,* Vol. 10, 1981, pp. 725–735.

[38] Lavy, T. L., Shepard, J. S., and Mattice, J. D., *Journal of Agricultural and Food Chemistry,* Vol. 28, 1980, pp. 626–630.

[39] Lavy, T. L., Walstad, J. D., Flynn, R. R., and Mattice, J. D., *Journal of Agricultural and Food Chemistry,* Vol. 30, 1982, pp. 375–381.

[40] Winterlin, W. L., Kilgore, W. W., Mourer, C. R., and Schoen, S. R., *Journal of Agricultural and Food Chemistry,* Vol. 32, 1984, pp. 664–672.

[41] Funckes, A. J., Hayes, G. R., and Hartwell, W. V., *Journal of Agricultural and Food Chemistry,* Vol. 11, 1963, pp. 455–457.

[42] Kolmodin-Hedman, B., Hoglund, S., Swenson, A., and Akerblom, M., *Archives of Toxicology,* Vol. 54, 1983, pp. 267–273.

[43] Franklin, C. A., Greenholgh, R., and Maibach, H. I. in *Human Welfare and the Environment,* J. Miyamoto, Ed., Pergamon Press, New York, 1983, pp. 221–226.

[44] Franklin, C. A., Muir, N. I., and Moody, R. P., *Toxicology Letters,* Vol. 33, 1986, pp. 127–136.

[45] Akerblom, M., Kolmodin-Hedman, B., and Hoglund, S. in *Human Welfare and the Environment,* J. Miyamoto, Ed., Pergamon Press, New York, 1983, pp. 227–232

[46] Rosenberg, N. M., Queen, R. M., and Stamper, J. H., *Bulletin of Environmental Contamination and Toxicology,* Vol. 35, 1985, pp. 68–72.

[47] Sell, C. R., Maitlen, J. C., and Aller, W. A., "Perspiration as an Important Physiological Pathway for the Elimination of 2,4-Dichlorophenoxyacetic Acid from the Human Body, 1982," presented at the American Chemical Society Meeting, Las Vegas, NV, 1 April, 1982.

[48] Peck, C. C., Lee, K., and Becker, C. E., *Journal of Pharmacokinetics and Biopharmaceutics,* Vol. 9, 1981, pp. 41–58.

[49] Nigg, H. N. and Stamper, J. H., *Archives of Environmental Contamination and Toxicology,* Vol. 12, 1983, pp. 477–482.

[50] Nigg, H. N., Stamper, J. H, and Queen, R. M., *Archives of Environmental Contamination and Toxicology,* Vol. 15, 1986, pp. 121–143.

[51] Dowdy, S. and Weardon, S. in *Statistics for Research,* Wiley, New York, 1983, p. 196.

Margaret Rucker,[1] *James Grieshop,*[2] *Alice Peters,*[3]
Heather Hansen,[4] *and Gordon Frankie*[5]

Pesticide Information and Attitudes Toward Chemical Protective Clothing Among Urban Pesticide Users

REFERENCE: Rucker, M., Grieshop, J., Peters, A., Hansen, H., and Frankie, G., "**Pesticide Information and Attitudes Toward Chemical Protective Clothing Among Urban Pesticide Users,**" *Performance of Protective Clothing: Second Symposium, ASTM STP 989,* S. Z. Mansdorf, R. Sager, and A. P. Nielsen, Eds., American Society for Testing and Materials, Philadelphia, 1988, pp. 788–795.

ABSTRACT: The objective of this project was to obtain data on the attitudes and practices of urban users of pesticides, especially as they related to protective clothing and use of pesticide labels for information on safety precautions. Results of a mail survey indicated that over one-fourth of the sample experienced health problems associated with pesticides and about 5% were serious enough to require professional help. The survey and followup interviews suggested that many consumers are disinclined to use any protective gear for home use of pesticides or give special care to clothing worn when applying pesticides. Reasons for these attitudes included perceived protection by government, little danger in low dosages, and the expenses and possible social ridicule involved in use of special gear. Problems in using pesticide lables for safety information included the high volume of badly organized information, technical jargon including the signal words, and spillage obliterating the instructions.

KEY WORDS: household pesticide application, protective clothing, household pesticide labels

Early concerns about pesticide safety focused on agricultural workers. More recently, attention has been directed toward the home users of pesticides and the potential for exposure resulting from household application. This interest has been fostered, at least in part, by surveys demonstrating the magnitude of household pesticide use; figures indicate that from 72% to over 97% of the households in various regions of the United States use pesticides in the home [1–6]. Furthermore, at least in some households, the rate of application is quite high. Finklea et al. [3] reported that one third of their respondents applied pesticides in the home at least once a week. Levenson and Frankie [7] reported a mean value for indoor use of ten times per year and a mean for outdoor use of seven times a year for their sample. In earlier work, these authors [8] found that about 7% of their respondents used pesticides as much as 90 to 100 times a year.

[1] Associate professor, Division of Textiles and Clothing, University of California, Davis, CA 95616.

[2] Lecturer and community education specialist. Department of Applied Behavioral Sciences and Cooperative Extension, University of California, Davis, CA 95616.

[3] Staff research assistant, Division of Textiles and Clothing, University of California, Davis, CA 95616.

[4] Graduate student, Department of Applied Behavioral Sciences, University of California, Davis, CA 95616.

[5] Professor, Department of Entomology, University of California, Davis, CA 94720.

The rate of home pesticide usage would not generate as much concern if the applicators were knowledgeable and careful in their use of these chemicals. However, several studies suggest they are neither. Savage et al. [5] found that household applicators often either did not know what pesticide they had used or were familiar with only the brand name. Finklea et al. [3] noted that safety precautions were ignored by many of their households. For example, 88% did not store pesticides in a locked area, 66% stored these chemicals within reach of small children, 54% left pesticides near food or medicine, and 66% neither wore protective gloves during application nor washed their hands after pesticide use. As these authors pointed out, industry and government agencies have attempted to reduce the probability of ill effects due to careless use of pesticides through restrictions on what is available to the general public. However, pesticides for home use are still not risk-free, especially some of those that must be diluted by the consumer before application. Hazards may be compounded when significant quantities of different pesticides are applied frequently and the need for minimum precautionary actions is ignored. As noted by Savage et al. [5], the survey data on extent of household pesticide use suggests that "the home environment in the United States may play a more significant role in human exposure to pesticides than previously thought."

The present project was designed to gain additional data on the attitudes and behavior of consumers who purchase pesticides for household use. Special attention was directed toward attitudes and practices regarding protective clothing and equipment and use of the pesticide label for advice regarding this type of precautionary action as well as other information on safe usage of the product. Two methods of data collection were employed in this project; a survey was designed to provide quantitative information on attitudes and practices and focused group interviews were used to provide more in-depth qualitative data to complement the survey information.

The Survey

Procedure

The survey data were collected through use of mailed questionnaires. Participants in this phase of the study were obtained by placing business reply postcards in stores that stocked pesticides for home use. The cards were made available for a five-month period, from the middle of April through the middle of September. The stores were located throughout a major metropolitan area in northern California. A total of 70 stores agreed to cooperate, including 22 grocery stores, 20 drug stores, 16 hardware stores, and 12 nurseries. One drug chain, one grocery chain, and one nursery refused to participate in the project.

Seven hundred eighty people returned the postcard and were sent an informational leaflet and a questionnaire. Of these, 415 returned the questionnaire, for a response rate of 53%. To obtain an indication of possible nonrespondent bias, responses were broken down by zip code and analyzed against a model of equal expected values from each of the areas. A chi-square (χ^2) test of goodness of fit indicated that the observed response rates were not significantly different from the equal expected values ($\chi^2 = 9.51$, $df = 25$). Although this finding indicates a lack of bias in terms of respondents from one type of neighborhood being better represented than respondents from another type, respondents may still differ from nonrespondents in other ways, such as level of concern about pesticide safety.

Results

Responses to demographic items on the questionnaire indicated that the sample was fairly well balanced regarding sex, 44% females versus 56% males. Ages of respondents ranged from 16 to 85 with a median of 49. The sample was generally well educated with a median value for years of education of 14 years and a range from 3 to 21 years. About half of the sample reported that family income fell in the $20 000 to $45 000 range with approximately equal numbers checking values above and below that range. The five main ethnic categories were all represented in the sample although most of the respondents, over 82%, indicated they were White.

The majority of respondents, over 85%, lived in a detached home. Similarly, about 85% reported owning the residence in which they lived. These latter values were not surprising given that contact with the respondents was made through stores that carried pesticides for home use. In their survey of adults from three metropolitan areas in the United States, Levenson and Frankie [7] found that significantly more personal use of outdoor pesticide was reported by owners compared to renters and by those in houses compared to those in apartments. Differences regarding inside use were not significant. It should be noted that these data do not imply that renters and apartment dwellers are at less risk with respect to pesticide exposure, since personal application is apt to be added to commercial application provided by landlords. For example, research by Wood et al. [9] on public housing residents indicated that, for cockroach control alone, their respondents spent 0.4% to 1% of their annual income to supplement the pest control program that was provided for them.

As shown in Table 1, two questions designed to assess the extent of health problems associated with household pesticides indicated that having and using these products is certainly not risk-free. Approximately 5% of the respondents reported that they had needed to call a poison control center or a medical service for a pesticide problem. In addition, over 27% said that they or someone in their family had experienced health problems associated with applying pesticides in the home or yard.

Table 2 summarizes responses to items about pesticide labels. As shown in this table, almost all of the respondents (over 99%) stated that they read the label before purchasing pesticides and again before use. This finding is consistent with the survey data reported by Bennett et al. [1]; 82% of their sample reported reading and following label directions. However, as pointed out by the authors, these figures are undoubtedly inflated. People who do not read and follow label directions are apt to be reluctant to admit such irresponsible behavior.

In the present survey, responses to more specific questions about opinions and practices regarding labels suggest that labels are not as effective an information provision system as they are intended to be. Only 62% of the respondents checked that they always read *and understand* the label; 26% reported that the label is too hard to read and 6% indicated that they read only parts of it. Some 27% felt that the label does not provide the information they need.

As shown in Table 3, close to one half of the sample reported that they normally wore some type of protective equipment or clothing or both when applying pesticides. The item

TABLE 1—*Health problems associated with use of pesticides.*

Question	Response	Cases	%
Have you ever had to call a poison control center or a medical service for a pesticide problem?	yes	21	5.1
	no	389	94.9
	total	410	
Have you or any member of your family experienced any health problems that might have been caused by using pesticides?	yes	113	27.6
	no	296	72.4
	total	409	

TABLE 2—*Responses to questions about pesticide labels.*

Question	Response	Cases	%
Do you read the pesticide label before purchasing?	yes	400	99.5
	no	2	0.5
	total	402	
Do you read the pesticide label before using?	yes	345	99.4
	no	2	0.6
	total	347	
I always read and understand the label	yes	252	61.9
	no	155	38.1
	total	407	
The label is too hard to read	yes	106	26.0
	no	301	74.0
	total	407	
I read only parts of the label	yes	24	5.9
	no	383	94.1
	total	407	
The label doesn't provide the information I need	yes	110	27.0
	no	297	73.0
	total	407	

listed most frequently was gloves, mentioned by about 24% of the respondents. Other than gloves, however, clothing was rarely listed as a precautionary measure.

In addition to the apparent lack of appreciation for clothing as a protective barrier between the pesticide and the applicator, the data in Table 3 also suggest some unawareness that clothing can serve as a carrier of pesticides. Several studies have indicated that pesticide-soiled clothing can be a source of secondary contamination of household members, especially when it is not separated from other clothing [10–13]. Answers to an opinion question on laundering pesticide-soiled clothing, "It is important to be as careful as possible when washing clothes that have pesticides on them," suggested that 92% of the respondents were concerned about proper care of such garments. However, some lack of concern about the possibility of secondary contamination is indicated by the fact that 23% of the respondents reported

TABLE 3—*Responses to questions about protective clothing and equipment.*

Question	Response	Cases	%
When applying pesticides, do you normally use protective clothing and equipment?	yes	194	47.4
	no	215	52.6
	total	409	
If yes, what do you normally wear?[a]	gloves	99	23.9
	mask	55	13.2
	goggles	23	5.5
	long sleeves	12	2.9
	glasses	9	2.2
	hat	6	1.4
After you apply pesticides, what do you do with clothing used?	nothing in particular	94	22.7
	wash it	307	74.0
	store it	5	1.2
	throw away	8	1.9
	total	414	

[a] Includes only items that were listed at least five times, and multiple responses; therefore, a total is not reported.

that they did "nothing in particular" with their clothing after applying pesticides. Some 74% said they washed the clothing, another 1% reported storing the clothing without laundering it, and 2% said they threw it away. Moreover, 17% of the respondents were unsure or disagreed with the statement, "People should not come into the house wearing clothes that have pesticides on them."

These data suggested a need for further investigation of the attitudes and behaviors of household pesticide applicators with respect to use and care of clothing worn during application, precautionary information that is available on pesticide products for home use, and how it is being interpreted. Therefore, a second phase of the project was designed to collect more in-depth information through focused group interviews with members of the questionnaire sample as well as people who had returned the postcard but had not yet received the questionnaire or leaflet. The postcard-only group was added to the study to check for any response biases that might occur as a result of exposure to the questionnaire and leaflet. Subsequent blind review of the transcripts indicated that responses to questions about pesticide labels and past use of protective clothing were not biased by exposure to the experimental materials. Therefore, the data sets were combined for content analysis.

The Focused Group Interview

Procedure

Before beginning the interviews, a set of approximately 100 labels of pesticide products for home use were obtained from the Office of Pesticide Information Coordination on the University of California, Davis campus. These labels were reviewed for safety information related to clothing. This review indicated that, with the exception of the recommendation to avoid getting pesticide on clothes, the majority of labels in each of the three toxicity categories did not mention clothing. In only a few cases, users were advised to wear rubber gloves or wash clothing after use.

Following this review, a focused group interview schedule and pictorial stimuli modeled after the Thematic Apperception Test (TAT) were developed. These instruments were designed to gather more information about household applicators' opinions and use of protective clothing, whether pesticide label information was related to these opinions and practices, and general problems that might be associated with the format and content of labels currently on the market. The marketing literature suggests that asking people to respond to pictorial stimuli is a useful means of assessing actual level of concern about an issue when direct questioning might produce exaggerated socially desirable responses. Furthermore, this technique is also less prone to experimenter bias and has proved useful in several other studies assessing respondents' awareness and concerns about clothing [14–16]. Focused group interviews are recommended for interpreting and expanding upon previously obtained quantitative results [17].

To obtain participants for the interviews, the postcards and questionnaires were reviewed for addresses that fell within the metropolitan area and were therefore within reasonable driving distance from the interview site. Those people for whom telephone numbers could be found were placed in a pool of potential participants and called regarding availability during the interview period. Of the 70 people who could be reached through this procedure, 24 were available and willing to cooperate. Assignment to one of five interview groups was based on times that were most convenient for the respondents. The groups ranged in size from three to seven people.

In advance of the interviews, participants were mailed three pictorial stimuli and asked to write responses to each. The stimuli were designed to represent a pesticide selection situation, a label inspection situation, and a pesticide application situation. These written responses were collected after each interview to provide additional insights on needs and concerns regarding protective clothing and pesticide labels.

Results

The interview transcripts and written responses highlighted a variety of problems related to label information and protective clothing. As expected, the interviews and pictorial techniques generated more willingness to admit to lack of knowledge and questionable practices than appeared in questionnaire responses.

One problem was that consumers were not familiar with signal words on labels and used other, often irrelevant, cues such as price to evaluate strength of the product and need for precautionary behaviors. The general feeling among interviewees was that toxicity levels should be stated on the label rather than implied through use of signal words.

Without a clear understanding of toxicity levels of the products they were using, a common assumption seemed to be that special clothing and equipment were not essential when applying household pesticides, because the government would not allow the sale of very toxic products for home use. Alternatively, some group members contended that protective gear was not needed for household applications since only small amounts of pesticides were used. Even when the label dictated use of protective clothing, some people failed to believe it was necessary. As one person wrote in response to the pictures, "They don't really expect me to wear rubber gloves mixing this, do they? They have to be kidding."

Another factor inhibiting use of special protective gear was anticipated reactions from neighbors. One participant expressed concern about looking ridiculous when he wore his baseball or rain hat, goggles, long-sleeved shirt and trousers, high top leather shoes, and gloves. He felt, "I look like I'm going to the North Pole or a man from outer space."

Expense was also mentioned as a problem in taking proper precautions. There was some feeling that most people do not have all the recommended clothing and equipment and requiring that kind of purchase in addition to paying for the pesticide "is asking for too much."

As might be expected from these attitudes, most of the interviewees reported use of common work or "knock-about" clothes for applying pesticides. Clothing was not given much thought. It was selected as one would for other household chores. Hats, when worn, were almost always baseball caps. One respondent noted that while broad brimmed hats might offer more protection from pesticides contacting the skin, they also reduced visibility. Therefore, if he needed to look up while spraying, he used a baseball cap; if he was applying along the ground, he wore a hat with a big brim. Shoes were usually leather and gloves sometimes were leather, too. Long pants and long-sleeved shirts were often worn, but this selection seemed more dependent on weather conditions than a conscious attempt to obtain protection from pesticides.

Additional questioning about pesticide labels was used to elicit problems that could affect transmission and use of label information in general, including but not limited to recommendations regarding protective clothing. More frequently mentioned general problems with some possible solutions included the following:

1. One factor making labels difficult to read is the form in which the information is presented. According to the respondents, many labels provide too much information, presentation as unbroken text makes it difficult to find what one wants, and the small type reduces readability, especially for older consumers. Respondents appreciated the desire of policy makers to make sure they were fully informed, but felt that some of

the information could be conveyed by means other than the label. Package inserts and label attachments were suggested as alternative ways of providing extra information. If only selected information was required on the label itself, the volume of information wold not seem so intimidating and a larger type face could be used to make it more readable. These suggestions are consistent with the model of a labeling system for toxic products proposed by Bettman et al. [18]. These authors recommend that essential usage instructions should go on labels while additional details should be placed on the inserts. In addition, major headings in large bold type could be used to facilitate finding different types of information.

2. Another factor related to readability of the label is the technical level of the instructions. As one respondent expressed it, it seems that instructions are not meant for people who do not garden as a rule. Another complimented one company for using layman's language and thought that other companies would be well advised to follow suit. The general feeling was that the language on a number of labels is more technical than necessary and could be simplified.

3. For liquid pesticides, the respondents reported problems with pesticide drips obliterating part or all of the label. Use of different materials for the labels or protective coatings on the labels could be helpful in solving this problem.

4. In terms of needs for additional information, a frequent request was for more explanations of the *how* and *why* for both usage and disposal, not just the *what* to do. Without the former types of information, participants reported that they either had difficulty in following the directions or decided not to follow them because they seemed ridiculous without further explanation. If there is no space for this type of information on the label, perhaps it could be added to the insert.

5. Another major concern was the lack of information on expiration dates and effects of age on the products. Respondents wanted to know whether a product changed after a certain time and what types of changes occurred at that point. They felt they should be informed as to whether the product became ineffective or more toxic with time. This problem was most troublesome for people who changed residences and "inherited" pesticides from previous occupants.

Conclusions

Although the sample for this study was drawn from only one geographical area and the qualitative data were obtained from a relatively small subset of respondents, consistent patterns of responses suggest several conclusions. First, it appears that home users of pesticides may not always follow the most appropriate procedures regarding use and care of protective gear due to various misperceptions and that label information is not always helpful in correcting these misperceptions. In addition, other factors inhibiting appropriate use of protective clothing and equipment may include unacceptable costs and anticipated responses of others to protective attire. These factors should be taken into account in designing products for the home use market. Furthermore, many pesticide labels need to be redesigned to present precautionary information in a more effective manner.

Acknowledgments

This research was supported in part by a grant from the Center for Consumer Research, University of California at Davis. Additional funding was provided by the L. J. and Mary C. Skaggs Foundation.

References

[1] Bennett, G. W. Runstrom, E. S., and Wieland, J. A., *Bulletin of the Entomological Society of America.*, Vol. 29, No. 1, Spring 1983, pp. 31–38.

[2] *Pesticide Usage Survey of Householders,* Quarterly Report No. 2, EPA Contract No. 68-02-1271, Colorado Epidemiologic Pesticide Studies Center, Colorado State University, Fort Collins, CO, 1974, pp. 14–23.

[3] Finklea, J. F., Keil, J. E., Sandifer, S. H., and Gadsden R. H., *The Journal of the South Carolina Medical Association*, Vol. 65, No. 2, Feb. 1969, pp. 31–33.

[4] Lande, S. S., *Public Health Report,* Vol. 90, 1975, pp. 25–28.

[5] Savage, E. P., Keefe, T. J., Wheeler, H. W., Mounce, L., Helwic, L., Applehans, F., Goes, E., Goes, T., Mihlan, G., Rench, J., and Taylor, D. K., *Archives of Environmental Health,* Vol. 36, No. 2, Nov./Dec. 1981, pp. 304–309.

[6] von Rumker, R., Lawless, E. W., Meiners, A. F., Lawrence, K. A., Kelson, G. L., and Horay, F., *Production, Distribution, Use and Environmental Impact Potential of Selected Pesticides,* Office of Pesticide Programs, U. S. Environmental Protection Agency, Washington, DC, 1974.

[7] Levenson, H. and Frankie G. W. in *Urban Entomology: Interdisciplinary Perspectives,* G. W. Frankie and C. S. Koehler, Eds., Praeger Press, New York, pp. 67–106.

[8] Frankie, G. W. and Levenson, H. in *Perspectives in Urban Entomology,* G. W. Frankie and C. S. Koehler, Eds., Academic Press, New York, pp. 359–399.

[9] Wood, F. E., Robinson, W. H., Kraft, S. K., and Zungoli, P. A., *Bulletin of the Entomological Society of America,* Vol. 27, No. 1, March 1981, pp. 9–13.

[10] Finely, E. L., Metcalfe, G. I., McDermott, F. G., Graves, J. B., Schilling, P. E., and Bonner, F. L., *Bulletin of Environmental Contamination and Toxicology,* Vol. 12, No. 3, Sept. 1974, pp. 268–274.

[11] Easley, C. B., Laughlin, J. M., Gold, R. E., and Tupy, D. R., *Archives of Environmental Contamination and Toxicology,* Vol. 12, No. 1, 1983, pp. 71–76.

[12] Bellin, J. S., *Occupational Health and Safety,* Vol. 50, No. 6, June 1981, pp. 39–42.

[13] Li, F. P., Lokich, J., Lapey, J., Neptune, W. B., and Wilkins, E. W., Jr., *Journal of the American Medical Association,* Vol. 240, No. 5, 4 Aug. 1978, p. 467.

[14] Rook, D. W., *Journal of Consumer Research,* Vol. 12, Dec. 1985, pp. 251–264.

[15] Damhorst, M. L., *Clothing and Textiles Research Journal,* Vol. 3, No. 2, Spring 1985, pp. 39–48.

[16] Rosencranz, M. L., *Clothing Concepts: A Social-Psychological Approach,* Macmillan, New York, 1972, p. 69.

[17] Churchill, G. A., Jr., *Marketing Research: Methodological Foundations,* Dryden Press, Chicago, 1983, pp. 179–184 and 186–188.

[18] Bettman, J. R., Payne, J. W., and Staelin, R., *Journal of Public Policy and Marketing,* Vol. 5, 1986, pp. 1–28.

Melville H. Litchfield[1]

A Review of the Requirements for Protective Clothing for Agricultural Workers in Hot Climates

REFERENCE: Litchfield, M. H., **"A Review of the Requirements for Protective Clothing for Agricultural Workers in Hot Climates,"** *Performance of Protective Clothing: Second Symposium, ASTM STP 989,* S. Z. Mansdorf, R. Sager, and A. P. Nielsen, Eds., American Society for Testing and Materials, Philadelphia, 1988, pp. 796–801.

ABSTRACT: The wearing of additional protective clothing has to be viewed in the overall context of recommendations for the safe and effective use of pesticides. In hot climates, the wearing of such protective clothing is more problematical than in temperate climates because of the heat and stress factors involved. Materials that make up protective clothing need to be comfortable to the wearer in a hot climate and also be protective to a variety of pesticide formulations. These materials should also be available at low cost for use in developing countries. The range of potentially acceptable materials, their availability, wearer acceptability, and performance in the field is reviewed in this paper.

Protective clothing should be designed according to the requirements of the situation, for example, for workers handling concentrated formulations or when spraying diluted solutions. More research into the problems in hot climates is needed, and the direction of this effort is discussed at the conclusion of this paper.

KEY WORDS: protective clothing, pesticide exposure, pesticide formulations, spraying, hot climates, agricultural workers, developing countries, clothing materials, designs, costs, comfort, hygiene, attitudes, training

This paper is given on behalf of GIFAP (Groupement International des Associations de Fabricants de Produits Agrochemiques) which is the international association of the national associations of agrochemical manufacturers. GIFAP's objective is to promote optimal food and fiber production worldwide through appropriate crop protection with agrochemicals and with the minimum of hazard to man, animals, and the environment. In the particular context of pesticide safety, it is pertinent to mention three guide-line booklets that GIFAP has produced. These are (1) *The Safe Handling of Pesticides During Their Formulation, Packing, Storage and Transport,* (2) *The Safe and Effective Use of Pesticides,* and (3) *Emergency Measures in Cases of Pesticide Poisoning.* More will be said about the Guideline on the safe and effective use of pesticides later in this paper that reviews the subject of protective clothing for agricultural operators in hot climates.

Basic Considerations

The discussion of the subject of the need for and type of protective clothing for pesticide operations is often confused because of lack of definition. The term protective clothing

---

[1] Product stewardship, ICI Plant Protection Division, Fernhurst, Haslemere, Surrey, GU27 3JE, England.

means different things to different people. Some describe normal work clothing as being protective, and in a sense it is, since it must afford a degree of protection to the bare skin. Others refer to protective clothing only when it covers every part of the body with impermeable material, including facial protection and a respirator. The fact is, the recommendation for protection should be always based upon the knowledge of the hazard of the product being used and the type of exposure encountered. In this context, also, a clear distinction should be made between protection required for handling the concentrated pesticide formulations and that for spraying the dilute solutions. However, it has to be recognized that however low the hazard or minimal the exposure, all pesticides should be handled with respect, and a basic minimum of protection of the body should be advised. The GIFAP guideline on the safe and effective use of pesticides states that even when no specific protective clothing is recommended on the product label, lightweight clothing covering as much of the body as possible should be worn. In practice, this refers to a long-sleeved upper garment, long trousers, footwear, and, sometimes, a hat and these garments should be worn when applying spray solutions. In addition, when pouring out and mixing a concentrated pesticide formulation the wearing of protective gloves and eye protection is usually the basic recommendation.

This is the baseline for the development of the subject, and it should be emphasized that there is plenty of support for this position from other sources that will be referred to shortly. Thus, any recommendation for protection over and above this minimum will be referred to as additional protective clothing. When is additional protective clothing to be worn? In developing the answer to this, some misunderstandings about the need for protection during pesticide use have to be clarified. First, it has to be emphasized that there are a large number of pesticide formulations that can be handled and sprayed by people taking adequate precautions to avoid contamination and wearing the basic minimum clothing as just described. A second point is that if contamination does occur, it is mainly via the skin, and thus most of the effort to prevent harm should be directed at skin protection. By comparison, inhalation, that is, exposure to the lungs, is usually a minor hazard for many types of pesticide application.

Additional protective clothing is required for the more hazardous formulations and for certain uses. The pesticide manufacturer will indicate on the product label when this is required and what is required.

Protection During Pesticide Use in Hot Climates

Although the principles just enumerated apply to the protection of operators in all situations, the special problems associated with hot and humid conditions are not often addressed. What do the various authorities say about this aspect and what is the overall assessment of the position at present?

The World Health Organization (WHO) document *Prevention, Diagnosis and Treatment of Pesticide Poisoning* [1] observes that additional protective clothing in tropical conditions may actually do more harm than good because of sweating, constant adjustment of equipment, irritation (physical and mental), and physical distress. Air permeable clothing such as cotton twill is advised even though it might afford less protection *per se*. In some cases, national dress can cover the body as effectively as overalls. To promote the wearing of extra clothing, the application of pesticides in the cooler hours of the day is advised.

A publication by the United Nations (UN) Economic and Social Commission for Asia and the Pacific [2] has rather more to say about protective clothing for hot climates. This also advises that pesticide applicators should select protective clothing that is the most comfortable to wear and to wear no more than is needed to do the job safely. Preferably, a pair of light durable cotton overalls should be worn, lacking this, then shirt and trousers worn to cover the full length of arms and legs and fastened at wrists and neck. For certain mixing and loading operations, a neoprene rubber apron is advocated, covering the front of the body from the chest to the ankles and wrapped around the sides of the legs. This publication also gives some practical advice on protective clothing such as types of gloves, boots, and eye protection.

The preceding advice appears to have been based mainly on accumulated experience rather than on hard facts. In more recent years, there has been a growing amount of basic data on the properties of protective clothing for pesticide users in hot climates. For example, the work of Davies et al. and Freed et al. [3,4] was based upon actual measurement of the retention or penetration of pesticide formulations on several fabrics either in laboratory tests or in field studies. They showed that there was considerable reduction in penetration of pesticides when the workers wore 100% cotton-denim coveralls compared to ordinary work clothing, which in this case included short-sleeved shirts and cotton trousers. Laboratory tests showed that cotton or cotton/polyester fabrics showed improved protection against a range of pesticide formulations after treatment with the fluoroaliphatic repellent, Scotchguard. Field studies verified these findings and there appeared to be no problems with the workers wearing the treated fabrics in hot conditions.

Similar conclusions have been reached in other field studies in the United States, for example, with the work of Waldron [5] and of Lavy and Mattice [6], that is, work clothing covering most of the body provides satisfactory protection in many cases. The results of these studies also give some insight into the attitude of operators towards the wearing of protective clothing and the fact that they will wear garments that are practical, comfortable, and inexpensive, but that cost can have overriding considerations compared to safety aspects.

Thus, the overall conclusion from these more detailed observations support the basic recommendation that lightweight clothing covering most of the body provides a significant degree of protection. There should not be too much difficulty for spray operators to adhere to this advice in hot climates nor for operators to wear gloves and eye protection for short periods when handling the more concentrated formulations. Provided that basic hygiene and operating precautions are observed, then there should be little problem with the use of many pesticide formulations in hot conditions.

Protective Clothing in Hot Climates

What about the situation when additional protective clothing is recommended and this has to be worn in hot climates? It is generally accepted that additional items such as rubber aprons, waterproof outer garments, face masks, etc., are very difficult to wear for any length of time when actively working. This is a dilemma of course when the need for additional protection is advocated. Therefore, more attention needs to be given to the study of materials to be used for making up protective clothing acceptable for wearing in hot climates. Other necessary considerations such as the design of such clothing, its durability, and wearer acceptability also require more investigation. What are the properties required for such materials? They are rather extensive, particularly when taking into account the needs for developing countries. Ideally, the materials should have most of the following qualities:

(a) protective against a range of pesticide formulations; aqueous, organic solvent, or oil based;
(b) lightweight;
(c) low price;
(d) readily washable but able to retain protective properties;

(*e*) durable;
(*f*) good air exchange to maintain wearer comfort;
(*g*) acceptable to wearer; and
(*h*) available for local manufacture.

This is a daunting challenge, but some inroads are being made to seek out and test materials in order to meet some or most of the requirements. Foremost in this respect is the work of DeJonge et al. [7] who have examined a range of materials for their protective properties, the effect of laundering, comfort in hot conditions, and user preferences. The materials examined included 100% cotton, cotton treated with Scotchguard, Tyvek (a 100% olefin spun-bonded nonwoven fabric), Crowntex (a laminate containing polypropylene), and Goretex (a three-layer structure consisting of an outer layer of rip-stop nylon and an inner layer of nylon tricot laminated to a film). With regard to protection *per se*, Tyvek, Goretex, and Crowntex were far more protective to an aqueous Guthion spray solution than cotton with or without Scotchguard treatment. Studies on thermal comfort indicated that Goretex and cotton chambray were the most acceptable materials. Laundering at temperatures from 38 to 60°C showed that a large percentage (75% and greater) of either captan or Guthion was removed in one wash from Goretex. In user preference trials with Tennessee agricultural workers, Goretex came out as highly acceptable. Thus, Goretex appears to be a front runner in many respects, the problem comes with its price, where it is far more expensive than other contenders. This is very important because one of the findings of the user preference study showed that when *all* the facts about a protective garment were known to the wearers, greater emphasis was placed on how much it costs than on its protective properties or even comfort.

DeJonge et al. have also confirmed that extra protection is afforded by a Scotchguard finish on cotton or polyester fabrics. This finish does not appear to detract from wearer acceptability for cotton and of course has some cost advantage over more sophisticated materials such as Goretex. The work from this group is giving basic information on many of the properties required for protective clothing in hot climates, and, since the research is ongoing, we can expect to see further leads from this quarter.

Factors Influencing the Wearing of Protective Clothing

Despite these leads there still remains the need to know how the materials will perform in actual field use in hot climates in developing countries and with a variety of pesticide formulations. Most of the experience that has been gained so far has been in the United States or by trying out the materials in more temperate conditions. It will only be after lengthy usage in practical conditions by being worn everyday and washed regularly that a proper assessment of their qualities can be made. There are several elements that go towards meeting this goal. One that has not been mentioned so far is the role of clothing manufacturers. It would seem essential to encourage one or more to have a more active involvement in this area. Their interest is essential for several reasons, above all in reducing the cost of materials to amounts affordable to agriculturists in many parts of the world. It would be pertinent to make a plea at this point for a better definition of the materials available. Studies have already shown, for example, that there can be significant differences in wetting or penetration for different weaves of cotton. When comparing the properties of the newer synthetic materials, it will be important to know their structural characteristics so that consistent comparisons can be made about their effectiveness.

The other important element is in the education and training of the users themselves. The efforts of producing materials and clothing can be wasted if they are not used correctly.

The advice to wear appropriate protective clothing is one aspect of a training and education program. It must be emphasized however, that first and foremost the education must concentrate on the correct and safe application of pesticides and avoidance of exposure. The wearing of appropriate protective clothing should not be an excuse for operators to neglect these basic precautions. In a similar context, application equipment should be maintained in good working order so that it is not necessary to have to wear additional protection to prevent contamination from leaks from this source.

Perhaps one of the most difficult features of the whole subject is the attitude of the wearer. There is a basic hurdle to overcome with people's views on what to do about their personal protection even when they know the facts. This aspect has already been touched upon in this paper for experiences in the United States. Elsewhere, Jeyaratnam [8] from studies in Sri Lanka found that many agricultural workers there had relevant knowledge of the hazards arising from pesticide use and realized the importance of undertaking necessary safety practices, particularly avoidance of skin contamination. However, despite this knowledge the relevant practices often were not followed. Jeyaratnam, quite rightly, advocates educational programs that also try to change attitudes rather than just extending knowledge *per se*.

Future Initiatives

The way forward should be undertaken pragmatically on a step-by-step basis. The first step is to identify the most appropriate materials for making up additional protective clothing that meet the three major requirements; protection, user comfort, and low cost. The protective clothing should be designed with practical considerations in mind and to cover those parts of the body where greatest exposure is expected according to the pesticide use.

Some moves have been made along these lines in the pesticide industry on an individual company basis or in collaboration with government bodies. However, in order to be able to provide more advice on this subject, GIFAP is mounting a program to build upon existing knowledge and experience. They key components of this program are:

(*a*) to determine suitable material(s) for additional protective clothing to be worn in hot climates,
(*b*) to determine which manufacturers are able to provide these materials and at prices acceptable to developing countries,
(*c*) to determine what facilities are available in developing countries for making up clothing from such materials, and
(*d*) to recommend designs for protective clothing for use in hot climates.

This program will be tackled by a special group from GIFAP, and the study will include field trials in appropriate locations.

This task is not being underestimated but is considered one more step in the research and development of appropriate protective clothing, in this case, in the particular context of agricultural operators working in hot climates. The new ideas and exchange of information shared at this conference will further help this endeavor.

References

[1] Plestina, R., *Prevention, Diagnosis and Treatment of Insecticide Poisoning,* World Health Organization, WHO Report No. WHO/VBC/84.889, 1984.
[2] Oudejans, J. H., *Agro-Pesticides: Their Management and Application,* United Nations Economic and Social Commission for Asia and The Pacific, Bangkok, Thailand, 1982.

[*3*] Davies, J. E., Enos, H. F., Barquet, A., Morgade, C., Peters, L. J., and Danauskas, J. X. in *Pesticide Residues and Exposure*, J. R. Plimmer, Ed., American Chemical Society Symposium Series No. 182, 1982, pp. 169–182.

[*4*] Freed, V. H., Davies, J. E., Peters, L. J., and Parveen, F., *Residue Reviews*, Vol. 75, 1980, pp. 159–167.

[*5*] Waldron, A. C. in *Dermal Exposure Related to Pesticide Use*, Honeycutt, Zweig, and Ragsdale, Eds., American Chemical Society Symposium Series No. 273, 1985, pp. 413–425.

[*6*] Lavy, T. L. and Mattice, J. D. in *Dermal Exposure Related to Pesticide Use*, Honeycutt, Zweig, and Ragsdale, Eds., American Chemical Society Symposium Series No. 273, 1985, pp. 163–173.

[*7*] DeJonge, J. O., Easter, E. P., Leonas, K. K., and King, R. M. in *Dermal Exposure Related to Pesticide Use*, Honeycutt, Zweig, and Ragsdale, Eds., American Chemical Society Symposium Series No. 273, 1985, pp. 403–411.

[*8*] Jeyaratnam, J. in *Education and Safe Handling in Pesticide Application*, van Heemstra and Tordoir, Eds., Elsevier, Amsterdam, Oxford, New York, 1982, pp. 23–30.

James I. Grieshop[1]

Protective Clothing and Equipment: Beliefs and Behavior of Pesticide Users in Ecuador

REFERENCE: Grieshop, J. I., **"Protective Clothing and Equipment: Beliefs and Behavior of Pesticide Users in Ecuador,"** *Performance of Protective Clothing: Second Symposium, ASTM STP 989,* S. Z. Mansdorf, R. Sager, and A. P. Nielsen, Eds., American Society for Testing and Materials, Philadelphia, 1988, pp. 802–809.

ABSTRACT: Although protective clothing and equipment are designed to provide protection to pesticide users, they are often not used by small farmers in developing countries such as Ecuador, South America. Protection is not used because of its high cost, inappropriateness, and the lack of availability to users who could be at great personal risk when using pesticides. Other important obstacles to their use are inaccurate "folk beliefs" about pesticides and pest management. These beliefs, although consistent with the Ecuadorean farmers' and nonfarmers' views that pesticide materials are not hazardous and that only weak individuals are at risk, are at odds with what is known about pesticides and the potential risks from misuse. Greater efforts must be made to understand pesticide users' beliefs and behaviors if pesticide safety education programs are to be effective and if users are going to be educated to use protective clothing and equipment.

KEY WORDS: accidents, protection, protective clothing, pesticides, hazardous materials, pest control, spraying, behavioral science, education, human behavior, attitudes, opinions, culture

Statistics on poisonings and deaths due to the use of agricultural pesticides indicate the worldwide nature of the problem [1–4]. Although many likely causes exist for this problem, the use of protective clothing and equipment is commonly promoted as a sure way to decrease accidents and personal risk [5,6]. However, far too little attention has been given to the real-life users of pesticides, especially in developing countries, and their use of pesticides and of protective clothing and equipment. Indeed, more attention has focused on researching and developing new clothing and equipment [7,8] and safer packaging [9] than on understanding the users' attitudes and behaviors. Such an understanding is crucial if effective and productive strategies for educating and persuading users are to be implemented. Such strategies are needed not only for increasing the safe use of pesticides in general, but also for increasing the use of protective equipment and clothing when pesticides are used [10]. A crucial question is: Are users, especially those in the developing countries, and protective equipment and clothing compatible?

More questions than answers exist for understanding attitudes and practices of small farmers and other users of pesticides in the developing countries. In countries such as Ecuador, South America, too little pertinent information exists, and too much of what is available is anecdotal [4]. The information gap must be narrowed. Ecuador is as useful a

[1] Lecturer and specialist, Department of Applied Behavioral Sciences, Cooperative Extension, University of California, Davis, CA 95616.

case in which to systematically investigate such attitudes and practices as any in Latin America.

The Ecuadorean Case

The Andes Mountains divide Ecuador and its population of 8 million into three distinct geographic and cultural regions. The hot and humid tropical coast on the west and the equally hot and humid Amazonian jungle on the east sandwich the temperate highlands. The population is more or less evenly divided between the coast and the highlands, between urban and rural areas, and between Spanish-speaking (60%) and indigenous-speaking groups (40%) represented by 20 distinct language groups. Annual per capita income is estimated at less than $1000 U.S. currency. Over 85% of all Ecuadoreans live in poverty, with the rural poor earning the equivalent of only $60 per year [11,12]. According to 1984 census data, over 45% of the population was involved in agriculture, many in very small scale agriculture. In the highlands area, most farmers till small acreages (2 hectares [5 acres] or less). Although larger parcels are more common in the coastal agricultural areas, small farmers are the rule there, also. Traditional agricultural practices have been rapidly changed, particularly in the past 25 years, by new agricultural technologies, practices, and inputs. Increasing amounts of chemical fertilizers and pesticides subsidized by government policy and resources [13] have found their way into use throughout the country. In the five-year period from 1978 to 1982, over 19 million kg of active ingredients of herbicides, fungicides, nematacides, and insecticides were imported [14]. (Ecuador has no facilities for manufacturing chemical pesticides.) Although the majority of pesticides are used in the coastal areas as part of banana production, even the smallest farmers in the remotest areas are now using pesticides.

The transfer of the pesticide technology to Ecuadorean agriculture has been rapid and widespread. However, the diffusion of the practices, techniques, and knowledge for understanding by users of the proper, safe, effective, responsible, and appropriate use of pesticide technology chemicals has been neither as rapid nor as universal. The development of the necessary infrastructure (that is, laws, regulations, governmental monitoring, education, and responsible private-sector involvement) is under- or undeveloped, thereby creating conditions ripe for greater personal risk. This situation is further compounded by low levels of education, functional literacy, and income of most direct users of pesticides. Although the literacy rate in Ecuador is estimated at 78% [11], the common measure of literacy is whether or not a person can read and write his name. If the user's ability to read is limited, then even the information, technical or safety, found on labels may be indecipherable. Low levels of income may also restrict the users' ability to afford recommended protective equipment. Finally, users of pesticides, such as the small farmers found throughout Ecuador and other developing countries, may labor with serious misconceptions, misinformation, and erroneous beliefs, thereby increasing their at-risk situation.

The study[2] was designed to investigate systematically the beliefs and behaviors of pesticide users throughout Ecuador, including farmers, farm workers, vendors, and "housewives." Special attention was also paid to the use and availability of the protective equipment and clothing.

---

[2] This study was part of a larger one conducted to determine the comprehension and acceptability of visual materials related to education programs on the safe use and management of pesticides among small farmers in Ecuador.

Procedure

In-depth ethnographic interviews, in Spanish, were conducted with 238 small farmers, farm workers, vendors, "housewives," technical experts, and students in agricultural schools. Both regular users (n = 111) and nonusers (infrequent use to total nonuse) (n = 127) of pesticides were interviewed. Respondents were interviewed in their work sites, in homes, community centers, places of work, and schools. Interviews took place in 8 of Ecuador's 23 provinces in both the highlands and the coast. Subjects were selected on the basis of contact through grower's associations, governmental ministries, and through private and volunteer education and development organizations. No attempt was made to randomly select interviewees. Responses to the interview questions were recorded in Spanish on a prepared form. Interviewers were two North Americans, fluent in Spanish, who lived in Ecuador. Field observations supplemented interview data, particularly in relation to use of protective equipment and clothing, the handling of pesticides, their application and disposal, and the purchasing behavior of the users.

Results

Results are presented in translated form and in relation to practices and use of protective equipment (Tables 1 and 2) and for respondents' beliefs, attitudes and opinions (Table 3).

Behaviors

Of particular interest were the practices followed by the 111 small farmers, farm workers, and housewives who used pesticides, in terms of how applied, what equipment was used, and reasons for not using the protective equipment and clothing. Respondents had a long history of pesticide use; the average number of years of use was slightly less than ten years, ranging from one to thirty years. Materials commonly used ranged from low- and moderate-toxicity organo-chlorines and organo-phosphates to highly toxic materials (for example, lindane, aldrin, dieldrin, parathion, and methyl parathion). The backpack sprayer was the predominant method for applying pesticides (91%). Application directly by hand (31%) was also extensively used. Six percent of the users indicated they regularly put pesticides directly in irrigation water. Observations confirmed these practices, and, if anything, tended to suggest a greater reliance on hand and water application than reported. Only a small minority of the respondents indicated they even had the basic protective equipment (Table 1). Although 21% of the respondents reported they had "special clothing" for applying pesticides, observations suggested that special clothing usually meant a piece of plastic worn on the back while applying pesticides. Clothing worn by applicators typically included pants, shirt, a poncho, leather shoes or sandals, and a felt hat. Often individuals would not wear shoes. In hotter climates, applicators would sometimes not wear shirts. Even though some pesticide material used was of low toxicity and the types of clothing used provided adequate protection, the fact that few individuals had multiple clothing changes created the risk of multiple exposure to residues. Reasons stated for not using equipment reflect realities of the marginal economic status of most small farmers and of the climate. To assess the availability of basic safety equipment at the local level, data were gathered from agricultural chemical stores in ten sites (villages, small towns, regional cities) throughout Ecuador (Table 2). Over 64% of the respondents indicated they purchased their pesticide materials from similar stores. Based upon these limited figures, a farmer seeking to buy the basic safety equipment would face a difficult search to do so. Prices were also a harsh reality; gloves,

TABLE 1—*Pesticide users' practices.*

METHOD OF APPLICATION ($n = 111$)		
	Yes	No
Backpack sprayer	101 (91%)	10 (9%)
By hand	34 (31%)	77 (69%)
By irrigation water	7 (6%)	104 (94%)
Other materials	8 (7%)	103 (93%)
PROTECTIVE EQUIPMENT AND CLOTHING OWNED/USED ($n = 102$)		
Goggles	16 (16%)	
Gloves	30 (27%)	
Mask	25 (25%)	
Boots	44 (43%)	
Special clothing	21 (21%)	
REASON FOR NONUSE OF PROTECTION CLOTHING/EQUIPMENT ($n = 94$)		
Don't own	22 (23%)	
Uncomfortable	16 (17%)	
Too expensive	24 (26%)	
Unnecessary	12 (13%)	
Other	43 (42%)	

of the type recommended for applying pesticides, cost the equivalent of $10.00, whereas the annual per capita income of most rural farmers was under $100 [*12*].

Observation of users, supplemented by observations of many other farmers from the highlands and the coast, reinforced the impression that the typical user of pesticides was neither adequately protected nor protecting herself/himself. Observations suggested that the claims for ownership and use of such protective equipment as masks, goggles, and gloves were exaggerated. The most commonly observed mask was a cloth bandana. Also, sightings of applicators using a backpack sprayer with protective goggles on top of the head and protective mask lowered around the neck were not uncommon. Scenes of farm workers mixing pesticides in large vats with a stick or by hand, with no protection of any type visible, were also common.

Beliefs and Opinions

Does behavior follow belief? Are the beliefs of users and their behaviors consistent? In order for us to test such questions, respondents were requested to indicate whether, for them personally, they believed statements related to pesticides were true or false or whether

TABLE 2—*Availability of protective equipment in local agrichemical stores* (n = *10*).

Equipment	No. of Stores Carrying Equipment (%)
Sprayer, backpack	5 (50%)
Goggles	4 (40%)
Gloves	6 (60%)
Mask, respirator	5 (50%)
Boots	1 (10%)
Special clothing	0 (...)

they were unsure. Statements were aimed at assessing respondents' beliefs about personal risk, risk to others, risk to animals, and risks to the environment from using pesticides. Further, the respondents' beliefs in terms of modes of poisoning by pesticides were also tested. (Results, in translated form, are reported in Table 3. Since relatively few indicated they were "Unsure," that category was merged with "False.") The majority of the 111 users and the 127 nonusers thought they ran great personal risk by using pesticides.

Responses to questions about modes of poisoning were of particular interest. Although the three major modes for poisoning by pesticides (that is, ingestion, olfactory, and dermal) are well documented in relevant literature [15], both users and nonusers expressed some surprising beliefs. In the case of poisoning by ingestion, users unanimously agreed with the statement that poisonings would occur by drinking a pesticide; nonusers agreed 98% of the time. In the case of poisoning by smelling, the rate of agreement was much less: 19% of the users thought smelling pesticides would not cause a poisoning and 24% of the nonusers were unsure or believed poisonings would not occur in this manner. Finally, in the case of dermal poisoning, over ⅓ of the users (36%) thought they were not at risk of being poisoned by touching pesticides, and nearly one half of nonusers (48%) expressed the same belief. This high incidence of "nonbelief" in dermal poisoning is sobering. It is also surprising in the sense that nonusers, as a group, had higher levels of formal education (average grade achievement for nonusers was 8, while the users averaged 5.9 grades), and presumably more knowledge about science, technology, and agriculture. The Chi-square statistic was calculated for each question category for users and nonusers. Table 3 illustrates the results. In general, nonusers expressed statistically significantly lower levels of accurate knowledge about pesticides and their effects than did users. As has already been indicated, the levels of knowledge of users were also low. With regard to knowledge and beliefs on modes of poisoning, the only difference between users and nonusers that approached significance ($p < 0.08$) was in reference to dermal poisoning.

As part of the interview process, additional views and beliefs of both users and nonusers were documented. A number of expressed beliefs illustrate the complexity and contradictory nature of the belief systems. For example, three terms were commonly used in Ecuador to refer to pesticides: remedies (*remedios*), cures (*curaciones*), and treatments (*tratamientos*). All have their reference to medical or healthful uses of pesticides. When asked about these uses, respondents often stated that since pesticides cured plant/pest problems, how could they be harmful to themselves? Other stated that they believed pesticides could cause poisonings, but only to individuals who were weak or allergic.

Discussion

The results of this study suggest that even if protective equipment and clothing were readily available and affordable, there is a high probability they would not be used by large segments of the target population because of discomfort, lack of appropriateness, cost and local custom. Discomfort in equatorial heat and humidity, even in the highlands, is a major consideration and obstacle to use. Most protective equipment and clothing are inappropriate for use in such conditions. Furthermore, the high cost of such equipment, given the economic realities of a poor country such as Ecuador, preclude their purchase by the vast majority of farmers. Custom cannot be overlooked. In the larger study mentioned earlier which was conducted to determine the comprehensibility of visual teaching materials, respondents would all too often state that the use of protective equipment and clothing "was not the custom" in their village or area. There were neither local models who used such equipment nor were there incentives for using such equipment.

TABLE 3—*Users' and nonusers' attitudes/knowledge of effects of pesticides.*

	Users ($n = 111$)		Nonusers ($n = 127$)		Chi-square	
	Yes	No/Not Sure	Yes	No/Not Sure		
Will pesticides make you sick?	77 (70%)	34 (30%)	90 (70%)	37 (30%)	0.012	...
Will pesticides make other people sick?	95 (86%)	16 (14%)	92 (72%)	35 (28%)	6.05	$p < 0.01$
Will pesticides harm farm animals?	107 (96%)	4 (4%)	103 (81%)	24 (19%)	11.91	$p < 0.001$
Will pesticides contaminate the soil?	78 (70%)	33 (30%)	61 (48%)	65 (52%)	11.16	$p < 0.0008$
Are pesticides the only effective way to control pests?	86 (78%)	25 (22%)	77 (60%)	50 (40%)	7.029	$p < 0.008$
Can you be poisoned by pesticides by						
drinking?	111 (100%)	0 (...)	124 (98%)	3 (2%)	1.097	...
smelling?	90 (81%)	21 (19%)	96 (76%)	31 (24%)	0.0749	...
touching?	71 (64%)	40 (36%)	66 (52%)	61 (48%)	3.015	...

Clearly, a number of obstacles to the regular use of equipment exist. Some of these are technical, some economic, and some cultural. But, in addition, the belief and attitude structures are significant barriers to change. For example, the all-too-common belief that a person does not run a risk of poisoning by touching a pesticide is real. It is a serious obstacle to persuading users to take the necessary precautions and use gloves and other equipment. This belief is reinforced by "folk beliefs," including those related to personal illness and wellness, namely, "only the weak and allergic are at risk from pesticides, and I am strong and well." Even if this "folk belief" were true, a serious problem would still exist. When 45% of all rural residents in Ecuador are estimated to be anemic and 40% are undernourished [12], rural residents are at greater risk. To assume the equipment will be used when available is naive. To assume that users will obey laws and regulations requiring the use of equipment when handling pesticides is also naive.[3] Anyone who wishes to change users' behavior has to first understand those users' beliefs and attitudes. To a great extent, users' behaviors and beliefs are consistent.

Although the aim may be to change the behavior regardless of the change in beliefs, in a situation such as that found in Ecuador and in other developing countries, the change will have to come about through education. Continuous educational and training programs can be offered to users, but until users are convinced that safety measures are important to them, such measures will be only haphazardly practiced. Therefore, the strategy has to include the analysis of the target users' behaviors, beliefs, and attitudes, with a keen eye to resistance points to the acceptance of the change. As done in this study, the users' view of the acceptability and appropriateness of protective equipment and measures, along with an analysis of the users' beliefs, must be important components of the strategizing process.

Some examples of this approach as applied to pesticide safety already exist [16,17]. Unfortunately, there are too few examples for a problem that is far too widespread. Much more attention has to be given to this area. Governmental agencies as well as the private sector must address these issues and pursue educational strategies. And, as part of those strategies, efforts to identify and analyze the range of obstacles and barriers to change through education must be a major component. The research and development for education for the protection of users must be as important as the research and development of protective equipment and clothing.

Acknowledgments

Fieldwork for this study was conducted while the author was a Fulbright Research Scholar in Ecuador in 1984 and 1985. The author also acknowledges the important field work of David Winter, Education Assistant with the World Wildlife Fund, Washington, DC and the support of Dr. Ray Smith from the Consortium for International Crop Protection.

References

[1] Almeida, W. F., "The Dangers and the Precautions," *World Health,* Aug./Sept., 1984, pp. 10–12.
[2] Copplestone, J. F., "A Global View of Pesticide Safety," *Pesticide Management and Insecticide Resistance,* Watson, D. L. and Brown, W. A., Eds., Academic Press, New York, 1977.
[3] Bull, D., *A Growing Problem: Pesticides and the Third World Poor,* Oxfam, Oxford, United Kingdom, 1982.

[3] In fact, Ecuador, in 1984, promulgated regulations that control the importation, sale, and use of agricultural pesticides. For the most part, these regulations are unenforceable because of the under-developed infrastructure.

[4] "The Estimation of Pesticide Poisoning: The Need for a Realistic Perspective," International Group of National Associations of Manufacturers of Agrochemical Products (GIFAP), Brussels, Belgium, April 1986.

[5] Oudejans, J. H., *Agro-Pesticides: Their Management and Application,* United Nations' Economic and Social Commission for Asia and the Pacific, Bangkok, Thailand, 1982.

[6] Granovsky, T., Howell, H. N., Jr., Heep, C. L., and Grieshop, J. I., *Training Program for Pesticide Users. Trainers Manual,* Consortium for International Crop Protection and the Agency for International Development, Berkeley, CA, 1985.

[7] Thornhill, E. W., "Maintenance and Repair of Spraying Equipment," *Tropical Pest Management,* Vol. 30, No. 3, 1984, pp. 266–281.

[8] Matthews, G. A. and Clayphon, J. E., "Safety Precautions for Pesticide Application in the Tropics," *Pest Articles and News Summaries,* (PANS), Vol. 19, No. 1, March 1973, pp. 1–12.

[9] Reynolds, R., "Report and Guideline for the Packaging and Storage of Pesticides," Report to the Food and Agriculture Organization of the United Nations, prepared under sponsorship of the U.S. Agency for International Development, Jan. 1982.

[10] Grieshop, J. I., "Strategies of Education, Enforcement and Engineering to Improve Pesticide Management and Safety," *Tropical Pest Management,* Vol. 30, No. 3, 1984, pp. 282–290.

[11] "Ecuador: An Agenda for Recovery and Sustained Growth," A World Bank Country Study, The World Bank, Washington, DC, 1984.

[12] "Ecuador: A Country Profile," U.S. Agency for International Development, prepared by Evaluation Technologies, Arlington, VA, March 1980.

[13] Repetto, R., "Paying the Price: Pesticide Subsidies in Developing Countries," Research Report No. 2, World Resources Institute, Washington, DC, Dec. 1985.

[14] Sevilla, R. and Perez de Sevilla, P., *Las Plaguicidas en el Ecuador: Mas Allá de una Simple Advertencia,* Fundación Natura, Quito, Ecuador, Jan. 1985.

[15] *An Agromedical Approach to Pesticide Management: Some Health and Environmental Considerations,* Davies, J., Freed, V., and Whittemore, F., Eds., U.S. Agency for International Development, Consortium for International Crop Protection, Berkeley, CA, 1982.

[16] "Guidelines for Emergency Measures in Cases of Pesticide Poisoning," International Group of National Associations of Manufacturers of Agrochemical Products (GIFAP), Brussels, Belgium, Nov. 1984.

[17] Rodriguez-Schneider, D. and Torrey, N., "Communications Support Project: Radio Campaign Design for Pesticide Safety in Bolivia," Institute for International Research, Washington, DC, Oct. 1985.

New Materials and Technologies

John Davies[1]

Conductive Clothing and Materials

REFERENCE: Davies, J., **"Conductive Clothing and Materials,"** *Performance of Protective Clothing: Second Symposium, ASTM STP 989,* S. Z. Mansdorf, R. Sager, and A. P. Nielsen, Eds., American Society for Testing and Materials, Philadelphia, 1988, pp. 813–831.

ABSTRACT: In the last few decades, many challenges have had to be met in protecting the human in industry against a large variety of hazards. In a number of areas in the world, maintenance work is required to be carried out on and in the vicinity of very high-voltage electrical equipment and in particular on overhead transmission lines while they are still energized.

To protect workers under these and similar situations, suits made from conducting materials are required. The early suits were considered very inefficient due to their inferior design and construction and poor electrical properties.

The practical functions of a Conductive Clothing Assembly extend far wider than the provision of adequate conductivity, flame retardancy, wear resistance, electrical resistance, current carrying capability, screening efficiency, ease of cleaning and maintenance, together with wearer acceptance and comfort, all have to be met.

This paper describes the initial investigations and development of conductive composite textile materials for a protective clothing assembly to combat the pick up from the electrostatic field in the immediate vicinity of 400-kV substation equipment and overhead line transmission towers giving a work environment where inductance buildup of electrical energy could occur to workers and subsequent earthing could result in a discharge above the threshold of feeling.

A review of the requirements of the new International Electrotechnical Commission Standard 78-15 Specification for Conductive Clothing for Live Working is also included in the paper together with a survey of the current work being undertaken in furthering the development of conductive composite materials.

KEY WORDS: protective clothing, conducting materials, Faraday suit, screening suits, electrostatic protection

The subject of the effect of human exposure to electrical energy is of particular biological and physiological interest. The influence of strong electrical fields to the human is now well represented in the literature [1–3]. Man is a very sensitive creature to electrical energy because of his highly developed nervous system. The impedance of the human body is predominantly resistive to low frequencies, however, capacitative effects become important at higher frequencies. There is always present an electrical field in any energized power line or high-voltage equipment and, at higher voltages, the field becomes more extended.

Working techniques developed in North America in the late sixties required linemen to work on power transmission lines while they were energized, the so-called "bare hand" or "live line" work system [4–6]. Various concerns were expressed about the exposure and in particular the biological effects of the magnitude and path of the electrical currents that could flow in the linemen's body before, during, and after contact with the high energy sources. Among a number of recommendations to facilitate the work system was the need

---

[1] Industrial consultant and vice-president, Institute of Quality Assurance, London, UK.

for operatives to be protected by properly designed metallic screens to reduce the induction of electrical currents in the workers to a negligible value.

Michael Faraday [7] had shown that there was no electrical field inside a cage made of conducting material regardless of the electrical charges that may exist on the exterior surface of the cage. While a man inside a total shield would be completely isolated from the electrical field, such an arrangement is obviously impractical where mobility and dexterity are essential to facilitate a safe and effective work operation. The challenge was given to develop a protective assembly to facilitate the screening effect of the Faraday cage, hence the Faraday suit came into being.

Developments in the United Kingdom

Early developments of the Faraday suit in the United Kingdom centered on investigations for a requirement for a clothing assembly to protect workers operating in the immediate vicinity of 400-kV overhead line transmission towers. This was later extended to cover any environment where an inductance build up from exposure to a high electrical field could occur to work personnel, where a subsequent earthing could result in a discharge above the threshold of feeling.

Enquiries made to the textile, woolen, and plastics industries for suitable conducting materials revealed a number of fashion and work wear fabrics containing aluminum foil. But, in every case, the foil was coated with lacquer to prevent the loss of the decorative or reflective aspect, and also, their electrical conductivity was poor. Woolen/metallic materials when examined gave spasmodic continuity figures that were due to the metallic fiber content being in short staple form and a pattern of conductivity could not be guaranteed. Conducting plastic compounds in tough flexible film form were also evaluated and, although suitable from a conductibe aspect, garments produced from these materials could not be considered comfortable.

A material used for radar targets containing twin stainless steel 0.09-mm-diameter solid wires woven into cotton on a 6.5 mm check appeared to offer good electrical properties. A suit designed and manufactured from this material shown in Fig. 1 had been successfully used on the first "bare hand" maintenance operation in Europe on a live 400-kV electrical conductor. The clothing assembly was, however, uncomfortable to wear, it was heavy and rigid and had to be lined to limit skin irritation. Articulation at the elbow and knee positions were very restrictive. One difficulty in manufacture was the need to sew the conductive foil into all seams to maintain continuity of the conducting screen.

The initial study was now extended to improve the performance and comfort of the live line clothing assembly as well as proximity clothing and to set standards of electrical and physical properties of a composite clothing assembly.

From the preliminary investigations, it became evident that in order to guarantee the conductivity and to meet wearer acceptability it would be necessary for a textile/metallic material to be woven specifically for this purpose. Experimental weaving was put in hand for seven fabrics using a 10-mm square and 30 by 40-mm square rectangular lattice a 12 denier/90 filament continuous stainless steel thread of the following specification:

Number of filaments in thread	90
Filament diameter	12 μm
Thread diameter	0.1 mm
Tensile strength	1.6 kg to break
Electrical resistance	0.8634 ohm/cm
Fusing current in free air	1.5 A

Solid steel conductor
X 100

FIG. 1—*Suit A assembly.*

TABLE 1—Comparison of conductive cloth with interwoven metallic thread and the properties of the 12/90 stainless steel thread.

Sample No.	Type of Material	Dimensions of Metallic Grid	Width of Cloth	Type of Weave	Resistivity, Ω/m^2	Properties of 12/90 Stainless-Steel Thread used to Produce Metallic Grid
D 58056/1	Denim	1-cm squares	20 cm	2 pick weft, single warp	5.36	number of strands in thread = 90
D 58046/2	polynosic rayon	1-cm squares	14 cm	2 pick weft, single warp	5.72	strand diameter = 12 μm
D 58046/3	nylon rayon blend	1-cm squares	16 cm	2 pick weft, single warp	7.10	thread diameter = 0.1 mm
D 58046/1	nylon	1-cm squares	32 cm	2 pick weft, single warp	7.31	tensile strength = 1.6 kg to break
D 58056/2	denim	1-cm squares	20 cm	single pick weft, single warp	9.51	electrical resistance = 0.8634 Ω/cm
D 58056/4	nylon	3 by 4 cm rectangles	30 cm	2 pick weft, single warp	19.8	fusing current in free air = 1.5 A
D 58056/5	nylon	3 by 4 cm rectangles	30 cm	single pick weft, single warp	44.7	

INCHES

Flexible steel conductor
X 100

FIG. 2—*Suit B assembly.*

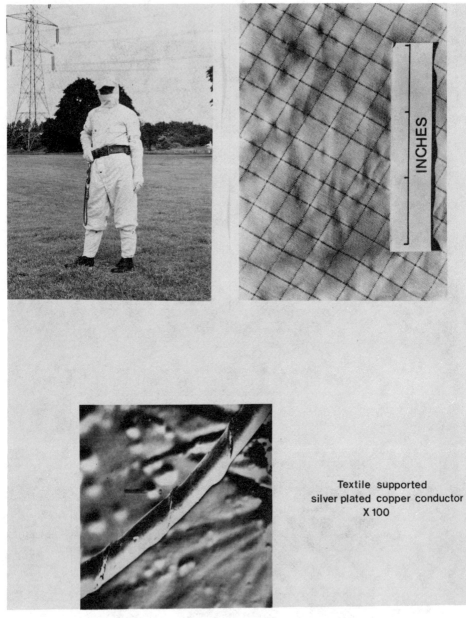

Textile supported
silver plated copper conductor
X 100

FIG. 3—*Suit C assembly.*

The base fabrics for these weaves included denim, nylon, rayon/nylon blend, and polynosic rayon, and the performance specification of these is shown in Table 1.

Production weaving of the cotton denim/stainless steel and polynosic rayon/stainless steel commenced, but the latter had to be abandoned because of weaving problems. In order to maintain continuity along the seams of any resultant garment, a 12-mm tape consisting of stainless steel warp yarn with nylon weft was also manufactured.

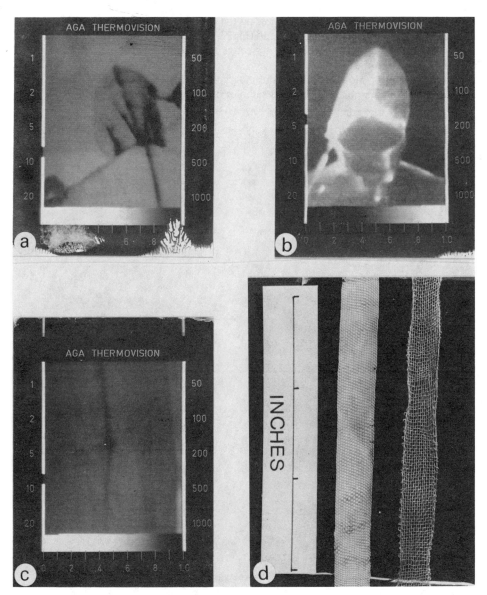

FIG. 4—*Examples of use of thermography on conducting/screening suits showing the effects of conducting tape sewn into seams of garment. (a) rear view of hood (black is hot), (b) front view of hood (white is hot), (c) rear seam showing spread (black is hot), and (d) examples of seam conducting tapes.*

The major problem arising from weaving of the cloth was the dissimilar nature of tensioning of the cotton yarn and steel thread producing a "cockle" as shown in Fig. 2. Therefore, an alternative to the stainless steel was desirable and a thread in the form of a 1% silver-plated copper strip wound round a 75-denier textured filament nylon support (Fig. 3) was used in experimental and production weaving with denim and nylon base fabrics and proved very successful from an electrical and cost basis. The cost of the silver-plated copper conductor was 25% that of the stainless steel.

Working garments were manufactured using the denim/stainless steel blend (Fig. 2, Suit B) and the nylon/silver-plated copper blend (Fig. 3, Suit C). They found general acceptance from the wearers, although there was a subjective response that small thermal "hot spots" were experienced in the seam areas.

To establish this opinion, an infrared thermography analysis of the clothing assembly was made. Figure 4 shows polaroid camera examples of this, for an a-c test of 6.5 A at 7.2 V

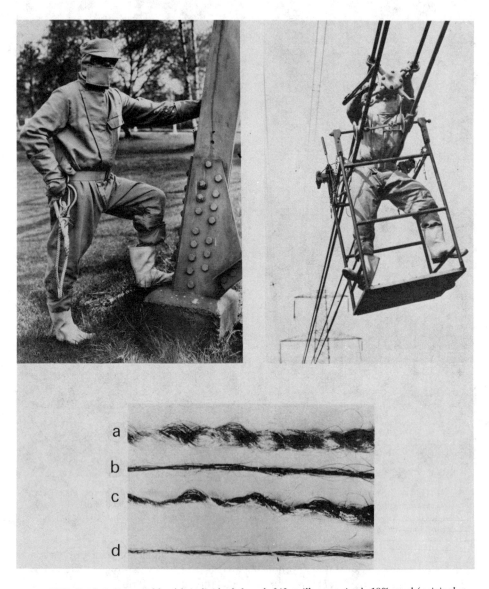

FIG. 5—*Suit D assembly; (a) individual thread, 2/2 twill as received, 18% steel (original magnification × 20); (b) same thread as (a), after removal of wool; (c) individual thread, 2/ 2 twill as received, 25% steel (original magnification × 20); (d) same thread as (c), after removal of wool.*

TABLE 2—*Physical properties of materials used for conducting/screening suits.*

Property	Performance and Suit Reference					Test Method
	A	B	C	D	D$_1$	
Base fabric	cotton	cotton	nylon	wool	wool	...
Conductor	solid	flexible	textile supported	staple	staple	...
Blend	10-mm mesh	10-mm mesh	10-mm mesh	75/25	82/12	ISO 1833
Weave	plain	denim	plain	2/2 twill	2/2 twill	...
Thread count						
warp	...	72	97	220	220	ISO 7211/4
weft	...	49	91	193	193	...
Weight, g/m^2	281	255	128	215	215	ISO 3801
Breaking load, N						
warp	...	1080	268	300	315	ISO 7211/2
weft	...	680	220	290	280	...
Abrasion resistance	...	25 000	50 500	23 000	23 500	BS *Handbook No. 11*

that gave a calibrated temperature rise of 1 to 5°C above ambient. Although not considered a serious defect, it was a comfort factor that could not be determined by electrical measurement and indicated the need to distribute the conducting media more evenly through the base fabric.

Where there is a low propensity for static electricity buildup, it has been the practice to blend with natural and man-made fibers an electrical conductor in the form of a staple fiber.

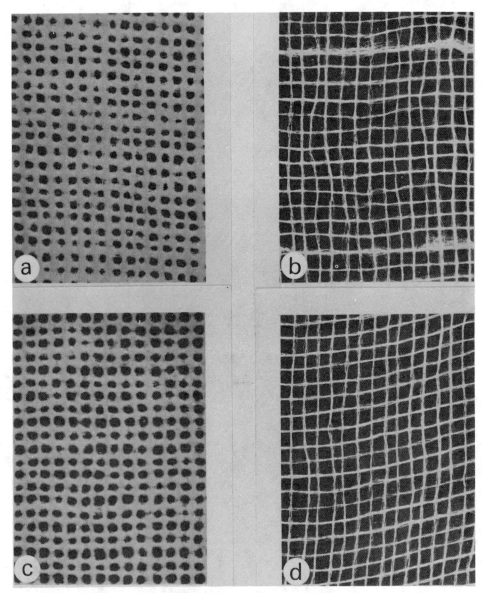

FIG. 6—*Examples of steel/wool fabrics before and after chemical removal of wool. (a) plain weave cloth as received, 18% steel (original magnification ×8); (b) same cloth as (a), after removal of wool; (c) plain weave cloth as received, 25% steel (original magnification ×8); (d) same cloth as (c), after removal of wool.*

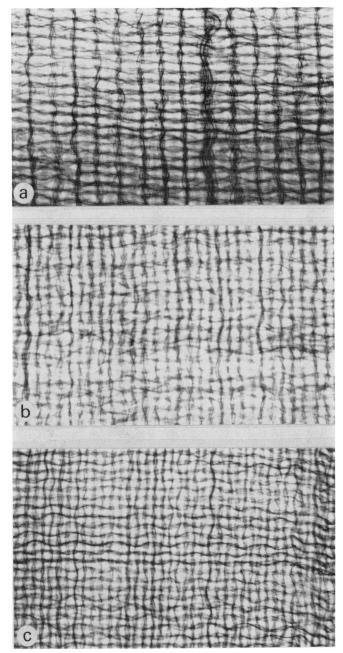

FIG. 7—*Radiographs of steel blended woven fabrics:* (a) *18% stainless steel/82% wool radiograph (original magnification* ×9), (b) *25% stainless steel/75% wool radiograph (original magnification* ×9), *and* (c) *25% stainless steel/75% aramid radiograph (original magnification* ×9).

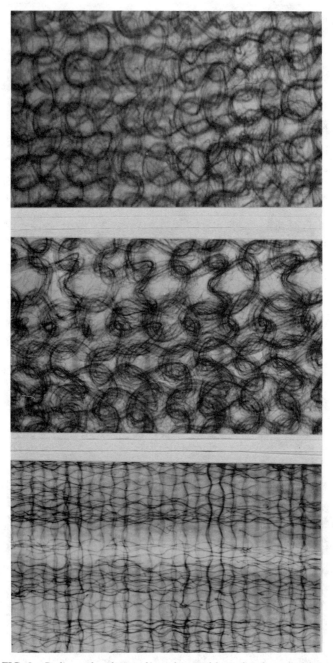

FIG. 8—*Radiographs of aramid/stainless steel knitted and woven fabrics.*

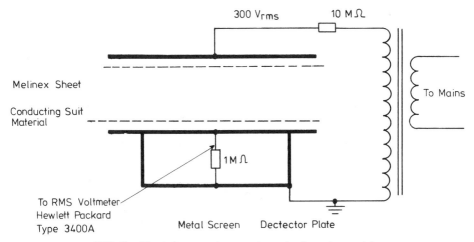

FIG. 9—*Circuit for screening text on conducting suit materials.*

Metallic fiber technology today makes available yarns spun from staple stainless steel film (75 to 150 mm long, 12 μm diameter) giving all the characteristics associated with spun natural or man-made fiber yarns. These steel yarns are also highly flexible, give good abrasion resistance, and can be blended as slivers with any type of yarn using the sandwich blending technique in spinning.

In conjunction with the International Wool Secretariat in England, a high ratio yarn 75/25% and 82/18% wool/steel blend, as shown in Fig. 5, was successfully spun and a cloth woven. In assessing the conductive media of these materials, it was possible to manufacture working garments, as shown in Fig. 5, without taping the seams, thus overcoming the "hot spot" problem.

Physical and Electrical Properties of Conducting Materials

The physical properties of the materials used for Suits B, C, and D using standard test methods are shown in Table 2 and compare favorably with the performance requirements expected from standard commercial work wear materials used for protective clothing in more conventional work areas.

The distribution and measurement of the wool/steel blend was evaluated using a chemical washing technique as follows. A known weight of cloth is heated in an aqueous sodium hydroxide until all traces of wool are removed. The skeletal remains of the cloth consist of

TABLE 3—*Screening properties of material fabrics current flow micro A/m^2.*

Material	Current Flow, $\mu A/m^2$
Nothing	30.5
Suit A	7.1
Suit B	7.4
Suit C	0.077
Suit D	0.006
Suit D_2	0.016
Aluminum foil	0.002

stainless steel fiber that should be washed, dried, and weighed. From these weights, the percentage steel content of the cloth can be determined.

Examples of results obtained using this technique are shown in Fig. 6.

Later, test and production weaves included aramid fiber as a base fabric, this however cannot be easily removed by chemical washing. Radiographs were taken by modifying a back reflection Laue camera. Samples are attached to the front of the film holder that was placed some 175 mm from the X-ray tube and exposed to the beam of an X-ray for 4 s. The resulting film is printed under a magnification ×9. Figures 7 and 8 show the steel distribution of aramid/metallic blended knitted and woven materials using this technique.

The effectiveness of the electrical screening can be found using the circuit arrangement shown in Fig. 9, where the separation between the plates is approximately 30 mm. Readings should be taken (1) with no material present, (2) with the materials earthed and laid on the melinex sheet above the guard ring and detector plate, and (3) with a sheet of aluminum foil in place of the material. Using the separation of 30 mm, an effective field of 10 kV/m is established, which approximates a typical electrical field encountered in the course of live line maintenance on a 400-kV system.

Table 3 shows a series of results using this technique.

Electrical Characteristics of Complete Suit Assemblies

In order to give protection, a Faraday suit must form a complete conducting surface surrounding the wearer and with only small openings. The suit may conveniently be made in one piece or two. In the latter case, the two parts must be physically connected together by using press studs or by having conductive tie ribbons on the two parts of the garment. For the feet, conductive footwear should be worn, and it is preferable to have conductive socks inside the footwear. Such socks may be knitted from the yarn used of the suit material. Similarly, conductive gloves are required. Both must be physically connected to the suit in such a manner to maintain a low resistance electrical path.

The material used for the suit should also have a flame-retarding property and should be lightweight. When a conducting suit is being used, the wearer, on first approaching the live line, must make an electrical contact and a very heavy charging current may flow for a short time through the terminal on the suit through which contact is made. It is therefore important to note that any difference of potential between parts of the clothing assembly would cause currents to flow through the wearer.

Electrical resistance of the conducting paths in the complete suit assemblies can be measured using the ohmic ranges of a Hewlett Packard Type 3490A multiammeter that provides a constant d-c current of 1 mA, with connections to the extremities being made using steel needles inserted at the measured positions indicated in Fig. 10. Initial measurements attempted with the suit stretched flat on a table and with a cardboard divider inserted in the suit showed that considerable variations in the resistance values rendered the results meaningless and that it was necessary to carry out the measurements while the suit was being worn. The results of the test are given in Fig. 10.

These preliminary measurements produced variations that indicated that the suit material needed to be extended in order to obtain reasonable stable results. However, when the suit was worn, movements of the arms and legs produced variable values for electrical resistance, hence the results given are with the wearer of the suit remaining still. From these investigations, a specified measurement of 25 Ω maximum could be obtained for each measured path for the suit C, D, and D_2 assemblies.

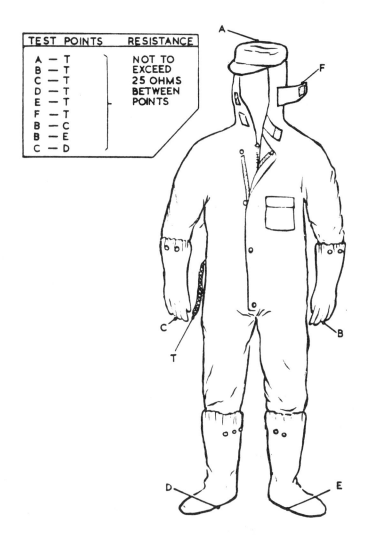

SUIT	TEST POINTS								
	A - T	B - T	C - T	D - T	E - T	F - T	B - C	B - E	C - D
A	300	95	100	100	160		240	55	150
B	11	98	200	38	8	511	200	130	145
C	8	17	25	7	16	11	27	34	14
D	12	12	9	12	15	21	18	13	14
D₁	9	10	10	6	8	21	23	22	15

Values in ohms

FIG. 10—*Results of resistance measurements on five conducting suits.*

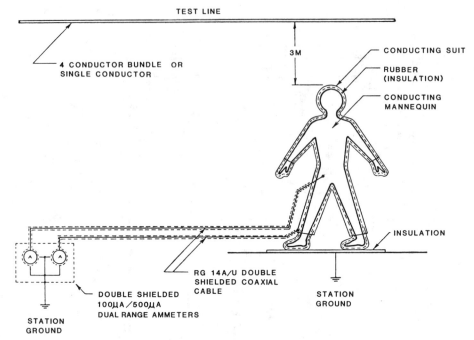

FIG. 11—*Induced current test circuit.*

Measurement of the Electrostatic Screen for Complete Suits

In order to measure the effectiveness of the conducting suit and to measure separately the induced current in the lineman and in the conducting suit, that is, the screening potential, the following test was applied.

TABLE 4—*Induced current test data.*

		Induced Current, μA			
		Suit		Manikin	
Conducting Suit	Test, kV, (60 Hz)	Face Open	Face Closed	Face Open	Face Closed
Aramid/25% Steel, one piece	120	260	260	3.5	>1.0
	200	410	420	7.0	>1.0
	300	660	660	11.5	>1.0
Wool/18% Steel, one piece	120	200	260	5.5	>1.0
	200	430	450	9.5	>1.0
	300	700	720	16.0	>1.0
Wool/25% Steel, one piece	120	260	240	3.5	>1.0
	200	430	450	7.0	>1.0
	300	680	680	11.5	>1.0
Wool 18%/Steel, two piece	120	210	210	5.5	>1.0
	200	360	350	10.0	>1.0
	300	550	560	17.0	>1.0

A rubber-insulated manikin (resistance = 2000 MΩ plus, height = 1.775 m, and weight = 80 kg clothed) in a suit was placed in a standing position directly under and 3 m away from a 500-kV conductor bundle as shown in Fig. 11. Currents were induced in the manikin and clothing assembly from line-applied test voltages of 120, 200, and 300 kV. At each test potential, the induced currents were measured using double-shielded coaxial cable to eliminate any influences from the high-voltage circuit on the instrumentation. The ammeters were dual range and shielded and housed in a grounded mobile measurement laboratory.

Four conducting suits were evaluated, and tests were made on the manikin with the face area of the suit either open or totally enclosed (results are shown on Table 4).

All suits gave good screening effect, and it was shown that current density on the body can be drastically reduced by keeping the face opening as small as possible.

Flammability Hazard from Making Line Connection

Each of the four conductive suits were further subjected to a test to evaluate the flammability hazard when making the line contact. This was accomplished by energizing the conducting suit at a line/ground potential of 133 kV and 60 Hz as shown in Fig. 12. Using the capacitive load of the suit with a ground clearance of 1.2 m, the cloth suit connecting tail was cut and a gap of 75 mm made. A test voltage of 133 kV was applied, and the current in the arc between the pieces of the self material measured 3.5 mA. The vigorous arc was sustained for a period of 3 min. The results of the conductive suit's charging current test are graphically represented in Fig. 13. All the suits tested were unaffected by this flammability test.

International Electro Technical Commission (IEC)

For many years, the subject of live line working, of which the Faraday suit has such an important part to play, has been the subject of informal international collaboration. At the beginning of 1970, the subject was taken up by the International Electro Technical Com-

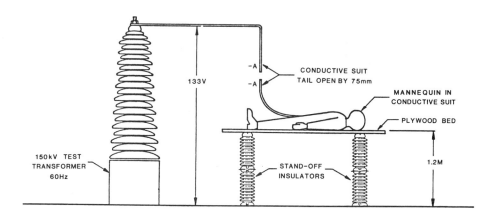

NOTE – LOCATION A–A WAS THE LOCATION OF AMMETER DURING CHARGING CURRENT TESTS AND THE ARC DURING FLAMMABILITY TESTS

FIG. 12—*Conducting suit charging current test circuit. Location A-A was the location of the ammeter during charging current tests and the arc during flammability tests.*

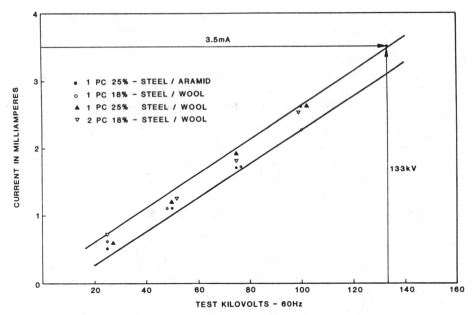

FIG. 13—*Conducting suit flammability test results.*

mission (IEC) and their Technical Committee 78 published a Specification for Conductive Clothing for Live Working No. 78/15 in 1986.

This specification is applicable to conductive clothing worn by electrical workers during live working (especially bare hand) at a voltage level up to 800 kV and is applicable to suit gloves and mitts.

In its terminology it refers to conductive material as material composed of metal and natural or synthetic threads closely woven. Its technical requirements specifies the following properties:

Flame retardancy.
Wear resistance.
Electrical resistance.
Current carrying capability.
Screening efficiency.
Cleaning requirements.
Spark discharge protection.

Test methods are given to cover both type tests, sampling tests, routine tests, and acceptance checks and tests.

The Faraday suit is now very much a reality.

Conclusion

As a result of the foregoing investigation, it was possible to produce a specification for clothing designed for the protection of personnel making contact with energized power lines operated at voltages up to and including 400 kV. This specification was based on the performance of the 25% stainless steel fiber/75% wool blend and the 25% stainless steel fiber/75% aramid fiber fabrics.

Acknowledgments

The author acknowledges with grateful thanks the technical contributions of Harry Haywood of Courtaulds, Dr. Parvez of Mehta International Wool Secretariate, and Jack Marshall of Marlin & Evans plc, in the development of materials. To Jack Simpson and the staff at Ontario Hydro Canada for supplying the data on high-voltage testing and to Ron Frost and the staff at the Central Electricity Generating Board, Scientific Services Department, for making available laboratory facilities and contributing to the general development of the project.

References

[1] Black, J. D., "Biological Effect of High Voltage Electric Fields," Health and Safety Division Ontario Hydro Canada, 1977.
[2] Cabanes, J., "Effects of Electric and Magnetic Fields on Living Organisms and in Particular Man, General review of the literature," R.G.E 1976-07, July 1978, pp. 19–26.
[3] Michaelson, S. M., "Health and Safety of High Voltage Transmisson Lines," Report No. UR-3490-1255, U. S. Energy Research and Development Administration, University of Rochester, Rochester, N.Y., 1976.
[4] Kouwenhaven, W. B., et al., "Body Currents in Live Line Working," IEEE Summer Power Meeting, Detroit, 1965 Institute of Electrical and Electronics Engineers, Inc., Paper No. TP65-73, 1965.
[5] Kouwenhaven, W. B., et al., "Medical Evaluation of Man Working in ac. Electro Fields," IEEE Summer Power Meeting, New Orleans, Institute of Electrical and Electronics Engineers, Inc., Paper No. 31-66-395, 1966.
[6] "Recommendations for Safety in Live Line Maintenance," IEEE Winter Power Meeting, New York, Institute of Electrical and Electronics Engineers, Inc., Paper No. 31, 1967, p. 67.
[7] Faraday, M., "Experimental Researches in Electricity," Dover Publications Inc., 1837, pp. 365–366.

David W. Jones[1] and Peter Watts[1]

Kinetics of Vapor Sorption by Latex-Bonded Carbon Particles

REFERENCE: Jones, D. W. and Watts, P., **"Kinetics of Vapor Sorption by Latex-Bonded Carbon Particles,"** *Performance of Protective Clothing: Second Symposium, ASTM STP 989,* S. Z. Mansdorf, R. Sager, and A. P. Nielsen, Eds., American Society for Testing and Materials, Philadelphia, 1988, pp. 832–846.

ABSTRACT: A study has been made of the sorption of isopropyl methylphosphonofluoridate vapor (IMPF) by an air-permeable protective material consisting of powdered activated carbon bonded to a fabric substrate. The interactions of IMPF with individual components of the system have been studied prior to examination of the composite material. The solubility and diffusivity coefficients of IMPF in films formed from two different acrylic latex binders have been determined by gravimetric sorption methods. Sorption of IMPF vapor by the activated carbon and the fabric substrate has been measured gravimetrically. Experimental materials employing different binders and carbon loadings have been prepared, and the uptake of IMPF vapor determined.

Vapor sorption by these materials has been studied under dynamic conditions of flow through the material, and breakthrough profiles have been obtained. A mathematical model of dynamic vapor sorption has been developed that accurately reproduces the experimentally observed results. This model divides the adsorbent into regions of different prevailing sorption kinetics. The major region has a sorption rate that is determined solely by mass transport of vapor. The remaining "slow" region shows sorption kinetics that are consistent with a slow surface deposition or displacement reaction as the rate determining step. Even though all of the carbon particles are coated with a film of binder, permeation through this film is not rate-controlling step. The "dynamic" sorption capacity of carbon incorporated into the material is largely uninfluenced by the presence of the latex binder.

The mathematical model is successful in predicting the performance of multiple-layer assemblies from the results observed on single layers, and this provides a verification of the approach to the analysis of the sorption kinetics. The model is useful in providing guidelines to show how the material performance may be improved, and indicating the research avenues that are most likely to lead to useful developments.

KEY WORDS: carbon, charcoal, kinetics, latex, permeability, diffusion, protective clothing, vapor, continuous-feed stirred tank reactor, sorption

Nomenclature

Terms Used in the General Text

C_p Concentration of penetrant dissolved in the polymer
C_v Concentration of penetrant in the vapor phase
D Diffusion coefficient of penetrant in the polymer
$\overline{D}$ Integral or mean diffusion coefficient, when D is concentration dependent

---

[1] Higher scientific officer and principal scientific officer, respectively, The Chemical Defence Establishment, Porton Down, Salisbury, Wilts, UK.

J Flux of penetrant through a polymer sheet at the steady state
ℓ Thickness of the polymer film
M_t Mass of penetrant absorbed by the polymer film at time t
M_∞ Mass of penetrant absorbed by the polymer film, in equilibrium with the surrounding vapor concentration
M_c Mass of adsorbate taken up by the carbon when in equilibrium with the surrounding vapor concentration
P Permeability coefficient, equal to the product of the diffusion and solubility coefficients
P_c Concentration-dependent permeability coefficient
S Solubility coefficient
S_c Concentration-dependent solubility coefficient

Terms Used in the Continuous-Feed Stirred Tank Reactor (CFSTR) Mathematical Model

B The normalized effluent from the CFSTR, that is, effluent/influent
C_0 Challenge vapor concentration
C_1 Vapor concentration within the CFSTR
G Fraction of the sorption sites that are fast
H Fraction of the flow passing through the holes
K_{12} Langmuir affinity coefficient
k_{13} Adsorption rate constant of the slow region
k_{31} Desorption rate constant of the slow region
k_{14} Adsorption rate constant of the very slow region
k_{41} Desorption rate constant of the very slow region
N Number of CFSTRs in sequence in the mathematical simulation
R Sorption capacity, when all sites are occupied
Z Term in the diffusion-controlled sorption model

British troops are currently issued with an air-permeable overgarment for protection against chemical warefare agents. This garment has two layers, the lower one consisting of a non-woven fabric coated with powdered activated carbon. During manufacture of this material, the carbon and an acrylic latex adhesive are mixed together in an aqueous slurry and sprayed onto the fabric substrate, which is then passed through an oven to dry the material and cure the latex binder. Examination of the surface by electron microscopy showed that the carbon particles are bonded to each other and to the substrate by bridges of polymer, and that all the visible carbon surfaces are covered by a film of polymer. The thickness of this film is variable, and may be estimated at ~50 nm over much of the surface, but with some regions where the coverage is much thicker than this [1]. The adsorbent is therefore a matrix of encapsulated carbon particles.

The purpose of the investigations described here was to gain a quantitative understanding of the processes of vapor sorption by this composite material, and in particular to assess the influence of the binder on the sorption kinetics. The experimental conditions were designed for this purpose, rather than being intended to correspond with any particular test method that is used for material evaluation, or to correspond with the conditions that may be anticipated during field use of the garment. The approach was to study the individual components and their interactions with isopropyl methylphosphonofluoridate vapor (IMPF), and, after the carbon, base fabric, and binders had been examined in isolation, then attention was focused on the composite material and its performance under dynamic conditions.

Experimental

Materials

Latex Binders—Two different acrylic latex binders were used in this study. These differed markedly from each other in their solubility and diffusivity properties with respect to IMPF (see a later section), and were therefore suitable for the purpose of probing the effects of the binder on the sorption processes.

1. "Primal B15" is manufactured by Rohm and Haas. This is a soft acrylic copolymer latex, of undisclosed composition. It is manufactured as an adhesive, and at one time was used in the commercial manufacture of the protective material.

2. Poly (*n*-butyl methacrylate) (PBMA) was prepared in the laboratory under clean surfactant-free conditions. It is a much harder polymer than B15, and as a carbon binder it was used only for these research purposes.

Carbon—Type RB, manufactured by Chemviron Ltd, was used in all of the material reported on here. It is a steam-activated coal-based carbon, with a specific surface area by the Brunauer, Emmett, and Teller (BET) method of 1500 $m^2 g^{-1}$ [1].

The Composite Materials—Experimental materials were made using the preceding carbon and binders, and a nylon/viscose/chloroprene non-woven base fabric. Four different material samples will be discussed here that all used the same ratio of carbon to binder, that is, 4:1 by dry weights, but differed in the type of binder and the carbon loading. The material specifications are summarized in Table 1.

IMPF—Isopropyl methylphosphonofluoridate was prepared at The Chemical Defence Establishment (CDE), and was of ≥96% purity.

$$H—\underset{\underset{CH_3}{|}}{\overset{\overset{CH_3}{|}}{C}}—O—\underset{\underset{CH_3}{|}}{\overset{\overset{F}{|}}{P}}=O$$

Apparatus and Procedures

Detailed descriptions of all of the apparatus and procedures have been given elsewhere [1], and so only brief outlines will be given here. All of the work was conducted at 30°C.

Gravimetric Measurement of Vapor Sorption—A CI Mk II microforce balance was used to measure the uptake of IMPF vapor by films of the latex binders, the carbon, and the base fabric. It was mounted in a "vacuum head" fitted with side arms to permit input and output of a stream of dry nitrogen that served as a carrier for the IMPF vapor. The nitrogen

TABLE 1—*Material specifications.*

Material Reference	Binder Type	Carbon Loading, g m^{-2}
A	Primal B15	21
B	Primal B15	33
C	Primal B15	56
D	PBMA	43

stream passed from the balance head to a flame ionization detector (FID), permitting continuous monitoring of the vapor concentration.

Dynamic Vapor Sorption by the Composite Material—A stream of dry nitrogen was passed through a 20-mm-diameter disc of the material to effect drying, and then a nitrogen stream containing IMPF vapor was passed through the material and subsequently to the FID, thereby providing a continuous record of the vapor concentration in the effluent stream. The experiment was continued until the effluent concentration was equal to the influent or "challenge" concentration, thereby producing a complete "breakthrough profile." The use of a dynamic method of this kind permits the precise quantification of the vapor concentrations and flow velocities, and is therefore particularly suitable for studies that aim to understand the sorption kinetics. An analogous condition of flow through the material occurs if wind pressure acts to force vapor through the air-permeable fabric layers. All of the experiments reported here were conducted at a linear flow velocity of 10.6 mm s^{-1}.

Approach to the Analysis of the Results

Evaluation of the Diffusion Coefficient of IMPF in the Polymers

The subject of diffusion of organic molecules in polymers has received a considerable amount of attention [2], and the mathematical treatment of the process of absorption by a film surrounded by vapor has been thoroughly investigated [3]. The initial rate of uptake of vapor is given by

$$\frac{M_t}{M_\infty} = \left(\frac{Dt}{\pi \ell^2}\right)^{1/2}$$

(1)

where

M_t = mass of penetrant absorbed at time t,
M_∞ = final equilibrium mass of absorbed penetrant,
ℓ = thickness of the film, and
D = diffusion coefficient of the penetrant in the polymer.

If a plot is made of M_t/M_∞ against $t^{1/2}$, then the initial slope of this sorption plot yields a value for D, via Eq 1. If the diffusion coefficient is a function of the concentration of dissolved penetrant, then this method yields an integral or mean value, usually denoted by $\overline{D}$ [3].

The solubility coefficient, S, of the penetrant in the polymer is obtained directly from the equilibrium uptake, and is defined by $S = C_p/C_v$, where C_p and C_v are the concentrations in the polymer and vapor phases, respectively. Since the same units are employed to express the concentration in the two phases, S is dimensionless.

If a polymer film is exposed to a vapor concentration, C_v, on one face, while concentration at the other face is maintained at a negligibly small value, then the steady-state flux, J, is given from Fick's first law by

$$J = \frac{DC_p}{\ell} = \frac{DSC_v}{\ell}$$

(2)

where C_p now represents the concentration in the immediate surface layer of polymer. If we introduce the term "permeability coefficient" P, defined by $P = DS$, then the flux is given by

$$J = \frac{PC_v}{\ell} \tag{3}$$

For concentration-dependent systems, the permeability coefficient is defined by $P_c = \overline{D}_c S_c$, where $\overline{D}_c$ is the integral diffusion coefficient and S_c is the solubility coefficient at the vapor concentration, C_v.

The CFSTR Approach to Dynamic Sorption Kinetics

The CFSTR Concept—Almost all of the published work on carbon sorption kinetics is concerned with filter beds of activated carbon, and the mathematical approach that has been traditionally adopted [4] is of little value for the present study of a very thin deposit of a heterogenous material.

The term, continuous-feed stirred tank reactor (CFSTR) is borrowed from chemical engineering. It is used here in an abstract sense to mean an entity that behaves mathematically in a manner that is analogous to this reactor vessel. The "abstract CFSTR" consists of a region of space into which the adsorbate vapor is admitted at a constant flow rate. Within the boundary of the CFSTR, mixing is complete and rapid, so that a uniform concentration exists throughout the interior. The effluent from the CFSTR has the same concentration as that existing in the interior, and the same volume flow rate as the input. In this study, it was shown that one layer of the material behaved analogously to one CFSTR in the mathematical model. But, this "CFSTR approach" has wider applications than the present one where the identification of a CFSTR with a physical entity was relatively straightforward. It was first applied to carbon sorption for the study of filter beds of granular charcoal [5], when the bed was modeled as a cascade of CFSTRs in sequence, and its use for the present purpose was a development from these studies.

CFSTR Models—The concept of the CFSTR as a region of space where a uniform concentration prevails may be used as the basis for considering a wide variety of systems, obeying differing kinetics and of varying degrees of complexity. The simplest model that was used in this study will be discussed first, followed by the more elaborate models that were required to provide a full explanation of the observed results.

The models all use a Langmuir sorption isotherm to relate the vapor concentration with the concentration of sorbate on the carbon. A further assumption of the Basic Model is that equilibrium between the adsorbed phase and the vapor phase is achieved effectively instantly, that is, that the sorption process is rapid in comparison with transport into the CFSTR. The equations describing this simplest system are

$$C_1 = \frac{C_2}{K_{12}Q} \tag{4}$$

$$C_A = C_2 + C_1 \tag{5}$$

$$Q = Q_0 - C_2 \tag{6}$$

$$\frac{dC_A}{dt} = k_F(C_0 - C_1) \tag{7}$$

where C_0 is the influent or challenge concentration, C_1 is the vapor concentration within the CFSTR, C_2 is the corresponding concentration of sorbate adsorbed onto the carbon, K_{12} is the Langmuir affinity coefficient, Q is the concentration of unoccupied sorption sites, Q_0 is the concentration of total sorption sites, C_A is the "total" concentration within the CFSTR, and k_F is given by

$$k_F = \frac{F}{V} \qquad (8)$$

where V is the void volume of the CFSTR and F is the volume flow rate.

The computer programs using the ISIS simulation package solve the simultaneous equations describing the system for the boundary condition $C_2 = 0$ at $t = 0$. The results are presented graphically as a normalized breakthrough curve, that is, a plot of C_1/C_0 (denoted by B) against time.

The Basic Model had partial success in reproducing the experimentally observed breakthrough profiles (see a later section), but it became clear that the simple model required some modification if a full description of all of the data were to be achieved. In many instances, immediate breakthrough was seen, with a non-zero effluent concentration at the start of the challenge of the material. This was contrary to the Basic Model, which invariably predicted that B should be zero at $t = 0$. This effect was modeled by proposing the existence of "holes," that is, that a fraction (H) of the flow effectively by-passes the CFSTR, and is then returned to the main flow after it has passed through the CFSTR. The H was typically ~ 0.05. The CFSTR with holes may be shown schematically by

The experimental results always showed a relatively long "tail" in the approach of B to unity when compared with simulations produced by the foregoing model. In order to reproduce this effect, it was necessary to propose the existence of a "slow" fraction of adsorbent, that is, that not all of the sites reached immediate equilibrium with the vapor concentration C_1 within the CFSTR. The assumption that the slow sites show a Langmuir sorption isotherm was retained with the same value for the affinity coefficient, so that the fast and slow regions differ only in their kinetics and not their equilibrium uptake. Different mathematical forms of the sorption rate dependence of the slow region may be proposed, corresponding with different rate-controlling sorption mechanisms.

The first mathematical form that was considered is an extension of the model that was used in the kinetic derivation of the Langmuir isotherm, and may be termed a Surface Deposition controlled model. The net rate of change of the concentration of sorbate on the slow sites, C_3, is given by

$$\frac{dC_3}{dt} = k_{13} C_1 Y - k_{31} C_3 \qquad (9)$$

where Y is the concentration of unoccupied slow sites. Following Langmuir [6], the affinity coefficient is regarded as being equal to the ratio of the rate constants for sorption and desorption, k_{13} and k_{31}, respectively, so that

$$k_{13} = k_{31} K_{12} \qquad (10)$$

Therefore, Eq 9, for Surface Deposition Control, may be written as

$$\frac{dC_3}{dt} = k_{13} C_1 Y - \frac{k_{13} C_3}{K_{12}} \qquad (11)$$

If diffusion through a barrier is the limiting process, then it may be shown (1) that the sorption rate for "diffusion control" is given by

$$Y \frac{dC_3}{dt} = ZC_1 Y - \frac{ZC_3}{K_{12}} \qquad (12)$$

where Z is a constant, proportional to the permeability coefficient. Comparison of Eqs 11 and 12 shows that there is a fundamental difference in the mathematical expression of the sorption rate for the two processes, since Eq 12 has a Y term on the left-hand side but Eq 11 does not. The different mathematical forms result in different shapes of the simulated breakthrough profiles, and comparison of these simulations with the experimental results therefore provides a means of determining which rate determining mechanism prevails (see a later section).

The model that was eventually adopted and provided an accurate description of all of the experimental results, incorporated three kinetic regions, that is, fast (instantaneous), slow, and very slow, with the latter two regions having the same mathematical form of the sorption rate, but differing in the value of their rate constants. This may be shown diagrammatically as

Although sorption onto the slow and very slow sites is represented as occurring directly from the vapor phase, the same mathematical form would result if sorption occurred first onto the fast sites followed by migration onto the other sites.

Results

Gravimetric Sorption Measurements

Vapor Sorption by the Carbon—The equilibrium uptake of IMPF by the carbon, M_c, is expressed as a fraction of its dry weight. The results at 30°C gave $M_c = 0.32$ at 1120 mg m^{-3} and $M_c = 0.58$ at 6700 mg m^{-3} vapor concentration. If these results are substituted into the Langmuir equation (Eq 4), then one has a pair of simultaneous equations, which are solved to give a value of the affinity coefficient $K_{12} = 7.64 \times 10^{-4}$ (mg m^{-3})$^{-1}$. This value, derived from the gravimetric measurements, will be used later in the computer simulations of dynamic sorption.

Vapor Sorption by the Fabric Substrate—The results on "bare fabric" showed that the contribution made by the substrate to the total of vapor sorption by the protective material was negligibly small, less than 3% in all cases [1].

Vapor Sorption by Latex Films

Primal B15—Repeat experiments at 30°C gave a mean value for the diffusion coefficient, D, of 12.8×10^{-14} m² s⁻¹, calculated via Eq 1. No dependence of D on the vapor concentration was observed over the range 3200 to 7250 mg m⁻³. The solubility coefficient was 4.8×10^3, giving a value for the permeability coefficient of $P = 6.1 \times 10^{-10}$ m² s⁻¹.

Poly(n-butyl methacrylate)—PBMA showed some dependence of both the diffusivity and solubility coefficients on the vapor concentration. At the lowest concentration studied, 2310 mg m⁻³, values of $\overline{D} = 1.1 \times 10^{-14}$ m² s⁻¹ and $S = 1.9 \times 10^3$ were observed. At a concentration of 4720 mg m⁻³, the corresponding values were $D = 1.4 \times 10^{-14}$ m² s⁻¹ and $S = 2.4 \times 10^3$. The values for the permeability coefficient were therefore 2.1×10^{-11} m² s⁻¹ and 3.4×10^{-11} m² s⁻¹ for the lower and higher vapor concentrations, respectively.

PBMA and Primal B15 therefore provide a suitable pair of materials with which to examine the influence of the binder permeability on the kinetics of vapor sorption, since the values of the permeability coefficients differ by approximately a factor of 20.

Dynamic Vapor Sorption by the Material

Curve Fitting– The experimentally observed breakthrough profiles are stored in the computer file and may be reproduced in the form of an asterisk (∗) on the graphical output of the simulated profile, thereby facilitating comparison of the simulation with experiment. The procedure was to repeat the simulations, changing the relevant parameters as deemed

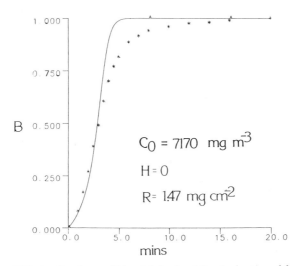

FIG. 1—*One layer of Material A, simulation by basic model.*

appropriate, until a fit of the computer-generated simulation curve to the experimental points was obtained. During this procedure, only certain of the parameters may legitimately be changed. Thus, C_0 and F (the "challenge" concentration and flow rate, respectively) are measured during the course of the experiment, and these values are used in the simulation. If repeat experiments on samples cut from the same material are considered, then any differences in the value of R (the total uptake if all sorption sites are occupied in milligrams IMPF per square centimetre of the fabric layers) must lie within the range that may represent local variations in the carbon coverage due to irregularity of spray, etc. If one considers materials with different charcoal loadings, but the same carbon/binder formulation, for example, Materials A, B, and C, then the parameters that describe the intensive properties of the materials (that is, the rate and affinity constants) should maintain the same values. In all of the simulations shown here, the affinity coefficient, K_{12}, was maintained at the value that had been derived from the gravimetric measurements (see the section on vapor sorption by the carbon).

Limitations of the Basic Model—Figure 1 shows an experimentally determined break-through profile of the Material A, together with a simulation curve produced by the Basic CFSTR Model, that is, on the assumption that equilibrium exists at all times between the vapor and adsorbed phases. Only the first part of the profile can be matched using this model, and we may take this as an indication that although a substantial number of fast sites are present, there is also a slow fraction that gives rise to the "tail" of the breakthrough profile.

Elaborating the Sorption Model—Having postulated the existence of a slow fraction of the total number of sorption sites, one must determine which mathematical description of the sorption kinetics of this fraction provides the best correspondence between the simulations and the observed results. The form of the sorption rate dependence given by Eq 11, corresponding with Surface Deposition as the rate-determining step, provides a relatively close fit of the simulation curve to the experimental points if it is assumed that 0.68 of the total sites are fast, with the remaining slow fraction having the desorption rate constant k_{31}

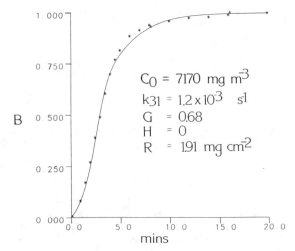

$$C_0 = 7170 \text{ mg m}^{-3}$$
$$k_{31} = 1.2 \times 10^3 \text{ s}^{-1}$$
$$G = 0.68$$
$$H = 0$$
$$R = 1.91 \text{ mg cm}^{-2}$$

FIG. 2—*One layer of Material A, simulation by surface deposition model.*

$= 1.2 \times 10^{-3}\,\text{s}^{-1}$ and with the adsorption rate constant $k_{13} = k_{31} \times K_{12}$ (see Fig. 2.) These values of the parameters were arrived at by repetition of the simulation until the best fit of the curve to these and other results was obtained.

In contrast, the model that used diffusion through a barrier as the rate-determining step, that is, with the sorption rate of the slow region described by Eq 12, proved to be incapable of producing a fit of the simulation curve with the experimental results. This is shown clearly in Fig. 3, where the previous results (that is, in Figs. 1 and 2) are compared with a simulation by the diffusion control model. The presence of a significant fraction of diffusion controlled sites always gives rise to a portion of the curve with a distinctly sigmoidal shape, as may be seen in Fig. 3 from ~5 min onwards. Whatever values of Z and G were adopted, the simulation curve always differed in shape from that found by experiment [1], and this finding is evidence against a diffusion process of any kind being the rate-determining step in sorption by the slow region. Therefore, the rate-determining process would appear to be neither diffusion through a film of binder, nor vapor diffusion through a depletion layer in the vicinity of the carbon particles.

In some, but not all, of the experimental results, the effluent concentration when extrapolated to zero time was not equal to zero. This is contrary to the predictions of the CFSTR model, and was accommodated by postulating the existence of holes, that is, a fraction (H) of the flow effectively bypassing the CFSTR and failing to make intimate contact with the adsorbent. Further improvement in the accuracy of the simulations was achieved by dividing the noninstantaneous fraction equally into slow and very slow regions. The model was now capable of reproducing accurately the breakthrough profiles of materials A, B, and C, when it was assumed that the single fabric layer corresponded with one CFSTR in the mathematical model. Figures 4, 5, and 6 show these simulations, with in each case the fast fraction, G, equal to 0.68 and the slow and very slow regions each occupying 0.16 of the total number of sorption sites. The values of the rate constants were maintained constant at $k_{31} = 3.5 \times 10^{-3}\,\text{s}^{-1}$ and $k_{41} = 0.7 \times 10^{-3}\,\text{s}^{-1}$. In addition to the results shown here for a linear flow rate of 10.6 mm s^{-1}, the model also proved satisfactory at 21.2 mm s^{-1} [1]. An important inference to be drawn from these results is that one layer of material may be modeled as one CFSTR, irrespective of the carbon loading within the range 21 to 56 g m^{-2}.

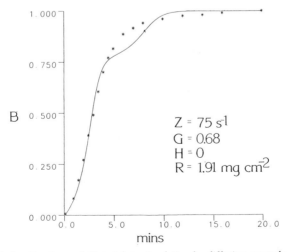

FIG. 3—*One layer of Material A, simulation by diffusion control model.*

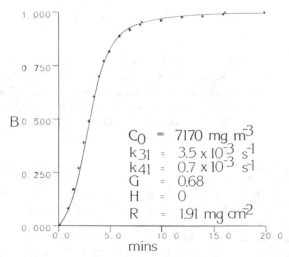

FIG. 4—*One layer of Material A, simulation by the fully developed model.*

Multiple Layers of Material—A stringent test of the equivalence of one material layer with one CFSTR lies in the behavior of multiple layers of material. If the adsorbent system is modeled as a number, N, of CFSTRs in sequence, then the outcome and characteristic shape of the simulation is markedly dependent on the value of N. The program that modeled a series of CFSTRs was confined to a single slow region, as in the simulation in Fig. 2, and the same values for k_{31} and G that had been deduced for a single layer were used for the simulations on multiple layers of material. A value of $H = 0.03$ was used, approximating to the mean value seen for single layers of Material A [1].

Figures 7 and 8 show the results for two and three layers of material, with two and three CFSTRs in the corresponding simulations. The agreement between the curves given by the mathematical model and the experiment is excellent. Figure 9 gives the same experimental points as shown in Fig. 8, but the computer simulation curve was generated by a model

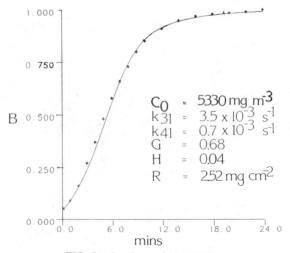

FIG. 5—*One layer of Material B.*

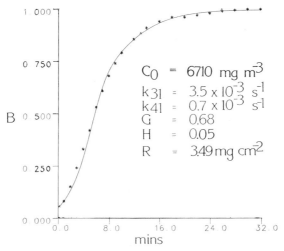

FIG. 6—*One layer of Material C.*

with only a single CFSTR, rather than the three used for the curve in Fig. 8. The poor correspondence between the simulation and experiment is immediately apparent, and this emphasizes the importance of assigning the correct number of CFSTRs to the mathematical model, this is, one CFSTR per fabric layer. The success of the model in predicting the behavior of multiple-layer assemblies from the results obtained on single layers (as exemplified by Figs. 7 and 8) is regarded as being forceful evidence for the soundness of the CFSTR approach to the analysis of sorption kinetics.

Inspection of Figs. 8 and 9 illustrates an important finding that emerges from these studies. If an adsorbent is divided into a series of separate layers then the breakthrough profile becomes progressively more sigmoid shaped as the number of layers is increased. The three-layer system shows an initial period when no vapor is detectable in the effluent, rather than

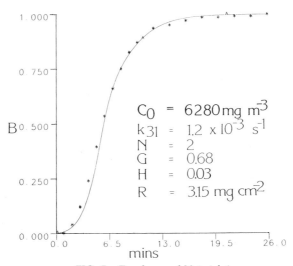

FIG. 7—*Two layers of Material A.*

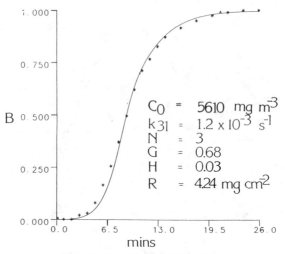

FIG. 8—*Three layers of Material A.*

the immediate breakthrough that is evident for the single-layer systems. Both the experimental results and the theoretically derived computer simulations are in agreement on this point. Improved protection by materials of this type may therefore be achieved without sacrifice of the air permeability property by increasing the number of adsorbent layers, without necessarily increasing the total amount of adsorbent present.

Material using PBMA Binder—Although the permeability coefficients of PBMA and Primal B15 differ by approximately a factor of 20, comparatively little difference was seen in the behavior of the composite materials using these binders. The same mathematical model that had been developed for the material with B15 binder was also sufficient for the PBMA materials, except that the fast fraction was reduced from 0.68 to 0.45 of the total. The slow and very slow fractions now each occupied 0.275 of the total sites, with the values

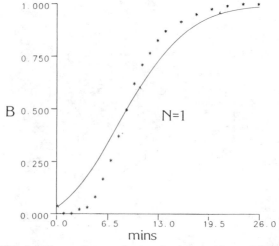

FIG. 9—*The previous results, but simulation with one CFSTR.*

of their rate constants, k_{31} and k_{41}, retained at their previous values. Figure 10 shows a typical breakthrough profile for material D, and equally close fits of the simulations with the results were also obtained at the higher linear flow rate of 21.2 mm s^{-1} [1].

Possible Sorption Mechanisms—The combined evidence of the preceding sections indicates very strongly that diffusion through the binder is not the rate-determining step for sorption by the slow regions. However, the results indicate that the polymer is implicated in the slow portion, since a definite difference was observed between the materials with B15 and PBMA binders.

Calculations based on the known diffusion coefficients indicate that equilibrium between the IMPF dissolved in the polymer and the vapor phase is achieved very rapidly [1]. The mathematical form of the sorption rate given by Eq 9 would therefore also apply to a mechanism where transfer from the polymer to the sorbent surface is the rate-controlling step, with the rate of forward transfer proportional to the number of unoccupied sites and the concentration of IMPF dissolved in the polymer. If displacement of polymer from the surface by the IMPF molecule must first occur, then this provides an explanation of the comparative slowness of the process. The statement that a fraction $(1 - G)$ of the sites are slow does not necessarily imply that this same fraction of the adsorbent surface is covered in polymer, but only that access to these sites is gained after a slow process. If migration through the internal pore network occurs after initial sorption at the particle surface, then the slow regions may represent sites accessible only after a displacement process of this kind.

It is recognized that the mechanism just proposed is conjectural, and considerable work would be required to assess it rigorously. However, some support is provided by studies on material using a carbon of lower mesoporosity, when it was observed that the results could be modeled only on the assumption that the fast fraction was of negligibly small extent [1].

Sorption Capacity of the Carbon "On the Cloth"—During all of the simulations, the value of the term, $\int_0^T (1 - B)dt$, was calculated. Where an accurate fit of the curve to the data

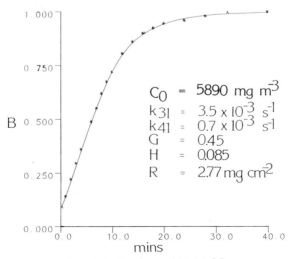

$$C_0 = 5890 \text{ mg m}^{-3}$$
$$k_{31} = 3.5 \times 10^{-3} \text{ s}^{-1}$$
$$k_{41} = 0.7 \times 10^{-3} \text{ s}^{-1}$$
$$G = 0.45$$
$$H = 0.085$$
$$R = 2.77 \text{ mg cm}^{-2}$$

FIG. 10—*One layer of Material D.*

was obtained, the total uptake of the material is then given by

$$\text{Mass Adsorbed} = FC_0 \int_0^T (1 - B)dt$$

The collected results for the four different materials showed that the sorption capacity of the carbon, as determined by this dynamic method, was the same as that obtained from the gravimetric measurements on the pure charcoal, within the limits of experimental error [1].

Conclusions

1. The polymer binder used in this material has remarkably little effect on the sorptive properties with regard to IMPF. Not only is the capacity of the adsorbent unaffected, but the presence of the binder has only a small effect on the sorption kinetics.

2. The CFSTR approach to the analysis of carbon sorption kinetics has been validated. This provides a powerful tool for the study both of protective clothing assemblies and carbon filters, and opens up the possibility of the analysis of relatively complex systems.

3. If further improvements in the performance of the protective garment are required, then attention should be focused on the structure of the ensemble. The inclusion of multiple adsorbent layers would result in very large increases in breakthrough times, and is much more likely to bring immediate benefit than the investigation of alternative adsorbent/binder systems. This conclusion is supported by the experimental results, and corroborated by the simulations produced from the mathematical model.

References

[1] Jones, D. W. "Vapour Sorption by Latex Bonded Carbon Particles," PhD thesis, Council for National Academic Awards, London, 1986.
[2] Diffusion in Polymers, J. Crank and G. S. Park, Eds., Academic Press, London, 1968.
[3] Crank, J., The Mathematics of Diffusion, 2nd ed., Clarendon Press, Oxford, 1975.
[4] Klotz, I. M., Chemical Reviews, Vol. 39, 1946, p. 241.
[5] Watts, P., to be published.
[6] Langmuir, I. Journal American Chemical Society, Vol. 40, 1918, p. 1368.

Jeffrey O. Stull,[1] Ruth A. Jamke,[2] and Mark G. Steckel[2]

Evaluating a New Material for Use in Totally Encapsulating Chemical Protective Suits

REFERENCE: Stull, J. O., Jamke, R. A., and Steckel, M. G., "**Evaluating a New Material for Use in Totally Encapsulating Chemical Protective Suits,**" *Performance of Protective Clothing: Second Symposium, ASTM STP 989,* S. Z. Mansdorf, R. Sager, and A. P. Nielsen, Eds., American Society for Testing and Materials, Philadelphia, 1988, pp. 847–861.

ABSTRACT: A totally encapsulating suit has been developed by the U.S. Coast Guard to provide a high level of protection for response personnel during chemical spills. The suit is fabricated by Chemical Fabrics Corporation (CHEMFAB) using a new material, Challenge 5100. Challenge 5100 is CHEMFAB's aramid-reinforced fluoroelastoplastic composite that has demonstrated excellent chemical resistance against a variety of chemical solvents. Other components of the suit include a Teflon FEP visor, Teflon TFE film gloves, a pressure sealing zipper, and various liquid-proof seam constructions. The Coast Guard and Chemical Fabrics Corporation are extensively testing these materials and components against selected chemicals to determine suit capabilities and limitations. The results of the testing program will be used to make suit use recommendations, as well as support further optimization of suit materials and construction.

KEY WORDS: totally encapsulating chemical protective suit, chemical resistance testing, permeation resistance, penetration resistance, protective clothing

The U.S. Coast Guard is mandated by the Clean Water Act of 1977 (as amended in 1978) and the Comprehensive Environmental Response Compensation and Liability Act (CERCLA) to respond to any chemical discharge into the waters of the United States. The Coast Guard also has the responsibility for inspecting and certifying marine chemical-carrying vessels. Both of these missions require appropriate personnel protection against a multitude of hazardous chemicals, especially those transported in bulk that are likely to be encountered in marine spills and during marine inspections. To aid spill response and monitoring, the Coast Guard developed its own Chemical Hazard Response Information System (CHRIS) [*1*], which now defines the properties, hazards, and response techniques for over 1100 chemicals.

As the Coast Guard's role in chemical spill response grew, it found that for many CHRIS chemicals, commercial chemical protective clothing either did not provide adequate protection, or had little chemical data available to judge its performance. As a consequence, a formal research and development project was established in 1978 to develop new chemical protective clothing and equipment that would satisfy Coast Guard requirements. Part of this project was directed toward developing a totally encapsulating chemical protective suit.

[1] Senior engineer, Texas Research International, 9063 Bee Caves Road, Austin, TX 78733.
[2] Chemical Fabrics Corporation, Merrimack, NH 03054.

The goals of this effort were to:

1. Select a material or group of materials that would provide broad protection against as many chemicals as possible that would eliminate the need for a large inventory of different chemical protective suits.
2. Reduce the problems of selecting a suit for mixtures and unknown chemicals.
3. Overcome a lack of standards for designing commercial suits by complete documentation of the suits' capabilities and limitations through thorough testing.

Early Work

When the Coast Guard began its research effort, the majority of chemical protective suits is used were constructed of butyl rubber with a polycarbonate visor. These suits were modified versions of those used by the U.S. Army for chemical warfare applications. An early Coast Guard study identified 400 CHRIS chemicals as requiring the use of a totally encapsulating protective garment and self-contained breathing apparatus for adequate protection [2]. From this same study, measurements of material-chemical permeation indicated that butyl rubber and polycarbonate were compatible with only 38% and 60% of these chemicals, respectively (for a 3-h period).

The Coast Guard then undertook the development of its own totally encapsulating chemical protective ensemble to include the selection of compatible materials and the development of a uniform suit design [3]. Several existing and state-of-the-art materials were screened by chemical resistance and physical property testing. This screening yielded two materials to supplement butyl rubber as garment materials in separate suits, that is, Viton/chlorobutyl laminate and chlorinated polyethylene (CPE). In addition, a Teflon fluorinated-ethylene propylene (FEP)/Surlyn laminate was chosen to replace polycarbonate as the visor material for all three suit materials.[3]

Each of the selected materials were subjected to extensive chemical resistance testing, including one-sided immersion testing against 160 representative CHRIS chemicals and permeation testing against 59 of those chemicals [3]. The immersion testing results indicated few chemical effects on the Teflon visor material, with Viton/chlorobutyl laminate moderately affected, and chlorinated polyethylene greatly affected. No chemicals permeated the FEP visor material within 3 h, but the Viton/chlorobutyl laminate and CPE exhibited breakthrough of 15 and 30 chemicals, respectively.

Despite the relative poor performance of CPE, it was retained in the Coast Guard's chemical protective suit "system" because of its resistance to inorganic acids and bases, and other chemicals with high spill frequencies. The Coast Guard Research and Development Center began to test the three garment and visor materials to further determine their resistance to other chemicals and mixtures under various conditions. Their research found a number of material failures and caused concern for using these materials, even though, the three materials collectively represented the most effective combination to provide broad chemical resistance [4].

Investigation of Alternative Materials

In 1984, the Coast Guard initiated a review of protective clothing materials to determine if new materials with greater chemical resistance could be identified. Selection criteria [5]

---

[3] Viton, Teflon, and Surlyn are registered trademarks of E. I. DuPont de Nemours and Company.

were divided into three areas:

1. Chemical resistance.
2. Physical properties.
3. Fabrication feasibility.

Chemical resistance performance was evaluated using ASTM Test Method for Resistance of Protective Clothing Materials to Permeation by Liquids and Gases (F 739-85) against a representative battery of chemicals specified in ASTM Standard Guide for Test Chemicals To Evaluate Protective Clothing Materials (F 1001-86) (Table 1). A 3-h period was specified to assess the compatibility of test chemicals. Physical property behavior was screened based on test methods and minimum performance levels established by Coast Guard engineering (Table 2). The performance levels were derived from physical property testing on existing chemical protective clothing materials that had demonstrated adequate material integrity and durability in actual field usage. Finally, the material supplier had to demonstrate their ability to fabricate strong, liquid-proof seams with the garment, visor, and closure tape materials. Testing in this area included seam penetration resistance to selected chemicals and measurement of seam strength.

In 1985, Chemical Fabrics Corporation introduced Challenge[4] 5100, a proprietary, aramid-reinforced fluoroelastoplastic composite. This material exhibited a high level of chemical resistance and possessed equal or better physical properties relative to the Coast Guard's originally selected materials. Tables 2 and 3 show a comparison of these materials' physical property and permeation properties results. The Coast Guard also adopted a Teflon FEP visor that facilitated suit fabrication while eliminating lamination difficulties inherent to the FEP/Surlyn composite. Additionally, a Teflon (TFE) film glove material was chosen for use in the Coast Guard Chemical Response Suit. The Coast Guard opted for Teflon components in the suit design, where possible, to provide a suit with improved uniformity in chemical resistance throughout the garment. The only two major non-Teflon components are the suit closure (a neoprene-brass pressure sealing zipper) and exhaust valves (nylon

TABLE 1—*Standard chemicals for evaluating protective clothing materials.*[a]

Chemical	Chemical Class
Acetone	Ketone
Acetonitrile	Nitrile
Carbon disulfide	Sulfur organic compound
Dichloromethane	Chlorinated hydrocarbon
Diethyl amine	Amine
Dimethylformamide	Amide
Ethyl acetate	Ester
Hexane	Aliphatic hydrocarbon
Methanol	Alcohol
Nitrobenzene	Nitrogen organic compound
Sodium hydroxide (50%)	Inorganic base
Sulfuric acid (93.1%)	Inorganic acid
Tetrachloroethylene	Chlorinated hydrocarbon (olefin)
Tetrahydrofuran	Heterocyclic ether
Toluene	Aromatic hydrocarbon

[a] These chemicals are recommended in ASTM F 1001-86.

[4] Registered trademark of Chemical Fabrics Corporation.

TABLE 2—Physical property requirements and data for candidate garment materials.

Property, units	Test Method	Coast Guard Requirement	Materials[a]			
			Chlorinated Polyethylene	Butyl Rubber	Viton/ Chlorobutyl	Challenge 5100
Weight, gm/m²	ASTM D 751-83	990 (max)	650	540	520	450
Thickness, mm	ASTM D 751-83	0.76 (max)	0.51	0.36	0.51	0.46
Tensile strength, kg/cm w = warp; f = fill	Fed. Std. 191A,5102	14.3 w (min) 14.3 f (min)	15.5 (w) 17.7 (f)	24.2 (w) 15.4 (f)	23.2 (w) 45.7 (f)	23.7 (w) 20.5 (f)
Tear strength, kg/cm	Fed. Std. 191A,5134	1.61 w (min) 1.61 w (min)	2.41 (w) 2.98 (f)	1.68 (w) 2.41 (f)	1.73 (w) 1.96 (f)	1.71 (w) 1.75 (f)
Hydrostatic resistance, kg/m²	Fed. Std. 191A,5512	0.284 (min)	0.284	0.462	0.547	0.445
Abrasion, grams lost H-18 wheel, 600 cycles	Fed. Std. 191A,5306	0.30 (max)	0.39	0.31	b	0.05
Stiffness—warp, cm	Fed. Std. 191A,5200	5.0 (max)	no data	no data	no data	4.5
Flammability ignition time, s	ASTM D 568-77	self-ext[c]	self-ext	self-ext	self-ext	Does not ignite
Low temperature bending	ASTM D 2136-84	pass at −32°C	pass	pass	pass	pass

[a] All materials have fabric supports; data for first three materials from Ref 3.
[b] Exposed fibers of the base material appeared after 600 cycles.
[c] Test method measures ignition properties of materials, however, Coast Guard requirement specifies that material is self-extinguishing.

TABLE 3—*Comparison of permeation results for chlorinated polyethylene, Viton/chlorobutyl laminate, and challenge 5100.*

Chemical	Breakthrough Time, min[a]		
	Chlorinated Polyethylene	Viton/Clorobutyl Laminate	Challenge 5100
Acetic Acid	no BT[b]	no BT	no BT
Acetone[c]	20-25	43-53	no BT
Acetonitrile[c]	80-85	90-105	no BT
Benzene	71-75	no BT	no BT
Carbon disulfide[c]	8-10	11-15	13-23
Dichloromethane[c]	15-25	25-36	35-45
Diethyl amine[c]		27-33	no BT
Diethyl ether		1-10	no BT
1,2-dichloroethane	15-25	no BT	no BT
Dimethyl formamide[c]		no BT	no BT
Dimethyl sulfoxide	no BT	no BT	no BT
Ethyl acetate[c]	60-70	20-40	no BT
Ethyl acrylate	14-32	34-45	no BT
Freon TF	no BT	no BT	no BT
Hexane[c]	no BT	no BT	no BT
Methanol[c]	no BT	no BT	no BT
Methyl ethyl ketone	28-35	25-40	no BT
Nitric acid (conc.)	no BT	no BT	no BT
Nitrobenzene[c]	60-70	170-180	no BT
Sodium hydroxide (50%)[c]	no BT	no BT	no BT
Styrene	60-70	no BT	no BT
Sulfuric acid (conc.)[c]	no BT	no BT	no BT
Tetrachloroethane	60-70	no BT	no BT
Tetrachloroethylene[c]			no BT
Tetrahydrofuran[c]	27-39	9-11	no BT
Trichloroethylene	10-15	25-30	108-143
Toluene[c]	69-75	no BT	no BT

[a] Breakthrough times determined using ASTM F 739-81. Blanks indicate the absence of data; breakthrough times are presented as ranges due to the imprecision in determining actual breakthrough time; breakthrough time is heavily dependent on the analytical sensitivity of the detector used.
[b] No BT denotes no breakthrough detected for a 3-h period.
[c] ASTM F 1001-86 chemicals.

and silicone rubber). Figure 1 displays the Coast Guard's Chemical Response Suit fabricated by Chemicals Fabrics Corporation.

Strategy for Testing Chemical Response Suit Materials

The selection of the Challenge 5100 and Teflon materials was based on limited data against a small number of representative chemicals. In order to support the development of a Challenge suit, and its use in the field, the Coast Guard initiated an extensive testing program that would document the performance of the overall suit, its materials, and components. This testing program encompasses an examination of all primary suit materials (garment, visor, and glove), critical suit seams, and suit components (closure and exhaust valves). Overall garment testing is being considered in a separate part of the program. The final goals of this test program are:

1. to integrate test data for assessing overall suit performance, and
2. to establish suit use recommendations against particular chemicals.

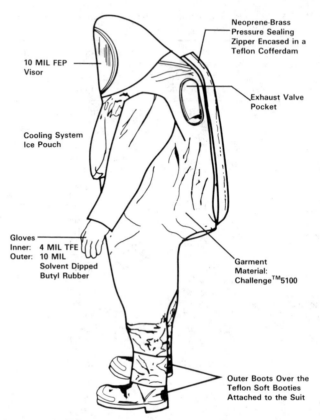

FIG. 1—*Coast Guard Chemical Response Suit (sketch showing the location and composition of each major material and component).*

Material performance is being further characterized in terms of chemical resistance to a larger set of chemicals under various conditions, and in terms of additional physical property or functional testing. In general, each material and component should be tested in the same fashion and against the same chemicals. Practically, this is difficult due to the enormous size of the test matrix. The Coast Guard adopted the philosophy of first testing the garment material against a large set of priority chemicals and then testing other primary materials and seams against a smaller subset of the priority chemicals. In this manner, material performance can be compared and judgements can be made on how to extend the testing of other materials to more chemicals. Table 4 provides this preliminary matrix of suit materials and components, types of testing, and chemical batteries. Eventually, predictive models will be necessary to overcome large testing demands and the problems of mixture exposure.

Garment Material Evaluation

Chemical Resistance Testing

The garment material comprises more than 75% of the total exposed surface area for the Coast Guard Chemical Response Suit. The Coast Guard started testing the Challenge composite against 115 priority liquid CHRIS chemicals using ASTM F 739-85 for measuring

TABLE 4—*Suit material and component test matrix.*

Material and Component	Type of Test	Test Chemical or Properties
Garment material	permeation[a]	115 priority liquids
		25 priority gases
		variable effects on ASTM F 1001-86
		chemicals (temperature, contact
		time, pressure, mixtures)
	strength	Tensile, tear, bursting
	resistance	abrasion, cut, puncture,
		UV light, ozone degradation
	other physical properties	stiffness, flammability,
		low temperature, performance
Creased garment material	permeation	ASTM F 1001-86 chemicals
Visor material	permeation	ASTM F 1001-86 chemicals
	strength	tensile, tear, bursting
	resistance	abrasion/clarity,
		UV light, ozone degradation
	other physical properties	light transmission,
		stiffness, flammability,
		low temperature performance
Creased visor material	permeation	ASTM F 1001-86 Chemicals
Inner glove material	permeation	ASTM F 1001-86 Chemicals
	strength	tensile, tear, bursting
	resistance	abrasion
	other physical properties	stiffness, flammability,
		low temperature performance
Outer glove material	degradation[b]	ASTM F 1001-86 chemicals
Critical suit seams	penetration[c]	water, MEK, HCl, hexane, toluene
	permeation	ASTM F 1001-86 chemicals
	strength	tensile, bursting, dead load
Suit closure (zipper)	penetration	water, MEK, HCl, hexane, toluene
	degradation	ASTM F 1001-86 chemicals
	strength	tensile, bursting, dead load

[a] Permeation resistance measured using ASTM F 739-81 over 3-h period.
[b] Degradation resistance measured using draft ASTM F23.30.03 method.
[c] Penetration resistance measured using ASTM F 903-84.

permeation resistance. These chemicals were chosen on the basis of the encapsulation requirement, spill frequency, and toxicity [6]. To date, eight chemicals have been found to permeate the garment material within a 3-h period; of these, three chemicals exhibit breakthrough in 1 h (see Table 5). The material is also being tested against chemical gases, and eventually will be evaluated against the 400+ CHRIS chemicals requiring encapsulation.

Investigation of Chemical Resistance Variables

Chemical resistance testing of the garment material also involves investigation of parameters expected to affect material performance. These parameters include contact time, internal suit pressure, temperature, and chemical mixture exposure. This testing takes advantage of earlier work performed by the Coast Guard R&D Center on Viton/chlorobutyl laminate and chlorinated polyethylene [7,8]. The results of the permeation testing at various temperatures are shown in Fig. 2 and demonstrate the expected relationship between breakthrough time and temperature—a decrease in breakthrough time at elevated temperatures. Although we have not yet developed a theoretical, predictive model for the permeation

TABLE 5—*Permeation testing results for Challenge 5100 (Teflon-coated Nomex). All tests conducted at 23 to 25°C.*

Chemical	CHRIS[a] Code	BT[b]	Perm[c] Rate	Det[d] Met'd	MDL[e] (ppm)	Source[f]
Acetaldehyde	AAD	>3 h		PID	ND	TRI
Acetic acid	AAC	>4 h		FID	35.46	R&DC
Acetic anhydride	ACA	>3 h		PID	0.57	TRI
Acetone	ACI	>3.5 h		FID	1.16	R&DC
Acetone cyanohydrin	ACY	N/A		PID	2.74	TRI
Acetonitrile	ATN	>4.5 h		FID	ND	R&DC
Acetophenone	ACP	>92 h		FID	ND	R&DC
Acetyl chloride	ACE	>3.1 h		PID	35.46	TRI
Acrolein	ARL	38 min		PID	0.06	TRI
Acrylic acid	ACR	>3 h		PID	0.86	TRI
Acrylonitrile	ACN	54 min	0.086	PID	0.46	TRI
Adiponitrile	ADN	>3.1 h		PID	0.3	TRI
Allyl alcohol	ALA	>14 h		PID	1.13	TRI
Allyl chloride	ALC	102 min	0.011	PID	0.16	TRI
Aniline	ANL	>3.3 h		PID	0.46	TRI
Benzene	BNZ	>3.2 h		PID	0.05	TRI
Benzyl chloride	BCL	>3.2 h		PID	0.11	TRI
Bromine	BRX	>3.3 h		PID	0.53	TRI
n-butyl acetate	BCN	>3 h		PID	0.25	TRI
n-butyl acrylate	BTC	>3 h		PID	0.22	TRI
n-butylamine	BAM	>3 h		PID	0.32	TRI
n-butyl alcohol	BAN	>15.6 h		PID	0.32	TRI
Butyraldehyde	BTR	>7.5 h		PID	0.29	TRI
Carbon disulfide	CBB	17.7 min	0.043	PID	0.05	TRI
Carbon tetrachloride	CBT	>3.0 h		PID	0.29	TRI
Chlordane, 85%	CDN	>3.4 h		PID	0.26	TRI
Chlorobenzene	CRB	>3 h		PID	0.20	TRI
Chloroform	CRF	>3.6 h		PID	0.19	TRI
Chloropicrin	CPL	>3.1 h		PID	1.80	TRI
Chlorosulfonic acid	CSA	>3.0 h		Ion Chr	0.50	TRI
Creosote	CCT	>18.1 h		PID	0.32	TRI
m-cresol	CRL	>4 h		PID	0.03	TRI
Crotonaldehyde	CTA	>3.1 h		PID	0.62	TRI
Cumene hydroperoxide	CMH	>3.5 h		PID	1.20	TRI
Cyclohexane	CHX	>3.4 h		PID	0.25	TRI
1,2-dibromoethane	EDB	>3.4 h		PID	0.10	TRI
1,2-dichloroethane	EDC	>5.7 h		PID	0.09	TRI
1,2-dichloroethyl ether	DEE	>3 h		PID	ND	TRI
Dichloromethane	DCM	46.8 min	0.023	PID	0.27	TRI
		37 min		ECD	0.03	R&DC
1,2-dichloropropane	DPP	>3.1 h		PID	0.31	TRI
1,3-dichloropropene	DPR	>3 h		PID	0.17	TRI
Diethanolamine	DEA	>3 h		PID	ND	TRI
Diethylamine	DEN	>4.5 h		FID	ND	R&DC
Dimethyl sulfate	DSF	N/A		PID	1.52	TRI
Disopropylamine	DIA	>11.2 h		PID	0.39	TRI
Dimethylformamide	DMF	>3.2 h		FID	ND	R&DC
1,4-dioxane	DOX	>3 h		PID	0.38	TRI
Di-n-propylamine	DNA	>3.4 h		PID	0.22	TRI
Epichlorohydrin	EPC	>3 h		PID	0.75	TRI
Ethion 4	ETO	>4.8 h		PID	0.03	TRI
Ethyl acetate	ETA	>4.3 h		FID	0.49	R&DC
Ethyl acrylate	EAC	>17 h		PID	1.72	TRI
Ethyl alcohol	EAL	>3 h		PID	2.86	TRI
Ethylamine (70%)	EAM	>3 h		PID	0.74	TRI

TABLE 5—*Continued.*

Chemical	CHRIS[a] Code	BT[b]	Perm[c] Rate	Det[d] Met'd	MDL[e] (ppm)	Source[f]
Ethyl benzene	ETB	>3 h		PID	0.14	TRI
Ethylenediamine	EDA	>3.2 h		PID	2.78	TRI
Ethylene glycol	EGL	>16.8 h		PID	2.63	TRI
Ethyl ether	EET	>3 h		PID	0.13	TRI
Formaldehyde (37%)	FMS	>3 h		PID	ND	TRI
Furfural	FFA	>1 h		PID	0.08	TRI
Gasoline	GAT	>14.9 h		PID	1.65	TRI
Glutaraldehyde (sol'n)	GTA	N/A		PID	0.43	TRI
Hexane	HXA	>5 h		PID	0.25	TRI
Hydrazine hydrate	HDZ	N/A		PID	0.90	TRI
Isopropyl alcohol	IPA	>3 h		PID	1.16	TRI
Isopropylamine	IPP	>3 h		PID	1.57	TRI
Malathion (50%)	MLT	>3.1 h		PID	1.03	TRI
Methyl acrylate	MAM	>3 h		PID	0.12	TRI
Methyl alcohol	MAL	>14.2 h		PID	4.07	TRI
Methyl ethyl ketone	MEK	>3 h		PID	0.65	TRI
Methyl isobutyl ketone	MIK	>3 h		PID	3.98	TRI
Methyl methacrylate	MMM	>3.1 h		PID	0.19	TRI
Methyl parathion	MPT	N/A		PID	0.15	TRI
Naled	NLD	>3.4 h		PID	N/A	TRI
Naphthalene	MLT	>13.2 h		PID	ND	TRI
Nitric acid	NAC	N/A		Ion Chr	0.20	TRI
Nitrobenzene	NTB	>3 h		PID	0.08	TRI
2-nitropropane	NPP	>3 h		PID	0.59	TRI
Oleum	OLM	>3.0 h		Ion Chr	0.20	TRI
Parathion	PTO	>3.0 h		PID	0.09	TRI
Petroleum ether		>3.4 h		PID	4.55	TRI
Phenol	PHN	>3 h		PID	0.03	TRI
Phosphoric acid	PAC	N/A		Ion Chr	0.50	TRI
Phosphorus oxychloride	PPO	>3 h		Ion Chr	0.50	TRI
Phosphorus trichloride	PPT	>3 h		Ion Chr	0.50	TRI
Propionic acid	PNA	>3 h		PID	0.31	TRI
n-propyl alcohol	PAL	>3 h		PID	0.76	TRI
n-propylamine	PRA	>10.2 h		PID	0.74	TRI
Propylene oxide	POX	137 min	0.024	PID	0.68	TRI
Silicon tetrachloride	STC	>3.0 h		Ion Chr	0.50	TRI
Sodium hydrosulfide	SHR	N/A		AA	0.50	TRI
Sodium hydroxide (50% aqueous)	CSS	>71 h		SE	ND	R&DC
Sodium hydroxide (50% aqueous)	CSS	>3.0 h		Ion Chr	0.50	TRI
Styrene	STR	>4 h		PID	0.05	TRI
Sulfur monochloride	SFM	N/A		Ion Chr	0.50	TRI
Sulfuric acid (conc.)	SFA	>72 h		Sulfate	ND	TRI
1,1,2,2-tetrachloroethane	TEO	>15.2 h		PID	0.23	TRI
Tetrachloroethylene	TTE	108 min		ECD	ND	R&DC
Tetrahydrofuran	THF	>5.5 h		FID	ND	R&DC
1,1,1-trichloroethane	TCE	>3 h		PID	0.60	TRI
Trichloroethylene	TCL	143 min	0.034	PID	0.07	TRI
Toluene	TOL	>3 h		PID	0.06	TRI
		>18.5 h		FID	0.69	TRI
o-Toluidine	TLI	>3.3 h		PID	0.43	R&DC
Toluene 2,4-diisocyanate	TDI	>3.3 h		PID	0.69	TRI
Turpentine	TPT	>3.6 h		PID	0.03	TRI
Vinyl acetate	VAM	74 min	0.055	PID	0.21	TRI
Vinylidene chloride	VCI	>3.0 h		PID	0.49	TRI

TABLE 5—*Continued.*

Chemical	CHRIS[a] Code	BT[b]	Perm[c] Rate	Det[d] Met'd	MDL[e] (ppm)	Source[f]
Xylenes	XLM	>3 h		PID	0.13	TRI
Xylenol	XYL	>3.3 h		PID	ND	TRI

[a] The CHRIS Code comes from the Coast Guard CHRIS list.
[b] BT = breakthrough time (>XH = time test run; nMin = BT in minutes for those compounds that did break through.)
[c] The permeation rate units are micrograms/square centimetre/hour.
[d] DET MET'D = detector used for determination of BT.
[e] MDL = minimum detection limit of the detector.
[f] SRC = source of data: TRI = Texas Research Institute; R&DC = Coast Guard results.

behavior of Challenge products, we observe an apparent inverse, linear relationship between temperature and log(breakthrough) for the limited data reported here. Splash testing with dichloromethane per Draft 3 of ASTM F 739-85, Procedure B, yields essentially the same breakthrough time as obtained when liquid remains in contact with the surface of Challenge 5100. This is not unexpected for this material, since the vapor pressure at the surface is the same, whether one has liquid or saturated vapor present.

Physical Property Testing

An original concern that the Teflon laminate may "microfracture" with use [9] is being investigated by a battery of physical property and chemical resistance testing. As a practice, most permeation testing is conducted with pristine material samples. Chemical Fabrics Corporation has devised a method for creasing samples as a preconditioning technique to determine if the chemical resistance of the material changes with physical abuse. Early tests with this method have shown no significant change in the permeation breakthrough time for dichloromethane and methyl ethyl ketone. Other physical property tests are being performed to determine how well the Challenge 5100 retains its characteristics following temperature changes and exposure to ultraviolet (UV) light and ozone.

Visor Material Evaluation

The critical performance requirements for visor materials in chemical protective clothing include:

1. High visible light transmittance and visual clarity.
2. Chemical permeation resistance.
3. Physical integrity and damage tolerance.

For screening purposes, light transmittance was measured with a visible light spectrophotometer. Chemical permeation resistance was measured against an aggressive chemical from the ASTM F 1001-86 battery. Physical integrity and damage tolerance were evaluated in terms of tear strength (trapezoid method) and stiffness (cantilever method). Stiffness is critical since it is related to the ease of film creasing that dramatically reduces visual clarity.

Of the commercially available perfluorinated polymer films, fluorinated ethylene-propylene (FEP) possesses the highest visible light transmittance per given thickness and was thus selected as the visor material. The preceding screening tests were employed to determine

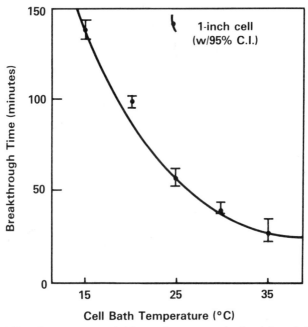

FIG. 2—*Effect of temperature on dichloromethane permeation breakthrough time on Teflon laminated Nomex.*

the optimum visor film thickness. Data on commercially available 5, 10, 14, and 20-mil FEP film are presented in Table 6. The data reveal that as film thickness increases, the chemical permeation (Fig. 2) and physical properties improve at the expense of light transmittance. Ten-mil FEP was selected since it provides adequate clarity and resistance to creasing while offering permeation resistance and tear strength consistent with the garment material (Tables 2 and 3). Additional physical properties of the FEP visor material are offered in Table 7.

The current Coast Guard test program includes further documentation of the chemical resistance properties of the FEP visor, including the ASTM F 1001-86 battery as well as specific chemicals that have exhibited breakthrough to Challenge. The effects of creasing on the permeation resistance of the 10-mil FEP will also be characterized.

TABLE 6—*Physical properties of FEP film visor candidates.*

Property, units	Test Method	5 Mil	10 Mil	14 Mil	20 Mil
Tear strength[a] trapezoid method, kg	Fed. Std. 191A-5136	19.6	47.3	62.0	44.4
Flexural rigidity[a] canti-lever method, mg-cm	ASTM D 1388-64	0.149	1.07	2.85	7.62
Light transmittance,[b] % visible	ASTM E 424-71	95.5	94.8	94.0	92.6
Permeation breakthrough time, min	ASTM F 739-81	see Fig. 3			

[a] Average of machine-direction and transverse-direction values.
[b] Average light transmittance from 390 to 876 nm; Perkin Elmer Lambda 4 spectrophotometer.

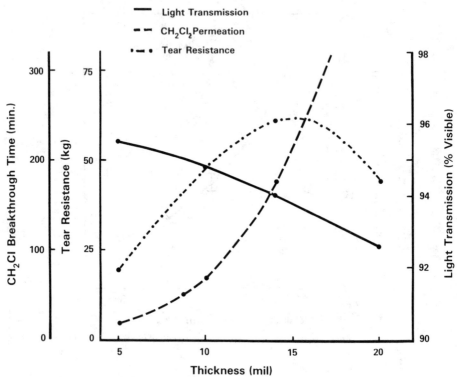

FIG. 3—*Effect of thickness on visor material tear strength, light transmission, and permeation characteristics.*

Evaluation of Secondary Suit Materials and Components

Glove Material Testing

The Coast Guard Chemical Response Suit is designed with Teflon (TFE) inner gloves and outer gloves of either butyl rubber or Viton. The inner glove consists of two simple hand silhouettes with a peripheral heat-sealed seam. The outer elastomer glove provides the shape to the composite glove, which dramatically improves dexterity. A strategy similar to that for testing the visor material is being used to evaluate the inner glove material (4-mil Teflon-TFE film). This includes testing the chemical permeation resistance of the glove material against the 15 chemical ASTM battery. Many of the same physical property tests used to evaluate the garment and visor materials have also been performed on the TFE film (Table 7). Lower physical properties of the glove material are believed to be acceptable due to the supporting nature of the glove. Vendor information is used to assess the performance of the outer glove, but the Coast Guard is verifying manufacturer recommendations with single-sided immersion testing against the ASTM F 1001-86 chemicals (using tentative ASTM Standard Method for Resistance of Protective Clothing Materials to Degradation by Liquids).

Suit Seam Testing

Critical seams of the Coast Guard Chemical Response Suit include:

1. Garment material to garment material.
2. Garment material to visor material.

TABLE 7—*Physical properties of visor and glove TFE films.*

Property, units	Test Method	Visor	Glove
Composition		fluorinated ethylene-propylene	polytetrafluoro-ethylene
Thickness, mm	ASTM D 374-79	0.254	0.102
Tear strength, kg[a] trapezoid	Fed. Std. 191A,5136	47.3	4.2
Abrasion resistance[b] taber, grams lost	ASTM D 3389-85	0.02	0.05
Flexural rigidity cantilever, mg-cm	ASTM D 1388-64	1.07×10^4	8.65×10^2
Low temperature bending, °C	ASTM D 2136-84	pass at $-40°C$	pass at $-40°C$
Flame resistance vertical	Fed. Std. 191A,5903		
char length, cm		3.6	3.8
after-flame, s		0	0
after-glow, s		0	0

[a] Average of machine-direction and transverse-direction valves.
[b] H-18 wheel, 600 cycles, 250-g weight.

3. Garment material to inner glove material.
4. Garment material to suit closure tape material.
5. Glove material to glove material.

Each of the seam constructions has been subjected to two types of seam strength tests: tensile and dead load stresses (Table 8). The ultimate tensile strength of each seam type generally reflects the tensile strength of the weakest base material, as opposed to the actual strength of the heat-sealed joint. In other words, the heat sealed seams are designed to be as strong or stronger than the base materials (with the possible exception of the glove-glove seam that exhibited both film and seam type failures). Dead load or creep testing was conducted to simulate the long term, but low stress conditions resulted from the positive pressure in a totally encapsulating suit. Representative seam stresses were calculated for two locations within the suit (torso and glove) based on an internal positive pressure of 5.8 mm Hg and on measured suit dimensions. Dead load testing was conducted at loads of 15 and 2.3 kg/cm representing stresses in the torso and gloves, respectively. No failures occurred in any of the seam configurations in 48 h under the preceding loading conditions.

TABLE 8—*Seam physical properties of Challenge and secondary components.*

Seam Type	Ultimate Tensile Strength, kg/cm	Dead Load,[a] kg/cm	Test[b] Duration, h
Challenge-Challenge	132	15	48+
Challenge-visor	25.3	15	48+
Challenge-closure[c]	129.5	15	48+
Challenge-glove	12.3	2.3	48+
Glove-glove	8.2	2.3	48+

[a] Dead loads were conducted at approximately 15 times the static seam stress resulting from normal suit positive pressure (5.8 mm Hg). Maximum interior dimensions of 25.9 cm radius in the suit torso and 7.4 cm radius in the glove yield stresses of 98.2 and 26.8 gm/cm, respectively.
[b] $n+$ indicates no failure in the time stated.
[c] Closure is a neoprene-brass pressure sealing zipper.

In addition, penetration testing using ASTM Test Method for Resistance of Protective Clothing Materials to Penetration by Liquids (F 903-84) of the first three seams has been conducted against a five chemical battery (Table 4) and permeation testing will be performed with the ASTM F 1001-86 chemicals. No penetration was noted for any seam-chemical combination [10]. This testing has already identified improvement of the garment material seam from a combined sewn and heat-sealed "T" seam to a heat-sealed lap seam.

Suit Component Testing and Considerations

Both the suit closure and exhaust valves are non-Teflon suit components. Their performance with respect to chemical exposure is difficult to assess since standard methods do not exist to measure component chemical resistance. Penetration testing of the suit zipper has been performed using a modified test cell against five selected chemicals with no evidence of penetration [10]. Sample suit zippers have also been subjected to zipper crosswise strength testing to determine tensile properties. Other physical property tests such as bursting strength are being considered for evaluating suit closure performance. The assessment of suit exhaust valve performance is being conducted by Lawrence Livermore National Laboratory in a separate study [11]. To protect both components, the suit has been designed with "splash" covers to reduce the likelihood of liquid chemical contact.

Integration of Test Data

The results from material chemical resistance and physical property testing must be related to overall suit performance in order to provide meaningful results to end-users. Physical property data are used to determine if materials and components possess sufficient integrity and resistance to physical and environmental abuse relative to evolving standards. Generally, each material should have similar physical property requirements, but these may differ based on the material's function. Such requirements should be set to reflect actual use conditions. While standards have been used in the past based on chemical warfare clothing material requirements, the Coast Guard is conducting new studies to better define which properties should be measured and what are reasonable requirements for those properties.

Using chemical resistance data to assess suit performance is a much more complex problem. Because dermal exposure limits do not exist, any permeation of hazardous chemicals through a protective garment is unacceptable. The problem arises in comparing material swatch testing against overall suit exposure to chemicals. In general, most permeation resistance testing represents "worst case" exposure, where the liquid or gaseous chemical is in constant contact with the material over the test period. This is not the usual case for field exposures during spill response and monitoring. Yet, many researchers recognize that certain variables (that is, temperature, chemical mixtures, etc.) can accelerate a chemical's effect on materials [5]. This combined with the inability to test any material-chemical combination under all conditions makes suit recommendations difficult.

The Coast Guard plans to use a 1-h criterion for permeation breakthrough time for initially recommending suit use against a particular chemical. One hour should provide a reasonable safety factor for all anticipated exposures. However, this rule should also be applied to all primary materials and components, that is, the recommendation is based on the performance of all primary materials. Mixtures must be addressed on a case-by-case basis and will require the use of field test kits or predictive models.

Conclusion

We have described an extensive suit material and component testing program to support the Coast Guard's use of Challenge in their Chemical Response Suit. This program represents a comprehensive approach for selecting materials and evaluating their performance for chemical spill response and clean-up. Moreover, this type of evaluation allows end-users to understand suit capabilities and limitations. The Coast Guard believes that the new material will provide protection for more chemicals than any one suit or combination of suits it now uses. Few chemical protective suits offer the same level of documentation. It appears, however, that all primary materials and components should be tested to identify weaknesses that might otherwise go undetected. Garment material performance alone does not provide a sufficient basis for making suit use recommendations.

References

[1] "Chemical Hazard Response Information System (CHRIS)," Coast Guard Commandant's Instruction M16465.12A, Government Printing Office, Washington, DC, Nov. 1984 (Stock No. 050-012-00215-1).
[2] Friel, J. V., McGoff, M. J., and Rodgers, S. J., "Material Development Study for a Hazardous Chemical Protective Clothing Outfit," Technical Report CG-D-58-80, MSA Research Corp., Evans City, PA, Aug. 1980 (AD A095 993).
[3] Stull, J. O., "Early Development of a Hazardous Chemical Protective Ensemble," Technical Report CG-D-24-86, U.S. Coast Guard, Washington, DC, Oct. 1986.
[4] Bentz, A. and Mann, V., "Critical Variables Regarding Permeability of Materials for Totally-Encapsulating Suits," 1st Scandanavian Symposium on Protective Clothing Against Chemicals, Copenhagen, Denmark, Nov. 1984.
[5] Stull, J. O., "Criteria for Selection of Candidate Chemical Protective Clothing Materials," *Journal of Industrial Fabrics*, Vol. 4, No. 2, 1985, pp. 13–22.
[6] Hendrick, M. S. and Billing, C. B., "Selection of Priority Chemicals for Permeation Testing and Hazardous Chemical Spill Detection and Analysis, Technical Report CG-D-22-86, U.S. Coast Guard R&D Center, Groton, CT, July 1986 (AD A172 370).
[7] Stull, J. O., et al., "A Comprehensive Materials Evaluation Program to Support the Development and Selection of Chemical Protective Clothing," *Proceedings*, 1986 Hazardous Material Spills Conference, St. Louis, April 1986.
[8] Man, V., et al., "Permeation of Protective Clothing Materials: 1. Comparison of Liquid, Liquid Splashes, and Vapors on Breakthrough Times," 1985 Pittsburg Conference on Analytical Chemistry and Applied Spectroscopy, Atlantic City, 1986.
[9] Weeks, R. W., Jr., and McLeod, M. J., "Permeation of Protective Garment Material by Liquid Halogenated Ethanes and a Polychlorinated Biphenyl," Technical Report DHHS 81-110, NIOSH, U. S. Government Printing Office, Washington, DC, Jan. 1981.
[10] Anderson, C., "Penetration and Degradation Tests of Selected Samples," Internal Coast Guard Technical Report, Anderson Associates, Groton, CT, Aug. 1986.
[11] Swearengen, P. M. and Stull, J. O., "Evaluating the Performance of One-Way Vent Valves used in the Construction of Totally-Encapsulating Chemical Protective Suits," 2nd International Symposium on the Performance of Protective Clothing, Tampa, FL, Jan. 1987.

Author Index

Subject Index

A

Acetylcholinesterase monitoring, 780
Aerosol spray
  permeability testing, 738–744
Airborne contaminants
  instrumentation for, 454, 477–479, 482
Airtight suit, 307–313
Allergens
  electron microscopy chemicals, 389–395
  skin cleaners, 345
Aluminized fabrics
  molten metal splashes, 137
American Association of Textile Chemists
    and Colorists
  test method 135–1978, R–85, 699, 707
American Industrial Hygiene Association,
    6
Asbestos
  gloves/mittens, 152, 155
ASTM Committee F23, 3–6
  activities and accomplishments, 3–6
  history of, 3–6
  organization of, 6
  service awards, 5
  subcommittees, 4–6
ASTM Standards
  D 471–79, 729
  D 568–77, 519
  D 737–75, 33–34, 123
  D 751–83, 850
  D 1059–76, 681
  D 1388–64, 34, 123
  D 1518–85, 123
  D 1682–64, 34
  D 1777–64, 34, 123
  D 1910–64, 123
  D 3775–79, 681
  D 3776–79, 681
  D 4108–82, 110–111, 114–119
  D 3786–80a, 34
  E 96–66, 33–34
  E 96–80, 317
  E 662, 143, 145
  F 739–81, 226, 236, 239, 252–256, 277,
    328, 853

F 739–85, 239–241, 244, 263, 267, 317,
    321, 519, 730, 733, 849, 856
F 903–84, 5, 853
F 955–85, 102
F 1001–84, 853
F 1001–86, 849, 857, 858
F 1052–87, 525
Audit scheme, 401
Autoradiograph
  chemical permeation tests, 287–288

B

Behavioral science
  protective clothing use, 804
Biological monitoring
  pesticides, 560–562, 767–768, 770, 780–
    783
Breakthrough times
  after contamination, 362–367, 369, 372–
    374
  data base, 407
  materials
    butyl, 219, 222–224, 369
    butyl-coated nylon, 320
    butyl rubber, 203–207, 264–265, 330–
      331
    Challenge 5100, 851
    chlorinated polyethylene, 851
    latex, 330–331
    natural rubber, 203–207, 264–265
    neoprene, 219, 222–224, 245–251,
      292, 330–331, 365–366, 369
    neoprene rubber, 203–207, 264–265
    nitrile, 245–251, 362–367,
    nitrile rubber, 203–207, 219, 222–224,
      264–265, 319, 330–331
    Nomex/Teflon, 369
    polyethylene, 203–207, 264–265
    polyvinyl alcohol, 203–207, 330–331
    polyvinylchloride, 203–207, 219, 222–
      224, 330–331, 523
    rubber sheeting, 735
    saran/chlorinated polyethylene, 523